Advances in Terahertz Source Technologies

Advances in Terahertz Source Technologies

edited by

**Gun-Sik Park, Masahiko Tani,
Jae-Sung Rieh and Sang Yoon Park**

JENNY STANFORD
PUBLISHING

Published by

Jenny Stanford Publishing Pte. Ltd.
101 Thomson Road
#06-01, United Square
Singapore 307591

Email: editorial@jennystanford.com
Web: www.jennystanford.com

British Library Cataloguing-in-Publication Data
A catalogue record for this book is available from the British Library.

Advances in Terahertz Source Technologies

Cover image: Hubble Space Telescope image of NGC 346 nebula with infant stars embedded, which radiates various spectra of electromagnetic waves, including the terahertz wave (Courtesy of NASA, ESA, Antonella Nota (STScI, ESA)).

ISBN 978-981-4968-89-8 (Hardcover)
ISBN 978-1-003-45967-5 (eBook)

Contents

Preface

Welcome to *Advances in Terahertz Source Technologies*, a comprehensive exploration of the exciting developments in the field of terahertz (THz) technology. This book is organized into three distinct parts, each focusing on specific applications and breakthroughs of THz sources within their respective domains: Photonics, Solid-State Electronics, and Vacuum Electronics.

Part I: Photonics

In the realm of Photonics, we shine a light on the limitless possibilities of THz technology. Here, we discover how the precision and control of lasers, optical parametric oscillators, and nonlinear optical processes have opened up new horizons. From medical imaging to security screening, the chapters in this section illuminate the vast array of applications where THz radiation is making a transformative impact.

Part II: Solid-State Electronics

Part II explores the world of Solid-State Electronics, where semiconductor electronic devices and novel materials are revolutionizing the THz landscape. Transistor circuits, semiconductor diodes, and plasmonic structures take center stage, driving advancements in imaging, spectroscopy, and communication. This section showcases how solid-state principles are transforming THz sources into compact, efficient, and versatile tools for a wide range of applications.

Part III: Vacuum Electronics

The final section, Vacuum Electronics, delves into the high-power domain of THz generation. Vacuum tubes, traveling wave tubes, gyrotrons, and other vacuum-based devices emerge as the powerhouses of the THz spectrum. From scientific research to industrial processes, this part highlights the pivotal role that vacuum electronics play in applications demanding high THz power levels, such as materials processing and plasma diagnostics.

As editors, we are profoundly grateful to the authors who have contributed their expertise, unveiling the diverse applications of THz sources within these three domains. Their dedication to advancing our understanding of terahertz technology is evident in each chapter.

For readers, whether you are a seasoned researcher, a student with a thirst for knowledge, or an enthusiast intrigued by the world of THz technology, we hope this book serves as a source of inspiration and enlightenment. Each part offers a unique perspective on the practical impact of terahertz sources in specific fields, illustrating how THz radiation is driving innovation, solving challenges, and opening new avenues of exploration.

This book represents a snapshot of the state of the art in THz technology, capturing the potential and promise of terahertz waves as we look forward to even greater breakthroughs in this dynamic and evolving field.

Thank you for joining us on this journey through *Advances in Terahertz Source Technologies*. We invite you to immerse yourself in the world of terahertz applications and to envision the boundless possibilities that lie ahead.

Part I

THz Photonic Sources

Chapter 1

THz Optical Parametric Generators and Oscillators

Kodo Kawase,[a] Kosuke Murate,[a] Hiroaki Minamide,[b] and Kouji Nawata[b]

[a] *Department of Electronics, Nagoya University, Furocho, Chikusa-ku, Nagoya 464-8603, Japan*
[b] *Tera-Photonics Research Team, RIKEN Center for Advanced Photonics, 519-1399 Aoba, Aramaki, Aoba-ku, Sendai 980-0845, Japan*
kawase@nuee.nagoya-u.ac.jp

We have long promoted research on the generation, detection, and application of THz waves using nonlinear optical effects at Tohoku University, RIKEN, and Nagoya University. We have tried various methods and crystals, but the optical parametric effects in $LiNbO_3$ crystals, whose potential we recognized early on, have the best performance, showing high output, high sensitivity, wide tunability, and high dynamic range as a system. In this chapter, we describe the basic and recent progress of injection-seeded THz parametric generation (is-TPG) and backward THz-wave parametric oscillation (BW-TPO), which are the most notable results of our studies.

Advances in Terahertz Source Technologies
Edited by Gun-Sik Park, Masahiko Tani, Jae-Sung Rieh, and Sang Yoon Park

ISBN 978-981-4968-89-8 (Hardcover), 978-1-003-45967-5 (eBook)
www.jennystanford.com

1.1 Injection-Seeded THz-Wave Parametric Generation Pumped by Subnanosecond Near-Infrared Pulses

(Hiroaki Minamide)

Injection-seeded terahertz-wave (THz-wave) parametric generation (is-TPG), invented in 2001 [1], is attractive owing to its simple configuration, room-temperature operation, and wide frequency tunability. Notably, the is-TPG has no resonator structure and generates THz waves from a lithium niobate (LN) bulk crystal pumped by a near-infrared pulse laser with weak injection seeding to the Stokes beam for actively promoting optical wavelength downconversion. This invention was based on previous fundamental research on phonon polariton scattering in LN crystals [2] and a THz-wave parametric oscillator (TPO) [3]. The is-TPG has several advantages. For example, owing to the absence of a Fabry–Pérot resonator, it simultaneously achieves a narrow linewidth of THz waves with the Fourier transform limit of the excitation light pulse and broadband continuous wavelength tunability with no mode hopping. This source is ideal for terahertz spectroscopic applications.

Phonon polariton scattering exhibits high gain in generating THz waves [4]. However, the output intensity has been an issue since its invention, with a maximum THz-wave output energy of 1 nJ and a maximum peak power of a few hundred milliwatts [5]. Therefore, THz-wave detection requires a cryogenically cooled silicon bolometer.

In theory, THz-wave parametric generation is based on stimulated Raman scattering (SRS) by polaritons, which is essential for optical parametric downconversion. However, our study revealed that stimulated Brillouin scattering (SBS) is simultaneously excited by nonlinear interactions with acoustic phonons. The SBS gain is approximately three orders of magnitude higher in the steady state [6]. Eventually, the SBS significantly suppressed wavelength conversion for THz-wave generation.

To improve the conversion efficiency while avoiding SBS effects we focused on a shorter pump beam pulse width because the SRS process has a superior time response compared to the

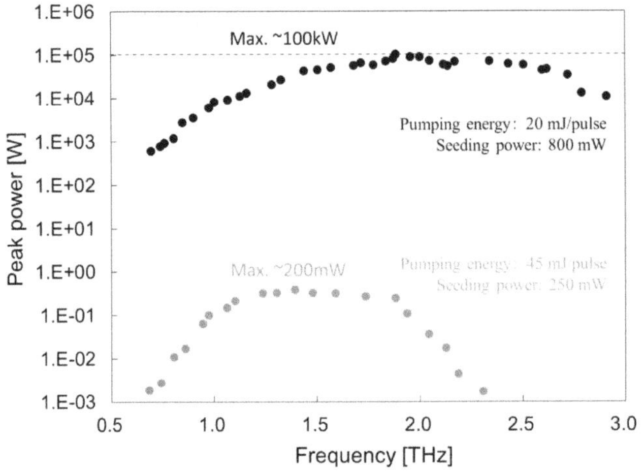

Figure 1.1 Terahertz-wave output from is-TPGs before and after suppressing SBS in an LN crystal.

relatively slow SBS process. Additionally, we propose the use of a subnanosecond pulse width over conventional nanosecond pulse excitation, considering that the spectral linewidth must be as narrow as possible. The proposed improvement can be implemented with minimal changes to the experimental structure of the conventional is-TPG. Furthermore, a single longitudinal-mode microchip laser with a wavelength of 1 μm can be used over a few hundred picoseconds.

The resulting THz-wave output from the is-TPG had a peak power of $\sim$100 kW, an enhancement of approximately five orders of magnitude over the conventional output, as shown in Fig. 1.1. The output is comparable in terms of spectral brightness to other THz-wave sources, mainly large or large-scale devices, and is one of the brightest THz-wave sources, especially in regions of $>$ 1 THz, with a brightness temperature of the order of 10^{18} K, even many years after its initial development. It can be used for various types of measurements, with numerous types of THz-wave detectors operating at room temperature. With array sensors based on recently developed, highly sensitive THz-wave detectors such as Fermi-level managed barrier (FMB) diodes, THz-wave imaging applications can be further expanded. This expansion is enabled because the THz-wave beam is widely irradiated on the object, enabling clear imaging even when the power density is reduced.

In recent studies, the mode-hop-free frequency-tunable range has been continuously realized over the ultra-broadband range of 0.8–4.7 THz [7]. Furthermore, a pulse repetition rate of 100 kHz, 1000 times higher than the conventional pulse repetition rate [8], was achieved. The improved average output power enables easy reception of measurements using slow-temporal-response detectors, such as pyrodetectors.

1.2 Highly Efficient THz-Wave Parametric Wavelength Conversion between Near-Infrared Light and THz Waves

(Hiroaki Minamide, Kouji Nawata)

The buildup time of SBS in LN crystals is estimated to be ~1.5 ns [9]. We show that subnanosecond pump pulses enable highly efficient wavelength conversion between near-infrared and THz waves. Our experimental setup of is-TPG was explained in previous reports [6, 10], except for the pump system. Several pump sources were used to test the various subnanosecond pulse widths. One pump laser was a Q-switched mode-locked Nd:YAG laser with a pulse width range of 35–200 ps (YG901C-10, Quantel Laser). The step-tuned pulse width was changed using etalons with different frequency bandwidths inside the cavity. The energy was regeneratively amplified to 35 mJ, at a pulse repetition rate of 10 Hz. The other pump lasers were microchip Nd:YAG lasers with different pulse widths of 700 and 850 ps. An output energy of up to 27 mJ was obtained by using a diode-pumped Nd:YAG amplifier.

A mode-hop-free external-cavity laser diode (λ-master 1040, SpectraQuest Lab. Inc.) and an optical amplifier were used for the injection-seeding laser. A continuous-wave output power of 350 mW was coupled to a 5 mol% MgO-doped LN with a pump beam. The frequency of the generated THz wave was 1.9 THz. The THz-wave output energy was measured using a calibrated pyrodetector.

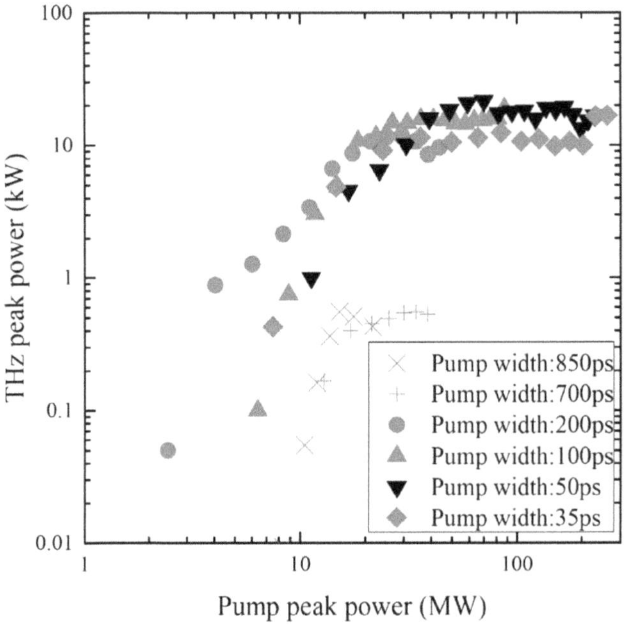

Figure 1.2 Peak power of THz-wave output as a function of the pump power with different pump widths of 35, 50, 100, 200, 700, and 850 ps.

Figure 1.2 shows the THz-wave peak power with different pump widths ranging from 35 to 850 ps. The maximum peak power of the THz-wave output was maintained at ~15 kW with a range of 10–20 kW below a pulse width of 200 ps. In contrast, the THz peak power with a pump width above 700 ps was saturated at ~0.5 kW at a pump power of 25 MW. The results indicate that the pump width affects the conversion efficiency.

Figure 1.3 shows the energy conversion efficiency as a function of pump width. The results show that the conversion efficiency changes significantly with the pulse width of the pump beam at ~1 ns. For subnanosecond pulse widths, the conversion efficiency is almost the same ($\sim 10^{-4}$), whereas for a pulse width of 15 ns, it degrades to the 10^{-8} range, which is consistent with the rise time of the SBS of ~1.5 ns. The experiment indicated that the conversion efficiency can be controlled by tuning the pulse width. In the region where the conversion efficiency is constant with a

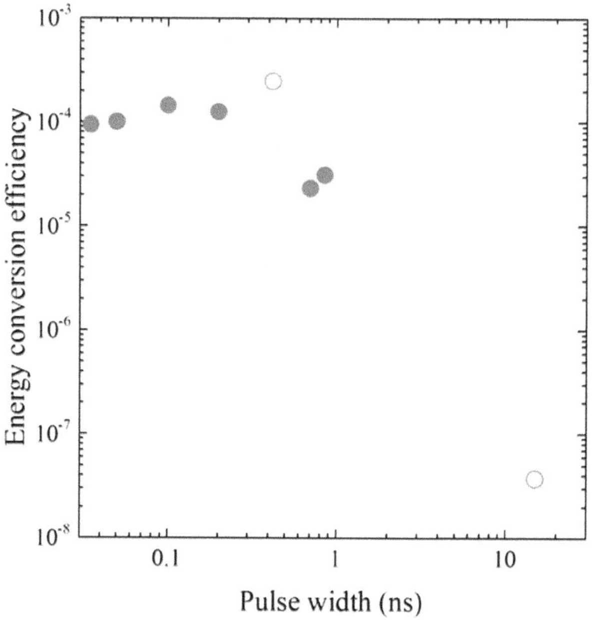

Figure 1.3 The energy conversion efficiency from the pump beam to THz waves at 1.9 THz as a function of the pump width. Closed circles represent the experimental data obtained in this study. The open circles represent the previous results with microchip laser pulses of 420 ps [6] and conventional Q-switched laser pulses of 15 ns [5].

subnanosecond pulse width, the peak power of the THz wave is almost constant; therefore, a higher pulse width of the pump beam leads to a higher energy of the THz wave. Subnanosecond pulses with pulse widths of ~1 ns are effective for THz-wave applications that require interaction with matter and energy.

1.3 Multi-Wavelength THz Parametric Generator

(Kodo Kawase, Kosuke Murate)

is-TPG has a narrow linewidth and generates high-brightness THz waves [11]; therefore, it can generate only one wavelength per pulse, limiting its application to spectroscopy. The first limitation is that the wavelength must be changed for each spectroscopic measurement. The second limitation is measurement uncertainty due to fluctuations in the output with each pulse; this is compensated for by data averaging. This chapter describes the generation of simultaneous multi-wavelength THz waves via is-TPG to overcome these limitations [12]. In multi-wavelength generation, the wavelength does not need to be changed, thus shortening the measurement time. Also, averaging is not required because the spectroscopic measurements are completed in one pulse, and the small fluctuation in each wavelength can be neglected. Thus, is-TPG enables real-time spectroscopy with a repetition rate equal to that of the excitation laser.

Multi-wavelength THz waves are generated by injecting multi-wavelength seed beams into the crystal and distributing the energy to each wavelength. For the multi-wavelength seed beams, we used multiple external-cavity diode lasers (ECDLs) with a semiconductor optical amplifier. Figure 1.4 shows the experimental setup for a real-time THz spectrometer using a multi-wavelength is-TPG. A microchip Nd:YAG laser (LE0600DPS; Hamamatsu Photonics, Shizuoka, Japan; duration: 650 ps; wavelength: 1,064.4 nm; repetition rate: 11 Hz; output: 0.7 mJ/pulse) was used as the pump source, and multiple ECDLs were utilized as multi-wavelength seed beam sources. The non-collinear phase-matching condition must be satisfied for all wavelengths to generate multi-wavelength THz waves. An achromatic optical setup was used with a grating and telescope system, in which the incidence angle of the seed beams into the crystal automatically satisfied the non-collinear phase-matching condition [13]. The emitted THz waves were focused onto the sample using lenses. The THz waves were detected based on nonlinear optical wavelength conversion [11]. Near-infrared (NIR) beam (or "detection Stokes beam") intensities were measured using a beam profiler or NIR camera (GS3-U3-41C6NIR-C; Point Grey Research Inc., Richmond, Canada). The detection Stokes beams were angle-dispersive, allowing each wavelength to be detected separately.

The five-wavelength generation/detection results are shown in Fig. 1.5a. Detection Stokes beams corresponding to THz waves of 1.34, 1.48, 1.68, 1.83, and 1.98 THz were observed. The wavelengths

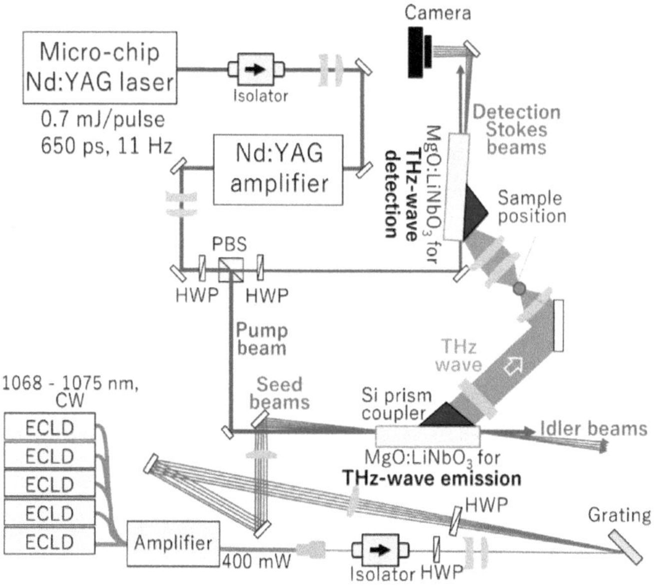

Figure 1.4 Experimental setup of the high-speed terahertz (THz) spectroscopic imaging system using a multi-wavelength injection-seeded THz parametric generator (is-TPG).

of the detection Stokes beams coincided with the wavelengths of the respective seed beams.

In general, THz detectors (such as pyroelectric detectors) detect wavelengths in groups, which limits their ability to obtain spectral information. In contrast, with the proposed approach, each wavelength can be spatially separated and detected. Notably, interference fringes appeared in the detection Stokes beams due to the cover glass of the sensor; however, this did not affect the measurements. Figure 1.5b shows the dynamic range of the spectroscopic system. THz-wave attenuators (TFA-4; Microtech Instruments Inc., Eugene, OR, USA) were inserted and attenuated down to −60 dB to measure the change in detected THz-wave intensity.

All five wavelengths decayed with THz-wave attenuation, and a dynamic range of about six orders of magnitude was obtained. The dynamic range of the single-wavelength generation/detection was more than seven orders of magnitude using the same pump intensity and detector, which is better than that of the multi-wavelength

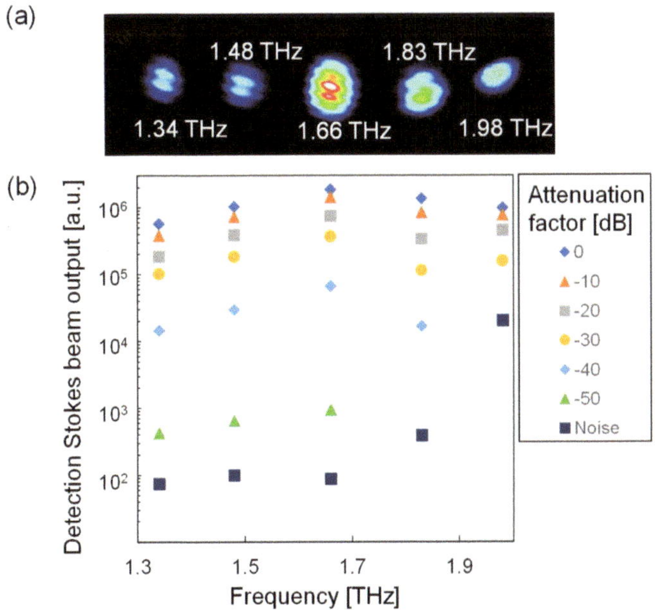

Figure 1.5 (a) Detection Stokes beams of five THz wavelengths captured by a complementary metal oxide semiconductor (CMOS) camera. (b) Dynamic range of the spectroscopic system.

configuration. This is because the energy is divided among the wavelengths in the multi-wavelength case, each of which is weaker than the single-wavelength case. The broadband THz parametric generation (TPG) that occurs in the high-frequency region (above 1.8 THz) effectively becomes noise that reduces the dynamic range in this area.

One-shot spectroscopic measurements of three saccharides were obtained using this system. Pellets (40% maltose, glucose, and lactose, and 60% polyethylene) were placed in the THz-wave optical path, and spectroscopic measurements were recorded. The central frequency of the absorption spectrum is approximately 1.62–2.02 THz for maltose, 1.44–2.09 THz for glucose, and 1.36–1.82 THz for lactose, according to a THz database [14] (blue line under each image in Fig. 1.6). Five frequencies, of 1.36, 1.48, 1.64, 1.82, and 2.00 THz, in the multi-wavelength is-TPG were selected to match these reference frequencies. The measurement results are shown in

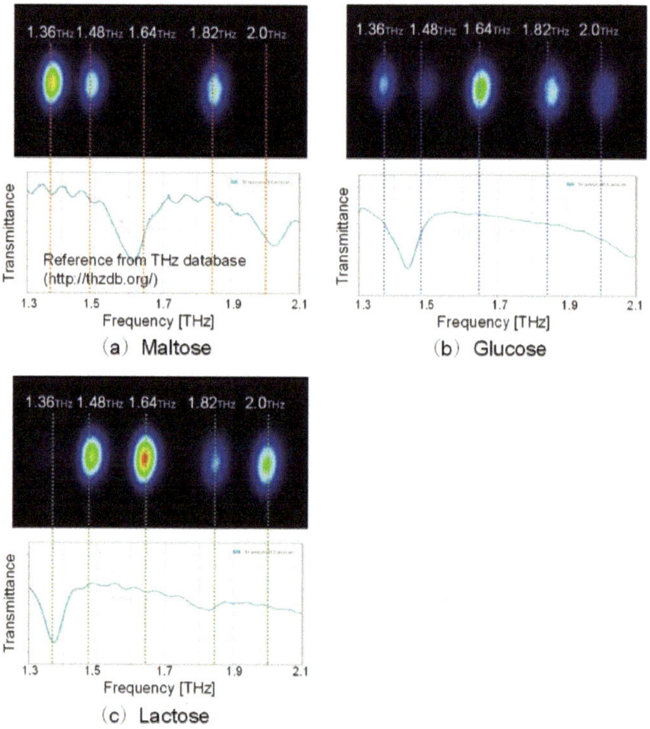

Figure 1.6 Results of real-time one-shot spectroscopy and the reference spectrum from a THz database (blue line under each image). Three saccharides were identified in real-time via is-TPG five-wavelength generation.

Fig. 1.6. The attenuation of the THz waves matched the absorption spectra of the reagents; thus, saccharides could be identified using one-shot multi-wavelength is-TPG. The measurements can be performed at a frame rate equal to the laser's repetition rate (11 Hz), thus enabling real-time reagent identification.

Finally, high-speed THz spectroscopic imaging was performed. Three saccharides, in the proportions measured by real-time spectroscopy, were placed in an express mail service (EMS) envelope, as shown in Fig. 1.7a. Multi-spectral images were obtained at five wavelengths (1.36, 1.48, 1.64, 1.83, and 2.00 THz) to observe the absorption peaks of the saccharides. Figure 1.6b shows the results for each saccharide, obtained by component spatial pattern analysis

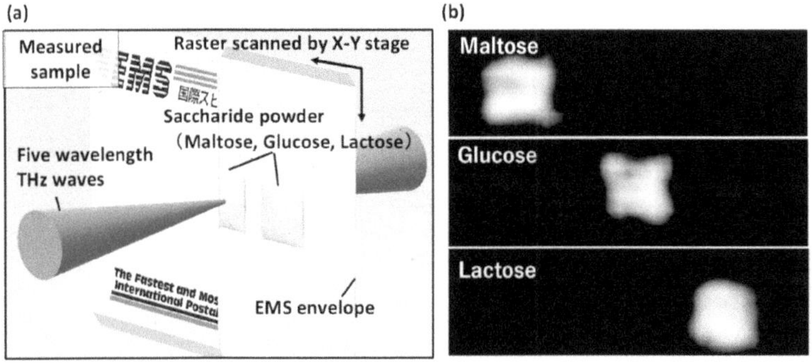

Figure 1.7 (a) Measured sample: three saccharide powders (maltose, glucose, and lactose) were placed in an express mail service envelope. (b) Multi-wavelength is-TPG spectroscopic imaging of the samples (image size: 48×16 pixels; step size: 1 mm).

of multi-spectral images [15]. The measurement time was shortened more than five-fold using this system.

1.4 Rapidly Wavelength-Switchable THz Parametric Generator

(Kodo Kawase, Kosuke Murate)

Multi-wavelength is-TPGs have some limitations, such as "parametric gain competition" among THz wavelengths and the need to match the number of ECDLs used for injection seeding with the number of THz wavelengths.

Thus, a real-time THz spectroscopic system involving rapid switching of the is-TPG wavelength was developed. In conventional is-TPGs, a commercially available ECDL is used as the seed source; however, this does not allow for rapid changes in the THz wavelength, resulting in slow tuning speeds. Therefore, an ECDL was developed, in which the wavelength could be rapidly switched by a digital micromirror device (DMD) [16, 17]. A DMD is a micro-electrical-mechanical device in which micromirrors are arranged in an array and each mirror can be independently controlled.

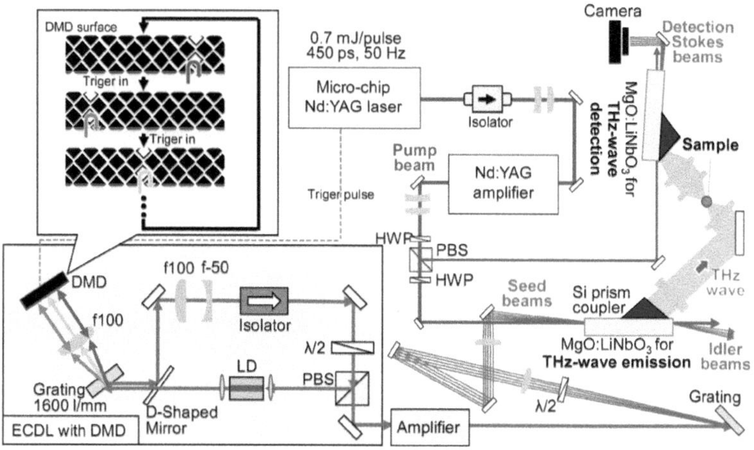

Figure 1.8 Experimental setup of a rapidly wavelength-switchable injection-seeded THz parametric generator using an external-cavity diode laser (ECDL) with a digital micromirror device (DMD).

Figure 1.8 shows the experimental setup of a rapidly tunable is-TPG that uses our developed ECDL (lower left part of the figure). A ring cavity was used as the resonator to suppress spatial hole burning and achieve stable oscillations. Furthermore, the ring cavity allowed for amplified spontaneous emission (ASE)-free high signal-to-noise output, as there was no ASE from the laser diode (OE-1040TA; Spectra Quest Lab, Inc., Chiba, Japan). Only the wavelength-selected beam can be output from the apparatus. In general, an ECDL selects a wavelength based on the angle of a mirror or diffraction grating; a DMD was used for this purpose in our system. The ASE from the laser diode was diffracted by the grating and focused onto the DMD. As shown in the upper left part of Fig. 1.8, when diffracted light is input into the DMD, the oscillation wavelength can be selected by tilting the micromirrors positioned according to the desired wavelength. The wavelength-selected beam was reflected downward slightly, into the plane of the diagram, and reflected by a D-shaped mirror to a different direction than that of the incident beam. Using a polarization beam splitter as the output mirror, the feedback rate was variable, such that the optimum laser power could be obtained.

The pump beam from the microchip laser and seed beam from the ECDL were input into the $LiNbO_3$ crystal to generate a THz-wave (linewidth < 4 GHz); 11 mJ of the pump beam was injected into the $LiNbO_3$ crystal for generation and 3 mJ was used for detection. The seed beam was amplified to 300 mW and injected into the generation crystal. The achromatic optical setup, consisting of a diffraction grating and pair of lenses in the seed beam path, ensures that the injection angle of the seed beam changes at each wavelength to automatically satisfy the phase-matching condition. The generated THz-wave was then input into the $LiNbO_3$ crystal after passing through the sample. During the detection process, the THz-wave was up-converted to a detection Stokes beam, which was captured by an NIR camera synchronized with the μ-chip laser. Information on all wavelengths can be obtained in one shot by adjusting the exposure time of the NIR camera according to the number of wavelengths.

Figure 1.9a shows the wavelength dependence of the power output from the developed ECDL. The tunable range was approximately 24 nm (1,065–1,089 nm) and had a flat output of 1,065~1,082 nm. The upper axis shows the corresponding frequency (in THz) when the ECDL was used as a seed source for the is-TPG. The ECDL was triggered by a signal from a function generator; stable wavelength-switching was confirmed up to 6.55 kHz.

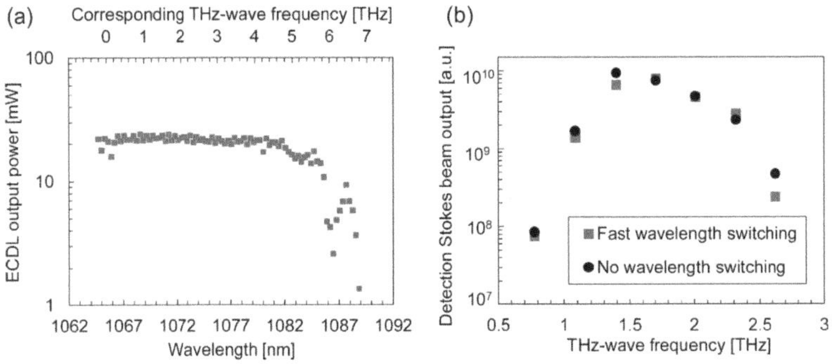

Figure 1.9 (a) Tunability of an ECDL using a DMD; single-mode oscillation was confirmed. The upper axis shows the corresponding frequency (in THz) when the laser is used as a seed source for an injection-seeded THz-wave parametric generator. (b) Frequency tunability of the is-TPG with (solid black line) and without (dashed orange line) fast wavelength-switching.

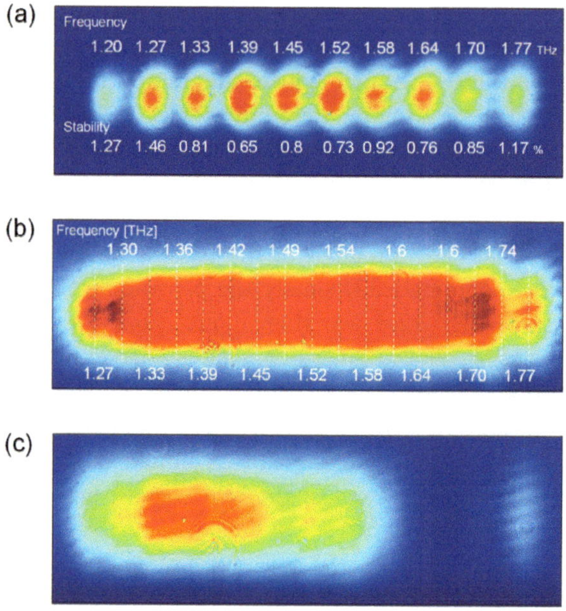

Figure 1.10 Detection Stokes beams when (a) switching among 10 wavelengths (1.20–1.77 THz) and (b) 17 wavelengths (1.27–1.77 THz). (c) Maltose was inserted into the THz-wave path while switching among the 17 wavelengths. The interference fringes shown in (b) and (c) were caused by the coverglass of the CMOS sensor of the near-infrared camera.

The tunability of the is-TPG with a developed ECDL was compared during fast switching among seven wavelengths at 50 Hz. Figure 1.9b shows the tunability of the is-TPG when tuning the THz wavelength in two ways. The dashed line shows the result of sweeping each wavelength by the conventional method using a commercial ECDL. The solid line shows the result of fast wavelength-switching; its tunable range was 0.78–2.62 THz, the same as that without fast switching.

The next demonstration involved switching among 10, and then 17, wavelengths. A 10-wavelength THz-wave (1.20–1.77 THz) was generated and detected, as shown in Fig. 1.10a. The average stability of the 10 wavelengths was 0.95%, which was almost the same as the 0.86% obtained for single-wavelength THz-wave generation using the same system. This indicates that fast wavelength-switching does not affect stability. Figure 1.10b shows the output in the case with 17 wavelengths. Due to the exposure time of the NIR camera used, 17 was the maximum number of wavelengths that could be detected in a single frame. Although the detection Stokes beams were not spatially separated by wavelength, we confirmed wavelength switching for these 17 wavelengths. In addition, Fig. 1.10c shows the results obtained when a reagent (maltose) was inserted into the THz-wave path. Spectral measurement was possible even when the detection Stokes beams were not spatially separated, because the THz waves were absorbed continuously according to the fingerprint spectrum. The absorption spectra were consistent with those of maltose, thus confirming that this system is capable of accurate spectral measurements.

1.5 Backward THz-Wave Parametric Oscillation

(Hiroaki Minamide, Kouji Nawata)

Backward optical parametric oscillation (BW-OPO) was proposed by Harris at Stafford University in 1966 [18]. Since then, several researchers have confirmed the backward optical parametric oscillation [19, 20]. A study was conducted on a more specific realization of the BW-OPO in the THz-wave region, wherein theoretical [21] and experimental investigations have been conducted [22, 23]. However, cases wherein BW-OPO has been realized simply by pumping a monochromatic laser into a nonlinear optical crystal, or cases wherein it has been realized as a wavelength-tunable source, have not been reported. The reasons for this are discussed as follows:

First, to realize the BW-OPO, special phase-matching conditions must be satisfied. In three-wave mixing, the momentum of either the idler or signal beam must exceed that of the pump beam. In this case, backward generation was realized.

Second, no phase-matching condition other than this process should simultaneously exist; if it does, the oscillation threshold of the BW-OPO must be minimum compared with the other nonlinear processes.

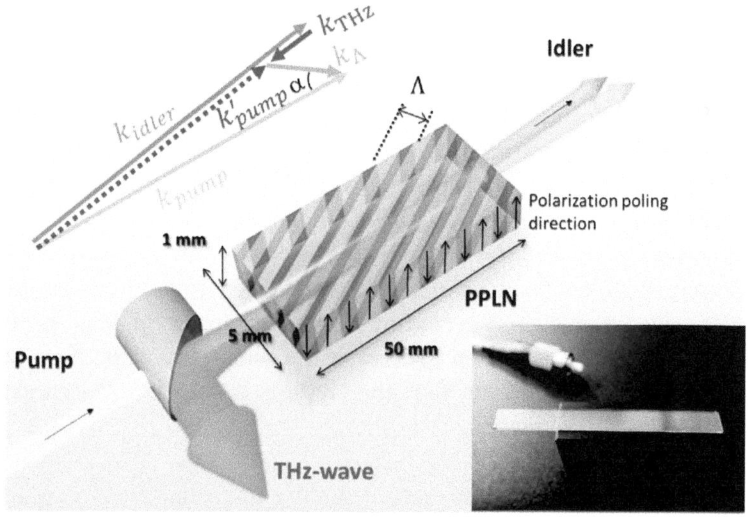

Figure 1.11 Optical setup of BW-TPO with vector momentum conservation and a photograph of a periodically poled LN. The parabolic mirror for THz-wave coupling has an aperture for passage of the pump beam. The left-upper figure is a phase-matching condition, while the right-bottom photograph is a PPLN.

We realized backward THz-wave parametric oscillation (BW-TPO) for the first time [24], which is ascribed to the discovery of a conversion efficiency enhancement in our previous is-TPG studies [6, 10, 25]. Hence, the threshold of the BW-TPO can be reduced. Notably, a few other phase-matching conditions, except for the established BW-TPO, were present, which was the second reason for this finding.

Figure 1.11 shows a schematic illustration of BW-TPO. The features of the BW-TPO are discussed. The proposed system is a simple optical system with no resonator structure and achieves terahertz-wave emission using a pump beam. The frequency of the system can be easily tuned by rotating the crystal or changing the incident angle of the pump beam. Furthermore, the proposed system is applicable to the sub-THz frequency region below 1 THz, which is beneficial for nondestructive testing. The cascaded TPO can be generated, exceeding the conversion efficiency limit determined using the Manley–Rowe relations, and quasi-collinear phase matching can be established, providing high conversion efficiency without any back conversion.

A major feature of BW-TPO is that a virtual pump beam vector can be constructed using the pump beam and lattice vectors of the PPLN crystal, as shown by the momentum vectors in Fig. 1.11. In this figure, the virtual pump beam, idler beam, and backward THz-wave vectors create a collinear phase-matching condition. The newly discovered condition can be modified by varying the angle between the pump beam and the lattice vector, which eventually changes the THz-wave frequency.

Figure 1.12 shows a plot of the THz-wave frequency and the angle between the pump beam and lattice vectors. The results of the experiment were consistent with those obtained using the new phase-matching condition model. The PPLN crystal has a rectangular aperture with a size of 1 mm × 5 mm, which limits its frequency tunability. However, continuous and wider frequency

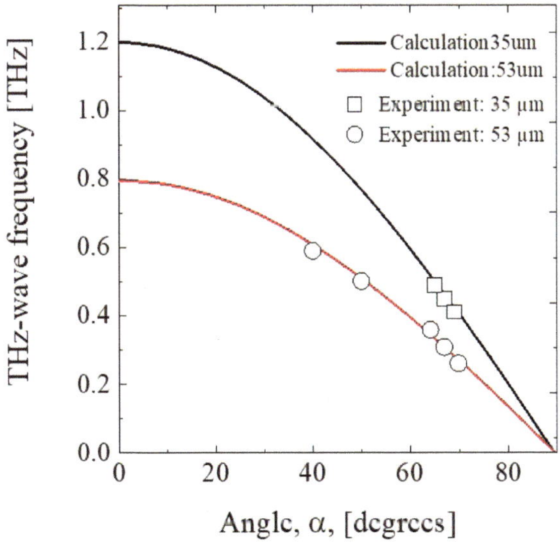

Figure 1.12 THz-wave frequency as a function of the angle between the pump beam and grating vectors, α. The solid lines were calculated. The plots are experimentally measured data. Squares and circles are for PPLN pitches of 35 and 53 μm, respectively.

tuning is possible using a disk-shaped PPLN that covers the entire sub-THz-wave region.

With respect to the output, BW-TPO is advantageous in terms of the realization of a cascaded OPO, as shown in Fig. 1.13, where the first-order idler beam and THz wave are generated from the

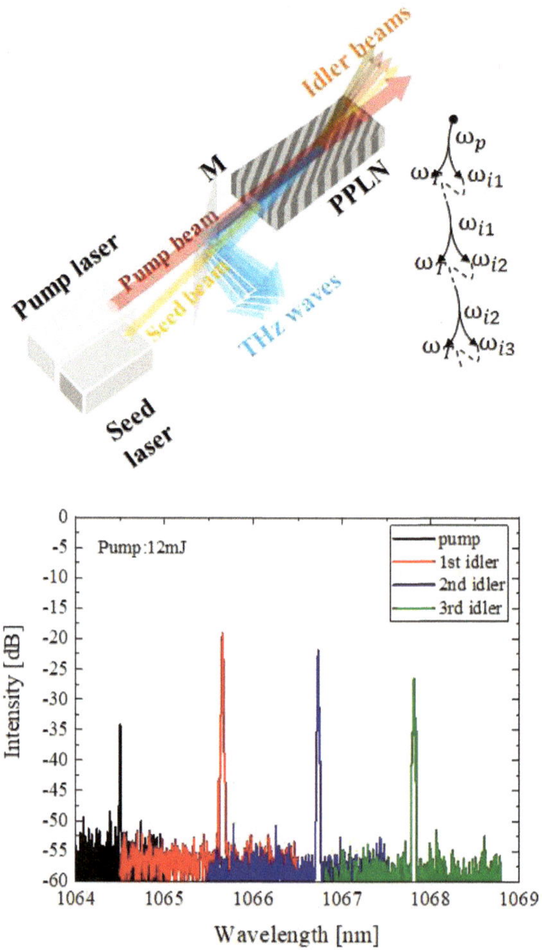

Figure 1.13 Schematic of cascaded backward THz-wave oscillation and experimental results of the wavelengths of idler beams generated during cascading.

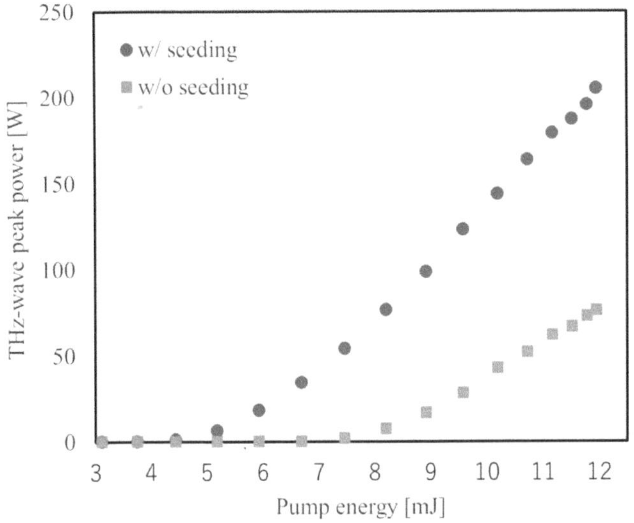

Figure 1.14 Experimental peak power of THz-wave output as a function of input pump pulse energy. Red circles represent results with injection seeding to the first idler beam, while orange squares represent results without injection seeding.

pump beam, the second-order idler beam, and the THz wave are sequentially generated from the first-order idler beam, and continuous optical parametric downconversion is similarly realized. This enables the THz-wave output to exceed the upper limit in a single conversion process, as determined using the Manley–Rowe relations, thus enabling high-output power generation. Furthermore, by externally injecting a seed beam into the first-order idler beam, the initial oscillation threshold can be reduced [26], creating conditions that effectively induce the cascade process occurring effectively. The experimental results are presented in Fig. 1.13. The experiment generated a third-order idler beam, with an input pump pulse energy of 12 mJ.

Figure 1.14 shows the experimental results of the THz-wave output as a function of the input energy of the pump beam. Highly efficient conversion occurred via injection seeding. A THz-wave output peak power exceeding 205 W was achieved at an excitation pulse energy of 12 mJ. The peak power was similar to the gyrotron-class output power [27]. With or without injection seeding, the oscillation threshold can be reduced by ~39% while substantially improving the slope efficiency, resulting in an output power enhancement of ~2.7 times.

In terms of applications, the developed high-power THz-wave source can expand the range of nondestructive inspection targets, despite the relatively large THz-wave absorption coefficient of the materials. Similarly, the THz-wave source can be applied to relatively thick samples and is expected to fulfill the needs of various users.

Next, a demonstration of THz-wave transmission imaging using the BW-TPO is presented. In the imaging system, the emitted THz waves were collimated, focused using a lens with a focal length of

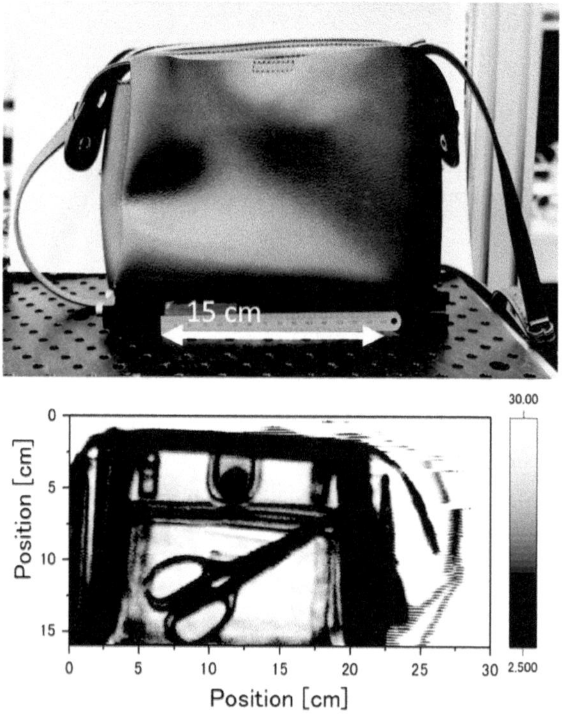

Figure 1.15 Imaging experiment using the output of a backward terahertz-wave parametric oscillator. The sample is an artificial leather bag with a middle partition. The results were obtained with scissors inside. The frequency of the irradiated THz wave was ~0.3 THz.

100 mm, and irradiated onto the sample. Another lens refocuses the THz waves transmitted through the sample onto an SBD detector (ACST). In this case, a lens with a focal length of 50 mm was placed 100 mm behind the sample to focus THz waves onto the detector. As the lens surface was not coated with an anti-reflective coating, only two lenses were used to minimize Fresnel loss.

Figure 1.15 shows the measurement results of scissors in a bag made of artificial leather. The bag contained several leather compartments inside and outside to hold the small objects. The THz wave penetrated the artificial leather well, and the inside of the bag could be clearly observed with high resolution. Consequently, the performance of BW-TPO is expected to be versatile for users.

References

1. K. Kawase, J. Shikata, K. Imai, and H. Ito, *Appl. Phys. Lett.* **78**, 2819–2821 (2001).
2. J. M. Yarborough, S. S. Sussman, H. E. Purhoff, R. H. Pantell and B. C. Johnson, *Appl. Phys. Lett.* **15**, 102–105 (1969).
3. K. Kawase, M. Sato, T. Taniuchi, and H. Ito, *Appl. Phys. Lett.* **68**, 2483–2485 (1996).
4. Y. Takida, J. Shikata, K. Nawata, Y. Tokizane, Z. Han, M. Koyama, T. Notake, S. Hayashi, and H. Minamide, *Phys. Rev. A* **93**, 4–7 (2016).
5. K. Kawase, H. Minamide, K. Imai, J. I. Shikata, and H. Ito, *Appl. Phys. Lett.* **80**, 195–197 (2002).
6. S. Hayashi, K. Nawata, T. Taira, J. Shikata, K. Kawase, and H. Minamide, *Sci. Rep.* **4**, 5045 (2014).
7. Y. Takida and H. Minamide, *Proc. SPIE*, **10210**, 102100W.
8. Y. Moriguchi, Y. Tokizane, Y. Takida, K. Nawata, T. Eno, S. Nagano, and H. Minamide, *Appl. Phys. Lett.* **113**, 121103 (2018).
9. A. de Bernabé, C. Prieto, and A. de Andrés, *J. Appl. Phys.* **79**, 143–148 (1996).
10. H. Minamide, S. Hayashi, K. Nawata, T. Taira, J. Shikata, and K. Kawase, *J. Infrared, Millimeter, Terahertz Waves* **35**, 25–37 (2014).
11. K. Kawase, J. Shikata, K. Imai, and H. Ito, *Appl. Phys. Lett.* **78**, 2819–2821 (2001).
12. K. Murate and K. Kawase *J. Appl. Phys.* **124**, 160901 (2018).

13. K. Kawase, M. Sato, T. Taniuchi, and H. Ito, *Appl. Phys. Lett.* **68**, 2483–2485 (1996).
14. THz-database. Available at: http://thzdb.org/.
15. K. Murate, S. Hayashi, and K. Kawase, *Appl. Phys. Express* **10**, 032401 (2017).
16. S. Mine, K. Kawase, and K. Murate, *Appl. Opt.* **60**, 1953–1957 (2021).
17. S. Mine, K. Kawase, and K. Murate, *Opt. Lett.* **46**, 2618–2621 (2021).
18. S. E. Harris, *Appl. Phys. Lett.* **9**, 114–116 (1966).
19. C. Canalias and V. Pasiskevicius, *Nat. Photonics* **1**, 459–463 (2007).
20. C. Liljestrand, A. Zukauskas, V. Pasiskevicius, and C. Canalias, *Opt. Lett.* **42**, 2435–2438 (2017).
21. Y. J. Ding and J. B. Khurgin, *Opt. Commun.* **148**, 105–109 (1998).
22. Y. J. Ding and W. Shi, *IEEE J. Sel. Top. Quantum Electron.* **12**, 352–359 (2006).
23. N. E. Yu, C. Jung, C. S. Kee, Y. L. Lee, B. A. Yu, D. K. Ko, and J. Lee, *Jpn. J. Appl. Phys.* **46**, 1501–1504 (2007).
24. K. Nawata, Y. Tokizane, Y. Takida, and H. Minamide, *Sci. Rep.* **9**, 726 (2019).
25. K. Nawata, S. Hayashi, H. Ishizuki, K. Murate, K. Imayama, Y. Takida, V. Yahia, T. Taira, K. Kawase, and H. Minamide, *IEEE Trans. Terahertz Sci. Technol.* **7**, 617–620 (2017).
26. Y. Takida, K. Nawata, and H. Minamide, *APL Photonics* **5**, 061301 (2020).
27. T. Idehara and S. P. Sabchevski, *J. Infrared, Millimeter, Terahertz Waves* **38**, 62–86 (2017).

Chapter 2

Terahertz Wave Emission with Photoconductive Antennas

Mona Jarrahi[a] **and Masahiko Tani**[b]

[a] *Electrical and Computer Engineering Department, University of California Los Angeles, 420 Westwood Plaza, Los Angeles, California, 90095, USA*

[b] *Research Center for Development of Far-Infrared Region, University of Fukui, 3-9-1 Bunkyo, Fukui 910-8507, Japan*

mjarrahi@ucla.edu, tani@fir.u-fukui.ac.jp

Photoconductive antennas are used for the generation and detection of pulsed and continuous-wave (CW) terahertz waves using photoconductive modulation of semiconductor carriers with pulsed and CW laser excitation, respectively. They are the most commonly used emitters and detectors for terahertz time-domain spectroscopy. In this chapter, the basic operation principles and properties of photoconductive antennas are described in relation to the material and structural design, as well as the pump laser. Improvement of the efficiency of photoconductive antennas is discussed with special attention to the enhancement effects enabled by plasmonic nanostructures.

Advances in Terahertz Source Technologies

Edited by Gun-Sik Park, Masahiko Tani, Jae-Sung Rieh, and Sang Yoon Park

ISBN 978-981-4968-89-8 (Hardcover), 978-1-003-45967-5 (eBook)

www.jennystanford.com

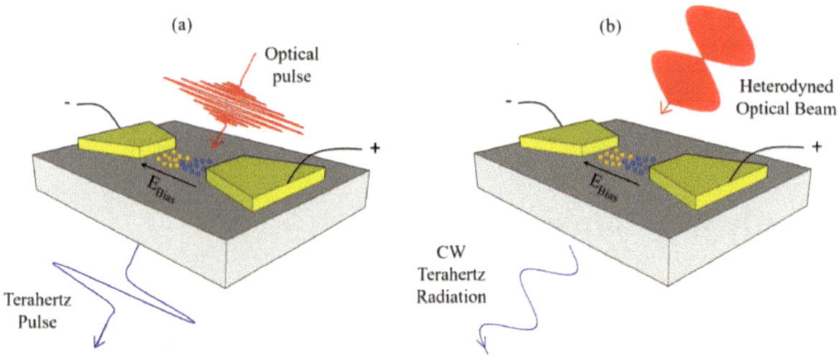

Figure 2.1 PCA operation principles for (a) pulsed and (b) CW terahertz generation. Photogenerated electrons/holes are shown in blue/yellow.

2.1 Operation Principles of Photoconductive Antennas

Since their first demonstration by David Auston [1], photoconductive antennas (PCAs) have emerged as the most commonly used terahertz sources for both pulsed and CW operations [1–19]. A PCA generally consists of a terahertz antenna fabricated on a photoconductive semiconductor substrate. In order to generate terahertz pulses, the PCA is illuminated with femtosecond optical pulses. When the antenna is biased, the induced electric field inside the semiconductor substrate drifts the photogenerated electron–hole pairs to feed the antenna with a pulsed photocurrent. This photocurrent generally has a sub-picosecond pulse width, corresponding to a broad terahertz frequency range. In order to generate CW terahertz waves, the PCA is illuminated with a heterodyning optical beam with a terahertz beat frequency. When the antenna is biased, the induced electric field inside the semiconductor substrate drifts the photogenerated electron–hole pairs to feed the antenna with a CW terahertz photocurrent. The operation principles of a PCA for pulsed and CW terahertz generation are illustrated in Figs. 2.1a,b.

2.2 Design Considerations of Photoconductive Antennas

The choices of photoconductive material, antenna structure, and pump laser are the most important design considerations for a PCA, which are discussed in the following sections.

2.2.1 Photoconductive Material

Carrier lifetime, mobility, and resistivity are the most important specifications of a photoconductive material that significantly impact the operation of a PCA-based terahertz source. While pulsed terahertz generation can be achieved by the use of both short- and long-carrier-lifetime photoconductors, efficient CW terahertz generation generally requires a sub-picosecond carrier lifetime. In the early days of the development of PCAs, ion-implanted silicon-on-sapphire (SOS) wafers were used as the photoconductive material [13]. Later, it was replaced with low-temperature grown GaAs (LT-GaAs), which is the most commonly used short-carrier-lifetime photoconductor for operation at optical wavelengths below 870 nm. LT-GaAs is grown at temperatures ranging from 200 to 400°C and annealed in an As ambient over a specific pressure condition to obtain a sub-picosecond recombination lifetime and high resistivity by introducing high-density deep-level defects [20, 21] and As precipitates [22]. ErAs:InGaAs and InAlAs/InGaAs multi-layer heterostructures are the most commonly used short-carrier-lifetime photoconductors for operation at optical wavelengths exceeding 1550 nm [23–27]. The introduced trap centers inside these photoconductors not only reduce the carrier recombination lifetime but also increase the dark resistivity of the photoconductor.

While the use of short-carrier-lifetime photoconductors ensures ultrafast carrier dynamics in PCAs, the quantum efficiency of PCAs utilizing short-carrier-lifetime photoconductors is degraded by the carrier mobility reduction as a result of excessive defect and trap centers in the photoconductor. As described in the following sections, the use of plasmonics-enhanced PCAs can eliminate the need for a short-carrier-lifetime photoconductor by mitigating the tradeoff between the high-quantum-efficiency and ultrafast operation of PCAs [28–32].

Apart from bulk semiconductors, 2D materials such as graphene and black phosphorus also offer broadband optical absorption and very high carrier mobility, which make them attractive for PCAs operating at a broad optical wavelength range [33–37]. However, the quantum efficiency of the demonstrated PCAs based on 2D materials has been limited by their low optical absorption due to the very small interaction length of the optical pump with these 2D materials [38–40].

2.2.2 Antenna Structure

Impedance, bandwidth, polarization, gain, and radiation pattern are the most important specifications of the antenna that significantly impact the operation of a PCA-based terahertz source. The complex impedance of the antenna and photoconductor should be chosen such that maximum power transfer is achieved from the optically induced photocurrent in the PCA active area to the antenna and free space, subsequently [41]. Since photoconductor contact electrodes exhibit capacitive parasitics, a positive imaginary antenna impedance is generally desired for maximum power efficiency. The geometry of the electrical bias line can be specifically chosen to provide the desired positive imaginary impedance [42].

Broadband antennas such as Hertzian dipole, bow-tie, log-periodic, or log spiral antennas are generally used for pulsed terahertz generation. In general, the antenna efficiency at lower frequencies increases with the antenna length or size. This tendency was well demonstrated in the report by Miyamaru et al. [43], where dipole-type photoconductive antennas with various geometrical parameters were investigated. For example, it was found that the peak terahertz signal amplitude was enhanced about 6 times when increasing the dipole antenna length from 20 to 200 μm, while the emission spectral peak was shifted from 0.5 to 0.2 THz. Narrowband, resonant antennas—such as resonant dipole, slot, and horn antennas—are better suited for CW terahertz generation. However, the use of narrowband antennas limits the frequency tuning range of CW terahertz sources. Ultimately, the bandwidth requirements of the application will determine the preferred choice of antenna. For example, while spectroscopic applications generally demand broad radiation bandwidths exceeding several THz, high data rate communications can be accomplished over much smaller bandwidths of a few tens of GHz.

The choice of polarization is also determined by the application requirements and the polarization sensitivity of the receiver/detector used in combination with the PCA-based terahertz source. While spiral antennas provide circular polarization, bow-tie, dipole, and horn antennas offer linear polarization. For simultaneous detection of two orthogonal polarization components, PCA detectors with three [44] or four contacts [45] can be used. The gain and radiation pattern of a PCA-based source are determined by not only the choice of antenna but also the silicon lens that is often used to extract terahertz radiation out of the PCA. Since electromagnetic radiation has a higher tendency to go to the substrate, which has a higher dielectric constant than air, placing a PCA on a silicon lens enables efficient extraction of radiation. The geometry of the utilized silicon lens, which can be in a collimating, hyper-hemispherical, and hemispherical shape would modify the divergence angle of the radiation according to the application requirements [46].

2.2.3 Pump Laser

Femtosecond lasers providing optical pulses with full width at half maximum (FWHM) pulse width of 200 fs or less are most commonly used for pumping PCA-based pulsed terahertz sources. While Ti:sapphire mode-locked lasers are generally used for operation at 800 nm wavelengths, Ytterbium(Yb) and Erbium(Er)-doped mode-locked fiber lasers are generally used for operation at 1 and 1.55 μm wavelength ranges, respectively. These mode-locked fiber lasers are more attractive for practical applications in field settings due to their smaller size, lower cost, and higher stability and their compatibility with fiber optic components.

By pumping PCAs with multi-mode laser diodes or chaos laser systems, multi-mode terahertz emissions are obtained as a replica of the multi-mode optical spectra. Note that the multi-mode terahertz spectra can be obtained through the same terahertz time-domain systems that typically use femtosecond pump lasers. The terahertz wave is generated by photomixing of the multiple modes with random phases (without mode-locking) through a PCA-based source. The generated terahertz waves are detected by a PCA-based detector, which measured the cross-correlation between the generated terahertz emission and the multi-mode optical pump incident on the detector. The details of this technique and its recent progress are described in Chapter 5, "Photomixing THz Sources," of this book.

On the other hand, two CW lasers with a terahertz frequency difference are generally used for pumping PCAs or other photomixing diodes, such as unitraveling carrier photodiodes [47], for the generation of CW terahertz waves. Polarization and spatial mode of the two lasers should be aligned to efficiently generate an optical pump with a terahertz beat frequency. The use of CW lasers, instead of femtosecond mode-locked lasers, enables realizing lower cost and more compact terahertz systems. For example, integrated frequency comb lasers, as well as dual distributed feedback (DFB), distributed Bragg reflector (DBR), and vertical-external-cavity surface-emitting Lasers (VeCSEL) have been demonstrated, providing the terahertz beat frequency for PCA-based CW terahertz sources [48–51].

2.2.4 Sub-bandgap excitation of LT-GaAs-based Photoconductive antennas

Photoconductance of LT-GaAs with sub-bandgap optical excitation was observed [52], and the operation of LT-GaAs PCAs with sub-bandgap excitation has been demonstrated [53, 54]. The sub-bandgap excitation of LT-GaAs PCAs benefits from the availability of cost-effective and stable telecommunication wavelength (> 1 μm) femtosecond laser sources. In addition, operation of LT-GaAs PCAs with sub-bandgap excitation does not suffer from the low bias voltage and large Johnson noise limitations observed for PCAs with low bandgap photoconductive substrates, such as InGaAs. It should be noted that the induced photoconductive signal with sub-bandgap excitation is much lower than that of above bandgap (e.g., > 1.43 eV) excitation: LT-GaAs photoconductivity at 1.55 μm excitation was reported to be about 10% of that at 800 nm excitation

[53]. The photoconductive efficiency of above [55] and sub-bandgap [54] excitations can be improved by using a TM-polarized optical pump (the polarization direction is orthogonal to the boundary of the photoconductive material and contact) compared with a TE-polarized optical pump (the polarization direction is parallel to the boundary of the photoconductive material and contact). A factor of 2 and 3 enhancement of the photoconductive signal was observed when using a TM-polarized optical pump, compared with a TE-polarized optical pump, for an LT-GaAs PCA with a gap of 5 and 2 μm, respectively, by sub-bandgap excitation at 1560 nm [54].

It was also found that the sub-bandgap photoconductivity can be considerably enhanced by using PCAs with narrower gaps. Kataoka et al. [56] reported that the photoconductive signal of an LT-GaAs dipole PCA used as a terahertz detector was enhanced about 10 times by reducing the gap from 5 to 1.5 μm under a 1560 nm femtosecond laser excitation with a 60 fs pulse width and a 50 MHz repetition rate. The probe power (P) dependence of the terahertz photoconductive signal was well fit to a power scaling of $P^{1.35}$. It is noteworthy that the spectral noise stayed almost constant with the probe power below 15 mW.

The photoconductivity of sub-bandgap excitation of LT-GaAs is explained by the two-photon absorption originating from the second-order nonlinear optical susceptibility (the χ^2 process) and the two-step absorption mediated with the mid-gap states due to defects in LT-GaAs [53]. It is difficult to separate the two processes since they are correlated with each other. However, considering the sub-quadratic probe power dependence observed by Kataoka et al. [56] and the saturation properties at a low probe power level of 20–30 mW [54], it is reasonable to conclude that the latter process is dominant.

LT-GaAs has 1–1.5% excess As above stoichiometry and the main type of defect in LT-GaAs is As-anti-site (As_{Ga}), whose density is estimated to be $\sim 10^{19}$ cm^{-3} for annealed samples [57]. Escaño et al. carried out theoretical investigation of As_{Ga} defect properties by using DFT calculations and found that mid-gap states originating from As_{Ga} are located almost in the middle of the GaAs bandgap ($\sim$0.79 eV from the valence band) [58].

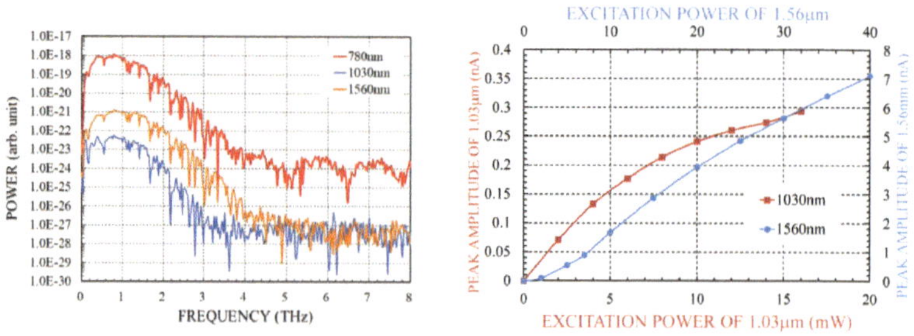

Figure 2.2 (a) Terahertz power spectra obtained by an LT-GaAs PCA-based source and detector at excitation wavelengths of 800, 1030, and 1560 nm. (b) Excitation power dependence at 1030 and 1560 nm wavelengths.

Recently, Kitahara et al. [59] reported the comparison of pulsed terahertz emission and detection at excitation wavelengths of 1030 and 1560 nm by using dipole PCAs fabricated on LT-GaAs substrates with a contact gap of 3.4 μm for both the emitter and detector [59]. Surprisingly, the peak terahertz signal at the 1560 nm excitation wavelength was about 5 times larger than that of the 1030 nm excitation wavelength under the same experimental conditions (16 mW pump and probe power levels, 32 V peak-to-peak bias voltage applied to the emitter). This indicates that the sub-bandgap emission and detection efficiency is 2.2 times larger at 1560 nm compared with 1030 nm. Figure 2.2a shows the observed terahertz power spectra for 1030 and 1560 nm excitations. The terahertz power spectrum obtained at 800 nm excitation wavelength under the same conditions is also shown for comparison. Figure 2.2b shows the excitation power dependence of the peak terahertz signal detected at 1030 and 1560 nm excitation wavelengths. The signal with the 1030 nm excitation saturates at a power of 15 mW, while the signal with the 1560 nm excitation does not show a significant saturation tendency even at 30 mW. This comparison indicates that a sub-bandgap excitation is mainly due to a two-step excitation mediated by the mid-gap states located in the middle of the GaAs bandgap due to the As_{Ga} defects. It is noteworthy that the terahertz power spectrum with the 800 nm excitation wavelength is larger than that of the 1560 nm excitation by 3 orders of magnitude but the noise level is also increased with the same order. Therefore, the spectral dynamic range obtained with the 1560 and 800 nm excitation wavelengths is almost the same.

2.3 Plasmonics-Enhanced Photoconductive Antennas

In order to achieve high optical-to-terahertz conversion efficiency, a large fraction of the photocarriers should drift from the photoconductive active region to the terahertz antenna in a sub-picosecond time scale. However, since the maximum carrier drift velocity is limited by scattering in the semiconductor lattice, only the photocarriers that are generated within approximately 100 nm distance from the antenna electrodes can have a sub-picosecond transit time to the antenna electrodes. Therefore, a small fraction of the photogenerated carriers contribute to terahertz generation, limiting the optical-to-terahertz conversion efficiency of conventional PCA sources.

To address this limitation, plasmonic light concentrators (Fig. 2.3a), plasmonic contact electrodes (Fig. 2.3b), plasmonic nanoantenna arrays (Fig. 2.3c), and plasmonic nanocavities (Fig. 2.3d) have been used to enhance the interaction between the optical pump and photoconductive material at the nanoscale and offer enhanced optical-to-terahertz conversion efficiencies [60, 61]. These optical-to-terahertz conversion enhancement techniques are discussed in the following sections.

2.3.1 PCAs Based on Plasmonic Light Concentrators

When an incident electromagnetic wave is coupled to surface plasmons—collective electron-plasma oscillations on the metal surface—surface plasmon waves can propagate along the dielectric-metal interface. To couple electromagnetic waves to surface plasmons, the electromagnetic wave vector along the metal surface should match the wave vector of the surface plasmons [62–65], which can be achieved by introducing sub-wavelength patterns in the metal structure. Hence, metallic nanostructures embedded in

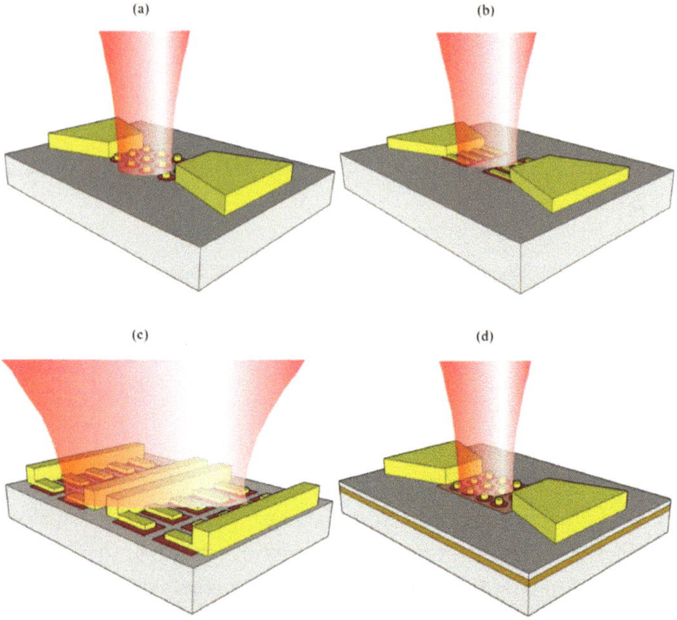

Figure 2.3 PCAs based (a) plasmonic light concentrators in the photoconductive active region, (b) plasmonic contact electrodes connected to the terahertz antenna, (c) plasmonic nanoantenna arrays, and (d) plasmonic nanocavities.

the photoconductor active area can be designed to concentrate light by exciting surface plasmon waves at the optical pump wavelength.

The use of plasmonic light concentrators is an effective way to improve the optical-to-terahertz conversion efficiency of PCAs [66–71]. By increasing the optical absorption in specific regions of the photoconductor, the number of photocarriers reaching the antenna electrodes in a sub-picosecond time scale can be increased, enhancing the optical-to-terahertz conversion efficiency.

Various PCA terahertz emitters with embedded plasmonic light concentrators have been demonstrated at different optical pump wavelengths using different photoconductive substrates. For example, 100% radiation power enhancement over 0.1–1.1 THz was reported [66] when using plasmonic nanorods in the active region of a PCA (Fig. 2.4a). Also, one order of magnitude terahertz field

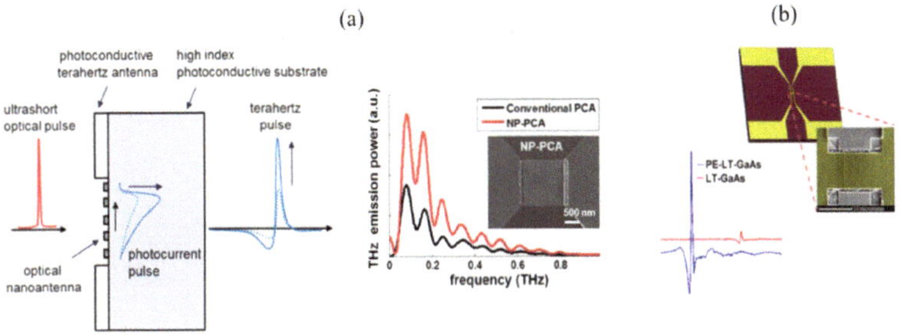

Figure 2.4 PCAs based on plasmonic light concentrators. (a) Schematic and SEM image of a PCA with plasmonic nanorods embedded in the photoconductor and its terahertz emission spectrum (red) compared with a similar PCA without plasmonic structures (black) [66]. (b) Schematic and SEM image of a PCA with a plasmonic nanoslit array and its radiated terahertz field (blue) compared with a similar PCA without plasmonic structures (red) [70].

enhancement was demonstrated [70] when embedding a plasmonic nanoslit array in a PCA active area (Fig. 2.4b).

2.3.2 PCAs Based on Plasmonic Contact Electrodes

The use of plasmonic contact electrodes is another effective way to improve the optical-to-terahertz conversion efficiency of PCAs. When the geometry of the contact electrodes is chosen to allow the excitation of surface plasmon waves at the optical pump wavelength, the optical pump beam coupled to the photoconductor will be tightly confined near the plasmonic contact electrodes. As a result, a large fraction of the photocarriers are generated in close proximity to the plasmonic contact electrodes. Since the plasmonic contact electrodes have a direct electrical connection to the terahertz antenna, the transport path length of the majority of the photocarriers to the terahertz antenna is significantly reduced compared to conventional PCAs without plasmonic contact electrodes, resulting in a significantly enhanced optical-to-terahertz conversion efficiency.

Various PCA terahertz emitters with different types of plasmonic contact electrodes have been demonstrated using different photo-conductors and terahertz antennas utilized for operation at different

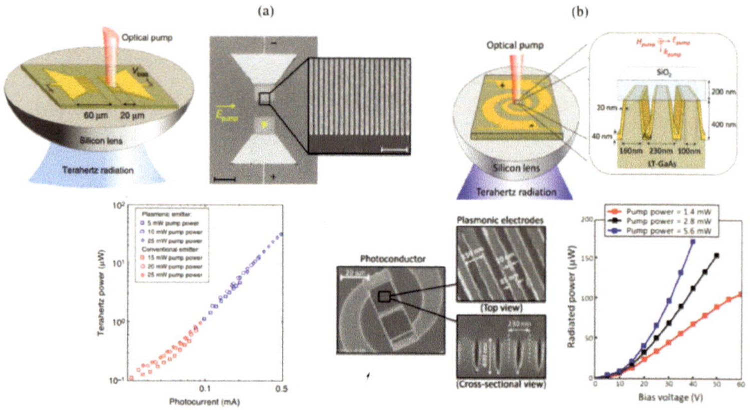

Figure 2.5 PCAs based on plasmonic contact electrodes. (a) Schematic and SEM image of a PCA with 2D plasmonic contact electrode gratings and its terahertz radiation power compared with a similar conventional PCA without any plasmonic structures [74]. (b) Schematic and SEM images of a PCA with 3D plasmonic contact electrode gratings and its terahertz radiation power at various optical pump power levels [77].

optical pump wavelengths and terahertz radiation bandwidths for both pulsed and CW operation [72–85]. For example, two- and three orders of magnitude enhancement in optical-to-terahertz conversion efficiency was demonstrated using PCAs with 2D and 3D plasmonic contact electrode gratings, respectively [74, 77], as shown in Figs. 2.5a,b.

2.3.3 PCAs Based on Plasmonic Nanoantenna Arrays

The use of plasmonic nanoantenna arrays is a very promising approach to enhance the radiation bandwidth and optical-to-terahertz conversion efficiency of PCAs, especially at high optical pump power levels.

Since nanoantenna arrays can be configured in a relatively large area and pumped with relatively large optical beam spot sizes, destructive thermal and carrier screening effects can be mitigated, enabling operation at high optical pump powers. In addition, the geometry of the nanoantenna arrays can be chosen to excite surface plasmon waves at the optical pump wavelength, enhancing

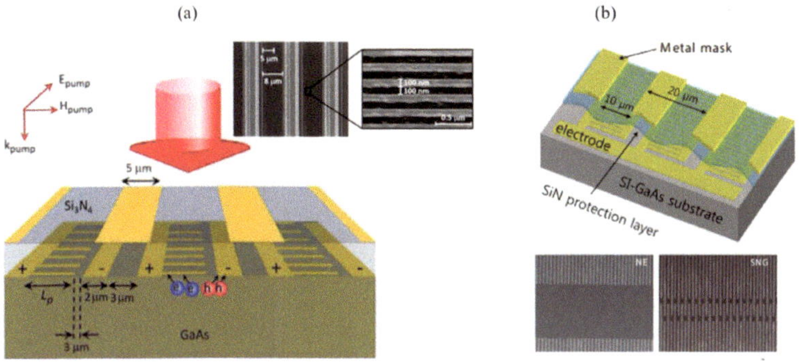

Figure 2.6 PCAs based on plasmonic nanoantenna arrays. (a) Schematic and SEM images of PCAs based on plasmonic nanoantenna arrays in the form of (a) uniform gap [86] and (b) shifted nanogap gratings [90].

optical-to-terahertz conversion efficiency by reducing the transport path length of a large fraction of the photogenerated carriers to the nanoantennas.

Several implementations of terahertz emitters based on plasmonic nanoantenna arrays have been demonstrated using different photoconductive substrates and nanoantenna structures, optimized for operation at various optical pump wavelengths [86–90]. For example, Figs. 2.6a,b show PCAs based on plasmonic nanoantenna arrays in the form of uniform gap [86] and shifted nanogap gratings [90], respectively. Benefitting from the large power handling capability of the PCAs based on plasmonic nanoantenna arrays, terahertz radiation powers exceeding 6 mW have been achieved [89]. In addition, the fabrication of plasmonic nanoantenna arrays on semiconductors with surface built-in electric fields has enabled high-efficiency optical-to-terahertz conversion without any need for a bias voltage [91–93].

2.3.4 PCAs Based on Plasmonic Nanocavities

The use of plasmonic nanocavities can further enhance the optical-to-terahertz conversion efficiency of PCAs by increasing the optical pump confinement and photocarrier concentration in close proximity to the terahertz radiating elements. For example, Fig. 2.7 shows a PCA based on a plasmonic nanocavity formed between a plasmonic

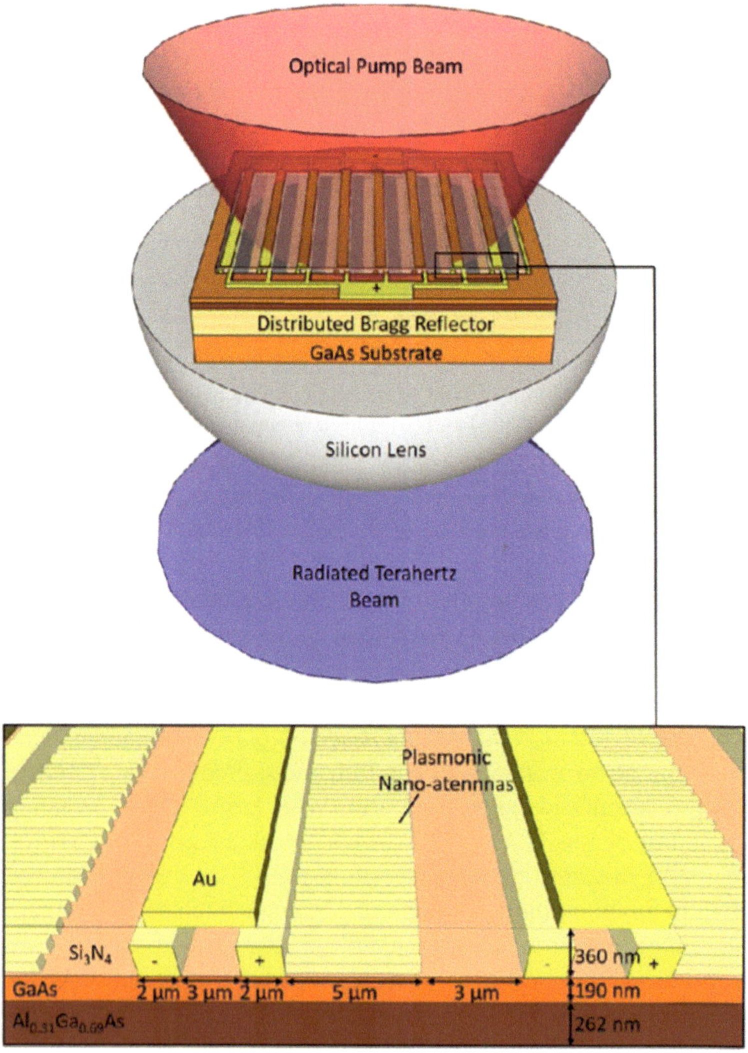

Figure 2.7 PCA based on a plasmonic nanocavity formed by a plasmonic nanoantenna array and a distributed Bragg reflector [94].

nanoantenna array and a DBR, offering an order of magnitude higher optical-to-terahertz conversion efficiency compared to a PCA with similar plasmonic nanoantennas fabricated on a substrate without the DBR [94].

2.4 Conclusion and Outlook

In concluding this chapter, we discuss the capabilities and limitations of PCAs and suggest future research directions for the development of higher-performance PCAs. Laser-driven terahertz emitting devices are characterized with (i) the efficiency of optical-to-terahertz power conversion, (ii) the terahertz output power, and (iii) the bandwidth or frequency tunability of the generated terahertz frequency spectrum. The efficiency of optical-to-terahertz power conversion has reached 7.5% by using three-dimensional plasmonic contact electrodes [77]. This is much higher than the efficiency of optical-to-terahertz power conversion observed for nonlinear optical terahertz emitters limited by the Manley–Rowe relations, where the upper limit of the quantum efficiency is determined by the energy ratio of the generated terahertz photon to the optical pump photon. The highest terahertz output power provided by PCA-based sources is limited by the damage threshold and saturation properties of the PCAs. The damage threshold of PCA-based terahertz sources is low compared to the terahertz emitters based on nonlinear optical processes since the pump optical power is absorbed in a thin layer of the photoconductive material. The saturation of terahertz emission from PCA-based sources originates from the screening of the bias electric field by the photoexcited carriers, and the available terahertz output power is limited by the static electromagnetic energy stored in the PCA under an applied bias field. Because of these properties, PCA-based terahertz sources are not suitable for the generation of a high terahertz peak field nor a high terahertz pulse energy by pumping it with a high optical pulse energy. However, by using large area and/or arrayed PCAs, it is possible to generate high terahertz power levels by preventing thermal breakdown and saturation effects. In the work of [75], 1.9 mW of 0.1–2 THz spectral power was obtained by pumping an array of logarithmic spiral PCAs with 320 mW optical power from a mode-locked Ti:sapphire laser operating at 800 nm wavelength.

The bandwidth of a PCA is not only limited by the carrier lifetime in the photoconductive material but also by the PCA parasitics and the onset time of photoconductance, which is determined by the pulse width of the pump laser. It should be noted that the ultrabroad band operation of PCAs based on compound semiconductors is limited by optical phonon absorptions, by which the PCA loses radiation power at the band of the optical phonons (Restrahlen band). To avoid optical phonon absorption, Si or Ge can be used as the photoconductive material. For example, a smooth ultrabroad band terahertz emission extending up to 70 THz was obtained by the excitation of a PCA implemented on an Au-implanted Ge substrate with 11 fs laser pulses with a central wavelength around 1100 nm [95].

Despite their limitations, PCAs will remain key components for terahertz imaging, spectroscopy, and sensing because of their high efficiency in the generation and detection of terahertz waves. Their efficiency is expected to be further enhanced through optimized photoconductive materials and structural designs, especially for the designs suitable for operation with the Yb- and Er-doped fiber lasers.

References

1. Auston, D. H., Cheung, K. P., and Smith, P. R. (1984). *Appl. Phys. Lett.*, **45**, 284–286.
2. Verghese, S., McIntosh, K. A., Calawa, S., Dinatale, W. F., Duerr, E. K., and Molvar, K. A. (1998). *Appl. Phys. Lett.*, **73**, 3824–3826.
3. Jepsen, P., Uhd, Jacobsen, R. H., and Keiding, S. R. (1996). *JOSA B*, **13**, 2424-2436.
4. Zhao, G., Schouten, R. N., van der Valk, N., Wenckebach, W. Th., and Planken, P. C. M. (2002). *Rev. Sci. Instrum.*, **73**, 1715–1719.
5. Shuji, M., Tani, M., and Sakai, K. (1997). *Appl. Phys. Lett.*, **70**, 559–561.
6. Mona J., and Lee, T. H. (2008). "High-power tunable terahertz generation based on photoconductive antenna arrays." *2008 IEEE MTT-S International Microwave Symposium Digest*. IEEE, 2008.
7. Peytavit, E., Lepilliet, S., Hindle, F., Coinon, C., Akalin, T., Ducournau, G., Mouret, G., and Lampin, J.-F. (2011). *Appl. Phys. Lett.*, **99**, 223508.
8. Roehle, H., Dietz, R. J. B., Hensel, H. J., Böttcher, J., Künzel, H., Stanze, D., Schell, M., and Sartorius, B. (2010). *Opt. Express*, **18**, 2296–2301.
9. Taylor, Z. D., Brown, E. R., Bjarnason, J. E., Hanson, M. P., and Gossard, A. C. (2006). *Opt. Lett.*, **31**, 1729–1731.

10. Jarrahi, M. (2009). *IEEE Photonics Technol. Lett.*, **21**, 830–832.
11. Sascha, P., Mittendorff, M., Lu, H., Weber, H. B., Winnerl, S., and Gossard, A. C. (2012). *Appl. Phys. Lett.*, **101**, 101105.
12. Sascha, P., Döhler, G. H., Malzer, S., Wang, L. J., and Gossard, A. C. (2011). *J. Appl. Phys.*, **109**, 061301.
13. Smith, P. R., Auston, D. H., and Nuss, M. C. (1988). *IEEE J. Quantum Electronics*, **24**, 255–260.
14. Piao, Z., Tani, M., and Sakai, K. (2000). *Jpn. J. Appl. Phys.*, **39**, 96–100.
15. Brown, E. R. (2003). *Terahertz Sensing Technology: Volume 1: Electronic Devices and Advanced Systems Technology*. Ed. Woolard, D. L., Loerop, W. R., and Shur M., THz generation by photomixing in ultrafast photoconductors. (World Scientific Publishing Co., Inc.), pp. 147–195.
16. Brown, E. R., Smith, F. W., and McIntosh, K. A. (1993). *J. Appl. Phys.*, **73**, 1480–1484.
17. Gregory, I. S., Baker, C., Tribe, W. R., Bradley, I. V., Evans, M. J., Linfield, E. H., Davies, A. G., Missous, M. (2005). *IEEE J. Quantum Electronics*, **41**, 717–728.
18. Yang, S.-H., and Jarrahi, M. (2020). *Opt. Photonics News*, **31**, 36–43.
19. Bjarnason, J. E., Chan, T. L. J., Lee, A. W. M., and Brown, E. R. (2004). *Appl. Phys. Lett.*, **85**, 3983–3985.
20. Warren, A. C., Katzenellenbogen, N., Grischkowsky, D., and Woodall, J. M. (1991). *Appl. Phys. Lett.*, **58**, 1512–1514.
21. Shen, Y. C., Upadhya, P. C., Beere, H. E., and Linfield, E. H. (2004). *Appl. Phys. Lett.*, **85**, 164–166.
22. Otsuka, N., Tasaki, Y., Yamada, T., and Suda, A. (2000). *J. Appl. Phys.*, **88**, 6016–6020.
23. Driscoll, D. C., Hanson, M. P., and Gossard, A. C. (2005). *Appl. Phys. Lett.*, **86**, 051908.
24. Ospald, F., Maryenko, D., von Klitzing, K., Driscoll, D. C., Hanson, M. P., Lu, H., Gossard, A. C., and Smet, J. H. (2008). *Appl. Phys. Lett.*, **92**, 131117.
25. Sukhotin, M., and Brown, E. R. (2002). *Appl. Phys. Lett.*, **82**, 3116–3118.
26. Dietz, R. J. B., Gerhard, M., Stanze, D., Koch, M., Sartorius, B., and Schell, M. (2011). *Opt. Express*, **19**, 25911–25917.
27. Dietz, R. J. B., Globisch, B., Gerhard, M., Velauthapillai, A., Stanze, D., Roehle, H., Koch, M., Göbel, T., and Schell, M. (2013). *Appl. Phys. Lett.*, **103**, 061103.
28. Yardimci, N. T., Turan, D., Cakmakyapan, S., Jarrahi, M. (2018). *Appl. Phys. Lett.*, **113**, 251102.

29. Lu, P.-K., Turan, D., and Jarrahi, M. (2020). *Opt. Express*, **28**, 26324–26335.
30. Turan, D., Yardimci, N. T., and Jarrahi, M. (2020). *Opt. Express*, **28**, 3835–3845.
31. Yardimci, N. T., Turan, D., and Jarrahi, M. (2021). *APL Photonics*, **6**, 080802.
32. Lu, P. K., Olvera, A. F., Turan, D., Seifert, T. S., Yardimci, N. T., Kampfrath, T., Preu, S., and Jarrahi, M. (2022). *Nanophotonics*, **11**, 2661–2691.
33. Luca, B., Schmitz, M., Engels, S., Dauber, J., Oellers, M., Haupt, F., Watanabe, K., Taniguchi. T., Beschoten, B., Stampfer, Ch. (2015). *Sci. Adv.*, **1**, e1500222.
34. Bolotin, K. I., Sikes, K. J., Jiang, Z., Klima, M., Fudenberga, G., Honec, J., Kim, P., Stormer, H. L. (2008). *Solid State Commun.*, **146**, 351–355.
35. Osman, B., Polat, E. O., Kakenov, N., and Kocabas, C. (2015). *Nature Commun.*, **6**, 6628.
36. Liu, C.-H., Chang, Y.-C., Norris, T. B., Zhong, Z. (2014). *Nature Nanotechnol.*, **9**, 273.
37. Chen, X., Lu, X., Deng, B., Sinai, O., Shao, Y., Li, C., Yuan, S, Tran, V., Watanabe, K., Taniguchi, T., Naveh, D., Yang. L., and Xia, F. (2017). *Nature Commun.*, **8**, 1672.
38. Prechtel, L., Song, L., Schuh, D., Ajayan, P., Wegscheider, W., and W. Holleitner, W. A. (2012). *Nat. Commun.*, **3**, 646.
39. Hunter, N., Mayorov, A. S., Wood, Ch. D., Russell, Ch., Li, L., Linfield, E. H., Davies, A. G., and Cunningham, J. E. (2015). *Nano Lett.*, **15**, 1591–1596.
40. Mittendorff, M., Suess, R. J., Leong, E., and Murphy, T. E. (2017). *Nano Lett.*, **17**, 5811–5816.
41. Berry, Ch. W., and Jarrahi, M. (2012). *J. Infrared, Millimeter, Terahertz Waves*, **33**, 1182–1189.
42. Duffy, S. M., Verghese S., McIntosh, A., Jackson, A., Gossard, A. C., Matsuura, S. (2001). *IEEE Trans. Microwave Theory Techniques*, **49**, 1032–1038.
43. Miyamaru, F., Saito, Y., Yamamoto, K., Furuya, T., Nishizawa, S., and Tani, M. (2010). *Appl. Phys. Lett.*, **96**, 211104.
44. Castro-Camus, E., Lloyd-Hughes, J., Fu, L., Tan, H. H., Jagadish, C., and Johnston, M. B. (2007). *Opt. Express*, **15**, 7047–7057.
45. Hirota, Y., Hattori, R., Tan, M., and Hangyo, M. (2006). *Opt. Express*, **14**, 4486–4493.
46. Van Rudd, J., and Mittleman, D. M. (2022). *J. Opt. Soc. Am. B*, **19**, 319–329.

47. Ishibashi, T., and Ito, H. (2020). *J. Appl. Phys.*, **127**, 031101.
48. Klehr, A., Fricke, J., Knauer, A., Erbert, G., Walther, M., Wilk, R., Mikulics, M., Koch, M. (2008). *IEEE J. Selected Topics Quantum Electronics*, **14**, 289–294.
49. Brenner, C., Hofmann, M., Scheller, M., Shakfa, M. K., Koch, M., Mayorga, I. C., Klehr, A., Erbert, G., and Tränkle, G. (2010). *Opt. Lett.*, **35**, 3859–3861.
50. Blin, S., Paquet, R., Myara, M., Chomet, B., Gratiet, L. L., Sellahi, M., Beaudoin, G., Sagnes, I., Ducournau, G., Latzel, P., Lampin, J.-F., Garnache, A. (2017). *IEEE J. Selected Topics Quantum Electronics*, **23**, 1–11.
51. Carpintero, G., Hisatake, S., de Felipe, D., Guzman, R., Nagatsuma, T., Keil, N. (2018). *Sci. Rep.*, **8**, 1–7.
52. Erlig, H., Wang, S., Azfar, T., Udupa, A., Fetterman, H. R., and Streit, D. C. (1999). *Electron. Lett.*, **35**, 173–174.
53. Tani, M., Lee, K.-Su, and Zhang, X.-C. (2000). *Appl. Phys. Lett.*, **77**, 1396–1398.
54. Estacio, E. S., Hibi, M., Saito, K., Que, Ch. T., Furuya, T., Miyamaru, F., Nishizawa, S., Yamamoto, K., Tani, M. (2013). *J. Infrared, Millimeter, Terahertz Waves*, **34**, 423–430.
55. Huggard, P. G., Shaw, C. J., Cluff, J. A., and Andrews, S. R. (1998). *Appl. Phys. Lett.*, **72**, 2069–2071.
56. Kataoka, T., Kajikawa, K., Kitagawa, J., Kadoya, Y., and Takemura, Y. (2020). *Appl. Phys. Lett.*, **97**, 201110.
57. Witt, G. L. (1993). *Mater. Sci. Eng.*, **B22**, 9–15.
58. Escaño, M. C., Nguyen, T. Q., Osanai, Y., Kasai, H., and Tani, M. (2019). *Mater. Res. Express*, **6**, 055914.
59. Kitahara, H., Furuya, T., Escaño, M. C., and Tani, M. (2022). *The International Conference on Infrared, Millimeter, and Terahertz Waves 2022 (IRMMW-THz)*, Paper No. Th-P-28.
60. Yardimci, N. T., and Jarrahi, M. (2018). *Small*, **14**, 1802437.
61. Jarrahi, M. (2015). *IEEE Trans. Terahertz Sci. Technol.*, **5**, 391–397.
62. Dionne, J. A., Sweatlock, L. A., Atwater, H. A., and Polman, A. (2005). *Phys. Rev. B*, **72**, 075405.
63. Pendry, J. B., Martin-Moreno, L., and Garcia-Vidal, F. J. (2004). *Science*, **305**, 847–848.
64. Ebbesen, T. W., Lezec, H. J., Ghaemi, H. F., Thio, T., and Wolff, P. A. (1998). *Nature*, **391**, 667–669.
65. Schuller, J. A., Barnard, E. S., Cai, W., Jun, Y. C., White, J. S., and Brongersma, M. L. (2010). *Nat. Mater.*, **9**, 193–204.

66. Park, S. G., Jin, K. H., Yi, M., Ye, J. C., Ahn, J., Jeong, K. H. (2012). ACS *Nano*, **6**, 2026–2031.
67. Park, S.-G., Choi, Y., Oh, Y.-J., Jeong, K.-H. (2012). *Opt. Express*, **20**, 25530–25535.
68. Jooshesh, A., Smith, L., Masnadi-Shirazi, M., Bahrami-Yekta, V., Tiedje, T., Darcie, T. E., and Gordon, R. (2014). *Opt. Express*, **22**, 27992–28001.
69. Burford, N. M., Evans, M. J., El-Shenawee, M. O. (2018). *IEEE Trans. Terahertz Sci. Technol.*, **8**, 237–247.
70. Jooshesh, A., Bahrami-Yekta, V., Zhang, J., Tiedje, T., Darcie, T. E., Gordon, R. (2015). *Nano Lett.*, **15**, 8306–8310.
71. Fesharaki, F., Jooshesh, A., Bahrami-Yekta, V., Mahtab, M., Tiedje, T., Darcie, T. E., Gordon, R. (2017). *ACS Photonics*, **4**, 1350–1354.
72. Hsieh, B-Y., and Jarrahi, M. (2011). *J. Appl. Phys.*, **109**, 084326.
73. Berry, C. W., Jarrahi, M. (2012). *New J. Phys.*, **14**, 105029.
74. Berry, C. W., Wang, N., Hashemi, M. R., Unlu, M., and Jarrahi, M. (2013). *Nat. Commun.*, **4**, 1622.
75. Berry, C. W., Hashemi, M. R., and Jarrahi, M. (2014). *Appl. Phys. Lett.*, **104**, 081122.
76. Yang, S.-H., and Jarrahi, M. (2013). *Opt. Lett.*, **38**, 3677–3679.
77. Yang, S.-H., Hashemi, M. R., Berry, C. W., and Jarrahi, M. (2014). *IEEE Trans. Terahertz Sci. Technol.*, **4**, 575–581.
78. Li, X., Yardimci, N. T., and Jarrahi, M. (2017). *AIP Adv.*, **7**, 115113.
79. Yang, S.-H., and Jarrahi, M. (2015). *Appl. Phys. Lett.*, **107**, 131111.
80. Berry, C. W., Hashemi, M. R., Preu, S., Lu, H., Gossard, A. C., and Jarrahi, M. (2014). *Appl. Phys. Lett.*, **105**, 011121.
81. Berry, C. W., Hashemi, M. R., Preu, S., Lu, H., Gossard, A. C., and Jarrahi, M. (2014). *Opt. Lett.*, **39**, 4522–4524.
82. Yang, S.-H., Watts, R., Li, X., Wang, N., Cojocaru, V., O'Gorman, J., Barry, L. P., and Jarrahi, M. (2015). *Opt. Express*, **23**, 31206–31215.
83. Yang, S.-H., and Jarrahi, M. (2015). *Opt. Express*, **23**, 28522–28530.
84. Yang, S.-H., Salas, R., Krivoy, E. M., Nair, H. P., Bank, S. R., and Jarrahi, M. (2016). *J. Infrared, Millimeter Terahertz Waves*, **37**, 640–648.
85. Huang, S.-W., Yang, J., Yang, S.-H., Yu, M., Kwong, D.-L., Zelevinsky, T., Jarrahi, M., and Wong, C. W. (2017). *Phys. Rev. X*, **7**, 041002.
86. Yardimci, N. T., Yang, S.-H., Berry, C. W., and Jarrahi, M. (2015). *IEEE Trans. Terahertz Sci. Technology*, **5**, 223–229.

87. Yardimci, N. T., Salas, R., Krivoy, E. M., Nair, H. P., Bank, S. R., and Jarrahi, M. (2015). *Opt. Express*, **23**, 32035–32043.
88. Turan, D., Garcia, S. C. C., Castro-Camus, E., and Jarrahi, M. (2017). *J. Infrared, Millimeter Terahertz Waves*, **38**, 1448–1456.
89. Yardimci, N. T., and Jarrahi, M. (2016). *Appl. Phys. Lett.*, **109**, 191103.
90. Moon, K., Lee, I. M., Shin, J. H., Lee, E. S., Kim, N., Lee, W. H., Ko H., Han, S. P., Park, K. H. (2015). *Sci. Rep.*, **5**, 13817.
91. Turan, D., Lu, P. K., Yardimci, N. T., Liu, Z., Luo, L., Park, J.-M., Nandi, U., Wang, J., Preu, S., and Jarrahi, M. (2021). *Nat. Commun.*, **12**, 4641.
92. Lu, P. K., Jiang, X., Zhao, Y., Turan, D., and Jarrahi, M. (2022). *Appl. Phys. Lett.*, **120**, 261107.
93. Lu, P. K., Turan, D., and Jarrahi, M. (2022). *Opt. Express*, **30**, 1584–1598.
94. Yardimci, N. T., Cakmakyapan, S., Hemmati, S., and Jarrahi, M. (2017). *Sci. Rep.*, **7**, 4166.
95. Singh, A., Pashkin, A., Winnerl, S., Welsch, M., Beckh, C., Sulzer, P., Leitenstorfer, A., Helm, M., and Schneider, H. (2020). *Light Sci. Appl.*, **9**, 30.

Chapter 3

Optical Rectification–Based Sources

Gyula Polónyi[a,b] and János Hebling[a,b,c]
[a] *HUN-REN - PTE High-Field Terahertz Research Group, University of Pécs, Hungary*
[b] *Szentágothai Research Centre, University of Pécs, 7624 Pécs, Hungary*
[c] *Institute of Physics, University of Pécs, 7624 Pécs, Hungary*
polonyi@fizika.ttk.pte.hu

Nowadays there are a few techniques based on ultrashort laser pulses, which are able to generate high energy and high-field THz pulses. These are large-area photoconductive antennas (LAPCA) [1], optical rectification (OR) in organic and inorganic nonlinear crystals [2–4], spintronics emitters [5], and dual-color laser-generated air plasma [6]. From among them OR is able to generate pulses with both larger than 1 mJ energy and higher than 10 MV/cm peak electric field, predestining them to use not only to drive and investigate dynamics in matter [7], but also as sources for compact, THz driven particle accelerators [8–11].

OR is a special form of the well-known nonlinear optical process, the difference frequency generation (DFG). However, in the case of OR, not two beams with well-separated spectra travel through the nonlinear optical material (NM), but an ultrashort pump pulse having a broad bandwidth. The OR process can be considered as the result of a huge number of DFG processes between the frequency components of the ultrashort pump pulse. In the

Advances in Terahertz Source Technologies
Edited by Gun-Sik Park, Masahiko Tani, Jae-Sung Rieh, and Sang Yoon Park

ISBN 978-981-4968-89-8 (Hardcover), 978-1-003-45967-5 (eBook)
www.jennystanford.com

absence of frequency dependence of both absorption, refraction, and nonlinearity, this process results in generation of an electromagnetic wave having a temporal shape proportional to the time derivative of the intensity of the ultrashort pump pulse [12], approximating a single-cycle THz pulse.

3.1 Phase Matching, Velocity Matching, Tilted Pulse Front

The generation of THz pulses with high efficiency demands the fulfillment of several conditions where, similarly to other nonlinear frequency conversion processes, the phase matching leads the major role. In the case of OR the phase matching is equivalent to a velocity matching. It requires the group velocity of the pump pulse to be equal to the phase velocity of the generated THz radiation [13]

$$v_{\mathrm{opt}}^{\mathrm{gr}} = v_{\mathrm{THz}}. \tag{3.1}$$

This is in accordance with the expectation, that in an ideal case, the optical pulse propagates with the same velocity as the THz pulse, thus the THz field is gradually amplified while propagating through the NM due to the constructive interference of the THz waves which are generated at different positions in it. This entails the equality of the corresponding indices

$$n_{\mathrm{opt}}^{\mathrm{gr}} = n_{\mathrm{THz}}. \tag{3.2}$$

Although, there are nonlinear materials like ZnTe, where the group refractive index matches well with the THz refractive index at the operational wavelength of the widespread used Ti:sapphire femtosecond lasers (800 nm), but strong limitations like multi-photon absorption (MPA) prohibit the efficient THz generation. Many materials like lithium niobate (LN) are better candidates owing to the much bigger nonlinear coefficient, which results in higher conversion efficiency at the same pump intensity and larger bandgap resulting in a less significant role of MPA [3], which allows for higher pump intensity with higher efficiency. However, the large mismatch between the group refractive index of the optical and the phase refractive index of THz beams does not enable efficient OR.

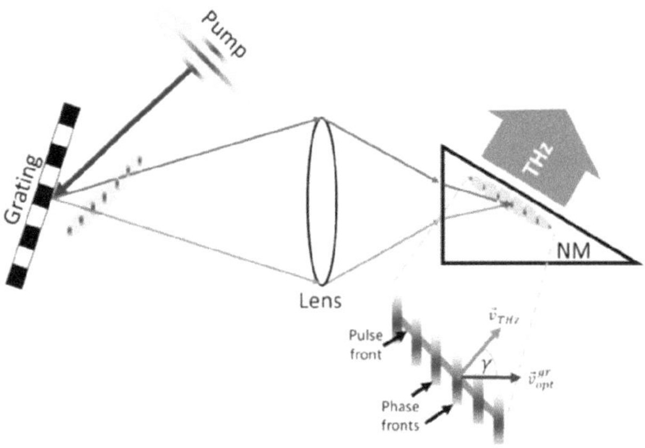

Figure 3.1 THz excitation setup based on tilted pulse front pumping geometry.

A qualitative improvement to overcome this problem was to tilt the pulse front (tilted pulse front: TPF), which was proposed in general for every NM where the group refractive index at optical is smaller than the phase refractive indices at THz frequencies ($n_{\text{opt}}^{\text{gr}}(\omega_{\text{opt}}) < n_{\text{THz}}(\omega_{\text{THz}})$) [14]. According to the enlarged part in Fig. 3.1, if there is an angle (γ) between the phase front and the pulse front, then although the pump pulse moves with a velocity $v_{\text{opt}}^{\text{gr}}$, the projection of this velocity in the propagation direction of the generated THz radiation is only $v_{\text{opt}}^{\text{gr}} \cdot \cos(\gamma)$. This should be equal with the phase velocity of the THz pulse, thus the velocity matching condition (Eq. 3.1) changes to a more general form

$$v_{\text{THz}} = v_{\text{opt}}^{\text{gr}} \cdot \cos(\gamma). \tag{3.3}$$

This pulse front tilt is introduced by a diffraction on a grating, then the beam spot on the grating is imaged into the nonlinear material by a lens or a telescope. This also changes the tangent of γ by the demagnification of the imaging system. Then, tangent γ decreases by n_{gr} as the pulse enters into the NM. It should be noted that the NM has to be used in wedged angle having the tilt angle γ. This technique has vastly broadened the scope of the available materials with various pump sources.

It is worth to present a simple method that provides the basis of comparing different nonlinear optical crystals for OR. Since OR is an intrapulse DFG, it is informative to consider the formula which describes the efficiency of DFG by long plane wave pulses in the absence of pump absorption and depletion, and it takes into account the effect of THz absorption [3, 15]. For phase-matched conditions the formula reads as

$$\eta_{\mathrm{THz}} = \frac{2\omega^2 d_{\mathrm{eff}}^2 L^2 I}{\epsilon_0 n_{\mathrm{optical}}^2 n_{\mathrm{THz}} c^3} \exp(-\alpha_{\mathrm{THz}} L/2) \cdot \frac{\sinh^2 \alpha_{\mathrm{THz}} \left(\frac{L}{4}\right)}{\alpha_{\mathrm{THz}} L/4}. \tag{3.4}$$

Here ω is the generated (THz) angular frequency, d_{eff} the effective nonlinear coefficient, I the intensity of the pump light, ϵ_0 the vacuum permittivity, c the velocity of light in vacuum, L the length of the nonlinear crystal, α_{THz} the intensity absorption coefficient for the THz radiation, and n_{opt} and n_{THz} the refraction indices. Two cases can be distinguished according to the absorption, where this formula simplifies to

$$\eta_{\mathrm{THz}} = \frac{2\omega^2 d_{\mathrm{eff}}^2 L^2 I}{\epsilon_0 n_{\mathrm{opt}}^2 n_{\mathrm{THz}} c^3} \tag{3.5}$$

in case of negligible absorption ($\alpha_{\mathrm{THz}} \cdot L \ll 1$) and

$$\eta_{\mathrm{THz}} = \frac{8\omega^2 d_{\mathrm{eff}}^2 L^2 I}{\epsilon_0 n_{\mathrm{opt}}^2 n_{\mathrm{THz}} c^3 \alpha_{\mathrm{THz}}^2} \tag{3.6}$$

in case of strong absorption ($\alpha_{\mathrm{THz}} \cdot L \gg 1$) [3]. From this, it is obvious that the selection of crystal length significantly greater than the penetration depth of the THz radiation $\alpha_{\mathrm{THz}}^{-1}$ is pointless. Only those THz photons that are produced within the region $L = \alpha_{\mathrm{THz}}^{-1}$ of the exit surface of the crystal can significantly contribute to the THz emission.

According to Eqs. (3.5)–(3.6), it is possible to introduce two figures of merits (FOMs) [3] of the nonlinear crystal used for OR defined by

$$\mathrm{FOM}_{\mathrm{NA}} \equiv \frac{d_{\mathrm{eff}}^2 L^2}{n_{\mathrm{opt}}^2 n_{\mathrm{THz}}}, \tag{3.7}$$

$$\mathrm{FOM}_{\mathrm{A}} \equiv \frac{4 d_{\mathrm{eff}}^2}{n_{\mathrm{opt}}^2 n_{\mathrm{THz}} \alpha_{\mathrm{THz}}^2}. \tag{3.8}$$

The $\mathrm{FOM}_{\mathrm{NA}}$ and $\mathrm{FOM}_{\mathrm{A}}$ values are measures of the pump-to-THz energy conversion efficiencies for optical rectification in weakly and strongly absorbing crystals, respectively. We propose to use FOM_A if α_{THz} is larger than 5 cm^{-1} and FOM_{NA} for smaller α_{THz} values [3].

Table 3.1 lists the most commonly used THz generator crystals in a comparable form. Here, d_{eff} is the nonlinear optical coefficient, from which one can predict the expected nonlinear response from the material, γ is the phase matching angle, f_{range} is the attainable frequency range based on experimental results, and f_{TO} is the frequency of the transversal-optical phonon resonance. This resonance at f_{TO} causes strong absorption and significant dispersion in the THz range close to it, which, in case of semiconductors and inorganic materials, limits the bandwidth. In contrast, the spectral width of the phonon resonances in organic materials is much smaller, thus THz generation can occur below and above these frequencies and the result is the presence of the corresponding spectral drops in the generated spectrum.

The first four materials are the commonly used semiconductors. Their d_{eff}-s and FOM values are the smallest compared to the other materials in the table. They require much less γ than LN. They only have middle/high frequency transversal-phonon absorption peaks, which provide bandwidth up to even 7 THz. The second four materials are the featured organic crystals with excellent d_{eff}-s and FOM-s. They are typically used in collinear geometry regardless of their phase matching angles. The appearance of the phonon absorption peaks around 1–2 THz does not limit the THz bandwidth, but distorts the spectrum. The last two items are the ferroelectric materials, LN and lithium-tantalate (LT). They have medium d_{eff} and FOM values. They require the biggest phase matching angles, which brings in difficulties. Although their phonon resonance frequency lies around 7–8 THz, they possess strong absorption (Fig. 3.7.) which limits the available THz frequencies below 2 THz.

3.2 Semiconductor-Based Sources

Since the optical and THz refractive indices in ZnTe match at the wavelength of Ti:sapphire lasers, it was a conventional approach to generate THz pulses with the combination of this pump source

Table 3.1 Collection of useful properties of various THz source materials

	$d_{\text{eff}}\ \left[\frac{\text{pm}}{\text{V}}\right]$	γ (@1 THz)	FOM $\left[\frac{\text{pm}^2\text{cm}^2}{\text{V}^2}\right]$	f_{range} [THz]	f_{TO} [THz]
GaP	24.8@ 1030 nm	6.6° @ 1030 nm	0.76	0.1–7	11
ZnTe	68.5@ 800 nm	20.6° @ 1030 nm	7.25	0.1–3	5.3
GaAs	65.6@ 800 nm	0.6° @ 1384 nm	3.63	0.1–3	8.1
CdTe	81.8@ 886 nm	0.9° @ 1049 nm	9.7	0.1–3	4.3
DAST	480@ 1.5 2 μm	10.7° @ 1500 nm	883	0.1–10	1.1
DSTMS	430@ 1.9 μm	5.5° @ 1500 nm	777	0.1–1	1.
				1.1–5	
				5.1–7	
OH1	570@ 1.3 2 μm	11.1° @ 1500 nm	1211	0.1–3	1.5
HMQ-TMS	292@ 633 nm	1.3° @ 1510 nm	356	0.1–5	1.7
				5.1–11	
				11.1–15	
PNPA				0.1–2	2.1
LN	168@ 1030 nm	64.6°	48	0.1–1	7.6
LT	161@ 1030 nm	70°	35	0.1–1	5.9

and crystal. The first proposal to increase the efficiency and the pulse energy was to expand the pump spot size and the crystal aperture [16]. Later, the TPF technique was applied to ZnTe as well, but the efficiency was still highly limited due to two-photon absorption (2PA). The significance of the restraining effect of 2PA was clearly shown in [17], where both ZnTe and GaP were pumped at 1 μm wavelength. Although the nonlinear coefficient is more than two times less in GaP than in ZnTe, but at this wavelength the dominant multiphoton absorption order is 3PA for GaP and 2PA for ZnTe, resulting in a 3-times higher conversion efficiency for GaP. At the first efficient use of TPF pumping in semiconductors, the pumping wavelength was long enough to eliminate 2PA in GaAs and led to a conversion efficiency of 0.05% [18]. With the elimination of even 3PA in ZnTe by using longer pump wavelength, as high as 0.7% conversion efficiency has been reported [19]. In order to overcome the low order MPA-s, longer wavelength has to be utilized, at which the velocity matching condition still requires to be fulfilled, for this, tilting the pulse front is necessary [3, 17]. Favorably the necessary tilt is less than half for semiconductors than for LN (Table 3.1.) at all wavelengths. In the ZnTe-related experiments two approaches were implemented for this, one with a conventional tilted pulse front pumping scheme using a prism shape crystal, and one where this conventional scheme was substituted by a novel, imaging-free construction, the so-called contact grating (CG) [20], where the grating structure was realized directly on the entrance surface of a plane-parallel crystal [4, 19]. In ZnTe, the available THz bandwidth is smaller than 3 THz, albeit the frequency of the lowest order transversal-optical phonon mode is 5.3 THz. The reason is the absorption peaks at 1.6 and 2.7 THz, which are related to the strong second-order phonon processes [21]. The effective nonlinear coefficient of ZnTe is 68.5 pm/V which is one of the highest among semiconductors. More details about multiphoton absorption and the contact grating technology can be found at the end of this subchapter.

There is a growing interest in GaP crystal as a NM for THz source, which is shown by the numerous articles, published recently [22–25]. The group velocity of GaP matches well with the phase velocity of the low-frequency THz pulses at 1 μm wavelength where commercially available Yb laser sources are available, therefore collinear phase matching is possible, which greatly simplify to work with it, moreover, the absence of 2PA at this wavelength allows the using of higher pump intensities. The conversion efficiency and the THz pulse power have increased by two orders of magnitude in the last ten years. These progressing results from GaP are partly due to the improvement of pump sources where the Yb-doped fiber technology was replaced by thin disk oscillators with MHz repetition rate. The other remarkable outcome is the broadband THz spectrum which spreads from 0.3 to 7 THz at relatively high power, since its transverse-optical phonon frequency is the highest (11 THz) amidst semiconductors, organic- and ferroelectric crystals (LN and LT), which leads to low dispersion and absorption in broader THz range. For this, GaP is ideal for spectroscopic purposes where the excessive pulse energy is less required, but the broad THz spectrum and the high repetition rate are advantageous.

GaAs just slightly lags behind ZnTe with its 65.6 pm/V effective nonlinear coefficient. It is not a coincidence that the highest reported value for THz pulse energy among semiconductors (17 μJ) was accomplished in this crystal by using longer wavelength and pulse front tilt [18]. Although the transversal-optical phonon absorption peak at 8.1 THz would let the available spectrum to spread much above 3 THz, but the tuning range is restricted by the cancellation of nonlinearity around 5.1 THz, see the corresponding valley of d_{eff} in Fig. 3.8. The bandgap of GaAs is 1.43 eV, which results that the 2PA edge of the crystal is at a relatively long wavelength of 870 nm, thus in order to eliminate at least three orders of multiphoton absorption, one would need a pump wavelength longer than 2600 nm.

CdTe shares the same value for bandgap with GaAs. However, the effective nonlinear coefficient of it is the highest among semiconductors, 81.8 pm/V, which holds promise to achieve high conversion efficiencies and pulse energies, but with limited spectral width, due to the fact that the transversal-optical phonon frequency is as low as 4.3 THz which is the lowest one among the semiconductors indicated in Table 3.1. Nevertheless, one could get an even broader spectrum from CdTe than from LN, of spreading up to 3 THz.

3.2.1 Contact Grating

A solution to omit the handicaps coming from the tilted pulse front pumping was suggested by Pálfalvi et al. [20]. The principle of operation of the semiconductor CG source is illustrated in Fig. 3.2a. A grating structure, responsible for tilting the pump pulse front, is formed on the entrance surface of the nonlinear material to substitute the entire TPF pumping scheme. In the setup shown here, the pump beam is perpendicularly incident on the CG and two symmetrically propagating diffraction orders ±1 are created to achieve the tilted pulse fronts [26, 27]. In these orders the pulse fronts are parallel with the incoming front, thus they are tilted by the angle of diffraction. The propagation direction, although different, hence the pulse fronts are tilted. The grating period was chosen so that the angle of diffraction would match the achievable γ. The generated THz beam propagates collinearly with the incident pump beam and leaves the crystal through the back surface, since the pulse fronts are parallel with the pumping direction in both orders. With this, a compact, monolithic, and alignment-free THz source can be realized. The collinear geometry is of great advantage for experiments while the plan-parallel shape of the NM ensures an excellent focusability of the generated THz beam.

The manufactured grating profile (see Fig. 3.2b) was closely fitted to the binary design profile, optimized for the highest diffraction efficiency [27]. The grating structure was achieved by a

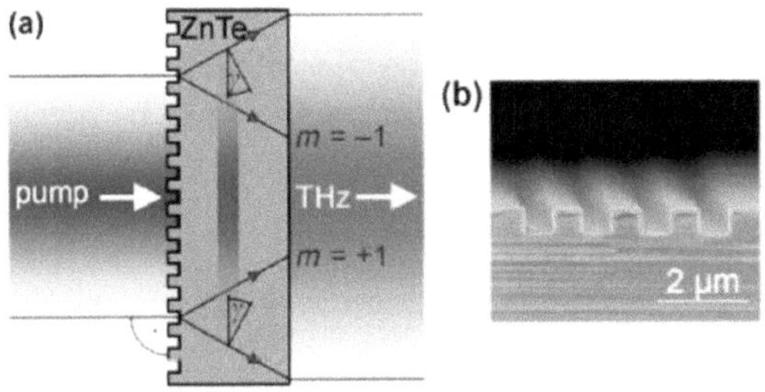

Figure 3.2 (a) Scheme of the ZnTe CG THz source with collinear geometry, utilizing the two diffraction orders ±1. The pulse front tilt angle is γ. (b) Scanning electron microscope micrograph of a cleaved test sample showing the grating profile [4].

combination of electron beam microlithography and dry (plasma) etching. Unfortunately, ZnTe inherently contains bubbles since its growth. These bubbles uncovered become pits in the surface causing defects of the grating profile. In this example, the defects were affecting about 20% of the total grating area. Nevertheless, the efficiency (0.3%) [4] demonstrated with the ZnTe CG source was as much as 6 times higher than the highest previously reported value from any semiconductor THz source and by the enormous two orders of magnitude higher than previously reported from a ZnTe source (0.0031%) [16].

3.2.2 Multiphoton Absorption

Due to its significant restraining effect on the applicable pump intensity, in case of semiconductors*, one should consider the multiphoton absorption process. Here, an electron makes a single transition from the valence band to the conduction band by absorbing more than one photon, simultaneously. This phenomenon is disadvantageous once, as the pumping photons are lost to optical rectification, and twice, which is even more important, these excited electrons in the conduction band are free carriers, which absorb the generated THz energy and restrain the THz output power to an almost constant value, as it was shown [19, 31]. Since the probability of the absorptive process occurring diminishes, as the photon order *N* is increased [32], by increasing *N* via utilizing longer pumping wavelengths, one can use more pumping intensity on OR. When one designs experiments, where higher order multiphoton absorption is occurring, one should consider the value of the corresponding multiphoton absorption coefficient, which can be measured by z-scan method [33]. It was recently shown that if the measuring

---

*It also occurs in organic and ferroelectric materials, but with less significant role. In organic materials, d_{eff} is large, thus, smaller pump intensity is sufficient. LN have large bandgap, for this, depending on the wavelength only 3PA (at 800 nm) or 4PA (at 1030 nm) is possible. Moreover, the corresponding absorption coefficient is much smaller due to the large bandgap [28]. Nevertheless, a detailed numerical investigation of the effects of 3PA was given in [29]. In addition, the effect of 3PA on the THz generation was shown experimentally: as the pump wavelength was increased, the corresponding change in the 3PA coefficient allowed for higher conversion efficiencies [30].

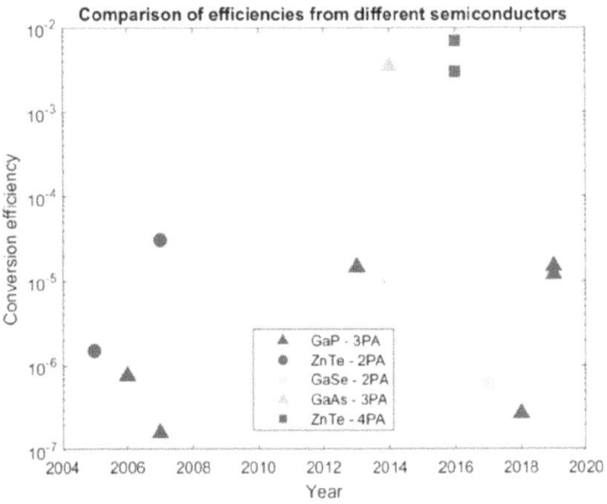

Figure 3.3 Conversion efficiency of GaP [23–25, 38–40], ZnTe [4, 16, 19, 41], GaSe [42], GaAs [18]. The colors refer to materials, the symbols refer to the dominating MPA.

pulse duration is not short enough, higher than the real absorption coefficient can be measured [34]. If it is unfeasible to carry out the measurement at the desired wavelength, various models can lead to estimated values [28, 32, 35].

Figure 3.3 shows the comparison of efficiencies from different semiconductors. The circles correspond to results where the dominant multiphoton absorption order was two, the triangles where it was three, and the squares where it was four, respectively. The differences between the results of the same symbols come from the various pumping conditions at the same material and wavelength. One can observe an outstanding jump in efficiency, which is independent from the material and pumping parameters. The highest energies and best efficiencies were achieved by eliminating the low order (2–3PA) MPA, by using longer pumping wavelengths, which points to the conclusion that pump source development is necessary in this direction. However, it was recently shown by a theoretical investigation through many orders of MPA, that if the conversion efficiency of the pump wavelength is also considered, a practical limit can be deduced, which determines the optimal pumping wavelength, like it is 2.06 μm and 3.85 μm in case of GaP and GaAs, respectively [36]. Besides this, the saturation of

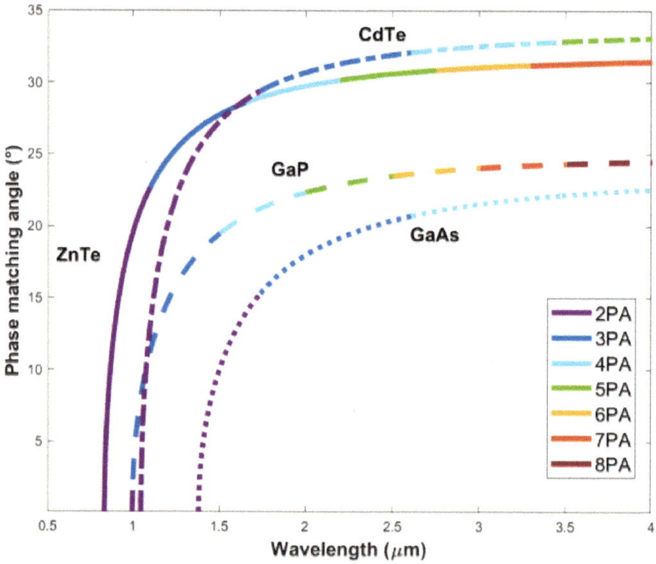

Figure 3.4 Necessary pulse front tilt for phase matching at 1 THz with various pump wavelengths for different semiconductors. The change in the color corresponds to the changes in the orders of MPA-s.

MPA was demonstrated at different wavelengths/orders, which was followed by a new rise of the THz pulse generation efficiency owing to the scattering of the free electrons to conduction band valleys with larger effective mass (THz absorption) [19, 31, 37].

As we saw, to overcome the MPA, longer pumping wavelengths have to be applied and one has to tilt the pulse front for efficient velocity matching. In Fig. 3.4. each curve presents the pulse front tilt angle necessary for phase matching at 1 THz versus the pumping wavelength in different materials. The change in the color corresponds to the change in the dominant multiphoton absorption order. Compared to LN, where the needed pulse front tilt angle, γ, is around 64°, here we see less than half of this value, which provides many advantages. It can be seen that above around 2 μm, the angle does not change significantly. GaAs and GaP provide the least angle and CdTe and ZnTe are the biggest ones. Moreover, GaAs and GaP possess the least material dispersion in the visible–near-IR range. All these factors make it easier to utilize them as THz sources.

3.3 Organic Crystal-Based Sources

Generally, organic crystals provide high laser-to-THz conversion efficiencies at room temperature and broad spectra (Table 3.1.), and with optimal pumping conditions mJ-level THz pulse energy can be achieved with them. They do not require pulse front tilting, for the plane-parallel crystals can be used in collinear geometry and the THz radiation is naturally collimated and aberration-free. This makes it possible to focus the beam by single optics to diffraction-limited spot size and to create high field strength even up to a few tens of MV/cm-s [43]. From the other side, their typical damage threshold is around 20 mJ/cm^2† for femtosecond pumping, while semiconductors can sustain 60 to 220 mJ/cm^2 (ZnTe, GaP, GaAs), and inorganic materials, like LN, even 2.5 J/cm^2. In this context, they can be viewed as optically sensitive THz source materials. Moreover, it is challenging to grow large crystals without defects, and they are expensive materials. They mostly need pump sources with a wavelength range around 1500 nm, that should be amplified to tens of mJ-s to have reasonably high THz energy and conversion efficiency, but these systems are not as widespread as the well-tried Ti:sapphire and Yb lasers. Although their THz spectrum is quite broad, it contains many absorption peaks which can affect the application possibilities.

The first organic crystal which was used as a THz emitter is DAST (4-N,N-dimethylamino-4′-N′-methylstilbazolium tosylate), which is composed of a cation that possesses a large molecular optical nonlinearity, and of an anion that is designed to force noncentrosymmetric packing in the crystalline phase. DAST crystal has large nonlinear coefficient, more than twice than that of LN. It is highly attractive for high-speed electro-optic applications and broadband THz-wave generation [45]. It has the highest damage threshold among organic crystals and it provided the best conversion efficiency when it was pumped by a high energy optical parametric amplifier (OPA) at 1950 nm [43]. Although, the spectrum

†An exotic exemption is DPFO with its >50 mJ/cm^2 which is close to the semiconductors, and it has an other interesting property, namely, its spectrum lies between two water absorption lines (1.35–1.65 THz) which makes it possible for long distance air propagation with little attenuation from ambient humidity [44].

of the generated THz pulses can go up to 10 THz, but it also has a strong absorption line at 1.1 THz where the transverse-optical phonon resonance lies (Table 3.1) and others at around 5, and 8.4 THz [46]. The absorption at 1.1 THz and the dispersion of the refractive index around this frequency result in roughly two different phase matching regions: for 0.2–1 THz, pumping wavelength from 1000 to 1200 nm is required. For frequencies above 1.5 THz, from 1200 to 1700 nm is necessary, which are around telecommunication wavelengths, where compact and low-cost fiber lasers are available [47].

DSTMS is structurally very similar to DAST, but here the counter-anion is slightly modified (from 4-methylbenzenesulfonate to 2,4,6-trimethylbenzenesulfonate). From the aspect of crystal growing, it is better than DAST, since its solubility in methanol is twice as large as DAST. Furthermore, its transverse-optical phonon absorption peak around 1 THz is half that of DAST [47]. The highest peak electric field (83 MV/cm) and the highest THz pulse energy (0.9 mJ) were demonstrated with it [48, 49]. Its THz spectrum spans up to 8 THz with only one moderate absorption peak at 1 THz, hence it is excellent for THz spectroscopic applications. The best pump wavelength range is, similar to DAST, around 1500 nm. With DSTMS, an attempt was made to overcome the limit of the small aperture size, by coherently combining the output of multiple, small crystals in a large-size partitioned THz emitter, with which GV/m level peak electric field was achieved [50].

OH1 (2-(3-(4-hydroxystyryl)-5,5-dimethylcyclohex-2-enylidene) malononitrile) is a non-ionic configurationally locked polyene crystal. It has the highest FOM, and has optimum velocity-matching between 1200 and 1460 nm for 0.3-2.5 THz. It has absorption lines at about 3 and 5 THz. It is well suited for linear and nonlinear THz spectroscopy applications, since it provides MV/cm level peak electric fields above 3% conversion efficiency [46].

Like DAST and DSTMS, HMQ-TMS ((2-(4-hydroxy-3-methoxystyryl)-1-methyl-quinolinium-2,4,6-trimethylbenzenesulfonate)) is also an ionic organic crystal with a quinolinium-based ionic-cation core structure and can be obtained in high-optical quality. Here the velocity matching is fulfilled at shorter wavelengths, compared to other organic materials. It can provide an extremely wide THz spectrum, but with certain absorption lines at 1.7, 5 (strong), 8, and 11 THz when it is pumped at 1500 nm with ultrashort pulses (65 fs) [51]. It is worth emphasizing that unlike the other organic materials, it can be efficiently pumped by Yb-based laser sources, too [52].

Since that demonstration of DAST as THz emitter, much attention has been focused on the development of organic crystals for THz generation like, DSTMS, and OH1. Several new THz generators have been developed; however, no new organic THz generator has significantly exceeded DAST. A promising candidate, PNPA (E)-(4-((4-nitrobenzylidene) amino)-N-phenylaniline) is the newest material on the list, and since it is a recent development, there is no available information on its many properties like, nonlinear coefficient, damage threshold or optical refractive index. Nevertheless, according to a comparative study, it outperforms DAST, and OH1 in terms of peak electric field and spectral width of THz spectrum when same pumping conditions have been applied [53].

Besides the presented organic crystals, one can find more demonstrations of THz generation by OR in other organic materials like BNA, 2A5NPP, DAP+NP-NP, EHPSI-4NBS, TMOAT, NMBA, and MBST, but we chose the crystals above, since they are more convincing and useful for the broadest range of THz science and application. Theoretical evaluations additionally show that the upper limits of nonlinear optical susceptibilities in organic crystals have not been reached yet [54]. The current direction of investigation is about to develop easier crystal growth procedures and the example of PNPA shows that it is still possible that new candidates will emerge as efficient organic THz sources.

3.4 Lithium Niobate–Based Sources

Both ferroelectric crystals, LT and LN have three important properties which are very advantageous from the point of view of effective THz generation: (i) according to Table 3.1. they have significantly larger nonlinear optical coefficients than semiconductors, (ii) they are dielectrics having much larger bandgap than semiconductors, and (iii) their damage threshold is much higher than all the other THz sources indicating that they are good candidates for high energy THz pulse generation. The advantage of the large nonlinear coefficient is evident according to Eq. (3.4). The large bandgap is beneficial since it means that for a given pump wavelength only higher-order MPA is possible (than for smaller bandgap material), thus higher pump intensity can be used, resulting higher THz generation efficiency as well, according to Eq. (3.4). However, for LT and LN the index of refraction on the THz range is more than two times larger than the group index on the visible–near infrared pump range leading to that velocity matching (Eq. 3.1) cannot be achieved by choosing appropriate pump laser wavelength.

OR of fs laser pulses was realized in LT as early as 1984. The focused fs pulses generated phonon-polaritons inside the LT crystal and their temporal shape was measured (also inside the LT crystal) by electro-optic sampling (EOS) [55]. Since velocity matching was not fulfilled, the generated phonon-polaritons composed a Cherenkov cone, and the authors called the generation process as electro-optic Cherenkov generation. The electric field strength measured inside the LT crystal was only 10 V/cm. Unfortunately, it is impossible to increase the THz energy and field strength in this arrangement by using larger pump energy and (in order to avoid the damage of the crystal) a larger pump spot size. The reason for this is that the spot size has to be significantly shorter than the wavelength of the generated THz radiation [55]. The propagation character determined by the Cherenkov cone (instead of the plane wave) is also very disadvantageous from an application point of view.

Introducing the TPF technique resulted in a large improvement in the THz pulse generation by ferroelectric crystal. Even in the first report, generation of THz pulses with 98 pJ energy and smaller than 40 mrad divergence was demonstrated by pumping congruent LN at 77 K with 800 nm pump pulses having 2.3 μJ energy [56]. Replacing the congruent LN for a stoichiometric one (which has smaller THz absorption [57]) resulted in 400 pJ THz energy, $\eta = 1.7 \times 10^{-2}\%$ optical-to-THz conversion efficiency and 6 kV/cm unfocused THz electric field outside the LN crystal [58]. In these experiments the setup illustrated in Fig. 3.1 was utilized. Using a pump laser with 0.5 mJ energy the scalability of the THz energy (up to 240 nJ) with the pump pulse energy, and an increased efficiency ($\eta = 5.0 \times 10^{-2}\%$) were demonstrated at room temperature [59]. Later, first of all by increasing the pump energy, but also by improving the imaging and diffraction system [20, 60–62], THz pulses with larger and larger energy were generated [63–65]. Up to now the highest reported THz pulse energy generated by OR in LN is 1.4 mJ, while the highest focused THz peak field is 6.3 MV/cm [65]. As high as 2% generation efficiency was achieved by using long pump pulse duration (680 fs), at cryogenic crystal temperature [66]. It was shown earlier, that it is advantageous to use at least a few hundred fs long Fourier-transform-limited pump pulses [60]. The reason for this is the large γ tilt angle and the correspondingly large $d\theta/d\lambda$ angular dispersion [67–69],

$$\tan(\gamma) = \frac{n}{n_g}\bar{\lambda}\frac{d\epsilon}{d\lambda} \tag{3.9}$$

which results in a large group-delay-dispersion (GDD, D) according to Eq. (3.10) [69, 70].

$$D = \frac{\lambda}{c}\left[n\left(\frac{d\epsilon}{d\lambda}\right)^2 - \frac{d^2n}{d\lambda^2}\right] = \frac{\lambda}{c}\left[\tan^2(\gamma)\frac{n_g^2}{n}\frac{1}{\bar{\lambda}^2} - \frac{d^2n}{d\lambda^2}\right] \tag{3.10}$$

In case of LN or LT, the first term in Eq. (3.10) is much larger than the second one. As a result of the large GDD, the temporal intensity profile of an ultrashort pump pulse evolves very fast inside the LN crystal, and the dispersion length is short [71]. As a result, the average pump pulse duration is much longer than the transform-limited one, and the average intensity is much lower than the peak, resulting in a reduced effective interaction length for THz generation. For LN the effective interaction length is about 6 and 1 mm for 600 and 100 fs pump pulses, respectively [60].

3.4.1 Limitations of TPF

Besides the limited interaction length connected to the large angular dispersion of the pump beam, two other disadvantageous properties of the conventional TPF setup (Fig. 3.1) have been recognized recently. These result in the restriction of upscaling of the generated THz energy and field strength by simply increasing the pumping energy and the pumped area. In summary, the following three properties of the conventional TFP setup limit the available THz pulse energy and field strength and degrade the THz beam quality:

i) **Reduced interaction length** of OR because of the GDD proportional to the square of the pump beam's angular dispersion (Eq. 3.10) and to $\tan^2(\gamma)$ (Eqs. 3.9 and 3.10).
ii) **Imaging errors in the presence of angular dispersion** result in significantly increased pump pulse duration at the edges of a large pump spot. The longer pulse means reduced pump intensity and reduced local pump-to-THz conversion efficiency. Furthermore, the generated THz pulse shape will vary along the cross section of the THz beam.
iii) **The prism shape of the LN crystal** also results in THz pulses with different temporal shapes across the THz beam cross section owing to the different generation lengths. It is unfeasible that such a bad quality, strongly asymmetric THz beam to be tightly focused, and thus the realization of high THz field strength cannot be achieved.

We notice that these limitations exist for semiconductor THz sources, too. However, since both three types of limitations increase with the γ pulse front tilt angle, these limitations are less severing for semiconductors in comparison to LN.

We also notice that since all limitations depend on the pump beam size, their effect is negligible even for LN THz sources for 100–200 fs pump pulse duration and less than 4 mm pump beam diameter. These restrictions allow generation of THz pulses with tens of μJ energy [60], and 1 MV/cm focused THz field [72]. Such THz pulses can be used in THz pump–probe measurements and for controlling material excitations [7].

However, a few possible applications of high energy THz pulses, for example, pumping of compact, THz-driven particle accelerators, require THz pulses with significantly larger energy and peak THz electric field strength [7, 9]. For this strong seek of extremely high-field THz pulses, over the past few years extensive effort has been made to reduce or even eliminate the above-mentioned limitations.

3.4.2 New Designs

According to Eq. (3.10), the effect of GDD can be reduced by decreasing the bandwidth of the pump pulse, via using pump pulses with longer transform limited pulse duration [60, 66] and/or by using nonlinear optical material (for example semiconductors) which need less tilt angle (see Fig. 3.4) [19]. A more radical solution is replacing the optical grating by an echelon one [72]. In such a setup, if the pump pulse duration is shorter than the delay between the adjacent beamlets introduced by reflection on the echelon grating, then the pump beam will not have angular dispersion since the beamlets propagate and diffract independently from each other, and GDD will not be present. In this case, the beamlets are without pulse front tilt, but in the whole beam an average tilt angle is realized. If this tilt has a γ angle satisfying Eq. (3.4), and the delay between the adjacent beamlets is much smaller compared to the temporal period of the THz radiation, then the THz pulse is generated efficiently, and the THz beam propagates perpendicular to the (average) tilted intensity front, similar to the conventional (continuous) TPF case. The advantage of such a setup is especially significant in the case of sub-hundred-fs pump pulse duration, when the pump bandwidth is large [72]. However, the excellent performance of such a setup was demonstrated for 280 fs pump pulse duration, too [73]. Using pump pulses with 0.4 mJ energy at 25 kHz repetition rate, generation of THz pulses with 1.3% pump-to-THz energy conversion efficiency was demonstrated. The peak electric field of the focused THz pulses was 400 kV/cm. Such a THz source can be applied very efficiently in THz pump-probe measurements. However, it contains the same prism-shaped LN crystal with $\sim 64^\circ$ wedge angle as the conventional TPF LN setup, which hinders upscaling of this system.

The CG THz source setup originally was proposed for LN THz source in order to get rid of the imaging system and make it possible to use plan-parallel LN crystal, thus eliminating limitations (ii) and (iii) [20]. However, detailed design of LN CG THz pulse sources indicated that (contrary to the case of semiconductor CG THz [26, 27]) the LN has to be wedged, although with more than three times smaller wedge angle than for the conventional TPF LN THz source [74, 75]. The reason for this is the required large pulse front tilt. This large pulse front tilt angle (which entails the consequently large groove density) also caused technical difficulties and resulted in an order of magnitude smaller pump-to-THz conversion efficiency than can be obtained by conventional TPF pumped LN source [76].

In order to reduce the needed groove density of the CG, a hybrid TPF excitation scheme was suggested [77]. In this case, the final pulse front tilt is created in two steps. First, an optical grating-imaging system creates one part of the needed tilt angle, then the CG (with a reduced groove density) increases the tilt angle to the needed value. In this case, the LN crystal has to be wedged with an about 30° wedge angle. Another improved hybrid TPF scheme was suggested recently [78]. Here the final pulse front tilt is created also in two steps, however, for the second step not a CG is used in first diffraction order on the input surface of the LN crystal, but an echelon structure is created as it is shown in Fig. 3.5. This structure is called "nonlinear echelon slab" (NLES).

The most important advantage of this TPF LN THz source setup compared to all existing and earlier theoretically designed tilted-pulse front schemes is, that it makes it possible to use a

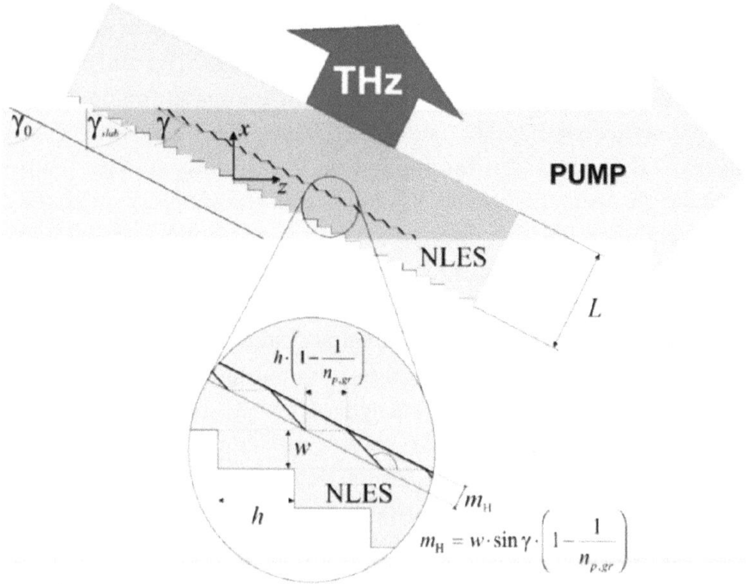

Figure 3.5 The alteration of the pump pulse front when passing through the input surface of the NLES [78].

plane-parallel LN structure if appropriate initial pulse front tilt is applied. This is essential in order to reach good THz beam quality. Furthermore, the NLES-based hybrid setup significantly reduces imaging errors compared to the conventional setup for the followings reason: as it is mentioned and also shown in Fig. 3.5, it is necessary to introduce a γ_0 pulse front tilt using a grating–imaging system before the pump beam hits the NLES. The γ_0 tilt angle has to be equal to the γ angle satisfying the velocity matching according to Eq. (3.3). Please note that this angle is significantly smaller than the tilt angle that has to be created by the grating–imaging system in the conventional TPF setup. The reason is that in the conventional setup when the pump beam enters into the LN the tilt angle decreases so that its tangent will be reduced by the $n_{\text{opt}}^{\text{gr}}$ group refractive index of the LN crystal. Hence, in the conventional setup, the grating–imaging system has to introduce much larger pulse front tilt than γ. As the magnified-out part of Fig. 3.5. shows, the tilt angle of the beamlets is reduced similarly in the hybrid NLES setup, too, but the average tilt of the whole cross section of the pump beam remains unchanged. In a proof-of-principle experiment the working of the hybrid NLES setup was demonstrated [79], but the THz generation efficiency was less than expected. The imperfect quality of the echelon structure was identified as the main reason for this. Although according to numerical calculations the construction of hybrid NLES setup makes possible to use pump beams with close to 2 cm diameter to produce THz pulses with 0.5 mJ energy [78], improved versions of it were suggested. These make it possible to increase the pump diameter by eliminating the imaging system, by applying a LN with a small wedge angle, and a slanted volume phase holographic grating, respectively [80, 81].

A very simple and compact scalable LN TPF setup, the reflective nonlinear slab (RNLS) and its modified version the RNLS with external structured reflector (RNLS-ESR) were proposed very recently [80, 82].

As it is illustrated in Fig. 3.6, they contain a plan-parallel LN slab with a structured reflector manufactured inside its back surface (RNLS), or a plan-parallel LN slab and an external structured reflector with an index-matching liquid (RNLS-ESR). The pump beam enters the nonlinear material (NM, LN or LT) perpendicular to its

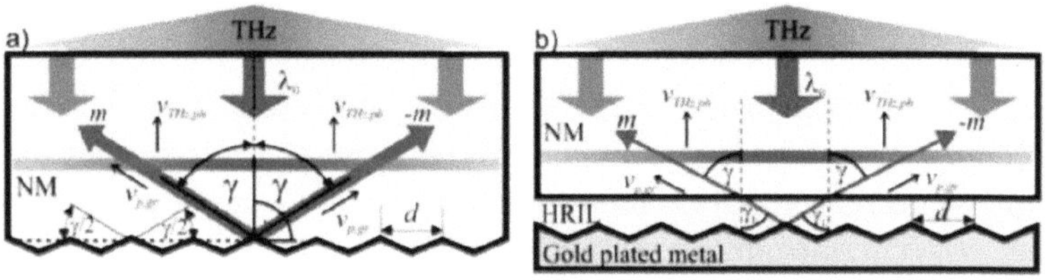

Figure 3.6 Schematic views of the RNLS and RNLS-ESR setups. NM: nonlinear material, LN or LT, HIRL: high refractive index liquid [82].

front surface, travels through the NM, then it is diffracted/reflected into the symmetric $\pm m^{\text{th}}$ orders/beams at the structured reflector. In these beams, the pulse fronts (red band parallel with the NM surfaces) are tilted. These overlapping tilted intensity fronts generate the THz pulse which propagates upwards and leaves the NM perpendicular to the NM surface. (In case of RNLS-ESR the back surface of NM is flat and the periodic structure is created on a separate metallic surface (ESR). For optical coupling, a high refractive index liquid (HRIL) is used between NM and ESR). Setups similar to the ones illustrated in Figs. 3.5 and 3.6 need creation of structure on the surface of LN (or other NM) with optical quality. Although this was not demonstrated yet for LN crystal structures planned for THz source, creation of structures on LN surface with optical quality (surface roughness with less than 10 nm rms) was demonstrated using micromachining with a diamond tool [83].

3.5 Dispersion of Refractive Index, Absorption and Nonlinear Coefficient

From the point of view of applications, like THz-driven particle acceleration, the THz pulse energy (and the related pump-to-THz conversion efficiency) and the peak electric field are the most important properties of THz pulses. Nevertheless, the temporal shape of the THz pulse also affects such applicability. For linear or nonlinear THz spectroscopy and similar applications, THz pulses with single-cycle shape and smooth and wide corresponding spectra are the best. As it was mentioned above, in the absence of frequency dependence of both absorption, refraction and nonlinearity of the NM, THz pulses with single-cycle THz pulse shape are generated, with a spectral width inversely proportional to the pump pulse duration. However, transversal-optical phonon-polaritons cause significant frequency-dependent absorption and refraction in the THz range. The nonlinear optical coefficient also has frequency dependence. Consequently, the pulse shape of THz pulses generated in semiconductors and LN contains a few oscillations when the average frequency is not much smaller than the f_{TO} frequency, but it approaches a single-cycle character for low frequency. Generation in organic crystals usually results in more irregular temporal shape of the electric field, since usually a broader spectrum is generated than the lowest f_{TO}.

As an example, Fig. 3.7 shows the frequency dependences of the index of refraction and absorption on the THz range for the three most frequently used crystals for THz generation: LN, GaP, and GaAs. The index of refraction and the absorption have the well-known resonant character around the TO phonon frequencies of f_{TO} = 7.6, 11, and 8.1 THz for LN, GaP and GaAs, respectively. Figure 3.7 depicts only the more important low frequency range of these

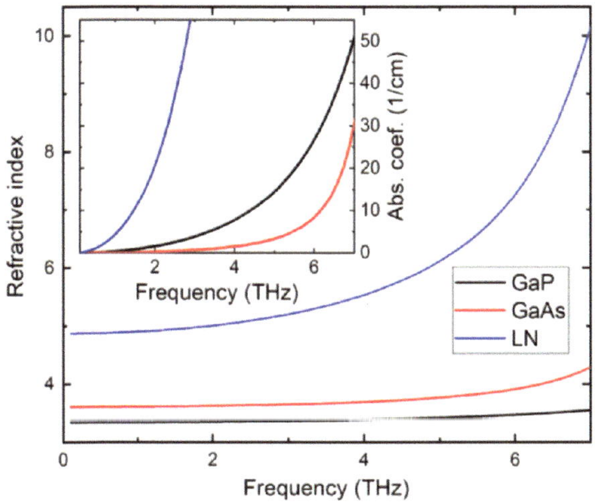

Figure 3.7 Refractive index and absorption coefficient (inset) of GaP, GaAs and LN up to 7 THz.

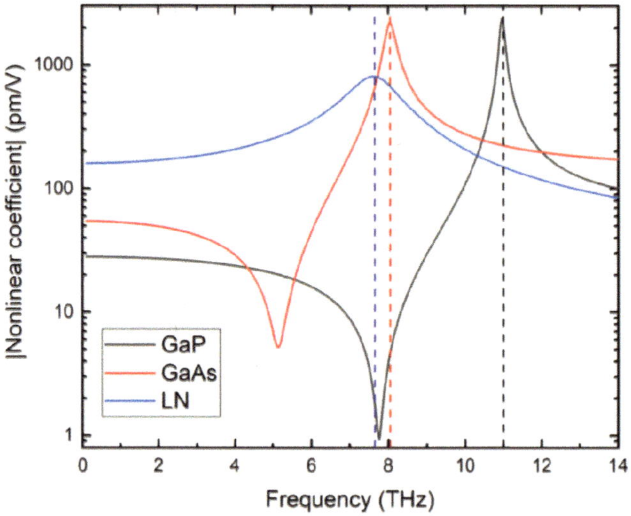

Figure 3.8 Frequency-dependent nonlinear coefficient of GaP, GaAs and LN up to 14 THz. The dashed lines indicate the transversal-optical phonon frequencies.

curves. It can be seen that approaching the resonance frequency both the absorption and the index of refraction increase more and more steeply. This limits the highest THz generation frequency to about 2.5 THz for LN at room temperature, and to about 4 THz at cryogenic temperature [47]. The THz absorption of both semi-insulating GaAs and GaP should allow generation even up to 7 THz. However, for most of the semiconductors the frequency dependence of the nonlinear optical coefficient means a limiting effect of the highest THz frequency.

Figure 3.8 displays the dispersion of the absolute value of the nonlinear coefficient at the THz frequency range for GaP, GaAs and LN. These curves were obtained from [84–86], after some small modification to fit d_{eff} to known values in the IR range, and in the low frequency THz range, with the help of Eq. (3.11). As can be seen, d_{eff} becomes zero at 5.1 and 7.8 THz, respectively for GaAs and GaP. The reason for this is that the nonlinearity at the THz range has electrical and ionic contributions. These have the same sign for LN, but the opposite sign for most of the semiconductors. Because of the opposite sign of the ionic and electronic contributions, they cancel each other at special frequencies, resulting that the nonlinearity changes sign around them. This effect limits the generated THz frequency to below 4.5 and 7.0 THz for GaAs and GaP, respectively.

Because of the dispersion of the refractive index on the THz range, there are a few possibilities to change the velocity-matched frequency, and with that the mean THz frequency, that is tuning of the THz pulses is possible. However, the mean THz frequency usually is not equal to the velocity-matched frequency, since similarly to the frequency dependence of the EOS sensitivity [87, 88], the frequency dependence of the THz absorption and the nonlinear coefficient, and the bandwidth of the pulp pulse also affect the mean frequency. Tuning of the THz frequency was demonstrated by tuning the pump wavelength [22], the pump pulse front tilt angle [58], the duration of the pump pulse [24], and the thickness of the generating GaP crystal [24]. A stronger control of the mean frequency, and even the bandwidth of the generated THz pulses can be achieved by using multi-pulse pumping [89, 90], periodically poled or periodically inverted nonlinear optical crystals [91–93], or both [94]. The THz pulse shape can be also controlled by spatial modulation of the pump pulse spot [95, 96].

In spectroscopic applications not only the bandwidth and mean frequency is important, but the signal-to-noise ratio and the dynamic range also play a crucial role. For a fixed measurement time, these properties can be increased by increasing the repetition rate of the THz pulse source. Although at the dawn of OR-based THz pulse sources around 2000, laser oscillators and regenerative amplifiers with tens of MHz and hundreds of kHz repetition rates, respectively were used as pump source [56, 97], later in order to make available higher energy pump pulses, an oscillator-amplifier system with 1 kHz [41] or even with only 10 Hz [65] repetition rate was used. However, very recently, aimed at the better suitability for spectroscopic applications and based on the huge improvement of modern thin-disc femtosecond lasers, a few groups intensively studied THz sources working at above 10 MHz repetition rate [22]. Very recently, the dynamic range of photoconductive antenna–based TDTS systems was approached, with which a dynamic range value of 90 dB was demonstrated at 3 THz with a Time Domain-Terahertz

Spectroscopy (TDTS) setup using CG GaP OR source and CG GaP EOS detector pair pumped by Yb laser working at 1.1 MHz repetition rate [98].

3.6 Models for THz Generation

As it was shown, Eq. (3.5), which describes the efficiency of difference frequency generation at perfect phase matching and in case of non-depleting pumping, and its derivatives, Eqs. (3.6) and (3.7), can be used to compare nonlinear materials' ability to generate THz pulses. It is important to note, that in these (and other OR related) equations, d_{eff} is not the nonlinear coefficient describing the frequency conversion (SHG or DFG) in the near-infrared, but the nonlinear coefficient, which can be determined by

$$d_{\text{eff}} = -\frac{n_p^4 r}{4}, \tag{3.11}$$

where n_p is the phase refractive index of the pump, and r is the electro-optic coefficient.

In Ref. 99 a theory was introduced to describe primally the generation of THz radiation in periodically poled LN. The differential equation introduced in this, became the basis of numerous theoretical researches [60, 78, 100, 101]. This model considered the phase unmatching, but not the dispersion of the pump pulse, which is important at short pump pulses and big angular dispersion. This was first taken into account in case of OR in reference [60]. Reference [101] considered first in case of TPF setups the shape of the NM prism and the effect of THz walk-off, but does not the effect of the THz pulse on the pump pulse. This latter effect of the THz pulses surpasses the effect of dispersion even if the conversion efficiency is of 1% magnitude. The reason for this is that the frequency of the pump is 200–300 times bigger than the frequency of the THz radiation, thus the photon conversion efficiency is 200–300%, that is cascading of the THz generation occurs [63, 102, 103]. The effect of the cascading, the self-phase modulation (SPM) and the group velocity dispersion (GVD-AD) produced by the angular dispersion (Eq. 3.10) (different extent for different pump pulse durations) is

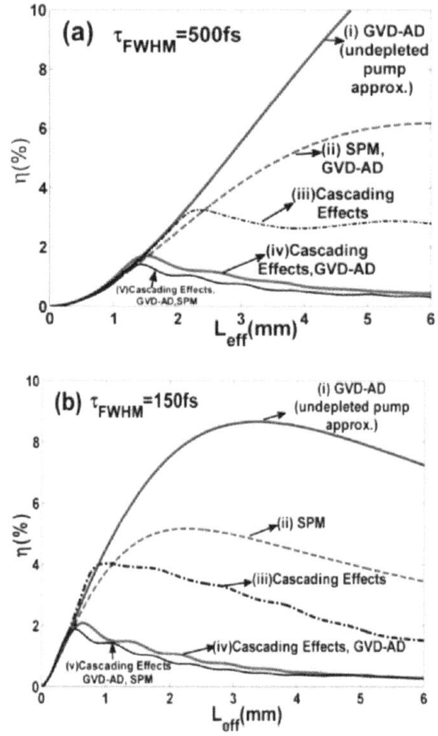

Figure 3.9 Conversion efficiency as a function of effective length is calculated by switching off various effects. Material dispersion and absorption are considered for all cases. Acronyms are resolved in the main text [106].

shown well in Fig. 3.9. for LN THz source. Based on this, it can be stated, that the results of models which do not consider the cascade effects can only be accepted, if the conversion efficiency is less than 1% [104]. The newest models [104, 105] calculate the THz waveform in 2D+1 or 3D+1 (space + time dimension) dimensions, and take into account the related diffraction of both the pump, and the generated THz pulse, which has a major role in the exact description of the pulse front tilt [68]. L. Wang pointed out that even the 2D+1 type models can give sufficiently accurate results for describing the generation of the THz pulses, increasing the number of the dimensions does not result in significantly different outcomes, only the run time grows crucially [105].

3.7 Summary

In summary, a tremendous improvement of the OR-based THz sources was witnessed in the last two decades. The reported energy of THz pulses generated by this method increased by 8 orders of magnitude from the 10 pJ to the 1 mJ level. The pump pulse to THz pulse energy conversion efficiency can be a few percent for high energy pumping. Three types of nonlinear optical materials are used: semiconductors (mostly ZnTe, GaAs and GaP), ferroelectric crystals (LN, LT), and organic crystals. Up to now LN produced the THz pulses with the largest energy (above 1 mJ), while organic crystals with the highest focused peak electric field (above 10 MV/cm). LN THz sources work efficiently up to 2 THz, semiconductors up to a few THz, while some organic crystal works even above 10 THz. The spectrum of organic crystal THz sources is usually modulated by absorption resonances, while the spectra of semiconductor and LN THz sources are smoother. These two types are suitable for single-cycle THz pulse generation at low mean frequency.

Further advancement of the OR-based THz sources is foreseen: Developing of new organic crystals with improved properties (e.g. better nonlinear susceptibility, higher damage threshold, with simple growing procedure) is expected. Application of TPF pumping to these materials can also result in enhanced THz generation. The technical improvement of new TPF setups (as was introduced in subchapter 3.4.2, and others) and their application in both semiconductor- and ferroelectric-based THz sources will result in more compact devices with excellent THz beam characteristics. Since the properties of the pump laser strongly determine the quality of the THz source, any progress in laser development will contribute. Besides the evolution of the presently used pump lasers, the emergence of ultrashort pulse CO_2 lasers is expected [107]. Utilizing them as THz pump sources can result in THz pulses with significantly higher energy [108]. Moreover, the further improvement in THz source development will lead to more advanced THz metrology and widens the scope of possible applications, such as THz pump–THz probe measurements, material control experiments, THz driven particle accelerations, among many others.

Acknowledgments

The authors would thank Gy. Tóth for his contribution to subchapter 3.6, Models for THz generation.

References

1. Ropagnol, X., Kovács, Z., Gilicze, B., Zhuldybina, M., Blanchard, F., Garcia-Rosas, C. M., Szatmári, S., Földes, I. B., and Ozaki, T. (2019), *New J. Phys.*, **21**, 113042.
2. Hauri, C. P., Ruchert, C., Vicario, C., and Ardana, F. (2011), *Appl. Phys. Lett.*, **99**, 161116.
3. Hebling, J., Yeh, K.-L., Hoffmann, M. C., Bartal, B., and Nelson, K. A. (2008), *J. Opt. Soc. Am. B*, **25**, B6.
4. Fülöp, J. A., Polónyi, Gy., Monoszlai, B., Andriukaitis, G., Balciunas, T., Pugzlys, A., Arthur, G., Baltuska, A., and Hebling, J. (2016), *Optica*, **3**, 1075.
5. Papaioannou, E. Th., and Beigang, R. (2021), *Nanophotonics*, **10**, 1243–1257.
6. Koulouklidis, A. D., Gollner, C., Shumakova, V., Fedorov, V. Yu., Pugžlys, A., Baltuška, A., and Tzortzakis, S. (2020), *Nat. Commun.*, **11**, 292.
7. Salén, P., Basini, M., Bonetti, S., Hebling, J., Krasilnikov, M., Nikitin, A. Y., Shamuilov, G., Tibai, Z., Zhaunerchyk, V., and Goryashko, V. (2019), *Phys. Rep.*, **836–837**, 1–74.
8. Pálfalvi, L., Fülöp, J. A., Tóth, Gy., and Hebling, J. (2014), *Phys. Rev. Spec. Top. Accel. Beams*, **17**, 031301.
9. Zhang, D., Fallahi, A., Hemmer, M., Ye, H., Fakhari, M., Hua, Y., Cankaya, H., Calendron, A.-L., Zapata, L. E., Matlis, N. H., and Kärtner, F. X. (2019), *Optica*, **6**, 872.
10. Hibberd, M. T., Healy, A. L., Lake, D. S., Georgiadis, V., Smith, E. J. H., Finlay, O. J., Pacey, T. H., Jones, J. K., Saveliev, Y., Walsh, D. A., Snedden, E. W., Appleby, R. B., Burt, G., Graham, D. M., and Jamison, S. P. (2020), *Nat. Photonics*, **14**, 755–759.
11. Liu, W., Sun, L., Yu, Z., Liu, Y., Jia, Q., Sun, B., and Xu, H. (2021), *Opt. Lett.*, **46**, 4398.
12. Wynne, K., and Carey, J. J. (2005), *Opt. Commun.*, **256**, 400–413.

13. Nahata, A., Weling, A. S., and Heinz, T. F. (1996), *Appl. Phys. Lett.*, **69**, 2321–2323.
14. Hebling, J., Almasi, G., Kozma, I., and Kuhl, J. (2002), *Opt. Express*, **10**, 1161.
15. Sutherland, R. L., McLean, D. G., and Kirkpatrick, S. (2003), *Handbook of Nonlinear Optics*. (Marcel Dekker).
16. Blanchard, F., Razzari, L., Bandulet, H. C., Sharma, G., Morandotti, R., Kieffer, J. C., Ozaki, T., Reid, M., Tiedje, H. F., Haugen, H. K., and Hegmann, F. A. (2007), *Opt. Express*, **15**, 13212.
17. Hoffmann, M. C., Yeh, K.-L., Hebling, J., and Nelson, K. A. (2007), *Opt. Express*, **15**, 11706.
18. Blanchard, F., Ropagnol, X., Hafez, H., Razavipour, H., Bolduc, M., Morandotti, R., Ozaki, T., and Cooke, D. G. (2014), *Opt. Lett.*, **39**, 4333.
19. Polónyi, Gy., Monoszlai, B., Gäumann, G., Rohwer, E. J., Andriukaitis, G., Balciunas, T., Pugzlys, A., Baltuska, A., Feurer, T., Hebling, J., and Fülöp, J. A. (2016), *Opt. Express*, **24**, 23872.
20. Pálfalvi, L., Fülöp, J. A., Almási, G., and Hebling, J. (2008), *Appl. Phys. Lett.*, **92**, 171107.
21. Schall, M., Walther, M., and Uhd Jepsen, P. (2001), *Phys. Rev. B*, **64**, 094301.
22. Aoki, K., Savolainen, J., and Havenith, M. (2017), *Appl. Phys. Lett.*, **110**, 201103.
23. Xu, J., Globisch, B., Hofer, C., Lilienfein, N., Butler, T., Karpowicz, N., and Pupeza, I. (2018), *J. Phys. B At. Mol. Opt. Phys.*, **51**, 154002.
24. Drs, J., Modsching, N., Paradis, C., Kränkel, C., Wittwer, V. J., Razskazovskaya, O., and Südmeyer, T. (2019), *J. Opt. Soc. Am. B*, **36**, 3039.
25. Meyer, F., Hekmat, N., Vogel, T., Omar, A., Mansourzadeh, S., Fobbe, F., Hoffmann, M., Wang, Y., and Saraceno, C. J. (2019), *Opt. Express*, **23**, 31313.
26. Bakunov, M. I., and Bodrov, S. B. (2014), *J. Opt. Soc. Am. B*, **31**, 2549.
27. Ollmann, Z., Fülöp, J. A., Hebling, J., and Almási, G. (2014), *Opt. Commun.*, **315**, 159–163.
28. Wherrett, B. S. (1984), *J. Opt. Soc. Am. B*, **1**, 67.
29. Zhong, S.-C., Zhai, Z.-H., Li, J., Zhu, L.-G., Li, J., Meng, K., Liu, Q., Du, L.-H., Zhao, J.-H., and Li, Z.-R. (2015), *Opt. Express*, **23**, 31313.
30. Wu, X., Carbajo, S., Ravi, K., Ahr, F., Cirmi, G., Zhou, Y., Mücke, O. D., and Kärtner, F. X. (2014), *Opt. Lett.*, **39**, 5403.

31. Ku, S. A., Tu, C. M., Chu, W.-C., Luo, C. W., Wu, K. H., Yabushita, A., Chi, C. C., and Kobayashi, T. (2013), *Opt. Express*, **21**, 13930.
32. Keldysh, L. (1965), *Sov. Phys. JETP*, **20**, 1307–1314.
33. Sheik-Bahae, M., Said, A. A., Wei, T.-H., Hagan, D. J., and Van Stryland, E. W. (1990), *IEEE J. Quantum Electron.*, **26**, 760–769.
34. Kawamori, T., Schunemann, P. G., Gruzdev, V., and Vodopyanov, K. L. (2022), *APL Photonics*, **7**, 086101.
35. Nathan, V., Mitra, S. S., and Guenther, A. H. (1985), *J. Opt. Soc. Am. B*, **2**, 294.
36. Mbithi, N., Toth, G., Tibai, Z., Benabdelghani, I., Nasi, L., Krizsan, G., Hebling, J., and Polónyi, G. (2022), *J. Opt. Soc. Am. B*, **39**.
37. Hebling, J., Hoffmann, M. C., Hwang, H. Y., Yeh, K.-L., and Nelson, K. A. (2010), *Phys. Rev. B*, **81**, 035201.
38. Chang, G., Divin, C. J., Liu, C.-H., Williamson, S. L., Galvanauskas, A., and Norris, T. B. (2006), *Opt. Express*, **14**, 7909.
39. Chang, G., Divin, C. J., Liu, C.-H., Williamson, S. L., Galvanauskas, A., and Norris, T. B. (2007), *Opt. Lett.*, **32**, 433.
40. Li, J., Chai, L., Shi, J., Liu, F., Liu, B., Xu, B., Hu, M., Li, Y., Xing, Q., Wang, C., Fedotov, A. B., and Zheltikov, A. M. (2013), *Laser Phys. Lett.*, **10**, 125404.
41. Löffler, T., Hahn, T., Thomson, M., Jacob, F., and Roskos, H. G. (2005), *Opt. Express*, **13**, 5353.
42. Bereznaya, S. A., Korotchenko, Z. V., Redkin, R. A., Sarkisov, S. Y., Tolbanov, O. P., Trukhin, V. N., Gorlenko, N. P., Sarkisov, Y. S., and Atuchin, V. V. (2017), *J. Opt.*, **19**, 115503.
43. Gollner, C., Shalaby, M., Brodeur, C., Astrauskas, I., Jutas, R., Constable, E., Bergen, L., Baltuška, A., and Pugžlys, A. (2021), *APL Photonics*, **6**, 046105.
44. Savoini, M., Huber, L., Cuppen, H., Abreu, E., Kubli, M., Neugebauer, M. J., Duan, Y., Beaud, P., Xu, J., Rasing, T., and Johnson, S. L. (2018), *ACS Photonics*, **5**, 671–677.
45. Jazbinsek, M., Mutter, L., and Gunter, P. (2008), *IEEE J. Sel. Top. Quantum Electron.*, **14**, 1298–1311.
46. Vicario, C., Jazbinsek, M., Ovchinnikov, A. V., Chefonov, O. V., Ashitkov, S. I., Agranat, M. B., and Hauri, C. P. (2015), *Opt. Express*, **23**, 4573.
47. Jazbinsek, M., Puc, U., Abina, A., and Zidansek, A. (2019), *Appl. Sci.*, **9**, 882.
48. Shalaby, M., and Hauri, C. P. (2015), *Nat. Commun.*, **6**, 5976.

49. Vicario, C., Ovchinnikov, A. V., Ashitkov, S. I., Agranat, M. B., Fortov, V. E., and Hauri, C. P. (2014), *Opt. Lett.*, **39**, 6632.
50. Vicario, C., Monoszlai, B., and Hauri, C. P. (2014), *Phys. Rev. Lett.*, **112**, 213901.
51. Vicario, C., Monoszlai, B., Jazbinsek, M., Lee, S.-H., Kwon, O.-P., and Hauri, C. P. (2015), *Sci. Rep.*, **5**, 14394.
52. Rovere, A., Jeong, Y.-G., Piccoli, R., Lee, S.-H., Lee, S.-C., Kwon, O.-P., Jazbinsek, M., Morandotti, R., and Razzari, L. (2018), *Opt. Express*, **26**, 2509.
53. Rader, C., Zaccardi, Z. B., Hang, S., Ho, Harrell, K. G., Petersen, P. K., Stephan, H., Michaelis, D. J., and Johnson, J. A. (2022), arXiv:2207.03382.
54. Bosshard, Ch., Bösch, M., Liakatas, I., Jäger, M., and Günter, P., Second-order nonlinear optical organic materials: recent developments, in *Nonlinear Optical Effects and Materials*, vol. 72, Günter, P., ed. (Springer Berlin Heidelberg) (2000), pp. 163–299. doi: 10.1007/978-3-540-49713-4_3.
55. Auston, D. H., Cheung, K. P., Valdmanis, J. A., and Kleinman, D. A. (1984), *Phys. Rev. Lett.*, **53**, 1555–1558.
56. Stepanov, A. G., Hebling, J., and Kuhl, J. (2003), *Appl. Phys. Lett.*, **83**, 3000–3002.
57. Pálfalvi, L., Hebling, J., Kuhl, J., Péter, Á., and Polgár, K. (2005), *J. Appl. Phys.*, **97**, 123505.
58. Hebling, J., Stepanov, A. G., Almási, G., Bartal, B., and Kuhl, J. (2004), *Appl. Phys. B Lasers Opt.*, **78**, 593–599.
59. Stepanov, A. G., Kuhl, J., Kozma, I. Z., Riedle, E., Almási, G., and Hebling, J. (2005), *Opt. Express*, **13**, 5762.
60. Fülöp, J. A., Pálfalvi, L., Almási, G., and Hebling, J. (2010), *Opt. Express*, **18**, 12311.
61. Kunitski, M., Richter, M., Thomson, M. D., Vredenborg, A., Wu, J., Jahnke, T., Schöffler, M., Schmidt-Böcking, H., Roskos, H. G., and Dörner, R. (2013), *Opt. Express*, **21**, 6826.
62. Tokodi, L., Hebling, J., and Pálfalvi, L. (2017), *J. Infrared Millim. Terahertz*, **38**, 22–32.
63. Yeh, K.-L., Hoffmann, M. C., Hebling, J., and Nelson, K. A. (2007), *Appl. Phys. Lett.*, **90**, 171121.
64. Fülöp, J. A., Ollmann, Z., Lombosi, Cs., Skrobol, C., Klingebiel, S., Pálfalvi, L., Krausz, F., Karsch, S., and Hebling, J. (2014), *Opt. Express*, **22**, 20155.

65. Zhang, B., Ma, Z., Ma, J., Wu, X., Ouyang, C., Kong, D., Hong, T., Wang, X., Yang, P., Chen, L., Li, Y., and Zhang, J. (2021), *Laser Photonics Rev.*, **15**, 2000295.

66. Huang, S.-W., Granados, E., Huang, W. R., Hong, K.-H., Zapata, L. E., and Kärtner, F. X. (2013), *Opt. Lett.*, **38**, 796.

67. Bor, Zs., and Rácz, B. (1985), *Opt. Commun.*, **54**, 165–170.

68. Martinez, O. E. (1986), *Opt. Commun.*, **59**, 229–232.

69. Hebling, J. (1996), *Opt. Quantum Electron.*, **28**, 1759–1763.

70. Martinez, O. E., Gordon, J. P., and Fork, R. L. (1984), *J. Opt. Soc. Am. A*, **1**, 1003.

71. Diels, J.-C., and Rudolph, W. (2006), *Ultrashort Laser Pulse Phenomena: Fundamentals, Techniques, and Applications on a Femtosecond Time Scale*. (Academic Press, an Imprint of Elsevier).

72. Ofori-Okai, B. K., Sivarajah, P., Huang, W. R., and Nelson, K. A. (2016), *Opt. Express*, **24**, 5057–5068.

73. Guiramand, L., Nkeck, J. E., Ropagnol, X., Ozaki, T., and Blanchard, F. (2022), *Photonics Res.*, **10**, 340.

74. Nagashima, K., and Kosuge, A. (2010), *Jpn. J. Appl. Phys.*, **49**, 122504.

75. Ollmann, Z., Hebling, J., and Almási, G. (2012), *Appl. Phys. B*, **108**, 821-826.

76. Tsubouchi, M., Nagashima, K., Yoshida, F., Ochi, Y., and Maruyama, M. (2014), *Opt. Lett.*, **39**, 5439.

77. Pálfalvi, L., Ollmann, Z., Tokodi, L., and Hebling, J. (2016), *Opt. Express*, **24**, 8156.

78. Pálfalvi, L., Tóth, G., Tokodi, L., Márton, Z., Fülöp, J. A., Almási, G., and Hebling, J. (2017), *Opt. Express*, **25**, 29560.

79. Nugraha, P. S., Krizsán, G., Lombosi, C., Pálfalvi, L., Tóth, G., Almási, G., Fülöp, J. A., and Hebling, J. (2019), *Opt. Lett.*, **44**, 1023.

80. Tóth, G., Pálfalvi, L., Tibai, Z., Tokodi, L., Fülöp, J. A., Márton, Z., Almási, G., and Hebling, J. (2019), *Opt. Express*, **27**, 30681.

81. Krizsán, G., Tibai, Z., Tóth, G., Nugraha, P. S., Almási, G., Hebling, J., and Fülöp, J. A. (2022), *Opt. Express*, **30**, 4434.

82. Krizsán, G., Tibai, Z., Hebling, J., Pálfalvi, L., Almási, G., and Tóth, G. (2020), *Opt. Express*, **28**, 34320.

83. Takigawa, R., Higurashi, E., Kawanishi, T., and Asano, T. (2014), *Opt. Express*, **22**, 27733.

84. Boyd, G. D., and Pollack, M. A. (1973), *Phys. Rev. B*, **7**, 5345–5359.

85. Faust, W. L., and Henry, C. H. (1966), *Phys. Rev. Lett.*, **17**, 1265–1268.
86. Dekorsy, T., Yakovlev, V. A., Seidel, W., Helm, M., and Keilmann, F. (2003), *Phys. Rev. Lett.*, **90**, 055508.
87. Wu, Q., and Zhang, X.-C. (1997), *Appl. Phys. Lett.*, **71**, 1285–1286.
88. Leitenstorfer, A., Hunsche, S., Shah, J., Nuss, M. C., and Knox, W. H. (1999), *Appl. Phys. Lett.*, **74**, 1516–1518.
89. Chen, Z., Zhou, X., Werley, C. A., and Nelson, K. A. (2011), *Appl. Phys. Lett.*, **99**, 071102.
90. Stummer, V., Flöry, T., Krizsán, G., Polónyi, G., Kaksis, E., Pugžlys, A., Hebling, J., Fülöp, J. A., and Baltuška, A. (2020), *Optica*, **7**, 1758.
91. Lee, Y.-S., Meade, T., Perlin, V., Winful, H., Norris, T. B., and Galvanauskas, A. (2000), *Appl. Phys. Lett.*, **76**, 2505–2507.
92. Lemery, F., Vinatier, T., Mayet, F., Aßmann, R., Baynard, E., Demailly, J., Dorda, U., Lucas, B., Pandey, A.-K., and Pittman, M. (2020), *Commun. Phys.*, **3**, 150.
93. Zhang, C., Avetisyan, Y., Glosser, A., Kawayama, I., Murakami, H., and Tonouchi, M. (2012), *Opt. Express*, **20**, 8784.
94. Jolly, S. W., Matlis, N. H., Ahr, F., Leroux, V., Eichner, T., Calendron, A.-L., Ishizuki, H., Taira, T., Kärtner, F. X., and Maier, A. R. (2019), *Nat. Commun.*, **10**, 2591.
95. Stepanov, A. G., Hebling, J., and Kuhl, J. (2004), *Opt. Express*, **12**, 4650.
96. Zhong, S.-C., Zhu, Y., Du, L.-H., Zhai, Z.-H., Li, J., Zhao, J.-H., Li, Z.-R., and Zhu, L.-G. (2017), *Opt. Express*, **25**, 17066.
97. Nagai, M., Tanaka, K., Ohtake, H., Bessho, T., Sugiura, T., Hirosumi, T., and Yoshida, M. (2004), *Appl. Phys. Lett.*, **85**, 3974–3976.
98. Cui, W., Awan, K. M., Huber, R., Dolgaleva, K., and Ménard, J. (2022), *Adv. Opt. Mater.*, **10**, 2101136.
99. Vodopyanov, K. L., (2006), *Opt. Express*, **14**, 2263.
100. Li, Y., Hu, X., Liu, F., Li, J., Xing, Q., Hu, M., Lu, C., and Wang, C. (2012), *J. Opt. Soc. Am. B*, **29**, 3114.
101. Bakunov, M. I., Bodrov, S. B., and Mashkovich, E. A. (2011), *J. Opt. Soc. Am. B*, **28**, 1724.
102. Cronin-Golomb, M. (2004), *Opt. Lett.*, **29**, 2046.
103. Jewariya, M., Nagai, M., and Tanaka, K. (2009), *J. Opt. Soc. Am. B*, **26**, A101.
104. Ravi, K., Huang, W. R., Carbajo, S., Nanni, E. A., Schimpf, D. N., Ippen, E. P., and Kärtner, Franz. X. (2015), *Opt. Express*, **23**, 5253.

105. Wang, L., Kroh, T., Matlis, N. H., and Kärtner, F. (2020), *J. Opt. Soc. Am. B*, **37**, 1000.
106. Ravi, K., Huang, W. R., Carbajo, S., Wu, X., and Kärtner, F. (2014), *Opt. Express*, **22**, 20239.
107. Polyanskiy, M. N., Pogorelsky, I. V., Babzien, M., and Palmer, M. A. (2020), *OSA Contin.*, **3**, 459.
108. Tóth, G., *to be published* (2023)

Chapter 4

Method of Terahertz Liquid Photonics

Yiwen E and X.-C. Zhang
The Institute of Optics,
University of Rochester, NY 14627, USA
zhangxc@rochester.edu

This chapter describes the method of extending technologies of terahertz (THz) air photonics to THz liquid photonics, using laser-induced micro-plasma with flowing liquid targets. Matter exists in distinct states. Among them, solids, gases, and plasmas have been used as sources of THz waves for decades. Exploring new THz sources could provide an alternative perspective in understanding fundamental physics. However, the current use of liquids, especially liquid water, as THz wave emitters was extremely limited until very recently. Broadband terahertz wave generation from liquid was first demonstrated in 2017. Since then, it has been widely studied with different liquid targets under various excitation geometries. The recent progress and challenges are summarized in this book chapter. It is reasonable to expect that liquids should have unique properties if they could be harnessed as THz sources.

Advances in Terahertz Source Technologies
Edited by Gun-Sik Park, Masahiko Tani, Jae-Sung Rieh, and Sang Yoon Park

ISBN 978-981-4968-89-8 (Hardcover), 978-1-003-45967-5 (eBook)
www.jennystanford.com

4.1 Background

Laser-induced plasma in air has been well studied since it is able to emit ultra-broadband terahertz (THz) pulses with a field strength over MV/cm. However, the widespread use of THz air photonics has been limited by the amount of laser energy required to implement the technique. This problem could be addressed by replacing the typical filament currently used (length of several mm to cm) with a micron-range long plasma, which is named "micro-plasma" [1]. A micro-plasma can be created by focusing a laser beam on ambient air through a high numerical-aperture (NA) objective. The minimum laser pulse energy required to generate THz waves from such micro-plasma could be many orders of magnitude smaller compared to that of a filament. This method has been first used in laser-induced THz air photonics.

Figure 4.1a shows the concept of THz emission (drawn as blue pulses) from a micro-plasma. In contrast to the traditional THz sources, the THz emission direction of the maximum energy is nearly perpendicular with respect to the optical propagation direction; therefore, it is spatially separated from the residual laser excitation. This sideways THz radiation is attributed to the steep ponderomotive potential at the focal plane, which accelerates the free electrons created by photoionization [2–4]. Figure 4.1b shows

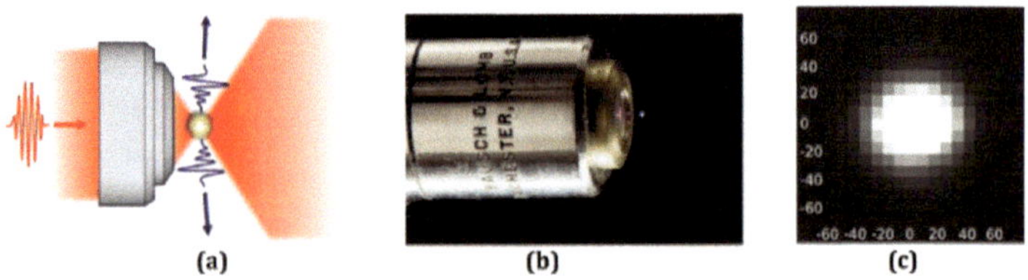

Figure 4.1 (a) Concept of THz generation from laser-induced micro-plasma. A high numerical-aperture (NA) objective focuses the laser beam. THz signal emitted from micro-plasma is spatially separated from the residual laser excitation. (b, c) Photos of the micro-plasma fluorescence. c is a magnified image of the micro-plasma (unit is μm), taken by an iCCD camera. Please note that the actual plasma spot's contribution to the THz wave is about one μm or less.

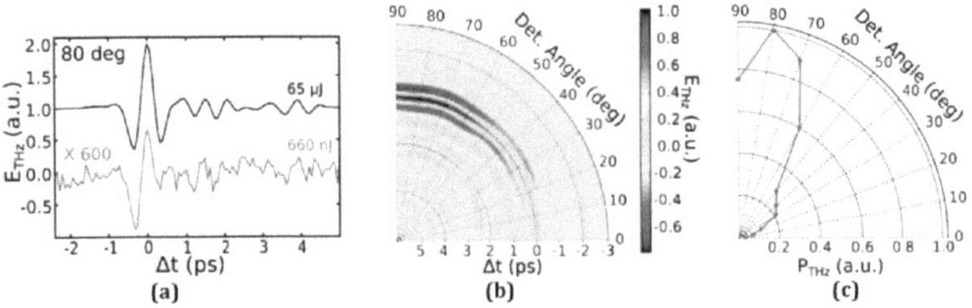

Figure 4.2 (a) The THz pulse generated from micro-plasma at 80° from the laser propagation direction. The THz signal generated by the 0.6 μJ optical pulse energy is observed. (b–c) Measured THz peak field and power as a function of the detection angle, respectively. The power is calculated as the time integral of the absolute square of the corresponding THz waveform.

a photo of the fluorescence spot at the focal point of a 0.85 NA microscope objective when a laser beam is focused (65 μJ pulse energy). The longitudinal and transverse dimensions of this micro-plasma are smaller than 30 μm in Fig. 4.1c, which are measured by an iCCD camera.

Figure 4.2a plots the measurement of THz time-domain waveforms generated from micro-plasma under different optical pulse energy for excitation. Figures 4.2b and 4.2c, respectively, show the THz peak field and power as a function of the detection angle, which is defined as the angle between the laser propagation direction and the optical axis of the THz detection optics. The figures show that, with one-color optical excitation, the THz emission is at maximum around 80° away from the laser propagation direction. THz wave signal with 0.6 μJ laser excitation energy is also observed, which is about 3 orders less than commonly used laser pulse energy in THz air photonics.

Since most liquids with polar molecules show a strong absorption at THz frequency, a thin target is the first choice to harvest THz emission from a liquid material. Therefore, the technology of producing micro-plasma is usually employed in the excitation of a thin target to avoid the contribution of THz emission from air plasma created near the liquid target.

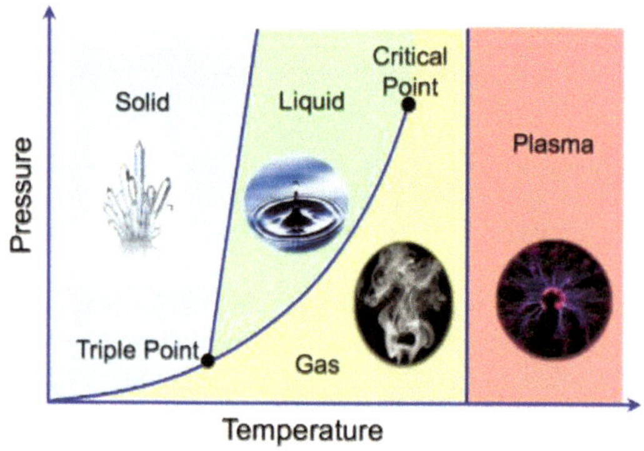

Figure 4.3 Three of four states, solid, gas, and plasma, have been used to generate THz waves. The use of liquids as THz sources is a real challenge.

4.2 Liquid for THz Source

The phase diagram of pressure vs. temperature for normal universal matters is shown in Fig. 4.3. Three of four states: solid, gas, and plasma have been well studied for THz wave generation, including the recent development of THz air photonics. However, until recently, the demonstration of THz wave generation from liquid sources was conspicuously absent. It is well known that water, the most common liquid on the earth, is a strong absorber at the THz frequency. Therefore, liquid water has historically been sworn off as a source for THz radiation. In 2017, broadband THz wave generation from a gravity-driven, free-flowing water film [5], and from bulk liquid in a cuvette [6] have been experimentally demonstrated through ionizing liquid media with intense laser pulses.

4.3 THz Wave Emission under Single-Color Optical Excitation in a Thin Water Film

It is well known in the THz community that water strongly absorbs far-infrared electromagnetic waves. Water vapor has many sharp absorption lines in the THz range resulting from molecular vibrational and rotational modes, and liquid water has a continuous

Table 4.1 Input photon numbers are needed in order to get one photon out after a water film with thickness d.

Water thickness d (mm)	THz photons in N_{in} (#)	THz photons out N_{out} (#)
1	3.6×10^9	1
0.1	9	1

absorption covering the THz regime, with its power absorption coefficient $\alpha = 220\ \text{cm}^{-1}$ at 1 THz [7]. To clearly show the strong absorption of liquid water, a simple calculation in the following is helpful. If we assume N_{out} and N_{in} are numbers of input and output THz photons, respectively, they follow Eqn. (4.1) regardless of the extra loss at the interface.

$$N_{out} = e^{-\alpha d} N_{in} \tag{4.1}$$

Table 4.1 lists the number of THz photons needed in order to get one THz photon out after passing through a water film with a certain thickness of d mm. For a 1 mm thick water film, it requires 3.6 billion THz photons to get one THz photon out. This calculation does not include the air/water interface reflection. To reduce the THz attenuation in water, using a thin water layer is the key. For example, while the thickness d decreases by one order, from 1 to 0.1 mm, the number of N_{in} drops eight orders, from 3.6 billion input THz photons to 9 input THz photons. Therefore, to get a measurable signal from a liquid target in a transmissive geometry, a target with a thickness of about 0.1 mm or less is required. On the other hand, to avoid the influence between pulses, a flowing liquid target with a certain flow rate refreshes the excitation area for each pulse. With a comparable molecular density to a solid, the fluidity of liquid presents its superiority.

The experiment setup of THz wave generation from a liquid target is as same as a general system for the study of THz emission from materials. For our case, the laser beam with mJ pulse energy, horizontal polarization, 800 nm center wavelength, and 1 kHz repetition rate, is focused into a thin water film. A 2 mm thick <110> oriented ZnTe crystal configured for electro-optical (EO) sampling [8] and a Golay detector are used for the THz wave detection. A liquid

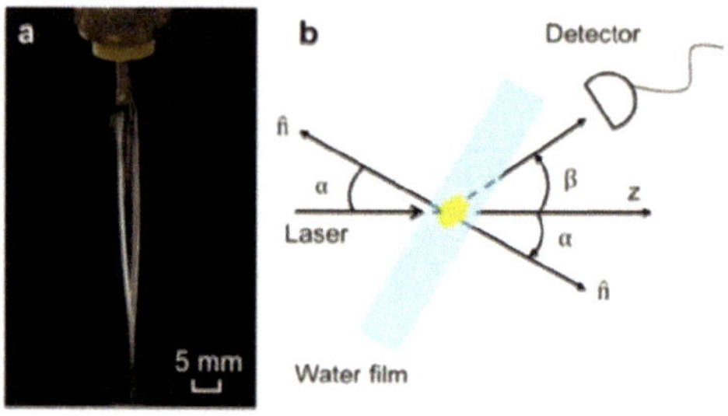

Figure 4.4 (**a**) Photo of a 120 μm thick water film formed by a water jet with a flat nozzle. (**b**) Illustration of incident angle α and detection angle β.

jet with a pressure of 30 psi is used to create a 5 mm wide, 120 μm-thick free-standing water film, shown in Fig. 4.4a. The thickness of the water film is measured by an auto-correlation system [9]. The relatively high flow rate of a jet benefits a stable thin film, which is crucial for obtaining intense and stable THz waves. Moreover, the noise level of the system with a flowing liquid target is usually limited by the flatness and stability of the target. Figure 4.4b is an illustration with an incident angle α for the laser beam and detection angle β for the detector, where $\hat{n}$ is the surface normal of the water film.

Figure 4.5 is the cross-sectional diagram of the THz wave generation process in a water film. Laser pulses ionize water molecules at the focus through multiphoton absorption and cascade ionization [10]. For simplicity, it is assumed that the focus of the laser is at the center of the target. At the ionized area, quasi-free electrons experience the ponderomotive force and move towards areas with lower electron density. Simultaneously, other particles are relatively stationary due to the relatively large mass. Since the electrons move slower than the envelope of the propagating laser pulse, the density of the ionized carriers is always kept identical in the forward direction. Therefore, a dipole is created along the laser-propagating direction (shown as the black arrow in Fig. 4.5), which radiates a broadband THz pulse. Due to the total internal reflection at the water/air interface, THz emission can be coupled out when $-24.6^\circ < \theta_t < +24.6^\circ$, where only a small portion (dashed area in Fig. 4.5) of THz emission can be collected.

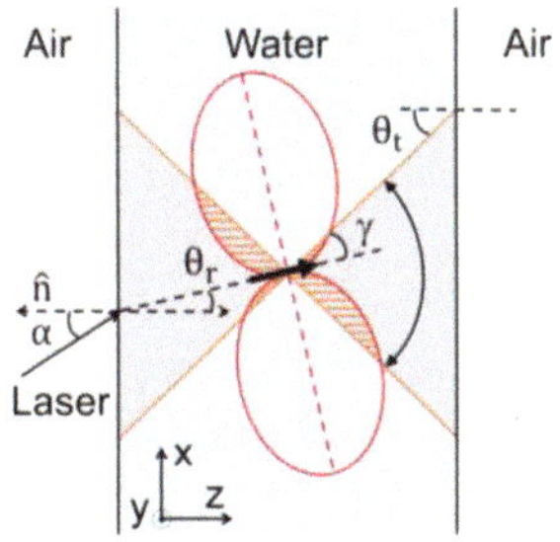

Figure 4.5 2D-cross section of a dipole model in a water film. Focused intense pulses are incident into the thin water film with an angle of α and ionize water at the focal point in the direction of the refracted laser beam.

Figure 4.6 plots the THz waveforms generated from the water film with two opposite incidence angles ($\alpha = \pm 65°$) and detected at the laser propagation direction ($\beta = 0°$). For normal incidence, $\alpha = 0°$. The sign of the angle indicates rotating the film in the opposite direction. The results show that the THz waveform keeps its amplitude with an opposite polarity when $\alpha = \pm 65°$. This observation further confirms that the direction of the dipole is along the laser propagation direction. Since the detector is in the forward direction of laser pulses, it can only detect the projection of dipoles perpendicular to the laser propagation direction. For these two cases (the opposite incidence angles), the projection of the dipole has opposite directions.

Figure 4.7 shows a simulated result of the THz wave intensity $I_{THz}(\alpha, \beta)$. In the simulation, the micro-plasma is considered as a point source emitting THz wave and is located at the center of the film for simplicity. Besides the forward (**F**) propagating signal, the signal propagating in the backward (**B**) is also expected. These two parts are separated by dashed lines in the plot and labeled separately. Due to the symmetric geometry of the model, intensity distributions for forward and backward propagating THz signals have the same patterns when the plasma is at the center of the water film. For a real case, the specific position of the focus should be considered. In other words, the THz intensity in either forward or backward direction could be optimized by the relative position of the plasma in the liquid target. The dashed lines also indicate the case

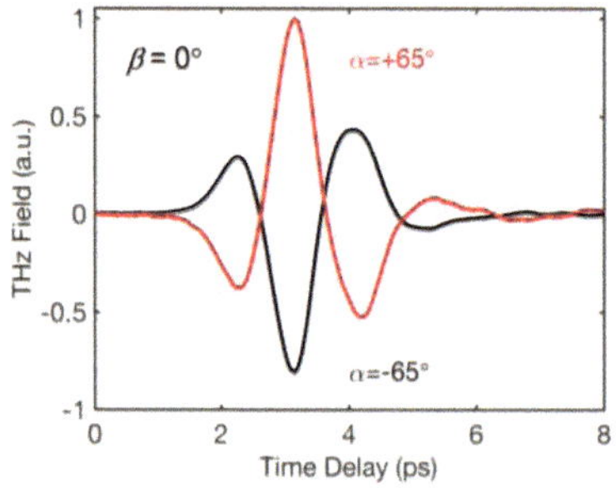

Figure 4.6 THz waveforms with opposite ± incident angles. The THz shows an opposite polarity with the same field strength when the incidence angle α changes its sign.

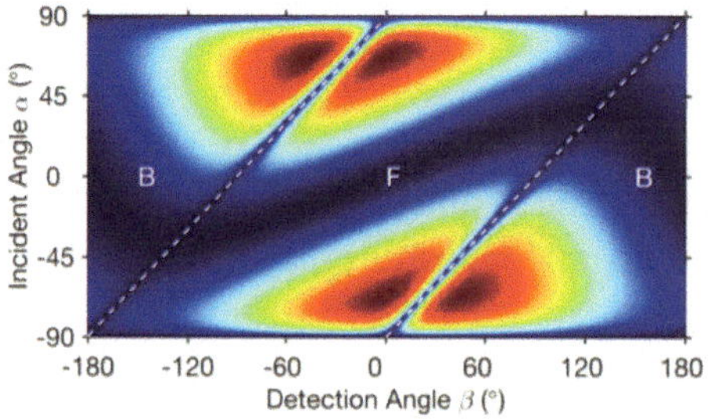

Figure 4.7 Simulation results of normalized intensity $I_{\text{THz}}(\alpha, \beta)$ using the dipole approximation. Labels "B" and "F" indicate the backward and forward propagating THz signals, respectively.

of $|\alpha-\beta| = 90°$, which means that the detector is put in the plane of the water film. If $|\alpha-\beta| > 90°$, THz waves propagate in the backward direction.

To verify the simulation result, THz signal versus α is measured when β is fixed at 0°. Both EO sampling and a Golay cell are used in the measurement. The corresponding results are plotted in Fig. 4.8, where the red solid line shows the simulation result. The EO sampling result (black squares) is calculated as the time integral of the absolute square of the corresponding THz waveform. Results from a Golay cell are plotted in as blue dots. As shown in the result, the optimal incidence angle of the laser beam is 65°, which is a combined result of the transmittance of the p-polarized excitation

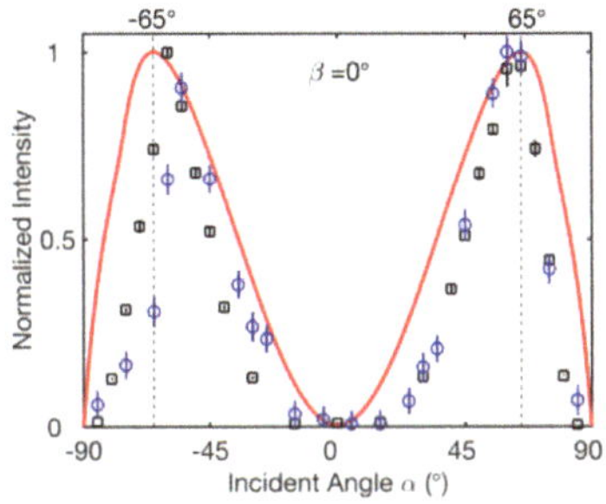

Figure 4.8 THz intensity dependence on an optical incident angle α when $\beta = 0°$. The black squares are the experiment data measured by EO sampling and the blue circles show the results from a Golay cell.

laser at the air/water interface and the dipole orientation direction. Note that only the forward THz signal can be measured when $\beta = 0°$.

4.4 THz Wave Emission under the Excitation of Asymmetric Optical Fields

The enhancement of THz wave radiation with an asymmetric excitation scheme has previously been observed when air plasmas act as THz sources [11]. The generation efficiency is enhanced by 3 orders in the electric field when two-color excitation is employed. To implement two-color excitation in the setup, a nonlinear crystal (BBO) is usually used to generate the second harmonic beam (2ω). Both the fundamental and second harmonic beams are focused to create the plasma, in which the optical fields are asymmetric since the relative phase between two beams. The schematic diagram of two-color excitation in water film is shown in Fig. 4.9a. An in-line phase compensator accurately controls the relative phase between two beams (ω and 2ω) [12]. The energy of 2ω pulses is about 10% of the entire excitation laser energy. Both ω and 2ω pulses have the same polarization. Subsequently, ω and 2ω laser pulses are co-focused into a 120 μm-thick water film by a 1 inch effective focal length parabolic mirror.

Figure 4.9b plots the THz waveforms generated by one-color (ω) and two-color (ω and 2ω) under the same total optical pulse energy.

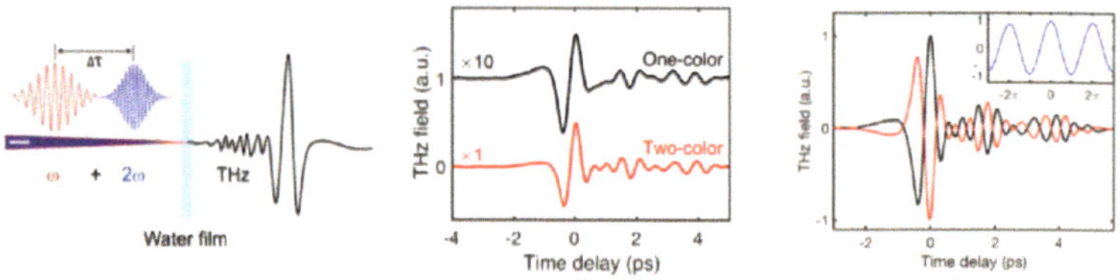

Figure 4.9 (a) Schematic diagram of the experimental setup for the asymmetric optical field excitation with two-color pulses. (b) THz field increases 10 times with two-color excitation than one-color excitation. (c) Two THz waveforms obtained by EO sampling when the relative phase between ω and 2ω pulses is varied by π. Inset, the peak of THz electric field as a function of the phase delay between ω and 2ω pulses.

A 10 times increase in field (100 times in energy) is obtained with the two-color excitation. Compared to the air plasma, the electrons ionized in liquid water are regarded as quasi-free electrons [13], since most electrons are excited to the conduction band rather than free space. Therefore, the THz wave generation process in water resembles this process in air. Phenomenologically, the transient photocurrent model can be used to explain the generation process in water and would predict the modulation of THz fields generated from a water film as well, which is experimentally confirmed, as shown in Fig. 4.9c. Specifically, Fig. 4.9c shows that the polarity of the THz electric field is completely flipped over by changing the relative phase ϕ between the fundamental and the second harmonic pulses by π. The inset of Fig. 4.9c plots the THz field as a function of optical phase delay between ω and 2ω pulses, which indicates that the polarity of the THz electric field is gradually changed with the optical phase delay.

An overall phase scan for THz wave emission from the water film is obtained by gradually adjusting the phase between ω and 2ω pulses at an attosecond-level accuracy while monitoring the THz energy with a Golay cell, as shown in Fig. 4.10. The noise floor is also shown in the figure, which is measured by blocking the THz wave while the optical excitation is kept. The modulated portion shows the phase modulation while the unmodulated portion remains blank at the bottom of the figure. By comparing the energy levels in the figure, the modulated and unmodulated components are estimated

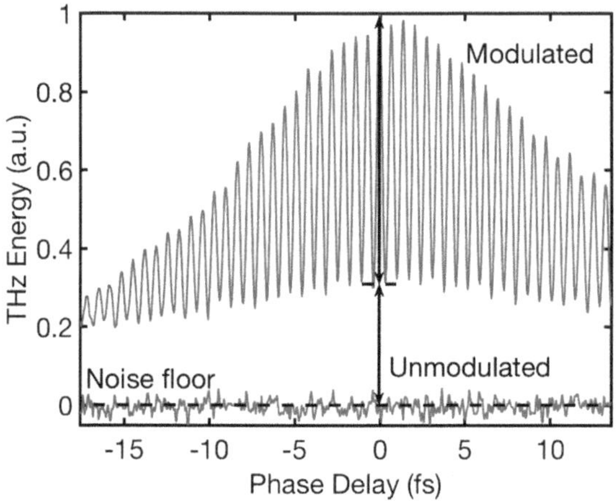

Figure 4.10 An overall phase scan for THz wave radiation from the water film obtained by gradually changing the phase between ω and 2ω pulses while monitoring the THz energy with a Golay cell.

to be 70% and 30%, respectively. The modulated and unmodulated THz waves relate to different generation processes in the plasma.

Figure 4.11 plots the THz energy dependence of different components on the total optical excitation energy. The unmodulated THz energy (red circles) shows a linear dependence on the laser pulse energy. For the modulated THz energy (blue dots), the modulation does not appear until the excitation pulse energy is beyond 200 μJ. Subsequently, the measurement matches a quadratic fitting above the threshold. The energy measurement from EOS (blue squares) is coincident with the modulated result from the Golay cell.

The two components show different mechanisms. The modulated component mainly comes from electron acceleration [14, 15] and the buildup of bremsstrahlung from electron-atom collisions [16]. The unmodulated component may arise from other processes, such as a spatial net charge distribution created by the ponderomotive force [2]. Since no threshold is observed for the unmodulated portion, the THz wave emission can be attributed to part of the broadband radiation from the combination of thermal bremsstrahlung from electrons and electron-ion recombination

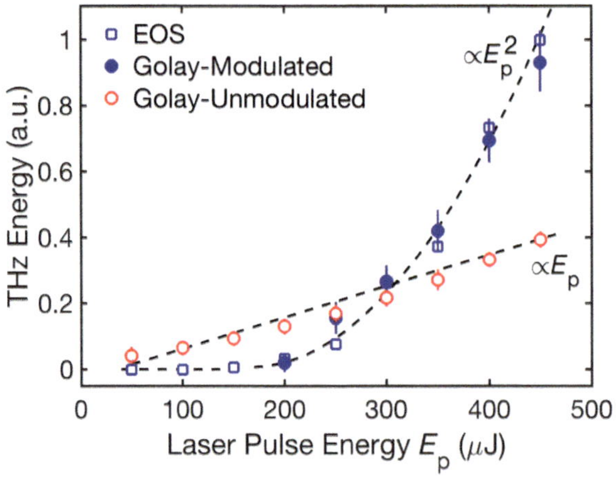

Figure 4.11 Normalized THz energy as a function of the total excitation laser pulse energy. Blue dots and red circles are modulated and unmodulated THz energy measured by the Golay cell, respectively.

[10]. Moreover, the energy dependence in Fig. 4.11 indicates that the ratio of the modulated to unmodulated THz energy increases with the laser pulse energy. The unmodulated component is stronger with weak excitation pulses while the modulated component will dominate if intense laser pulses are used due to the quadratic feature.

4.5 THz Emission from Waterlines

Due to the total internal reflection at the liquid/air interface, only a small fraction of total THz energy can be collected from a liquid film. Based on the dipole model, using a line to replace the film, theoretically, coupling up to 5 times or more THz power out is expected. In the experiment, a stronger THz signal emitted from a waterline than a water film is observed. The waterline with a diameter of a few hundreds of microns can be easily produced by using syringe needles. Figure 4.12 shows a photo of 8 syringe needles with inner diameters of 60, 90, 160, 210, 260, 330, 410, and 510 μm (corresponding to their gauge number from G34 to

Figure 4.12 Eight (8) syringe needles. The distinct color indicates the different gauge number (G#). Lime green needle (G34) has 60 μm inner diameter, which is the smallest among the collection.

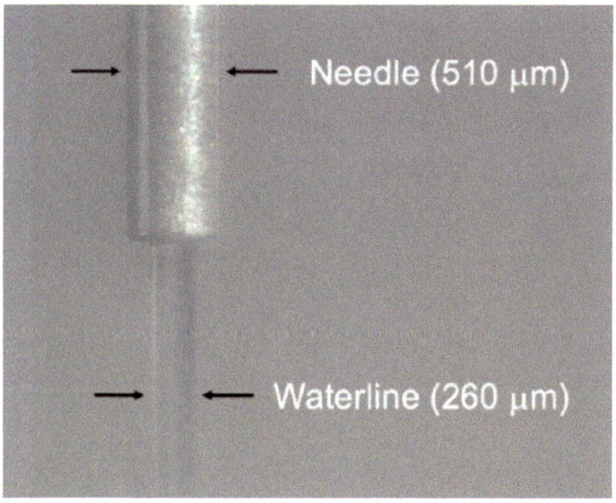

Figure 4.13 Photo of a flowing 260 μm waterline produced by a syringe needle (G27). For a laminar flow, the diameter is determined by the inner diameter of the needle.

G21), respectively. Figure 4.13 is a photo of 260 μm waterline from a syringe needle (G27).

Figure 4.14a shows the measured THz peak field emitted from a 260 μm diameter waterline crossed the laser focal spot. When x-position equals the zero, the laser beam is focused at the center of the water line. Figure 4.14b plots the THz temporal waveform at the optimal position (± 90 μm from the center of the waterline). The opposite polarity at ± 90 μm supports the dipole model discussed

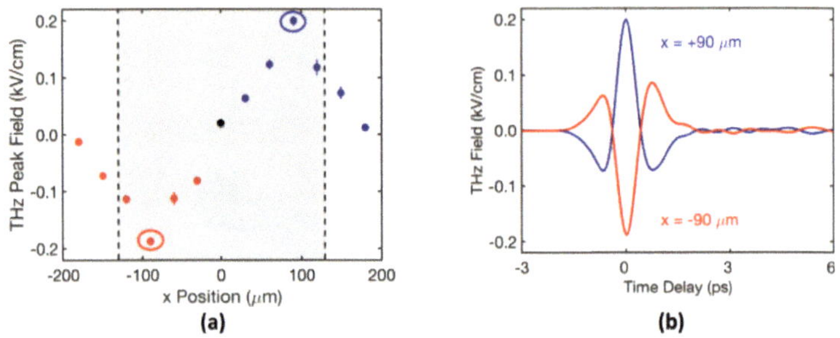

Figure 4.14 (**a**) Peak field and (**b**) temporal waveform of THz emission when a waterline is crossed the laser focal spot.

previously, where the incident angles of the optical beam have opposite signs in these two cases.

4.6 Summary of Results of THz Wave Generation from Liquid Water

The following shows the key observations and confirmations from current investigations from three research groups from the University of Rochester (USA), Capital Normal University (China) and ITMO University (Russia).

4.6.1 Key Observations

- THz wave emission from water is stronger than the one from air under the comparable laser condition for excitation.*
- Excitation with subpicosecond optical pulse duration (300 ~500 fs) in water is better than femtosecond laser pulse duration (<100 fs).
- The brightest white fluorescence from water does not offer the strongest THz signal at the same time.

*The experimental observation showed that the optimal laser pulse duration for the strongest THz emission from water depends on the thickness of the water film. For example, a 120 μm thick film, the pulse duration is 0.3 ps for maximizing the THz signal. The thicker the film, the longer optical pulse duration is preferred.

- Water vapor mist ejected by the laser excitation follows in both forward and backward directions.

4.6.2 Other Confirmations

- THz wave emission pattern under the single-color optical excitation agrees with the dipole model.
- THz wave emission is very sensitive to the position of laser focus to the liquid target.
- THz wave polarization (s-, p-) is independent of the optical polarization of the excitation beam under the normal incident angle but strongly depends on the optical polarization under the oblique incident angle.

4.7 THz Wave Generation from Liquid Metal

Solid metals were the first targets to facilitate the generation of THz waves via femtosecond laser excitation [17]. Recently, new reports on using solid metals with ultra-intense laser pulses have been released [18]. It is believed that the strength of the THz field by using these methods would set new records worldwide. However, under excitation with ultrahigh optical energy, the solid target is completely destroyed after one laser shot. The single-shot emission limits its further application. To solve this, we use liquid metals as the target instead of their solid phase. Based on our study of THz emission from liquid water, we replace the water with liquid metals to create a flowing target. The well-known liquid metal is mercury. However, considering the safety in the laboratory, the gallium-based alloys, which are lower both in their vapor pressure at room temperature and toxicity, are being used as a replacement in various applications. Table 4.2 lists the melting point, surface tension, density, viscosity, and ionization energy of four liquid metals. The corresponding properties of water are used as a reference in the second column.

Among these materials, one could find that gallium (Ga) is the best candidate. Ga has its melting temperature at 30°C, which is easy to implement in the lab. The surface tension of liquid gallium

Table 4.2 Melting point for selected samples of the liquid metals

Material	Water	Galinstan*	Gallium	Phosphorus	Indium
Melting point	0°C (23.00°F)	-19.0°C (-2.00°F)	29.8°C (85.60°F)	44.0°C (111.00°F)	156.6°C (313.88°F)
Surface tension	0.073 N/m (at 20°C)	0.718 N/m (at 20°C)	0.724 N/m (at 40°C)	0.698 N/m (at 50°C)	0.558 N/m (at 170°C)
Density	0.998 g/cm^3 (at 20°C)	6.440 g/cm^3 (at 20°C)	6.080 g/cm^3 (at 53°C)	1.740 g/cm^3 (at 50°C)	7.020 g/cm^3 (at 170°C)
Viscosity	0.0010 Pa·s (at 20°C)	0.0024 Pa·s (at 20°C)	0.0019 Pa·s (at 53°C)	0.0016 Pa·s (at 50°C)	0.0019 Pa·s (at 170°C)
Ionization energy	9.50 eV		5.98 eV	10.64 eV	6.08 eV

*Galinstan: (62–95% Gallium, 5–22% Indium, 0–16% Tin by weight).

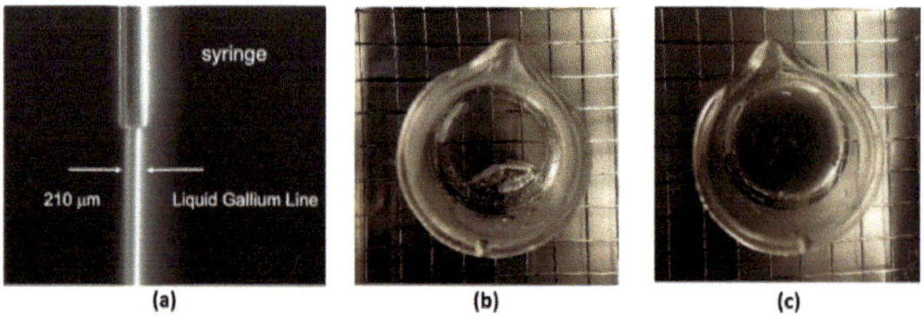

Figure 4.15 (**a**) Photo of a flowing liquid gallium line ($d = 210$ µm) from a syringe. The high surface tension of Ga (10 times greater than water) makes an excellent surface quality. (**b**) Adding liquid gallium to water and breaking Ga into mixed solution in an ultrasonic cleaner. (**c**) Mixed Ga-H_2O solution is obtained.

at 30°C is 0.708 N/m, which is 10 times higher than that of liquid water (0.0712 N/m). Such a high surface tension forms a high-quality flowing liquid line, as shown in Fig. 4.15a. A liquid metal line with its diameter comparable to the laser focal spot is possible. Figures 4.15b,c show that Ga can be made into a milky solution with water by using an ultrasonic cleaner. Under the excitation of laser pulses, Ga_2O_3 can be formed during oxidization, which may cause pipe clog during the experiment. For other liquid metals, special

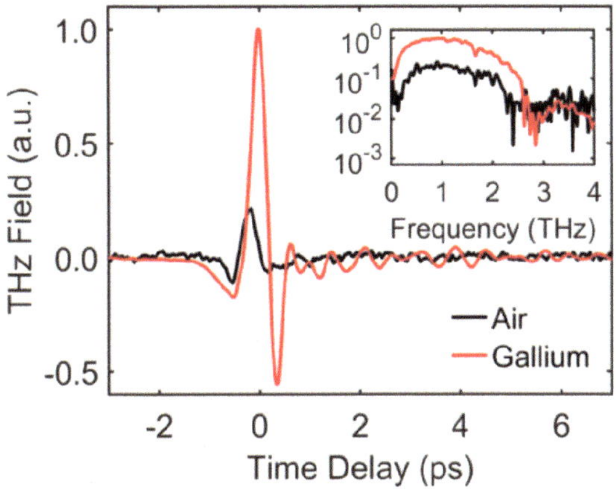

Figure 4.16 THz waveforms generated from air plasma and liquid gallium line, respectively. The corresponding comparison in the frequency domain is shown in the inset.

cautions are needed for handling indium and phosphorus due to safety concerns. For example, Indium compounds could be toxic, and some phosphorus co-products are flammable.

The THz signals excited from single-color air and liquid gallium line are shown in Fig. 4.16. Compared to the maximum signal generated by an air plasma source under single-color excitation and with 370 fs pulse duration, the liquid gallium source has a higher energy and peak field strength with a maximum conversion efficiency of 2.5×10^{-7}. Figure 4.17 shows the THz peak field dependence on the position of the line across the focus (moving in the *x*-axis). As shown in the plot, translation along the *x*-axis and keeping the detection static shows two nearly identical peaks corresponding to angles that are opposite in sign, which is similar to the observations of the waterline. Since liquid metal is opaque for both THz and optical waves, the optimized position is near the surface of the target. The z position (in the laser-propagating direction) is optimized before each measurement. We note that the *x*-axis is perpendicular to the optical pulse propagation direction. The optimum positions are reported as $\pm$ 85 μm.

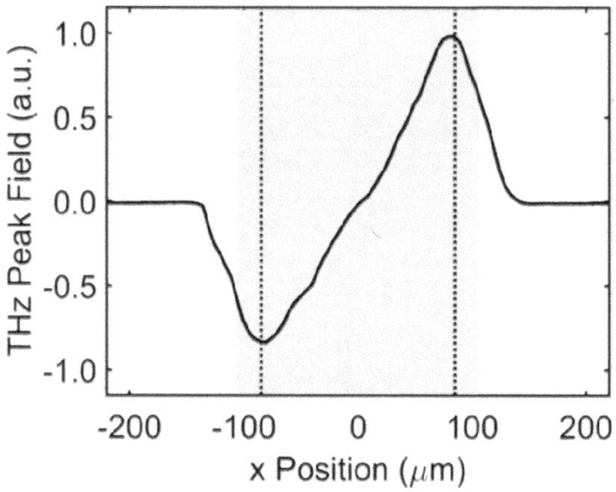

Figure 4.17 Normalized THz field strength as a function of the *x*-axis position from liquid gallium line. The dashed black line represents the *x*-axis position of the maximum THz fields for liquid gallium. The width of the gray area stands for the diameter of the liquid gallium line.

4.8 THz Wave Generation from Liquids with Nanoparticles

Using liquids to carry nanoparticles for THz wave generation may offer many new opportunities, such as gold (Au) nanoparticles in water, and dye liquids. We experimentally investigated the possibility of enhancing THz radiation by adding Au nanoparticles into the liquid target. The resonant wavelength of nanoparticles is around the central wavelength (800 nm) for excitation. The absorption at 800 nm decreases obviously with declining concentration indicating that the resonant absorption intensity is strongly dependent on the concentration. However, the THz radiations generated from the Au nanoparticle solution are insensitive to the concentration. Compared with the THz radiation generated from the deionized water, THz waveforms from the nanoparticle solution are almost the same. Although the THz radiation is not enhanced by the localized surface plasmon of Au nanoparticles, the experiment provides worthy information that the concentration of nanoparticles within the order of 10^8 to 10^{11} is not a sensitive factor to THz

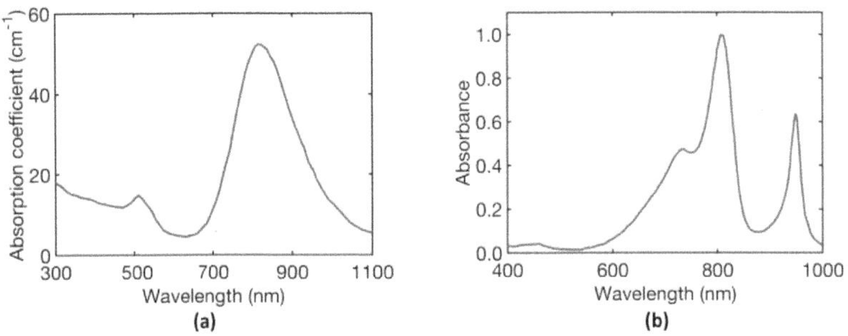

Figure 4.18 (**a**) The absorbance spectrum of Au nanoparticle solution. (**b**) The absorbance spectrum of a dye solution (Product Code: NIR818D from Exciton Company). Ti:sapphire laser has a turning range from 700 to 900 nm, with the central wavelength at 800 nm, it can be turned for resonance or non-resonance excitation.

wave generation. It is an important reference for investigating the possibility of enhancing THz wave emission from liquid by nanoparticles.

Organic dye materials have been used for dye lasers in the 70s and 80s [19]. We found that some dyes have strong absorption at the laser wavelength. Using selected dyes as carriers of nanoparticles with strong resonance at laser wavelength may increase the absorption of the optical absorption. Figures 4.18a and 4.18b plot the absorbance of Au nanoparticles in liquid water (Nano Composix) and a dye in methanol and water (Exciton company, product code NIR818D), respectively. With a femtosecond Ti:sapphire laser at a center wavelength of 800 nm, both liquids have its peak absorption around 800 nm. However, the liquid with Au nanoparticles has the plasmonic effect. It is unclear if such plasmonic resonance effect will contribute to THz wave generation. A measurement of THz emission by tuning laser wavelength from resonance to non-resonance excitation for comparison will give us more physics about the influence of optical absorption. This topic provides a new direction for the study of laser-liquid interaction. Measuring properties of THz wave emission from different liquids also helps us better understand the physics of laser-liquid interaction.

4.9 THz Wave Emission from Liquid Nitrogen

Liquid nitrogen with nonpolar molecules hardly absorbs THz waves. It will be an interesting topic to study THz wave generation from liquid nitrogen. However, the surface tension of liquid nitrogen at atmospheric pressure is about one order smaller than that of water. This low surface tension of liquid nitrogen makes it extremely difficult to form a stable nitrogen line in the lab. Moreover, the pressure in a storage dewar has a higher pressure than the atmospheric pressure leading to a higher boiling point in the dewar.

Figure 4.19a shows a photo of a nitrogen stream right after a liquid nitrogen tank. Gas bubbles and liquid droplets get mixed in the stream and scatter the visible beam. We have built a unique device with extra thermal isolation, similar to the design for the cryogenic storage dewar, to generate a liquid nitrogen line, as shown in Fig. 4.19b. A stable and optically transparent liquid nitrogen line is obtained in the ambient environment, as shown in Fig. 4.19c. The device to make a liquid nitrogen line is using gravity-driven and free-flowing method to replace pressure pumping. The vaporization creates an insulating layer which keeps the liquid from boiling rapidly. Then, a steady liquid line is formed.

(a) (b) (c)

Figure 4.19 (**a**) Photo of a nitrogen stream from a liquid nitrogen tank. This nitrogen stream cannot be used for the experiment since both the gas bubbles and liquid droplets in the stream scatter the optical beam. (**b**) A specially designed device is capable to produce a stable and uniform free-flow liquid nitrogen line. (**c**) Photo of a free-flow liquid nitrogen line in the ambient environment.

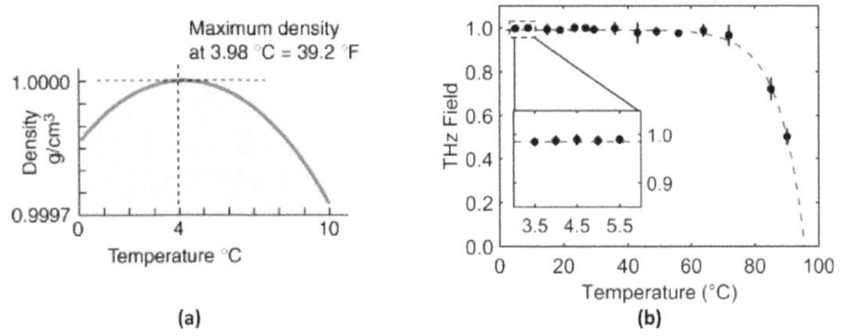

Figure 4.20 (**a**) Temperature-dependent water density (from Wikipedia). The density peaks at 3.98°C. (**b**) Preliminary test of THz emission from water from 2 to 90°C. In order to measure the density change near 4°C, an SNR > 10,000 and temperature better than 0.1°C is needed.

4.10 Density Singularity of Water at 4°C

It is well known that liquid water has the highest density at 3.98°C. We need to improve the system SNR to better than beyond 10,000 in order to measure the density singularity of water at 4°C. Figure 4.20a shows the water density vs temperature in the range of 0 to 10°C. The change in the density over this range is about 0.003%. The preliminary measurement of the THz signal from water at different temperatures is plotted in Fig. 4.20b. In order to see this 4°C effect, assuming the THz emission is linearly related to the target density, then the control step of water temperature should be less than 0.1°C, and the SNR of the THz field measurement should be better than 10,000. This requires us to further improve our liquid-target THz system to measure a small change in the THz field. A fast drop of the THz signal at 75°C in Fig. 4.20b might be due to the water vapor generated by the laser beam. At 75°C or higher, each laser pulse generates a large amount of water vapor which was not observed at lower temperatures. The reason is unclear at this moment.

4.11 Molecular Orientation and Alignment

It is expected that if the molecules are spatially oriented or aligned, then the THz emission from these aligned/oriented molecules may show its strength and polarization depending on the degree of molecules' orientation and alignment. We would like to study the dynamics of the molecular orientation and alignment by using double-optical-excitation in the femtosecond time frame. Double-optical-excitation uses two optical pump pulses: the first pump pulse to ionize in liquid, and the molecules in the ionization area are oriented and aligned by the THz field generated by the first pump pulse, then the second time-delayed pump pulse will generate a new THz field from the previous plasma which is created by the first pump pulse. This study will help us to understand the kinetic role of ionized water molecules during the THz wave generation, especially its dynamic process.

To understand the role of laser-induced micro-plasma kinetics in liquids, we examine the THz signal by using the double-optical-excitation method. Figure 4.21 is the schematic diagram of the setup. The water film is excited by two optical pulses (1 and 2). Only the THz signal generated by pulse 1 is measured by the detector, which is selected by the chopper frequency with a lock-in amplifier. We check the THz signal change by pulse 2, pre- or pro- excitation. For simplicity, the detector is in the laser propagation direction ($\beta = 0°$) in the preliminary measurement.

Figure 4.22 plots the preliminary data of the THz peak field vs the time delay $\Delta\tau$ measured by EO sampling. Both enhanced amplitude (K) and temporal window (T) depend on the energy ratio of pulse 1 and pulse 2. This plot should be redone with a THz energy detector (Golay cell) in order to see the true energy change. THz field should

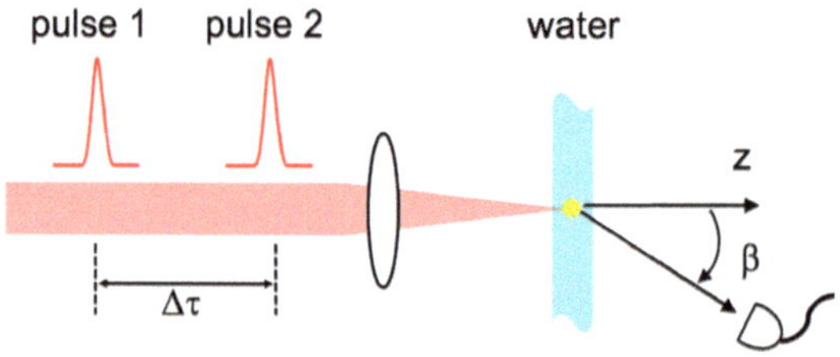

Figure 4.21 Schematic diagram of double-pulse excitation setup. Main pulse (pulse 2) as a pump pulse is amplitude modulated, and the detector only measures the THz signal generated by pulse 1.

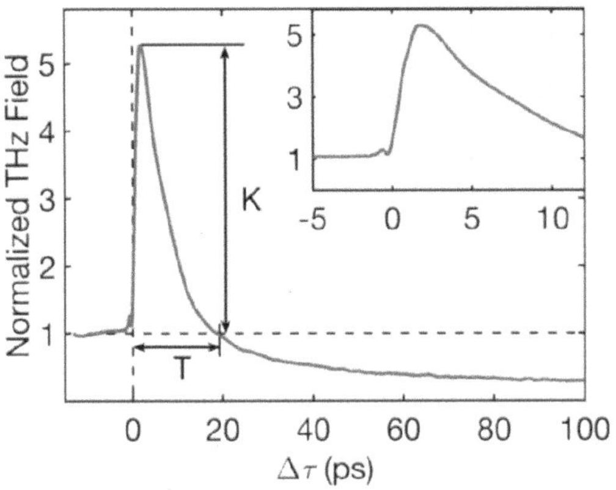

Figure 4.22 Measured THz peak field vs relative timing $\Delta\tau$ between two pulses by EO sampling. K refers to the amplitude enhancement. T refers to the time existing enhancement. Inset zooms the plot near the peak at zero time delay ($t = 0$).

align or orient the molecules in the pre-plasma created by pulse 2, then we measure the enhanced THz field with pulse 1.

Surprisingly even though s-polarized optical pulses generate a weak THz signal from the water line, the enhancement of THz waves with s-polarized optical pulses is stronger than that by using p-polarized optical pulses, as shown in Fig. 4.23. The enhanced THz signal (red curve) shows that the s–p arrangement is better than the p–p arrangement. Possibly, molecular alignment is one of the explanations.

4.12 Magnetic Fluids

Magnetic fluids can at large be categorized into a ferrofluid or a magnetorheological (MR) fluid. A ferrofluid is composed of nanoscale ferromagnetic particles, typically magnetite (Fe_3O_4) or hematite (Fe_2O_3), colloidally suspended in carrier fluids, such as organic solvents or water. The ferromagnetic nanoparticles are coated with a surfactant (see soap, detergent) to prevent their agglomeration (due to van der Waals and magnetic forces). The physical change from liquid to solid can be done by applying an

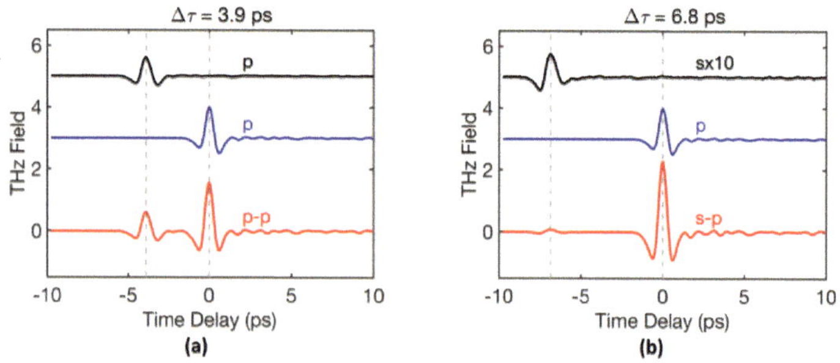

Figure 4.23 THz signals enhanced by a preformed plasma. (**a**) THz signals individually generated by the p-polarized pre-pump and main pump are plotted as the top and middle lines. The bottom line shows the THz signal generated by two beams with a certain time delay $\Delta\tau$. (**b**) Similar results are plotted when the pre-pump is s-polarized. The enhancement is greater while the pre-pump is s-polarized. All the signals are normalized by the signal from the p-polarized main pump. The possible reason for the enhancement is the molecules in the pre-plasma are aligned and oriented by the pre-pump.

external magnetic field. A ferrofluid becomes strongly polarized in the presence of a magnetic field and forms chains of particles locally [20]. A video of ferrofluid in the proximity of a magnet can be found in Ref. [21].

The MR fluid is a magnetically sensitive magnetorheological fluid whose viscosity changes in milliseconds when exposed to a magnetic field. The particle size lies within the μm scale. The magnetic field creates sharp and stiff structure of iron particles with magnetorheological fluid as compared to ferrofluid.

The majority of efforts having been invested in magnetic fluids concentrates on exploiting the change in their viscosity in a magnetic field, and thus the applied force to materials in contact. Typical applications stemming from these studies include damping mechanisms in automobile brakes and speaker voice coils, and optics surface polishing.

While there have also been some scattered investigations on the optical properties of magnetic fluids [22], the generation of terahertz (THz) wave from magnetic fluids, nonetheless, has not been reported before. A prospective advantage lying beneath these

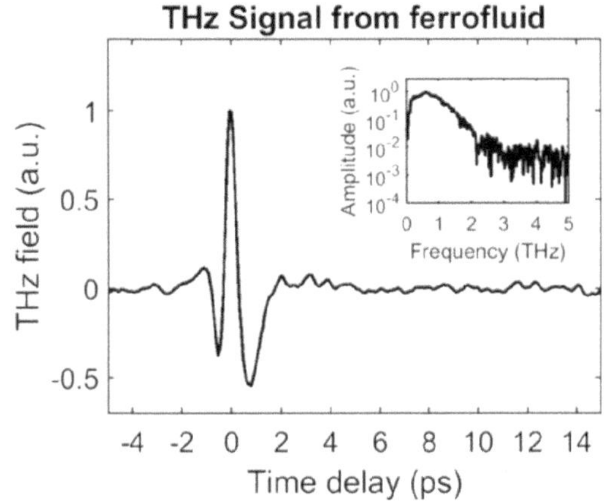

Figure 4.24 A THz waveform generated from ferrofluid. The inset shows the corresponding spectra.

novel THz emitter candidates is the extra degree of freedom offered by the material response to an external magnetic field. The magnetic field modifies the relative arrangements and proximity of the magnetic particles, enabling the construction of chains of particles of arbitrary orientations largely dictated by the magnetic field. These adjustable formations could bring forth real-time, efficient tunability in THz emission.

Figure 4.24 shows the THz signal emitted from ferrofluid (Ferrotec EMG900) measured with a THz-TDS. The ferrofluid is circulated in a form of a jet with a diameter of 1.18 mm at the point of excitation. The liquid jet optically excited with a single-color, 800 nm femtosecond pulse train of a 195 mW average power. Inset is the spectrum of the terahertz pulse. Note the absence of an obvious absorption dip at 1.7 THz which corresponds to the use of a hydrocarbon carrier solvent in the ferrofluid.

4.13 Future Perspective

It is very vital to study the mechanism of ionization in liquids, especially in liquid water because of its importance in various applications, such as water disinfection, radiotherapy, liquid-cooled systems, etc. During the ionization process, multiple ultrafast processes are involved. Among them, proton transfer and electron-initiated reactions have been studied for decades. THz generation from ionized liquid provides an alternative perspective to investigate the process of laser-liquid interaction. Some of the unknowns and challenges are listed as follows:

(1) What are the roles of electrons (free or hydrated) and ions in the THz emission process? Which particle plays the most important role?
(2) How can we control the properties (THz emission bandwidth, THz central frequency, the strength of THz peak field) of THz emission from the ionized liquid?
(3) How does the THz emission depend on the ultrafast processes (optical pulse duration and wavelength)?
(4) Hydrated electrons can be excited by chemical reactions. Can we generate THz without laser-induced plasma generation? What's the role of hydrated electrons in the THz emission process?
(5) Liquid metal and liquid water present different properties in optical transparency, conductivity, etc. What are the roles of these properties in the THz emission process?

4.14 Summary

THz Liquids Photonics is considered as an interdisciplinary and transformative research topic in the THz wave community. THz liquid photonics under extreme conditions include ultrafast laser pulse application (< 50 fs), ultralow temperature operation (< 2.17 K), ferrofluids, and liquid metals. A successful investigation in THz generation from liquids will complete the last piece of the matter-phase puzzle for THz sources. New science and technology will be developed by fully characterizing THz emission in liquids to support new THz wave science, technology, and applications. The further exploration of THz photonics in liquids is certainly bright. . .

Funding

This research of the University of Rochester is supported by the Air Force Office of Scientific Research (FA9550-21-1-0389) and National Science Foundation (ECCS-2152081).

Acknowledgments

The project of THz liquid photonics has an international collaboration with Liangliang Zhang, and Cunlin Zhang from Capital Normal University of China, with Anton Tcypkin, and Sergey Kozlov from ITMO University of Russia. The authors would like to thank ShingYiu Fu, Qi Jin, Kareem Garriga Francis, Fang Ling, Yuqi Cao, Jianming Dai, and Evgenia Ponomareva.

References

1. F. Buccheri and X.-C. Zhang, Terahertz emission from laser-induced microplasma in ambient air, *Optica*, 4, 366 (2015).
2. H. Hamster, A. Sullivan, S. Gordon, W. White, and R. W. Falcone, Subpicosecond, electromagnetic pulses from intense laser-plasma interaction, *Phys. Rev. Lett.* 71(17), 2725–2728 (1993).
3. H. Hamster, A. Sullivan, S. Gordon, and R. W. Falcone, Short-pulse terahertz radiation from high-intensity-laser-produced plasmas, *Phys. Rev. E* 49(1), 671–677 (1994).
4. C. D'Amico, A. Houard, M. Franco, B. Prade, A. Mysyrowicz, A. Couairon, and V. T. Tikhonchuk, Conical forward THz emission from femtosecond-laser-beam filamentation in air, *Phys. Rev. Lett.* 98(23), 235002 (2007).
5. Qi Jin, E. Yiwen, K. Williams, J. Dai, and X.-C. Zhang, Observation of broadband terahertz wave generation from liquid water, *Appl. Phys. Lett.* 111(7), 071103 (2017).
6. I. Dey, K. Jana, V. Yu Fedorov, A. D. Koulouklidis, A. Mondal, M. Shaikh, D. Sarkar, A. D. Lad, S. Tzortzakis, A. Couairon, and G. R. Kumar, Highly efficient broadband terahertz generation from ultrashort laser filamentation in liquids, *Nat. Commun.* 8(1), 1184 (2017).

7. C. Rønne, L. Thrane, P.-O. Åstrand, A. Wallqvist, K. V. Mikkelsen, and S. R. Keiding, Investigation of the temperature dependence of dielectric relaxation in liquid water by THz reflection spectroscopy and molecular dynamics simulation, *J. Chem. Phys.* 107(14), 5319–5331 (1997).
8. Q. Wu and X.-C. Zhang, Free-space electro-optic sampling of terahertz beams, *Appl. Phys. Lett.* 67(24), 3523 (1995).
9. T. Wang, P. Klarskov, and P. U. Jepsen, Ultrabroadband THz time-domain spectroscopy of a free-flowing water film, *IEEE Trans. Terahertz Sci. Technol.* 4(4), 425 (2014).
10. P. K. Kennedy, D. X. Hammer, and B. A. Rockwell, Laser-induced breakdown in aqueous media, *Prog. Quantum Electron* 21(3), 155 (1997).
11. X. Xie, J. Dai, and X. C. Zhang, Coherent control of THz wave generation in ambient air, *Phys. Rev. Lett.* 96(7), 075005 (2006).
12. J. Dai, N. Karpowicz, and X.-C. Zhang, Coherent polarization control of terahertz waves generated from two-color laser-induced gas plasma, *Phys. Rev. Lett.* 103(2), 023001 (2009).
13. J. Noack and A. Vogel, Laser-induced plasma formation in water at nanosecond to femtosecond time scales: calculation of thresholds, absorption coefficients, and energy density, *IEEE J. Quantum Electron* 35(8), 1156 (1999).
14. K. Y. Kim, J. H. Glownia, A. J. Taylor, and G. Rodriguez, Terahertz emission from ultrafast ionizing air in symmetry-broken laser fields, *Opt. Express* 15(8), 4577 (2007).
15. K.-Y. Kim, A. Taylor, J. H. Glownia, and G. Rodriguez, Coherent control of terahertz supercontinuum generation in ultrafast laser–gas interactions, *Nat. Photonics* 2(10), 605 (2008).
16. K.-Y. Kim, Generation of coherent terahertz radiation in ultrafast laser-gas interactions, *Phys. Plasmas* 16(5), 056706 (2009).
17. W. P. Leemans, C. G. R. Geddes, J. Faure, C. Tóth, J. V. Tilborg, C. B. Schroeder, E. Esarey, G. Fubiani, D. Auerbach, and B. Marcelis, Observation of terahertz emission from a laser-plasma accelerated electron bunch crossing a plasma-vacuum boundary, *Phys. Rev. Lett.* 91(7), 074802 (2003).
18. G. Liao, Y. Li, H. Liu, G. G. Scott, D. Neely, Y. Zhang, B. Zhu, Z. Zhang, C. Armstrong, and E. Zemaityte, Multimillijoule coherent terahertz bursts from picosecond laser-irradiated metal foils, *Proc. Natl. Acad. Sci.* 116(10), 3994–3999 (2019).

19. F. J. Duarte, P. Kelley, L. W. Hillman, and P. F. Liao, *Dye Laser Principles: with Applications.* (Academic Press, 1990).
20. A. O. Ivanov and A. Zubarev. Chain formation and phase separation in ferrofluids: the influence on viscous properties. *Materials* 13(18), 3956 (2020).
21. Ferrofluid: MAGRON/Ferrozone: Republic of Korea. MAGRON/ Ferrozone, www.ferrozone.co.kr/.
22. S. Chen, et al. Tunable optical and magneto-optical properties of ferrofluid in the terahertz regime. *Opt. Express* 22(6), 6313–6321 (2014).

Chapter 5

Photomixing THz Sources

Osamu Morikawa[a] and Fumiyoshi Kuwashima[b]
[a]*Chair of Liberal Arts, Japan Coast Guard Academy,*
Wakabacho 5-1, Kure, Hiroshima 737-8512, Japan
[b]*Department of Electrical and Electronic Engineering,*
Fukui University of Technology, 3-6-1 Gakuen, Fukui 910-8505, Japan
morikawa@jcga.ac.jp, f7_kuwashima@outlook.jp

The first THz photonic sources were the photoconductive antennas (PCAs) irradiated by femtosecond pulse lasers [1, 2], which are expensive instruments. It is also feasible for the PCAs to generate continuous-wave (CW) THz radiations from laser power modulation due to optical beat in beams from low-cost CW lasers.

5.1 Generation of CW THz Radiation Using Photomixing

The optical beat in CW laser beams can be generated by mixing two or more spectral components (photomixing). This is realized with a few methods: (a) superposing two single-mode laser beams; (b) constructing a laser holding two oscillating modes (dual-mode laser); and (c) using a multimode laser, such as a multimode laser diode (MLD). Before we describe the irradiation sources, we first

Advances in Terahertz Source Technologies
Edited by Gun-Sik Park, Masahiko Tani, Jae-Sung Rieh, and Sang Yoon Park

ISBN 978-981-4968-89-8 (Hardcover), 978-1-003-45967-5 (eBook)
www.jennystanford.com

introduce the devices for photomixing THz sources and the THz radiation power.

5.1.1 Devices for Photomixing THz Sources and THz Radiation Powers

As a conversion device from the CW laser power modulation to the THz radiation, PCAs are utilized in most experiments (for example in [3]), although some groups have reported THz generation based on difference frequency generation using a quasi-CW beam (laser pulses with the long duration time of around 10 ns [4, 5]). The advantages of the PCAs are broad tunability, which typically ranges from 0.1 to 3 THz, and the integration that is possible with diode and fiber technologies [6]. In particular, PCAs with interdigitated electrode structures are often adopted for the CW laser irradiation to enhance the radiation efficiency [3, 6–8], since the THz radiation power is lower than that by the pulse lasers as explained as follows [9]: one pulse with a power p and a continuous n-pulse train each with a power p/n have the same average powers (Fig. 5.1), where the pulse widths are taken as equal and then the duty ratio of the pulse and CW lasers are $1/n$ and 1, respectively. These cause THz radiation with a power of p^2 and $n \times (p/n)^2 = p^2/n$, respectively, because the radiation power is roughly proportional to the square of the radiated electric field, which, in turn, is roughly proportional to modulation in the irradiating beam power [10]. Thus, the THz radiation power from a PCA is inversely proportional to the duty

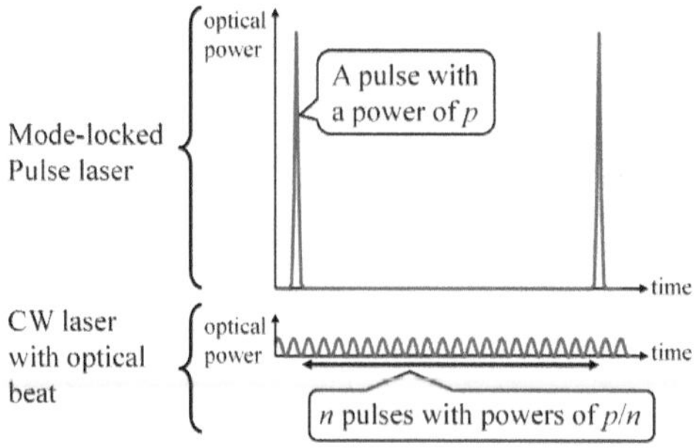

Figure 5.1 The pulses in laser beams.

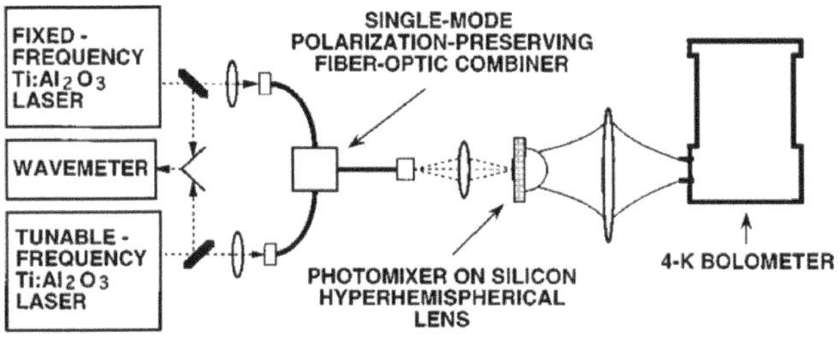

Figure 5.2 Schematic diagram of a photomixing THz source using two single-mode lasers. In the figure, "PHOTOMIXER" denotes PCA [3].

ratio of the irradiating laser beam. The radiation power by a CW laser is typically around 1/10000 times as small as that by a pulse laser. The challenge for CW systems is to achieve sufficient THz power for applications and the THz power of a few microwatts has been reported using LT-GaAs PCAs in CW operation [6].

In the early stages of the photomixing THz sources, the lasers with a wavelength of around 800 nm were employed (Fig. 5.2) since they can excite the PCA with a photoconductive layer of low-temperature-grown GaAs [3, 11, 12]. To combine 1.55 μm photonics and optoelectronics technologies developed for the telecommunication industry with the terahertz technologies, Sukhotin et al. used a PCA with a photoconductive layer of ErAs/InGaAs on InP substrate and generated THz radiation using 1.55 μm LDs [13]. About a decade ago, devices other than the PCAs have been proposed to raise the CW THz radiation power, such as uni-traveling-carrier photodiodes (UTC-PDs) [14], the waveguide integrated photodiodes with a thin absorbing layer or a PIN photodiode [15, 16].

5.1.2 Generation of THz Radiation Using Superposed Two Single-Mode Laser Beams (Two-Beam Photomixing)

A laser beam with two spectral components can be obtained by superposing two single-mode-laser beams (hereafter referred to as two-beam photomixing, Fig. 5.2) [3, 11, 13, 17, 18]. The use of wavelength-tunable lasers enables the frequency tunability of the generated THz radiation, which is the main advantage of this technique [6]. Since the wavelength change of 2.1 nm around

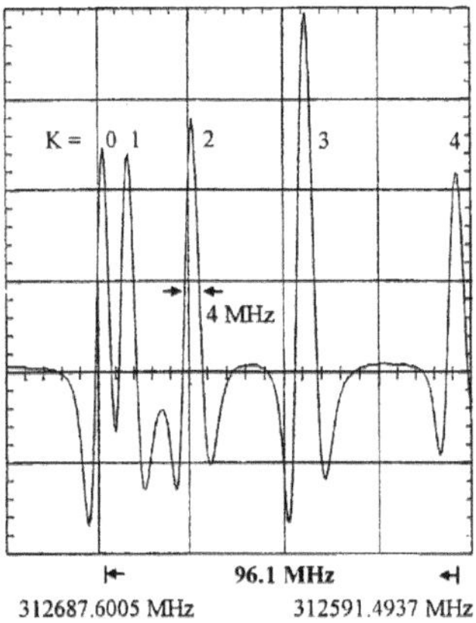

Figure 5.3 Acetonitrile absorption lines measured by the two-beam photomixing THz source and an InSb bolometer [21].

the wavelength of 800 nm corresponds to the optical frequency change of 1 THz, the tunable range of a CW Ti:Sapphire laser [3] or a diode laser [11] corresponds to more than several THz, which exceeds the bandwidth of typical PCAs. On the other hand, the main disadvantage to overcome in this technique is its low conversion efficiency as mentioned in Subsection 5.1.1. Its optical-to-THz conversion efficiency is 10^{-6}–10^{-5}, and the typical output power is in the microwatt range [6].

Using the tunability of the THz radiation frequency, this radiation source can be combined with a power detector such as a Golay cell or a hot-electron bolometer to construct a THz spectrometer [19, 20]. The spectral linewidth of the THz radiation can be as narrow as several tens of kHz using lasers with external cavities and it can be applied to high-resolution spectroscopy such as that for gas molecules (Fig. 5.3) [21].

5.1.3 Generation of THz Radiation by Photomixing Using a Dual-Mode Laser

The two-beam photomixing technique requires sophisticated techniques to precisely superpose the two laser beams [3, 13]. Instead, one can use a single dual-mode laser.

There are some methods to construct the dual-mode laser. Lee et al. proposed a combination of a Fabry-Perot (F-P) external cavity, lenses, and a grating [22]. The lens and grating were adjusted for two of the F-P modes to selectively oscillate. Hyodo et al. constructed a dual-mode laser based on a combination of laser medium with a rather narrow bandwidth of laser gain ($Nd{:}YVO_4$) and a laser cavity with a length of 3.0 mm and observed two modes oscillated in the laser gain region with the fixed optical frequency separation of 101 GHz [23]. Gu et al. constructed a distributed-Bragg-reflector laser diode (DBR-LD) with a specially designed DBR, which holds two modes with a fixed optical frequency separation of 163.5 GHz [24]. Lotem et al. proposed a construction with separately tunable wavelengths, where the output beam of an LD was divided into two and reflected back to the LD by using two gratings [25]. Wang et al. proposed a construction with an antireflection-coated LD (AR-coated LD), a grating, lenses, and a V-shaped mirror, which enables tunability of the optical frequency separation by sliding the V-shaped mirror [26]. Gu et al. applied this laser to the PCA irradiation and observed the THz radiation generation in the frequency region from 0.25 to 1.75 THz [27].

In the case of dual-mode lasers with a single optical cavity, the optical frequencies of the two modes fluctuate in the same direction when the fluctuation is caused through the optical cavity, such as electrical, thermal, mechanical, or other effects. Then the frequency fluctuation of the frequency difference, optical beat, or THz radiation is reduced compared to those of the two optical modes (common-mode-noise rejection effect) [6]. In Refs. [23] and [24], while the optical linewidths were estimated to be 100 kHz and 240 MHz, the observed linewidths of the electromagnetic radiations from the PCAs (101 and 163.5 GHz) were reduced to 430 Hz and 130 MHz, respectively.

5.1.4 Generation of THz Radiation by Photomixing Using a Multimode Laser

Multimode lasers can also be used as the irradiation source for PCAs. Like dual-mode lasers, multimode lasers do not require sophisticated techniques for the superposition of two beams. Unlike the dual-mode lasers, the multimode lasers have simple structures, which reduce their cost, and are commercially available.

The first experiment was reported by Tani et al. [12]. They employed an MLD with the optical spectrum consisting of discrete lines with equal intervals (longitudinal-mode spacing, LMS). The observed THz radiation spectrum consisted of discrete lines at the LMS frequency (52 GHz) and its multiples (hereafter referred to as a comb structure) below 0.5 THz.

The obtained THz radiation shows a broad spectrum and can be a candidate for a radiation source of spectroscopy. Since spectroscopic data such as transmittance are obtained by taking the ratios of the spectra with and without samples, the radiation spectrum should not be a discrete comb but with continuous components for spectroscopy. Morikawa et al. employed a broad-area LD (BLD), an MLD with a broad emission region (typically 50 μm or more). While an MLD with a narrow emission region (for example, 3 μm) oscillates with a single transverse mode and has a discrete optical spectrum, the BLD oscillates with multiple transverse modes and has a quasi-continuous optical spectrum with much more spectral lines. They pointed out that the optical components with common transverse modes or different transverse modes are mixed to contribute to the comb or to the continuous components in the THz spectrum, respectively [28]. They observed the appearance of the continuous components by defocusing the laser beam on the photoconductive gap of the PCA, although they are almost absent when the laser beam was tightly focused (Fig. 5.4). However, some MLDs are sometimes unstable due to intermittent mode hopping. Such MLDs can be stabilized by feeding back a part of the output beam to the MLD itself [29]. The details are described in Section 5.3.

To obtain a continuous THz radiation spectrum, one can also employ optical sources other than BLD, such as amplified spontaneous emission noise from an Er-doped fiber amplifier [30], an AR-coated LD with an external cavity that consists of a grating, lenses, and a flat mirror [31], or a superluminescent diode (SLD) [32].

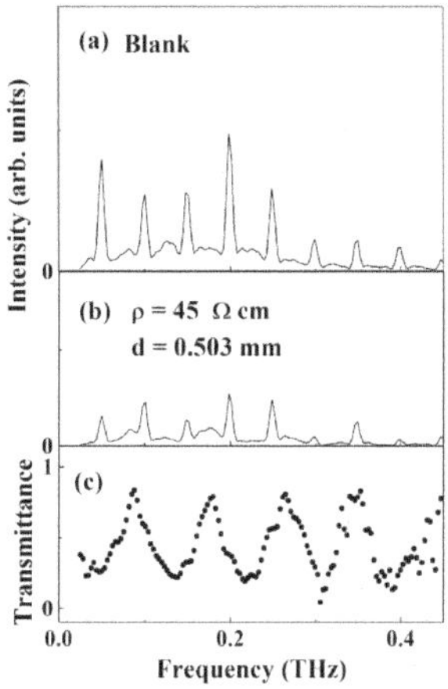

Figure 5.4 Sub-THz radiation spectra and transmittance of a Si wafer. The inset in (a) shows spectra obtained by tight focusing (upper curve) and defocusing (lower curve). Spectra of the radiation (a) before and (b) after inserting the Si wafer sample measured with defocusing. Transmittance obtained from (a) and (b) is shown in (c) [28].

5.2 Photomixing THz Sources Combined with Coherent Detection

The optically generated CW THz radiation can be measured by detecting the electric field using, for example, a PCA irradiated by the laser beam split from the one irradiating the emitter, like the conventional THz time-domain spectroscopic system (THz-TDS) [33]. Such systems can be used to obtain phase information of the THz radiation, like the conventional THz-TDS. However, the signal is weak as explained as follows: the photocurrent signal induced in the detector PCA is proportional to both the THz electric field and laser power. Since the THz radiation electric field is roughly proportional to the laser power modulation amplitude [10], the obtained signal is roughly proportional to the square of the laser power, like the THz radiation power from an emitter PCA, as explained in Subsection 5.1.1. Then the obtained signal is also inversely proportional to the duty ratio of the pump laser and is weak compared with that of a pulse laser.

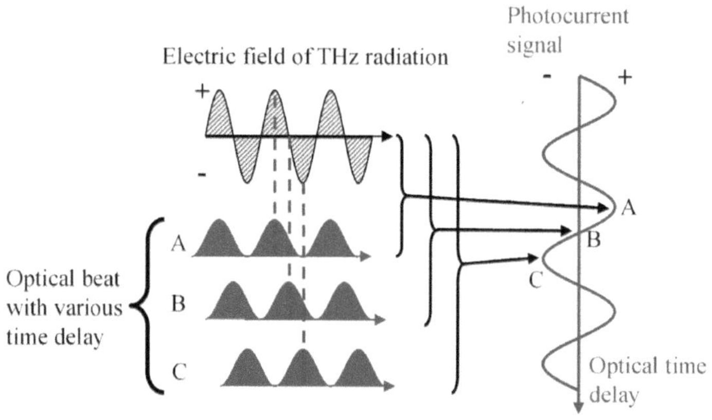

Figure 5.5 Timing of the THz radiation electric field and optical beat at the detector PCA, and generation of the photocurrent signal in the detector PCA.

5.2.1 Coherent Detection System Using Superposed Two Single-Mode Laser Beams

By employing a coherent detector such as a PCA [8, 34] or an electro-optic detector [35], we can obtain phase information of the THz radiation generated with CW lasers as well as amplitude information, as in the case of the conventional THz-TDS. The experimental setup is the same as that of the conventional THz-TDS except for the irradiating laser, where the coincidence timing of the THz radiation and optical beat at the detector is varied by scanning the optical time delay [34]. When both the laser power modulation and THz radiations are sinusoidal, the timing A, B, and C in Fig. 5.5 causes positive, zero, and negative photocurrent, respectively. One can obtain a sinusoidal photocurrent signal by scanning the time delay while monitoring the photocurrent. This photocurrent signal includes both amplitude and phase information of the THz radiation and can be used to obtain the refractive index and absorption coefficient of a sample [34]. This coherent detection system can also be applied to imaging [7, 36]. However, for the spectroscopy using this system, one has to scan the time delay at each frequency or at each step of changing the optical frequency difference between the two single-mode lasers [34]. This procedure is not required if an electro-optic phase modulator is combined with a lock-in amplifier as follows [37]: when the optical phase of one of the two laser beams to be mixed is shifted by φ with an electro-optical phase modulator, the optical beat phase is shifted by φ [10]. By modulating the optical beat phase of the beam irradiating the detector in a saw-tooth form (linear lines with 2π jumps), the phase of the sinusoidal photocurrent signal in Fig. 5.5 is modulated linearly, resulting in an AC photocurrent signal. By detecting this AC signal using a lock-in amplifier, taking the saw-tooth signal as the reference, one can measure both the amplitude and phase information of the THz signal without scanning the time delay [37, 38].

5.2.2 Cross-Correlation Spectroscopic System (CCS)

Although the CW single-mode lasers are less expensive than the pulse lasers, the spectroscopic system described in Subsection 5.2.1 requires two single-mode lasers (and an electro-optic modulator [37, 38]). One can construct a low-cost and simple THz-TDS system using a single MLD, where the setup is the same as that of the conventional THz-TDS except that the MLD is used instead of the pulse laser (hereafter referred to as cross-correlation spectroscopy, CCS) [39, 40].

In both the CCS and conventional THz-TDS, the detector PCAs produce photocurrents proportional to both the THz-radiation electric fields and laser powers and act as multipliers. The time-domain signal obtained by each of those systems is a convolution of the laser power autocorrelation and the instrumental function of the optics in the THz radiation path, the emitter PCA, and the detector PCA. In the case of the CCS, although the laser power modulation differs randomly from scan to scan, the laser power autocorrelation is stable since the laser optical spectrum is stable. Then, the complex transmittance of a sample is obtained by comparing the Fourier components of the time-domain signals with and without the sample, like the conventional THz-TDS [33, 39]. Although the CCS can produce signals equivalent to those by the conventional THz-TDS, it suffered from a low signal-to-noise ratio (SNR). This is because of the following: (a) the CCS utilizes the CW laser source and the signals are weak as depicted at the beginning of Section 5.2; (b) the signal spectrum consists of discrete lines and the spectral

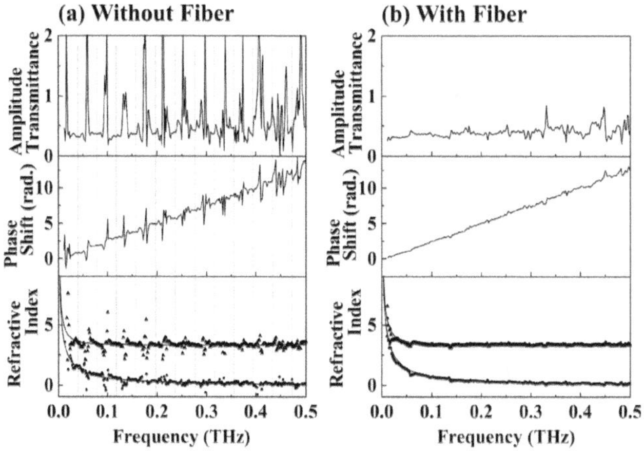

Figure 5.6 Amplitude transmittance, phase shift, and complex refractive index of an *n*-type Si wafer. For data in (a), the BLD beam was directly used, and in (b), the BLD beam was used through a single-mode optical fiber. Broken lines denote the positions of the discrete peaks of the comb structure. The symbols ▲ and • in the bottom figures denote the experimentally obtained real and imaginary parts of the refractive index, respectively, and the solid lines are the calculations using the Drude model [41].

components are weak at frequencies other than the line positions; and (c) the signal spectrum is limited to the low-frequency region below about 0.5 THz.

To enhance the SNR, some groups have tried to make the signal spectrum continuous. Unlike the THz radiation spectrum shown in Fig. 5.4, the signal spectrum obtained by the CCS with the BLD cannot be continuous by defocusing. Morikawa et al. noted that the continuous components are caused from the optical beat with the different-transverse-modes combination, or the different-beam-pattern combination. They utilized a single-mode fiber optics [41] or spatial filter [42] to unify the beam pattern and obtained the signal spectrum with the continuous components and applied it to the spectroscopy (Fig. 5.6), where noise suppression at frequencies other than the peak positions was clearly observed. Furthermore, when measuring thick samples, artifacts were found to appear in the complex refractive indices. These can be removed by using trimmed window functions in the Fourier transformation of the time-domain signals [43].

The continuous CCS signal spectrum can also be obtained by utilizing the irradiation source other than the BLD. Brenner et al. have constructed an external-cavity laser with a continuous broad spectrum, which consisted of an AR-coated LD, a grating, lenses, and a flat mirror [31]. They obtained a continuous signal spectrum from several tens GHz to about 600 GHz. Molter et al. have proposed the photomixing of two beams from two non-AR-coated LDs each with feeding-back optics [44]. In one of these, a single oscillating mode hopped repetitively in sequence and the time-averaged spectrum was discrete with an equal interval of LMS, while in the other a single oscillating mode was modulated without mode hopping and the time-averaged spectrum was a single line with the width of LMS. These two beams were mixed to cause the optical beat and the CCS signal with a continuous spectrum from 100 GHz to 1 THz. Moler et al. have also proposed employing a superluminescent diode (SLD) for the CCS light source [32]. Their CCS system showed a smooth spectrum from several tens GHz to about 1.5 THz and was applied to the spectroscopy of organic samples [32] and to thickness measurement of a coating with a thickness of several tens of μm [45].

To improve the SNR, high-efficiency emitters such as the UTC-PD, the waveguide integrated photodiodes with a thin absorbing layer or a PIN photodiode have been developed for 1.5 μm lasers [15, 16] as mentioned in Subsection 5.1.1. Simultaneously, high-efficiency detectors such as the PCA with the InGaAs/InAlAs multilayered structure and interdigitated finger electrodes have also been developed for 1.5 μm lasers [15, 16]. These were applied to the coherent detection system using two-beam photomixing [46], and broad bandwidth up to 3.5 THz and high SNR of 105 dB was reported [16]. These emitter and detector devices were applied to the CCS system, and broad bandwidth up to 1.8 THz and high SNR of 60 dB were reported although the signal spectrum was discrete [47]. Recently, Deumer et al. have reported the coherent detection system using two-beam photomixing with the broad bandwidth of 4.5 THz and the high dynamic range of 112 dB using a PIN-photodiode emitter and an InGaAs:Fe-based PCA detector [48]. By applying these devices to the CCS system, its specification may also be improved. On the other hand, Rehn et al. have tried to enhance the dynamic range of the CCS system by reducing the duty ratio of the CW LD [49], since the CCS signal is inversely proportional to the duty ratio, as mentioned at the beginning of Section 5.2. Although they did not observe the enhancement of the dynamic range, they successfully expanded the signal bandwidth from about 550 GHz to about 1 THz.

Higher frequency components with a broad spectrum of the CCS signal can be obtained by the combination of the MLD and the two-beam photomixing technique. In Refs. [50–52], the THz radiation sources using this technique have been reported, showing the appearance of the higher frequency components with discrete spectra, and were applied to imaging or thickness measurement. Morikawa et al. employed the BLD in the two-beam photomixing and observed the appearance of a continuous spectrum component in a higher frequency region, which was suitable for spectroscopy [53, 54].

5.3 Stable CW THz Wave Generation and Detection Using Laser Chaos

To make a stable, low-cost, and compact CW THz source with a broad spectrum, a chaotically oscillating laser diode (COLD) can be employed. This section is devoted to the description of the chaotically oscillating lasers and their application to THz-TDS. The high stability of optical beats in chaotically oscillating lasers is verified and is compared to that of a free-running continuous-wave laser using a highly efficient plasmonic PCA. The COLDs have a great potential for the realization of low-cost, stable, and compact THz-TDS.

5.3.1 Laser Chaos

COLD is constructed by feeding back a part of the output beam with an external mirror and can be used for the generation of a stable and broad-spectrum THz radiation. Although femtosecond optical pulses from mode-locked lasers are frequently used to excite voltage-biased PCAs to generate THz radiation, such systems suffer from the high costs of the lasers. To reduce the system cost, conventional free-running CW multimode laser diodes (FRCW-MLDs) can be also used to excite the PCAs. However, the generated THz radiation is sometimes unstable due to intermittent mode hopping. Instead, COLDs can be employed as the irradiation source for the PCAs to generate stable and broad-spectrum CW THz radiation at a low cost [55–60]. The COLD shows more stable optical spectra in the time scale of down to 0.1 ms than that of the FRCW-MLD. In this subsection, observation results indicating the stability of optical beats in COLD are described.

5.3.1.1 Time evolution of variables

Since ancient days, the word "chaos" has been used in various scenes. The physical term "chaos" has been first defined by Y. Ueda in observing behaviors of analog computers [61]. This behavior was characterized by a trace in the phase space based on the variables describing the system (Broken Egg Attractor), which was found on 27 November 1961. Then, this phenomenon was separately found by Lorentz [62] and named chaos.

The minimal condition for the onset of chaos in a dissipative system is the presence of at least three degrees of freedom, which was proven by Ruelle and Takens [63].

The chaotic oscillation of lasers was first predicted by H. Haken. He pointed out the mathematical equivalence between the Lorenz and Laser Maxwell Born equations in 1975 [64].

The single-mode laser equations based on the semiclassical theory are written as [65],

$$\dot{E}(t) = -\kappa(1 + i\Delta)E - igP \tag{5.1}$$

$$\dot{P}(t) = -\gamma_{-}\perp(1 + i\delta)P + \mathrm{ig}ED \tag{5.2}$$

$$\dot{D}(t) = -\gamma_{\parallel}(D_0 - D) + 2ig(E^*P - EP^*), \tag{5.3}$$

where E is the electric field, P is the atomic polarization, D is the population inversion, D_0 is the unsaturated population inversion, κ is the cavity damping constant, $\gamma_{\perp}$ is the transverse relaxation constant, and $\gamma_{\parallel}$ is the longitudinal relaxation constant. Here we introduce two detuning parameters defined by

$$\Delta = (\omega_c - \omega_L)/\kappa \tag{5.4}$$

$$\delta = (\omega_O - \omega_L)/\gamma_\perp, \tag{5.5}$$

where ω_c is the cavity frequency, ω_L is the laser frequency, and ω_O is the atom center frequency. In the case of a high-Q laser, ω_L is close to ω_c.

5.3.1.2 Classification of lasers

The single-mode laser is a typical dissipative dynamical system with three degrees of freedom (E, P, and D) as shown in Eq. (5.1)–(5.3). The conditions for the onset of chaos is that $\delta = \Delta = 0$ (perfectly tuned laser), which was theoretically predicted by H, Haken [64]. In Ref. 64, the conditions for chaotic behaviors under the bad cavity condition ($\kappa > \gamma_\perp + \gamma_{//}$) were derived. Although this condition is difficult to be realized in real laser systems, Weiss and King first observed chaotic emission from 3.39 μm He-Ne laser under this condition in 1982 [66].

From the viewpoint of laser dynamics, lasers are classified by decay rates of three variables, electric field (E), population inversion (D), and polarization (P), which are κ, $\gamma_{//}$, and $\gamma_\perp$, respectively [67].

If one of the relaxation rates is larger than the others, the corresponding variable relaxes rapidly and follows the other two variables adiabatically. Thus, degrees of freedom (number of differential equations) are reduced from three to two (adiabatic elimination). Lasers in which only the polarization relaxes rapidly are called class B lasers. In these lasers the energy alternates between E and P, resulting in relaxation oscillation of output power. Class B lasers can exhibit chaotic behavior when one external force is applied to it, for example, modulation of cavity loss, or injection of an external light signal. Optical delayed feedback can also be the external force, which is the special one that increases the degrees of freedom to infinity and is described later. Laser diodes (LD) are also classified into the class B laser and sensitive to external force and are easy to show chaotic oscillations. The principles of these processes have been reviewed in detail by Ohtsubo [68].

Lasers with D and P decaying faster than E are called class A lasers. The laser equations for the class A lasers can be reduced to one (E). In order to generate chaos in the class A laser, the degree of freedom must be increased at least by two.

5.3.1.3 Effects of delayed feedback

A delayed feedback of light to the laser cavity is a powerful and simple method to yield chaotic behaviors by increasing the degrees of freedom to infinity. It was theoretically predicted by K. Ikeda [69, 70], and experimentally observed using a single-mode class A laser, which has one degree of freedom [71–73]. Only by adding optical feedback, a single-mode He-Ne laser (a class A laser) showed power fluctuation with modulation depth of around 10% and time scale of around 10 ms [74].

The chaotic behavior is characterized by the Lyapunov exponent (λ), which describes the divergence of the nearest trajectories [75]. By estimating the Lyapunov exponents, Kuwashima et al. proved chaotic oscillation in the above-mentioned single-mode class A laser with optical delayed feedback [71–74].

5.3.2 Application of Laser Chaos to Generation of THz Radiations

5.3.2.1 Merits of LDs as an irradiation source for THz radiation generation

The COLDs oscillate with multi-longitudinal modes. The output beam includes the optical beat and can be used as the irradiation source of PCAs for the generation of THz radiation.

The characteristics of the PCA-irradiating LDs required to make a compact, cheap, and stable THz wave system are as follows:

1. Multi-longitudinal-mode oscillation under their rated current operation
2. Transversally single-mode, which enables tight focusing to small photoconductive gaps of PCAs
3. Low cost
4. High power, more than 80 mW

When the driving current is well above the threshold, transversally single-mode LDs often oscillate with a longitudinally single mode. Moreover, small differences between individual FRCW-MLDs result in different THz-TDS signals even if the THz-TDSs employ LDs with the same model number. From the viewpoint of the optical spectrum stability, the FRCW-MLDs sometimes have different characters with instability due to intermittent mode hopping, even if those are ones with the same model number. These problems can be removed by adopting the COLD construction, where the output beam is partly fed back to LDs, which can be a low-cost one, for example, LDs for CD-R drives. The COLDs show multimode oscillation under high driving current with optical spectra stable in the time scale of down to 0.1 ms, which can be used to construct stable THz-TDSs [57–60]. Kuwashima et al. have tried several LDs, for example, GH0781JA2, RLD78PPY6, or DL-7140-213X. They are commercially available through internet procurement with the costs of about several dollars.

5.3.2.2 Optical spectra of laser chaos

Optical spectra of an LD (DL-7140-213X) are shown in Fig. 5.7 (FRCW-MLD) and 5.8 (COLD). The differences between Figs. 5.7 and 5.8 are caused by only 5%-optical feedback of the output beam with an external mirror in the COLD construction. In the case of the FRCW-MLD, several lines are observed when changing the current from I_{op} = 60 to 80 mA, as indicated by arrows in Figs. 5.7a,b,c. However, when I_{op} is increased up to 90 mA, the FRCW-MLD starts to oscillate with a single mode due to gain-narrowing (Fig. 5.7d),

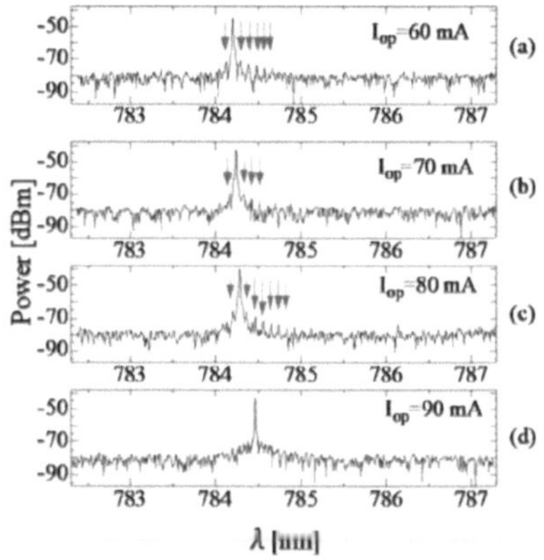

Figure 5.7 The observed optical spectra of the FRCW-MLD (DL-7140-213X) at operating currents (I_{op}) of (a) 60, (b) 70, (c) 80, and (d) 90 mA [29].

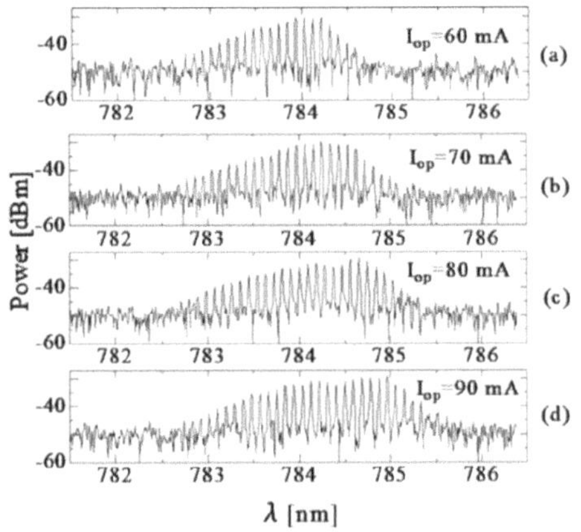

Figure 5.8 The observed optical spectra of the COLD (DL-7140-213X) at operating currents (I_{op}) of (a) 60, (b) 70, (c) 80, and (d) 90 mA [29].

indicating that the optical beat disappears. In addition, all the optical spectra in Fig. 5.7 are unstable in the time scale of around 1s or more due to intermittent mode hopping, although they may be stable in the time scale less than 1s. This long-term instability is thought to be caused by relatively slow fluctuations in the ambient temperature or in the injected current. On the other hand, in the case of COLD, many longitudinal modes from 782.7 to around 785 nm are observed in all the optical spectra in Fig. 5.8 because the optical feedback prevents the gain-narrowing. They also show stability in the time scale of 0.1 ms and above because the relatively fast disturbance due to the optical feedback reduces the influence of the slow fluctuations in the ambient temperature and injected current [29]. In Fig. 5.8, all the spectral lines shift in the same direction and the frequency intervals or the optical beat frequencies do not change.

5.3.2.3 Generated THz waves

Figure 5.9 shows the experimental setup of the THz-TDS with the COLD, which is the same as the conventional one except that the femtosecond pulse laser is replaced by the COLD. An LD (780 nm, Rohm, RLD78PPY6) is used to excite the emitter and detector PCAs. The operation current of the laser (I_{op}) is set to 120 mA. The PCAs

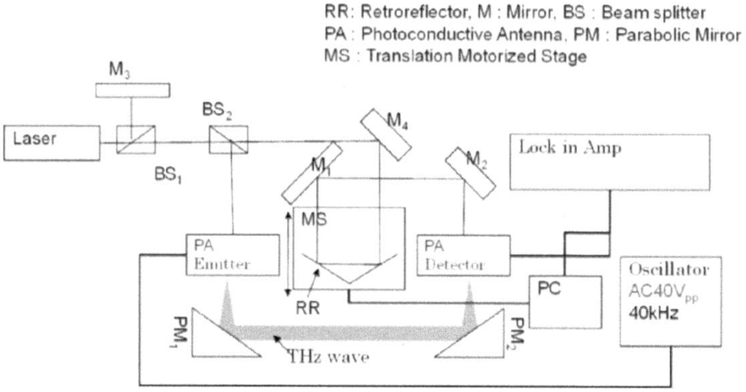

Figure 5.9 Experimental setup for THz-TDS system in which the conventional fs laser is replaced by the COLD [57].

are log spiral type and the emitter PCA is applied with an ac voltage of 40 V_{pp} with the frequency of 40 kHz for lock-in detection [57]. Here, the effective reflectivity of the external mirror is defined as

$$R_{3(\text{eff})} = R_3 R_{BS1}^2. \tag{5.6}$$

where R_{BS1} is reflectivity of beam splitter 1 (BS1) and R_3 is the reflectivity of external mirror (M_3).

The obtained time-domain signals are shown in Fig. 5.10 for $R_{3(\text{eff})}$ = (a) 29.3%, and (b) 0%. The noise level is shown in Fig. 5.10c, where the emitter PCA is not irradiated by the COLD. The signal in Fig. 5.10a is reproducible while that in Fig. 5.10b is not, indicating the stability of the THz-TDS is improved by using the COLD. The signal amplitude in Fig. 5.10a is about 5 times as large as that in Fig. 5.10b, indicating the magnitude of the THz radiation is enhanced by using the COLD.

Fourier spectra of the time-domain signals in Fig. 5.10 are shown in Fig. 5.11. In the case of FRCW-MLD, the signal spectrum is limited below 0.5 THz (Fig. 5.11b). In the case of COLD with $R_{3(\text{eff})} = 29.3\%$, the signal spectrum shows spectral lines up to about 1 THz. In addition, the signal spectrum by the COLD (Fig. 5.11a) shows an amplitude about several times as large as that by the FRCW-MLD (Fig. 5.11b).

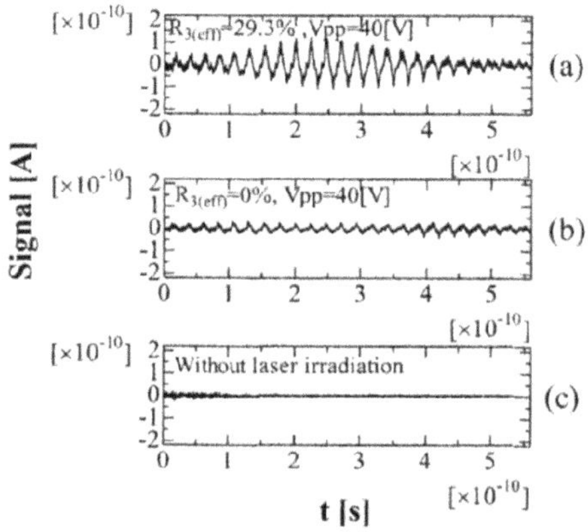

Figure 5.10 Time-domain signals with $R_{3(\text{eff})}$ = (a) 29.3% (COLD), (b) 0% (FRCW-MLD) and (c) without laser irradiation to the emitter PC [57].

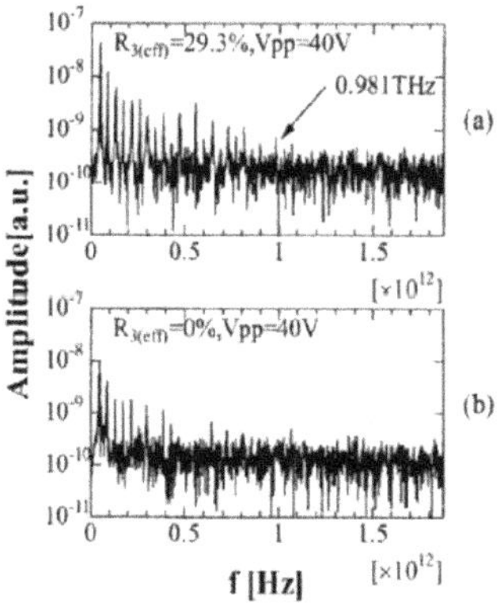

Figure 5.11 Fourier spectrum of the time-domain signals in Fig. 5.10 [57].

5.3.2.4 Simple stabilization mechanism

Simple stabilization mechanisms are shown in Fig. 5.12. For simplicity, we consider a 2-mode laser as the FRCW-MLD. If alternating mode hopping occurs as shown in the leftmost figure in Fig. 5.12, THz radiation is not generated. If simultaneous mode hopping occurs as shown in the middle figure in Fig. 5.12, THz radiation can be generated when the 2 modes simultaneously oscillate. On the

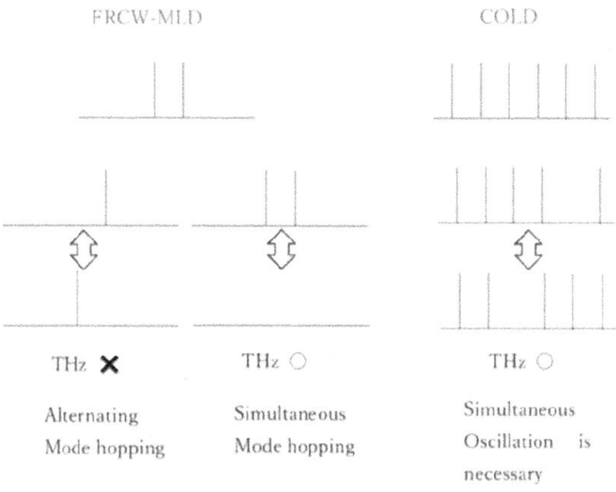

Figure 5.12 Stabilization mechanisms. The schematic diagrams represent instantaneous optical spectra.

other hand, in the COLD, many modes can simultaneously oscillate (right diagram in Fig. 5.12) and the simultaneously oscillating mode number cannot probably be one or zero even if mode hopping occurs. The experimental observation shows that the COLD produces much more stable optical beats than those by the FRCW-MLD as described in the following section, suggesting the case as shown in Fig. 5.12.

5.3.2.5 Stability of optical beats in laser chaos

Since the photocurrent induced in a PCA is proportional to both the irradiating laser power and THz radiation electric field, the PCA can be used as a mixer in the THz range. If the PCA is irradiated by a THz radiation from a local oscillator, the optical beat with a THz frequency in the laser beam irradiating the PCA is down-converted to a low-frequency signal, which can be used for monitoring the stability of the optical beats (Fig. 5.13). In Fig. 5.13, the PCA is the plasmonic photomixer, which has a specially designed photoconductive gap to enhance the signal photocurrent [76–81]. The local oscillator is a frequency multiplier chain producing 87 GHz radiation. The optical beat component to be monitored is ~88 (= 44 × 2)-GHz one in an LD (780 nm, Sharp, DL-7140-213X), where 44 GHz is the successive longitudinal-mode spacing in the LD

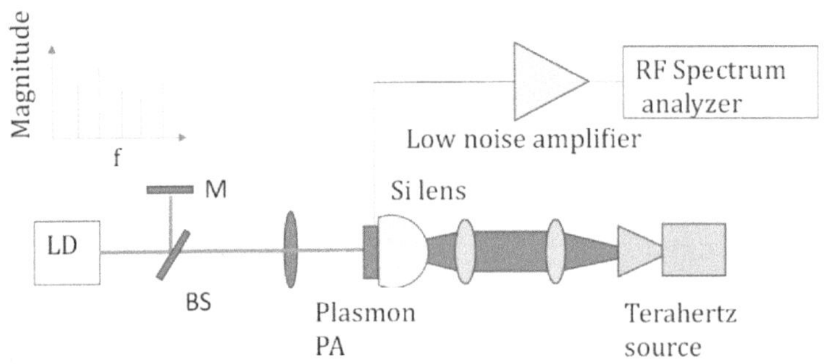

Figure 5.13 Experimental setup for monitoring the stability of the optical beats. By adding and removing the mirror M, the laser is changed from an FRCW-MLD to a COLD and back [29].

spectrum. It is down-converted to an IF signal with a frequency of ~1 GHz, which is monitored by an RF spectrum analyzer after an ~60 dB amplification using two low noise amplifiers. Considering the high stability of the 87 GHz frequency multiplier chain, the instabilities in the optical beats should be directly translated to instabilities in the down-converted IF signal observed by the RF spectrum analyzer.

This system is applied to the comparison of the optical beat stability between the FRCW-MLD and COLD. First, the LD is operated without an external mirror under an operating current of $I_{op} <$ 90 mA. The spectrum of the FRCW shows a comb structure with the line spacing of ~44 GHz for $I_{op} < 80$ mA. The resolution of the optical spectrum analyzer is 0.01 nm, which corresponds to ~5 GHz. Next, the output power is fed back into the laser via an external mirror, M (Reflectivity, R), to obtain a COLD [71–74]. In these experiments, the effective reflectivity of the external mirror is set to $R_{(eff)} = 6.3\%$. The spectrum of the obtained COLD has a comb structure for $I_{op} <$ 90 mA.

The IF signals of the FRCW-MLD and COLD observed by the RF spectrum analyzer are shown in Figs. 5.14 and 5.15, respectively. Only the broad peak around 1 GHz is the IF signal and other spectral components are noise. The scan time and span of the RF spectrum analyzer are set to 500 ms and 700 MHz, respectively. The number of data points in this RF frequency range is set to 601, corresponding

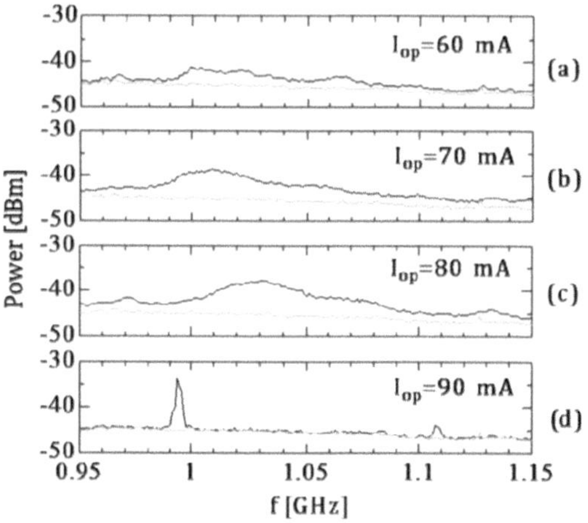

Figure 5.14 RF spectra of the IF signal obtained by the system in Fig. 5.13 with the FRCW-MLD. Operation Currents of LD (I_{op}) are (a) 60, (b) 70, (c) 80, and (d) 90 mA, respectively. The red line is the noise level [29].

to a scan time of 0.83 ms for each RF frequency point. The operating current, I_{op}, is set to 60, 70, 80, and 90 mA, corresponding to the laser irradiation powers of 20.30, 26.61, 30.78, and 33.71 mW, respectively.

In the case of the FRCW-MLD, the observed IF signal is weak at the I_{op} of 60 mA and it is hard to identify a clear RF peak due to the instability of the optical beat (Fig. 5.14a). With increasing the I_{op} to 70 mA, the IF signal level increases, and a peak is observed at ~1.01 GHz (Fig. 5.14b). The RF peak is observed when increasing the I_{op} up to 80 mA (Fig. 5.14c). However, the RF peak disappears when I_{op} is increased to 90 mA (Fig. 5.14d) since the laser starts operating with a longitudinally single mode. The sharp peaks observed in Fig. 5.14d around 1 GHz are caused by the radio signals picked up by the broadband logarithmic spiral antenna of the photomixer. On the other hand, in the case of the COLD, the observed IF signal is stable in both power and frequency over the entire operating current range of 60–90 mA, corresponding to the laser irradiation powers of 21.50, 27.33, 31.42, and 34.22 mW, respectively (Fig. 5.15). They are reproducible throughout the experiment time of several hours. Furthermore, a similar signal is obtained when the system is

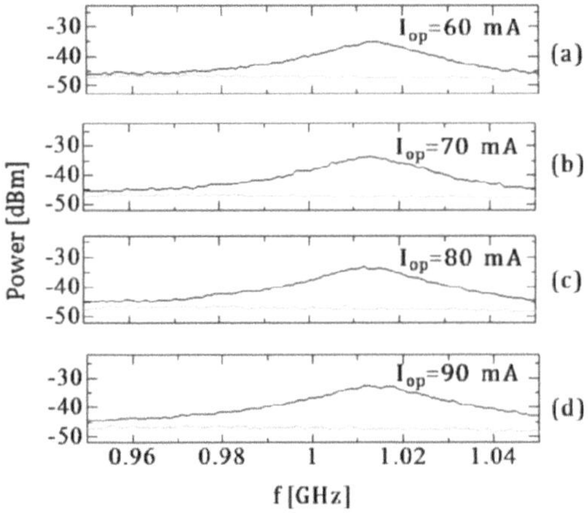

Figure 5.15 RF spectra for the IF signal obtained by the system in Fig. 5.13 with the COLD. Operation Currents of LD (I_{op}) are (a) 60, (b) 70, (c) 80, and (d) 90 mA, respectively. The red line is the noise level [29].

switched on again after several days without alignment. From these results, we can conclude that the 88 GHz optical beats from the COLD are stable over the measurement time scale of 500 ms for each scan and several hours or several days between scans.

The IF signals are observed to be reproducible in the whole range of the RF spectrum analyzer scan speed up to 50 ms (our own highest speed). Since one frequency point corresponds to about 0.1 ms (50 ms /data points 601), the stability of optical beats in COLD is verified in the time scale down to 0.1 ms.

Since the IF signal is proportional to the optical beat amplitude, the magnitude information of the optical beat can be obtained from the peak height in Figs. 5.14 and 5.15. The IF signals by the COLD clearly show higher peaks than those by the FRCW-MLD, indicating that the optical beat amplitudes in the COLD beam are larger than those in the FRCW-MLD beam [29].

In summary, the stability of the optical beats from an FRCW-MLD and COLD are directly measured by mixing the optical beats with a millimeter-wave radiation through a plasmonic PCA and monitoring the down-converted IF signals on an RF spectrum analyzer. In the case of the COLD, the optical beats are highly stable in the entire laser operating current range. On the other hand, in the case of the FRCW-MLD, the optical beats are not stable because of the intermittent mode hopping in the laser. This stability of the COLD enables the system to produce reproducible THz signals for several days without alignments. These studies show the great potential of COLDs for low-cost, stable, and compact THz-TDS systems [29].

5.3.3 Further Challenges

As described above, laser chaos can be a promising approach to stable THz systems. The stabilization mechanism is based on the disturbance caused by the delayed optical feedback, which supports the stable simultaneity of laser modes and stale optical beat. However, in the case of a FRCW-MLD, the oscillations of different laser modes may occur alternately [82], resulting in a low efficiency generation of the optical beat. Probably individual modes in the COLD can be treated as if they were individual lasers and their synchronization may be explained by the well-known chaos synchronization theory between individual lasers with mutual coupling. Although the general conditions for chaotic synchronization with one-way coupling have been recently investigated [83–85], those with mutual coupling are still under investigation, which will be helpful for understanding the behavior of the COLDs and FRCW-MLDs.

References

1. G. Mourou, C. V. Stancampiano, A. Antonetti, and A. Orszag. (1981). *Appl. Phys. Lett.*, **39**, 295–296.
2. D. H. Auston, K. P. Cheung, and P. R. Smith. (1984). *Appl. Phys. Lett.*, **45**, 284–286.
3. E. R. Brown, K. A. McIntosh, K. B. Nichols, and C. L. Dennis. (1995). *Appl. Phys. Lett.*, **66**, 285–287.
4. P. Zhao, S. Ragam, Y. J. Ding, and I. B. Zotova. (2011). *Appl. Phys. Lett.*, **98**, 131106.
5. T. Tanabe, K. Suto, and J. Nishizawa. (2003). *Appl. Phys. Lett.*, **83**, 237–239.
6. R. Safian, G. Ghazi, and N. Mohammadian. (2019). *Opt. Eng.*, **58**, 110901.

7. I. S. Gregory, W. R. Tribe, C. Baker, B. E. Cole, M. J. Evans, L. Spencer, M. Pepper, and M. Missous. (2005). *Appl. Phys. Lett.*, **86**, 204104.
8. S. Verghese, K. A. McIntosh, S. Calawa, W. F. Dinatale, E. K. Duerr, and K. A. Molvar. (1998). *Appl. Phys. Lett.*, **73**, 3824–3826.
9. O. Morikawa, M. Tonouchi, M. Tani, K. Sakai, and M. Hangyo. (1999). *Jpn. J. Appl. Phys.*, **38** part 1, 1388–1389.
10. M. Tani, O. Morikawa, S. Matsuura, and M. Hangyo. (2005). *Semicond. Sci. Technol.*, **20**, S151–S163.
11. K. A. McIntosh, E. R. Brown, K. B. Nichols, O. B. McMahon, W. F. DiNatale, and T. M. Lyszczarz. (1995). *Appl. Phys. Lett.*, **67**, 3844–3846.
12. M. Tani, S. Matsuura, K. Sakai, and M. Hangyo. (1997). *IEEE Microw. Guided Wave Lett.*, **7**, 282–284.
13. M. Sukhotin, E. R. Brown, A. C. Gossard, D. Driscoll, M. Hanson, P. Maker, and R. Muller. (2003). *Appl. Phys. Lett.*, **82**, 3116–3118.
14. S. Preu, G. H. Döhler, S. Malzer, L. J. Wang, and A. C. Gossard. (2011). *J. Appl. Phys.*, **109**, 061301.
15. B. Sartorius, D. Stanze, T. Göbel, D. Schmidt, and M. Schell. (2012). *J. Infrared Milli. Terahz. Waves.*, **33**, 405–417.
16. T. Göbel, D. Stanze, B. Globisch, R. J. Dietz, H. Roehle, and M. Schell. (2013). *Opt. Lett.*, **38**, 4197–4199.
17. K. A. McIntosh, E. R. Brown, K. B. Nichols, O. B. McMahon, W. F. DiNatale, and T. M. Lyszczarz. (1996). *Appl. Phys. Lett.*, **69**, 3632–3634.
18. S. Verghese, K. A. McIntosh, and E. R. Brown. (1997). *IEEE Trans. Microwave Theory Tech.*, **45**, 1301–1309.
19. E. R. Brown, J. E. Bjarnason, T. L. J. Chan, A. W. M. Lee, and M. A. Celis. (2004). *Appl. Phys. Lett.*, **84**, 3438–3440.
20. T. L. J. Chan, J. E. Bjarnason, A. W. M. Lee, M. A. Celis, and E. R. Brown. (2004). *Appl. Phys. Lett.*, **85**, 2523–2525.
21. P. Chen, G. A. Blake, M. C. Gaidis, E. R. Brown, K. A. McIntosh, S. Y. Chou, M. I. Nathan, and F. Williamson. (1997). *Appl. Phys. Lett.*, **71**, 1601–1603.
22. K.-S. Lee and C. Shu. (1997). *IEEE J. Quantum Electron.*, **33**, 1832–1838.
23. M. Hyodo, M. Tani, S. Matsuura, N. Onodera, and K. Sakai. (1996). *Electron. Lett.*, **32**, 1589–1591.
24. P. Gu, M. Tani, M. Hyodo, K. Sakai, and T. Hidaka. (1998). *Jpn. J. Appl. Phys.*, **37** Part 2, L976–L978.
25. H. Lotem, Z. Pan, and M. Dagenais. (1993). *Appl. Opt.*, **32**, 5270–5273.
26. C.-L. Wang and C.-L. Pan. (1995). *Opt. Lett.*, **20**, 1292–1294.

27. P. Gu, F. Chang, M. Tani, K. Sakai, and C.-L. Pan. (1999). *Jpn. J. Appl. Phys.*, **38** part 2, L1246–L1248.
28. O. Morikawa, M. Tonouchi, and M. Hangyo. (1999). *Appl. Phys. Lett.*, **75**, 3772–3774.
29. F. Kuwashima, M. Jarrahi, S. Cakmakyapan, O. Morikawa, T. Shirao, K. Iwao, K. Kurihara, H. Kitahara, T. Furuya, K. Wada, M. Nakajima, and M. Tani. (2020). *Opt. Express*, **28**, 24833–24844.
30. T. Nagatsuma, T. Kumashiro, Y. Fujimoto, K. Taniguchi, K. Ajito, N. Kukutsu, T. Furuta, A. Wakatsuki, and Y. Kado. (2009). *34th International Conference on Infrared, Millimeter, and Terahertz Waves* (IRMMW-THz 2009). M3C05.0088.
31. C. Brenner, M. Hofmann, M. Scheller, M. K. Shakfa, M. Koch, I. C. Mayorga, A. Klehr, G. Erbert, and G. Tränkle. (2010). *Opt. Lett.*, **35**, 3859–3861.
32. D. Molter, M. Kolano, and G. von Freymann. (2019). *Opt. Express*, **27**, 12659–12665.
33. M. Hangyo, T. Nagashima, and S. Nashima. (2002). *Meas. Sci. Technol.*, **13**, 1727–1738.
34. G. Mouret, S. Matton, R. Bocquet, D. Bigourd, F. Hindle, A. Cuisset, J. F. Lampin, and D. Lippens. (2006). *Appl. Phys. Lett.*, **88**, 181105.
35. A. Nahata, J. T. Yardley, and T. F. Heinz. (1999). *Appl. Phys. Lett.*, **75**, 2524–2526.
36. I. S. Gregory, W. R. Tribe, B. E. Cole, C. Baker, M. J. Evans, I. V. Bradley, E. H. Linfield, A. G. Davies, and M. Missous. (2004). *Electron. Lett.*, **40**, 143–145.
37. T. Göbel, D. Schoenherr, C. Sydlo, M. Feiginov, P. Meissner, and H. L. Hartnagel. (2009). *Electron. Lett.*, **45**, 65–66.
38. A. M. Sinyukov, Z. Liu, Y. L. Hor, K. Su, R. B. Barat, D. E. Gary, Z.-H. Michalopoulou, I. Zorych, J. F. Federici, and D. Zimdars. (2008). *Opt. Lett.*, **33**, 1593–1595.
39. O. Morikawa, M. Tonouchi, and M. Hangyo. (2000). *Appl. Phys. Lett.*, **76**, 1519–1521.
40. M. Scheller and M. Koch. (2009). *Opt. Express*, **17**, 17723–17733.
41. O. Morikawa, M. Fujita, and M. Hangyo. (2004). *Appl. Phys. Lett.*, **85**, 881–883.
42. O. Morikawa, M. Fujita, K. Takano, and M. Hangyo. (2011). *J. Appl. Phys.*, **110**, 063107.
43. O. Morikawa, D. Hamada, T. Ozturk, K. Yamamoto, K. Kurihara, F. Kuwashima, and M. Tani. (2021). *J. Opt. Soc. Am. B*, **38**, 1386–1391.

44. D. Molter, A. Wagner, S. Weber, J. Jonuscheit, and R. Beigang. (2011). *Opt. Express*, **19**, 5290–5296.
45. D. Molter, J. Klier, S. Weber, M. Kolano, J. Jonuscheit, and G. von Freymann. (2021). *Appl. Phys. Rev.*, **8**, 021311.
46. B. Sartorius, M. Schlak, D. Stanze, H. Roehle, H. Künzel, D. Schmidt, H.-G. Bach, R. Kunkel, and M. Schell. (2009). *Opt. Express*, **17**, 15001–15007.
47. R. B. Kohlhaas, A. Rehn, S. Nellen, M. Koch, M. Schell, R. J. B. Dietz, and J. C. Balzer. (2017). *Opt. Express*, **25**, 12851–12859.
48. M. Deumer, S. Breuer, R. Kohlhaas, S. Nellen, L. Liebermeister, S. Lauck, M. Schell, and B. Globisch. (2021). *Opt. Express*, **29**, 41819–41826.
49. A. Rehn, M. Mikerov, S. Preu, M. Koch, and J. C. Balzer. (2018). *Opt. Express*, **26**, 32758–32764.
50. I. S. Gregory, W. R. Tribe, M. J. Evans, T. D. Drysdale, D. R. S. Cumming, and M. Missous. (2005). *Appl. Phys. Lett.*, **87**, 034106.
51. M. Scheller, K. Baaske, and M. Koch. (2010). *Appl. Phys. Lett.*, **96**, 151112.
52. F. Kuwashima, T. Shirao, M. Tani, K. Kurihara, K. Yamamoto, M. Hangyo, T. Nagashima, and H. Iwasawa. (2012). In: *Int. Symp. Frontiers in THz Technology (FTT)*, 2012. Pos1.3.
53. O. Morikawa and M. Hangyo. (2019). *Jpn. J. Appl. Phys.*, **58**, 052006.
54. O. Morikawa and M. Hangyo. (2020). *Jpn. J. Appl. Phys.*, **59**, 082003.
55. F. Kuwashima, T. Shirao, M. Tani, K. Kurihara, K. Yamamoto, T. Nagashima, and M. Hangyo. (2014). In: *Proc. Int. Symp. Nonlinear Theory and Its Applications*, Luzern, Switzerland, September 14–18, paper C3L–E4.
56. F. Kuwashima, T. Shirao, T. Kishibata, T. Okuyama, Y. Akamine, M. Tani, K. Kurihara, K. Yamamoto, T. Nagashima, and M. Hangyo. (2014). In: *Int. Conf. Infrared, Millimeter and Terahertz Waves, Tucson, Arizona*, September 14–18, paper W5-P24.2.
57. F. Kuwashima, T. Shirao, M. Tani, K. Kurihara, K. Yamamoto, M. Hangyo, T. Nagashima, and H. Iwasawa. (2014). In: JPS. Conf. Proc. 1, 015018.
58. F. Kuwashima, T. Shirao, M. Tani, K. Kurihara, K. Yamamoto, M. Hangyo, T. Nagashima, and H. Iwasawa. (2012). In: *Int. Conf. Infrared, Millimeter and Terahertz Waves, Wollongong, Australia*, September 24–28, paper Thu-Pos-2.
59. F. Kuwashima, Takei, K., Fujii, T., M. Tani, K. Kurihara, M. Hangyo , T. Nagashima, and H. Iwasawa. (2011). In: *Int. Conf. Infrared, Millimeter and Terahertz Waves, Houston, Texas*, October 2–7, paper W5.13.

60. F. Kuwashima, S. Taniguchi, K. Nonaka, M. Hangyo, and H. Iwasawa, IRMMW-THz2010, Rome, Italy, 2010, paper Tu-P.63.
61. Y. Ueda: *The Road to Chaos*, with the assistance of R. H. Abraham and H. B. Stewart, The Science Frontier Express Series, Aerial Press, Inc. POB 1360, Santa Cruz, CA 95061, 1991. 2 ed., enlarged 2001 (Kyoto University Repository: http://hdl.handle.net/2433/71101).
62. E. N. Lorenz. (1963). *J. Atmospheric Sci.*, **20**, 130–141.
63. D. Ruelle and Takens. (1971). *Commun. Math. Phys.*, **20**, 167.
64. H. Haken. (1975). *Phys. Lett. A*, **53**, 77.
65. F. Kuwashima, I. Kitazima, and H. Iwasawa. (2001). *Jpn. J. Appl. Phys.*, **40**, 601–608.
66. C. O. Weiss and H. King. (1982). *Opt. Commun.*, **44**, 59.
67. F. T. Arecchi, G. L. Lippi, G. P. Pucciooni, and J. R. Tredicce. (1984). *Opt. Commun.*, **51**, 308–314.
68. J. Ohtsubo. (2017). *Springer International Publishing*, ISBN 978-3-319-56137-0.
69. K. Ikeda and M. Matsumoto. (1986). *J. Stat. Phys.*, **44**, 955–983.
70. K. Ikeda and K. Matsumoto. (1987). *Phys. D*, **29**, 223–235.
71. F. Kuwashima, I. Kitazima, and H. Iwasawa. (1998). *Jpn. J. Appl. Phys.*, **37**, L325–L328.
72. F. Kuwashima, T. Ichhikawa, I. Kitazima, and H. Iwasawa. (1999). *Jpn. J. Appl. Phys.*, **38**, 6321–6326.
73. F. Kuwashima and H. Iwasawa. (2007). *Jpn. J. Appl. Phys.*, **46**, 1526–1527.
74. F. Kuwashima, I. Kitazima, and H. Iwasawa. (2001). *Jpn. J. Appl. Phys.*, **40**, 601–608.
75. A. Wolf, J. B. Swift, H. L. Swineey, and J. A. Vastano. (1985). *Phys. D*, **16**, 285–317.
76. N. Wang, S. Cakmakyapan, Y. J. Lin, H. Javadi, and M. Jarrahi. (2019). *Nature Astronomy*, DOI: 10.1038/s41550-019-0828-6.
77. N. Wang and M. Jarrahi. (2016). In: *Int. Symp. Antennas and Propagation Symposium*, Fajardo, Puerto Rico, June 26–July 1, pp. 1189–1190.
78. N. Wang and M. Jarrahi. (2016). In: *Conf. Lasers and Electro-Optics*, San Jose, CA, June 5–10, pp. JTh2A-59.
79. N. Wang, H. Javadi, and M. Jarrahi (2016). In: *IEEE MTT-S International Microwave Symposium*, San Francisco, CA, May 22–27, pp. 1–3.
80. N. Wang, H. Javadi, and M. Jarrahi. (2016). In: *European Conf. on Antennas and Propagation, Davos*, Switzerland, April 11–15, pp. 1–3.

81. N. Wang and M. Jarrahi. (2013). *J. Infrared, Millimeter and Terahertz Wave*, **34**, 519–528.
82. A. M. Yacomotti, L. Furfaro, X. Hachair, F. Pedaci, M. Giudici, J. Tredicce, J. Javaloyes, and S. Balle. (2004). *Phys. Rev. A*, **69**, 053816.
83. A. Uchida, K. Higa, T. Shiba, S. Yoshimori, F. Kuwashima, and H. Iwasawa. (2003). *Phys. Rev. E*, **68**, 016215-1.
84. K. Umeno. (2000). *Jpn. J. Appl. Phys.*, **39**, 1442–1456.
85. M. Shintani and K. Umeno. (2018). *Prog. Theor. Exp. Phys.*, 013A01.

Chapter 6

Spintronic THz Emitters

Evangelos Papaioannou,[a,b] **Garik Torosyan,**[c] **and René Beigang**[d]

[a]*Institute of Physics, Martin-Luther University Halle-Wittenberg, 06120 Halle, Germany*
[b]*Department of Physics, Aristotle University of Thessaloniki, Thessaloniki 54124, Greece*
[c]*Photonic Center Kaiserslautern, Kaiserslautern 67663, Germany*
[d]*Department of Physics, University of Kaiserslautern, Kaiserslautern 67663, Germany*
evangelos.papaioannou@physik.uni-halle.de

The THz frequency range coincides with many low-energy modes in all phases of matter, i.e., plasma, gases, liquids, and solids that render it an excellent candidate for basic research. Furthermore, THz radiation can serve not only as a probe but also to control excitations like magnons and phonons. In terms of applications, THz imaging and sensing have recently gained considerable scientific attention. Applications in biomedicine, security, military, quality control and inspection such as in pharmaceutical and circuitry manufacturing, have been enabled in recent years due to the development of versatile THz sources based on photonic technologies. Despite this progress in THz technology, there is still a need for stronger, more efficient and broad bandwidth THz sources. To this end, the field of ultrafast spintronics with a focus on THz spintronics is a novel direction that has high potential to be an innovative part in the

Advances in Terahertz Source Technologies
Edited by Gun-Sik Park, Masahiko Tani, Jae-Sung Rieh, and Sang Yoon Park

ISBN 978-981-4968-89-8 (Hardcover), 978-1-003-45967-5 (eBook)
www.jennystanford.com

development of new THz sources and even to overcome existing THz technologies with respect to bandwidth. This chapter focuses on experimental approaches of ultrafast spintronics to generate broadband THz electromagnetic pulses. First, we will introduce the field of THz spintronics and then we will analyse the physics of the spin-to-charge-current conversion leading to the subsequent THz emission. We will present experimental configurations to generate and detect spintronic THz pulses and finally we will give an overview of strategies to engineer intensity and bandwidth of the emitted THz signal.

6.1 Introduction

The field of THz spintronics is a novel direction in the research field of nanomagnetism and spintronics that combines magnetism with optical physics and ultrafast photonics. The experimental scheme of the field involves the use of femtosecond laser pulses to trigger ultrafast spin and charge dynamics in thin films composed of ferromagnetic and non-magnetic thin layers where the non-magnetic layer features a strong spin-orbit coupling. The results obtained so far in this field can also be used to realize a new type of THz emitter based on the spintronic properties of pairs of thin magnetic and non-magnetic layers excited by fs laser pulses. The technological and scientific key challenges of THz spintronic emitters are to increase their efficiency and to shape the frequency bandwidth. To achieve this the control of the source of the radiation, namely the transport of the ultrafast spin current is required. The generation, detection, efficiency and future perspectives of THz emitters are addressed, and the state of the art of efficient emission in terms of materials, geometrical stack, interface quality and patterning is presented.

The physical mechanism of generation of THz radiation with spintronic emitters is based on the inverse spin Hall effect [1] and appears in multilayer heterostructures that consist of ferromagnetic (FM) and non-magnetic (NM), usually heavy metal, layers. When illuminated by ultrafast femtosecond (fs) laser pulses, spin-polarized electrons are excited in the FM layer and subsequently diffuse as a spin current into the NM layer. Due to the inverse spin Hall effect (ISHE) the spin current is converted into a transient transverse charge current in the NM layer resulting in THz emission [1]. This new source of THz radiation is an emerging topic subject to intensive research. The efficiency of such emitters is in some cases comparable with established types of THz sources (for example with nonlinear crystals) [2]. The engineering of THz emission is the main technological and scientific challenge and is currently the target of many research activities. In this chapter, we highlight and examine the different strategies that have been followed in order to explore the properties of the emitted THz signal, like THz amplitude and bandwidth, e.g. different material compositions of FM/NM systems with a variety of thicknesses, spintronic emitters in different geometrical stacking order using different layers and patterned structures, different substrates, interface materials and quality of interfaces and dependence on the excitation wavelength. First we present the radiation mechanism and the experimental way to measure the emitted THz radiation. We explore the current trends to engineer the properties of the THz emission and we discuss the future challenges to integrate spintronic emitters in THz devices. We finally explore the potential of the emitter to extend the THz field and widen its applications. This chapter only summarizes experimental results from the last couple of years and does not review theoretical models and calculations of the spin transport in magnetic heterostructures. In addition, we focus on the mechanism of THz generation from the ISHE and we will not deal in depth with other novel sources of THz radiation originating from other mechanisms like anomalous Hall effect or Rashba interfaces, since both are much less efficient as THz sources compared to the ISHE mechanism.

6.2 Spin-to-Charge Conversion Mechanism Responsible for THz Radiation

Commonly, optically generated THz radiation driven by femtosecond laser pulses is based on either transient photocurrents in semiconductors or by utilizing nonlinear optical responses of bound electrons in nonlinear crystals [3–6]. Equation (6.1) summarizes the different physical mechanisms of optically induced THz emission: it expresses the far-field THz electric field amplitude E_{THz}, that is experimentally measured in the time-domain as [3, 4]:

$$\mathbf{E}_{\text{THz}} \propto \frac{\partial \mathbf{J}_c}{\partial t} = \frac{\partial}{\partial t}\left[\mathbf{J}_{\text{ph}}(t) + \frac{\partial \mathbf{P}(t)}{\partial t} + \nabla \times \mathbf{M}(t)\right] \tag{6.1}$$

$\mathbf{E}_{\text{THz}}$ is proportional to the time-derivative of the induced local charge current $\mathbf{J}_{\text{c}}$. As Eq. (6.1) reveals the source of the THz radiation is either the time-varying photocurrent $\mathbf{J}_{\text{ph}}(t)$ in semiconductors (photoconductive switches/antennas) or the time-varying polarization $\mathbf{P}(t)$, in electro-optical crystals (nonlinear optical methods). Further, THz radiation can be obtained from free electrons in vacuum either from a bunch of relativistic electrons or from periodically undulated electron beams [3, 4]. Besides the aforementioned physical principles for the THz radiation, which are known as electric dipole emission, the last term of Eq. (6.1) defines that a time-varying magnetization on the ultrafast, (sub)-picosecond timescale can also act as a THz source, the so-called magnetic dipole emission. Such emission was first observed from optically excited magnetic metallic structures during the ultrafast demagnetization process of magnetic layers [7, 8]. This type of radiation is weaker due to its magnetic dipole character.

For all these methods there are limitations concerning the generated intensities and/or the available bandwidth. The use of photoconductive antennas (PCAs) is, in general, limited to low-power excitation sources and, as a consequence, to low intensities of the generated THz pulses. The bandwidth is usually governed by material properties (intrinsic phonon absorption in the materials used) and typically ends around 8 THz. In addition, the fabrication of PCAs requires an advanced technology to obtain, for example, the correct structuring which significantly determines the desired properties. Nonlinear methods are, in principle, well suited for THz generation. However, in order to achieve phase matching only very short nonlinear crystals in combination with ultrashort laser pulses can be used to obtain a decent bandwidth. This in turn limits the intensity of the generated THz pulses. The use of high-power excitation sources is limited by the damage threshold of the nonlinear crystals.

Spintronic THz emitters (STEs) define a novel and different type of radiation source. For THz generation in spintronic THz emitters the physical mechanism is based on the excitation of a spin current and the inverse spin Hall effect (ISHE), Fig. 6.1. In order to understand the evolution of the THz amplitude from a spintronic emitter, the different effects that take part in the emission process need to be clarified step for step. Similar to Eq. (6.1), the far-field THz electric field amplitude E_{THz}, that is experimentally measured, is proportional to the time-derivative of the local charge current j_{c} that is integrated over the thickness dz of the non-magnetic layer corresponding to the summation over all emitting dipoles:

$$E_{\mathrm{THz}} \propto \frac{\partial \int j_{\mathrm{c}} dz}{\partial t} \tag{6.2}$$

Different from the aforementioned sources of $\mathbf{J}_{\mathrm{c}}$ in Eq. (6.1), here the charge current j_{c} is induced by the inverse spin Hall effect and is proportional to the spin-current j_{s}. In detail, a femtosecond laser pulse excites electrons in the FM-layer from quasi-localized states (d-like states) beneath the Fermi level to more mobile states (sp-like states) above the Fermi level. The excitation and transport properties of the excited electrons such as the lifetime and velocity depend on the energy and the spin type (majority or minority). Majority spins are more mobile than minority spins since more majority carriers are available near the Fermi level. The imbalance in the number of majority and minority spins gives rise to an ultrafast spin current $\mathbf{j}_{\mathrm{s}} = \mathbf{j}_{\uparrow} - \mathbf{j}_{\downarrow}$. Subsequently, the spin current $\mathbf{j}_{\mathrm{s}}$ propagates across the FM/NM interface in the non-magnetic layer through a super diffusive process [1, 9]. This longitudinal spin-polarized current $\mathbf{j}_{\mathrm{s}}$ is then converted into a transient transverse charge current $\mathbf{j}_{\mathrm{c}}$ due to the ISHE in the NM layer. The spin current induces an effective charge current perpendicular to itself and the magnetization axis according to:

$$\mathbf{j}_{\mathrm{c}} = \Theta_{\mathrm{SH}}\, \mathbf{j}_{\mathrm{s}} \times \frac{\mathbf{M}}{|\mathbf{M}|}, \tag{6.3}$$

where the spin Hall angle is denoted by Θ_{SH}. Examples of the direction of the vectors $\mathbf{j}_{\mathrm{s}}$, $\mathbf{j}_{\mathrm{c}}$ and $\mathbf{M}$ are shown in Fig. 6.1. The

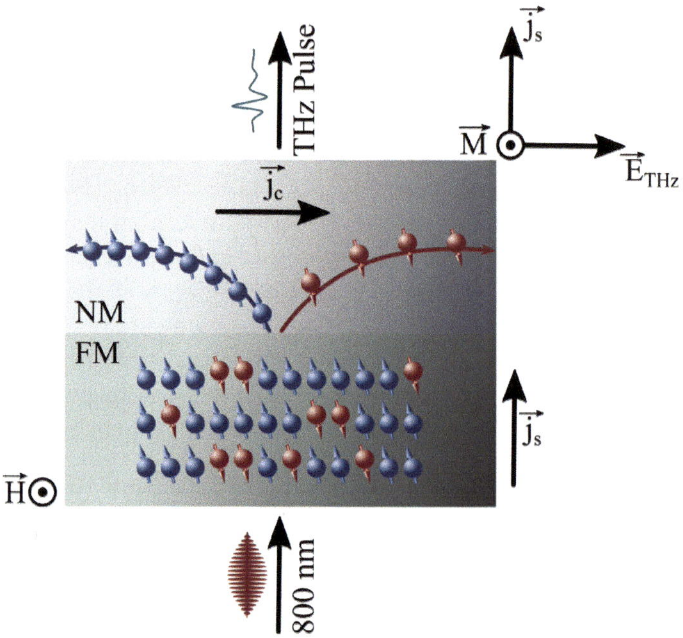

Figure 6.1 Graphical representation of the THz emission from ferromagnetic (FM)/non-magnetic (NM) heavy metal heterostructures after fs-laser excitation of the spin system. The fs-laser pulse pumps energy into the system and excites majority and minority spins which travel to the interface and form the injected spin current denoted as $\mathbf{j}_S$. Majority and minority spins travel in opposite directions after reaching the FM/NM interface due to the inverse spin-Hall effect. The imbalance in the number majority and minority spin in the NM-layer results in a transverse charge current $\mathbf{j}_C$. The latter gives rise to a THz transient emission according to Eq. 6.2. The applied external magnetic field **H**, in this representation points out of the plane forcing the Magnetization **M**, to be parallel to it. The polarization of the THz field is perpendicular to the direction of the magnetic field and, hence to the magnetization direction. The relative directions of $\mathbf{j}_S$, $\mathbf{j}_C$, $\mathbf{E}_{THz}$, **H** and **M** are highlighted in the side of the diagrams. These directions are independent of the pump pulse polarization.

transient charge current generates a short terahertz pulse according to Eq. (6.2) that propagates perpendicular to the electrical current. The emission is dominated by the electric dipole emission driven by a time varying ISHE-type electric current.

Theoretically, the origin of the THz radiation that is the evolution of the hot-carrier distribution after the femtosecond laser excitation can be captured by the Boltzmann transport theory. The latter has been proven to be an adequate tool to simulate excited carrier dynamics in metallic structures on the nanoscale and the reader is referred to references [1, 2, 9–12], for further reading since this chapter focuses mainly on the experimental findings of the field.

Experimental examples of THz pulses and their corresponding bandwidths are shown in Fig. 6.2. The pulses have been obtained with THz Time Domain Spectroscopy (THz-TDS) detection scheme. In the left column a photoconductive antenna (PCA) is used as a detector, in the right column electro-optical sampling (EOS) detection is applied. The THz pulses are emitted from a Fe(2 nm)/Pt(3 nm) bilayer grown on MgO substrate with the Pt layer facing the detector. The shape of the experimentally obtained pulses and their corresponding spectra depend on the experimental THz setup and is a convolution of emitter and detector response as Fig. 6.2 reveals. The measured bandwidth, for example for the TDS setup in Fig. 6.2 (left column), reaches 8 THz [13], which is mainly limited by the detector response of a low temperature (LT)-GaAs photoconductive antenna used in this case. On the other side, bandwidths with full width at half-maximum (FWHM) up to 30 THz measured with electro-optical sampling is shown in Fig. 6.2 (right column) for the same sample, using a 10 μm ZnTe detector and faster excitation pulses.

In the end of this section, it is worth mentioning that the origin of the THz emission after fs-laser excitation is difficult to disentangle between ISHE-type and laser-driven transient demagnetization in the magnetic layers. Recently, Zhang et al. [14] successfully separated and measured the weak THz emission during the demagnetization process, suggesting the implementation of this type of weak THz emission for ultrafast magnetometry on a pico- and sub-pico timescales. Presently, new spin-to-charge-conversion mechanisms have been proposed as sources of THz radiation. One is the THz emission from Rashba type interfaces [15, 16] where the generation of THz waves takes place at interfaces between two non-magnetic materials due to the inverse Rashba Edelstein effect. In a similar experimental procedure, the fs-laser pulse induces

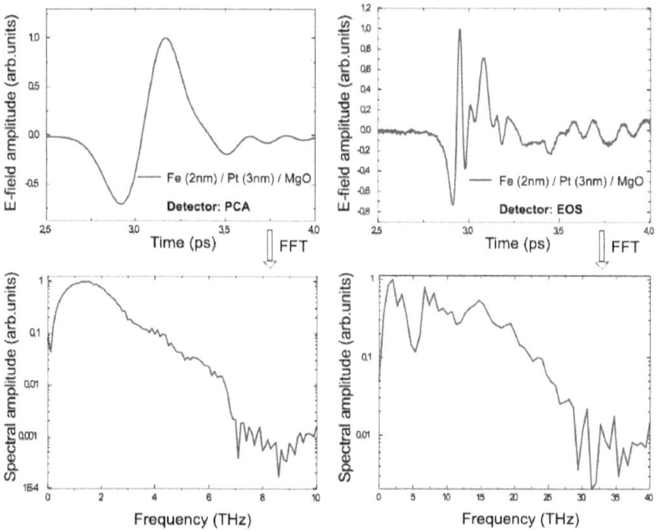

Figure 6.2 Emitted pulses from a Fe (2 nm)/ Pt (3 nm) bilayer grown on MgO substrate. The pulses were detected with THz time domain spectroscopy (THz-TDS). The laser pulse impinges first to the substrate and the signal is detected from the Pt-layer (Pt faces the detector). In the left column, the pulse is recorded by using a fs-excitation laser pulse length of 22 fs and a photoconductive antenna (PCA) as a detector. In the Right Column the pulse is recorded by using excitation pulse length of 10 fs and electro-optical sampling (EOS) technique is used for detection. The lower panels are the Fast Fourier Transformation (FFT) of the time traces of the pulses. The electro-optical sampling combined with an appropriate detector (ZnTe) reveals that STEs as Fe/Pt provide bandwidths more than 30 THz. The EOS measurement was performed in the lab of Prof. T. Kampfrath, FU Berlin.

a non-equilibrium electron flow in FM/Ag/Bi heterostructures. A femtosecond spin current pulse is launched in the ferromagnetic $Co_{20}Fe_{60}B_{20}$ layer and drives terahertz transients at a Rashba interface between two nonmagnetic layers, Ag and Bi. In contrast to the THz emission in spintronic emitters via the inverse spin-Hall effect, the inverse Rashba Edelstein effect transforms a nonzero spin density induced by the spin current injection into a charge current carried by interfacial states. Another alternative mechanism for generating THz emission from an ultrathin FM layer via the anomalous Hall effect was recently proposed [17]. The process involves a single FM layer and the generation of backflow superdiffusive currents at the dielectric/FM/dielectric interfaces and subsequent conversion of the charge current to transverse current due to the anomalous Hall effect. The THz generation is suggested to be mainly caused by the non-thermal superdiffusive current near the two FM/dielectric interfaces.

6.3 Experimental Detection of THz Emission

Detection of the emitted THz pulses from spintronic emitters can be accomplished with well-established THz timed domain spectroscopy (THZ-TDS) systems [18]. Details about different detection schemes can be found in other chapters of this book. A specific example is presented in Fig. 6.3. The system is driven by a femtosecond laser delivering sub-100 fs optical pulses at a repetition rate of usually a few MHz with an average output power of typically 600 mW. The laser beam is split into a pump and probe beam by a 90:10 beam-splitter. The stronger part is led through a mechanical computer-controlled delay line to pump the THz emitter, and the weaker part is used to gate the detector, a photoconductive switch with a dipole antenna of specific dimensions. In our example here we refer to a 20 μm dipole antenna. In a classical (standard) THz setup both the emitter and the detector operate with photoconductive antennas (PCA), whereas in the case of spintronic emitters the PCA emitter is substituted by the spintronic sample, which is placed in a weak magnetic (a few mT) field perpendicular to the direction of the pump beam and usually in the direction of the easy axis of the magnetic layer in order to achieve saturation. The direction of the magnetic field determines the polarization of the THz field, which is perpendicular to the direction of the magnetic field. Changing the direction of the magnetic field into the opposite direction changes the phase of the detected THz waveforms by 180° (Fig. 6.3b)). In this way, by changing the orientation of the magnetic field the polarization of the generated THz radiation can be changed easily. In combination with a polarization-sensitive detector this can be used to modulate the generated THz radiation (see, e.g., Ref. [19])

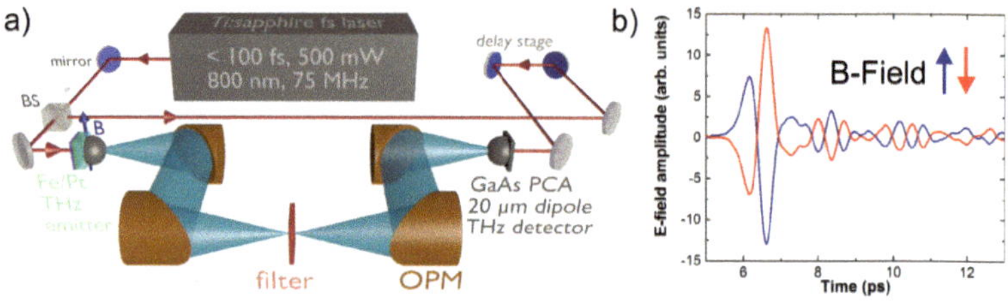

Figure 6.3 (a) An example of a terahertz time-domain spectroscopy (THz TDS) experimental set-up modified for spintronic emitters. (b) THz pulses for two opposite directions of the magnetic field. Reproduced under the terms of a Creative Commons Attribution 4.0 Licence [20] Copyright 2018, The Authors, published by Springer Nature.

The optical pump beam, which is sharply focused onto the sample at normal incidence by an aspherical short-focus lens, excites spin-polarized electrons in the magnetic layer (Fe), which give rise to a spin current, which in turn excites a transverse transient electric current in the Pt layer. The latter results in THz pulse generation of sub-picosecond duration being emitted forward and backwards into free space in the form of a strongly divergent beam.

Therefore, a hyperhemispherical Si-lens was attached at the back side of the sample, able to collect the divergent beam in the form of a cone and direct it further. With the lens attached an enhancement factor of up to 30 in electric field amplitude has been reported [20]. After that, the beam is collimated with an off-axis parabolic mirror and sent to an identical parabolic mirror in the reversed configuration. The latter focuses the beam to the second Si-lens which finally focuses the beam through the GaAs substrate of the detector PCA onto the dipole gap for detection. In this way, the THz optical system images the point source of THz wave on the emitter surface onto the gap of the detector PCA and ensures an efficient transfer of the emitted THz emission from its source to the detector.

In addition to having two collimated beam paths, this 4-mirror THz optics also allows for an intermediate focus of the THz beam aimed at imaging applications (Fig. 6.3a)).

In such a configuration, the THz beam path is determined by the silicon lens on the emitter, the parabolic mirrors and the silicon lens on the photoconducting antenna of the detector. The alignment of these components is not changed during an exchange of the spintronic emitter. If in addition the position of the pump beam focus remains constant, the spintronic emitter can easily be exchanged without changing the beam path, as the lateral position of the focus on the emitter is not critical assuming a homogeneous lateral layer structure. This is important for comparative studies of different spintronic emitter designs. The frequency response of the photoconductive dipole antenna with 20 μm dipole length limits the observable bandwidth. With this dipole length, a maximum detector response of around 1 THz can be expected with a reduction to 50% at 330 GHz and 2 THz. The 10% values are at 100 GHz and 3 THz [21]. In fact, the detected signal is strongly influenced by the detector response. The frequency response of the PCA detector above 3 THz is very flat and at 8 THz a strong phonon resonance in GaAs, which is used as substrate material for the photoconductive antenna, causes strong absorption of the THz radiation. Above 8 THz almost no THz radiation can be detected. Alternatively, THZ-TDS can utilize free space electro-optical sampling (EOS) for the detection of THz radiation from spintronic emitters [2, 22, 23]. In order to be able to detect frequencies well above 10 THz very thin ZnTe crystals are used as a detector in combination with extremely short laser pulses. The ease of implementation and the possibility to detect higher bandwidths as shown in Fig. 6.2 render this technique also popular for the investigation of the efficiency of the magnetic heterostructures.

6.4 Strategies to Engineer Intensity and Bandwidth of THz Signal

Currently, the main goal of the research on spintronic emitters is the engineering of THz emission and different strategies have been followed in order to explore the THz amplitude and bandwidth of the signal. In Fig. 6.4, we construct a road map with the most important physical parameters, that can influence the efficiency of the spintronic THz-emitters towards higher signal strengths

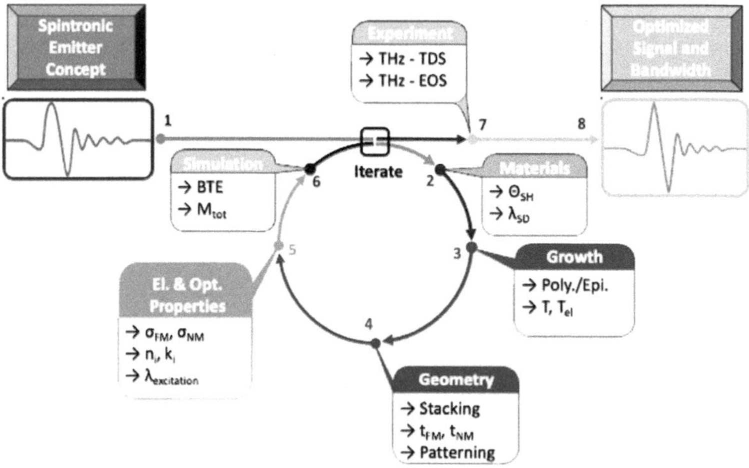

Figure 6.4 Road map to efficient spintronic THz-emitters with high signal strength and broad bandwidth. Here, Θ_{SH} is the spin-Hall-angle, λ_{SD} is the spin-diffusion-length, T and τ_{el} are the interface transmission parameter and the elastic scattering lifetime, respectively. t_{FM}, t_{NM}, σ_{FM} and σ_{FM} are the FM and NM layer thickness and electrical conductance, respectively. n_i and κ_i are the index of refraction and the absorption coefficient of all involved layers (metals and insulators) in the THz range, respectively. $\lambda_{excitation}$ is the wavelength of the laser excitation. BTE refers to the Boltzmann transport equation, M_{tot} is the transfer matrix used to describe the propagation of the generated THz wave throughout the total layer stack. THz-TDS detection schemes are defined in the graph as THz-TDS for the time domain spectroscopy using PCA as detectors and THz-EOS for using detectors based on free space electro-optical sampling. Figure modified from [24].

and broader bandwidths. Figure 6.4 summarizes all the important factors that can affect the temporal and spatial evolution of the spin current inside the metallic layers, by taking into account the generation and optical propagation of the THz wave, and forecasts the THz-pulse shapes and spectra, by taking into account the electron scattering lifetime and the interfacial spin current transport. Parameters that are included in Fig. 6.4 and play a decisive role in the emission are: the spin-Hall-angle Θ_{SH}, the spin-diffusion-length λ_{SD}, the interface transmission parameter T, and the inelastic scattering lifetime τ_{el}. Moreover, included in the road map are effects originating from the excitation wavelength, from geometrical stacking order, from the FM and NM layer thickness t_{FM}, t_{NM} and electrical conductance σ_{FM}, σ_{NM}, and from the index of refraction and the absorption coefficient of all involved layers (metals and insulators) in the THz-range n_i and κ_i. Theoretical calculations and simulation of the spin current transport with the Boltzmann equation and of the optical path with the use of the transfer matrix for the total layer stack M_{tot} are also necessary tools for the understanding and the evaluation of the research progress.

The following sections will analyse the road map by reviewing the research efforts on different material compositions, geometrical factors, interfaces and nanopatterning. We will present key works and summarise the unique advantages and the promises for applications of spintronic emitters to the development of THz technology.

6.4.1 Material Dependence

A variety of material combinations of ferromagnetic/non-magnetic layer systems have been so far explored as spintronic emitters. We first focus on the choice of the magnetic layer. Direct comparison between the 3d metals like Fe, Co and Ni, in structures like FM(3 nm)/Pt(3 nm) [2] and in FM(2 nm)/Pt(5 nm) [25] have revealed that Ni yields the lowest signal with Fe having a slightly higher signal than Co. The comparison of ferromagnetic alloys like $Ni_{89}Fe_{19}$, $Co_{70}Fe_{30}$, $Co_{40}Fe_{40}B_{20}$ and $Co_{20}Fe_{60}B_{20}$/Pt (3 nm) [2] showed the prominent role of the CoFeB alloy. Sasaki et al. [26] varied the composition of Co and Fe in the CoFeB alloy. They found that the THz emission in $Ta/(Co_xFe_{1-x})_{80}B_{20}$ is enhanced at compositions of approximately $x = 0.1$–0.3 which show a maximum saturation magnetization. Modification of the magnetic layer with ferrimagnetic gadolinium and terbium-iron alloys was achieved by Schneider et al. [27, 28]. Both systems Gd_xFe_{1-x}/Pt and Tb_xFe_{1-x}/Pt present the highest THz emission for small rare-earth content. However, the Gd_xFe_{1-x} exhibit up to 17 times higher amplitudes. A strong THz output was observed from compensated ferrimagnetic $Co_{1-x}Gd_x$(7 nm)/Pt(6 nm) bilayers. The THz signal decreases as the Gd fraction increases. At the compensation point for $x = 26$, the almost zero net magnetization was not strongly correlated to the emitted THz signal [22]. The replacement of the ferromagnetic layer with $DyCo_5$, $Gd_{24}Fe_{76}$, Fe_3O_4 and FeRh showed that they exhibit lower THz emission with respect to CoFeB layer [29]. Furthermore, Co-based Heusler alloys and in particular Co_2MnSi (CMS) were investigated as spin injector in CMS/Pt bilayers [30]. The achievement of the highly ordered B2 structural phase after annealing led to up to two times higher THz emission compared to CoFe/Pt bilayers. The effect was attributed to the small conductivity and the comparable to FeCo spin current injection of the CMS samples.

In addition, a variety of NM layers have been studied so far for THz emission. The efficiency of spin-to-charge current conversion is quantified by the spin Hall angle and the spin current relaxation length of the NM layer. The THz signal amplitude seems to scale with the intrinsic spin Hall conductivity of the used NM layer [2], see Fig. 6.5. The THz emission was even used to estimate the relative spin Hall angle [23, 31]. The prominent choice at the moment is Pt since it has provided so far the highest signal compared to W [23], $Cu_{80}Ir_{20}$ [23], Ta [31], Ru [31, 32], Ir [31], Pd[32], and $Pt_{38}Mn_{62}$ [2].

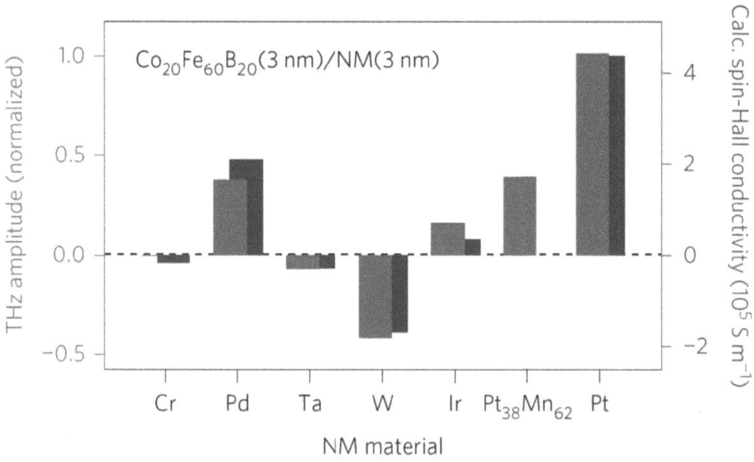

Figure 6.5 Spin Hall angle and spin Hall conductivity for the non-magnetic (NM) layer. Pt has so far the highest efficiency for THz emission. Reproduced with permission from [2]. Copyright © 2016, Springer Nature Limited.

Interestingly, NM layers with opposite sign of the spin Hall angle compared to Pt give rise to THz signals with opposite polarities and subsequently confirm the ISHE origin of the THz emission.

The spin-to-charge-conversion mechanism driven by the spin Seebeck effect was additionally probed in insulating magnetic/NM interfaces like YIG/$Cu_{1-x}Ir_x$ [33] and YIG/Pt [34] showing, however, much lower efficiency of the THz emission compared to the metallic magnetic layers.

According to experiments performed up to now and described above, bilayers composed of CoFeB/Pt and Fe/Pt provide the highest THz signal and they are the most prominent material choices so far. Other classes of magnetic layers have been also studied, such as compensated ferrimagnets, Heusler alloys, and insulating ferrimagnets like YIG; however, their efficiency is low. Besides the FM layer material investigations, different NM layers have been also studied with Pt being the prominent choice. Novel capping materials as spin-to-charge converters like at ferromagnetic/semiconductor interface [35], and topological insulators with large spin Hall angle like BiSe [36] used in combination with FM layers as in Bi_2Se_3/Co interface [37] can promote large spin injection efficiency, and pave the way for future investigations on materials.

6.4.2 Thickness Dependence

Various thickness dependence studies on the efficiency of the THz emission have revealed the critical role of the thicknesses of the individual layers. Studies on the variation of the thickness of the NM-layer and by keeping constant the thickness of the FM layer have shown the existence of an optimum thickness around 2–3 nm [20]. This range of thicknesses holds for different NM-layers like Pt [20, 23, 31], W [23, 31] where the maximum signal is obtained. By fitting the NM-thickness dependence of the THz amplitude, the spin current diffusion length could be extracted [20, 23]. Surprisingly, the latter was found to agree with values extracted from GHz spin pumping experiments although the excitation of the spin current is performed in different time- and energy scales. The physical mechanism behind this similarity is currently under ongoing research [38].

In the same way, for a constant NM-layer thickness there is an optimum FM layer thickness in the same range of about 2–3 nm [20, 23, 31]. When the ferromagnetic layer is too thin, then either the loss of magnetic order due to heating effects or the reorientation of the magnetization due to monolayer thickness cause a drastic reduction of the THz signal. The theory of the effect of changing layer thickness was first developed in Ref. [2] for detection of THz signals from the NM-layer side. A further expansion of the model was achieved in Ref. [20] and it is expressed by Eq. (6.4). The equation holds when the fs-laser illuminates first the NM-layer side and the detection is performed after the THz-signal has travelled through the substrate (see also Ref. [20]):

$$E_{\mathrm{THz}} \propto \Theta_{\mathrm{SH}} \frac{P_{\mathrm{abs}}}{d_{\mathrm{Fe}} + d_{\mathrm{Pt}}} \cdot \tanh\left(\frac{d_{\mathrm{Fe}} - d_0}{2\lambda_{\mathrm{pol}}}\right) \cdot \frac{1}{n_{\mathrm{air}} + n_{\mathrm{Sub}} + Z_0 \cdot (\sigma_{\mathrm{Fe}} d_{\mathrm{Fe}} + \sigma_{\mathrm{Pt}} d_{\mathrm{Pt}})} \cdot \tanh\left(\frac{d_{\mathrm{Pt}}}{2\lambda_{\mathrm{Pt}}}\right) \cdot e^{-(d_{\mathrm{Fe}} + d_{\mathrm{Pt}})/s_{\mathrm{THz}}}, \tag{6.4}$$

where n_{air}, n_{Sub} and Z_0 are the index of refraction of air, the index of refraction of the substrate at THz frequencies and the impedance of vacuum, respectively. Equation (6.4) takes into account all successive effects that take place, after the laser pulse impinges on the bilayer. In particular, the first term contains the spin Hall angle, and the second term accounts for the absorption of the femtosecond laser pulse in the metal layers. As only spin-polarized electrons within a certain distance from the boundary between FM and NM will reach the NM layer, only a fraction of the measured total absorbed power contributes to the generated THz signal. This fraction scales with the inverse of the total metal layer thickness $1/(d_{\mathrm{Fe}} + d_{\mathrm{Pt}})$. The third term describes the generation and diffusion of the generated spin current flowing in FM towards the interface with NM; The possibility that extra thin FM layers can lose their ferromagnetic properties below a certain thickness can be captured by introducing the term d_0 in Eq. (6.4). Below this critical thickness it is considered that the flow (if any) of spin current in FM does not reach the NM layer. Above this critical thickness, the generated spin polarization saturates with a characteristic constant λ_{pol}. The fourth and fifth term, the tangent hyperbolic function divided by the total impedance, refer to the spin accumulation in Pt, which is responsible for the strength of the THz radiation and it depends on the finite diffusion length λ_{Pt} of the spin current in NM layer. The symbols σ_{Fe} and σ_{Pt} correspond to the electrical conductivities of the layers. This part of the equation also includes the shunting effect of the parallel connection of the resistances of the individual FM and NM layers. The last term describes the attenuation of the THz radiation during propagation through the metal layers (with s_{THz} as an effective inverse attenuation coefficient of THz radiation in the two metal layers). The last term is required especially when the sample is excited from the outer Pt-side whereas the THz radiation is collected behind the substrate, thus propagating through both metal layers, see Ref. [20]. In the case of small losses, the attenuation can be taken into account by this single exponential factor. The fourth factor in Eq. (6.4) also accounts for the multiple reflections of the THz pulse at the metal/dielectric interfaces [2, 20]. All terms together describe the layer thickness dependence of the measured THz amplitudes.

In summary, the behaviour of the thickness dependence of the THz signal from spintronic emitters is established. In general, the optimal thickness of the NM layer depends in particular on the spin diffusion length. For thicknesses larger than the spin diffusion length the signal is reduced. For very thin magnetic layers the change of the magnetic order also reduces the THz emission. Future studies on the thickness dependence will aim to study the effect in complex multilayer structures and to reveal additional factors that can influence the thickness dependence like the interfacial spin memory loss [39]. Of large interest are the investigations and the comparison on spin current diffusion length for ultrafast fs-laser excitation and microwaves (GHz) excitation schemes.

6.4.3 Wavelength Dependence

The majority of the experiments up to now with spintronic THz emitters are performed at 800 nm excitation wavelength using femtosecond Ti:sapphire lasers. Would different excitation energies of the incoming fs-laser pulses influence the spin current dynamics

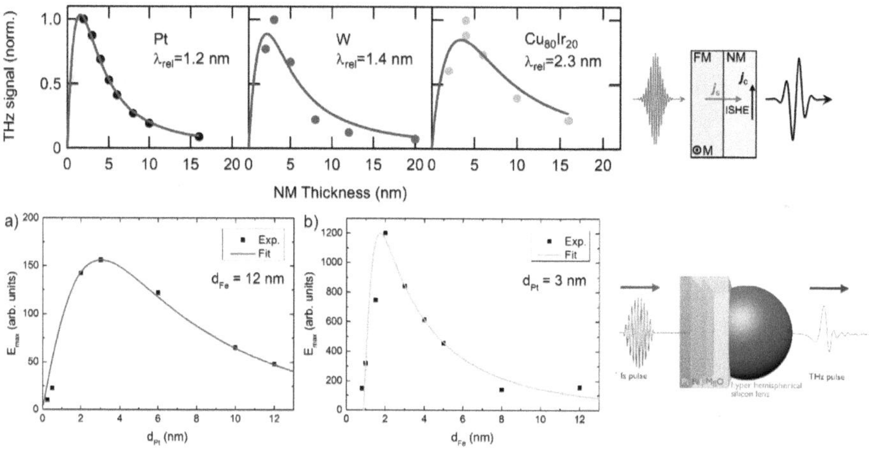

Figure 6.6 (Upper panel) Non-magnetic layer thickness dependence of THz emission for different NM layers like Pt, W, CuIr in CoFeB/NM bilayers [23], and the geometrical arrangement of the experiment. Reproduced under the terms of a Creative Commons Attribution 3.0 Licence [23], 2018 IOP Publishing Ltd. (Lower panel) (a) Pt thickness dependence of the THz field amplitude for a constant Fe thickness of 12 nm. (b) Fe thickness dependence of the THz field amplitude for a constant Pt thickness of 3 nm [20]. The geometrical arrangement is also depicted to the right of the graph. Reproduced under a Creative Commons Attribution 4.0 International License [20], The Authors 2018, published by Springer Nature. Figure modified from [24]

and subsequently the THz-emission? Recent investigations have tried to answer this question using different excitation wavelengths. Papaioannou et al. [40] used an optimized spintronic bilayer structure of 2 nm Fe and 3 nm Pt grown on 500 μm MgO substrate to show that the emitter is just as effective as a THz radiation source when excited either at $\lambda = 800$ nm or at $\lambda = 1550$ nm by ultrafast laser pulses from a fs fibre laser (pulse width 100 fs, repetition rate 100 MHz). Even with low incident power levels, the Fe/Pt spintronic emitter exhibits efficient generation of THz radiation at both excitation wavelengths. It should be mentioned that there is a linear dependence of the generated THz amplitude on pump power in the low power excitation regime for both pump wavelengths (see Fig. 6.7). At higher pump powers there will be a deviation from the linear behaviour as also shown in

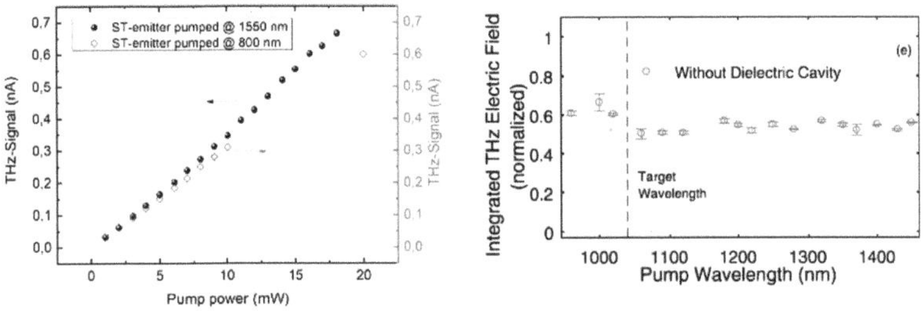

Figure 6.7 Excitation wavelength dependence of the THz emission. (Left panel) Maximum THz-E-field as a function of pump power of the laser recorded at 1550 nm (black squares) and 800 nm pump wavelength, $\lambda_{excitation}$. Figure modified from Ref. [40], Copyright 2018, IEEE. (Right panel) THz-E-field, emitted from a W/CoFeB/Pt sample as a function of the pump wavelength while the energy, focus diameter, and duration of the pump pulses were kept constant. Reprinted from Ref. [41] with the permission of AIP Publishing, Copyright 2019.

Ref. [25]. Herapath et al. [41] also found that the efficiency of THz generation is essentially flat for excitation by 150 fs pulses with centre wavelengths ranging from 900 to 1500 nm, using a CoFeB ferromagnetic layer between adjacent non-magnetic W and Pt layers. From both experiments, it seems that the crucial factor is the amount of energy that is deposited by the pump pulse in the electronic system and not the details of the involved optical transitions. By further probing the THz emission of Fe/Pt bilayers at $\lambda = 400$ nm [42], the spintronic THz emission efficiency of an optimized spintronic emitter was equal strong as with probing at 800 nm and 1550 nm. So it still remains independent of the optical pump wavelength. In addition, the efficiency is highly tunable with optical pump power [42]. The reason behind the observed effect is a consequence of the fact that the out-of-equilibrium transport is done not only by the electrons directly excited by the laser, but also by all the electrons that are excited at intermediate energies due to the scattering of the first-generation electrons. Considering that the electron–electron scattering lifetimes sharply decrease with the energy of the excited electron, the direct impact of high energetic electrons on the transport is expected to be minor at high-energy excitation. Instead, the most important impact is that by scattering with another electron, the electrons will lose energy, they will go down to an intermediate energy and transfer that energy to secondary electrons. This results in an intermediate energy electrons multiplication, which is very similar to a direct excitation by a laser at a lower frequency. Even with lower energies now, they will have longer lifetimes and they will contribute to the transport more importantly. If the system is excited with very low-frequency photons, other effects can become important, changing the qualitative scenario mentioned. In this case, one has to take into account that the spin diffusion requires asymmetry not only between the two spin channels but also between electrons and holes.

This was recently investigated by V. Mag-usara et al. [43]. They have shown that THz emission from a metallic spintronic heterostructure, such as the Fe/Pt bilayer on MgO substrate, is not totally independent of the optical excitation wavelength due to optical absorptance and spin-filtering. While the Fe/Pt bilayer is indeed a versatile THz source as its performance remains fairly the same over a wide range of pump wavelengths, it actually exhibits a slight enhancement of the spintronic THz emission in the 1200 nm to 1800 nm pump wavelength range and its THz generation efficiency continuously decreases when the excitation wavelength goes further beyond 2200 nm. The observed influence of the optical pump wavelength on the THz emission which, at wavelengths longer than 2500 nm, is inconsistent with the absorptance, led them to infer a $\approx$0.35 eV threshold pump photon energy for effective spintronic terahertz generation in the Fe/Pt bilayer. The threshold can be ascribed to the onset of significant spin-filtering in the Fe/Pt bilayer. The experimental and theoretical results are consistent with information based on first-principles calculations on the energy-dependent spin transport. Therefore, in studying the wavelength-dependence of THz radiation from a metallic spintronic heterostructure by THz emission spectroscopy, they were able to probe the optically excited spin transport scenario as well as demonstrate the influence of both optical pump absorptance and energy-dependent spin transport on the spintronic THz emission process.

The fact that efficient generation of THz radiation from spintronic structures is possible with lasers over a wide wavelength range will have important consequences for the use of such emitters in future applications. For example, the efficient excitation at 1550 nm wavelength allows the immediate integration of such spintronic emitters in THz systems driven by relatively low-cost and compact fs fibre lasers, without the need for frequency conversion. In the future, it will be crucial for the development of efficient spintronic THz emitters to understand the mechanisms of spin current generation and transport at lower and at higher excitation energies also from a theoretical point of view.

6.4.4 Interface Dependence

The transfer of a spin current from a FM to a NM layer (that is the source of THz emission) is a highly interface-sensitive effect. It depends on the structural properties of both layers and on possible interlayers between FM and NM layers. Both effects have not been studied in great detail so far. However, only few investigations have tried to correlate the structural quality of the interface with the THz signal strength and spectrum. A direct comparison of an epitaxial Fe(3 nm)/Pt(3 nm) bilayer with a signal-optimized polycrystalline CoFeB/Pt structure with the same layer thicknesses revealed a comparable THz signal strength [2]. In contrast, a significant increase in signal amplitude between Fe/Pt emitters grown epitaxially on MgO (100) substrates compared to polycrystalline emitters grown on sapphire substrates was reported by Torosyan et al. [20]. Similarly, the better crystal quality of a CoFeB layer, controlled by the annealing temperature, has significantly enhanced THz emission intensity [44].

Nenno et al. [13] have investigated in detail the performance of spintronic terahertz emitters by modifying the interface quality and its defect density. In particular, the presence of defect density was correlated with the elastic electron-defect scattering lifetime τ_{el} in the FM and NM layers and the interface transmission T for spin-polarized, non-equilibrium electrons.

A decreased defect density increases the electron-defect scattering lifetime and this results in a longer-lasting and stronger spin current pulse 6.8. Accordingly, a significant enhancement of the THz signal amplitude and a shift of the spectrum towards lower THz frequencies [13] was observed, see Fig. 6.8. Furthermore, besides

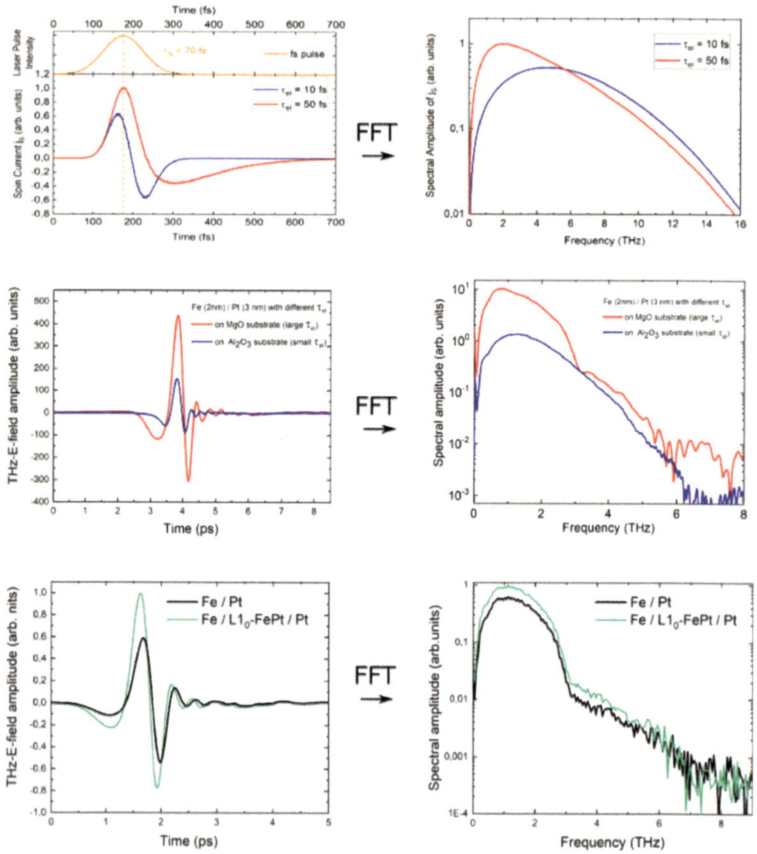

Figure 6.8 Dependence of the THz emission on the structural quality of the interface. (Upper panel, left) Numerical simulations of the Boltzmann transport equation for an Fe/Pt spintronic emitter for different elastic scattering lifetimes τ_{el}.The temporal evolution of the laser pulse intensity with a length of 70 fs is also shown on top. (Upper panel, right) Spectral amplitude of the simulated spin current (and consequently of the charge current) inside Pt layer obtained by the Fourier transform of the pulses on the left. (Middle panel, left) Experimental THz-E-field amplitudes and (right) corresponding spectral amplitudes for two Fe (2 nm)/ Pt(3 nm) bilayer with different τ_{el} as detected after travelling the optical beam path of a THz-TDS setup[13]. (Lower panel, left) Comparison of experimental obtained THz pulses and (right) corresponding spectral amplitudes between a Fe/Pt and a Fe/$L1_0$-FePt (2 nm)/Pt heterostructures with different interface transmissions T [46]. Figure modified from [24].

the defect density, the presence of the parameter of the interface transmission T plays an important role. The latter is correlated to the ability of the interface to transfer hot carriers into the NM layer. It was shown [13] that the interface transmission influences the spectral amplitude of the emitted THz field but conserves the composition of the spectrum, see Fig. 6.8.

The decisive role of the microstructural properties in the THz emission at the FM/NM was further probed by alloying at Co–Pt and Fe–Pt interface. Li et al. [45] showed that high interfacial roughness between the Co–Pt interface led to a decrease of the THz emission. The introduction of an Co_xPt_{1-x} alloy as a spacer at the Co–Pt interface showed that the intermixing amplified the THz emission. Maximum amplification by a factor of 4.2 was achieved for a $Co_{25}Pt_{75}$(1 nm thick) interlayer. Possible explanation could be the higher flux of spin currents into Pt due to reduction of spin decoherence at the interface caused by the presence of the alloy. Scheuer et al. [46] observed large enhancement in THz emission amplitude and bandwidth in Fe/$L1_0$-FePt/Pt which on the other hand, is dependent on the extent of alloying on either side of the interface, see Fig. 6.8, lower panel. It was concluded that the presence of the $L1_0$-FePt alloy of 2 nm thickness promotes the THz emission by improving the interface transmission T and enhancing the spin asymmetry between the Pt 5d band and the Fe 3d band [46].

The influence of an NM interlayer on the THz emission has been also studied. Such interlayers may be useful to realize multiple stacks of spintronic emitters. One prominent choice as interlayer is Cu, due to its negligible spin Hall angle and its large spin diffusion length. The strength of the THz signal amplitude was found to decrease with increasing Cu-layer thickness [40], either exponentially [23] or linearly [47]. Interestingly, the spin current relaxation length obtained from thickness dependence studies was revealed to be larger than for metals with sizeable spin Hall effect but, on the other hand, small compared to DC spin-diffusion lengths, that in Cu are of the order of 100 nm [23]. The use of a semiconducting material such as ZnO as interlayer has strongly suppressed the THz signal [47]. The use of insulating interlayers like MgO also strongly suppresses the THz signal. However, the suppression of spin current through MgO can be an advantage in multilayers, as we will see in the next section of geometry dependence.

Up to now the understanding and the structural engineering of THz intensity and spectral bandwidth of spintronic emitters remains an uncharted territory. First efforts have revealed the defect engineering as a factor that can be utilized in future to shape the THz emission. Non-magnetic interlayers like Cu, ZnO, MgO lead to a reduction of the emitted radiation in FM/Interlayer/NM structures. On the other side FM/NM alloyed interfaces may open a new direction in designing future STEs due to their ability to enhance the THz radiation.

6.4.5 Stack Geometry Dependence

The geometrical arrangement of the layers can play a decisive role for the THz emission. Seifert et al. [2] used instead of a FM/NM bilayer a trilayer in the form of NM1/FM/NM2 layers. The special feature of the trilayers is that the NM1 and NM2 layers have opposite spin Hall angles, like W and Pt. In such a way the spin Hall currents in W and Pt flow in the same direction and radiate in phase and thus enhance the THz amplitude. A large improvement of the signal was achieved in multilayers made of Fe/Pt/MgO [25, 42]. The MgO layer as an insulating layer hinders the flow of the spin current. The spin current can flow from each Fe layer only to its neighbouring Pt layer and generate THz radiation. Since the transverse charge currents are almost in phase due to the small thickness of the NM and FM layers compared to the THz wavelength, the THz signal from each layer is added constructively and boosts the emission. The maximum signal was observed for 3 repetitions of Fe/Pt/MgO multilayers while above 3 the detected THz signal dropped [42] due to the increased absorption of THz radiation caused by the additional MgO interlayers (see Fig. 6.9).

Hibberd et al. [48] followed another route to manipulate the THz radiation. They altered not the layer stack but the magnetic field pattern that is applied to the spintronic source. The goal was to manipulate the magnetic state of the ferromagnetic layer and therefore the polarization profile of the THz radiation. When the spintronic source was placed between two magnets of opposing

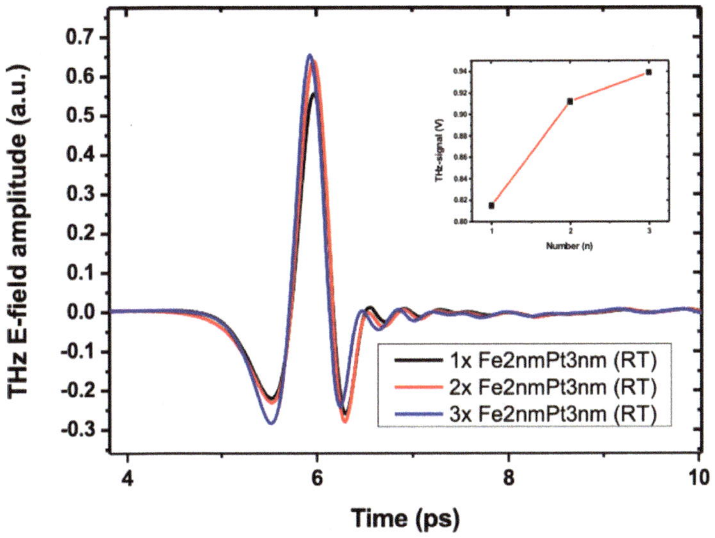

Figure 6.9 Emitted THz pulses from [Fe (2 nm)/Pt (3 nm)/MgO (3 nm)] $\times n$, $n = 1, 2, 3$ multilayers. The inset presents the maximum pulse amplitude with respect to the number of repetitions n.

polarity, THz radiation with a quadrupole-like polarization was generated. When focused, it resulted in longitudinal THz electric field amplitudes twice as much as that of the transverse signals achieved with linearly polarized THz radiation.

Ogasawara et al. [49] used synthetic magnets to manipulate the THz radiation. In particular, they prepared NM1(Ta)/CoFeB(FM)/Ir(NM2)/CoFeB(FM)/Ta(NM2) synthetic magnetic multilayers where the Ir spacer had different thicknesses. In such stacking THz emission is expected when the magnetizations of the CoFeB layers are anti-parallel. The authors observed THz emission under no applied magnetic field for samples with antiferromagnetic coupling and oscillations of the signal with respect to Ir thicknesses. In addition, Li et al. [50], showed as a proof-of-concept the use a magnetic tunnel junction (MTJ) as a THz source. They used a MTJ multi-stack of Ta/Ru/pinned FM stack/MgO/free FM stack/Ta/Ru. The source of the THz radiation due to the ISHE was the free FM layer/Ta interface while the pinned FM layer controlled the magnetic state of the free FM layer and so the THz signal strength and polarization. A different kind of multilayered structure was examined by Fix et al. [51] composed of ferrimagnetic (FIM) rare earth-3d transition metal alloys. In particular, the structure had the form of Pt(3 nm)/$Gd_{10}Fe_{90}$(3 nm)/W(3 nm)/$Gd_{30}Fe_{70}$(3 nm)/Pt(3 nm). The combination of NM-layers with opposite spin Hall angle (Pt,W) with FIM-layers of GdFe with suitable film thickness and composition was utilized to control the terahertz high and low emission state by temperature. Temperature was able to switch the relative alignment of the Fe moments in the different GdFe layers and thus the strength of the THz emission.

Another approach was presented by Feng et al. [52] where the performance of the spintronic THz emitter was improved by utilizing optics. In particular, the work utilized metal (NM1/FM/NM2)-dielectric photonic crystal structure where the metal Pt(1.8 nm)/Fe(1.8 nm)/W(1.8 nm) served as the spintronic emitter and the dielectric interlayers was SiO_2, while the maximum number of layer repeats in the photonic crystal was 3. The goal was to suppress the reflection and transmission of laser light simultaneously aiming to maximize the laser field strength in the metal layers. As a result, the maximization of the laser energy absorption in the metallic emitter improved the conversion efficiency by about 1.7 times compared to a NM1/FM/NM2 single THz emitter. Similarly, one year later Herapath et al. [41] utilized a dielectric cavity composed of TiO_2 and SiO_2, which was attached on a W/CoFeB/Pt emitter resulting in an enhancement factor of 2 in the THz emission. Chen et al. [53] used a cascade of two spintronic emitters to produce circularly polarized terahertz waves. They used the residual transmitted optical pump power after the first emitter to excite the second stage emitter. Equal amplitudes were achieved by adjusting the losses of the pump laser in the first emitter. The arrangement of the applied magnetic field directions on both emitters was such that perpendicular electric-field directions were achieved for the terahertz beams generated in the two stages. By adjusting the refractive index difference of low-pressure air between optical and terahertz frequency ranges, a phase difference between the terahertz waves was introduced in such a way that the mixing of the waves could lead to arbitrary control of the terahertz polarization-shaping, including also the chirality.

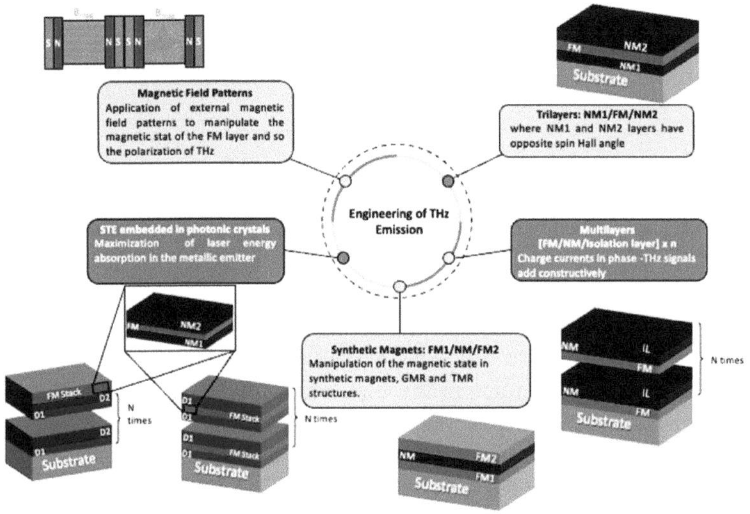

Figure 6.10 Graphical representation of geometrical stacking order that can influence the intensity, bandwidth and the polarization state of the emitted THz signals. FM denotes the ferromagnetic layer, NM the non-magnetic layer, IL the isolation layer, STE is the spintronic emitter, D1,2 the dielectric layers, INT the interlayer in synthetic magnetic structures. Figure modified from [24].

The different strategies on geometrical stacking order are summarized in Fig.6.10. Investigations carried out so far show that trilayer structures NM1/FM/NM2 with the non-metallic layers of opposite spin Hall angle are proven to be the most efficient spintronic emitters. However, the structural quality of very thin NM is not always guaranteed. The studies on optically assisted THz generation using photonic crystals have just started. The idea of manipulating the magnetization direction of the FM layer, either by creating external magnetic field patterns or manipulating the magnetic state of the magnetic layers like in synthetic magnets, is very effective and interesting from a physics and technological point of view. The future investigation of the stack geometry would definitely include manipulation of the magnetic state of the FM layer in different giant magnetoresistance (GMR) and tunnel magnetoresistance (TMR) structures. Application of field patterns and multilayers structure can be very useful for enhancing and shaping the THz emission. After all, the concept of modifying the stack geometry by means of geometrical and magnetic order holds big promises for future designs of tunable THz emitters.

6.5 Future Perspectives of THz STEs

THz spintronic emitters are certainly a new direction in physics with a huge potential for technological applications. They define the new field of ultrafast terahertz optospintronics. The emitters have many advantages: They are low cost, ultra-broadband, easy to use and robust. The performance of the emitters is comparable to state-of-the-art terahertz sources like the ones routinely used to cover the range from 0.3 to 8 THz, e.g. non-linear optical crystals such as ZnTe (110) and GaP or high-performance photoconductive switches [2]. The challenge nowadays is to generate THz broadband radiation with sufficient power. In this direction, Seifert et al. [54] explored the upscaling capabilities of metallic spintronic stacks in order to achieve intense THz sources. They used large area W/CoFeB/Pt stacks of 5.6 nm thickness.

The emitted THz pulse was found to reach energies of 5 nJ with peak field values of up to 300 kVcm^{-1} when excited with a laser pulse energy of 5.5 mJ. This corresponds to a conversion efficiency of 10^{-6}. Similar conversion efficiencies were reported for high repetition rate systems with repetition rates of 100 MHz and moderate average powers (up to 500 mW) using Fe/Pt bilayers [40]. The estimated absolute average power of the generated THz pulses was almost 50 μW for a pump power of 500 mW using 800 nm excitation wavelength. Taking into account a THz pulse length of approximately 1 ps this corresponds to a comparable conversion efficiency.

Currently, the downscaling through nanopatterning of the emitters is under investigation. In one of the first investigations on patterned emitters by Yang et. al. [25] Fe/Pt heterostructures were patterned in stripes of 5 μm width and a spacing of 5 μm and measurements were performed with the magnetic field along and perpendicular to the stripes. Anisotropic emission was observed with higher intensity for the perpendicular configuration and a blue shift in frequency compared to the parallel configuration. Further, modulation of THz radiation in terms of magnitude, bandwidth and centre frequency was achieved based on W/CoFeB/Pt stripe patterns [55]. The patterns had dimensions of 5 μm width and 5 μm spacing. By changing the angle between the applied magnetic field and the stripe direction, modulation of the THz emission was measured. Interestingly, Jin et al. [55], attributed this effect to nanoconfinement of the charge current due to nanopatterning. For a specific configuration of stripe and magnetic field at 90 degrees, the spatial separation of the photo-induced positive and negative charges accumulates and then creates a non-equilibrium electric potential $U(t)$ at the edge of the stripe. The transient potential creates an opposite built-in electric field. The latter induces a back flowing current, changes the effective length of the charge flowing and so alters the THz emission. Song et al. [56] fabricated CoFeB/Pt patterned as rectangles and orthogonal structures of 20 μm width and various lengths L, of 20, 40, 80, 160 and 320 μm, while the separation between the elements was 10 μm. They observed a reshaping of the THz wavefront, which indicates that the pattern structure influences the temporal distribution of the transient dynamics of the charge current. The FWHM decreased by decreasing L pointing to the backflow effect as mentioned before, while the centre frequency increases by decreasing L as a result of the antenna effect of the microstructures, where it is expected that shorter antennas will lead to higher oscillation frequencies. Microfabricated spintronic emitters were also studied by Wu et al. [57]. Fe(3 nm)/Pt(3 nm) bilayers were patterned by optical lithography in the form of stripes of constant separation of 510 μm and a variable width of 240, 300, 400 and 490 μm. The THz waveform was modulated by the different stripe widths. As the stripe width is reduced, a second peak in the spectra appears, showing that the bandwidth can be increased in such microstructures. Wu et al. interpreted the results in terms of a simplified multi-slit interference model, which can be a useful tool for designing future patterned emitters. Another approach in patterning was followed by Nandi et al. [58] where the spintronic trilayer of W(2 nm)/CoFeB(1.8 nm)/Pt(2 nm) was inserted in the gap of two antenna types: a slotline (with spacing of 25 μm) and an H-type dipole antenna (with length of 200 μm and gap size of 10 μm). Such antennas are frequently used to improve the radiation efficiency in the state-of-the-art semiconductor-based photoconductive THz emitters and receivers. The coupling between the spintronic emitter and the antennas led to a significant increase in the THz signal compared to plain continuous film emitters. The incorporation of spintronic emitters with already existing THz technologies can drive the on-chip application of this new class of THz emitters based on ultrafast spin and charge currents.

In conclusion, the first studies in nanopatterned emitters revealed the ability to modify the THz emission, rendering nanostructuring as a key ingredient for future applications of THz emitters. Meanwhile, the first implementation of spintronic emitters in real applications has been realized. Arrays of W/Fe/Pt were proposed to form a THz near-field microscope capable of illuminating an object at an extreme near field [59]. Guo et al. [60] examined the possibility to integrate W/CoFeB/Pt heterostructures in laser terahertz emission microscopy (LTEM) technique. Müller et al. [61] have demonstrated efficient coupling of THz-pulses emitted from W/CoFeB/Pt trilayer to the junction of a scanning tunnelling microscope, that has the potential to enable spatiotemporal imaging with femtosecond temporal and sub-nm spatial resolution.

Certainly, future investigations will deal with THz emission from giant magnetoresistance (GMR) and tunnel magnetoresistance (TMR) structures, where the potential for THz manipulation by defining the magnetic states of the layers is large. Open challenges for the coming years are the use of spintronic structures as detectors of THz radiation, the exploration of appropriate materials, interfaces and nanopatterns, the electrical detection of the ultra-short current pulses, and the use of emission for microscopy and imaging in the near- and far-field.

6.6 Conclusion

Spintronic THz emitters can cover a frequency range from 0,1 to well above 30 THz without phonon absorption of the emitter material. The latter renders them superior to all the current solid-state emitters. The big challenge nowadays is to generate broadband THz radiation with sufficient power by an appropriate combination of materials, multilayer stacks, growth quality and nanopatterning. The use of such sources in THz-imaging has just launched. The physics of the optospintronic phenomena can shed light on many light-induced magnetic interactions. The field is still in its infancy and the research on the design and control of THz waves emitted from spintronic structures has brilliant perspectives.

Acknowledgment

Marco Rahm, Dominik Sokoluk and Laura Scheuer (TU Kaiserslautern) are acknowledged for their long-standing collaboration. Prof. Georg Schmidt (MLU-Halle Wittenberg) is acknowledged for scientific support. Oliver Gueckstock and Prof. Tobias Kampfrath (FU Berlin) are acknowledged for their lab and scientific support in obtaining the data of Fig. 6.2 with the EOS setup.

References

1. T. Kampfrath, M. Battiato, P. Maldonado, G. Eilers, J. Nötzold, S. Mährlein, V. Zbarsky, F. Freimuth, Y. Mokrousov, S. Blügel, M. Wolf, I. Radu, P. M. Oppeneer, and M. Münzenberg, Terahertz spin current pulses controlled by magnetic heterostructures, *Nature Nanotechnology* **8**, 256 (2013).
2. T. Seifert, S. Jaiswal, U. Martens, L. Hannegan, J. Braun, P. Maldonado, F. Freimuth, A. Kronenberg, J. Henrizi, I. Radu, E. Beaurepaire, Y. Mokrousov, P. M. Oppeneer, M. Jourdan, G. Jakob, D. Turchinovich, L. M. Hayden, M. Wolf, M. Münzenberg, M. Kläui, and T. Kampfrath, Efficient metallic spintronic emitters of ultrabroadband terahertz radiation, *Nature Photonics* **10**, 483–488 (2016).
3. B. Ferguson and X.-C. Zhang, Materials for terahertz science and technology, *Nature Materials* **1**, 26–33 (2002).
4. R. A. Lewis, A review of terahertz sources, *Journal of Physics D: Applied Physics* **47**, 374001 (2014).
5. D. M. Mittleman, Perspective: Terahertz science and technology, *Journal of Applied Physics* **122**, 230901 (2017).

6. M. Tonouchi, Cutting-edge terahertz technology, *Nature Photonics* **1**, 97–105 (2007).
7. E. Beaurepaire, G. M. Turner, S. M. Harrel, M. C. Beard, J.-Y. Bigot, and C. A. Schmuttenmaer, Coherent terahertz emission from ferromagnetic films excited by femtosecond laser pulses, *Applied Physics Letters* **84**, 3465–3467 (2004).
8. D. J. Hilton, R. D. Averitt, C. A. Meserole, G. L. Fisher, D. J. Funk, J. D. Thompson, and A. J. Taylor, Terahertz emission via ultrashort-pulse excitation of magnetic metal films, *Optics Letters* **29**, 1805 (2004).
9. M. Battiato, K. Carva, and P. M. Oppeneer, Superdiffusive spin transport as a mechanism of ultrafast demagnetization, *Physical Review Letters* **105**, 027203–4 (2010).
10. J. Hurst, P.-A. Hervieux, and G. Manfredi, Spin current generation by ultrafast laser pulses in ferromagnetic nickel films, *Physical Review B* **97**, 014424–5 (2018).
11. G. Manfredi and P. A. Hervieux, Finite-size and nonlinear effects on the ultrafast electron transport in thin metal films, *Physical Review B* **72**, 155421–13 (2005).
12. D. M. Nenno, R. Binder, and H. C. Schneider, Simulation of hot-carrier dynamics and terahertz emission in laser-excited metallic bilayers, *Physical Review Applied* **11**, 054083 (2019).
13. D. M. Nenno, L. Scheuer, D. Sokoluk, S. Keller, G. Torosyan, A. Brodyanski, J. Lösch, M. Battiato, M. Rahm, R. H. Binder, H. C. Schneider, R. Beigang, and E. T. Papaioannou, Modification of spintronic terahertz emitter performance through defect engineering, *Scientific Reports* **9**, 13348 (2019).
14. W. Zhang, P. Maldonado, Z. Jin, T. S. Seifert, J. Arabski, G. Schmerber, E. Beaurepaire, M. Bonn, T. Kampfrath, P. M. Oppeneer, and D. Turchinovich, Ultrafast terahertz magnetometry, *Nature Communications* **11**, 4247 (2020).
15. C. Zhou, Y. P. Liu, Z. Wang, S. J. Ma, M. W. Jia, R. Q. Wu, L. Zhou, W. Zhang, M. K. Liu, Y. Z. Wu, and J. Qi, Broadband terahertz generation via the interface inverse Rashba–Edelstein effect, *Physical Review Letters* **121**, 086801 (2018).
16. M. B. Jungfleisch, Q. Zhang, W. Zhang, J. E. Pearson, R. D. Schaller, H. Wen, and A. Hoffmann, Control of terahertz emission by ultrafast spin-charge current conversion at Rashba interfaces, *Physical Review Letters* **120**, 207207 (2018).

17. Q. Zhang, Z. Luo, H. Li, Y. Yang, X. Zhang, and Y. Wu, Terahertz emission from anomalous Hall effect in a single-layer ferromagnet, *Physical Review Applied* **12**, 054027 (2019).
18. C. Fattinger and D. Grischkowsky, Terahertz beams, *Applied Physics Letters* **54**, 490–492 (1989).
19. O. Gueckstock, L. Nádvorník, T. S. Seifert, M. Borchert, G. Jakob, G. Schmidt, G. Woltersdorf, M. Kläui, M. Wolf, and T. Kampfrath, Modulating the polarization of broadband terahertz pulses from a spintronic emitter at rates up to 10 kHz, *Optica* **8**, 1013–1019 (2021).
20. G. Torosyan, S. Keller, L. Scheuer, R. Beigang, and E. T. Papaioannou, Optimized spintronic terahertz emitters based on epitaxial grown Fe/Pt layer structures, *Scientific Reports* **8**, 1311 (2018).
21. P. U. Jepsen, R. H. Jacobsen, and S. R. Keiding, Generation and detection of terahertz pulses from biased semiconductor antennas, *Journal of the Optical Society of America B* **13**, 2424–2436 (1996).
22. M. Chen, R. Mishra, Y. Wu, K. Lee, and H. Yang, Terahertz emission from compensated magnetic heterostructures, *Advanced Optical Materials* **6**, 1800430 (2018).
23. T. S. Seifert, N. M. Tran, O. Gueckstock, S. M. Rouzegar, L. Nadvornik, S. Jaiswal, G. Jakob, V. V. Temnov, M. Münzenberg, M. Wolf, M. Kläui, and T. Kampfrath, Terahertz spectroscopy for all-optical spintronic characterization of the spin-Hall-effect metals Pt, W and $Cu_{80}Ir_{20}$, *Journal of Physics D: Applied Physics* **51**, 364003 (2018).
24. E. T. Papaioannou and R. Beigang, THz spintronic emitters: a review on achievements and future challenges, *Nanophotonics* **10**, 1243–1257 (2021).
25. D. Yang, J. Liang, C. Zhou, L. Sun, R. Zheng, S. Luo, Y. Wu, and J. Qi, Powerful and tunable THz emitters based on the Fe/Pt magnetic heterostructure, *Advanced Optical Materials* **4**, 1944–1949 (2016).
26. Y. Sasaki, Y. Kota, S. Iihama, K. Z. Suzuki, A. Sakuma, and S. Mizukami, Effect of Co and Fe stoichiometry on terahertz emission from $Ta/(Co_xFe_{1-x})_{80}B_{20}/MgO$ thin films, *Physical Review B* **100**, 140406 (2019).
27. R. Schneider, M. Fix, R. Heming, S. Michaelis de Vasconcellos, M. Albrecht, and R. Bratschitsch, Magnetic-field-dependent THz emission of spintronic TbFe/Pt layers, *ACS Photonics* **5**, 3936–3942 (2018).
28. R. Schneider, M. Fix, J. Bensmann, S. Michaelis de Vasconcellos, M. Albrecht, and R. Bratschitsch, Spintronic GdFe/Pt THz emitters, *Applied Physics Letters* **115**, 152401 (2019).

29. T. Seifert, U. Martens, S. Günther, M. A. W. Schoen, F. Radu, X. Z. Chen, I. Lucas, R. Ramos, M. H. Aguirre, P. A. Algarabel, A. Anadón, H. S. Körner, J. Walowski, C. Back, M. R. Ibarra, L. Morellón, E. Saitoh, M. Wolf, C. Song, K. Uchida, M. Münzenberg, I. Radu, and T. Kampfrath, Terahertz spin currents and inverse spin Hall effect in thin-film heterostructures containing complex magnetic compounds, *SPIN* **07**, 1740010 (2017).
30. Y. Sasaki, Y. Takahashi, and S. Kasai, Laser-induced terahertz emission in Co2MnSi/Pt structure, *Applied Physics Express* **13**, 093003 (2020).
31. Y. Wu, M. Elyasi, X. Qiu, M. Chen, Y. Liu, L. Ke, and H. Yang, High-performance THz emitters based on ferromagnetic/nonmagnetic heterostructures, *Advanced Materials* **29**, 1603031 (2017).
32. S. Zhang, Z. Jin, Z. Zhu, W. Zhu, Z. Zhang, G. Ma, and J. Yao, Bursts of efficient terahertz radiation with saturation effect from metal-based ferromagnetic heterostructures, *Journal of Physics D: Applied Physics* **51**, 034001 (2017).
33. J. Cramer, T. Seifert, A. Kronenberg, F. Fuhrmann, G. Jakob, M. Jourdan, T. Kampfrath, and M. Kläui, Complex terahertz and direct current inverse spin Hall effect in YIG/$Cu_{1-x}Ir_x$ Bilayers Across a Wide Concentration Range, *Nano Letters* **18**, 1064–1069 (2018).
34. T. S. Seifert, S. Jaiswal, J. Barker, S. T. Weber, I. Razdolski, J. Cramer, O. Gueckstock, S. F. Maehrlein, L. Nadvornik, S. Watanabe, C. Ciccarelli, A. Melnikov, G. Jakob, M. Münzenberg, S. T. B. Goennenwein, G. Woltersdorf, B. Rethfeld, P. W. Brouwer, M. Wolf, M. Kläui, and T. Kampfrath, Femtosecond formation dynamics of the spin Seebeck effect revealed by terahertz spectroscopy, *Nature Communications* **9**, 2899 (2018).
35. L. Cheng, X. Wang, W. Yang, J. Chai, M. Yang, M. Chen, Y. Wu, X. Chen, D. Chi, K. E. J. Goh, J.-X. Zhu, H. Sun, S. Wang, J. C. W. Song, M. Battiato, H. Yang, and E. E. M. Chia, Far out-of-equilibrium spin populations trigger giant spin injection into atomically thin MoS_2, *Nature Physics* **15**, 347–351 (2019).
36. N. H. D. Khang, Y. Ueda, and P. N. Hai, A conductive topological insulator with large spin Hall effect for ultralow power spin–orbit torque switching, *Nature Materials* **17**, 808–813 (2018).
37. X. Wang, L. Cheng, D. Zhu, Y. Wu, M. Chen, Y. Wang, D. Zhao, C. B. Boothroyd, Y. M. Lam, J.-X. Zhu, M. Battiato, J. C. W. Song, H. Yang, and E. E. M. Chia, Ultrafast spin-to-charge conversion at the surface of topological insulator thin films, *Advanced Materials* **30**, 1802356 (2018).
38. L. Scheuer, M. Ruhwedel, G. Torosyan, R. Beigang, G. Schmidt, and E. T. Papaioannou, Correlation of interface transmission in THz spintronic

emitters with spin mixing conductance in spin pumping experiments, in H.-J. M. Drouhin, J.-E. Wegrowe, and M. Razeghi (Editors), *Spintronics XIII*, volume 11470, pp. 64–71, International Society for Optics and Photonics (SPIE, 2020).

39. H. S. Qiu, K. Kato, K. Hirota, N. Sarukura, M. Yoshimura, and M. Nakajima, Layer thickness dependence of the terahertz emission based on spin current in ferromagnetic heterostructures, *Optics Express* **26**, 15247–15254 (2018).
40. E. T. Papaioannou, G. Torosyan, S. Keller, L. Scheuer, M. Battiato, V. K. Mag-Usara, J. L'huillier, M. Tani, and R. Beigang, Efficient terahertz generation using Fe/Pt spintronic emitters pumped at different wavelengths, *IEEE Transactions on Magnetics* **54**, 1–5 (2018).
41. R. I. Herapath, S. M. Hornett, T. S. Seifert, G. Jakob, M. Kläui, J. Bertolotti, T. Kampfrath, and E. Hendry, Impact of pump wavelength on terahertz emission of a cavity-enhanced spintronic trilayer, *Applied Physics Letters* **114**, 041107 (2019).
42. R. Beigang, E. T. Papaioannou, L. Scheuer, S. Keller, G. Torosyan, M. Rahm, D. Sokoluk, M. Talara, Y. Oda, H. Kitahara, J. Afalla, V. K. Mag-usara, and M. Tani, Efficient terahertz generation using Fe/Pt spintronic emitters pumped at different wavelengths, in L. P. Sadwick and T. Yang (Editors), *Terahertz, RF, Millimeter, and Submillimeter-Wave Technology and Applications XII*, volume 10917, pp. 74–80, International Society for Optics and Photonics (SPIE, 2019).
43. V. K. Mag-usara, M. C. Escano, C. E. Petoukhoff, G. Torosyan, L. Scheuer, J. Madeó, J. Afalla, M. L. Talara, J. E. Muldera, H. Kitahara, D. R. Bacon, M. Nakajima, K. Dani, E. T. Papaioannou, R. Beigang, and M. Tani, Optimum excitation wavelength and photon energy threshold for spintronic terahertz emission from Fe/Pt bilayer, *iScience* **25**, 104615 (2022).
44. Y. Sasaki, K. Z. Suzuki, and S. Mizukami, Annealing effect on laser pulse-induced THz wave emission in Ta/CoFeB/MgO films, *Applied Physics Letters* **111**, 102401 (2017).
45. G. Li, R. Medapalli, R. V. Mikhaylovskiy, F. E. Spada, T. Rasing, E. E. Fullerton, and A. V. Kimel, THz emission from Co/Pt bilayers with varied roughness, crystal structure, and interface intermixing, *Physical Review Materials* **3**, 084415 (2019).
46. L. Scheuer, M. Ruhwedel, D. Karfaridis, I. G. Vasileiadis, D. Sokoluk, G. Torosyan, G. Vourlias, G. P. Dimitrakopoulos, M. Rahm, T. Kehagias, R. Beigang, and E. T. Papaioannou, THz emission from Fe/Pt spintronic emitters with Fe/$L1_0$-FePt/Pt alloyed interface, *iScience* **25**, 104319 (2022).

47. G. Li, R. V. Mikhaylovskiy, K. A. Grishunin, J. D. Costa, T. Rasing, and A. V. Kimel, Laser induced THz emission from femtosecond photocurrents in Co/ZnO/Pt and Co/Cu/Pt multilayers, *Journal of Physics D: Applied Physics* **51**, 134001 (2018).
48. M. T. Hibberd, D. S. Lake, N. A. B. Johansson, T. Thomson, S. P. Jamison, and D. M. Graham, Magnetic-field tailoring of the terahertz polarization emitted from a spintronic source, *Applied Physics Letters* **114**, 031101 (2019).
49. Y. Ogasawara, Y. Sasaki, S. Iihama, A. Kamimaki, K. Z. Suzuki, and S. Mizukami, Laser-induced terahertz emission from layered synthetic magnets, *Applied Physics Express* **13**, 063001 (2020).
50. J. Li, Z. Jin, B. Song, S. Zhang, C. Guo, C. Wan, X. Han, Z. Cheng, C. Zhang, X. Lin, G. Ma, and J. Yao, Magnetic-field-free terahertz emission from a magnetic tunneling junction, *Japanese Journal of Applied Physics* **58**, 090913 (2019).
51. M. Fix, R. Schneider, J. Bensmann, S. Michaelis de Vasconcellos, R. Bratschitsch, and M. Albrecht, Thermomagnetic control of spintronic THz emission enabled by ferrimagnets, *Applied Physics Letters* **116**, 012402 (2020).
52. Z. Feng, R. Yu, Y. Zhou, H. Lu, W. Tan, H. Deng, Q. Liu, Z. Zhai, L. Zhu, J. Cai, B. Miao, and H. Ding, Highly efficient spintronic terahertz emitter enabled by metal–dielectric photonic crystal, *Advanced Optical Materials* **6**, 1800965 (2018).
53. X. Chen, X. Wu, S. Shan, F. Guo, D. Kong, C. Wang, T. Nie, C. Pandey, L. Wen, W. Zhao, C. Ruan, J. Miao, Y. Li, and L. Wang, Generation and manipulation of chiral broadband terahertz waves from cascade spintronic terahertz emitters, *Applied Physics Letters* **115**, 221104 (2019).
54. T. Seifert, S. Jaiswal, M. Sajadi, G. Jakob, S. Winnerl, M. Wolf, M. Kläui, and T. Kampfrath, Ultrabroadband single-cycle terahertz pulses with peak fields of 300 kV/cm from a metallic spintronic emitter, *Applied Physics Letters* **110**, 252402 (2017).
55. Z. Jin, S. Zhang, W. Zhu, Q. Li, W. Zhang, Z. Zhang, S. Lou, Y. Dai, X. Lin, G. Ma, and J. Yao, Terahertz radiation modulated by confinement of picosecond current based on patterned ferromagnetic heterostructures, *physica status solidi (RRL) Rapid Research Letters* **13**, 1900057 (2019).
56. B. Song, Y. Song, S. Zhang, K. Jin, W. Zhu, Q. Li, Z. Zhang, X. Lin, Y. Dai, X. Yan, G. Ma, Z. Jin, and J. Yao, Controlling terahertz radiation with subwavelength blocky patterned CoFeB/Pt heterostructures, *Applied Physics Express* **12**, 122003 (2019).

57. W. Wu, S. Lendinez, M. Taghipour Kaffash, R. D. Schaller, H. Wen, and M. B. Jungfleisch, Modification of terahertz emission spectrum using microfabricated spintronic emitters, *Journal of Applied Physics* **128**, 103902 (2020).
58. U. Nandi, M. S. Abdelaziz, S. Jaiswal, G. Jakob, O. Gueckstock, S. M. Rouzegar, T. S. Seifert, M. Kläui, T. Kampfrath, and S. Preu, Antenna-coupled spintronic terahertz emitters driven by a 1550 nm femtosecond laser oscillator, *Applied Physics Letters* **115**, 022405 (2019).
59. S.-C. Chen, Z. Feng, J. Li, W. Tan, L.-H. Du, J. Cai, Y. Ma, K. He, H. Ding, Z.-H. Zhai, Z.-R. Li, C.-W. Qiu, X.-C. Zhang, and L.-G. Zhu, Ghost spintronic THz-emitter-array microscope, *Light: Science & Applications* **9**, 99 (2020).
60. F. Guo, C. Pandey, C. Wang, T. Nie, L. Wen, W. Zhao, J. Miao, L. Wang, and X. Wu, Generation of highly efficient terahertz radiation in ferromagnetic heterostructures and its application in spintronic terahertz emission microscopy (STEM), *OSA Continuum* **3**, 893–902 (2020).
61. M. Müller, N. Martín Sabanés, T. Kampfrath, and M. Wolf, Phase-resolved detection of ultrabroadband THz pulses inside a scanning tunneling microscope junction, *ACS Photonics* **7**, 2046–2055 (2020).

Chapter 7

Terahertz Frequency Comb

Takeshi Yasui
Institute of Post-LED Photonics, Tokushima University, Japan
yasui.takeshi@tokushima-u.ac.jp

Preface

Three techniques for terahertz (THz) frequency metrology based on frequency comb, namely, a THz-comb-referenced spectrum analyzer, a continuously tunable, single-frequency continuous-wave (CW)-THz generator, and dual-THz-comb spectroscopy, are reviewed. Since the frequency comb enables to coherently link the frequency among microwave, optical, and THz regions, it is possible to establish the THz frequency metrology traceable to time of the SI base units. Using a THz-comb-referenced spectrum analyzer based on a stable THz comb generated in a photoconductive antenna for THz detection, the absolute frequency of CW test sources in the sub-THz and THz regions was determined at a precision of 10^{-11}. Also, a continuously tunable, single-frequency CW-THz generator

Advances in Terahertz Source Technologies
Edited by Gun-Sik Park, Masahiko Tani, Jae-Sung Rieh, and Sang Yoon Park

ISBN 978-981-4968-89-8 (Hardcover), 978-1-003-45967-5 (eBook)
www.jennystanford.com

was demonstrated by photomixing of an accurately tunable CW laser and a tightly fixed CW laser in the optical frequency region, phase-locked to two independent optical combs. Furthermore, dual-THz-comb spectroscopy gives the precise frequency scale for broadband spectroscopy, which is secured by a rubidium frequency standard. These three techniques will open the door for the establishment of frequency metrology in the THz region.

7.1 Introduction

Terahertz (THz) electromagnetic waves, lying at the boundary between optical and electrical waves, have emerged as a new innovative mode for sensing, spectroscopy, communication, and other applications [1, 2]. Along with recent progress in THz technology, the requirements of THz metrology have increased in various applications [3]. In particular, THz frequency metrology is important because frequency is a fundamental physical quantity of electromagnetic waves [4]. For example, when the THz wave is used as a carrier wave for next-generation wireless communications (Beyond 5G or 6G), its wireless carrier frequency should be highly accurate and stable in order to secure necessary and sufficient bandwidth for broadband communication without interferences with other applications, such as astronomy or sensing. Also, the THz frequency metrology will play an important role in the frequency calibration of various types of commercial THz instruments, such as sources, detectors, and systems. For example, the precisely calibrated THz spectrometer increases identification power in spectroscopic applications based on THz spectral fingerprints. However, in contrast to electrical and optical regions, reliable frequency metrology has not yet been established in the THz region due to the fact that source and detector technologies are still immature in this frequency region. Such a "THz gap of frequency metrology" could hinder the growth of various THz applications. If reliable THz frequency metrology could be established based on the national frequency standard, the scope of THz applications would be greatly expanded as a result of the high reliability achieved.

An ideal approach for establishing reliable THz frequency metrology is to use a transition frequency in an atom for a frequency standard. For example, a frequency standard at 9.2 GHz was established based on the hyperfine transition in the cesium atom, which is the basis of the definition of time in the International System of Units (SI). In the THz region, a scheme for establishing a frequency standard based on three-photon coherent population trapping in stored ions has been proposed [5]; however, realizing this scheme will be challenging in practice. Another practical approach for frequency metrology is to transfer the frequency uncertainty of other electromagnetic wave regions, where reliable frequency standards have already been established, to the THz region. To achieve this, frequency linking between two electromagnetic wave regions is required. Recently, an optical frequency comb was used to distribute the frequency uncertainty of a microwave frequency standard to the optical region directly [6–9]. Furthermore, the concept of a frequency comb has been extended to the THz region [10]. These techniques show promise for realizing reliable frequency metrology in the THz region [4].

In this chapter, we first describe coherent frequency linking among the microwave, optical, and THz regions in Section 7.2, achieved by frequency comb techniques [4]. Then, three techniques for THz frequency metrology based on a frequency comb are reviewed: THz-comb-referenced spectrum analyzer [11–15], optical-comb-referenced frequency synthesizer [16–18], and dual-THz-comb spectroscopy [10, 19–24]. A THz-comb-referenced spectrum analyzer for precise frequency measurement of CW THz waves is presented in Section 7.3. Section 7.4 describes the demonstration of an optical-comb-referenced frequency synthesizer using photomixing of two optical frequency synthesizers phase-locked to a microwave frequency standard via optical comb. In Section 7.5, dual-THz-comb spectroscopy is described, in which each mode of the THz comb is used as a frequency marker for a broadband THz spectrum. Section 7.6 summarizes those works and discusses some goals of future research.

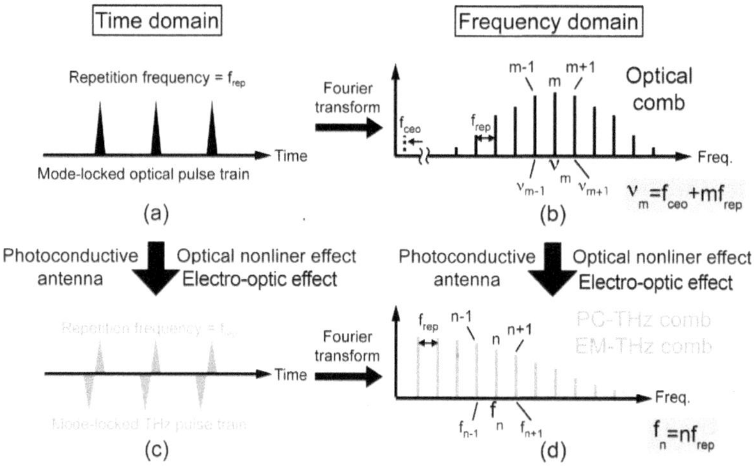

Figure 7.1 Behavior of an optical comb and a THz comb in the time and frequency domains.

7.2 Coherent Link of Frequency Using Frequency Comb

Figure 7.1 illustrates the behavior of an optical comb and a THz comb in the time and frequency domains. A femtosecond mode-locked (fs-ML) laser generates a sequence of ultrashort optical pulses that are essentially copies of the same pulse separated by an interval equal to the inverse of a repetition frequency f_{rep}. The highly stable fs-ML pulse train is synthesized by a series of regular frequency spikes separated by f_{rep} in the optical frequency domain. This structure is referred to as an optical frequency comb or optical comb. Since the frequency comb structure can be used as a precision frequency ruler in the optical region, the fs-ML-laser-based optical comb has received a lot of interest as a powerful metrological tool capable of covering the optical region by virtue of precise laser stabilization [6–9]. The concept of the frequency comb has been extended to the THz region by combining fs-ML pulse trains with a photoconductive antenna (PCA) [10]. When a PCA for THz generation is irradiated by the fs-ML optical pulse train, a free-space-propagating, mode-locked THz pulse train is radiated from the PCA.

This THz pulse train is synthesized by a regular comb of frequency spikes of electromagnetic waves in the THz frequency domain; namely, an electromagnetic THz comb (EM-THz comb). On the other hand, when the fs-ML optical pulse train is incident on a PCA used for THz detection, the instantaneous generation of photocarriers is repeated in the PCA in synchronization with the optical pulse train. The resulting mode-locked THz pulse train of photocarriers constructs a frequency comb structure of photocarriers in the THz region; namely, a photocarrier THz comb (PC-THz comb). The EM-THz comb propagates in free space whereas the PC-THz comb is stored in PCA. In this way, the optical comb is down-converted to the THz region without any change in its frequency spacing f_{rep}. The resulting THz comb is a harmonic frequency comb of f_{rep} without any frequency offset, composed of a fundamental component and a series of harmonic components of f_{rep}. This is a big difference compared with an optical frequency comb having a carrier-envelope offset frequency (f_{ceo}). Since the THz comb is considered to be a set of several thousand or several tens of thousands of narrow-linewidth, single-mode CW-THz waves which are phase-locked with each other, they provide attractive features for frequency metrology, such as simplicity, broadband selectivity, high spectral purity, and absolute frequency calibration. Therefore, if f_{rep} can be well stabilized by laser control, the THz combs can be used as precise frequency rulers in the THz region. The optical comb can be also used for precise frequency rulers in the optical region by stabilizing f_{rep} and f_{ceo}.

Figure 7.2 shows the concept of THz frequency metrology based on the frequency comb. Cesium (Cs) atomic clock in the microwave region (frequency = 9.2 GHz) has been widely used as standards of time and frequency. However, it has been difficult to transfer the frequency uncertainty of the Cs atomic clock to the optical region due to the large frequency gap ($\sim 10^5$) between the microwave and optical regions. Although such a gap has been bridged by a frequency chain [25], it is a rather bulky and complicated apparatus. Furthermore, the frequency uncertainty deteriorates while passing through many intermediate oscillators in the chain. Recently, an optical comb has emerged as a powerful tool for frequency linking between the microwave and optical regions directly, without losing

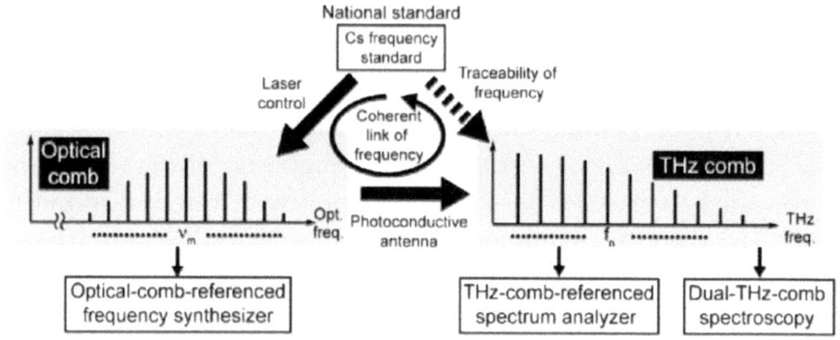

Figure 7.2 THz frequency metrology based on the frequency comb.

the frequency uncertainty, by achieving precise laser stabilization of f_{rep} and f_{ceo} with an atomic clock [6–9]. Furthermore, the frequency comb has been extended to the THz region using a PCA, as shown in Fig. 7.1 [10]. Here, the most important point is that frequency linking among the microwave, optical, and THz regions is based on a coherent process, involving laser control and a photoconductive antenna. The resulting coherent frequency linking enables us to share the same frequency uncertainty in three different regions of the electromagnetic spectrum. Therefore, based on a THz comb or optical comb, one can construct a THz frequency metrology system that is directly connected to a frequency standard in the microwave region. In other words, THz frequency metrology traceable to the SI base unit of time can be established.

7.3 THz-Comb-Referenced Spectrum Analyzer [11–15]

A spectrum analyzer is a fundamental frequency measurement instrument widely used for radio frequency (RF), microwave, millimeter, and even optical waves. However, it is still difficult to use in the THz region, although steady efforts are being made to extend its frequency range. The electrical heterodyne method with a superconductor-insulator-superconductor mixer [26] or a hot-electron-bolometer mixer [27] enables frequency measurement of CW waves in the sub-THz and THz regions. Conversely, the optical interferometric method can be used as an optical spectrum analyzer in the THz region. However, those methods often require cryogenic cooling of the mixer or detector to suppress thermal noise, which is a major obstacle to practical use. Recently, a new type of spectrum analyzer based on a harmonic mixing technique with a frequency comb has been proposed and developed. This spectrum analyzer can measure the absolute frequency and spectral shape of CW-THz waves in real time without the need for cryogenic cooling.

The proposed THz spectrum analyzer is based on a heterodyne technique as shown in Fig. 7.3a. In this heterodyne technique, the PC-THz comb is used as a local oscillator with known multiple frequencies covering from the sub-THz to THz regions. Combining PCA and PC-THz comb enables heterodyne mixing covering from the sub-THz to THz regions without the need for cryogenic cooling. Since each mode of PC-THz comb has a low-power local oscillator, moderate power is required for a measured CW-THz wave to get the beat signal at a good signal-to-noise ratio. Consider a PCA detector triggered by an fs-ML laser light with a repetition frequency f_{rep}. Figure 7.3b illustrates the corresponding spectral behaviors in THz and RF regions. The optical mode-locked pulse train emitted from the fs-ML laser constructs an optical frequency comb in the frequency domain, whose mode spacing is equal to a repetition frequency f_{rep} (see Fig. 7.1). When the PCA is triggered by such an

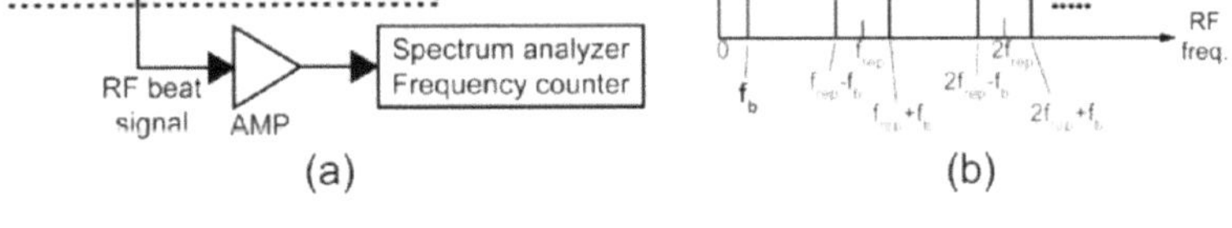

Figure 7.3 Principle of operation for THz-comb-referenced spectrum analyzer. (a) Schematic drawing and (b) the corresponding spectral behavior in THz and RF regions.

optical pulse train, the PC-THz comb is induced in the PCA. Next, consider what happens when a measured CW-THz wave (frequency $= f_{\mathrm{THz}}$) is incident on a PCA detector having the PC-THz comb with f_{rep}. The photoconductive mixing process in the PCA generates a group of beat signals between the CW-THz wave and the PC-THz comb in the RF region. Focus on a beat signal at the lowest frequency ($= f_{\mathrm{b}}$), namely the f_{b} beat signal. Since this f_{b} beat signal is generated by mixing the CW-THz wave (frequency $= f_{\mathrm{THz}}$) with the m-th mode of the PC-THz comb (frequency $= mf_{\mathrm{rep}}$) nearest in frequency to the CW-THz wave, f_{THz} is given by

$$f_{\mathrm{THz}} = mf_{\mathrm{rep}} \pm f_{\mathrm{b}}. \tag{7.1}$$

Since f_{rep} and f_{b} can be measured by RF frequency instruments, f_{THz} can be determined if the value of m and the sign of f_{b} are measured. To this end, the repetition frequency is changed from f_{rep} to $f_{\mathrm{rep}} + \delta f_{\mathrm{rep}}$ by adjustment of the laser cavity length with the laser control system. This results in a change of the beat frequency to $f_{\mathrm{b}} + \delta f_{\mathrm{b}}$. Since $|\delta f_{\mathrm{b}}|$ is equal to $|m\delta f_{\mathrm{rep}}|$, the value of m is determined by

$$m = \frac{|\delta f_{\mathrm{b}}|}{|\delta f_{\mathrm{rep}}|}. \tag{7.2}$$

Since $\delta f_{\mathrm{b}}/\delta f_{\mathrm{rep}}$ and f_{b} have opposite signs, the absolute frequency of the measured CW-THz wave can be determined by measuring f_{rep}, f_{b}, δf_{rep}, and δf_{b} as follows:

$$f_{\mathrm{THz}} = mf_{\mathrm{rep}} + f_{\mathrm{b}} \left\langle \frac{\delta f_{\mathrm{b}}}{\delta f_{\mathrm{rep}}} < 0 \right\rangle \tag{7.3}$$

and

$$f_{\mathrm{THz}} = mf_{\mathrm{rep}} - f_{\mathrm{b}} \left\langle \frac{\delta f_{\mathrm{b}}}{\delta f_{\mathrm{rep}}} > 0 \right\rangle. \tag{7.4}$$

Figure 7.4 shows a schematic diagram of the experimental setup. The THz-comb-referenced spectrum analyzer was composed of a home-built, fs-ML Er-doped fiber laser (center wavelength = 1550 nm, pulse duration = 40 fs, and f_{rep} = 56,122,206 Hz), a PCA for THz detection, and RF frequency measurement instruments. The repetition frequency f_{rep} was stabilized using a laser control system referenced to a rubidium (Rb) atomic clock (frequency = 10 MHz,

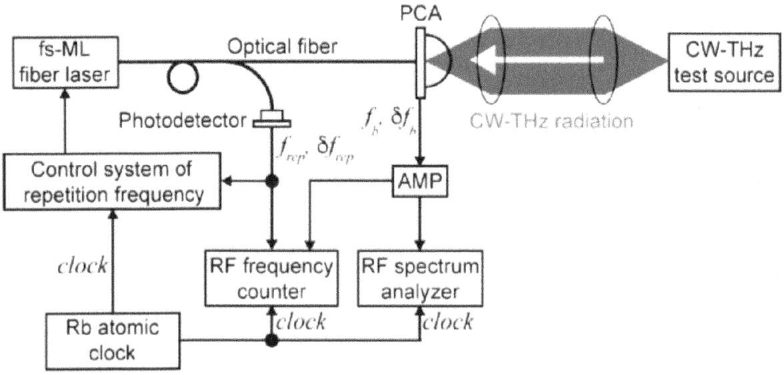

Figure 7.4 Experimental setup of THz-comb-referenced spectrum analyzer. fs-ML fiber laser: femtosecond mode-locked fiber laser; PCA: photoconductive antenna for THz detection; AMP: current preamplifier.

accuracy $= 5 \times 10^{-11}$, stability $= 2 \times 10^{-11}$ at 1 s). The output of the fiber laser was delivered to a bowtie-shaped, low-temperature-grown InGaAs/InAlAs PCA for a 1550 nm laser light by an optical fiber. This results in the generation of the PC-THz comb in the PCA equivalent to the Rb atomic clock. The CW-THz wave from a test source propagated in free space through a pair of THz lenses and was then incident on the PCA. Photoconductive heterodyne mixing between the CW-THz wave and the PC-THz comb in the PCA generates a group of beat signals in the RF region. The f_b beat signal is amplified by a high-gain current preamplifier (AMP, bandwidth $= 10$ MHz and sensitivity $= 10^5$ V/A) and was measured with an RF spectrum analyzer (frequency range $= 100$ Hz $\sim$3 GHz) and an RF frequency counter (frequency range $<$225 MHz) to determine its spectral shape and center frequency. A portion of the laser light is detected with a photodetector, and its repetition frequency ($= f_{rep}$) is measured using the RF frequency counter. The RF spectrum analyzer and RF frequency counter are synchronized to the Rb atomic clock.

The first test source (output power $= 4$ dBm $= 2.5$ mW, frequency range $= 75$–110 GHz, and linewidth $<$0.6 Hz) is achieved by a combination of an active frequency multiplier chain (multiplication factor $= 6$) and a microwave frequency synthesizer (frequency $=$ 12.5–18.33 GHz), referred to as an active frequency multiplier chain

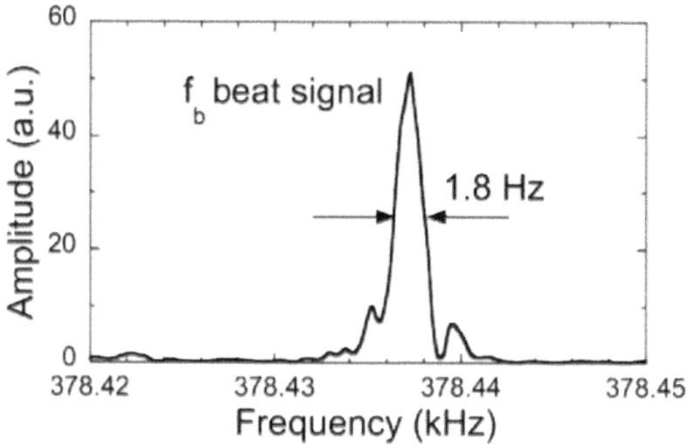

Figure 7.5 RF spectrum of f_b beat signal for the AFMC source measured by the RF spectrum analyzer.

(AFMC) source. Since the frequency synthesizer is synchronized to the Rb atomic clock, the output frequency of this AFMC source is accurate and stable. We first evaluated the spectral linewidth of the f_b beat signal when the output frequency of the AFMC source is set to be 100 GHz. Figure 7.5 shows a linear-scale spectrum of this signal, measured by the RF spectrum analyzer [resolution bandwidth (RBW) = 1 Hz and sweep time = 690 ms]. The resulting linewidth of the beat signal was 1.8 Hz. This result indicates that each mode of the PC-THz comb has sufficiently narrow linewidth to perform the frequency measurement with high precision. On the other hand, the signal-to-noise ratio (SNR) of the f_b beat signal reached 40 dB (not shown). From this SNR and the output power of the test source (namely +4 dBm = 2.5 mW), the detection limit of the THz power is estimated to be −36 dBm, or 250 nW. A comparison of SNR among the f_b beat signals obtained by three different PCAs and electro-optic sampling is given in detail elsewhere [12]. To determine the absolute frequency of the test source, it is necessary to measure the deviation of f_b before and after changing f_{rep}. The initial f_{rep} and f_b values were measured to be 56,122,206.03 Hz and 356,156 Hz as indicated as the left spectrum in Fig. 7.6 (RBW = 1 kHz and sweep time = 902 ms). Then, the frequency f_{rep} was changed by δf_{rep} (= +25 Hz) using the laser control system. This resulted in a change of the beat frequency by δf_b of +44,549 Hz (see the right spectrum

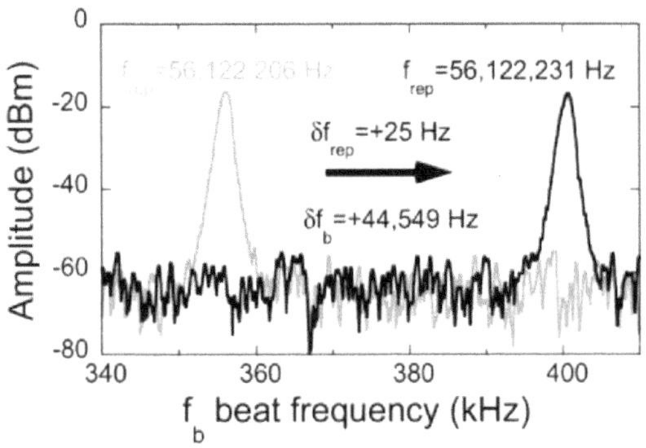

Figure 7.6 RF spectrum of f_b beat signal before and after changing f_{rep} by 25 Hz.

in Fig. 7.6). Since $|\delta f_b|$ is equal to $|m\delta f_{rep}|$, m is determined as

$$m = \frac{|\delta f_b|}{|\delta f_{rep}|} = \frac{|+44,549|}{|+25|} \approx 1,782. \tag{7.5}$$

Since the sign of $\delta f_b/\delta f_{rep}$ (positive in this case) is opposite to that of f_b, the value of f_{THz} was determined as follows:

$$\begin{aligned} f_{THz} &= mf_{rep} - f_b = 1,782 \times 56,122,206.03 - 356,156 \\ &= 100,009,414,988.9 \text{ Hz}. \end{aligned} \tag{7.6}$$

Since the actual set frequency of the AFMC source was 100,009,414,988.9 Hz from the output frequency of the microwave synthesizer, the error between the set and measured frequencies was only 0.56 Hz. When the precision of frequency measurement is defined as a ratio of the error to f_{THz}, the corresponding precision was 5.6×10^{-12}. To evaluate the precision of frequency measurement in the available frequency range of the AFMC source, we determined the absolute frequency of the source while tuning its output frequency from 75 to 110 GHz at 5 GHz intervals. The resulting precision for eight different measurement frequencies was shown in Fig. 7.7, in which the precision is defined as a ratio of the frequency error to the set frequency of the test source, A mean precision of 2.4×10^{-11} was obtained for this source, which is limited by the performance of the Rb atomic clock used. The frequency precision will be further improved if we use a frequency standard having better uncertainty, such as Cs atomic clock.

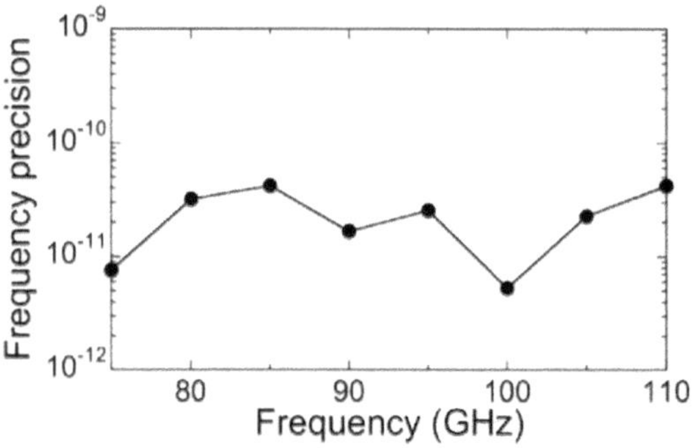

Figure 7.7 Frequency precision of the f_b beat signal of the AFMC source measured by the RF frequency counter.

The second test source was achieved by photomixing of two CW near-infrared lasers at adjacent wavelengths with a uni-traveling-carrier photodiode (UTC-PD) [28]. The two CW lasers were external cavity wavelength-tunable laser diodes with an emission wavelength of 1550 nm and operated in free-running mode (spectral linewidth $\leq$ 100 kHz, frequency fluctuation $<$ 100 MHz/hour). The optical frequency difference between them was set to be approximately 120 GHz. The outputs of the lasers were combined using a fiber coupler, amplified with an Er-doped fiber amplifier, and then photomixed by an F-band UTC-PD (available frequency = 90–140 GHz) equipped with a horn antenna. The output power was set to be 100 μW at a frequency of 120 GHz. We used the THz-comb-referenced spectrum analyzer for the monitoring of this photomixing source. The resulting spectrum of the beat signal is shown in Fig. 7.8 (RBW = 1 kHz, sweep time = 13 ms). The detailed spectral behavior of the beat signals is shown as a movie in elsewhere [12]. In contrast to the stable AFMC source, the beat frequency of the photomixing source exhibits large fluctuations within a spectral window of 0.8 MHz. This is because the two CW lasers used for the photomixing are operated in a free-running mode without any frequency control.

Although the frequencies of the above test sources were in the low-frequency THz region due to the limitation of our available CW-THz sources, it is natural to believe that this THz spectrum analyzer can be extended to the higher frequency THz region. This

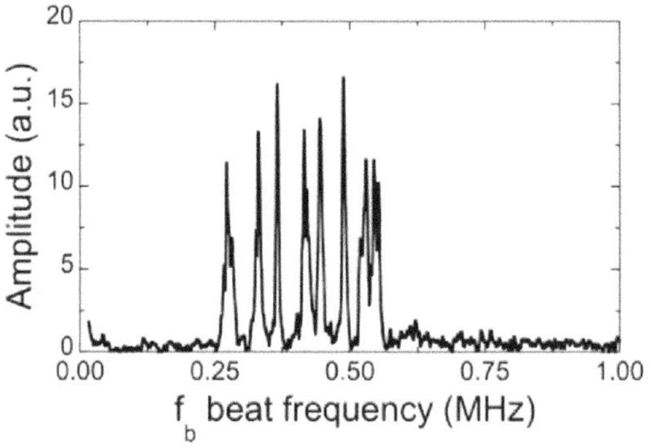

Figure 7.8 RF spectrum of f_b beat signal for the photomixing source measured by the RF spectrum analyzer.

is because the spectral bandwidth of the THz spectrum analyzer is limited mainly by the spectral sensitivity of the PCA. For example, the spectral characteristics of the LT-GaAs-PCA as a THz detector have been investigated using THz time-domain spectroscopy (THz-TDS), for which a bandwidth over 100 THz was achieved [29]. Therefore, the THz-comb-referenced spectrum analyzer can cover the entire THz spectral region.

7.4 Optical-Comb-Referenced Frequency Synthesizer [16–18]

Accurate, stable, and/or tunable single-frequency signal generation is an essential technique to achieve the THz frequency synthesizer. One promising method for widely tunable CW-THz sources is the photomixing of two CW near-infrared lasers of adjacent wavelengths with a photomixer, such as UTC-PD [28] or PCA [30]. However, if the two CW lasers are operated in the free-running mode, it is difficult to generate the accurate and stable frequency in the CW-THz wave as shown in Fig. 7.8. One attractive frequency reference for simultaneous control of these two CW lasers is an optical frequency comb [31–33]. Accurate, stable, phase-locked CW-THz wave has been discretely generated at four different frequencies (0.30, 0.56, 0.84, and 1.1 THz) by photomixing of two CW lasers locked to a single optical comb based on a mode-locked Ti:Sapphire laser [31]. Furthermore, the output frequency was tuned continuously by scanning the frequency interval of the optical comb while locking the CW lasers to the comb; however, the range of continuous tuning was limited to several kilohertz. In the case where the two CW lasers share the same optical comb, when scanning the frequency interval of the comb, the optical frequencies of the two CW lasers change simultaneously. This common-mode change cancels most of the optical frequency change in the two CW lasers. As a result, the continuous tuning range of the CW-THz wave is much smaller than that of the optical frequency in the CW lasers. One promising approach to increase the continuous tuning range while maintaining excellent frequency uncertainty is the photomixing of two independent optical frequency synthesizers (OFSs) phase-locked to a frequency standard.

An OFS is realized by phase-locking a tunable single-frequency CW laser to one of the optical comb modes [34]. When f_{rep}, f_{ceo}, and the beat frequency between the CW laser and one of the comb modes are fully phase-locked to the frequency standard, the optical frequency of the OFS can be determined at the uncertainty of the frequency standard via the optical comb. Furthermore, the optical frequency can be widely and continuously tuned by scanning f_{rep} while maintaining the phase-locking. Therefore, if the outputs from a tunable OFS and a fixed OFS are optically heterodyned with a photomixer, the generated CW-THz radiation is widely and continuously tunable while its frequency is always determined, referenced to the frequency standard.

Let us consider the optical frequency f_{ofs1} of a fixed OFS (OFS1), composed of a CW laser (CWL1) and an optical frequency comb (OFC1), as shown in the upper part of Fig. 7.9. The f_{ofs1} is represented by

$$f_{\text{ofs1}} = f_{\text{ceo1}} + m_1 f_{\text{rep1}} + f_{\text{beat1}}, \qquad (7.7)$$

where f_{ceo1} and f_{rep1} are the carrier-envelope offset frequency and repetition rate of OFC1, m_1 is the mode number of the OFC1 to which CWL1 is phase-locked, and f_{beat1} is the beat frequency between CWL1 and the m_1-th mode. In the fixed OFS, f_{ceo1}, f_{rep1}, m_1, and f_{beat1} are all stabilized at fixed values by precise laser control of

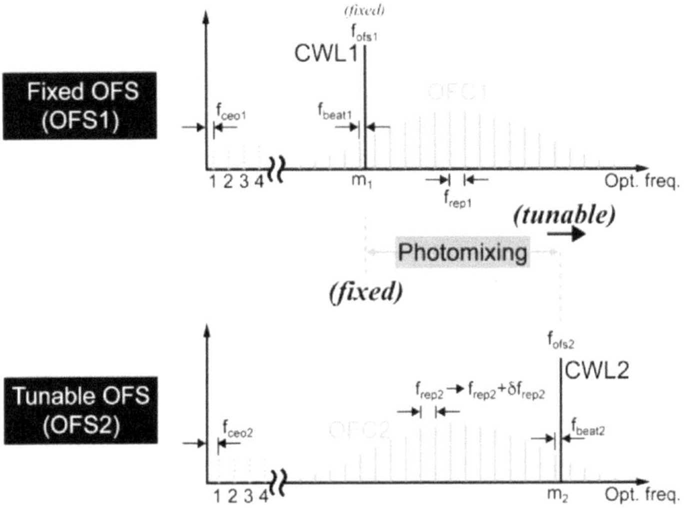

Figure 7.9 Principle of optical-comb-referenced frequency synthesizer based on photomixing of two OFSs.

OFC1 referenced to a frequency standard. Therefore, f_{ofs1} can be determined at the uncertainty of the frequency standard. On the other hand, the optical frequency f_{ofs2} of a tunable OFS (OFS2), composed of another CW laser (CWL2) and another fiber comb (OFC2), as shown in the lower part of Fig. 7.9, is given by

$$f_{ofs2} = f_{ceo2} + m_2 f_{rep2} + f_{beat2}, \tag{7.8}$$

where f_{ceo2} and f_{rep2} are the carrier-envelope offset frequency and repetition rate of OFC2, m_2 is the mode number of the OFC2 to which CWL2 is phase-locked, and f_{beat2} is the beat frequency between CWL2 and the m_2-th mode. If the comb spacing is changed from f_{rep2} to $f_{rep2} + \delta f_{rep2}$ while keeping f_{ceo2} at a fixed value, all the modes of OFC2 expand in a manner similar to the bellows of an accordion, and thus, the frequency of the m_2-th mode changes by the product of m_2 and δf_{rep2}. Therefore, f_{ofs2} can also be tuned continuously by $m_2 \delta f_{rep2}$ if CWL2 remains phase-locked to the m_2-th comb mode. In the tunable OFS, f_{ceo2}, m_2, and f_{beat2} are fixed at certain values while f_{rep2} is tuned precisely, via the precise laser control referenced to the frequency standard. Even though OFS2 is used for continuous tuning, its absolute frequency f_{ofs2} can be determined within the uncertainty of the frequency standard by simply monitoring f_{rep2} in real time with a frequency counter.

When a THz synthesizer is realized by photomixing OFS1 and OFS2 with a photomixer, its output frequency ($= f_{\text{THz}}$) is given by

$$\begin{aligned} f_{\text{THz}} &= |f_{\text{ofs2}} - f_{\text{ofs1}}| \\ &= \left|\left(f_{\text{ceo2}} + m_2 f_{\text{rep2}} + f_{\text{beat2}}\right) - \left(f_{\text{ceo1}} + m_1 f_{\text{rep1}} + f_{\text{beat1}}\right)\right|. \end{aligned} \tag{7.9}$$

Here, it is important to emphasize that the absolute value of f_{THz} can be determined at the uncertainty of the frequency standard by measuring f_{ceo1} (fixed), f_{ceo2} (fixed), f_{rep1} (fixed), f_{rep2} (variable), m_1 (fixed), m_2 (fixed), f_{beat1} (fixed), and f_{beat2} (fixed). This is the main advantage over traditional THz frequency synthesizers. The continuous tuning range of f_{THz} (δf_{THz}) is represented as

$$\delta f_{\text{THz}} = \delta f_{\text{ofs2}} = m_2 \delta f_{\text{rep2}}. \tag{7.10}$$

The value of m_2 reaches 3,880,000 when the optical frequency of CWL2 is 194 THz (corresponding wavelength = 1550 nm) and f_{rep2} is 50 MHz. For example, when $\delta f_{\text{rep2}} = 500$ kHz, which is 1% of f_{rep2}, δf_{THz} could reach about 2 THz from Eq. 7.10. Therefore, the proposed method can also be used for widely and continuously tunable THz synthesizers.

The experimental setup of the optical-comb-referenced frequency synthesizer is shown in the upper part of Fig. 7.10. We first constructed two OFSs operating at a wavelength of 1542 nm for photomixing. The fixed OFS (OFS1) was composed of a distributed feedback fiber laser (CWL1) and a custom-built optical comb including a mode-locked Er-doped fiber laser, namely, a fiber comb (OFC1). On the other hand, the tunable OFS (OFS2) was composed of an external cavity laser diode (CWL2) and another fiber comb (OFC2). The frequencies f_{ceo1}, f_{ceo2}, f_{rep1}, f_{rep2}, f_{beat1}, and f_{beat2} were all phase-locked to the microwave frequency reference synthesized from a hydrogen maser linked to coordinated universal time (UTC), operated by the National Metrology Institute of Japan (UTC-NMIJ). Details of the phase-locking process in the OFSs are given elsewhere [34]. The mode numbers m_1 and m_2 were selected depending upon the frequency of the generated CW-THz radiation. An optical wavemeter was used to determine m_1 and m_2. The

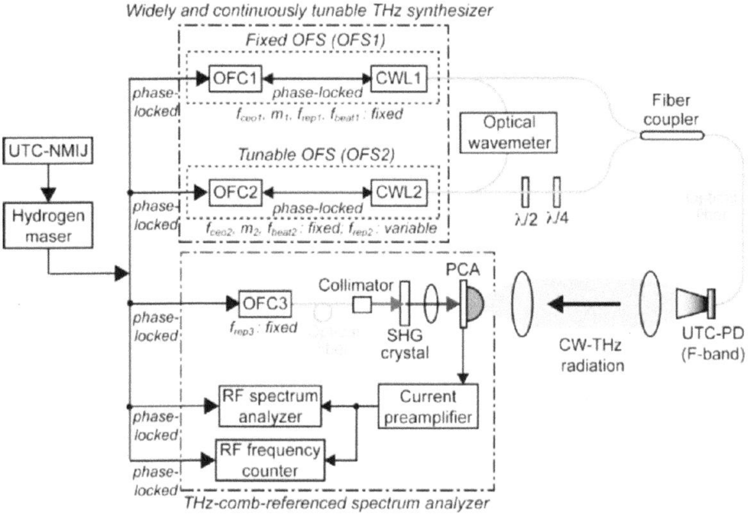

Figure 7.10 Experimental setup of optical-comb-referenced frequency synthesizer. OFC1, OFC2, and OFC3: fiber combs; CWL1 and CWL2: CW near-infrared lasers; $\lambda/2$: half-wave plate; $\lambda/4$: quarter-wave plate; UTC-PD: uni-traveling-carrier photodiode for photomixing; PCA: photoconductive antenna for THz detection; SHG crystal: second-harmonic-generation crystal; UTC-NMIJ: coordinated universal time operated by the National Metrology Institute of Japan.

continuous tuning range of δf_{ofs2} was achieved up to 990 GHz by expanding the tuning range of CWL2 using a combination of a piezoelectric (PZT) actuator and a linear translation stage. After adjusting the polarization overlap with a half-wave plate ($\lambda/2$) and a quarter-wave plate ($\lambda/4$), the output beams from OFS1 ($f_{\mathrm{ofs1}} =$ 194.4 THz, power = 7.3 mW) and OFS2 ($f_{\mathrm{ofs2}} =$ 194.3 THz, power = 9.4 mW) were combined with a fiber coupler and then optically heterodyned with a fiber-coupled, F-band UTC-PD (frequency range = 90–140 GHz) equipped with a horn antenna. We estimated from the observed photocurrent value (7 mA) and the set bias voltage (−2.5 V) of the UTC-PD that the average power of generated CW-THz radiations was around 250 μW in the F-band. We here used an F-band UTC-PD as a photomixer for actual proof of the proposed method; however, it should be emphasized that it is possible to easily use another UTC-PD or PCA in place of the F-band UTC-PD if THz synthesizers with higher frequency and/or broader tunability are required.

To evaluate the spectral characteristics of the THz synthesizer, we used a THz-comb-referenced spectrum analyzer [11–15] composed of a third fiber comb whose frequency f_{rep} was stabilized (OFC3 with $f_{\text{rep3}} = 56{,}122{,}639$ Hz), a bow-tie-shaped, low-temperature-grown GaAs PCA for THz detection, and RF frequency instruments, as shown in the lower part of Fig. 7.10. The frequency f_{rep3} and RF frequency instruments were synchronized to the microwave frequency reference. The free-space-propagating CW-THz wave passing through a pair of THz lenses was made incident on the PCA. The PCA was triggered by second-harmonic-generation (SHG) light (center wavelength = 775 nm, average power = 10 mW) from OFC3. This resulted in the generation of the PC-THz comb of OFC3 in the PCA. Photoconductive heterodyne mixing between the CW-THz radiation and the PC-THz comb in the PCA generated beat signals in the RF region. The beat signal of the current from the PCA was amplified by a high-gain current preamplifier (bandwidth = 10 MHz, sensitivity = 10^5 V/A) and was measured with an RF spectrum analyzer and frequency counter. Details of the THz-comb-referenced spectrum analyzer are given in Section 7.3. The spectral behavior and frequency instability of the CW-THz wave were measured with the THz-comb-referenced spectrum analyzer. On the other hand, the absolute frequency of the CW-THz radiation was determined based on Eq. 7.9.

To evaluate the spectral characteristics in the THz frequency synthesizer, we first generated a frequency-locked CW-THz wave at upper and lower frequency limits within the F-band, ranging from 90 to 140 GHz. Figures 7.11a,b show spectra of the f_b beat signal for CW-THz radiation around 91.97 and 140.0 GHz, respectively. The lower horizontal axes give the frequency scale (f_{beat3}) measured with the RF spectrum analyzer. One can clearly confirm that the CW-THz radiation spectra have similar Gaussian-like shapes at these two frequencies. The linewidth of the CW-THz radiation was 599 kHz at 91.97 GHz and 631 kHz at 140.0 GHz when a Gaussian function was fitted to the spectral shape by regression analysis based on the Levenberg–Marquardt algorithm. If the combination of narrower-linewidth CW lasers with fast feedback control is employed for

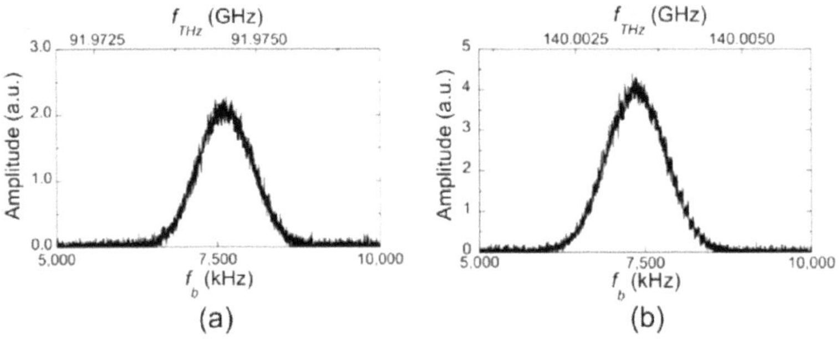

Figure 7.11 Spectra of f_b beat signal for CW-THz wave at (1) 91.97 GHz and (b) 140.0 GHz.

the OFSs, the linewidth of the CW-THz radiation will be further decreased.

We also evaluated the frequency instability of the CW-THz radiation set at 132.0 GHz by measuring the frequency fluctuation of the beat signal (f_{beat3}) with the RF frequency counter. Figure 7.12 shows the fluctuation of the beat frequency, represented by the Allan standard deviation σ_A [35], with respect to various gate times τ. A power law relationship in the form $\sigma_A = 1/\tau$ between gate time τ and frequency fluctuation σ_A was found. The inverse proportionality between them clearly indicates that the output frequency of the THz frequency synthesizer was phased-locked to the hydrogen maser. Here, we consider whether the frequency fluctuation of the THz

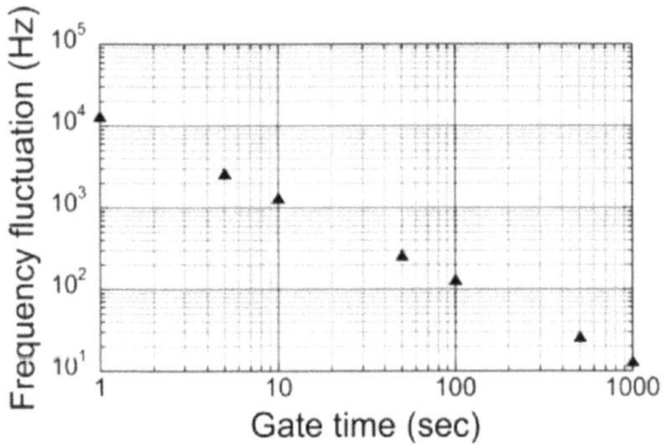

Figure 7.12 Frequency fluctuation of CW-THz wave at 132.0 GHz with respect to gate time.

synthesizer is consistent with that of the OFS. The THz frequency synthesizer had a frequency fluctuation of about 10 kHz at a gate time of 1 s (see Fig. 7.12). Conversely, since the frequency uncertainties of f_{ceo1} (or f_{ceo2}), f_{rep1} (or f_{rep2}), and f_{beat1} (or f_{beat2}) in OFS1 (or OFS2) were, respectively 10^{-17}, 10^{-12}, and 10^{-13} at an averaging time of 1 s [34], the frequency fluctuation in the OFS was estimated to be a few hundreds of Hz from the uncertainty of $m_1 f_{rep1}$ (or $m_2 f_{rep2}$), which is mainly due to the uncertainty of the frequency synthesis of f_{rep1} (or f_{rep2}). We consider that the discrepancy of frequency fluctuation between THz and optical synthesizers is mainly due to electrical noise caused in the amplification processes of the considerably weak beat signal in the THz-comb-referenced spectrum analyzer.

To demonstrate the utility of this THz frequency synthesizer in gas spectroscopy, we performed broadband THz spectroscopy of low-pressure gas molecules. The tuning range of the THz synthesizer was expanded over 120 GHz by using a broadband UTC-PD (freq. = 0.2–1.5 THz). We selected gas-phase acetonitrile (CH_3CN) as a sample gas. Since CH_3CN is a symmetric-top molecule with a rotational constant, B, of 9.199 GHz and a centrifugal distortion constant, D_{JK}, of 17.74 MHz [36], the frequencies of rotational transitions are given by

$$\nu = 2B\,(J+1) - 2D_{JK}K^2\,(J+1), \tag{7.11}$$

where J and K are rotational quantum numbers. From this equation, the molecule displays two characteristic features in its THz spectrum. The first term in Eq. 7.11 indicates that many manifolds of absorption lines regularly spaced by $2B$ (= 18.398 GHz) appear. The second term indicates that each manifold includes a series of closely spaced absorption lines of decreasing strength due to $2D_{JK}$. In this way, CH_3CN gas indicates the widths of characteristic spectral structures with both GHz and MHz orders. Figure 7.13a shows the transmittance spectrum of the CH_3CN gas at 20 Pa within a frequency range from 600 to 720 GHz at a spectral resolution of 5 MHz. Seven periodic manifolds of rotational transitions were clearly confirmed with a constant frequency spacing, which was exactly equal to $2B$ (= 18.398 GHz) in Eq. 7.11. From a comparison with the

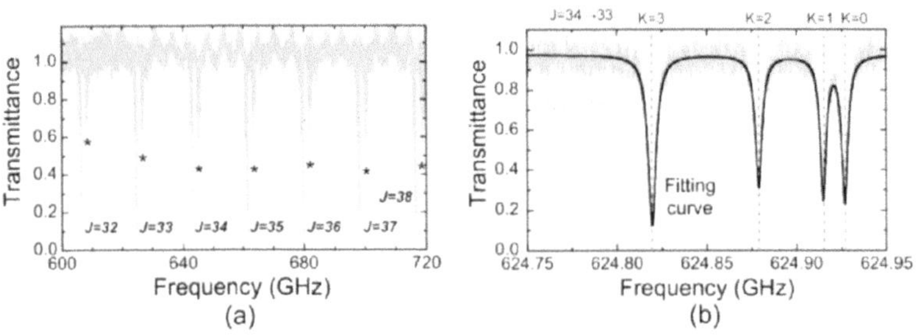

Figure 7.13 (a) Broadband transmittance spectrum of CH_3CH gas at 20 Pa (spectral range = 120 GHz, spectral resolution = 5 MHz). (b) High-precision transmittance spectrum of CH_3CH gas at 20 Pa (spectral range = 200 MHz, spectral resolution = 520 kHz). The black line indicates the curve fitting result with multiple Lorentzian functions.

literature values in the JPL database [37], we assigned them to $J = 32$ to $J = 38$, respectively [see text in Fig. 7.13a]. Satellite manifolds appearing at the higher frequency side of the main manifolds [see asterisks in Fig. 7.13a] were due to the rotational transition of the vibrationally excited molecules. In this way, we confirmed that the widths of the spectral structures with GHz order were characterized by the rotational constant B in the first term of Eq. 7.11. To confirm the spectral features of MHz order characterized by the centrifugal distortion constant D_{JK} in the second term of Eq. 7.11, we fine-tuned f_{THz} from 624.75 to 627.95 GHz across a part of the transition $J = 34 \rightarrow 33$. Figure 7.13b shows the transmittance spectrum of CH_3CN at 20 Pa with a spectral resolution of 520 kHz. For comparison, literature values of CH_3CN in the JPL database [37] are indicated by the dashed lines in Fig. 7.13b. Four sharp absorption lines with a linewidth of a few MHz appeared with a frequency spacing of a few tens of MHz. We determined the center frequencies of these four absorption lines by performing curve fitting analysis with multiple Lorentzian functions. From a comparison with the literature values in the JPL database [37], these peaks could be assigned to $K = 0$, 1, 2, and 3 in the transition $J = 34 \rightarrow 33$. The frequency discrepancy of the determined values from the literature values 52.6 ± 77.5 kHz for $K = 0$, 1, 2, and 3, corresponding to a spectral accuracy of 8.42×10^{-8} at 625 GHz.

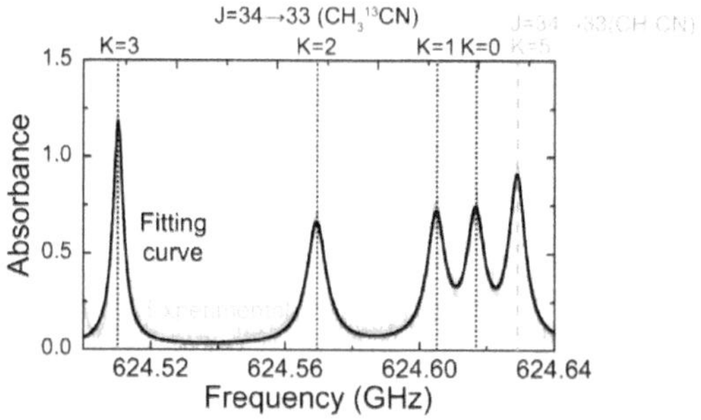

Figure 7.14 High-precision absorbance spectrum of the mixed gas of CH_3CN and $CH_3^{13}CN$ gas (spectral range = 140 MHz, spectral resolution = 520 kHz). The black line indicates the curve fitting result with multiple Lorentzian functions.

We further demonstrated THz spectroscopy of a mixture gas of CH_3CN and its isotope ($CH_3^{13}CN$). Because $CH_3^{13}CN$ is a symmetric-top molecule with a slightly different B value (= 9.194 GHz) and D_{JK} value (= 17.67 MHz) from those of CH_3CN (B = 9.199 GHz, D_{JK} = 17.74 MHz) [36], the absorption lines appear in the THz frequency with a slight frequency offset. We mixed the CH_3CN and $CH_3^{13}CN$ at an unknown mixture ratio in a gas cell. To observe the spectral features characterized by D_{JK} in these two molecules, we precisely tuned f_{THz} from 624.50 to 624.64 GHz across a part of the J = 34→33 transition. Figure 7.14 shows the absorbance spectrum of the mixed $CH_3CN/CH_3^{13}CN$ gas with a spectral resolution of 520 kHz. For comparison, the literature values in the JPL database [37] are also indicated by the dashed line for CH_3CN and the dotted lines for $CH_3^{13}CN$. A comparison between them allowed us to assign one transition of CH_3CN (J = 34→33, K = 5) and four transitions of $CH_3^{13}CN$ (J = 34→33, K = 0, 1, 2, and 3). In this way, the difference in the spectral features between CH_3CN and $CH_3^{13}CN$ was clearly observed. From the quantitative analysis of this spectrum, the mixture ratio of the gases was estimated to be 1:2.66. This demonstration indicates the high potential of this approach for the qualitative analysis of multiple gas molecules.

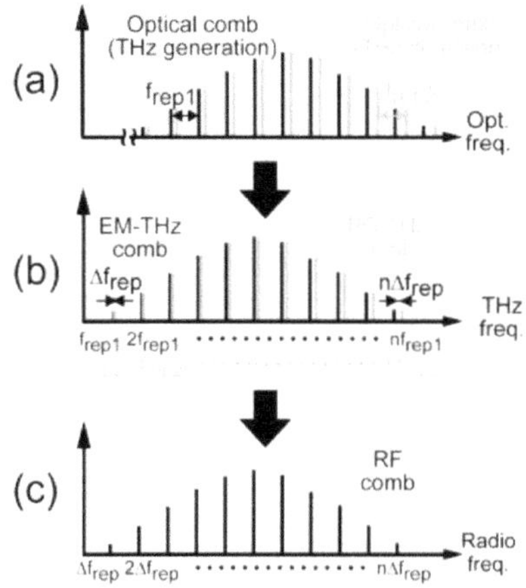

Figure 7.15 Principle of operation for dual-THz-comb spectroscopy in the frequency domain. Spectral behaviors in (a) optical, (b) THz, and (c) radio-frequency (RF) regions.

7.5 Dual-THz-Comb Spectroscopy [10, 19–24]

THz time-domain spectroscopy (THz-TDS) [38] is a popular spectroscopic technique for obtaining a broadband THz spectrum. In this method, marking of a frequency scale in the THz spectrum is performed by a mechanically-scanning time delay with a motor-driven translation stage. However, poor positioning precision and a limited stroke length of the translation stage hinder the achievement of high spectral resolution and accuracy in THz-TDS. This sometimes leads to inconsistency of frequency scales in the THz spectrum among different THz-TDS systems. Coherent linking of the frequency with a frequency comb, described in Section 7.2, has the potential to achieve the ultimate THz spectrometer with high accuracy, high resolution, and a broad spectral range because the THz comb can work as a THz frequency ruler traceable to a frequency standard. To read out the THz comb correctly, dual-THz-comb spectroscopy has been developed [10, 19–24].

The principle of operation in dual-THz-comb spectroscopy can be described as multi-frequency-heterodyning in frequency domain [10]. Let us consider THz generation from a PCA emitter excited by an fs-ML laser (pump laser; repetition frequency = f_{rep1}) and THz detection using a PCA detector gated by another fs-ML laser (probe laser; repetition frequency = f_{rep2}). Figure 7.15 illustrates spectral behaviors in (a) optical, (b) THz, and (c) RF regions. In the frequency domain, since the generation of the mode-locked THz pulse train by the PCA emitter can be considered as an ultra-wideband demodulation of an optical frequency comb, the optical comb is down-converted to the THz region without any change to the frequency spacing. The resulting THz comb of electromagnetic wave, namely EM-THz comb, is a harmonic frequency comb without a frequency offset, composed of a fundamental component (freq. = f_{rep1}) and a series of harmonic components (freq. = $2f_{\text{rep1}}$, $3f_{\text{rep1}}, \ldots, nf_{\text{rep1}}$) of a repetition frequency. Next, we consider what happens when the repetition frequency of the probe laser (= $f_{\text{rep2}} = f_{\text{rep1}} + \Delta f_{\text{rep}}$) is slightly detuned from that of the pump laser (= f_{rep1}) by a certain frequency offset (Δf_{rep}). Instantaneous photoconductive gating by the probe laser induces a frequency comb of photocarrier, namely PC-THz comb, having a different frequency spacing [freq. = $f_{\text{rep2}}, 2f_{\text{rep2}}, 3f_{\text{rep2}}, \ldots, nf_{\text{rep2}} = (f_{\text{rep1}} + \Delta f_{\text{rep}}), 2(f_{\text{rep1}} + \Delta f_{\text{rep}}), 3(f_{\text{rep1}} + \Delta f_{\text{rep}}), \ldots, n(f_{\text{rep1}} + \Delta f_{\text{rep}})$] in the PCA detector, which is also existed in THz frequency region. Under this condition, it is possible to detect the EM-THz comb as a result of the photoconductive process occurring between the EM-THz and PC-THz combs, giving rise to the multi-frequency-heterodyning effect. This results in the generation of a secondary frequency comb in the RF region, termed the RF comb (freq. = $\Delta f_{\text{rep}}, 2\Delta f_{\text{rep}}, 3\Delta f_{\text{rep}}, \ldots, n\Delta f_{\text{rep}}$). Since the RF comb is a replica of the EM-THz comb only downscaled by $f_{\text{rep1}}/\Delta f_{\text{rep}}$ in frequency, one can utilize the THz comb easily via direct observation of the RF comb using an RF spectrum analyzer and calibration of the frequency scale.

The principle of operation in dual-THz-comb spectroscopy can be also described as asynchronous-optical-sampling (ASOPS) in the time domain. A THz-TDS employing mechanical time-delay scanning measures the temporal waveform of a portion of a single THz pulse, as shown in Fig. 7.16a. Taking the Fourier transformation (FT) of the waveform gives the continuous spectrum of the broadband THz wave without comb structure, as shown in Fig. 7.16b. On the other hand, if the time window is greatly expanded to cover more than

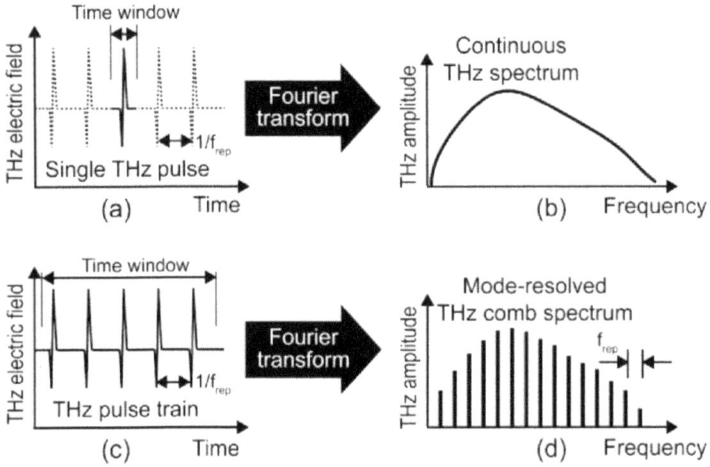

Figure 7.16 Principle of operation for dual-THz-comb spectroscopy in the time domain. (a) Temporal waveform of a single THz pulse and (b) corresponding THz continuous spectrum. (c) Temporal waveform of THz pulse train and (d) corresponding mode-resolved THz comb spectrum.

one repetition period, one can measure the temporal waveform of consecutive THz pulses, that is to say, a THz pulse train, as shown in Fig. 7.16c. Taking the FT of this will give the fine-structured spectrum of the THz comb because the FT of the periodic THz pulses imprints a frequency modulation on the continuous THz spectrum, as shown in Fig. 7.16d. However, in a conventional THz-TDS system, it is not practical to measure such a temporal waveform of the THz pulse train because it requires mechanical scanning over a long time delay. Instead, we focus on ASOPS using two fs-ML lasers with slightly mismatched repetition frequencies as shown in Fig. 7.17a [39–44]. Since ASOPS enables us to expand the picosecond time scale of THz pulse up to the microsecond scale based on the principle of ASOPS, the resulting slow-downed temporal waveform can be directly measured by a standard oscilloscope without the need for the mechanical time-delay scanning [see Fig. 7.17b]. One can therefore arbitrarily adjust the observed time window of the THz pulse temporal waveform by changing the time scale of the oscilloscope. In other words, it is possible to extend the time window

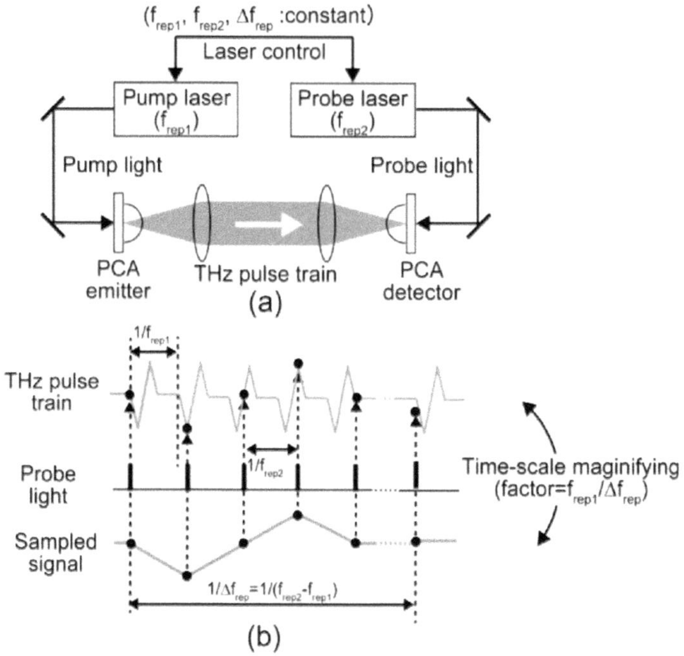

Figure 7.17 Principle of operation for asynchronous optical sampling (ASOPS). (a) Block diagram and (b) corresponding timing chart.

to more than one pulse period due to its non-mechanical nature of the time-delay scanning.

Figure 7.18 shows an experimental setup of ASOPS-based dual-THz-comb spectroscopy consisting of dual fs-ML Er-fiber lasers (center wavelength = 1550 nm, pulse duration = 50 fs, mean power = 500 mW) and a delay-stage-free THz-TDS setup including a pair of PCAs. The individual repetition frequencies of the two lasers (f_{rep1} = 250,000,000 Hz and f_{rep2} = 250,000,050 Hz) and the frequency offset between them ($\Delta f_{\text{rep}} = f_{\text{rep2}} - f_{\text{rep1}} = 50$ Hz) were stabilized by two independent laser control systems referenced to a rubidium (Rb) atomic clock (frequency = 10 MHz, accuracy = 5×10^{-11}, and instability = 2×10^{-11} at 1 s). After wavelength conversion of the two laser beams by SHG crystals, pulsed THz radiation was emitted by a dipole-shaped, low-temperature-grown, GaAs PCA (PCA1, bias voltage = 20 V) triggered by pump light

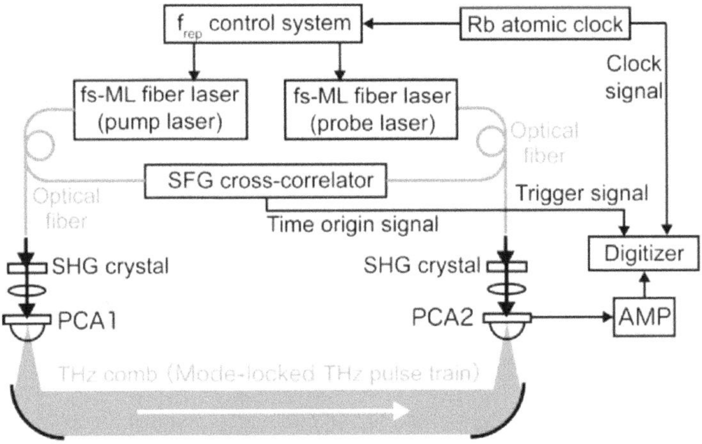

Figure 7.18 Experimental setup of ASOPS-based dual-THz-comb spectroscopy. fs-ML fiber laser: femtosecond mode-locked fiber laser; SFG cross-correlator: sum-frequency-generation cross-correlator; SHG crystal: second-harmonic-generation crystals; PCA1: dipole-shaped LT-GaAs photoconductive antenna for THz generation; PCA2: dipole-shaped LT-GaAs photoconductive antenna for THz detection; AMP: current preamplifier.

(power = 18 mW), and was then detected by another dipole-shaped LT-GaAs-PCA (PCA2) triggered by probe light (power = 9 mW). The optical path in which the THz beam propagated was purged with dry nitrogen gas to avoid absorption by atmospheric moisture (not shown). Portions of the output light from the two lasers were fed into a sum-frequency-generation (SFG) cross-correlator. The resulting SFG signal was used to generate a time origin signal in the ASOPS measurement. After amplification with a current preamplifier (AMP, bandwidth = 1 MHz, gain = 4×10^6 V/A), the temporal waveform of the output signal from PCA2 was acquired with a digitizer (sampling rate = 2×10^6 Sample/s, resolution = 20 bit) by using the SFG cross-correlator signal as a trigger signal. Then, the time scale of the observed signal was calibrated by a temporal magnification factor of $f_{rep1}/\Delta f_{rep}$ (= 250,000,000/50 = 5,000,000). Finally, the mode-resolved amplitude spectrum of the THz comb was obtained by FT of the temporal waveform of the THz pulse train.

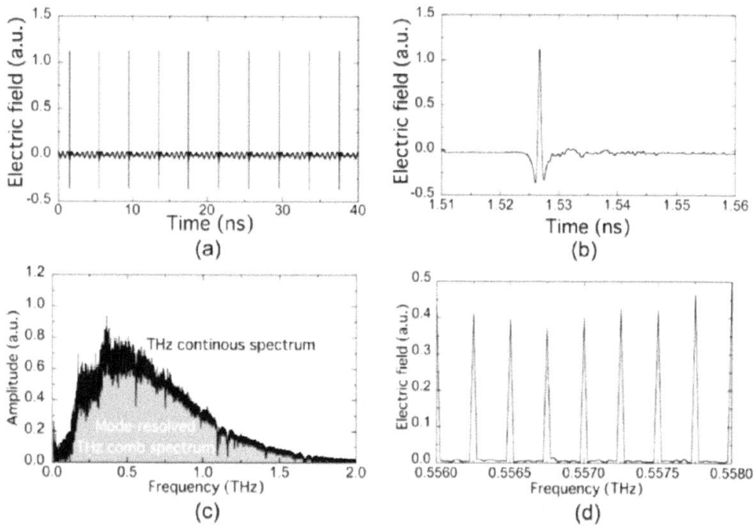

Figure 7.19 Temporal waveform of electric field in (a) 10 consecutive THz pulses (time window = 40 ns) and (b) the first THz pulse (time window = 50 ps). (c) Comparison of amplitude spectrum between mode-resolved THz comb spectrum and THz continuous spectrum. (d) Mode-resolved amplitude spectrum of THz comb around 0.557 THz.

We acquired the temporal waveform of 10 consecutive THz pulses by setting the time window size of 40 ns and a sampling interval of 100 fs as shown in Fig. 7.19a. Ten consecutive THz pulses are clearly observed at an interval of 4 ns. Figure 7.19b shows a magnified temporal waveform of the first THz pulse. The transient evolution of the pulsed THz radiation in the picosecond region was also observed clearly. The gray graph in Fig. 7.19c shows the mode-resolved amplitude spectrum of the THz comb obtained by taking the FT of the temporal waveform of the THz pulse train in Fig. 7.19a. For comparison, the black line in Fig. 7.19c shows a continuous spectrum of the THz amplitude obtained by taking the FT of the temporal waveform of the single THz pulse (time window = 4 ns). The spectral envelope of the mode-resolved THz comb spectrum is in good agreement with the spectral shape of the THz continuous spectrum. The mode-resolved THz comb spectrum is composed of a series of frequency spikes regularly separated by f_{rep1}. To observe its detailed structure, we expanded the spectral region around 0.5570 THz, as shown in Fig. 7.19d. The THz comb modes have a frequency spacing of 250 MHz and a linewidth of 25 MHz. The frequency spacing was equal to f_{rep1}, whereas the linewidth was consistent with the reciprocal of the temporal window in Fig. 7.19a (= 40 ns).

To evaluate the spectroscopic potential of the THz comb, we recorded the transmission spectra of several pharmaceutical tablets because they display sharp spectral fingerprints in the THz region resulting from their crystal structure. For the sake of this demonstration, we prepared two different sugar tablets (D-maltose and lactose powders mixed with polyethylene powder) and one reference tablet (polyethylene powder) with identical shapes (diameter = 10 mm and thickness = 1 mm). Sugar content by weight was 25% for the D-maltose tablet and 50% for the lactose tablet, respectively. The gray graphs in Figs. 7.20a,b show mode-resolved amplitude spectra of THz comb after passing through the D-maltose tablet and the lactose tablet, respectively. For comparison,

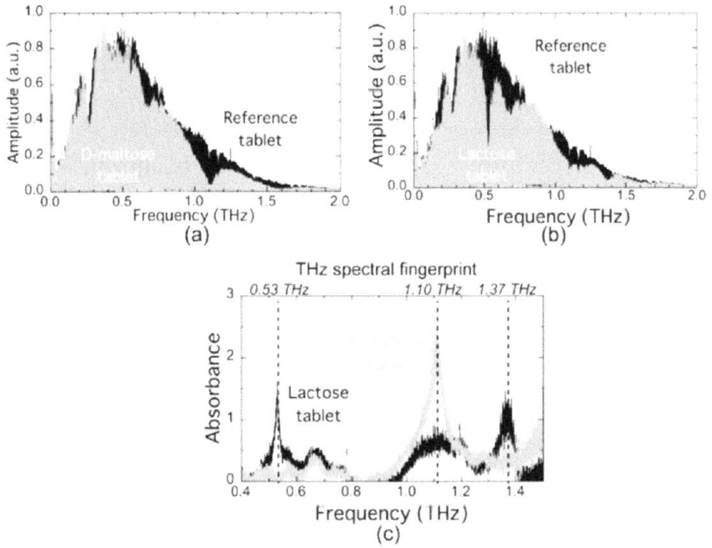

Figure 7.20 Mode-resolved amplitude spectra of THz comb after passing through (a) D-maltose tablet and (b) lactose tablet. (c) Comparison of absorbance spectra between D-maltose and lactose tablets.

the amplitude spectrum of the THz comb after passing through the reference tablet was also indicated as the black graph in both. Spectral dips clearly appeared in both spectra. Figure 7.20c shows the absorbance spectra of the two sugar tablets. Each absorption spectrum was calculated from the envelope of power spectra of the THz comb for the sugar tablet and the reference tablet. The two sugar tablets exhibited characteristic spectral fingerprints of THz absorption: 0.53 and 1.37 THz for the lactose and 1.10 THz for the D-maltose, in good agreement with the value reported in the literature [45]. In this way, it is possible to classify pharmaceutical tablets based on their spectral fingerprints appearing in THz comb even though they have a similar appearance.

To evaluate the potential of THz comb spectroscopy for gas analysis, the absorption spectrum of the rotational transition $1_{10} \leftarrow 1_{01}$ at 0.557 THz of water vapor was measured. A sample gas, a mixture of water vapor (partial pressure = 10 Pa) and nitrogen (partial pressure = 320 Pa), was introduced into the low-pressure gas cell. The sample has an expected pressure-broadening linewidth of 23 MHz full-width at half-maximum (FWHM) from the self-broadening of water vapor and the collision-broadening induced by nitrogen [46] and an expected Doppler-broadening linewidth of 0.026 MHz [47]. Figure 7.21a shows the mode-resolved amplitude spectrum of the THz comb expanded around 0.557 THz. The comb modes had a frequency gap of 250 MHz and a linewidth of 25 MHz. The amplitude spectrum did not indicate the spectral shape of the absorption line due to the excessively coarse distribution of the comb modes compared with the absorption linewidth of the low-pressure water vapor. Next, we demonstrated incremental sweeping of the comb modes across the absorption line at 0.557 THz by changing the repetition frequencies of both lasers. Incremental increases of f_{rep1} and f_{rep2} (= $f_{\text{rep1}} + \Delta f_{\text{rep}}$) by 11,220.8 Hz were repeated ten times while keeping Δf_{rep} at 50 Hz. The resulting overlaid spectra of the comb modes are shown in Fig. 7.21b. In this demonstration, a single shift of f_{rep1} by 0.004488% resulted in sweeping of the comb modes by 10% of their interval (= 25 MHz) due to the large number of THz comb modes (m = 2,228 at 0.557 THz). The frequency gaps between the comb modes in Fig. 7.21a were filled by interleaving additional frequency marks.

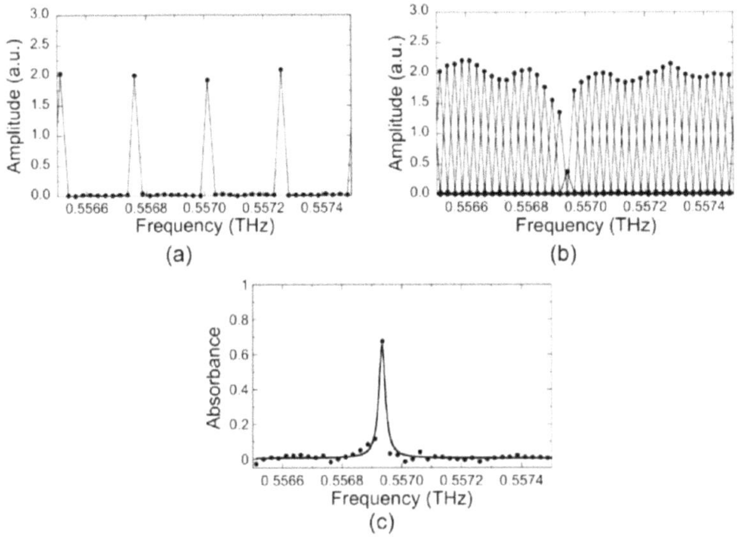

Figure 7.21 Mode-resolved amplitude spectra of (a) standard THz comb and (b) gap-less THz comb around 0.557 THz after passing through low-pressure water vapor contained in the gas cell. (c) Absorbance spectrum of the rotational transition $1_{10} \leftarrow 1_{01}$ in water vapor.

In other words, a spectrally interleaved or gap-less THz comb was successfully achieved. A sharp spectral dip clearly appeared at the position of the water absorption line of $1_{10} \leftarrow 1_{01}$. The absorption was extracted from the power spectrum by taking the peak amplitude of each mode normalized with a reference spectrum obtained under identical conditions, as shown in Fig. 7.21c. To determine the spectral linewidth, we fitted a Lorentzian function to the measured spectral profile because the spectral shape in the THz region arises from the pressure broadening rather than Doppler broadening (= 0.026 MHz) [47] at the present experimental condition. As a result, the spectral linewidth was determined to be 24 MHz, indicated by a solid line in Fig. 7.21c, and this value is consistent with the expected pressure-broadening linewidth (= 23 MHz). This result clearly indicated that the increased spectral sampling density in the gap-less THz comb enhanced the spectral resolution of this gas spectroscopy to the level of the linewidth of the comb mode (= 25 MHz) rather than the gap size (= 250 MHz) for unswept measurements.

7.6 Conclusions and Future Trends

We described the concept of THz frequency comb, together with three concrete THz frequency metrology techniques embodying this concept. The most important point in this approach is that the coherent frequency linking established by the frequency comb enables us to realize THz frequency metrology traceable to the SI base unit of time. Also, the fiber-based techniques used here, including fiber lasers and optical fibers, have the advantage of being portable, alignment-free, robust, and flexible.

First, the THz-comb-referenced spectrum analyzer was achieved by photoconductive mixing of a CW-THz wave with a PC-THz comb. Based on the stable PC-THz comb generated in a PCA, the absolute frequency of the CW-THz wave was determined at a precision of 10^{-11}, which is limited by the performance of the Rb atomic clock used. At the same time, the detailed spectral behavior of the CW-THz waves was monitored in real time. Although the two-step measurement of $f_{\rm rep}$ and $f_{\rm b}$ is required in Fig. 7.3, we also proposed a method for real-time monitoring of the absolute frequency of CW-THz radiation involving temporally parallel, i.e., simultaneous, measurement of two pairs of beat frequencies and laser repetition frequencies based on dual THz combs of photocarriers (PC-THz combs) with different frequency spacings [13]. To demonstrate this method, THz-comb-referenced spectrum analyzers were constructed with a dual configuration based on dual fs-ML lasers. Regardless of the presence or absence of frequency control in the PC-THz combs, a frequency precision of 10^{-11} was achieved at a measurement rate of 100 Hz. Furthermore, to eliminate the need for a precisely stabilized fs-ML laser, we further demonstrated the real-time monitoring of the absolute frequency of CW-THz wave with a free-running single fs-ML laser [14] or a free-running single-cavity dual-wavelength fs-ML laser [15]. These approaches will reinforce the versatility of the THz-comb-referenced spectrum analyzer.

Second, a widely and continuously tunable THz frequency synthesizer traceable to the hydrogen maser linked to UTC-NMIJ was demonstrated by the photomixing of two independent OFSs. A combination of dual OFSs and the photomixing technique achieved a frequency uncertainty of 10^{-12} in the THz frequency range. Furthermore, photomixing of a tunable OFS and a fixed one enabled us to extend the continuous tuning range of the CW-THz radiation up to 120 GHz while maintaining unprecedented frequency uncertainty. The effectiveness of the THz synthesizer was successfully demonstrated in low-pressure spectroscopy of gas molecules.

Third, we have described dual-THz-comb spectroscopy. Since THz comb has spectral characteristics of both narrow-linewidth and broadband spectrum, the advantages of THz time-domain spectroscopy and THz frequency-domain spectroscopy (namely, broad spectral coverage and high spectral resolution) can be combined based on THz comb. Furthermore, since the uncertainty of the frequency axis is traceable to the frequency standard via coherent frequency linking, the universality of the measurement is guaranteed. In this way, dual-THz-comb spectroscopy can achieve broad spectral coverage, high spectral resolution, and high spectral accuracy at the same time. Although the dual-THz-comb spectroscopy has high performance, its use is often hampered by the requirement for dual f_{rep}-stabilized fs-ML lasers because of being large, complicated, and expensive. We are conducting research and development to solve this practical problem. An adaptive sampling method was developed to eliminate the technical requirement of "frequency stabilization" [22]. When dual free-running fs-ML lasers are used for ASOPS, the time-scale magnification factor ($= f_{rep1}/\Delta f_{rep}$) is temporally fluctuated due to frequency fluctuation of f_{rep1} and Δf_{rep}; this leads to non-linearity of the time scale in the acquired RF pulse train. In the adaptive sampling method, an adaptive sampling clock that reflects the fluctuation of the time-scale magnification factor is generated and is used to correct the non-linearity of the time scale in real time. By using this method, higher performance is realized in dual-THz-comb spectroscopy with dual free-running fs-ML lasers than in that with dual f_{rep}-stabilized fs-ML lasers. Next, the need for "dual fs-ML lasers" can be resolved by using a single-cavity dual-wavelength dual-optical-comb fiber laser that can generate dual optical combs with slightly different wavelengths from a single cavity [23]. The dual-wavelength dual optical combs have a slightly different f_{rep} due to the refractive index dispersion of the cavity fiber. Furthermore, since they share a common cavity, f_{rep} fluctuations behave similarly; cancelling out by common-mode fluctuation enables to make Δf_{rep} stable without laser control. As a result, the performance of dual-THz-comb spectroscopy with the single-cavity dual-wavelength dual-optical-comb fiber laser is comparable to that with dual f_{rep}-stabilized fiber lasers. Furthermore, by combining the adaptive sampling method with the single-cavity dual-wavelength dual-optical-comb fiber laser, it is possible to achieve both versatility and high performance [24].

Our approach for THz frequency metrology enables us to treat the frequency in the THz region as a universal quantity based on the national frequency standard, in the same manner as in the optical and electrical regions. Therefore, the so-called THz gap of the THz frequency metrology will be essentially solved. The establishment of THz frequency metrology will cause a ripple effect in various fields of THz technology and applications. For example, to achieve universal identification power in spectroscopic applications, the uncertainty of the frequency scale in the THz spectrum should be certified by referencing to the SI base unit of time, that is, a microwave frequency standard. Our THz frequency metrology will play an important role in the frequency calibration and traceability of various types of commercial THz instruments, such as sources, detectors, and systems. The absolute frequency of CW-THz sources can be accurately determined by THz spectrum analyzers, whereas the frequency scale of a THz spectrometer can be precisely calibrated by THz frequency synthesizers. The precisely calibrated THz spectrometer will increase the identification power in spectroscopic applications based on THz spectral fingerprints. Also, the highly spectroscopic performance made possible by using a THz comb will contribute to the development of precise THz spectrometers and preparation of a precise spectral database atlas. Furthermore, when THz waves are used as carrier waves for next-generation wireless communications (Beyond 5G or 6G), the wireless carrier frequency

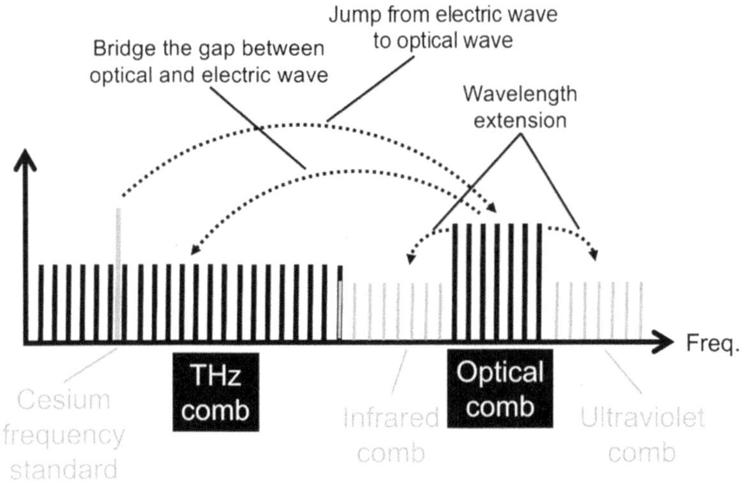

Figure 7.22 Seamless frequency linking with frequency comb.

should be highly accurate and stable in order to secure the necessary and sufficient bandwidth for broadband communication without interference with other applications, such as astronomy or sensing. A THz frequency synthesizer can be used to generate an accurate and stable wireless carrier frequency. In this way, THz frequency metrology will spread to various THz applications based on a high-reliability frequency scale.

Finally, I would like to mention the essence of coherent frequency linking that makes full use of frequency combs. This concept makes it possible to go beyond the framework of the THz frequency band and share the basic physical quantity of electromagnetic waves, namely frequency, with common uncertainty across a wide range of frequencies in electromagnetic waves (see Fig. 7.22). Currently, the frequency standard is the cesium frequency standard in the microwave band (frequency = 9.2 GHz). The optical comb has made it possible to bring the frequency uncertainty of the microwave band directly into the optical region via laser control regardless of a large frequency difference over 10^5. Recently, by using wavelength conversion technology such as nonlinear optical effect, it has become possible to expand the frequency certainty of the optical comb to the ultraviolet region and the infrared region. Furthermore, the THz comb introduced makes it possible to bridge the RF region and the optical region. In this way, the frequency comb technology has extremely high wavelength diversity, and these wavelength regions can be covered without the gap by the common concept of the frequency comb. More importantly, since all frequency combs in each wavelength region are based on a common frequency standard, it is possible to discuss frequencies with "same uncertainty" beyond the boundaries of each electromagnetic wave region seamlessly. This is the essence of frequency combs including optical combs and THz combs.

References

1. D. M. Mittleman, *Sensing with THz Radiation* (Springer, Berlin, 2003).
2. M. Tonouchi, Cutting-edge terahertz technology, *Nature Photon.* **1**, 97–105 (2007).
3. T. Kleine-Ostmann, T. Schrader, M. Bieler, U. Siegner, C. Monte, B. Gutschwager, J. Hollandt, R. Müller, G. Ulm, I. Pupeza, and M. Koch, THz metrology, Frequenz, **62**, 135–146 (2008).
4. T. Yasui, S. Yokoyama, H. Inaba, K. Minoshima, T. Nagatsuma, and T. Araki, Terahertz frequency metrology based on frequency comb, *IEEE J. Sel. Top. Quantum Electron.* **17**, 191–201 (2011).
5. C. Champenois, G. Hagel, M. Houssin, M. Knoop, C. Zumsteg, and F. Vedel, Terahertz frequency standard based on three-photon coherent population trapping, *Phys. Rev. Lett.* **99**, 013001 (2007).
6. T. Udem, J. Reichert, R. Holzwarth, and T. W. Hänsch, Accurate measurement of large optical frequency differences with a mode-locked laser, *Opt. Lett.* **24**, 881–883 (1999).
7. M. Niering, R. Holzwarth, J. Reichert, P. Pokasov, T. Udem, M. Weitz, T. W. Hänsch, P. Lemonde, G. Santarelli, M. Abgrall, P. Laurent, C. Salomon, and A. Clairon, Measurement of the hydrogen 1S–2S transition frequency by phase coherent comparison with a microwave cesium fountain clock, *Phys. Rev. Lett.* **84**, 5496–5499 (2000).
8. D. J. Jones, S. A. Diddams, J. K. Ranka, A. Stentz, R. S. Windeler, J. L. Hall, and S. T. Cundiff, Carrier-envelope phase control of femtosecond mode-locked lasers and direct optical frequency synthesis, *Science* **288**, 635–639 (2000).

9. T. Udem, R. Holzwarth, and T. W. Hänsch, Optical frequency metrology, *Nature* **416**, 233–237 (2002).
10. T. Yasui, Y. Kabetani, E. Saneyoshi, S. Yokoyama, and T. Araki, Terahertz frequency comb by multi-frequency-heterodyning photoconductive detection for high-accuracy, high-resolution terahertz spectroscopy, *Appl. Phys. Lett.* **88**, 241104 (2006).
11. S. Yokoyama, R. Nakamura, M. Nose, T. Araki, and T. Yasui, Terahertz spectrum analyzer based on a terahertz frequency comb, *Opt. Express* **16**, 13052–13061 (2008).
12. T. Yasui, R. Nakamura, K. Kawamoto, A. Ihara, Y. Fujimoto, S. Yokoyama, H. Inaba, K. Minoshima, T. Nagatsuma, and T. Araki, Real-time monitoring of continuous-wave terahertz radiation using a fiber-based, terahertz-comb-referenced spectrum analyzer, *Opt. Express* **17**, 17034–17043 (2009).
13. T. Yasui, K. Hayashi, R. Ichikawa, H. Cahyadi, Y.-D. Hsieh, Y. Mizutani, H. Yamamoto, T. Iwata, H. Inaba, and K. Minoshima, Real-time absolute frequency measurement of continuous-wave terahertz radiation based on dual terahertz combs of photocarriers with different frequency spacings, *Opt. Express* **23**, 11367–11377 (2015).
14. T. Minamikawa, K. Hayashi, T. Mizuguchi, Y.-D. Hsieh, D. G. Abdelsalam, Y. Mizutani, H. Yamamoto, T. Iwata, and T. Yasui, Real-time determination of absolute frequency in continuous-wave terahertz radiation with a photocarrier terahertz frequency comb induced by an unstabilized femtosecond laser, *J. Infrared Millim. Terahertz Waves* **37**, 473–485 (2016).
15. G. Hu, T. Mizuguchi, X. Zhao, T. Minamikawa, T. Mizuno, Y. Yang, C. Li, M. Bai, Z. Zheng, and T. Yasui, Measurement of absolute frequency of continuous-wave terahertz radiation in real time using a free-running, dual-wavelength mode-locked, erbium-doped fibre laser, *Sci. Rep.* **7**, 42082 (2017).
16. T. Yasui, H. Takahashi, Y. Iwamoto, H. Inaba, and K Minoshima, Continuously tunable, phase-locked, continuous-wave terahertz generator based on photomixing of two continuous-wave lasers locked to two independent optical combs, *J. Appl. Phys.* **107**, 033111 (2010).
17. T. Yasui, H. Takahashi, K. Kawamoto, Y. Iwamoto, K. Arai, T. Araki, H. Inaba, and K. Minoshima, Widely and continuously tunable terahertz synthesizer traceable to a microwave frequency standard, *Opt. Express* **19**, 4428–4437 (2011).
18. Y.-D. Hsieh, H. Kimura, K. Hayashi, T. Minamikawa, Y. Mizutani, H. Yamamoto, T. Iwata, H. Inaba, K. Minoshima, F. Hindle, and T. Yasui,

Terahertz frequency-domain spectroscopy of low-pressure acetonitrile gas by a photomixing terahertz synthesizer referenced to dual optical frequency combs, *J. Infrared Millim. Terahertz Waves* **37**, 903–915 (2016).

19. T. Yasui, M. Nose, A. Ihara, K. Kawamoto, S. Yokoyama, H. Inaba, K. Minoshima, and T. Araki, Fiber-based, hybrid terahertz spectrometer using dual fiber combs, *Opt. Lett.* **35**, 1689–1691 (2010).
20. Y.-D. Hsieh, Y. Iyonaga, Y. Sakaguchi, S. Yokoyama, H. Inaba, K. Minoshima, F. Hindle, Y. Takahashi, M. Yoshimura, Y. Mori, T. Araki, and T. Yasui, Terahertz comb spectroscopy traceable to microwave frequency standard, *IEEE Trans.THz Sci. Technol.* **3**, 322–330 (2013).
21. Y.-D. Hsieh, Y. Iyonaga, Y. Sakaguchi, S. Yokoyama, H. Inaba, K. Minoshima, F. Hindle, T. Araki, and T. Yasui, Spectrally interleaved, comb-mode-resolved spectroscopy using swept dual terahertz combs, *Sci. Rep.* **4**, 3816 (2014).
22. T. Yasui, R. Ichikawa, Y.-D. Hsieh, K. Hayashi, H. Cahyadi, F. Hindle, Y. Sakaguchi, T. Iwata, Y. Mizutani, H. Yamamoto, K. Minoshima, and H. Inaba, Adaptive sampling dual terahertz comb spectroscopy using dual free-running femtosecond lasers, *Sci. Rep.* **5**, 10786 (2015).
23. G. Hu, T. Mizuguchi, R. Oe, K. Nitta, X. Zhao, T. Minamikawa, T. Li, Z. Zheng, and T. Yasui, Dual terahertz comb spectroscopy with a single free-running fibre laser, *Sci. Rep.* **8**, 11155 (2018).
24. J. Chen, K. Nitta, X. Zhao, T. Mizuno, T. Minamikawa, F. Hindle, Z. Zheng, and T. Yasui, Adaptive-sampling near-Doppler-limited terahertz dual-comb spectroscopy with a free-running single-cavity fiber laser, *Adv. Photon.* **2**, 036004 (2020).
25. H. Schnatz, B. Lipphardt, J. Helmcke, F. Riehle, and G. Zinner, First phase-coherent frequency measurement of visible radiation, *Phys. Rev. Lett.* **76**, 18–21 (1996).
26. S. Kohjiro, K. Kikuchi, M. Maezawa, T. Furuta, A. Wakatsuki, H. Ito, N. Shimizu, T. Nagatsuma, and Y. Kado, A 0.2–0.5 THz single-band heterodyne receiver based on a photonic local oscillator and a superconductor-insulator-superconductor mixer, *Appl. Phys. Lett.* **93**, 093508 (2008).
27. J. J. A. Baselmans, M. Hajenius, J.R. Gao, T. M. Klapwijk, P. A. J. de Korte, B. Voronov, and G. Gol'tsman, Doubling of sensitivity and bandwidth in phonon cooled hot electron bolometer mixers, *Appl. Phys. Lett.* **84**, 1958–1960 (2004).

28. T. Nagatsuma, H. Ito, and T. Ishibashi, High-power RF photodiodes and their applications, *Laser Photon. Rev.* **3**, 123–137 (2009).
29. I. Katayama, R. Akai, M. Bito, H. Shimosato, K. Miyamoto, H. Ito, and M. Ashida, Ultrabroadband terahertz generation using 4-N,N-dimethylamino-4/-N/-methyl-stilbazolium tosylate single crystals, *Appl. Phys. Lett.* **97**, 021105 (2010).
30. S. Matsuura, M. Tani, and K. Sakai, Generation of coherent terahertz radiation by photomixing in dipole photoconductive antennas, *Appl. Phys. Lett.* **70**, 559–561 (1997).
31. Q. Quraishi, M. Griebel, T. Kleine-Ostmann, and R. Bratschitsch, Generation of phase-locked and tunable continuous-wave radiation in the terahertz regime, *Opt. Lett.* **30**, 3231–3233 (2005).
32. G. Mouret, F. Hindle, A. Cuisset, C. Yang, R. Bocquet, M. Lours, and D. Rovera, THz photomixing synthesizer based on a fiber frequency comb, *Opt. Express* **17**, 22031–22040 (2009).
33. F. Hindle, G. Mouret, S. Eliet, M. Guinet, A. Cuisset, R. Bocquet, T. Yasui, and D. Rovera, Widely tunable THz synthesizer, *Appl. Phys. B* **104**, 763–768 (2011).
34. H. Takahashi, Y. Nakajima, H. Inaba, and K. Minoshima, Ultra-broad absolute-frequency tunable light source locked to a fiber-based frequency comb, *Technical Digest of Conference on Lasers and Electro-Optics 2009*, Optical Society of America, Washington DC, paper CTuK4 (2009).
35. D. W. Allan, Statistics of atomic frequency standards, *Proc. IEEE* **54**, 221–230 (1966).
36. J. C. Pearson and H. S. P. Müller, The submillimeter wave spectrum of isotopic methyl cyanide, *Astrophys. J.* **471**, 1067–1072 (1996).
37. H. M. Pickett, R. L. Poynter, E. A. Cohen, M. L. Delitsky, J. C. Pearson, and H. S. P. Muller, Submillimeter, millimeter, and microwave spectral line catalog, *J. Quant. Spectrosc. Radiat. Transf.* **60**, 883–890 (1998).
38. D. H. Auston and K. P. Cheung, Coherent time-domain far-infrared spectroscopy, *J. Opt. Soc. Am. B* **2**, 606–612 (1985).
39. P. A. Elzinga, R. J. Kneisler, F. E. Lytle, Y. Jiang, G. B. King, and N. M. Laurendeau, Pump/probe method for fast analysis of visible spectral signatures utilizing asynchronous optical sampling, *Appl. Opt.* **26**, 4303–4309 (1987).
40. Y. Takagi and S. Adachi, Subpicosecond optical sampling spectrometer using asynchronous tunable mode-locked lasers, *Rev. Sci. Instrum.* **70**, 2218 (1999).

41. C. Janke, M. Först, M. Nagel, H. Kurz, and A. Bartels, Asynchronous optical sampling for high-speed characterization of integrated resonant terahertz sensors, *Opt. Lett.* **30**, 1405–1407 (2005).
42. T. Yasui, E. Saneyoshi, and T. Araki, Asynchronous optical sampling terahertz time-domain spectroscopy for ultrahigh spectral resolution and rapid data acquisition, *Appl. Phys. Lett.* **87**, 061101 (2005).
43. T. Yasui, K. Kawamoto, Y.-D. Hsieh, Y. Sakaguchi, M. Jewariya, H. Inaba, K. Minoshima, F. Hindle, and T. Araki, Enhancement of spectral resolution and accuracy in asynchronous-optical-sampling terahertz time-domain spectroscopy for low-pressure gas-phase analysis, *Opt. Express* **20**, 15071–15078 (2012).
44. Y.-D. Hsieh, S. Nakamura, D. G. Abdelsalam, T. Minamikawa, Y. Mizutani, H. Yamamoto, T. Iwata, F. Hindle, and T. Yasui, Dynamic terahertz spectroscopy of gas molecules mixed with unwanted aerosol under atmospheric pressure using fibre-based asynchronous-optical-sampling terahertz time-domain spectroscopy, *Sci. Rep.* **6**, 28114 (2016).
45. M. Schirmer, M. Fujio, M. Minami, J. Miura, T. Araki, and T. Yasui, Biomedical applications of a real-time terahertz color scanner, *Biomed. Opt. Express* **1**, 354–366 (2010).
46. T. Seta, H. Hoshina, Y. Kasai, I. Hosako, C. Otani, S. Loßow, J. Urban, M. Ekström, P. Eriksson, and D. Murtagh, Pressure broadening coefficients of the water vapour lines at 556.936 and 752.033 GHz, *J. Quantum Spectrosc. Radiat. Transfer* **109**, 144–150 (2008).
47. J. M. Hollas, *Modern Spectroscopy* (John Wiley & Sons, Chichester, 2003).

Part II

THz Solid-State Electronic Sources

Chapter 8

High-Efficiency THz Oscillators

Yu Ye and Qun Jane Gu
Electrical and Computer Engineering,
University of California, Davis,
One Shields Avenue, Davis, California 95616, USA
leavestc@gmail.com, jgu@ucdavis.edu

Recently, terahertz (THz) and sub-terahertz (sub-THz) oscillators have received an increasingly intensified attraction from both academic research and the consumer electronics market. As the oscillating frequency approaches the f_{max} of the advanced silicon/III–V semiconductor processes, achieving high power efficiency while maintaining a decent output power is essential and challenging for THz oscillators. And with the electrical wavelength scaling down, more distributed effects as well as high-frequency parasitic make the "active" device itself no longer seen as a "lump" device but more as a mixed model with EM and semiconductor device effects. Similarly, the boundary of the so-called "active core" and "passive network" which is clearly defined in low frequencies becomes obscure as frequency increases. Hence, there is a need to have a systematic design methodology, which synthesizes the individual terms, i.e., "active transistor," "active core," and "passive network," together to have a global optimization. In this chapter, we share a holistic optimization method of designing a THz oscillator with a design flow

Advances in Terahertz Source Technologies
Edited by Gun-Sik Park, Masahiko Tani, Jae-Sung Rieh, and Sang Yoon Park

ISBN 978-981-4968-89-8 (Hardcover), 978-1-003-45967-5 (eBook)
www.jennystanford.com

provided for optimization iterations. A design example is followed to demonstrate both good power efficiency and high output power for a sub-THz oscillator by using such a methodology.

8.1 Introduction

Terahertz (THz) and sub-terahertz (sub-THz) frequencies beyond 100 GHz provide a large bandwidth that can be implemented for a number of applications such as high-data-rate wired and wireless communication systems, imaging systems, high-resolution radars for precise position and movement measurements, as well as chemical sensing using rotational spectroscopy [1–6].

Also, with the continuous scaling of fabrication process nodes, silicon-based integrated circuit (IC) processes are continuously evolving with advanced nanometer-scale transistors at f_t and f_{max} beyond 300 GHz [7–9]. Hence, silicon-based semiconductor technology, such as CMOS technology, is considered a potential platform to provide a low-cost and high-yield solution for the sub-THz/THz ICs.

In general, based on the output frequency relation to the oscillation frequency, oscillators typically can be characterized in two types, fundamental-based [10–11] and harmonic-based [12, 13].

8.1.1 Fundamental Oscillators

The fundamental-based sources have a larger output power but with the operating frequency constrained by the device f_t/f_{max}. Several sub-THz/THz fundamental oscillators have been investigated to push up the frequency with advanced silicon-based processes [10, 11, 14–24]. In 2013, Ref. [11] demonstrated a fundamental 240 GHz oscillator by adopting IBM CMOS 32 nm SOI process, achieving −7 dBm output power while consuming 13.3 mW DC power, leading to 1.5% DC-to-RF efficiency. With a SiGe HBT process, in the same year, Ref. [10] demonstrated a 218–245 GHz fundamental VCO with −3.6 dBm output power and 0.81% efficiency. By using a 130 nm SiGe BiCMOS process, in 2017, Ref. [20] achieved a 4.8 dBm output power with 11.7% DC-to-RF efficiency at 175 GHz while in 2021, Ref. [19] achieved a 160 GHz fundamental oscillator with a 6.53 dBm output power and 13.4% DC-to-RF efficiency. Also in 2020, Ref. [21] published a 251 GHz fundamental oscillator with 1.35% efficiency in a 65 nm CMOS process. And in 2017, Ref. [22] achieved a 15.3% efficiency for a fundamental oscillator at 195 GHz with 6.5 dBm output power on a 55 nm SiGe process. In 2016, Ref. [24] presented a 177 GHz fundamental oscillator with a 9.4% DC-to-RF efficiency while generating 3.4 dBm output power in a 65 nm CMOS process. Moreover, by optimizing the bias condition, it demonstrated a record 25.9% DC-to-RF efficiency with 0.66 dBm output power. Several fundamental oscillator-based transmitters have also been presented [16–18]. In [16], with a 40 nm CMOS process, a 120 GHz FSK transmitter with −1 dBm output power and 7.22% efficiency has been demonstrated. Ref. [17] presented a 135 GHz ASK transmitter with −8.7 dBm output power and 0.73% efficiency. With a 65 nm CMOS process, Ref. [18] demonstrated a 108 GHz ASK transmitter with −2.3 dBm output power, 21.6 mW power consumption and 2.7% efficiency.

8.1.2 Harmonic Oscillators

On the other hand, harmonic-based sources, such as N-push VCOs [25-29], harmonically coupled oscillators [30, 31], linear superposition [32], and multiplier-chain [33–36], can exceed the operating frequency higher than the device f_{max}. The main drawbacks are the limited output power and poor power efficiency. In 2008, Ref. [32] generated a 324 GHz signal with −46 dBm output power with 12 mW DC consumption by using a linear superposition technique on a 90 nm CMOS process. In 2014, Ref. [28] presented a multi-core second-order harmonic coupled oscillator working at 256 GHz with 4.1 dBm output power and 227 mW DC power consumption with 1.14% efficiency on a 65 nm CMOS process. In the same year, Ref. [31] demonstrated a 293 GHz third-order harmonic coupled VCO on a 65 nm CMOS process with a maximum −2.74 dBm output power and 2.76% efficiency. Also in 2014, Ref. [36] presented a 240 GHz oscillator, which employs an 80 GHz oscillator with a frequency tripler. The transmitter generated 0 dBm output power with 220 mW DC power consumption. In 2015, Ref. [25] presented a 239 GHz push-push VCO on a 65 nm CMOS process, with −4.8 dBm output power and 1.45% DC-to-RF efficiency. In the same year, In Ref. [26], a 206–220 GHz push-push VCO with −1.6 dBm output power and 2.1% maximum DC-to-RF efficiency was demonstrated. In 2016, at 296 GHz, Ref. [37] achieved a third harmonic oscillator with 5.4 dBm output power and 5.15% DC-to-RF efficiency. Also, in 2017, Ref. [38] presented a 5.6 dBm output power with 4.6% DC-to-RF efficiency at 215 GHz. In 2018, Ref. [40] achieved a 2.95% DC-to-RF efficiency with a 3 dBm output power at 225 GHz in a 65 nm CMOS process. In 2020, Ref. [39] showed a 229 GHz oscillator with 3.4 dBm output power and 1.16% DC-to-RF efficiency by using a 65 nm CMOS process.

8.2 Challenges

As for silicon-based IC processes, despite the above-mentioned advancements, the low breakdown voltage and the relatively small effective f_{max} of active devices and the various loss mechanisms of silicon processes make the high-power and high energy-efficiency circuits at high frequencies continue to be a challenge. Moreover, the transconductance (g_{m}) of a transistor drops with frequencies, especially at the frequency close to device f_{max} [8, 9].

Figure 8.1 shows the state-of-the-art (SoA) of recently published silicon-based THz/sub-THz oscillators' output power versus the frequencies. The trade-off between the output power and oscillation frequency has been a main challenge. A large-size transistor has advantages in terms of higher output power, which presents higher parasitic capacitance and thus reduces the oscillation frequency. Therefore, silicon-based sub-THz/THz oscillators tend to adopt smaller-size transistors. As a result, as shown in Fig. 8.1, normally the power induced by a single oscillator stays still below 1 mW. Because of such a trade-off, there is a trend that shows a decreasing power performance once frequency goes higher. Also, to compare fundamental and harmonic-based THz/sub-THz oscillators, it is no surprise to see that fundamental oscillators provide better output power performance compared to harmonic ones, while harmonic

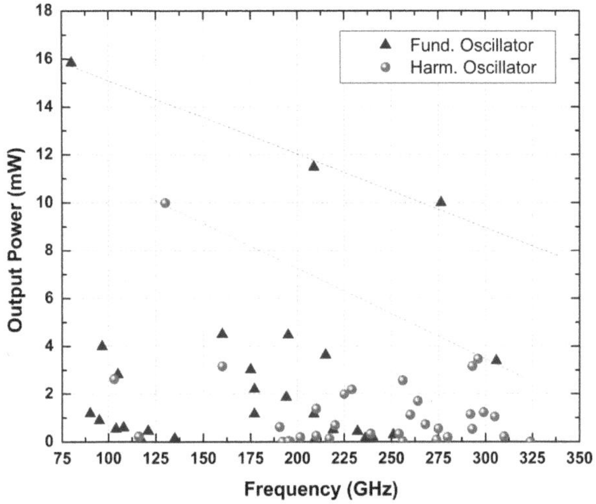

Figure 8.1 Recent SOAs of silicon-based signal sources above 80 GHz, output power vs. frequency [8–11, 14–32], and [37–71].

oscillators have the capabilities to exceed the transistor f_t/f_{max} limitations.

From another characterization aspect, Fig. 8.2 shows the SoA of recently published THz/sub-THz oscillators DC-to-RF efficiency versus the operating frequencies. Due to the nature of the semiconductor process that (a) the g_m of a transistor drops and (b) the loss of passive circuit increases with frequencies, the DC-to-RF efficiency of sub-THz/THz oscillators drops across the frequencies, especially when the oscillator operates at or above the half of the f_t and f_{max} of the process. Meanwhile, it is observed that fundamental oscillators can achieve better efficiency than harmonic ones. Many of the SoA fundamental designs achieved higher than 5% DC-to-RF efficiency, whereas most SoA harmonic ones observed lower efficiency.

Figure 8.3 shows the SoA of recently published THz/sub-THz oscillators' DC-to-RF efficiency versus the output power. There is also a trade-off between output power and DC-to-RF efficiency. Since THz/sub-THz oscillators tend to adopt smaller sizes of transistors to minimize the large parasitics. Normally to achieve decent output power, many techniques have been used such as

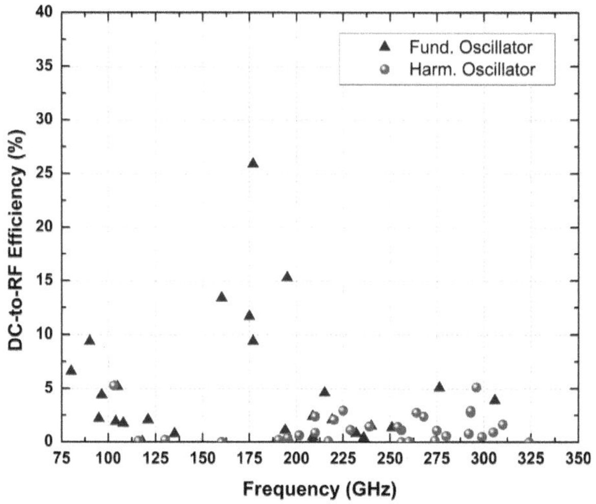

Figure 8.2 Recent SOAs of silicon-based signal sources above 80 GHz, DC-to-RF efficiency vs. frequency [8–11, 14–32], and [37–71].

post-oscillator amplifiers/buffers [15] or power combining from multiple oscillators [39], etc. However, such methods will either introduce more DC power consumption or increase the combining loss, hence, decreasing the DC-to-RF efficiency of the whole oscillator.

As observed from Figs. 8.1–8.3, the trade-off among output power, DC-to-RF efficiency and operating frequency would be a main obstacle for the THz/sub-THz oscillators design. The designer needs to balance the performance among these three specs for good results.

8.3 Design and Optimization Flow

Although the Si-based semiconductor processes keep on evolving and the nodes are keeping scaling down, currently the f_t and f_{max} are still within several hundreds of GHz, which makes the oscillator over 100 GHz on Si-based semiconductors not trivial, especially when the output power and DC-to-RF efficiency are considered. Due to many of the challenges as aforementioned, designers need to

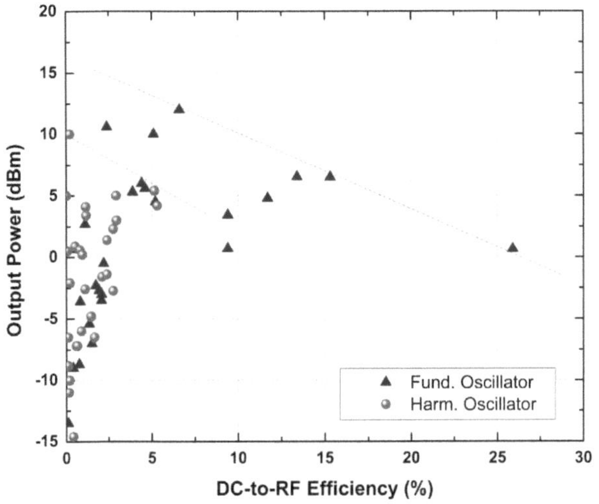

Figure 8.3 Recent SOAs of silicon-based signal sources above 80 GHz, DC-to-RF efficiency vs. Output power [8–11, 14–32], and [37–71].

carefully optimize the designs to save or gain the treasurable power or efficiency.

Normally, an oscillator design can be separated into two parts: (a) the oscillation part of circuits and (b) the output part of circuits. The oscillation part of circuits majorly contains the active transistors for oscillation and its related circuits or components. Typically, the design of such parts started from transistor size selection, layout, modeling, and optimization. Many techniques, such as g_m-boosted technique [72, 73], have been applied to improve performance. And the output part of circuits is used to deliver the signal generated from the oscillation part to other designs, such as antenna, PA, mixer, etc. The output part of the circuit can be a pure passive output matching network, or output buffers, or even some power combining circuits to combine several oscillator cores, etc. There are also many techniques, such as impedance boosting [54], that have been investigated to optimize the output part of circuits.

Since sub-THz/THz oscillators are very sensitive to frequency, output power and efficiency, every improvement in output power or DC-to-RF efficiency is crucial. And the separated optimization on the oscillation part or the output part of circuits might not be able to

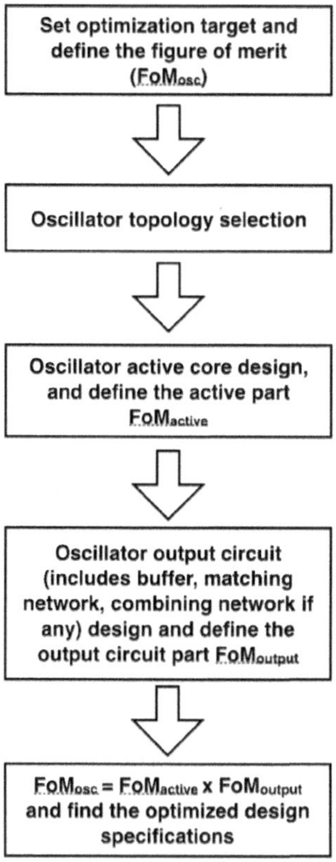

Figure 8.4 Design and optimization flow.

simultaneously achieve the optimal performance with regard to the whole oscillator. Hence, a holistic optimization method is proposed as shown in Fig. 8.4. The purpose of this method is to take both the oscillation and output parts into account for a holistic oscillator optimization. As shown in the design flow in Fig. 8.4, first, designers need to define a figure of merit (FoM) to include the most important specifications that they care about the most. Further, for different oscillator topologies, analytical equivalent circuits are used to model oscillation and output parts of circuits. Hence, the pre-defined FoM can be represented by using such equivalent circuits and linked to one or several oscillator design parameters, such as transistor W/L, etc. Then an optimized FoM can be found with certain oscillator parameters. Followed by the analytical evaluation, a pre-layout and post-layout simulation is needed to prove the analysis and finalize the designs.

8.4 Design Example

Due to the large signal attenuation and the low detection sensitivity in the frequency range above 100 GHz, high-power and high-efficiency generation from signal generators is preferred. Harmonic oscillators have the advantage of high signal frequency, but at the cost of low output power and poor power efficiency. Some other methods, such as power combining and power amplification, have been implemented to generate higher power levels but at the expense of lower efficiency due to the lossy silicon substrate, poor g_m performance and use of larger chip areas. This example uses a fundamental oscillator with the transformer-based impedance optimization technique [24] and [54].

8.4.1 Optimization Target

The two key performances of THz/sub-THz oscillators are the output power and DC-to-RF efficiency. To better quantify the performances, a FoM_{osc} is defined as

$$\text{FoM}_{\text{osc}} = \eta_{\text{osc}} P_{\text{out,osc}}, \tag{8.1}$$

where $P_{\text{out,osc}}$ is the generated output power and η_{osc} is the DC-to-RF efficiency, which are defined as

$$P_{\text{out,osc}} = P_{\text{active,osc}} H(R_\text{L}), \tag{8.2}$$

$$\eta_{\text{osc}} = \frac{P_{\text{out,osc}}}{P_{\text{DC,osc}}} = \frac{P_{\text{active,osc}}}{P_{\text{DC,osc}}} H(R_\text{L}) = \eta_{\text{active,osc}} H(R_\text{L}), \tag{8.3}$$

where $P_{\text{active,osc}}$ is the power generated from the active devices and $\eta_{\text{active,osc}}$ is the DC to RF efficiency of the active part, and $H(R_\text{L})$ is the insertion loss of the output impedance transformation network. The $P_{\text{active,osc}}$ and $\eta_{\text{active,osc}}$ are defined as

$$P_{\text{active,osc}} = 2 \cdot \frac{V_\text{p}^2}{2R_\text{L}} = \frac{V_\text{P}^2}{R_\text{L}}, \tag{8.4}$$

$$\eta_{\text{active,osc}} = \frac{P_{\text{active,osc}}}{2I_{\text{d}}V_{\text{DD}}} = \frac{V_{\text{p}}^2}{\mu_{\text{n}}C_{\text{ox}}\frac{W}{L}\left(V_{\text{gs,ave}} - V_{\text{th}}\right)^2 V_{\text{DD}}R_{\text{L}}}, \quad (8.5)$$

where the parameters V_{P} is the single-ended output signal voltage amplitude, μ_{n} is the electron mobility, C_{ox} is the unit gate oxide capacitance, W/L is the core transistor channel width to length ratio, I_{d} is the dc current consumption of single core transistor, V_{DD} is the dc power supply, $V_{\text{gs,ave}}$ is the average gate source voltage of the core transistor, V_{th} is the threshold voltage of the core transistor and R_{L} is the load resistance. A FoM_{A} for the active device performance is defined as

$$\text{FoM}_{\text{A}} = \eta_{\text{active,osc}}P_{\text{active,osc}}, \quad (8.6)$$

The relationship between FoM_{osc} and FoM_{A} is defined as

$$\text{FoM}_{\text{osc}} = \text{FoM}_{\text{A}}H^2(R_{\text{L}}). \quad (8.7)$$

To achieve the highest FoM_{osc}, both $P_{\text{out,osc}}$ and η_{osc} are preferred to be large. However, the design tradeoffs prevent the objectives to be achieved at the same time. A larger output power prefers a small load resistance R_{L}, which therefore requires a larger device width W with the fixed minimum device length. A larger device size leads to a smaller device output impedance. Therefore, it causes more power consumed by the active device and leads to efficiency reduction. Moreover, the load-matching network performance needs to be considered. The larger the impedance transformation ratio, the larger the insertion loss, $H(R_{\text{L}})$, which also degrades the efficiency. However, accurate optimization requires quantitative analysis, which is the focus of this example to provide a holistic design and optimization methodology. The following of this example starts from the active part for optimum $P_{\text{active,osc}}$, $\eta_{\text{active,osc}}$ and FoM_{A}, and then combined with the insertion loss, $H(R_{\text{L}})$, of the output impedance transformation network to target the best FoM_{osc} for the whole oscillator.

8.4.2 Core Transistor Optimization

The equivalent oscillator topology is plotted in Fig. 8.5, where C_{T} and R_{T} are adopted to represent the total parasitic capacitance and resistance of the transistor, respectively. To simplify the analysis, half

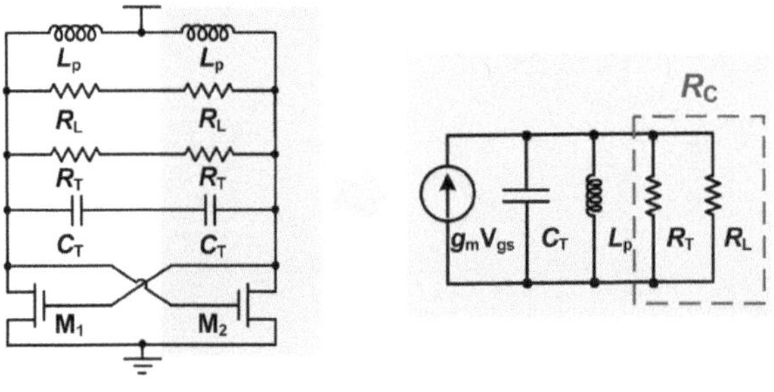

Figure 8.5 Topology of (a) cross-coupled oscillator with transistor's parasitic, C_T and R_T and (b) half of the equivalent circuit.

of the equivalent circuit is used, shown in Fig. 8.5. By extracting from the different sizes of transistors, the simulated peak g_m (g_{mp}), C_T, $1/R_T$ and I_d are shown in Fig. 8.6. The values of g_{mp}, I_d and C_T are proportional to W/L. However, due to the parasitic effect, R_T is not inversely proportional to the W/L. With a larger transistor size, R_T drops faster than the inverse proportion rate of W/L. To illustrate it more clearly, an effective peak g_m ($g_{mp,eff}$) is defined as

$$g_{mp,eff} = g_{mp} - 1/R_T. \tag{8.8}$$

As shown in Fig. 8.7, simulated $g_{mp,eff}$ is not proportional to W/L. With the larger size of the transistor, the $g_{mp,eff}$ grows slower than the transistor's g_{mp}, which can be presented as

$$g_{mp,eff} = m(W/L) \cdot \frac{W}{L}, \tag{8.9}$$

where $m(W/L)$ decreases with a larger design and accounts for the additional relation between $g_{mp,eff}$ and the device size. As for a certain silicon-based process, the transistor length is typically selected as the process minimum value to achieve the highest possible f_T/f_{max}. For example, for a 65 nm CMOS process, normally L is selected as 60 nm which reaches the length limitation. As shown in Fig. 8.7, because of R_T, since $g_{mp,eff}$ is not proportional to W/L but grows much slower with a larger transistor size due to parasitics, the variable $m(W/L)$ is used to illustrate this behavior.

As shown in Fig. 8.5, $g_m R_C \geq 1$ is required to start the oscillation, where R_C equals $R_T//R_L$. When the oscillation is stable, the power

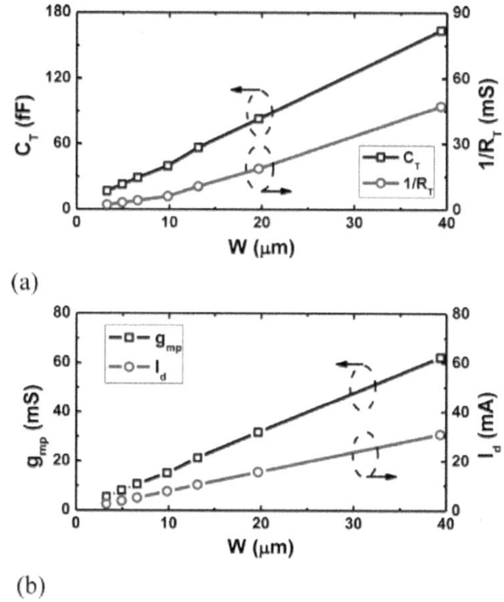

Figure 8.6 Simulated (a) C_T and R_T and (b) g_{mp} and I_d versus device width W.

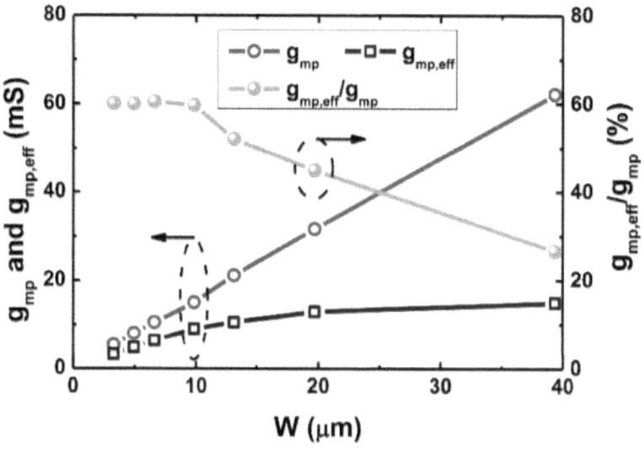

Figure 8.7 Simulated g_{mp}, $g_{mp,eff}$ and the ratio of g_{mp} and $g_{mp,eff}$ versus device width W.

generated by the transistor equals the power consumed by the resistors, including both the load resistance and circuit parasitic resistance. Hence, the average $g_m R_C$, characterized by $g_{m,ave} R_C$, over one period equals 1,

$$g_{m,ave}\left(R_T // R_{L,opt}\right) = 1, \tag{8.10}$$

where $R_{\mathrm{L,opt}}$ is defined as the optimized load resistor which extracts the maximum output power from the oscillator for a given device size. For a given size device, the dc power consumption with the structure configuration in Fig. 8.5 is relatively constant with a fixed dc power supply, therefore the $R_{\mathrm{L,opt}}$ is also the optimized load resistance with the best DC-to-RF efficiency. From the transistor side, $g_{\mathrm{m,ave}}$ can also be represented by device parameters as

$$g_{\mathrm{m,ave}} = \mu_{\mathrm{n}} C_{\mathrm{ox}} \frac{W}{L} (V_{\mathrm{X}} - V_{\mathrm{th}}), \tag{8.11}$$

where V_{X} is the gate voltage for which the transistor achieves the $g_{\mathrm{m,ave}}$. To start the oscillation, the load resistance should be larger than $1/g_{\mathrm{mp,eff}}$. Assume $R_{\mathrm{L,opt}}$ is proportional to $1/g_{\mathrm{mp,eff}}$, where α is a fitting factor and larger than 1 to meet the initial oscillation condition.

$$R_{\mathrm{L,opt}} = \frac{\alpha}{g_{\mathrm{mp,eff}}} = \frac{\alpha}{m(W/L)} \frac{L}{W}, \tag{8.12}$$

Substituting Eqs. (8.9), (8.10) and (8.12) into Eq. (8.11), V_{X} is written as

$$\begin{aligned} V_{\mathrm{X}} &= V_{\mathrm{th}} + \frac{1}{\mu_{\mathrm{n}} C_{\mathrm{ox}}} \cdot \frac{g_{\mathrm{mp}} L}{W} - \frac{\alpha - 1}{\mu_{\mathrm{n}} C_{\mathrm{ox}}} \cdot \frac{g_{\mathrm{mp,eff}} L}{W} \\ &= A - \frac{\alpha - 1}{\mu_{\mathrm{n}} C_{\mathrm{ox}}} \cdot \frac{g_{\mathrm{mp,eff}} L}{W}. \end{aligned} \tag{8.13}$$

Because g_{mp} is proportional to the W/L, the terms of $V_{\mathrm{th}} + g_{\mathrm{mp}} L/(\mu_{\mathrm{n}} C_{\mathrm{ox}} W)$ in Eq. (8.13) are constant with different device sizes and combined as A. Assuming the minimum gate voltage $V_{\mathrm{gs,min}}$ proportional to V_{X},

$$V_{\mathrm{gs,min}} = \beta V_{\mathrm{X}}, \tag{8.14}$$

where β is a fitting factor. Hence, V_{p} as in Eq. (8.4) can be expressed as

$$V_{\mathrm{p}} = V_{\mathrm{gs,ave}} - V_{\mathrm{gs,min}} = V_{\mathrm{gs,ave}} - \beta \left(A - \frac{(\alpha - 1)}{\mu_{\mathrm{n}} C_{\mathrm{ox}}} m\,(W/L) \right), \tag{8.15}$$

where for the cross-coupled oscillator, the average gate voltage $V_{\mathrm{gs,ave}}$ equals the dc power supply V_{DD}. So $P_{\mathrm{active,osc}}$, $\eta_{\mathrm{active,osc}}$ and

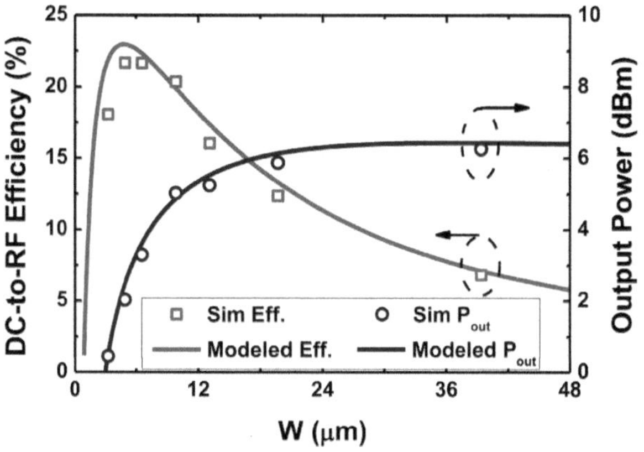

Figure 8.8 Simulated and analytical results based on the model of DC-to-RF efficiency and output power of the oscillator versus device width W.

FoM_A as defined in Eqs. (8.4–8.6) can be derived as

$$P_{\text{active,osc}} = \frac{m\,(W/L)}{\alpha}\left[V_{\text{DD}} - \beta\left(A - \frac{(\alpha-1)}{\mu_n C_{\text{ox}}}m\,(W/L)\right)\right]^2 \cdot \frac{W}{L}. \tag{8.16}$$

$$\eta_{\text{active,osc}} = \frac{m\,(W/L)\left[V_{\text{DD}} - \beta\left(A - \frac{(\alpha-1)}{\mu_n C_{\text{ox}}}m\,(W/L)\right)\right]^2}{\alpha\mu_n C_{\text{ox}}\,(V_{\text{DD}} - V_{\text{th}})^2\, V_{\text{DD}}}, \tag{8.17}$$

$$\text{FoM}_A = \frac{m\,(W/L)^2\left[V_{\text{DD}} - \beta\left(A - \frac{(\alpha-1)}{\mu_n C_{\text{ox}}}m\,(W/L)\right)\right]^4}{\alpha^2\mu_n C_{\text{ox}}\,(V_{\text{DD}} - V_{\text{th}})^2\, V_{\text{DD}}} \cdot \frac{W}{L}. \tag{8.18}$$

The first term in Eq. (8.18) decreases and the second term in Eq. (8.18) increases with a larger W/L. Hence, there exists a highest FoM_A at an optimum W/L. In the design, all the active devices maintain a minimum transistor length $L = 60$ nm. Based on simulation fitting, $\alpha = 1.75$ and $\beta = 0.25$. Figure 8.8 presents and compares the output power and DC-to-RF efficiency from model analytical results and simulation results. Both results match well and show that a larger-size transistor yields a larger output power and a lower power efficiency.

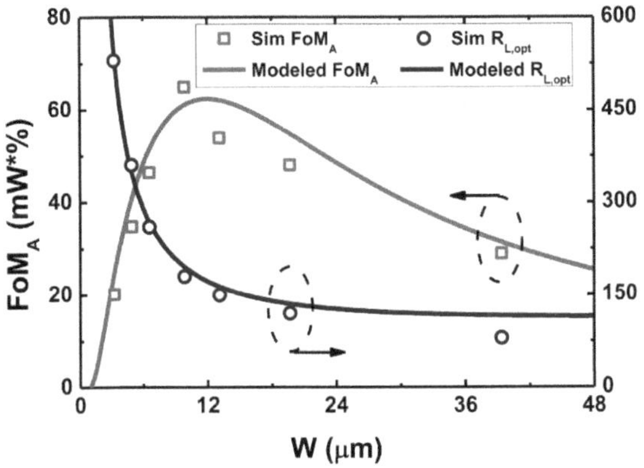

Figure 8.9 Simulated optimized R_L and FoM_A of the oscillator versus device width W.

The analytically derived FoM_A versus device width W is plotted in Fig. 8.9, which also matches well with the simulated results. The best FoM_A appears when the transistor width is around 9.6 to 12 μm. Besides, Fig. 8.9 also plots the required $R_{L,opt}$ versus device width W. When the transistor goes smaller, the required $R_{L,opt}$ grows larger, leading to smaller output power. In addition, high resistance such as more than 400 Ω is difficult to be matched to a standard 50 Ω load. While at the best FoM_A, the optimized load is as large as 150 to 200 Ω, which is also three to four times larger than the standard 50 Ω. So normally the 50 Ω load, such as antennas, is hard to directly connect to the core transistor-pair.

8.4.3 Transformer-Based Impedance Optimization

To solve the impedance matching issue, instead of directly extracting power from the core transistor pair, the proposed cross-coupled oscillator uses a transformer-based output network to realize the impedance conversion, as shown in Fig. 8.10a. The transformer is also used as the tuning inductor for the LC tank. As shown in Fig. 8.10b, the primary and secondary coil inductance of the transformer are L_1 and L_2, and mutual inductance and coupling coefficient are M and k, respectively. By using Kirchhoff laws to analyze the circuit in Fig. 8.10b, the input impedance, Z_1, seen from

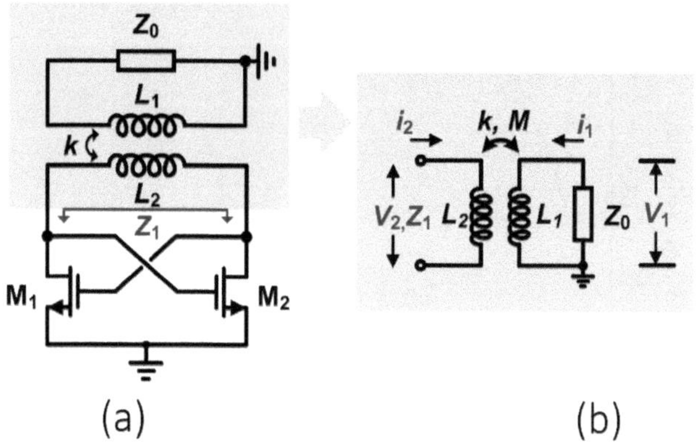

Figure 8.10 Schematic of the proposed oscillator of (a) topology of the oscillator with transformer-based impedance boosting network, and (b) equivalent circuit of the transformer-based impedance boosting network with the load.

V_2 is derived as

$$Z_1 = \frac{V_2}{I_2} = J\,\omega\left(L_2 - \frac{\omega^2 M^2 L_1}{\omega^2 L_1^2 + Z_0^2}\right) + \frac{\omega^2 M^2}{\omega^2 L_1^2 + Z_0^2} Z_0. \qquad (8.19)$$

So the L_s and R_s as shown in Fig. 8.11 are expressed as

$$L_s = \frac{1}{2}\left(L_2 - \frac{\omega^2 M^2 L_1}{\omega^2 L_1^2 + Z_0^2}\right), \qquad (8.20)$$

$$R_s = \frac{1}{2}\frac{\omega^2 M^2}{\omega^2 L_1^2 + Z_0^2} Z_0, \qquad (8.21)$$

$$Q = \frac{\omega L_s}{R_s}, \qquad (8.22)$$

where Q is the Q-factor of L_s. Hence, the equivalent parallel L_p and R_p shown in Fig. 8.11 are derived as

$$L_p \approx L_s = \frac{L_2}{2}\left(1 - \frac{k^2 \omega^2 L_1^2}{\omega^2 L_1^2 + Z_0^2}\right), \qquad (8.23)$$

$$R_p = Q^2 R_s = \frac{1}{2} \cdot \frac{L_2}{k^2 L_1} \cdot \frac{\left((\omega^2 L_1^2 + Z_0^2) - k^2 \omega^2 L_1^2\right)^2}{Z_0^2\left(\omega^2 L_1^2 + Z_0^2\right)} \cdot Z_0. \qquad (8.24)$$

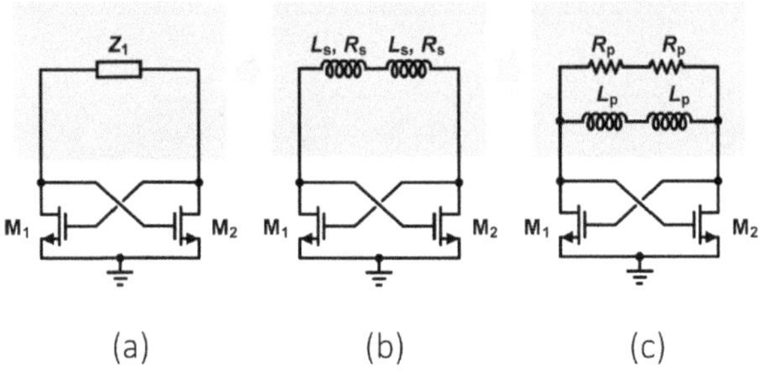

Figure 8.11 Equivalent circuit of the proposed oscillator with transformer-based load represented (a) by Z_1, (b) by L_s and R_s, and (c) by L_p and R_p.

As shown in Eq. (8.24), compared with directly loading the oscillator with an impedance Z_0, this approach boosts the impedance by $L_2(\omega^2 L_1^2 + Z_0^2)/(2k^2 L_1 Z_0^2)$ times, defined as the boosting index. By choosing a proper value for each parameter, the boosting index can be larger than one to convert 50 Ω to the designated larger value.

The insertion loss is a key performance figure of the transformer. A useful choice of the figure of merit for the transformer is the minimum insertion loss ($IL_{\min}$) [74], which is defined as the insertion loss when the transformer is properly matched with its input and output circuitry. The $IL_{\min}$ is expressed as [74–77],

$$I L_{\min}\,(\text{dB}) = -10\log\left[1+2\left(X-\sqrt{X^2+X}\right)\right], \tag{8.25}$$

where $x = 1/(k^2 Q_p Q_s)$, k is the coupling coefficient, Q_p and Q_s are the Q-factor of the primary and secondary coil inductors, respectively. Assuming $Q_p = Q_s = Q_T$ and $kQ_T >> 1$, the term $IL_{\min}$ is approximately simplified as $IL_{\min}(\text{dB}){=}{-}10\log(1{-}2/(kQ_T))$, which means smaller k or Q_T results in a larger insertion loss. For the well-matched circuit, the insertion loss of the output transformation network, $H(R_L)$, as defined in Eq. (8.2), equals the minimum insertion loss $IL_{\min}$, which can be expressed as

$$H\,(R_L) = I L_{\min} \approx 1 - \frac{2}{kQ_T}. \tag{8.26}$$

As shown in Eq. (8.24), to get a better boosting index, k needs to be small and the ratio of L_2/L_1 needs to be large. Due to these

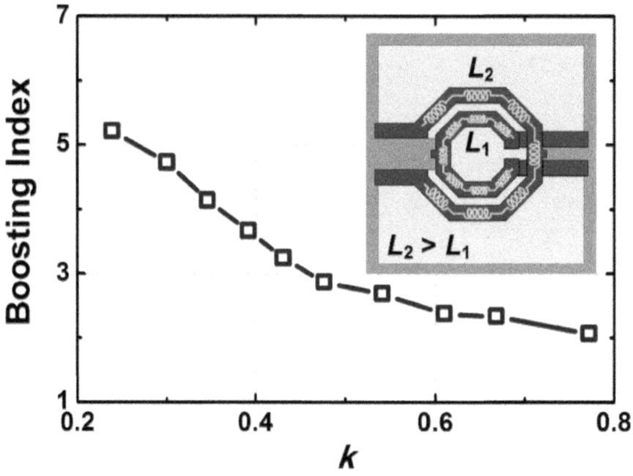

Figure 8.12 Top view of the transformer, where the outside coil is L_2 and inner coil is L_1, and the simulated boosting index of the transformer versus different coupling coefficient k.

two main restrictions, the topology of the transformer is designed as shown in Fig. 8.12, in which the radius of L_2 is larger than L_1, making the inductance of L_2 larger than L_1. Also, the horizontally weak coupling generates a small k. However, a smaller k enlarges the transformer insertion loss as illustrated in Eq. (8.26). Figure 8.12 shows the simulated boosting index and Fig. 8.13a shows the transformer insertion loss versus the coupling coefficient k. The most advantage to implement the proposed transformer-based output network is that it realizes large boosting index while only scarifies small power degradation. For example, this transformer-based output network can convert 50 to 200 Ω, which means the boosting index is 4, while with less than 1.5 dB simulated insertion loss. Second, the proposed matching network is based on a transformer which also constructs the LC tank to remove additional components for chip area reduction. For impedance matching, the R_{p} needs to be equaled as $R_{\mathrm{L,opt}}$ shown in Eq. (8.12) as

$$R_{\mathrm{p}} = R_{\mathrm{L,opt}} \tag{8.27}$$

Substituting Eqs. (8.24) and (8.27) into Eq. (8.26), the $H(R_{\mathrm{L}})$ with the relation of W/L is derived as

$$H\left(R_{\mathrm{L}}\right) = 1 - \frac{2C}{Q_{\mathrm{T}}}\sqrt{R_{\mathrm{L,opt}}} = 1 - \frac{2C}{Q_{\mathrm{T}}}\sqrt{\frac{\alpha L}{m\left(W/L\right)W}}, \tag{8.28}$$

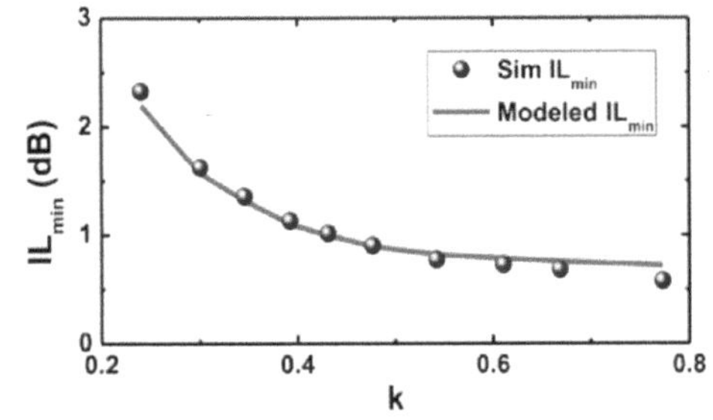

(a)

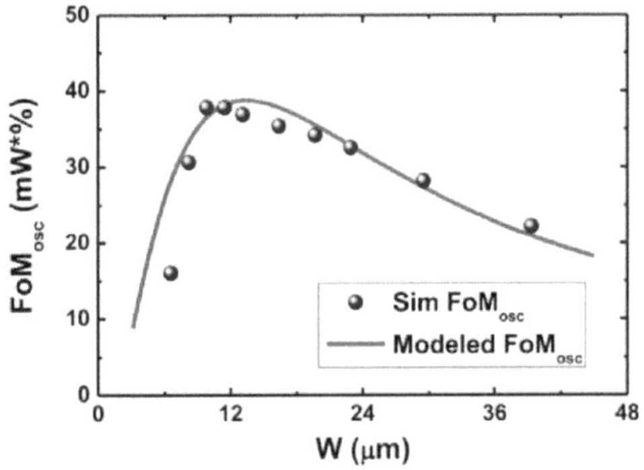

(b)

Figure 8.13 Simulated and analytical results of (a) $IL_{\min}$ of the transformer versus different coupling coefficient k and (b) $\mathrm{FoM}_{\mathrm{osc}}$

where $C = (2L_1 Z_0/(L_2(\omega^2 L_1^2 + Z_0^2)))^{1/2}$, $m(W/L)$ and α are defined in Eq. (8.12). Hence, the $\mathrm{FoM}_{\mathrm{osc}}$ that was defined in Eq. (8.1) is written as

$$\mathrm{FoM}_{\mathrm{osc}} = \mathrm{FoM}_{\mathrm{A}} \left(1 - \frac{2C}{Q_{\mathrm{T}}} \sqrt{\frac{\alpha L}{m\,(W/L)\,W}}\right). \tag{8.29}$$

In Eq. (8.29), the $\mathrm{FoM}_{\mathrm{osc}}$ is written as a formula with the parameter of W/L and plotted versus device width W in Fig. 8.13b with a comparison of the simulated $\mathrm{FoM}_{\mathrm{osc}}$. As shown in Fig. 8.13b, to get the best $\mathrm{FoM}_{\mathrm{osc}}$, by fixing the transistor length as its minimum value of 60 nm, the results show the optimized transistor width is around 8.4–12 μm. By a trade-off between the boosting index and insertion loss, a transformer with inter diameters of 22/34 μm

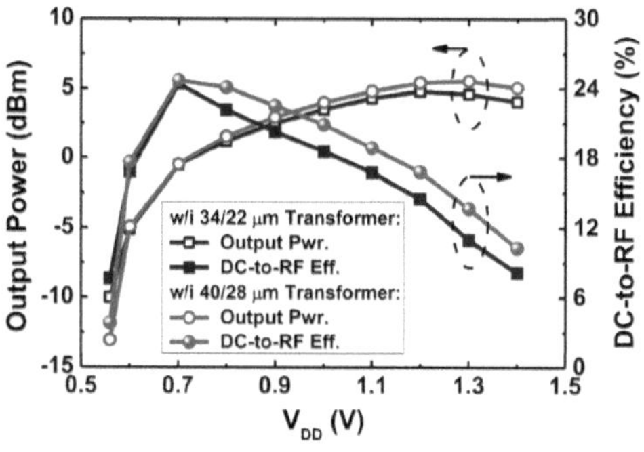

Figure 8.14 Simulated output power and DC-to-RF efficiency of the proposed oscillator with different size of the impedance boosting transformer.

and 28/40 μm for the primary/secondary coil can be selected for optimization. Figure 8.14 shows the simulated efficiency and output power of the oscillator with the different sizes of transformers. The two oscillators oscillate at 165 and 177 GHz with 28/40 and 22/34 μm diameter transformers, respectively. At 1 V power supply, both simulation results achieve better than 15% DC-to-RF efficiency with better than 3.5 dBm output power.

8.5 Conclusion

With the scaling of silicon-based integrated circuits technologies, silicon-based sub-THz/THz circuits and systems have been attracting increasing interest from both academia and industry. As one of the key components in the sub-THz/THz systems, sub-THz/THz oscillators are sensitive to operating frequency, output power and efficiency, and every step of improvement on output power or DC-to-RF efficiency is plausible and brings closer to their deployment in real life applications. This chapter demonstrates an example to use a holistic optimization method to achieve very good sub-THz signal generation by taking into consideration of both the oscillator active device and passive output component. The holistic design methodology and philosophy should be able to extend to broader domains and complex systems, such as package/antenna circuit co-design, channel-aware-based system optimization.

Acknowledgement

The authors would like to thank the National Science Foundation for partial funding support.

References

1. B. Yu, et al., High-efficiency micromachined sub-THz channels for low-cost interconnect for planar integrated circuits, *IEEE Trans. Microw. Theory Techn.*, vol. 64, no. 1, pp. 96–105, Jan. 2016.
2. Q. J. Gu, THz interconnect: the last centimeter communication, *IEEE Commun. Mag.*, vol. 53, no. 4, pp. 206–215, Apr. 2015.
3. Q. Jane Gu, Sub-THz/THz interconnect, complement to electrical and optical interconnects, *Fall 2020 IEEE Solid-State Circuits Magazine.*
4. A. Singh, et al., A D-band radio-on-glass module for spectrally-efficient and low-cost wireless backhaul, *IEEE Radio Freq. Integr. Circuits Symp. (RFIC)*, pp. 99–102, Aug. 2020.
5. B. P. Ginsburg, et al., A 160 GHz pulsed radar transceiver in 65 nm CMOS, *IEEE J. Solid-State Circuits*, vol. 49, no. 4, pp. 984–995, Apr. 2014.
6. N. Sharma et al., 85-to-127 GHz CMOS transmitter for rotational spectroscopy, *Proc. IEEE Custom Integr. Circuits Conf.*, pp. 1–4, Sep. 2014.
7. R. Guillaume, et al., Energy efficient distributed-oscillators at 134 and 202 GHz with phase-noise optimization through body-bias control in 28 nm CMOS FDSOI technology, *IEEE Radio Freq. Integr. Circuits Symp. (RFIC)*, pp. 156–159, Jun. 2017.
8. Y. Tousi and E. Afshari, A high-power and scalable 2-D phased array for terahertz CMOS integrated systems, *IEEE J. Solid-State Circuits* , vol. 50, no. 2, pp. 597–609, Feb. 2015.
9. O. Momeni and E. Afshari, High power terahertz and millimeter-wave oscillator design: a systematic approach, *IEEE J. Solid-State Circuits*, vol. 46, no. 3, pp. 583–597, Mar. 2011.
10. S. P. Voinigescu, et al., A study of SiGe HBT signal sources in the 220–330-GHz range, *IEEE J. Solid-State Circuits*, vol. 48, no. 9, pp. 2011–2021, Sep. 2013.
11. N. Landsberg and E. Socher, 240 GHz and 272 GHz fundamental VCOs using 32 nm CMOS technology, *IEEE Trans. Microw. Theory Techn.*, vol. 61, no. 12, pp. 4461–4471, Dec. 2013.

12. K. Sengupta and A. Hajimiri, A 0.28 THz power-generation and beam-steering array in CMOS based on distributed active radiators, *IEEE J. Solid-State Circuits*, vol. 47, no. 12, pp. 3013–3031, Dec. 2012.
13. C.–H. L, et al., A 37.5 mW 8 dBm-EIRP 15.5o-HPBW 338 GHz terahertz transmitter using SoP heterogeneous system integration, *IEEE Trans. Microw. Theory Techn.*, vol. 63, no. 2, pp. 470–480, Feb. 2015.
14. W. Volkaerts, M. Steyaert, and P. Reynaert, 118 GHz fundamental VCO with 7.8% tuning range in 65 nm CMOS, *IEEE RFIC Symp. Dig.*, June 2011.
15. Z. Wang, et al., A CMOS 210 GHz fundamental transceiver with OOK modulation, *IEEE J. Solid-State Circuits*, vol. 49, no. 3, pp. 564–580, Mar. 2014.
16. W. Volkaerts, et al., An FSK plastic waveguide communication link in 40 nm CMOS, *ISSCC Dig. Tech. Papers*, pp. 177–178, Feb. 2015.
17. N. Ono, et al., 135 GHz 98 mW 10 Gbps ASK transmitter and receiver chipset in 40 nm CMOS, *IEEE Symp. VLSI Circuits*, pp. 50–51, June 2012.
18. N. Weissman, et al., A packaged 106–110 GHz Bi-directional 10 Gbps 0.11 pJ/bit/cm CMOS transceiver, *IEEE Int'l Microw. Symp. Dig.*, May 2015.
19. X. Li, W. Chen, Y. Wang, and Z. Feng, A 160 GHz high output power and high DC-to-RF efficiency fundamental oscillator in a 130 nm SiGe BiCMOS process, *2020 50th European Microwave Conference*, pp. 1159–1162, Jan 2021.
20. H. Khatibi, S. Khiyabani, and E. Afshari, An efficient high-power fundamental oscillator above fmax/2: a systematic design, *IEEE Trans. Microw. Theory Techn.*, vol. 65, no. 11, pp. 4176–4189, Nov. 2017.
21. T. D. Nguyen, J.-P. Hong, A high fundamental frequency sub-THz CMOS oscillator with a capacitive load reduction circuit, *Microwave Theory Tech. IEEE Trans.*, vol. 68, no. 7, pp. 2655–2667, 2020.
22. H. Khatibi, et al., A 195 GHz single-transistor fundamental VCO with 15.3% DC-to-RF efficiency, 4.5 mW output power, phase noise FoM of –197 dBc/Hz and 1.1% tuning range in a 55 nm SiGe process, *IEEE Radio Freq. Integr. Circuits Symp. (RFIC)*, pp. 152–155, Jun. 2017.
23. H. T. Kwon, D. Nguyen, and J. P. Hong, A 219-GHz fundamental oscillator with 0.5 mW peak output power and 2.08% DC-to-RF efficiency in a 65 nm CMOS, *IEEE MTT-S Int. Microw. Symp. Dig.*, pp. 1–3, May 2016.
24. Y. Ye, B. Yu, and Q. J. Gu, A 165-GHz transmitter with 10.6% Peak DC-to-RF efficiency and 0.68 pJ/b energy efficiency in 65 nm bulk CMOS, *IEEE Trans. Microw. Theory Techn.*, vol. 64, no. 12, pp. 4573–4584, Dec. 2016.

25. H. Koo, et al., Design and Analysis of 239 GHz CMOS push-push transformer-based VCO with high efficiency and wide tuning range, *IEEE Trans. Circuits Syst. I, Reg. Papers*, vol. 62, no. 7, pp. 1883–1893, July 2015.
26. P.-H. Chiang, et al., A 206∼220 GHz CMOS VCO using body-bias technique for frequency tuning, *IEEE Int'l Microw. Symp. Dig.*, May 2015.
27. P.-Y. Chiang, O. Momeni, and P. Heydari, A 200 GHz inductively tuned VCO with –7 dBm output power in 130 nm SiGe BiCMOS, *IEEE Trans. Microw. Theory Tech.*, vol. 61, no. 10, pp. 3666–3673, Oct. 2013.
28. M. Adnan and E. Afshari, A 247-to-263.5 GHz VCO with 2.6 mW peak output power and 1.14% DC-to-RF efficiency in 65 nm bulk CMOS, *ISSCC Dig. Tech. Papers*, pp. 262–263, Feb. 2014.
29. R. Han and E. Afshari, 260 GHz broadband source with 1.1 mW continuous-wave radiated power and EIRP of 15.7 dBm in 65 nm CMOS, *ISSCC Dig. Tech. Papers*, pp. 138–139, Feb. 2013.
30. B. Khamaisi, S. Jameson, and E. Socher, A 210–227 GHz transmitter with integrated on-chip antenna in 90 nm CMOS technology, *IEEE Trans. THz Sci. Techn.*, vol. 3, no. 2, pp. 141–150, Mar. 2013.
31. S. Jameson, et al., High efficiency 293 GHz radiating source in 65 nm CMOS, *IEEE Microw. Wireless Compon. Lett.*, vol. 24, no. 7, pp. 463–465, Jul. 2014.
32. D. Huang, et al., Terahertz CMOS frequency generator using linear superposition technique, *IEEE J. Solid-State Circuits*, vol. 43, no. 12, pp. 2730–2738, Dec. 2008.
33. U. R. Pfeiffer, Silicon CMOS/SiGe transceiver circuits for THz applications, *Proc. IEEE 12th Topical Meeting Silicon Monolithic Integr. Circuits RF Syst.*, pp. 159–162, Jan. 2012.
34. Y. Ye, et al., A high efficiency E-band CMOS frequency doubler with a compensated transformer-based balun for matching enhancement, *IEEE Microw. Wireless Compon. Lett.*, vol. 26, no. 1, pp. 40–42, Jan 2016.
35. M. Abbasi, et al., Single-chip frequency multiplier chains for millimeter-wave signal generation, *IEEE Trans. Microw. Theory. Tech.*, vol. 57, no. 12, pp. 3134–3141, Dec. 2009.
36. S. Kang, S. V. Thyagarajan, and A. M. Niknejad, A 240 GHz wideband QPSK transmitter in 65 nm CMOS, *IEEE RFIC Symp. Dig.*, June 2014.
37. S. Jameson, E. Halpern, and E. Socher, A 300 GHz wirelessly locked 2 × 3 array radiating 5.4 dBm with 5.1% DC-to-RF efficiency in 65 nm CMOS, *2016 IEEE International Solid-State Circuits Conference (ISSCC)*, pp. 348–349, 2016.

38. R. Kananizadeh and O. Momeni, High-power and high-efficiency millimeter-wave harmonic oscillator design, exploiting harmonic positive feedback in CMOS, *IEEE Trans. Microw. Theory. Tech.*, vol. 65, no. 10, pp. 3922–3936, Oct. 2017.
39. H. Jalili and O. Momeni, A 230-GHz high-power and wideband coupled standing wave VCO in 65 nm CMOS, *IEEE J. Solid-State Circuits*, vol. 55, no. 3, pp. 547–556, March 2020.
40. A. Nikpaik, et al., A 219-to-231 GHz frequency-multiplier-based VCO with ~3% peak DC-to-RF efficiency in 65 nm CMOS, *IEEE J. Solid-State Circuits*, vol. 53, no. 2, pp. 389–403, Feb. 2018.
41. F. Ahmed, et al., 0.3-THz SiGe-based high-efficiency push–push VCOs with > 1 mW peak output power employing common-mode impedance enhancement, *IEEE Trans. Microw. Theory. Tech.*, vol. 66, no. 3, pp. 1384–1398, March 2018.
42. H. Rücker, et al., A 0.13 um SiGe BiCMOS technology featuring of 240/330 GHz and gate delays below 3 ps, *IEEE J. Solid-State Circuits*, vol. 45, no. 9, pp. 1678–1686, Sep. 2010.
43. Y. M. Tousi, et al., A 283-to-296 GHz VCO with 0.76 mW peak output power in 65 nm CMOS, *ISSCC Dig. Tech. Papers*, pp. 286–288, Feb. 2011.
44. H. Koo, C.-Y. Kim, and S. Hong, A G-band standing-wave push-push VCO using a transmission-line resonator, *IEEE Trans. Microw. Theory Tech.*, vol. 63, no. 3, pp. 1036–1045, Mar. 2015.
45. B. Khamaisi and E. Socher, A 209–233 GHz frequency source in 90 nm CMOS technology, *IEEE Microw. Wireless Compon. Lett.*, vol. 22, no. 5, pp. 260–262, May 2012.
46. N. Deferm and P. Reynaert, A 120 GHz 10 Gb/s phase-modulating transmitter in 65 nm LP CMOS, *2011 IEEE Int. Solid-State Circuits Conf. Dig. Tech. Papers*, pp. 290–292, Feb. 2011.
47. Y. Yang, et al., A 155 GHz 20 Gbit/s QPSK transceiver in 45 nm CMOS, *IEEE RFIC Symp.*, pp. 365–368, Jun. 2014.
48. J. Park, S. Kang, S. Thyagarajan, E. Alon, and A. M. Niknejad, A 260 GHz fully integrated CMOS transceiver forwireless chip-to-chip communication, *IEEE VLSI Circuits Symp.*, pp. 48–49, Jun. 2012.
49. S. Jameson, E. Halpern, and E. Socher, Sub-Harmonic Wireless Locking of a THz Radiating On-chip Source in 65 nm CMOS, *10th European Microwave Integrated Circuit Conference (EuMIC)*, pp. 105–107, 2015.
50. S. Jameson, E. Halpern, and E. Socher, A 300 GHz wirelessly locked 2 × 3 array radiating 5.4 dBm with 5.1% DC-to-RF efficiency in 65 nm CMOS, *ISSCC Dig. Tech. Papers*, pp. 348–349, Feb. 2016.

51. R. Kananizadeh and O. Momeni, A 190.5 ghz mode-switching VCO with 20.7% continuous tuning range and maximum power of −2.1 dBm in 0.13 μm BiCMOS, *ISSCC Dig. Tech. Papers*, pp. 46–47, Feb. 2016.
52. K. K. Tokgoz, et al., A 56 Gb/s W-band CMOS wireless transceiver, *ISSCC Dig. Tech. Papers*, pp. 242–243, Feb. 2016.
53. K. Katayama, et al., A 300 GHz 40 nm CMOS transmitter with 32-QAM 17.5 G/s/ch capability over 6 channels, *ISSCC Dig. Tech. Papers*, pp. 342–343, Feb. 2016.
54. Y. Ye, B. Yu, and Q. J. Gu, A 165 GHz OOK transmitter with 10.6% peak DC-to-RF efficiency in 65 nm bulk CMOS, *IEEE MTT-S Int. Microwave Symposium (IMS)*, May 2016.
55. K. Wu and M. Hella, A 103 GHz voltage controlled oscillator with 28% tuning range and 4.2 dBm peak output power using SiGe BiCMOS technology, *IEEE MTT-S Int. Microwave Symposium*, pp. 606–609, 2018.
56. J. Yoo, H. Son, J. Kim, D. Kim, and J.-S. Rieh, A 90 GHz high DC-to-RF efficiency VCO with multi-way transformers in 65 nm CMOS, *IEEE Inter-National Symposium Radio-Frequency Integration Technol.*, pp. 25–27, 2020.
57. M. Adnan and E. Afshari, A 105 GHz VCO with 9.5% tuning range and 2.8 mW peak output power in a 65 nm bulk CMOS process, *IEEE Trans. Microw. Theory Techn.*, vol. 62, no. 4, pp. 753–762, April 2014.
58. D. R. Utomo, D. Park, S. Lee, and J. Hong, High-power 268 GHz push-push transformer-based oscillator with capacitive degeneration, *IEEE Microw. Wireless Compon. Lett.*, vol. 28, no. 7, pp. 612–614, July 2018.
59. H. Koo, et al., A 254 GHz CMOS transmitter with VCO-Q modulation, *IEEE Microw. Wireless Compon. Lett.*, vol. 26, no. 6, pp. 458–460, June 2016.
60. C. S. Hsieh and J. Y. Liu, A low phase noise 210 GHz triple-push ring oscillator in 90 nm CMOS, *IEEE Trans. Microw. Theory Techn.*, vol. 66, no. 4, pp. 1983–1997, April 2018.
61. S. Khiyabani, H. Khatibi, and E. Afshari, A compact 275 GHz harmonic VCO with −2.6 dBm output power in a 130 nm SiGe process, *2019 IEEE Custom Integrated Circuits Conference (CICC)*, pp. 1–4, 2019.
62. D. R. Utomo, D.-W. Park, J.-P. Hong, and S.-G. Lee, A 264 GHz, 2.3 dBm push-push transformer-based hartley oscillator, *IEEE Microw. Wireless Compon. Lett. (Early Access)*, May 2021.
63. X. Liu and H. C. Luong, A fully integrated 0.27 THz injection-locked frequency synthesizer with frequency-tracking loop in 65 nm CMOS, *IEEE J. Solid-State Circuits*, vol. 55, no. 4, pp. 1051–1063, April 2020.

64. X. Liu and H. C. Luong, A 170-GHz 23.7% tuning-range CMOS injection-locked LO generator with third-harmonic enhancement, *IEEE Trans. Microw. Theory Techn.*, vol. 68, no. 7, pp. 2668–2678, July 2020.
65. S. Kang, J. Chien, and A. M. Niknejad, A W-band low-noise PLL with a fundamental VCO in SiGe for millimeter-wave applications, *IEEE Trans. Microw. Theory Techn.*, vol. 62, no. 10, pp. 2390–2404, Oct. 2014.
66. N. Pohl, H. Rein, T. Musch, K. Aufinger, and J. Hausner, SiGe Bipolar VCO With Ultra-Wide Tuning Range at 80 GHz Center Frequency, *IEEE J. Solid-State Circuits*, vol. 44, no. 10, pp. 2655–2662, Oct. 2009.
67. D. R. Utomo, et al., A 293/440 GHz push-push double feedback oscillators with 5.0/−3.9 dBm output power and 2.9/0.6% DC-to-RF efficiency in 65 nm CMOS, *2020 IEEE Symposium on VLSI Circuits*, pp. 1–2, 2020.
68. T. D. Nguyen and J.-P. Hong, igh power and high frequency CMOS oscillator with source-to-drain coupling and capacitive load reduction circuit, *IEEE Access*, vol. 8, pp. 138584–138594, 2020.
69. J. Yun, et al., A 280 GHz 10 dBm signal source based on InP HBT technology, *IEEE Microw. Wireless Compon. Lett.*, vol. 27, no. 2, pp. 159–161, Feb. 2017.
70. J. Yun, D. Yoon, H. Kim, and J. S. Rieh, 300 GHz InP HBT oscillators based on common-base cross-coupled topology, *IEEE Trans. Microw. Theory Techn.*, vol. 62, no. 12, pp. 3053–3064, Dec. 2014.
71. D. Yoon, J. Yun, and J.-S. Rieh, A 310–340 GHz coupled-line voltage-controlled oscillator based on 0.25 μm InP HBT technology, *IEEE Trans. THz Sci. Technol.*, vol. 5, no. 4, pp. 652–654, Jul. 2015.
72. T.-P. Wang and S.-Y. Wang, Frequency-tuning negative-conductance boosted structure and applications for low-voltage low-power wide tuning-range VCO, *IEEE Trans. Very Large Scale Integr. (VLSI) Syst.*, vol. 23, no. 6, pp. 1137–1144, Jun. 2015.
73. R. Levinger, O. Katz, J. Vovnoboy, R. B. Yishay, and D. Elad, A K-band low phase noise and high gain G_m boosted Colpitts VCO for 76–81 GHz FMCW radar applications, *IEEE Int. Microwave Symposium (IMS)*, May 2016.
74. K. T. Ng, et al., Substrate effects in monolithic RF transformers on silicon, *IEEE Trans. Microw. Theory Techn.*, vol. 50, no. 1, pp. 377–383, Jan. 2002.
75. K.-J. Kim, T. H. Lim, and K. H. Ahn, The novel high efficiency on chip transformers for the CMOS power amplifier, *Proceedings of the 2009 12th International Symposium on Integrated Circuits, ISIC'09*, pp. 401–404, 2009.

76. O. E. Gharniti, et al., Modeling and characterization of on-chip transformers for silicon RFIC, *IEEE Trans. Microw. Theory Techn.*, vol. 55, no. 4, pp. 607–615, Apr. 2007.

77. L. F. Tiemeijer, et al., Analysis, design, modeling, and characterization of low-loss scalable on-chip transformers, *IEEE Trans. Microw. Theory Techn.*, vol. 61, no. 7, pp. 2545–2557, July 2013.

Chapter 9

Resonant Tunneling Diode (RTD) THz Sources

Safumi Suzuki and Masahiro Asada
Department of Electrical and Electronic Engineering,
Tokyo Institute of Technology,
O-okayama 2-12-1, Meguro-ku, Tokyo 152-8552, Japan
safumi@ee.e.titech.ac.jp

Traditionally, terahertz (THz) applications have been developed using photonic-based systems with THz time-domain spectroscopy, parametric light sources, and single traveling carrier photodiodes. However, compact THz sources using semiconductor electronic/optical devices have recently been developed [1], and a highly compact THz application system is expected to be realized in the near future. Among such electronic THz sources, resonant tunneling diodes (RTDs) have long been known for their high-frequency operation. Initially, a high-frequency oscillator was developed using waveguide-type resonators, and in 1991, 712 GHz fundamental oscillation was achieved [2]. Research on planar integration of the oscillator circuit was then conducted, and a 650 GHz fundamental oscillation was obtained [3]. Subsequently, further efforts to develop RTD oscillators have been conducted. Since the frequency was updated in 2009 [4], it has been continuously improved [5–12], and a fundamental oscillation frequency of approximately 2 THz has

Advances in Terahertz Source Technologies
Edited by Gun-Sik Park, Masahiko Tani, Jae-Sung Rieh, and Sang Yoon Park

ISBN 978-981-4968-89-8 (Hardcover), 978-1-003-45967-5 (eBook)
www.jennystanford.com

been achieved, which is the highest oscillation frequency among room-temperature electronic single oscillators [12]. A milliwatt-class output power was also obtained at approximately 1 THz with a large-scale oscillator array [13]. Alongside a high-frequency and high output power, various characteristics, including wide frequency-tuning and high-frequency modulation, have been developed. Utilizing these characteristics, a wide range of applications, such as wireless communication, imaging, radar, sensing, and analysis, have been conducted using RTD devices. This chapter summarizes the characteristics and functions of RTD oscillators and their applications.

9.1 Introduction

The THz band, which has a frequency of approximately 0.1 to 10 THz, has various applications, such as imaging, chemistry and biotechnology analysis, and communications [14]. Compact solid-state THz sources are important for realizing such applications in general use, and various THz sources have been studied using both optical and electronic devices.

Figure 9.1 shows the current state of various compact semiconductor THz sources that can be integrated on a semiconductor chip and operated without the need for very low-temperature conditions.

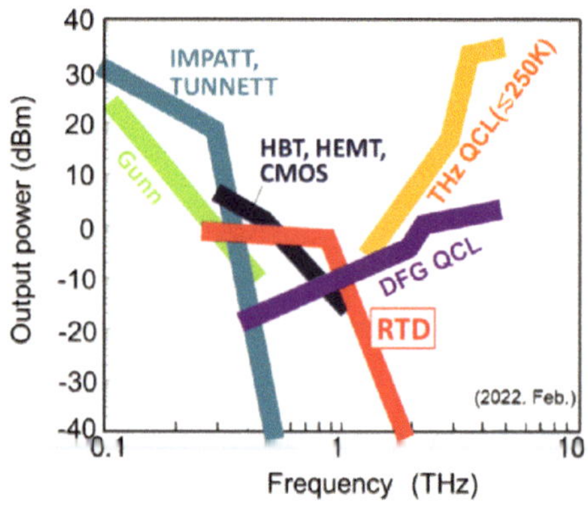

Figure 9.1 Current state of compact semiconductor THz signal sources.

Regarding optical devices, a quantum cascade laser (QCL) has been studied, exhibiting a continuously increasing operating temperature [15–17]. Recently, difference frequency generation (DFG) using mid-infrared QCLs has been reported [18, 19]. This device was integrated into a single chip, and THz waves were obtained by a DC power supply under room-temperature conditions. Regarding electronic devices, diodes; such as the impact ionization avalanche transit-time (IMPATT) diode, tunneling transit-time (TUNNETT) diode, and Gunn diode [20–22]; and transistors; such as the heterojunction bipolar transistor (HBT), high electron mobility transistor (HEMT), and complementary metal-oxide semiconductor (CMOS) [23–25]; have been studied as THz sources. Recently, the operating frequency of Si/SiGe circuits has increased remarkably, owing to the maturation of transistor characterization and circuit design techniques.

Resonant tunnel diodes (RTDs) are also candidates for room-temperature THz sources [26]. Previously, an oscillation of up to 1.98 THz was obtained at room temperature [12], and structures for high output power have been studied [13]. Research on several applications, such as wireless communication and sensing, has continuously developed. This chapter reviews recent progress in RTD oscillators and related applications. In Section 9.2, the operating principle of RTD oscillators and attempts to achieve high-frequency, high-power, and high functionality are described. Section 9.3 presents THz applications, including wireless communications, imaging, radar, and material analysis using RTD devices. Finally, Section 9.4 provides a summary of this chapter.

9.2 Characteristics of RTD Oscillators

9.2.1 Structure and Oscillation Principle

Generally, the epitaxial structures of RTDs used for THz oscillators are epigrown on a semi-insulating InP substrate and have an AlAs/InGaAs double barrier (or triple barrier) structure, as shown in Fig. 9.2a. Figure 9.2b shows the current-voltage characteristics and corresponding band structure of the conduction band. A quantum level is formed in the quantum well sandwiched by the

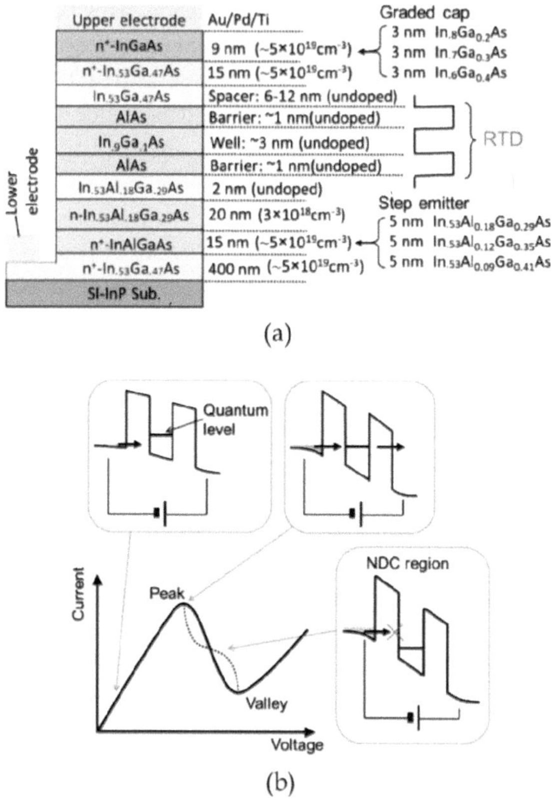

Figure 9.2 (a) RTD layer structure, and (b) current-voltage characteristics.

AlAs barriers. As the bias voltage increases, the energy of the electrons of the emitter matches the quantum level, and a tunneling current flows. When the bias voltage is further applied, the level drops below the bottom of the emitter conduction band edge, the electrons are unable to tunnel, and the current decreases. As a result, differential negative conductance (NDC) characteristics are obtained in the current-voltage characteristics, which are a gain for the THz oscillation. It should be noted that under the oscillation condition, owing to the nonlinearity of the current-voltage characteristics, self-bias current flows by rectification of its own oscillation signal, and the observed current-voltage characteristics in the NDC region change as shown by the dashed line in Fig. 9.2b [27]. The peak valley current ratio is approximately 2–4 in the RTDs used in THz oscillators. As the electrons move at a very high speed from the emitter to the collector and the response of the RTD is very fast, it is possible to form an oscillator in the THz frequency band by combining it with a resonator.

To date, high current densities of up to 50 mA/μm^2 have been obtained with a very thin barrier (0.9 nm) and well (2.5 nm) structure [11]. As a very high current flows to the diode and oscillator circuit and causes thermal destruction, the heat dissipation design for the oscillator structure is important. As well as InP-based RTDs, GaN-based RTDs have been developed recently, achieving stable operation at room temperature and a current density of more than 10 mA/μm^2 [28]. GaN is a well-known wide-bandgap material with high breakdown voltage. Depending on the optimization of the structure, it is possible to obtain current-voltage characteristics suitable for high-power operation.

The structure of the RTD oscillator, integrated with a slot antenna, and its equivalent circuit are shown in Figs. 9.3a,b, respectively. This is a typical oscillator structure, and the basic operation of the RTD oscillator can be explained using this structure. The RTD mesa is integrated into the slot antenna, and a metal-insulator-metal (MIM) capacitor is formed beside the antenna. As the MIM capacitor becomes a short circuit at a high frequency and an open circuit with a DC voltage, it is possible to form a slot antenna in the THz range and apply a DC voltage to the RTD simultaneously. The slot antenna acts as both a resonator and a radiator. A stabilization resistor is formed on the outside of the MIM in parallel with the RTD to cancel the NDC in the DC circuit. The RTD is represented

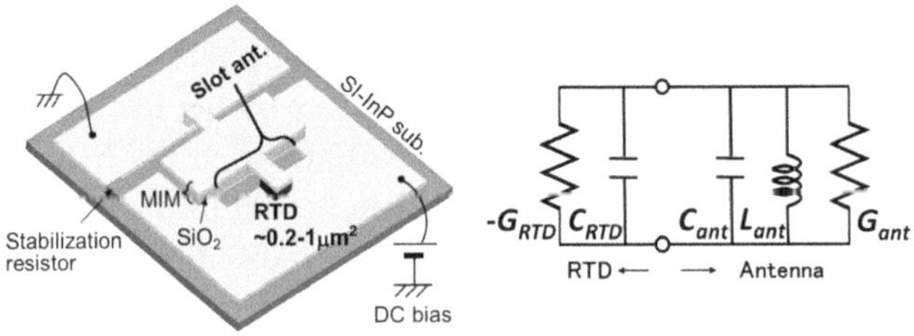

Figure 9.3 (a) RTD oscillator structure, and (b) its equivalent circuit.

as an equivalent circuit of the parallel connection between the NDC ($-G_{RTD}$) and the capacitance (C_{RTD}). The C_{RTD} consists of the depletion layer capacitance and additional capacitance owing to the electron transit-time. The antenna consists of an LC component (L_{ant} and C_{ant}) and conductance (G_{ant}) owing to the radiation and conduction losses. An oscillation is obtained when the NDC of the RTD exceeds the antenna loss, and the oscillation frequency is determined by the LC resonance of the antenna with the added RTD capacitance. Therefore, the oscillation frequency can be changed by changing the RTD area or antenna length. As the output is radiated toward the substrate side, owing to the high dielectric constant of the InP substrate, the device is mounted on a Si lens with a dielectric constant similar to that of InP, and the output is extracted from the lens.

Further, in the oscillator, the direct current and the high-frequency circuit are essentially separated by the MIM capacitor, but there is a drawback that the manufacturing process of the device becomes complicated. Therefore, a simplified structure has been proposed and demonstrated in which the MIM is removed and stabilizing resistors are placed at both ends of the slot [29]. This simple structure can employ another type of resonator; an oscillator integrated with a split-ring resonator, which is used in metamaterials [30]. These simple structures have another advantage in that the variation in characteristics is reduced owing to the reduction in the fabrication processes. A massive array can be formed by utilizing this small variation.

9.2.2 Toward High-Frequency and High-Power Oscillation

As mentioned previously, THz oscillations are obtained by canceling the antenna loss by the NDC. However, because the absolute value of the NDC decreases with frequency owing to the electron transit-time, there is an oscillation limit at which the NDC cannot compensate for antenna loss. Therefore, to obtain high-frequency oscillations, it is necessary to reduce the transit time and antenna loss. To date, the transit time has been reduced by thinning the quantum well and barrier layers, optimizing factors such as the depletion layer thickness, and the conductor loss has been

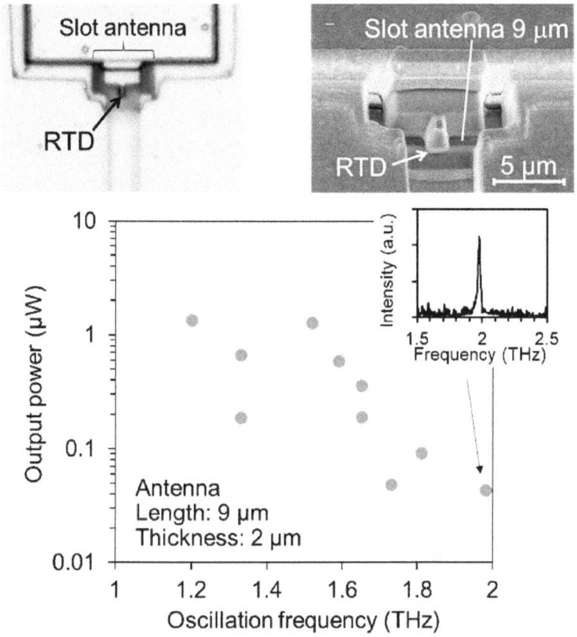

Figure 9.4 (a) Microphotograph and (b) SEM image of the 1.98 THz RTD oscillator, and (c) output power as a function of frequency and oscillation spectrum.

reduced by thickening the antenna electrode. Owing to these improvements, an oscillation frequency of 1.98 THz was obtained, which is the highest fundamental oscillation frequency among room-temperature semiconductor single oscillators (Fig. 9.4) [12, 31]. A cavity-type resonator can further reduce the conductor loss for high-frequency oscillations [31, 32], and oscillations up to approximately 3 THz are expected using the structure. High output power is also possible using the cavity-type resonator, as will be described subsequently. In addition, harmonic oscillators are effective for obtaining high frequencies. A 1.5 THz harmonic oscillation was achieved using the third harmonic component from three coupled 500 GHz oscillators [33].

The output of the RTD oscillator is maximized when the matching condition is satisfied, and the radiation conductance of the antenna is half the absolute value of the NDC [26]. Essentially, the center-fed slot antenna has a small radiation conductance, and thus, it is difficult to satisfy the matching condition. In an offset-fed structure in which the RTD was shifted from the center of the slot, the radiation conductance was high, the matching condition was satisfied, and an output of several 100 μW was obtained [34]. To further increase the output power, a rectangular cavity resonator structure was proposed and investigated. The cavity resonator has low loss and inductance characteristics that can increase the area of the RTD while maintaining the frequency, and an output power exceeding 1 mW is theoretically expected in the 1 THz range [35, 36]. A power combination using an oscillator array is also effective for high output power, and 0.73 mW was obtained at 1 THz by a large-scale 89-element array [13]. In this case, because the array elements were not synchronized due to the absence of a coupling structure between the arrayed elements, the oscillation spectrum became a multi-peak. Although the multi-peak spectrum is not suitable for communication applications, it is suitable for imaging applications where interference fringes become a problem. When the number of arrays is small, a coherent power combination is possible via mutual injection-locking with a coupling structure. An oscillator with two RTDs integrated on a coplanar line showed an output of over 1 mW at approximately 300 GHz [37]. Oscillators with two RTDs integrated on a ring-slot antenna or patch antenna have been reported, and these devices show coupled output power radiation with a good radiation pattern [38, 39]. A multi-element array with a coupling structure has been studied, and preliminary analyses and experiments for the operating modes in an arrayed oscillator have previously been conducted [40].

9.2.3 Functionality

In THz applications, such as wireless communications and imaging radar, various functions, such as high-speed modulation characteristics, narrow line width, and frequency-tuning characteristics, are required. The functionality of the RTD oscillator and efforts to add new functions are introduced in this section.

As described previously, a THz oscillation is obtained when biased to the NDC region, and the output is maximized approximately at the center of the NDC region. Therefore, intensity modulation is possible using bias voltage. The modulation frequency is limited by the MIM capacitor; however, a modulation cutoff frequency of approximately 30 GHz has already been achieved by optimizing the MIM capacitor size [41]. In addition to this intensity-modulation characteristic, a frequency-tuning characteristic can be obtained with a bias voltage because it changes the capacitance. The tuning frequency range was approximately 1–5% of the oscillation frequency. The tuning range was widened by the integration of the varactor diode, and a wide tuning range of 320 GHz was obtained for the 4-element array with different tuning ranges [42]. The spectral line width of the RTD oscillator was approximately 10 MHz in full width at half maximum and was determined by the phase noise originating from the shot noise of the flowing current in the RTD [43]. The spectral line width can be narrowed by a phase-locked loop technique, and a narrow line width of 1 Hz or less has been obtained [44, 45].

Injection-locking, which occurs when an external signal is injected into an oscillator, is a well-known phenomenon in the case of electrical oscillators. Injection-locking in RTD oscillators has been theoretically and experimentally investigated [46, 47]. Attempts have been made to add new functions utilizing this injection-locking phenomenon. The first attempt is the phase control of the oscillator. In the injection-locking state, the phase of the oscillator is determined by the frequency difference between the injection signal and free-running frequencies. By changing the free-running oscillation frequency by the bias voltage, the oscillator phase can be changed from $-\pi/2$ to $\pi/2$, and the experimental results are in agreement with the theoretical expectation [48]. If this phase control technique is applied to arrayed oscillators, a phased array can be realized. The second attempt is highly sensitive signal detection using the injection-locking phenomenon. The RTD can perform square-law detection using the second-order nonlinear component included in the current-voltage characteristics. However, when a large external signal is injected in the oscillation state and injection-locking occurs, square-law detection changes to synchronous detection, and a high responsivity is achieved [49, 50]. These high-responsivity characteristics have been utilized in high-capacity wireless communication. The third attempt is frequency comb generation. In the self-injection-locking state owing to external feedback, the frequency-tuning characteristic with bias voltage becomes discontinuous, and a comb-shaped spectrum appears at the discontinuity point [51]. This phenomenon is well-known in laser device technology and is called passive mode-locking. A novel function can be realized by combining photonic technology.

Many oscillators that integrate other antennas and resonators instead of slot antennas have been reported. Vivaldi [5] and tapered-slot antennas [52, 53] are integrated on thin substrates, which enable output radiation to the air without using a Si lens. This type of structure can be coupled to a photonic crystal waveguide. Features including external high-Q resonators, duplexers, high-directional antennas, and coupling with hollow waveguides have been reported using THz band photonic crystals, and even higher functionality has been added by integration with RTD oscillators [54]. Similarly, integrated patch antennas do not require a Si lens for output power radiation [37, 55, 56]. In addition, the integrated radial line-slot antenna is capable of circularly polarized radiation or optical vortex radiation with high directivity [57]. A new type of packaging using a small metal parabolic mirror was also developed for high-directive radiation without a Si lens [58].

9.3 Applications of RTD Oscillators

9.3.1 Wireless Communication

Utilizing the high-frequency oscillation characteristics and high functionality of the RTD oscillator, described in Section 9.2, wireless communication, imaging, radar, and analysis have been studied. Recently, the 5th generation mobile communication (5G) service has been implemented, and there is active research and development for the next generation (Beyond5G/6G) mobile communication service. The use of communications in the THz band is highlighted in Beyond 5G/6G white papers from various companies and institutions, and expectations are very high [59]. Therefore, a wireless communication experiment using RTD devices is introduced in this section.

As the output of the RTD oscillator changes with the bias voltage, the amplitude shift keying (ASK) or on-off keying (OOK) modulation schemes can be easily implemented by superimposing a modulation signal on the bias voltage, enabling simple and large-capacity THz wireless communication. In a typical THz wireless communication system using an RTD device, the RTD oscillator is intensity-modulated with the signal from a pulse pattern generator (PPG), and the modulated THz signal is received and demodulated by a Schottky barrier diode (SBD) receiver. The demodulated signal is amplified by a low-noise amplifier, and the data transmission quality is evaluated by an oscilloscope or a bit error rate tester.

Utilizing the high-frequency characteristics of RTDs, communication experiments have been conducted over a wide frequency range up to approximately 800 GHz. Recently, the transmission rate has rapidly increased owing to the optimization of modulation circuits. Using an RTD oscillator with a frequency of 650 GHz, an output power of 60 μW, and an SBD receiver, a transmission data rate of 44 Gbps was demonstrated with an error rate of (5×10^{-4}), which is below the forward error correction (FEC) limit [60]. Data transmission using frequency and polarization multiplexing techniques has also been achieved using a device chip that integrates four RTD oscillators with orthogonal polarization and two oscillation frequencies of 500 and 800 GHz (Fig. 9.5). In each of the two frequencies and polarizations, 28 Gbps transmissions were obtained, which is below the FEC limit [61]. In the 300 GHz band, data transmission capacities of 30 Gbps (error-free) and 56 Gbps (below the FEC limit) have been realized for communication using RTD devices for both the transmitter and receiver [49]. A combination of a UTC-PD transmitter and an RTD receiver succeeded in transmitting quadrature amplitude phase modulation (QAM) signals at 60 Gbps (below the FEC limit) [62] and pulse amplitude modulation (PAM) at 48 Gbps (below the FEC limit) [63]. In addition, 1 Gbps data transmission using circularly polarized waves by a radial line-slot array (RLSA) integrated device has also been reported [57]. Circularly polarized waves are useful for THz communication in mobile terminals because a communication link

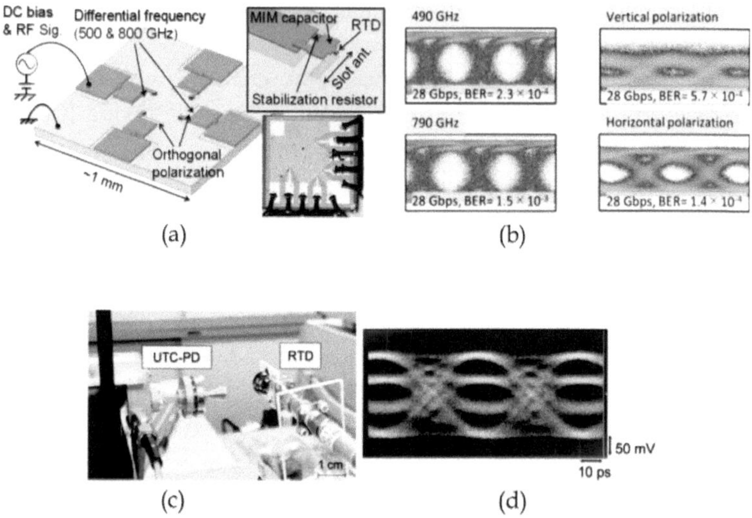

Figure 9.5 (a) Schematic device structure and microphotograph of RTD device for FDM and PDM communication, and (b) the obtained eye diagrams. Reproduced from Ref. [61] by permission of IEEE, © 2017. (c) Experimental setup of communication with UTC-PD transmitter and RTD receiver, and (d) the eye diagram of PAM4 data transmission.

can be established even when the transmitter and receiver are rotating.

9.3.2 Imaging and Radar

Recently, incidents using cutlery and combustibles have occurred in closed spaces, such as onboard trains, and the number of cases where public safety is threatened is increasing. Therefore, it is necessary to identify these factors in the future. In these applications, the use of transparent and high-resolution THz waves is expected, and RTD oscillators that can generate high frequencies with short wavelengths are suitable for these applications. If these applications are extended to radar imaging combined with distance measurements, the accuracy can be further improved [64]. In this section, various types of imaging and radar systems are introduced, that utilize the high-speed modulation characteristics of the RTD oscillator.

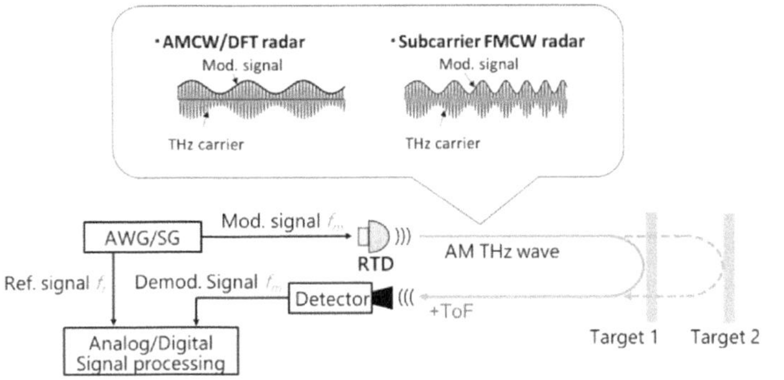

Figure 9.6 Measurement setup for various radar systems using an amplitude-modulated RTD oscillator.

The first imaging using an RTD oscillator began with transparent imaging in the 300 GHz range. An image was obtained by mechanical 2D scanning of a coin in a plastic case [65]. Line scanning of oscillators arranged in a one-dimensional RTD oscillator array has been developed, which enables the imaging of objects placed on a high-speed conveyor belt. In addition, simple encoders using a 500 GHz RTD oscillator have also been developed [66].

Next, various radar systems are introduced. A schematic of the various radar systems using an RTD device is shown in Fig. 9.6. The first is an amplitude-modulated continuous-wave (AMCW) radar [67]. The distance measurement principle of the AMCW radar is as follows: the output of the RTD device is intensity-modulated with a sine wave, the THz wave reflected by the object is received and demodulated by the receiver, and the time of flight (ToF) of the THz wave is obtained from the phase difference between the demodulated signal and the reference signal. The distance uncertainty can be avoided by using two modulation frequencies, and fine distance measurements were achieved with an error of 0.06 mm (standard deviation) when using two frequencies of 5 and 5.5 GHz [68]. Furthermore, a 3D image can be obtained by combining the AMCW-type radar and a 2D scan of the stage on which the target is placed [69]. In the future, if the THz beam is swept by a Galvano mirror or a phased array oscillator, obtaining a 3D image without a stage scan may be possible.

In the aforementioned AMCW method, the measurement is limited when there is only one reflection from the object's surface. However, when there are two reflectors, such as clothes and the body surface, they cannot be distinguished. It is possible to separate the two reflectors using the frequency-modulated continuous-wave (FMCW) scheme. Unfortunately, perfect continuous frequency tuning is difficult to achieve in RTD oscillators because of difficulties in avoiding the external feedback effect. Therefore, a new type of radar called a subcarrier FMCW system is proposed, in which the subcarrier frequency changes with time [70]. In this method, the output of the RTD is modulated by the frequency-chirped signal, the modulated THz signal is reflected by the target, and the returned signal is received and demodulated. Then, the demodulated signal and the reference signal are mixed, and the frequency beat of the demodulated and reference signals provides the ToF in the same way as the normal FMCW. By using an RTD oscillator with a frequency of 511 GHz and an output of about 10 μW modulated by a chirp signal of 3.5–10.5 GHz, the two targets were measured separately with a small distance error of approximately 1 mm, as shown in Fig. 9.7. The measurement time for obtaining the distance at one point in the FMCW experiment was approximately 1 min, but this was dramatically reduced to 4 ms by improving the data processing [71]. The measurement time can be further reduced by using a high-speed field-programmable gate array (FPGA). In addition, the subcarrier FMCW requires a high-quality chirp signal for modulation; however, as in the wavelength-sweep-type optical coherence tomography (OCT) method, the distance information can be obtained by discretely changing the modulation frequency and Fourier transformation. This is called the discrete Fourier transform (DFT) radar and was used to obtain a fine distance measurement with an error of approximately 1 mm [72], which is similar to the subcarrier FMCW method.

9.3.3 Analytics

Finally, applications in analytics are presented. The RTD oscillator can operate in a wide frequency range from 100 GHz or less to 2 THz, and if multiple oscillators in each frequency band are used using

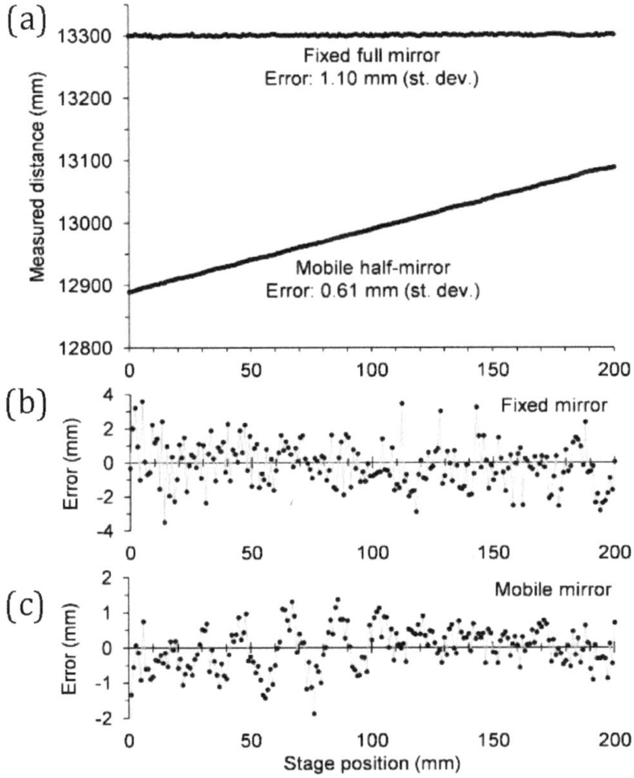

Figure 9.7 (a) Measured distances and errors of (b) fixed and (c) mobile half-mirrors by a subcarrier FMCW radar system.

the voltage-controlled oscillator (VCO) characteristics, a wide-band spectrum can be obtained and the substance can be identified. A system has been constructed that is capable of continuous frequency sweeping in the range of 400 to 900 GHz using seven RTD VCOs and the absorbance of the sample has been measured [73]. Allopurinol, which has a steep absorption band near 700 GHz, was used as the sample, and the absorbance measured using the RTD oscillator was approximately the same as the measurement result with THz time-domain spectroscopy (TDS). It is expected that instantaneous spectral measurement will be possible if a system that sweeps bias at high-speed is constructed. It has also been shown that the refractive index of a substance can be sensed from changes in the oscillation spectrum by using an RTD oscillator coupled to a photonic crystal resonator [74].

9.4 Summary

The characteristics and functions of the RTD oscillator and its applications were introduced. The application of the THz band has yet to be realized, but specific goals, such as applications for Beyond 5G/6G, have emerged. Consequently, research and development of THz sources is likely to intensify in the near future. RTD devices have the advantages of a compact size and high-frequency operation. However, the key limitation of RTD device performance for general use across several applications is the high output power characteristics, which remain a major challenge to overcome. The methods and preliminary studies for achieving high output power in RTD devices were introduced in this chapter. It is expected that continuous efforts to improve RTD devices will inspire future developments in THz technology.

References

1. Sengupta, K., Nagatsuma, T., Mittleman, D. M. (2018). Terahertz integrated electronic and hybrid electronic–photonic systems, *Nature Electronics*, **1**, pp. 622–635.
2. Brown, E. R., Sonderstrom, J. R., Parker, C. D., Mahoney, L. J., Molvar, K. M., McGill, T. C. (1991). Oscillations up to 712 GHz in InAs/AISb resonant-tunneling diodes, *Appl. Phys. Lett.*, **58**, pp. 2291–2293.
3. Reddy, M., Martin, S. C., Molnar, A. C., Muller, R. E., Smith, R. P., Siegel, P. H., Mondry, M. J., Rodwell, M. J. W., Kroemer, H., Allen Jr., S. J. (1997). Monolithic Schottky-collector resonant tunnel diode oscillator arrays to 650 GHz, *IEEE Electron Device Lett.*, **18**, pp. 218–221.
4. Suzuki, S., Asada, M., Teranishi, A., Sugiyama, H., Yokoyama, H. (2010). Fundamental oscillation of resonant tunneling diodes above 1 THz at room temperature, *Appl. Phys. Lett.*, **97**, 242102.
5. Feiginov, M., Sydlo, C., Cojocari, O., Meissner, P. (2011). Resonant-tunnelling-diode oscillators operating at frequencies above 1.1 THz, *Appl. Phys. Lett.*, **99**, 233506.

6. Kanaya, H., Shibayama, H., Sogabe, R., Suzuki, S., Asada, M. (2012). Fundamental oscillation up to 1.31 THz in resonant tunneling diodes with thin well and barriers, *Appl. Phys. Express*, **5**, 124101.
7. Koyama, Y., Sekiguchi, R., Ouchi, T. (2013). T. Oscillations up to 1.40 THz from resonant-tunneling-diode-based oscillators with integrated patch antennas, *Appl. Phys. Express*, **6**, 064102.
8. Kanaya, H., Sogabe, R., Maekawa, T., Suzuki, S., Asada, M. (2014). Fundamental oscillation up to 1.42 THz in resonant tunneling diodes by optimized collector spacer thickness, *J. Infrared, Millimeter, and THz Waves*, **35**, pp. 425–431.
9. Feiginov, M., Kanaya, H., Suzuki, S., Asada, M. (2014). Operation of resonant-tunneling diodes with strong back injection from the collector at frequencies up to 1.46 THz, *Appl. Phys. Lett.*, **104**, 243509.
10. Maekawa, T., Kanya, H., Suzki, S., Asada, M. (2014). Frequency increase in terahertz oscillation of resonant tunnelling diode up to 1.55 THz by reduced slot-antenna length, *Electron. Lett.*, **50**, pp. 1214–1216.
11. Maekawa, T., Kanaya, H., Suzuki, S., Asada, M. (2016). Oscillation up to 1.92 THz in resonant tunneling diode by reduced conduction loss, *Appl. Phys. Express*, **9**, 024101.
12. Izumi, R., Suzuki, S., Asada, M. (2017). 1.98 THz resonant-tunneling-diode oscillator with reduced conduction loss by thick antenna electrode, *42nd Int. Conf. Infrared, Millimeter, and THz Waves (IRMMW-THz)*, Cancun, Mexico, MA3.1, doi: 10.1109/IRMMW-THz.2017.8066877.
13. Kasagi, K., Suzuki, S., Asada, M. (2019). Large-scale array of resonant-tunneling-diode terahertz oscillator for high output power at 1 THz, *J. Appl. Phys.*, **125**, 151601.
14. Song, H.-J., Nagatsuma, T. (2015). (eds.) *Handbook of Terahertz Technologies: Devices and Applications*, Pan Stanford Publishing: Singapore.
15. Köhler, R., Tredicucci, R., Beltram, F., Beere, H. E., Linfeld, E. H., Davies, A. G., Ritchie, D. A., Iotti, R. C., Rossi, F. (2002). Terahertz semiconductor heterostructure laser, *Nature*, **417**, pp. 156–159.
16. Williams, B. S. (2007). Terahertz quantum-cascade lasers, *Nat. Photonics*, **1**, pp. 517–525.
17. Khalatpour, A., Paulsen, A. K., Deimert, C., Wasilewski, Z. R., Hu, Q. (2021). High-power portable terahertz laser systems, *Nat. Photonics*, **15**, pp. 16–20.
18. Fujita, K., Jung, S., Jiang, Y., Kim, J. H., Nakanishi, A., Ito, A., Hitaka, M., Edamura, T., Belkin, M. A. (2018). Recent progress in terahertz difference-frequency quantum cascade laser sources, *Nanophotonics*, **7**, pp. 1795–1817.

19. Fujita, K., Hayashi, S., Ito, A., Dougakiuchi, T., Hirata, M., Nakanishi, A. (2022). Broadly tunable lens-coupled quantum cascade laser sources in the sub-THz to THz frequency range. published online in *Photonic Research*, doi: 10.1364/PRJ.443819.
20. Nishizawa, J., Plotka, P., Kurabayashi, T., Makabe, H. (2008). 706 GHz GaAs CW fundamental-mode TUNNETT diodes fabricated with molecular layer epitaxy, *Phys. Status Solidi (C)*, **5**, pp. 2802–2804.
21. Eisele, H., Kamoua, R. (2004). Submillimeter-wave InP Gunn devices, *IEEE Trans. Microw. Theory Technol.*, **52**, pp. 2371–2378.
22. Eisele, H. (2010). 480 GHz oscillator with an InP Gunn device, *Electron. Lett.*, **46**, pp. 422–423.
23. Urtega, M., Griffith, Z., Seo, M., Hacker, J., Rodwell, M. J. W. (2017). InP HBT technologies for THz integrated circuits, *Proc. IEEE*, **105**, pp. 1051–1067.
24. Pfeiffer, U. R., Zao, Y., Grzyb, J., Hadi, R. A., Sarmah, N., Förster, W., Rücker, H., Heinemann, B. (2014). A 0.53 THz reconfigurable source module with up to 1 mW radiated power for diffuse illumination in terahertz imaging applications, *IEEE J. Solid State Circuits*, **49**, pp. 2938–2950.
25. Hu, Z., Mehmet, M., Han, R. (2018). High-power radiation at 1 THz in silicon: A fully scalable array using a multi-functional radiating mesh structure, *IEEE J. Solid State Circuits*, **53**, pp. 1313–1327.
26. Asada, M., Suzuki, S. (2021). Terahertz emitter using resonant-tunneling diode and applications, *Sensors*, **21**, 1384.
27. Asada, M., Suzuki, S. (2015). Theoretical analysis of external feedback effect on oscillation characteristics of resonant-tunneling-diode terahertz oscillators, *Jpn. J. Appl. Phys.*, **54**, 070309.
28. Zhang, H., Xue, J., Sun, Z., Li, L., Yao, J., Liu, F., Yang, X., Wu, G., Li, Z., Fu, Y., Liu, Z., Zhang, J., Hao, Y. (2021). 1039 kA/cm^2 peak tunneling current density in GaN-based resonant tunneling diode with a peak-to-valley current ratio of 1.23 at room temperature on sapphire substrate, *Appl. Phys. Lett.*, **119**, 153506.
29. Mai, T. V., Suzuki, Y., Yu, X., Suzuki, S., Asada, M. (2020). Structure-simplified resonant-tunneling-diode terahertz oscillator without metal-insulator-metal capacitors, *J. Infrared, Millimeter, Terahertz Waves*, **41**, pp. 1498–1507.
30. Yu, X., Suzuki, Y., Mai, T. V., Suzuki, S., Asada, M. (2021). Highly efficient resonant tunneling diode terahertz oscillator with a split ring resonator, *IEEE Electron Device Lett.*, **42**, pp. 982–985.

31. Izumi, R., Sato, T., Suzuki, S., Asada, M. (2019). Resonant-tunneling-diode terahertz oscillator with a cylindrical cavity for high-frequency oscillation, *AIP Adv.*, **9**, 085020.
32. Bezhko, M., Suzuki, S., Asada, M. (2020). Frequency increase in resonant-tunneling diode cavity-type terahertz oscillator by simulation-based structure optimization, *Jpn. J. Appl. Phys.*, **59**, 032004.
33. Lee, J., Kim, M., Yang, K. (2016). 1.52 THz RTD triple-push oscillator with a μW-level output power, *IEEE Trans. THz Sci. Technol.*, **6**, pp. 336–340.
34. Shiraishi, M., Shibayama, H., Ishigaki, K., Suzuki, S., Asada, M., Sugiyama, H., Yokoyama, H. (2011). High output power (~400 μW) oscillators at around 550 GHz using resonant tunneling diodes with graded emitters and thin barriers, *Appl. Phys. Express*, **4**, 064101.
35. Kobayashi, K., Suzuki, S., Han, F., Tanaka, H., Fujikata, H., Asada, M. (2020). Analysis of a high-power resonant-tunneling-diode terahertz oscillator integrated with a rectangular cavity resonator, *Jpn. J. Appl. Phys.*, **59**, 050907.
36. Han. F., Kobayashi, K., Suzuki, S., Tanaka, H., Fujikata, H., Asada, M. (2021). Impedance matching in high-power resonant-tunneling-diode terahertz oscillators integrated with rectangular-cavity resonator, *IEICE Trans. Electronics*, **E104.C**, pp. 398–402.
37. Al-Khalidi, A., Alharbi, K. H., Wang, J., Morariu, R., Wang, L., Khalid, A., Figueiredo, J. M. L., Wasige, E. (2020). Resonant tunneling diode terahertz sources with up to 1 mW output power in the J-band, *IEEE Trans. THz Sci. Technol.*, **10**, pp. 150–157.
38. Iwamatsu, S., Nishida, Y., Fujita, M., Nagatsuma, T. (2021). Terahertz coherent oscillator integrated with slot-ring antenna using two resonant tunneling diodes, *Appl. Phys. Express*, **14**, 034001.
39. Ourednik, P., Hackl, T., Spudat, C., Nguyen, T. D., Feiginov, M. (2022). Double-resonant-tunneling-diode patch-antenna oscillators, *Appl. Phys. Lett.*, **119**, 263509.
40. Mai, T. V., Namba, T., Suzuki, Y., Suzuki, S., Asada, M. (2021). Resonant-tunneling-diode oscillator array with zigzag arrangement for terahertz power combination, *9th Russia-Japan-USA-Europe Symposium on Fundamental & Applied Problems of Terahertz Devices & Technologies (RJUSE-TeraTech)*, Sendai, Japan, TH-1-5.
41. Ikeda, Y., Kitagawa, S., Okada, K., Suzuki, S., Asada, M. (2015). Direct intensity modulation of resonant-tunneling-diode terahertz oscillator up to ~30 GHz, *IEICE Electron. Express*, **12**, 2014116.

42. Kitagawa, S., Suzuki, S., Asada, M. (2016). Wide frequency-tunable resonant tunneling diode terahertz oscillators using varactor diodes, *Electron. Lett.*, **52**, pp. 479–481.

43. Karashima, K., Yokoyama, R., Shiraishi, M., Suzuki, S., Aoki, S., Asada, M. (2010). Measurement of oscillation frequency and spectral linewidth of sub-terahertz InP-based resonant tunneling diode oscillators using Ni–InP Schottky barrier diode, *Jpn. J. Appl. Phys.*, **49**, 020208.

44. Ogino, K., Suzuki, S., Asada, M. (2017). Spectral narrowing of a varactor-integrated resonant-tunneling-diode terahertz oscillator by phase-locked loop, *J. Infrared, Millimeter, and THz Waves*, **38**, pp. 1477–1486.

45. Ogino, K., Suzuki, S., Asada M. (2018). Phase locking and frequency tuning of resonant-tunneling-diode terahertz oscillators, *IEICE Trans. Electron.*, **E101-C**, pp. 183–185.

46. Asada, M. (2020). Theoretical analysis of subharmonic injection locking in resonant-tunneling-diode terahertz oscillators, *Jpn. J. Appl. Phys.*, **59**, 018001.

47. Arzi, K., Suzuki, S., Rennings, A., Erni, D., Weimann, N., Asada, M., Prost, W. (2020). Subharmonic injection locking for phase and frequency control of RTD-based THz oscillator, *IEEE Trans. THz Sci. Technol.*, **10**, pp. 221–224.

48. Suzuki, Y., Mai, T. V., Yu, X., Suzuki, S., Asada, M. (2021). Phase control of injection-locked RTD terahertz oscillator, *46th Int. Conf. Infrared, Millimeter, and THz Waves* (*IRMMW-THz*), Chengdu, China, TH-AM-5-2, doi: 10.1109/IRMMW-THz50926.2021.9566843.

49. Nishida, Y., Nishigami, N., Diebold, S., Kim, J., Fujita, M., Nagatsuma, T. (2019). Terahertz coherent receiver using a single resonant tunneling diode, *Sci. Rep.*, **9**, 18125.

50. Takida, Y., Suzuki, S., Asada, M., Minamide, H. (2020). Sensitive terahertz-wave detector response originated by negative differential conductance of resonant-tunneling-diode oscillator, *Appl. Phys. Lett.*, **117**, 021107.

51. Hiraoka, T., Arikawa, T., Yasuda, H., Inose, Y., Sekine, N., Hosako, I., Ito, H., Tanaka, K. (2021). Injection locking and noise reduction of resonant tunneling diode terahertz oscillator, *APL Photonics*, **6**, 021301.

52. Urayama, K., Aoki, S., Suzuki, S., Asada, M., Sugiyama, H., Yokoyama, H. (2009). Sub-terahertz resonant tunneling diode oscillators integrated with tapered slot antennas for horizontal radiation, *Appl. Phys. Express*, **2**, 044501.

53. Yu, X., Kim, Y. J., Fujita, M., Nagatsuma, T. (2019). Efficient mode converter to deep-subwavelength region with photonic-crystal waveguide platform for terahertz applications, *Opt. Express*, **27**, pp. 28707–28721.
54. Fujita, M., Nagatsuma, T. (2016). Photonic crystal technology for terahertz system integration, *Proc. SPIE 9856, Terahertz Physics, Devices, and Systems X: Advanced Applications in Industry and Defense*, 98560P.
55. Okada, K., Kasagi, K., Oshima, N., Suzuki, S., Asada, M. (2015). Resonant-tunneling-diode terahertz oscillator using patch antenna integrated on slot resonator for power radiation, *IEEE Trans. THz Sci. Technol.*, **5**, pp. 613–618.
56. Lee, J., Lim, M., Lee, J. (2021). 692 GHz high-efficiency InP-based fundamental RTD oscillator, *IEEE Trans. Terahertz Sci. Tech.*, **11**, pp. 716–719.
57. Horikawa, H., Chen, Y., Koike, T., Suzuki, S., Asada, M. (2018). Resonant-tunneling-diode terahertz oscillator integrated with a radial line slot antenna for circularly polarized wave radiation, *Semicond. Sci. Technol.*, **33**, 114005.
58. Tsuruda, K., Nishida, Y., Mukai, T., Terumoto, K., Miyamae, Y., Oku, Y., Nakahara, K. (2020). Development of practical terahertz packages for resonant tunneling diode oscillators and detectors, *IEEE Int. Symp. Radio-Frequency Integration Technology (RFIT)*, pp. 234–236.
59. NTT DOCOMO (2022). *White Paper 5G Evolution and 6G*, https://www.nttdocomo.co.jp/english/corporate/technology/whitepaper_6g/, etc.
60. Asada, M., Suzuki, S. (2018). Terahertz oscillators using resonant tunneling diodes, *Asia-Pacific Microwave Conf.* (*APMC*), Kyoto, Japan, pp. 521–523, doi: 10.23919/APMC.2018.8617175.
61. Oshima, N., Hashimoto, K., Suzuki, S., Asada, M. (2017). Terahertz wireless data transmission with frequency and polarization division multiplexing using resonant-tunneling-diode oscillators, *IEEE Trans. THz Sci. Technol.*, **7**, pp. 593–598.
62. Yamamoto, T., Nishigami, N., Nishida, Y., Fujita, M., Nagatsuma, T. (2020). *IEICE Conferences Archives*, C-14-3 (in Japanese).
63. Oshiro, A., Nishigami, N., Yamamoto, T., Nishida, Y., Webber, J., Fujita, M., Nagatsuma, T. (2021). PAM4 48 Gbit/s wireless communication using a resonant tunneling diode in the 300 GHz band, *IEICE Electronics Express*, **19**, 20210494.
64. Cooper, B. K., Dengler, J. R., Llombart, N., Thomas, B., Chattopadhyay, G., Siegel, H. P. (2011). THz imaging radar for standoff personnel screening, *IEEE Trans. Terahertz Sci. Tech.*, **1**, pp. 169–182.

65. Miyamoto, T., Yamaguchi, A., Mukai, T. (2016). Terahertz imaging system with resonant tunneling diodes, *Jpn. J. Appl. Phys.*, **55**, 032201.
66. Yamashita, G., Tsujita, W., Tsutada, H., Ma, R.,Wang, P., Orlik, P. V., Suzuki, S., Dobroiu, A., Asada, M. (2019). Terahertz polarimetric sensing for linear encoder based on resonant-tunneling-diode and CFRP polarizing, *44th Int. Conf. Infrared, Millimeter, and THz Waves* (*IRMMW-THz*), Paris, France, 4429649, doi: 10.1109/IRMMW-THz.2019.8873859.
67. Dobroiu, A., Wakasugi, R., Shirakawa, Y., Suzuki, S., Asada M. (2018). Absolute and precise terahertz-wave radar based on an amplitude-modulated resonant-tunneling-diode oscillator, *MDPI Photonics*, **5**, 52.
68. Dobroiu, A., Wakasugi, R., Shirakawa, Y., Suzuki, S., Asada, M. (2020). Amplitude-modulated continuous-wave radar in the terahertz range using lock-in phase measurement, *Meas. Sci. Technol.*, **31**, 105001.
69. Dobroiu, A., Asama, K., Suzuki, S., Asada, M., Ito, H. (2021). Three-dimensional terahertz imaging using an amplitude-modulated resonant-tunneling-diode oscillator, *46th Int. Conf. Infrared, Millimeter, and THz Waves* (*IRMMW-THz*), Chengdu, China, MO-AM-4-4, doi: 10.1109/IRMMW-THz50926.2021.9567337.
70. Dobroiu, A., Shirakawa, Y., Suzuki, S., Asada, M., Ito, H. (2020). Subcarrier frequency-modulated continuous-wave radar in the terahertz range based on a resonant-tunneling-diode oscillator, *Sensors*, **20**, 6848.
71. Ito, J., Dobroiu, A., Suzuki, S., Asada, M., Ito, H. (2021). Real-time distance measurement using a subcarrier FMCW radar based on a terahertz-wave resonant-tunneling-diode oscillator, *46th Int. Conf. Infrared, Millimeter, and THz Waves* (*IRMMW-THz*), Chengdu, China, WE-AM-4-2, doi: 10.1109/IRMMW-THz50926.2021.9567210.
72. Konno, H., Dobroiu, A., Suzuki, S., Asada, M., Ito, H. (2021). Discrete Fourier transform radar in the terahertz-wave range based on a resonant-tunneling-diode oscillator, *Sensors*, **21**, 4367.
73. Kitagawa, S., Mizuno, M., Saito, S., Ogino, K., Suzuki, S., Asada, M. (2017). Frequency-tunable resonant-tunneling-diode oscillators applied to absorbance measurement, *Jpn. J. Appl. Phys.*, **56**, 058002.
74. Okamoto, K., Tsuruda, K., Diebold, S., Hisatake, S., Fujita, M., Nagatsuma, T. (2017). Terahertz sensor using photonic crystal cavity and resonant tunneling diodes, *J. Infrared Millimeter and THz Waves*, **38**, pp. 1085–1097.

Chapter 10

Plasmon-Based THz Oscillators

Taiichi Otsuji
Research Institute of Electrical Communication,
Tohoku University, 2-1-1 Katahira,
Aoba-ku Sendai, Miyagi 9808577, Japan
taiichi.otsuji.e8@tohoku.ac.jp

In this chapter, the theory and experiments for plasmon-based THz oscillators are reviewed. Two-dimensional (2D) plasmons in submicrometer-scaled transistors like field-effect transistors (FETs) and/or high-electron-mobility transistors (HEMTs) have attracted considerable attention due to their nature of promoting emission and detection of electromagnetic radiation in the THz range. First, the theory of hydrodynamics and instabilities of 2D plasmons are described. Second, experimental studies for massive 2D plasmons in InGaAs-based, GaAs-based, and GaN-based compound semiconductor heterostructures as well as massless 2D plasmons in graphene are described. In particular, graphene Dirac plasmons open a pathway towards the realization of intense THz laser transistors operating at room temperatures with a dry-cell battery. Third, future trends and technological subjects are addressed and their solutions are discussed.

Advances in Terahertz Source Technologies
Edited by Gun-Sik Park, Masahiko Tani, Jae-Sung Rieh, and Sang Yoon Park

ISBN 978-981-4968-89-8 (Hardcover), 978-1-003-45967-5 (eBook)
www.jennystanford.com

10.1 Introduction

In the research of modern high-speed electronics and broadband photonics, the development of compact, tunable and coherent sources operating at terahertz (THz) frequencies is one of the technological bottlenecks [1]. Two-dimensional (2D) plasmons in submicrometer-scaled transistors like FETs and/or HEMTs have attracted considerable attention due to their nature of promoting emission and detection of electromagnetic radiation in the THz range. The channel of a transistor can act as a resonator for 2D plasmons, the quanta of the charge density waves of collectively excited 2D electrons. The plasmon mode frequencies are determined by the resonator dimensions and the density of 2D electrons, easily reaching the THz range in submicrometer-scaled transistors. Therefore, various submicrometer-scaled transistor structures supporting 2D plasmons were intensively studied as possible candidates for solid-state far-infrared (FIR) and THz sources [2–22]. The mechanisms of plasmon excitation/emission can be divided into two types: (i) incoherent and (ii) coherent type. The first is related to thermal excitation of broadband nonresonant plasmons by hot electrons [3, 4, 7, 8], whereas the second is related to the plasmon instability mechanisms like Dyakonov–Shur Doppler-shift-type model [6, 10, 13–16], Ryzhii–Satou–Shur electron-transit-type model [11–13, 18, 22, 23], and/or Cherenkov plasmonic-boom-type model [2, 9, 20] in which coherent plasmons can be excited either by hot electrons or by optical phonon emission under near ballistic electron motion [23].

In this chapter, the theory and experiments for plasmon-based THz oscillators are reviewed. First, the theory of hydrodynamics and instabilities of 2D plasmons are described. Second, experimental studies for massive 2D plasmons in compound semiconductor heterostructures as well as massless 2D plasmons in graphene are described. Third, future trends and technological subjects and their solutions are addressed.

10.2 Theory

10.2.1 Hydrodynamics of 2D Plasmons

The 2D plasma-wave motions in massive semiconductor materials can be described by the following hydrodynamic Euler equation and the current continuity equation [6]:

$$m\left(\frac{\partial \boldsymbol{u}}{\partial t}+\boldsymbol{u}\frac{\partial \boldsymbol{u}}{\partial \boldsymbol{r}}\right)=-e\frac{\partial V}{\partial \boldsymbol{r}}-\frac{\boldsymbol{u}}{m\tau}, \tag{10.1}$$

$$\frac{\partial n}{\partial t}+\frac{\partial}{\partial \boldsymbol{r}}(n\boldsymbol{u})=0, \tag{10.2}$$

where m is the electron effective mass, $\mathbf{r}$ is an arbitrary in-plane vector, $\mathbf{u}$ $(\mathbf{r}, t)$ is the in-plane electron spatiotemporal local velocity, τ is the electron momentum relaxation time, $V(\mathbf{r}, t)$ is the local potential at $\mathbf{r}$, and $n(\mathbf{r}, t)$ is the spatiotemporal local density of electrons. The 1st term of the right-hand in (10.1) is the Coulomb force and the 2nd term is the Drude friction. 2D electron channels in HEMTs consist of gated and ungated regions (Fig. 10.1). The ungated 2D plasmon receives the in-plane longitudinal Coulomb force so that it holds a square root dependence of dispersion relation with respect to the wave vector which is identical to that for general surface plasmons. The gated 2D plasmon receives transverse Coulomb force via a gate capacitor which is far stronger than the in-plane force due to the geometrical situation so that it holds a linear dispersion. In a simple case of gradual-channel approximation with infinite channel width (perpendicular to the source-drain direction), the 2D plasmon dynamics are deduced to one-dimensional systems [6].

The plasmon group velocity s is given by $s=\sqrt{eV_0/m}$ where V_0 is the gate swing voltage [6]. Assuming V_0 of an order of 1 V and m of an order of 0.1 m_0 (m_0 is the electron rest mass in vacuum) for InP-based heterostructure HEMTs, s becomes an order of 1×10^6 m/s which is at least two orders of magnitude higher than the electron drift velocity of any compound semiconductors. Thus, when we consider a submicrometer-gate-length HEMT, the fundamental mode of gated 2D plasmons stays at frequency in the THz range. This is the main advantage for use in plasmon resonant modes that can

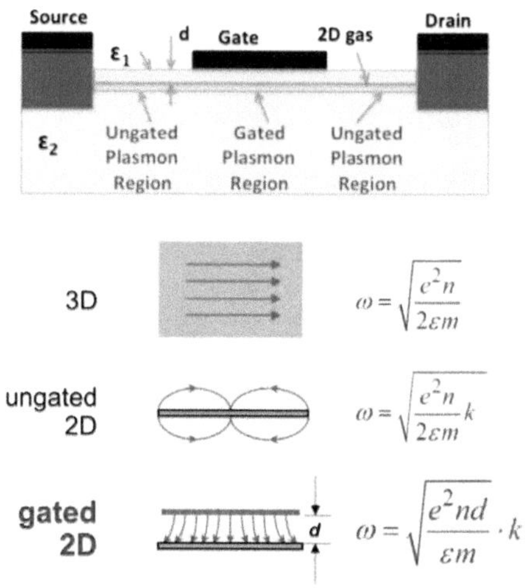

$$\omega = \sqrt{\frac{e^2 n}{2\varepsilon m}}$$

$$\omega = \sqrt{\frac{e^2 n}{2\varepsilon m} k}$$

$$\omega = \sqrt{\frac{e^2 nd}{\varepsilon m}} \cdot k$$

Figure 10.1 Plasmons in a HEMT (top) and dispersion relations of 3D, ungated 2D, and gated 2D plasmons (bottom).

operate in the frequencies far beyond the transit frequency limit of transistors.

Since the Q factor of the plasmon resonance is defined by the ratio of the carrier momentum relaxation time (mean lifetime) to the period of plasmon resonant frequency; $Q = s\tau/L$, as superior as electron transport property with as long as momentum relaxation time is the key factor to generate coherent, intense THz oscillation emission originated from the plasmon resonance. In this regard, graphene a carbon monoatomic honeycomb lattice sheet is the best-suited material [24–29]. Due to its exotic conical band structure holding linear and symmetric band dispersion around the Brillouin zone edges the conduction electrons and valence holes behave as if they were massless Dirac fermions [24, 25]. The momentum relaxation time in a high-quality graphene stays on the orders of picoseconds to 10s of picoseconds even at room temperatures, which are about two-three orders of magnitudes longer than any existing massive semiconductor materials [24]. The dispersion relations of graphene Dirac plasmons are well formulated already, which differ from those for massive semiconductors, but hold the gate tunability of the plasmon mode frequencies [26–29].

10.2.2 Dyakonov–Shur Doppler-Shift-Type Instability

When a single-gate HEMT is situated in source-terminated and drain-opened configuration with dc potential at drain terminal with respect to the source terminal, the drain end of the channel becomes depleted so that the drain-side impedance is mainly given by the depletion capacitance and takes a high value at high frequencies. In such a case, Doppler-shift effect occurs on the plasma-wave propagation/reflection at the drain boundary under a unidirectional dc-current flow (from drain to source), promoting the Dyakonov–Shur (DS) instability [6]. Consider the case in which plasmons are excited in a HEMT with a constant dc drain bias causing a background constant dc electron drift flow with velocity v_d and the gate length L is shorter than the coherent length of electrons. The plasma-wave-originated local displacement current δj_p is given by the product of the perturbation of the local electron charge density $e\delta n$ and the plasma-wave velocity. The forward (backward) component $\delta \vec{j}_p$ $(\delta \overleftarrow{j}_p)$ traveling to (from) the drain boundary is given by: $\delta \vec{j}_p = e\delta\vec{n} \cdot (s + v_d)$, $\delta \overleftarrow{j}_p = e\delta\overleftarrow{n} \cdot (s - v_d)$. Since the open-drain boundary conserves the current before and after the reflection, $|\delta \vec{j}_p| = |\delta \overleftarrow{j}_p|$, $\delta\overleftarrow{n} = (s + v_d)/(s - v_d) \cdot \delta\vec{n} > \delta\vec{n}$. This increment of the electron charge density $\delta n = (\delta\overleftarrow{n} - \delta\vec{n})$ directly reflects the increment of the gate potential δV_g via gate capacitor C: $\delta V_g = e\delta n \cdot C$. Since the source-terminated boundary gives a lossless reflection (reflection coefficient is -1) the gate potential becomes infinite after infinitesimal repetition of plasma-wave reflections, leading to the DS instability. When the plasmons are excited by the incoming THz radiation with angular frequency ω, $e^{-i\omega t}$, the effect of the instability is characterized by the imaginary part of ω, ω'', as

$$\omega'' = \frac{s^2 - v_d^2}{Ls} \ln\left|\frac{s + v_d}{s - v_d}\right|. \tag{10.3}$$

When $\omega'' > 0$, the system becomes unstable, giving rise to the condition of the DS instability. The plasma-wave increment in units of $s/2L$, a dimensionless plasma-wave increment $2\omega'' L/s$ is plotted as a function of the Mach number s/v_d [6]. In reality with finite τ value, the Drude loss factor should consider to obtain an overall gain, which is shown as a threshold level in Fig. 10.2 [19]. So far

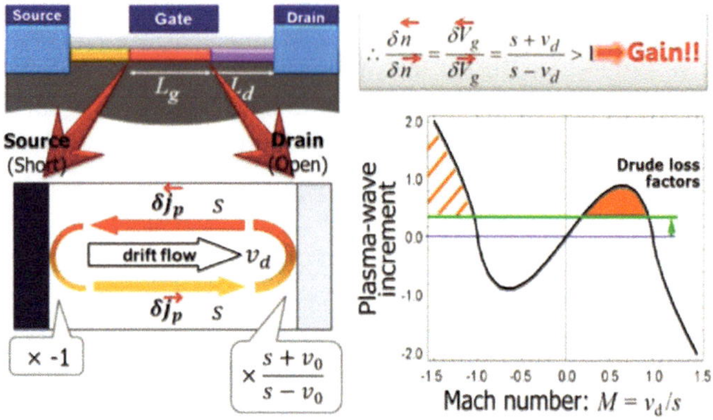

Figure 10.2 Dyakonov–Shur type instability in a 2D electron channel under source-terminated and drain-opened boundaries with dc drift velocity vd [6]. In reality with finite τ value, the Drude loss factor should consider to obtain an overall gain, which is shown as a threshold level. After Ref. [19].

DS-instability-driven voltage-tunable millimeter-wave to THz-wave emission has been observed at low and room temperatures from GaAs-, InP-, and GaN-based HEMTs [10, 13–19].

10.2.3 Ryzhii–Satou–Shur Electron-Transit-Type Instability

When the channel pinch-off is insufficient and the drain terminal is not open but yet conductive at THz frequencies, the plasmons are effectively "absorbed" or overdamped in the high-field gate-drain region. Hence their reflection is insufficient to promote necessary positive feedback for the occurrence of the DS instability. At large drain-source voltages, the THz conductivity of this region would be rather high due to relatively high value of the electron drift velocity v_{gd} in the high-field gate-drain region. In this case v_{gd} becomes much higher than that in the intrinsic channel region v_d. Note that the electrons propagating in the high-field gate-drain region induce the ac current in the gated channel and the drain contact. One can find that the electron ac concentration as a function of the coordinate varies as $n_\omega(x) = n_\omega|_{x=L_g} e^{-i\omega(x-Lg)/v_{gd}}$ [11, 12]. As a result, the ac current induced in the gated channel is presented. Its frequency dependence is directly reflected by the electron-transit time τ_{dd} at the gate-drain region $\tau_{gd} = L_d/v_{gd}$ where L_d is the length of the gate-drain region, which may contribute to promoting the plasma-wave instability [11–13]. In this case, the instability condition is given by the following inequality using the imaginary part of the plasma-wave current:

$$Im(\omega) = \omega'' \simeq -\frac{\nu}{2} + \gamma_{DS} + \gamma_{TT} > 0, \quad (10.4)$$

$$\gamma_{DS} \simeq \frac{v_d}{Lg}, \quad (10.5)$$

$$\gamma_{RSS} \simeq -\frac{v_{gd}}{L_g} r \cos(\omega_n \tau_{gd}/2) J_0(\omega_n \tau_{gd}/2), \quad (10.6)$$

where ν is the electron collision frequency, r is a phenomenological parameter ($r \leq 1$), and J_0 is the 0th Bessel function, and $n = 1, 2, 3, \ldots$ is the plasmon mode index [11, 12]. Here, γ_{DS} and γ_{RSS} are the *DS*-instability index and the transit-time-driven RSS-instability index, respectively. γ_{RSS} can take both positive and negative values depending on $\omega_n \tau_{gd}$. The contributions to the plasma-wave instability growth rate of the *DS* and *RSS* mechanism, ($\gamma_{DS}/2\pi$ and $\gamma_{RSS}/2\pi$) in a HEMT with typical geometric and material parameters are plotted in Fig. 10.3 [19] as functions of the fundamental plasmon mode frequency [12]. Due to the nature of transit-time-driven mechanism the *RSS* instability is sensitive to the plasmon mode frequency.

10.2.4 Cherenkov Plasmonic-Boom-Type Instability

Cherenkov-type radiation is well known as an electromagnetic radiation that occurs when a charged particle passes through a dielectric medium at a speed greater than the phase velocity of light in the medium [2, 9, 19]. If electrons (charged particles) travel surpassing the phase velocity of the plasma field, the field becomes unstable leading to radiation emission of plasmons. If there is an antenna structure that efficiently couple the non-radiative plasmons to radiative photons one can get an electromagnetic radiation originated from the plasmon instability. In this regard, we call it as Cherenkov-type plasmon instability, and also as plasmonic-boom-type instability.

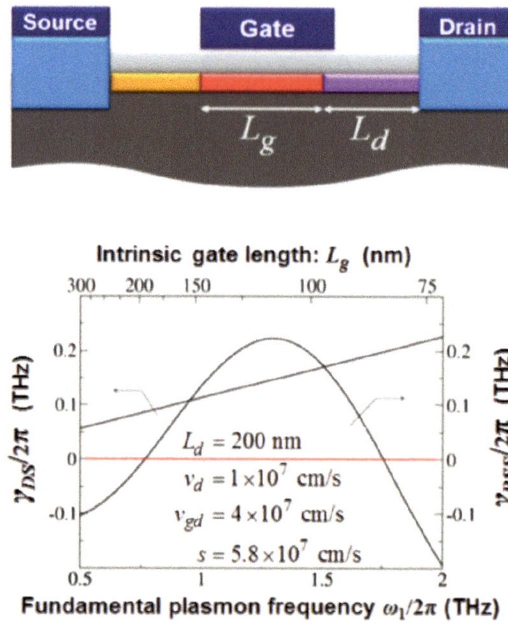

Figure 10.3 Contributions to the plasma-wave instability growth rate of the DS and RSS mechanism, ($\gamma_{DS}/2\pi$ and $\gamma_{RSS}/2\pi$) in a HEMT with typical geometric and material parameters as functions of the fundamental plasmon mode frequency. Due to the nature of transit-time driven mechanism the RSS instability is sensitive to the plasmon mode frequency. After Ref. [12, 19].

The plasmonic boom is similar to the supersonic boom. When the velocity of an aircraft exceeds that of the sonic wave, it promotes the supersonic boom generating an acoustic shockwave. Likewise, when the electron drift velocity exceeds the plasma velocity, it promotes the plasmonic boom generating a plasma shockwave [25]. If the structure along which the shockwave propagates serves a resonant nature, only the plasma resonant frequency components of the original broadband shockwave could survive whereas anti-resonant components attenuate totally as shown in Fig. 10.4. As a result, the plasmonic boom may also give rise to emissions of coherent plasmon oscillations in the plasmonic cavity. This plasmonic-boom-type instability was theoretically studied and resonant frequencies in the

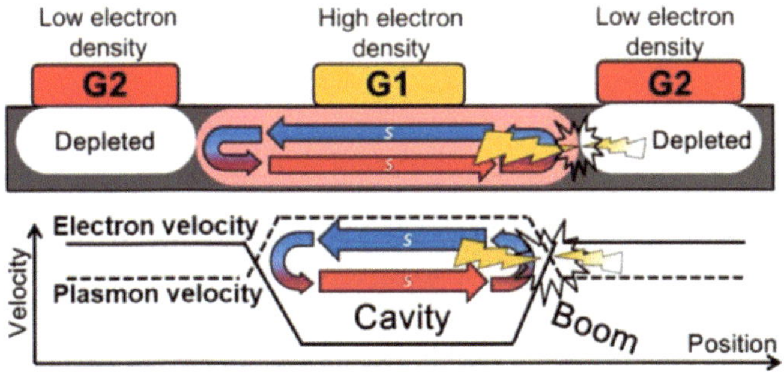

Figure 10.4 Plasmonic boom instability in a FET structure with dual gates (G1 and G2) structure.

THz region are predicted in sub-micrometer transistor structures [2, 9, 20].

In reality, under normal conditions of electron densities of the order of 10^{11} to 10^{12} cm^{-2} in the channel, the electron drift velocity stays around 10^5 to 10^6 cm/s whereas the plasmon velocity approaches around 10^7 to 10^8 cm/s. Therefore, it is difficult to make the condition to promote the plasmonic boom instability in transistor channels where electron drift velocity exceeds the plasmon velocity. To make it possible, electrostatic spatial electron density modulation by the gate voltage application is a good solution. When the electrons are depleted in a portion (like depleted drain-side ungated region) by setting the gate voltage to or below the threshold level whereas the main gated channel is highly doped, the current continuity and energy conservation conditions forth electrons travel slower (faster) in doped (depleted) regions and forth plasmons propagate faster (slower) in doped (depleted) regions simultaneously. Thus, the boundary between the doped and depleted region may see the situation that electrons travel with increasing drift velocity surpassing the plasmon velocity as shown in Fig. 10.4.

10.2.5 Coupling between Plasmons and Photons

2D plasmon itself is a non-radiative mode so that a coupling mechanism to efficiently transfer the energy between photons and plasmons is needed to work for radiation emission devices like plasmonic oscillators. A single-gate metal electrode works as a narrow-band dipole antenna to couple the THz photons with plasmons in the channel underneath the gate electrode. A metal-wired grating coupler structure is frequently utilized as a broadband antenna to yield THz electromagnetic-wave emission [3, 4]. If the single gate is replaced with a grating-finger type gate [5], a plural number of plasmonic cavities (whose electron density is modulated by the bias voltage of the grating-finger gate) are electrostatically coupled (via inter-finger region with less-electron-density) in a spatially distributed configuration. In this case constant dc channel current, which is generated by the applied dc drain bias, gives rise to periodic electron velocity modulation over the channel. This may promote the RSS instability [11, 12].

The dual grating-gate (DGG) structure provides an improved confinement of the 2D electrons into the plasmon cavities independent of the tuned gate biases, resulting in more intense resonant plasma excitation with higher quality factors (see Fig. 10.5a) [16, 18, 19, 30, 31]. The DGGs can alternately modulate the 2D electron densities to periodically distribute the plasmonic cavities (~100 nm width in microns distance) along the channel by applying large fraction of gate biases for sub-grating G1 and G2 [12]. Under pertinent drain-source dc bias conditions, dc electron drift flows may promote the plasmon instability, resulting in self-oscillation with characteristic frequencies in the THz regime. The 2D plasmons are originally longitudinal evanescent mode, but can be converted to radiation emission of freely propagating electromagnetic waves via DGG antenna structure (see Fig. 10.5a). Figure 10.5b depicts the potential distribution and equivalent circuit model for the RSS instability under a constant dc drain bias condition [13]. The 2D electron channel consists of a periodic series of highly confined 2D plasmon cavity section underneath the gate finger G1 and depleted section underneath the gate finger G2.

An asymmetric DGG (ADGG) structure in which the left-hand space and right-hand space between the adjacent gate fingers are asymmetric is proposed [31]. A wider (narrower) space gives a higher (lower) characteristic impedance to the boundary of the plasmon cavity underneath a gate finger, so that the ADGG structure

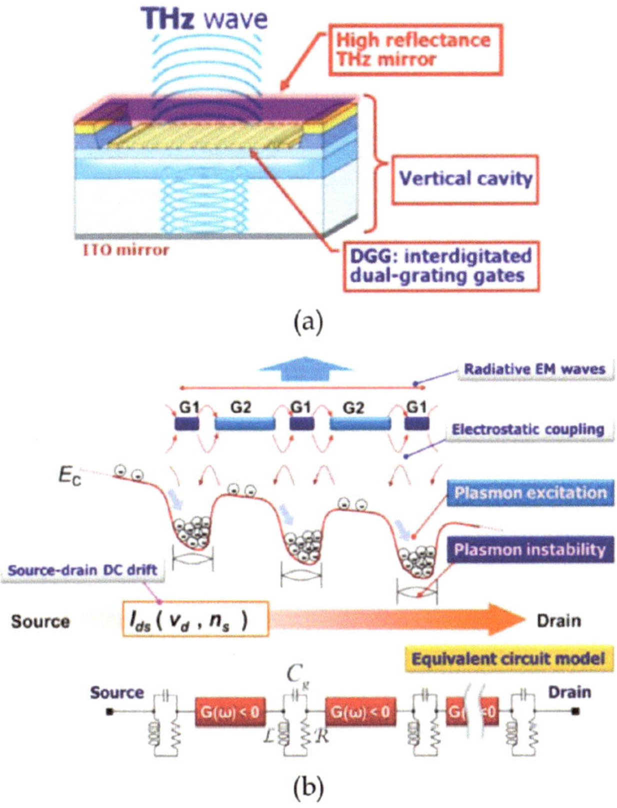

Figure 10.5 DGG-HEMT-type THz emitter with a vertical cavity. (a) structure, and (b) band diagram and equivalent circuit.

causes asymmetric plasmon cavity boundaries, which may promote the DS instability as well as the RSS instability [21, 22].

10.3 Experiments

10.3.1 AlGaN/GaN Single-Gate HEMT

AlGaN/GaN single-gate HEMTs were utilized to investigate tunable THz radiation emission originated from the plasmon instability [17]. The devices were based on AlGaN/GaN heterostructures grown

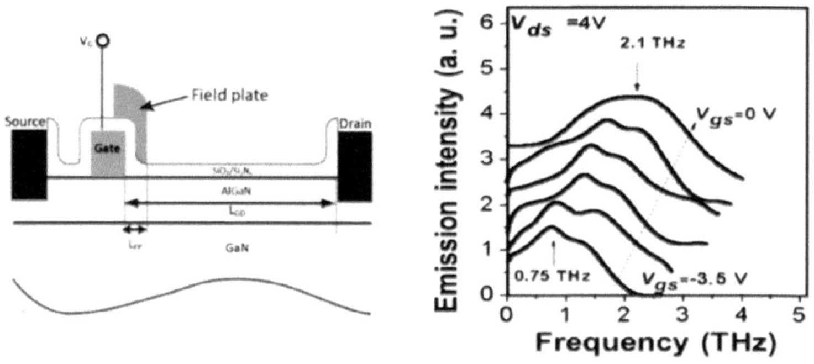

(a) (b)

Figure 10.6 THz emission from a current-driven AlGaN/GaN HEMT. (a) Device cross-sectional structure. (b) THz emission spectra under dc-current flow with drain-source bias V_{ds} of 4 V. After Ref. [17].

by metal-organic chemical-vapor deposition (MOCVD) method, and featured by a field plate structure that is popular in GaN-based HEMT's to help suppress the crystal defects-oriented gate leakage and relax lowering the breakdown voltages. The gate length, that determines the plasmon mode frequencies, were 150 (250) nm for sample 1 (2). The field-effect mobility of electrons in the channel was estimated as $\sim$1,500 cm^2/Vs.

The THz emission spectra were measured using a Fourier transform far-infrared spectrometer (JASCO FARIS-I) with a 4.2 K-cooled Si composite bolometer (Infrared Lab.), demonstrating room-temperature tunable THz emission over 0.75 to 2.1 THz peak frequency range responding to the gate-voltage-tuned electron density modulation in the channel (see Fig. 10.6). The radiation emission exhibited threshold-like behavior with respect to the applied drain-source voltage, manifesting the occurrence of current-driven plasmon instability. The plasmonic cavity of the gated channel region holds asymmetric boundaries (high impedance at the depleted drain-side and low impedance at the heavily doped source side), indicating the DS instability as the major mechanism of the instability.

The point of discussion is rather broad spectral width of emission with Q factors of around 0.5$\sim$1.5, and rather weak emission intensity that was not characterized well. These are attributed by the short momentum relaxation time of 2D electrons in the GaN channel. Replacement of the channel material with far superior electron transport properties is mandatory to realize coherent THz oscillation radiation from the device.

10.3.2 InGaAs/InAlAs/InP Dual-Grating-Gate HEMT

In our early works, we designed, fabricated, and measured several different types of DGG-HEMTs utilizing InGaAs/InAlAs/InP heterostructure material systems including symmetric DGG, asymmetric DGG, asymmetric and chirped DGG, as well as double-deck symmetric DGG structures [16, 18, 19]. These ideas to modify the DGG structure are as follows.

(i) Asymmetric DGG causes asymmetric plasmon cavity boundaries so that it may promote the Dyakonov–Shur Doppler-shift-type instability.
(ii) Chirped DGG can spatially modulate the plasma resonant frequencies depending on the chirped grating-finger size. To promote the current-driven plasmon instability one has to apply a dc drain-source bias voltage resulting in linear slope of the potential and electron densities along the channel, resulting in broadening of the plasmon resonant frequencies. The chirped DGG can align the plasmon resonant frequencies over the channel region, which helps improve the *Q* factor of THz radiation emission.
(iii) Double-deck DGG can replace the metal DGG electrode with heterojunction semiconducting materials. This help match the characteristic impedance of the DGG antenna structure to that of the free space. This help increase the coupling efficiency between non-radiative plasmons and radiative photons.

Those improvements reflected on the measured THz emission spectra in a certain amount. The mechanisms to promote the instability were identified to be a mixture of the Dyakonov–Shur Doppler-shift-type and Ryzhii–Satou–Shur electron-transit type. However, no one has yet succeeded in coherent single-mode THz emission so far [18, 19], and the major part of the emission was dominated by hot-plasmon-originated spontaneous broadband emission.

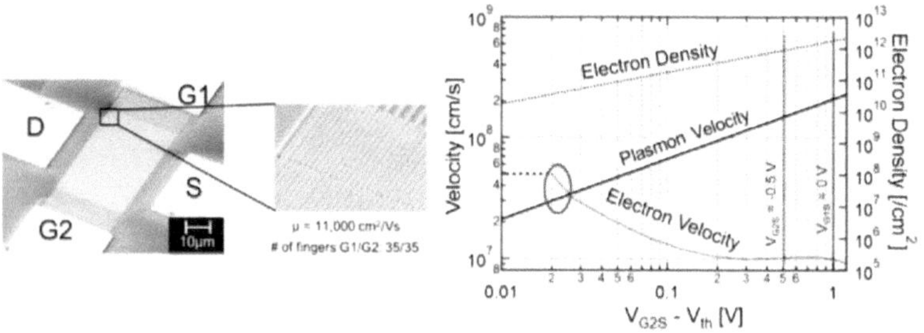

(a) (b)

Figure 10.7 A fabricated InGaAs/InAlAs/InP DGG-HEMT. (a) Scanning electron microscopic images. (b) Extracted electron drift velocity and plasmon velocity as functions of Gate-2 swing voltage $V_{G2S}-V_{th}$ at $V_{DS}=1$ V, $V_{G1S}=0$ V. It was confirmed that when V_{G2S} is close to the threshold level electron velocity exceeds the plasmon velocity. After Ref. [32].

Recently we succeeded in experimental demonstration of THz emission in an InGaAs-based DGG HEMT structure promoted by the plasmonic boom instability [32]. We designed and fabricated an InGaAs-based DGG-HEMT (Fig. 10.7a). The gate length was 150 nm, and the distance of each gate electrodes was 150 nm. The mechanism of generation of the plasmonic boom instability in a DGG-HEMT is similar to the case shown in Fig. 10.4. The DGG structure modulates the electron drift velocity and plasma velocity in a complementary manner. One gate electrode G1 (the other gate G2) is biased high (low), working as plasmonic cavities (high-field, depleted regions) wherein the plasma velocity is higher (lower) than the electron drift velocity. The plasmonic boom is generated at the boundaries where the electron drift velocity exceeds the plasma velocity as shown in Fig. 10.7b which is only given when Gate-2 bias voltage V_{G2S} is close to the threshold voltage V_{th}. We compared the THz emissions when $V = -1.0$ V (close to V_{th}) and -0.5 V (far beyond V_{th}).

The THz radiation emission from the DGG-HEMT device was measured under photomixed dual-CW laser irradiation with different frequencies around the plasmon mode frequencies at different HEMT bias conditions by using a Fourier transform IR (FTIR) spectrometer and a 4.2 K-cooled Si bolometer. The plasmonic boom instability generates a shockwave with rather broadband spectrum.

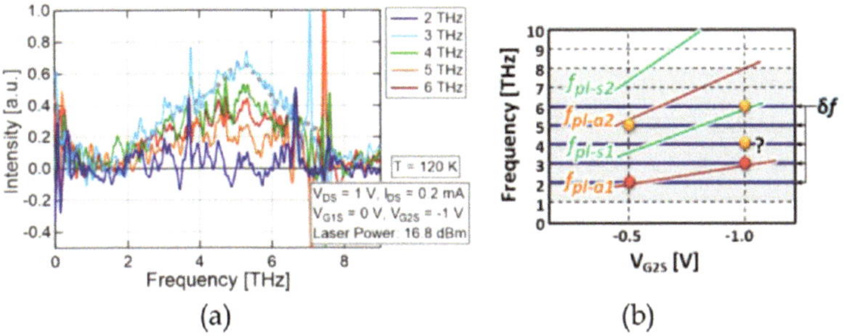

Figure 10.8 THz emission from a current-driven InGaAs/InAlAs/InP DGG-HEMT under IR photomixed irradiation with difference frequency δf of 2, 3, 4, 5, and 6 THz. (a) THz emission spectra beyond the black-body radiation at 120 K. Only the conditions of $\delta f = 3$ THz gives a distinctive enhancement of THz emission with a mono peak at $\sim$ 5.5 THz. (b) Observed emission spectral peak frequencies (indicated with colored circles) versus plasmon mode frequencies (lines) for $V_{G2S} = -0.5$ V and -1.0 V. $f_{pl\_a1(s1)}$ and $f_{pl\_a2(s2)}$ are the 1st and 2nd plasmon mode frequencies under asymmetric (symmetric) boundaries, respectively. The photomixed difference frequencies (δf) are shown with horizontal blue lines. The shaded area ranging from 1 to 8 THz shows the available bandwidth of the FTIR measurement setup. After Ref. [32].

The photomixed dual-CW-IR laser beams generate coherent spectral photocurrent around the frequency of δf so that the broad plasmonic ac current could be injection-locked to the photogenerated δf component. We set δf at 2, 3, 4, 5, and 6 THz. The device was cooled at 120 K in order to reduce the background thermal noise. The measured radiation spectra from the device showed distinctive emissions beyond the black-body radiation, which was promoted by δf-dependent coherent plasmons (see Fig. 10.8). The results suggest the occurrence of the plasmonic boom instability stimulated by the DC-current flow in the 2D channel under pertinent DC bias voltages [32].

10.3.3 Graphene-Channel Dual-Grating-Gate FET

We investigated dc-current-driven Dirac-plasmonic instabilities in high mobility graphene metamaterials that combine the advantage of an efficient tunable absorber, emitter and amplifier at room temperatures [21, 22]. Plasmon modes in our devices are excited in monolayer graphene on hexagonal boron nitride (hBN) with a periodic DGG structure positioned above the graphene sheet (see Figs. 10.9a,b and c). The DGG structure modulates the incoming electromagnetic wave and defines the plasmonic wave vectors. The samples were fabricated as graphene-channel FETs (GFETs) with structures featuring an interdigitated DGG. The plasmonic cavities are formed below the gates electrodes and designed with symmetric or asymmetric boundaries and electron mobilities around 50.000 cm^2/Vs at room temperatures.

DGG-bias-dependent electron density modulation causes spatial complementary modulation of the plasmon and drift velocities. This may cause Dyakonov–Shur Doppler-shift (DS) type, Ryzhii–Satou–Shur electron-transit (ET) type, and/or Cherenkov plasmonic boom (PB)-type instabilities under pertinent cavity boundaries and fraction of the electron drift velocity and the plasmon velocity. When the instability-driven gain surpasses the Drude loss, the system yields the net gain, resulting in plasmon self-oscillation at the resonant frequencies. The DGG works as a broadband antenna that can convert the non-radiative plasma oscillations to radiative THz waves. This, in turn, enables spontaneous THz emission of radiation as well as coherent light amplification of stimulated emission to the incident THz waves.

We examined three samples: two asymmetric DGG (ADGG) types (ADGG1 and ADGG2) [21, 22] and one symmetric DGG type (SDGG) alignment as summarized in Table 10.1 and as shown in 10.9b,c [22]. With the applied gate voltages V_{g1} and V_{g2} each device supports the formation of two different plasmonic cavities (types C1 and C2) below the fingers of Gate 1 and Gate 2 in the DGG device structure. Terahertz TDS was employed to measure the changes in the THz pulses transmitted through the graphene plasmonic cavities of type C1 (C2) when sweeping V_{g1} (V_{g2}) and keeping the voltage on the other gate electrode constant at the charge neutral point (CNP) $V_{g2} = V_{CNP2}$ ($V_{g1} = V_{CNP1}$) (see Figs. 10.9d,e). The drain-to-source voltage (V_d) dependent measurements were also conducted at a constant V_g. The transmission coefficient at a given V_g and V_d is referred to as T while T_{CNP} is the transmission coefficient at $V_g = V_{CNP}$. The measured extinction spectra $(1-T/T_{CNP})$ at

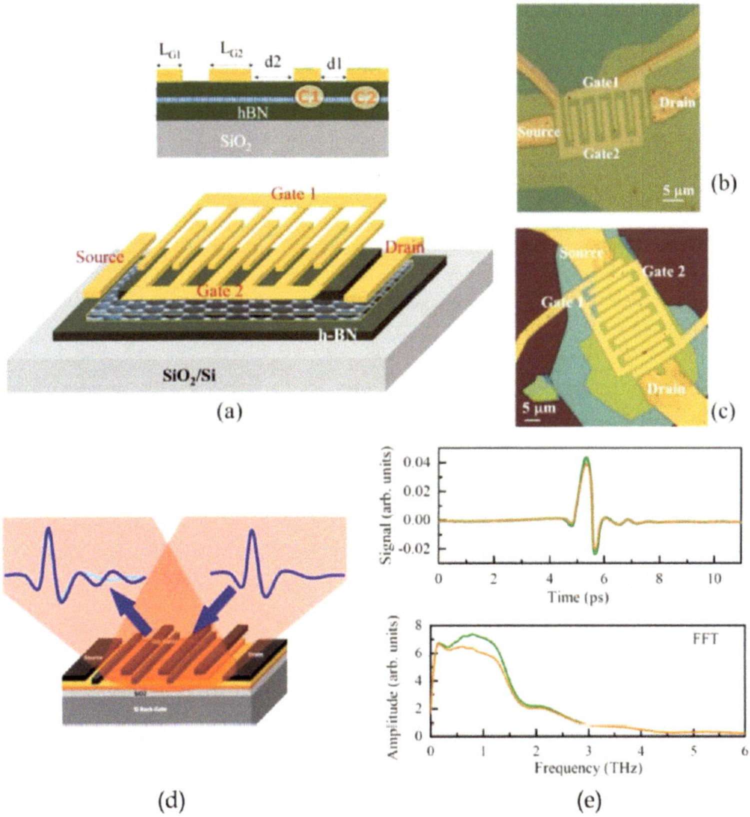

Figure 10.9 DGG-GFET samples for THz time-domain spectroscopic measurement. (a) Schematic illustrating of the hBN/graphene/hBN heterostructure asymmetric DGG-GFET. (b) Optical image of an asymmetric DGG-GFET (ADGG-1). (c) Optical image of the symmetric DGG-GFET (S-DGG). (d) Experimental geometry of incident THz pulse and transmitted/reflected pulse in an oblique angle. (e) Measured temporal waveforms (upper) and their Fourier spectra (lower). Orange: incident, green: output. After Ref. [22].

$V_d = 0$ V exhibited similar tendencies of polarization-sensitive resonant absorptions among the three samples as shown in Fig. 10.10 [21, 22].

Next, we conducted the TDS measurement under application of various non-zero V_d voltages. The observed extinction spectra

Table 10.1 Samples specifications [22]

Device	ADGG-1		ADGG-2		SDGG
Cavity	C1	C2	C1	C2	C1
Biased cavity length (μm)	0.75	1.50	0.5	1.0	2.0
Total channel length (μm)	24.0		26.5		38.5
$d1$ and $d2$ (μm)*	0.5 and 1.0		0.5 and 2.0		1.0
Channel width (μm)	1.325		4.9		3.2
CNP (V)	−0.12		+0.15		−0.10

*$d1$ and $d2$ are spaces between one Gate 1 finger and the right-hand-side and left-hand-side adjacent Gate 2 finger, respectively, as shown in Fig. 10.9a.

are plotted in Fig. 10.11 [22]. When V_d increases, the absorption peaks weakened with red shifting, all the samples approaching perfect transparency over the measured entire frequency range (at $V_d = 370$ mV for C2 of ADGG1). Further increase in V_d for only ADGG samples gives rise to negative absorption peak, i.e., resonant amplification, appears in the extinction spectra with a noticeable blue shift, reaching the maximal gain (9%), which is far beyond the interband-transition-limited quantum efficiency (2.3%).

In summary, we explored current-driven plasmon dynamics in monolayer graphene active metamaterials. Frequency tunable THz light amplification up to 9% gain at RT by current-driven plasmon instabilities produced in an ADGG-GFET structure has been successfully demonstrated [21, 22].

10.4 Future Subjects and Prospects

As described in the previous subsections, 2D plasmons in massive compound semiconductor heterostructures tend to promote current-driven instability and showed asymptotic behavior for coherent THz self-oscillation. However, due to insufficient electron transport properties of those 2D plasmons, hot-plasmon-originated incoherent broadband emission was dominated in the emission spectra so that perfect coherent THz oscillation was yet to be obtained. However, recent progress on observation of clear enhancement of THz emission promoted by the plasmonic boom-instability encourages us to drive forward.

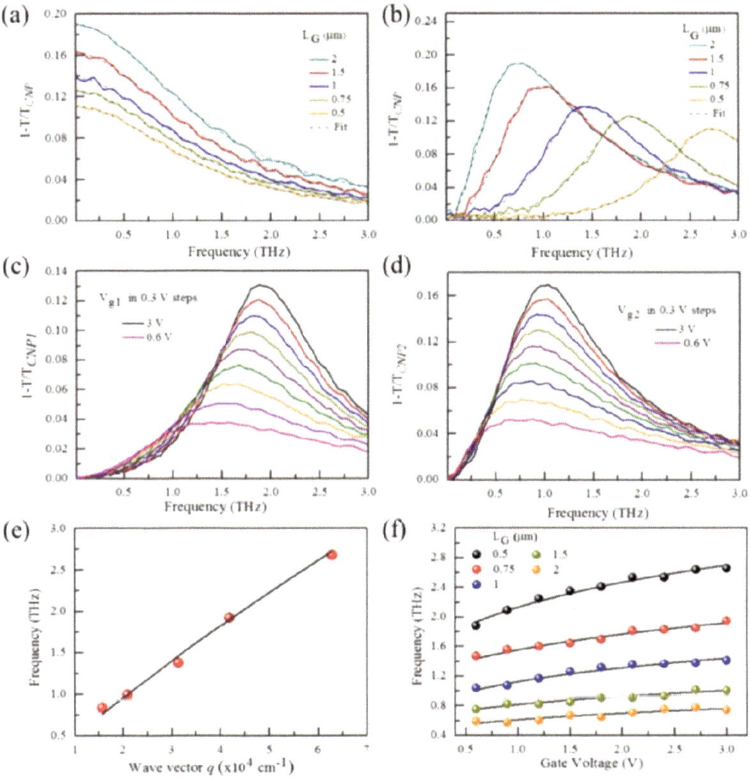

Figure 10.10 Gate voltage and length dependent extinction spectra of the graphene structures for $V_d = 0$ V. Plasmonic cavity length (L_G) dependent extinction spectra of the three devices with incident light polarized parallel (a) and perpendicular (b) to the DGG fingers. The measured line shape is well reproduced by a Drude model fit for parallel polarization (dashed line in (a)) and a damped oscillator model fit for perpendicular polarization (dashed line in (b)). Measured gate voltage-dependent transmission spectra $1-T/T_{CNP}$ of the asymmetric device A-DGG1 with incident light polarized perpendicular to the DGG fingers, with biased cavities C1 when sweeping V_{g1} while the voltage on the other gate electrode is kept constant at $V_{g2} = V_{CNP2}$ (c) and biased cavities C2 when sweeping V_{g2} while the voltage on the other gate electrode is kept constant at $V_{g1} = V_{CNP1}$ (d). Scaling laws of graphene plasmon resonance frequency in the three devices as a function of wave vector $q = \pi/L_G$ (e) and gate voltage (f). The solid lines in (e) and (f) are fits to data using a standard damped oscillator model. After Ref. [22].

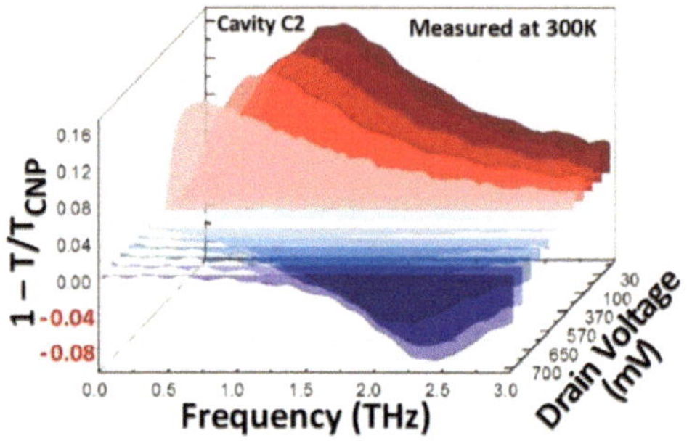

Figure 10.11 Drain bias dependent extinction spectra of the graphene structures. Spectra measured in cavity C2 of device A-DGG1 for fixed $V_{g2} - V_{CNP2} = 3$ V and $V_{g1} = V_{CNP1}$ when varying V_d. After Ref. [22].

Graphene Dirac plasmons, on the contrary, made it possible in DGG-GFET structure to realize coherent light amplification of stimulated plasmon emission of THz radiation even at room temperatures thanks to the extraordinary superior transport properties of graphene Dirac fermions and plasmons. The amplification gain approached 9% by monolayer graphene, which is four times higher than the quantum mechanical limit that is given when THz photons directly interact with graphene carriers of Dirac fermions. This suggests that the amplification gain could be multiplied by the number of graphene layers in the channel. In this case, non-Bernal stacked multiple layers of graphene is needed, which is synthesized by the thermal decomposition of a C-face hexagonal SiC substrate as long as the gate stack works for electrostatic charge doping weakened with increasing the number of layers by the charge screening effect of the doped charge at topmost graphene layers [33].

The measured frequency dependence of the extinction peaks perfectly traces the theory with sharp gain thresholds. Considering the extracted plasmon and drift velocity relations, a mixture of DS and RSS instabilities may cause such phenomena [21]. However there still exists quantitative discrepancies of the threshold drain-bias condition that changes the graphene from a lossy medium to a gain medium where electron drift velocity is lower than the plasmon velocity [21, 22]. A new phenomenological modeling was given to qualitatively reproduce the occurrence of THz amplification without

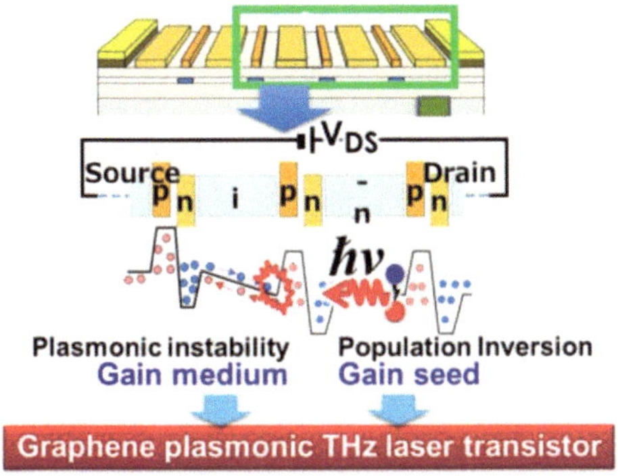

Figure 10.12 A possible structure for graphene plasmonic THz laser transistor.

need for any plasmon instabilities [21, 22]. Quantitative perfect interpretation of observed phenomena needs further study which will be a future subject.

Besides, we theoretically predicted the possibility of graphene THz laser transistors under optical and/or current-injection pumping [34–38]. So far, we succeeded in world-first single-mode THz emission and broadband THz amplified spontaneous emission (ASE) in a distributed-feedback (DFB) cavity dual-gate GFET structure at 100K [39]. The lasing threshold temperature was low up to 100K and its THz emission intensity was very weak (~0.1 μW for single-mode emission at 5.2 THz and ~10 μW for ASE). To dramatically increase both the lasing threshold temperature and the laser output intensity, introduction of the graphene Dirac plasmons is the key, we believe. In a single DGG-GFET structure, accommodating a gain-seed section of spontaneous THz emission under carrier population inversion conditions and a plasmonic amplifier section that promotes the plasmon instability (see Fig. 10.12) would be a possible solution [40]. This is a future subject.

10.5 Conclusion

The theory and experiments for plasmon-based THz oscillators were reviewed. First, theory of hydrodynamics and instabilities of 2D plasmons were described. Second, experimental studies for massive 2D plasmons in compound semiconductor heterostructures as well as massless 2D plasmons in graphene were described. 2D plasmons in massive compound semiconductor heterostructures are still on the way toward coherent THz oscillation. Recent progress on observation of clear enhancement of THz emission promoted by the plasmonic boom-instability encourages us to go forward. Graphene Dirac plasmons, on the other hand, made it possible in DGG-GFET structure to realize coherent light amplification of stimulated plasmon emission of THz radiation even at room temperatures with a maximal gain of 9% which is four times higher than the quantum mechanical limit when THz photons directly interact with graphene electrons. To realize room-temperature, intense THz lasing operation with dry-cell battery, graphene plasmonic laser transistor would be a promising solution, which will become in reality in the near future.

Acknowledgments

The author thanks Victor Ryzhii, Akira Satou, Stephane A. Boubanga-Tombet, Takayuki Watanabe, Deepika Yadav, Tomotaka Hosotani, Maxim Ryzhii, Vladimir Mitin, Michael S. Shu, Alexander A. Dubinov, Vyacheslav V. Popov, Vladimir Ya Aleshkin, Dmitry Svintsov, Wojciech Knap, Dmytro B. But, Ilya V. Gorbenko, Valentin Kachorovskii, Abdel El Fatimy, Nina Diakonova, Frederic Teppe, Dominique Coquillat and Yahya M. Meziani for their contributions. He also thanks Michel Dyakonov for his valuable discussion and encouragements. The works done by the author was supported by JSPS KAKENHI #23000008, 16H06361, 20K20349, and No. 21H04546, Japan, and JSPS-RFBR Bilateral Joint-Research Program #120204801, Japan and Russia.

References

1. Tonouchi, M. (2007). Cutting-edge terahertz technology, *Nature Photon.*, **1**, pp. 97–105.

2. Krasheninnikov, M., and Chaplik, A. (1980). Instabilities of two-dimensional plasma waves, *Sov. Phys. JETP*, **52**, pp. 279–282.
3. Tsui, D. C., Gornik, E., and Logan, R.A. (1980). Far infrared emission from plasma oscillations of Si inversion layers, *Solid State Comm.*, **35**, pp. 875–877.
4. Hopfel, R., Lindemann, G., Gornik, E., Stangl, G., Gossard, A. C., and Wiegmann, W. (1982). Cyclotron and plasmon emission from two-dimensional electrons in GaAs, *Surf. Sci.*, **113**, pp. 118–123.
5. Wilkinson, R. J., Ager, C. D., Duffield, T., Hughes, H. P., Hasko, D. G., Armed, H., Frost, J. E. F., Peacock, D. C., Ritchie, D. A., Jones, A. C., Whitehouse, C. R., and Apsley, N. (1992). Plasmon excitation and self-coupling in a bi-periodically modulated two-dimensional electron gas, *J. Appl. Phys.*, **71**, pp. 6049–6061.
6. Dyakonov, M., and Shur, M. (1993). Shallow water analogy for a ballistic field effect transistor: new mechanism of plasma wave generation by dc current, *Phys. Rev. Lett.*, **71**, pp. 2465–2468.
7. Hirakawa, K., Yamanaka, K., Grayson, M., and Tsui, D. C. (1995). Far-infrared emission spectroscopy of hot two-dimensional plasmons in Al0.3Ga0.7As/GaAs heterojunctions, *Appl. Phys. Lett.*, **67**, pp. 2326–2328.
8. Matov, O., Meshkov, O., Polischuk, O., and Popov, V. V. (1997). Generation of submillimeter electromagnetic radiation from two-dimensional plasma waves in a semiconductor heterostructure with metal grating, *Physica A*, **241**, pp. 409–413.
9. Mikhailov, S. (1998). Plasma instability and amplification of electromagnetic waves in low- dimensional electron systems, *Phys. Rev. B*, **58**, pp. 1517–1532.
10. Knap, W., Lusakowski, J., Parenty, T., Bollaert, S., Cappy, A., Popov, V. V., and Shur, M. S. (2004). Terahertz emission by plasma waves in 60 nm gate high electron mobility transistors, *Appl. Phys. Lett.*, **84**, pp. 2331–2333.
11. Ryzhii, V., Satou, A., and Shur, M. S. (2005). Transit-time mechanism of plasma instability in high electron mobility transistors, *Phys. Status Solidi (A)*, **202**, pp. R113–R115.
12. Ryzhii, V., Satou, A., and Shur, M. S. (2005). Plasma instability and terahertz generation in HEMTs due to electron transit-time effect, *IEICE Trans. Electron.*, **E89-C**, pp. 1012–1019.
13. Ryzhii, V., Satou, A., Ryzhii, M., Otsuji, T., and Shur, M. S. (2008). Mechanism of self-excitation of terahertz plasma oscillations in periodically double-gated electron channels, *J. Phys.: Condens. Matter*, **20**, p. 384207.

14. Lusakowski, J., Knap, W., Dyakonova, N., and Varani, L. (2005). Voltage tunable terahertz emission from a ballistic nanometer InGaAs/InAlAs transistor, *J. Appl. Phys.*, **97**, p. 064307.
15. Dyakonova, N., Teppe, F., Lusakowski, J., Knap, W., Levinshtein, M., Dmitriev, A. P., Shur, M. S., Bollaert, S., and Cappy, A. (2005). Magnetic field effect on the terahertz emission from nanometer InGaAs/AlInAs high electron mobility transistors, *J. Appl. Phys.*, **97**, p. 114313.
16. Otsuji, T., Meziani, Y. M., Hanabe, M., Ishibashi, T., Uno, T., and Sano, E. (2006). Grating-bicoupled plasmon-resonant terahertz emitter fabricated with GaAs-based heterostructure material systems, *Appl. Phys. Lett.*, **89**, p. 263502.
17. El Fatimy, A., Dyakonova, N., Meziani, Y., Otsuji, T., Knap, W., Vandenbrouk, S., Madjour, K. K., Théron, D., Gaquiere, C., Poisson, M. A., Delage, S., Pristawko, P., and Skierbiszewski, C. (2010). AlGaN/GaN high electron mobility transistors as a voltage-tunable room temperature terahertz sources, *J. Appl. Phys.*, **107**, p. 024504.
18. Otsuji, T., Watanabe, T., Boubanga-Tombet, S. A., Satou, A., Knap, W. M., Popov, V. V., Ryzhii, M., and Ryzhii, V. (2011). Emission and detection of terahertz radiation using two-dimensional electrons in III-V semiconductors and graphene, *IEEE Trans. Terahertz Sci. Technol.*, **3**, pp. 63–71.
19. Otsuji, T., Watanabe, T., Boubanga-Tombet, S. A., Satou A., Ryzhii, V., Popov, V. V., and Knap, W. (2014). Emission and detection of terahertz radiation using two-dimensional plasmons in semiconductor nanoheterostructures for nondestructive evaluations, *Opt. Eng.*, **53**, p. 031206.
20. Aizin, G., Mikalopas, J., and Shur, M. (2016). Current-driven plasmonic boom instability in three- dimensional gated periodic ballistic nanostructures, *Phys. Rev. B*, **93**, p. 195315.
21. Boubanga-Tombet, S., Knap, W., Yadav, Y., Satou, A., But, D. B., Popov, V. V., Gorbenko, I. V., Kachorovskii, V., and Otsuji, T. (2020). Room temperature amplification of terahertz radiation by grating-gate graphene structures, *Phys. Rev. X*, **10**, p. 031004.
22. Boubanga-Tombet, S., Knap, W., Yadav, Y., Satou, A., But, D. B., Popov, V. V., Gorbenko, I. V., Kachorovskii, V., and Otsuji, T. (2021). Paving the way for tunable graphene plasmonic THz amplifiers, *Front. Phys.*, **7**, p. 726806.
23. Ryzhij, V., Bannov, N. A., and Fedirko, V. A. (1984). Ballistic and quasi-ballistic transport in semiconductor structures, *Sov. Phys. Semicond.*, **18**, pp. 481–491.

24. Novoselov, K. S., Geim, A. K., Morozov, S. V., Jiang, D., Zhang, Y. et al. (2004). Electric field effect in atomically thin carbon films, *Science*, **306**, pp. 666–669.
25. Geim, A. K., and Novoselov, K. S. (2007). The rise of graphene, *Nat. Mater.*, **6**, pp. 183–191.
26. Grigorenko, A. N., Polini, M., and Novoselov, K. S. (2012). Graphene plasmonics, *Nature Photon.*, **6**, pp. 749–758.
27. Hartmann, R. R., Kono, J., and Portnoi, M. E. (2014). Terahertz science and technology of carbon nanomaterials, *Nanotechnol.*, **25**, p. 322001.
28. Tredicucci, A., and Vitiello, M. S. (2014). Device concepts for graphene-based terahertz photonics, *IEEE J. Selected Topics in Quantum Electron.*, **20**, p. 8500109.
29. Li, Y., Tantiwanichapan, K., Swan, A. K., and Paiella, R. (2020). Graphene plasmonic devices for terahertz optoelectronics, *Nanophoton.*, **9**, pp. 1901–1920.
30. Otsuji, T., Hanabe, M., Nishimura, T., and Sano, E. (2006). A grating-bicoupled plasma-wave photomixer with resonant-cavity enhanced structure, *Opt. Express*, **14**, pp. 4815–4825.
31. Popov, V. V., Fateev, D. V., Otsuji, T., Meziani, Y. M., Coquillat, D., and Knap, W. (2011). Plasmonic terahertz detection by a double-grating-gate field-effect transistor structure with an asymmetric unit cell, *Appl. Phys. Lett.*, **99**, p. 243504.
32. Hosotani, T., Satou, A., and Otsuji, T. (2021). Terahertz emission in an InGaAs-based dual-grating-gate high-electron-mobility transistor plasmonic photomixer, *Appl. Phys. Exp.*, **14**, p. 14 051001.
33. Ryzhii, M., Ryzhii, V., Otsuji, T., Mitin, V., and Shur, M. S. (2010). Electrically induced n-i-p junctions in multiple graphene layer structures, *Phys. Rev. B*, **82**, p. 075419.
34. Ryzhii, V., Ryzhii, M., and Otsuji, T. (2007). Negative dynamic conductivity in optically pumped graphene, *J. Appl. Phys.*, **101**, p. 083114.
35. Ryzhii, M., and Ryzhii, V. (2007). Injection and population inversion in electrically induced p-n junction in graphene with split gates, *Jpn. J. Appl. Phys.*, **46**, pp. L151–L153.
36. Ryzhii, V., Ryzhii, M., Mitin, V., and Otsuji, T. (2011). Toward the creation of terahertz graphene injection laser, *J. Appl. Phys.*, **110**, p. 094503.
37. Satou, A., Ryzhii, V., Kurita, Y., and Otsuji, T. (2013). Threshold of terahertz population inversion and negative dynamic conductivity in graphene under pulse photoexcitation, *J. Appl. Phys.*, **113**, p. 143108.

38. Otsuji, T., Boubanga-Tombet, S., Satou, A., Ryzhii, M., and Ryzhii, V. (2013). Terahertz-wave generation using graphene-toward new types of terahertz lasers, *IEEE J. Selected Topics in Quantum Electron.*, **19**, p. 8400209.
39. Yadav, D., Tamamushi, G., Watanabe, T., Mitsushio, J., Tobah, Y., Sugawara, K., Dubinov, A. A., Satou, A., Ryzhii, M., Ryzhii, V., and Otsuji, T. (2018). Terahertz light- emitting graphene-channel transistor toward single-mode lasing, *Nanophoton.*, **7**, pp. 741–752.
40. Otsuji, T., Boubanga-Tombet, S. A., Satou A., Yadav, D., Fukidome, H., Watanabe T., Suemitsu T., Dubinov, A. A., Popov, V. V., Knap, W., Kachorovskii, V., Narahara, K., Ryzhii, M., Mitin, V., Shur, M. S., and Ryzhii, V. (2022). Graphene-based plasmonic metamaterial for terahertz laser transistors, *Nanophoton.*, **11**, pp. 1677–1696.

Chapter 11

Beamforming THz Transmitters

Sanggeun Jeon
School of Electrical Engineering, Korea University, Seoul 02841, Korea
sgjeon@korea.ac.kr

This chapter presents the implementation of THz beamforming transmitters. First, as a key beamforming component, the THz phase shifters in various topologies are discussed. Two passive topologies (reflective-type and switched-type phase shifters) and an active topology (a vector-sum phase shifter) are described and compared. Then, introduced are the beamforming THz transmitters implemented in various architectures and technologies. As an early work of an integrated beamforming transmitter, a 200 GHz CMOS beamforming array based on the distributed active radiators is described. An integrated 1 × 4 phased-array transmitter integrating a 20 GHz frequency synthesizer using a 0.13 μm SiGe BiCMOS technology is introduced. Finally, a CMOS eight-element beamforming transmitter operating at 370–410 GHz is described.

11.1 Introduction

The THz band attracts considerable attention due to the unprecedentedly wideband spectrum available for numerous applications.

Advances in Terahertz Source Technologies
Edited by Gun-Sik Park, Masahiko Tani, Jae-Sung Rieh, and Sang Yoon Park

ISBN 978-981-4968-89-8 (Hardcover), 978-1-003-45967-5 (eBook)
www.jennystanford.com

Ultra-fast wireless communication, high-resolution imaging, low-cost medical diagnostics, spectroscopy, and remote sensing are envisioned with an aid of the wideband THz spectrum [1–3]. However, the practical use of THz spectrum for such applications has not been fully arrived yet and there are still many hurdles to overcome in the field. One of the critical challenges is the high free space path loss (FSPL) and high atmospheric absorption in the THz band. Being proportional to the frequency squared, the FSPL at 300 GHz, for example, reaches as high as 102 dB over 10 m, thus limiting the link budget severely. The absorption by oxygen or moisture included in the atmosphere is significantly increased in the THz band and hence makes the link budget even worse. The high path loss and tight link budget obviously diminish the link range of THz transceivers and thus impedes the use of THz for practical applications.

A common way of extending the link range to a reasonable distance enough for practical use is to increase the effective isotropically radiated power (EIRP) of transmitters. As expressed as a product of the transmit power and antenna gain, the EIRP can be increased by boosting either of the two factors. First, boosting of the transmit power available from solid-state devices is heavily dependent on the device technology. As the scaling progresses, the silicon-based transistors achieve the f_{max} exceeding 450 GHz [4]. Unfortunately, the device scaling lowers the operating voltage at the same time and thus severely limits the output power available from the device. The compound-semiconductor transistors such as InP HBT and HEMT exhibit a higher f_{max} and higher operation voltage [5] than the silicon counterpart. However, the output power of a power amplifier is still limited to a range of 13–24 dBm at the WR-3.4 band (220–320 GHz) [6, 7].

Another approach to increase the EIRP is to boost the antenna gain. Obviously, a high antenna gain can be obtained by adopting a high aperture size. A horn or parabola antenna achieves a high gain exceeding 26 or 55 dBi at the WR-3.4 band [8, 9]. Unfortunately, such a single-element antenna with a high gain would require a large form factor; moreover, it allows for only a very narrow fixed point-to-point link. In this regard, a beamforming array can be considered as a promising way of increasing the antenna gain effectively and

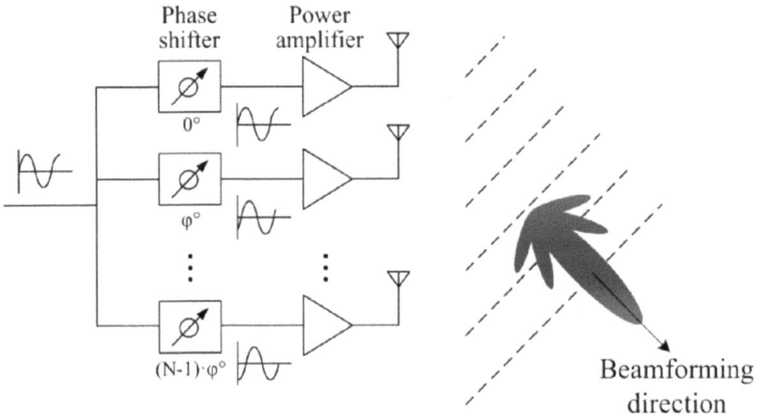

Figure 11.1 Structure of a conventional beamforming transmitter.

hence the EIRP to extend the limited link range. By combining multiple transmit elements with an antenna array as shown in Fig. 11.1, the EIRP is increased at the expense of a beamwidth. In principle, if N transmit array elements are combined coherently, the EIRP is effectively boosted by a factor of 20log$N$ dB. In addition, the beamforming array provides a functionality of beam steering by controlling the signal phase of each element. Therefore, the beaming array overcomes the limitation of the fixed point-to-point link possessed by the single-element transmitters and thus can be used more flexibly for various applications.

In this chapter, the structures, building blocks, and implementation of the beamforming THz transmitters are introduced. First, as a key building block, THz phase shifters are presented and discussed. Then, several implementation examples of the integrated beamforming THz transmitters are discussed. The THz frequencies considered in this chapter are the WR-3.4 band or higher.

11.2 THz Phase Shifters

A conventional beamforming transmitter consists of an array of transmit elements, each requiring a phase shifter shown in Fig. 11.1. The phase shifter introduces a variable phase shift between two adjacent elements, so that coherent signal combining is fulfilled in the radiating wave at a given direction. Unfortunately, the design of a phase shifter becomes challenging as the frequency increases to the THz band because of high component loss and parasitic effects. There are several types of integrated THz phase shifters reported so far [10–13]. A reflective-type phase shifter (RTPS) and a switched-type phase shifter (STPS) are two typical passive topologies that offer the advantages of no or low dc power consumption and high linearity. However, the passive phase shifters would suffer from high insertion loss and limited phase-shift range or resolution. On the other hand, a vector-sum phase shifter (VSPS) is a typical active topology. Therefore, the VSPS would achieve low insertion loss or even gain with a continuous 360° range at the expense of dc power consumption and linearity. The implementation of each topology at the THz band is described as follows.

11.2.1 Reflective-Type Phase Shifters (RTPS)

An RTPS consists of a 90° hybrid and two reflective loads with equal reactance of X_L as shown in Fig. 11.2. The input signal at port 1 is divided into two parts at ports 2 and 3. Each signal is reflected by the load with a reflection coefficient Γ_L and then is combined coherently at the output of port 4. By tuning X_L, the phase of Γ_L is varied and accordingly the phase of the output signal is shifted. A benefit of this topology is that a continuous phase shift can be achieved with a simple structure and a single tuning voltage if X_L is implemented

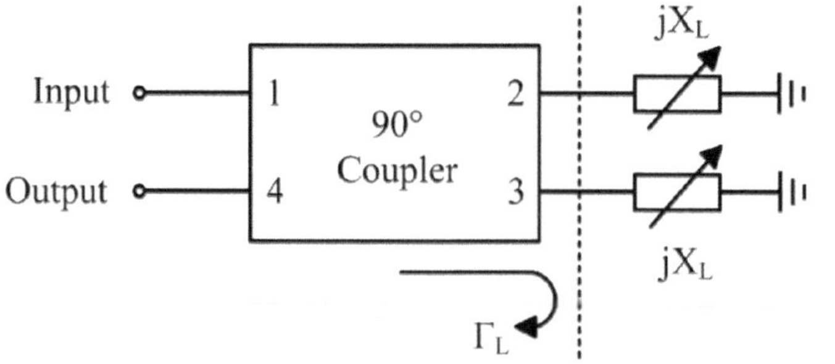

Figure 11.2 Simplified structure of an RTPS.

with a varactor. No dc power consumption and high linearity are other benefits. In addition, the input and output impedances are decently matched regardless of X_L. Nonetheless, an RTPS would suffer from high loss due to the passive nature. Also, the phase-shift range is limited by the tuning range of X_L and thus often fails to fulfill a full 360°.

A WR-3.4-band RTPS was demonstrated using a 50 nm InGaAs mHEMT technology with f_T and f_{max} of 380 and 600 GHz, respectively [10]. A schematic and a chip micrograph of the RTPS are shown in Fig. 11.3. The 90° hybrid is implemented in a Lange coupler which offers lower insertion loss than a branch-line coupler at the operation frequency. The reflective load is implemented by a varactor diode which is a single-finger transistor with drain and source tied to each other. The gate-to-source capacitance offers a capacitance tuning ratio (C_{max}/C_{min}) of 2.2. The varactor control voltage (V_{ctrl}) is applied through a series resistor and a bypass capacitor. The impedance matching between the hybrid and the reflective load is realized by a T-network consisting of three transmission-line sections and RF-short capacitances. The input and output reflection coefficients of the matching network were optimally determined, such that the RTPS performance such as a phase-shift range, bandwidth, insertion loss, and phase and amplitude errors are optimized together considering unavoidable trade-offs.

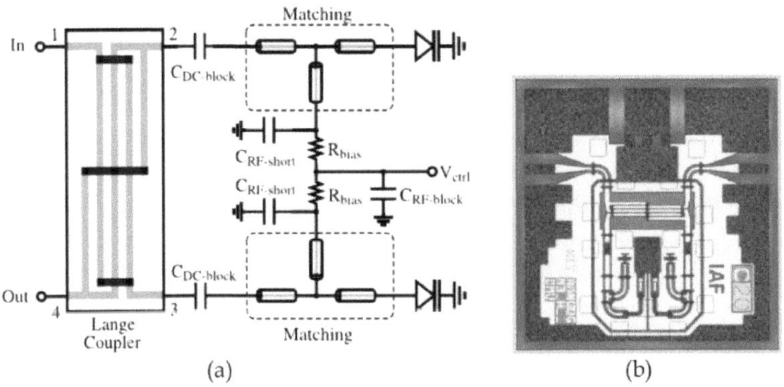

Figure 11.3 Schematic (a) and chip micrograph (size: 0.5 × 0.5 mm^2) (b) of a WR-3.4-band RTPS [10], reproduced by permission of © IEEE.

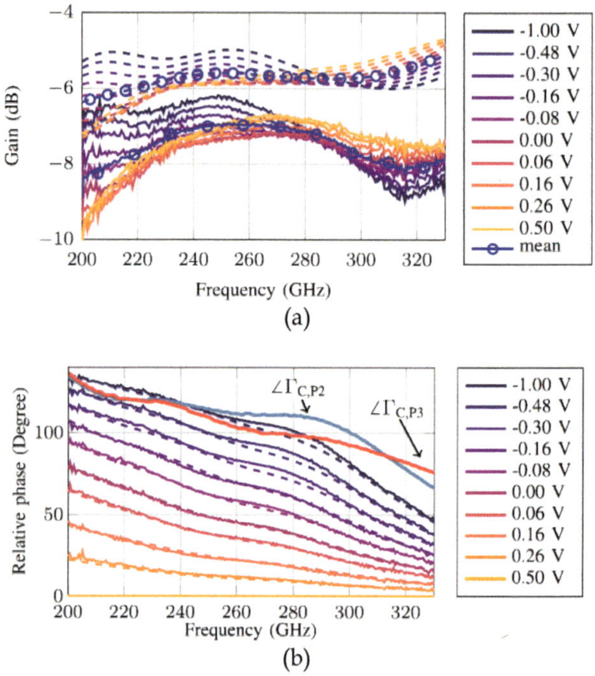

Figure 11.4 Measured (solid lines) and simulated (dashed lines) performance of a WR-3.4 RTPS [10]: (a) gain, (b) relative phase shift.

The gain and relative phase of the RTPS with V_{ctrl} varied from −1.0 to 1.5 V are shown in Fig. 11.4a,b, respectively. The average insertion loss exhibits 7 ± 0.7 dB over the entire WR-3.4 band. The amplitude variation over the different phase states is less than ± 0.6 dB at the center frequency. The maximum value of the relative phase-shift ranges between 135° at 200 GHz and 100° at 280 GHz with an almost linear decrease in between. The input and output matching are below −10 dB over the entire WR-3.4 band. The root-mean-squared (RMS) amplitude and phase errors are below 0.67 dB and 5.6°, respectively.

Although the RTPS in [10] shows a low amplitude variation, decent port matching, low amplitude and phase errors over a wide bandwidth, it suffers from a narrow phase-shift range and high loss. A simply way of extending the range and compensating the

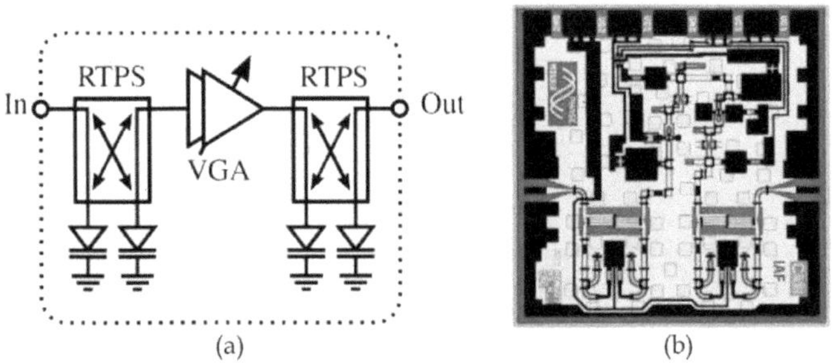

Figure 11.5 Schematic (a) and chip micrograph (size: 0.75 × 0.75 mm^2) (b) of a WR-3.4-band RTPS [11], reproduced by permission of © IEEE.

loss is to connect multiple RTPS units and an amplifier in cascade, as demonstrated in [11]. A schematic and a chip micrograph of the WR-3.4 RTPS are shown in Fig. 11.5. Two RTPS units are cascaded to extend the phase-shift range. Each unit employs a similar reflective load structure to [10], resulting in a phase shift less than 135°. To compensate the high loss from the cascaded RTPS units, a variable-gain amplifier (VGA) is inserted in between the units. This arrangement is advantageous in that the input and output impedances of the VGA are buffered by the RTPS units at both ends. Therefore, the overall RTPS maintains decent impedance matching performance. The VGA in a two-stage cascode configuration provides a peak gain of 9 dB over a 3 dB bandwidth from 180 to 280 GHz. Therefore, the measured gain of the overall RTPS was improved to a peak of −0.9 dB with a 3 dB bandwidth from 218 to 268 GHz, as shown in Fig. 11.6a. The relative phase shift, shown in Fig. 11.6b, ranges between 247° and 205° at the lower and upper boundary of the 3 dB bandwidth. The phase-shift range is extended to almost double to that of each RTPS unit. The RMS phase and amplitude errors are below 10.4° and 1.6 dB, respectively, over the 3 dB bandwidth.

11.2.2 Switched-Type Phase Shifters (STPS)

A conventional structure of an STPS is shown in Fig. 11.7. By switching the signal between the reference path and delay path, a desired phase shift of $\Delta\varphi$ is realized. The reference-delay section

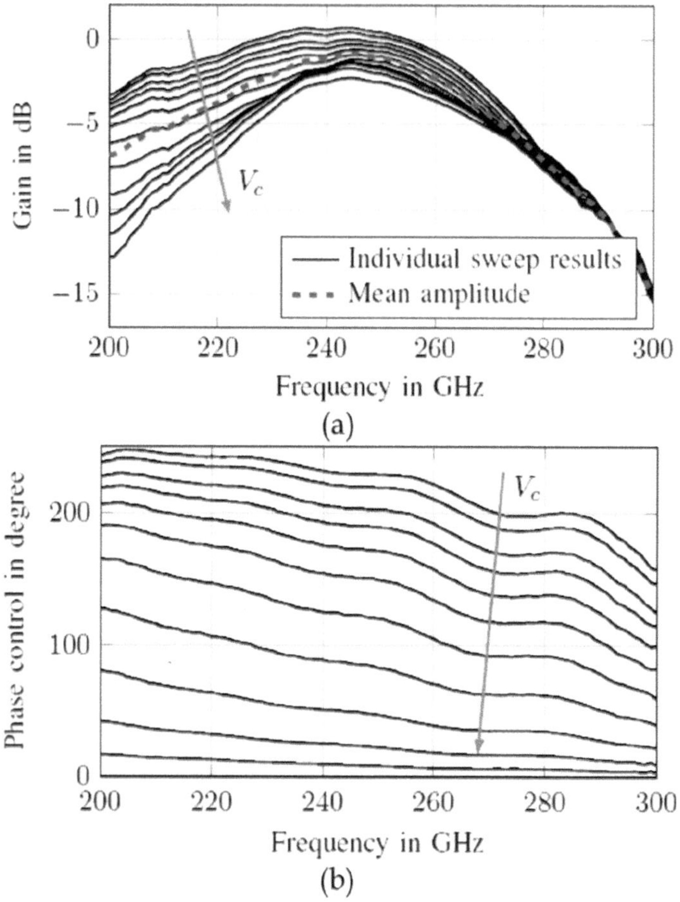

Figure 11.6 Measured performance of the WR-3.4-band RTPS [11]: (a) gain, (b) relative phase shift, reproduced by permission of © IEEE.

can be replaced by other phase-shift networks such as a low-pass-high-pass section and lumped LC section, etc. To increase the range and (or) resolution of the phase shift, multiple phase-shift units can be connected in cascade. An STPS is beneficial in that a discrete phase shift is conveniently achieved with no need of an additional digital-to-analog converter (DAC). No dc power consumption and high linearity are additional merits of the passive structure just like an RTPS. However, high insertion loss is a critical downside of an

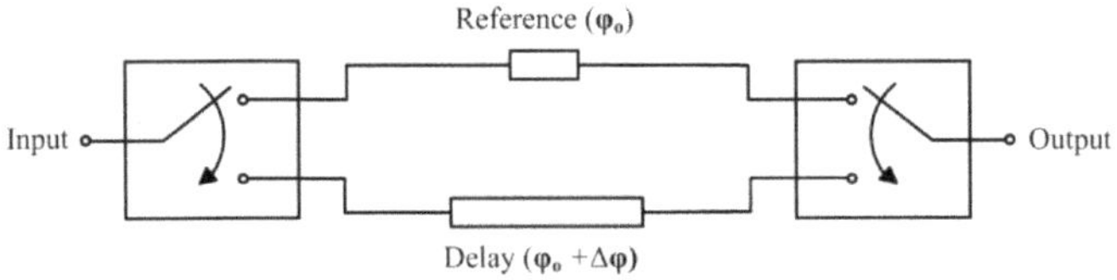

Figure 11.7 Conventional structure of an STPS.

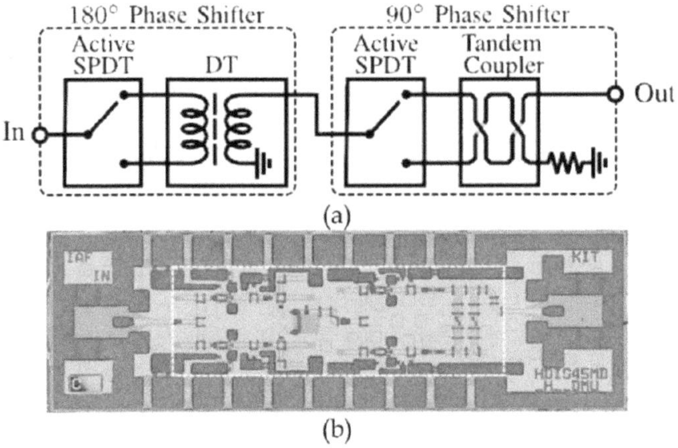

Figure 11.8 Schematic (a) and chip micrograph (size: 0.82 × 0.26 mm^2) (b) of a WR-3.4-band 2-bit STPS [12], reproduced by permission of © IEEE.

STPS. As the number of phase-shift units increases, the high loss is exacerbated. Moreover, as the frequency increases toward THz, the switch loss and bandwidth are significantly degraded. This would make the conventional STPS relatively less favorable in the THz band.

In [12], a modified STPS structure employing an active switch was demonstrated at the WR-3.4 band using a 35 nm InGaAs mHEMT technology. The high loss of a passive switch at the THz frequencies is compensated by the active switch. A schematic and a chip micrograph of the WR-3.4-band 2-bit STPS are shown in Fig. 11.8. The active single-pole double-throw (SPDT) switch is designed in a series-shunt type with input and output matching networks shown in Fig. 11.9a. When the signal path operates in a through mode, the series common-gate (CG) device is biased

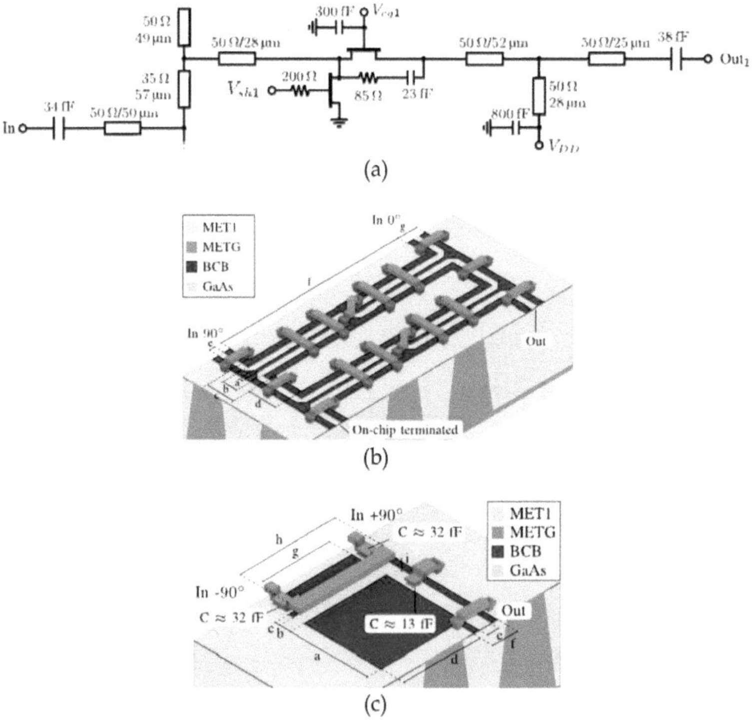

Figure 11.9 (a) Schematic of the active SPDT switch, (b) 90° tandem coupler, (c) 180° distributed transformer [12], reproduced by permission of © IEEE.

for maximum gain whereas the shunt device is off. On the other hand, in an isolated mode, the shunt device is turned on with a low on-resistance whereas the CG device is off. The gain and isolation are 5.3 and 25 dB, respectively, at 250 GHz. The 90° and 180° phase-shift units are implemented using an edge-coupled hybrid tandem coupler [14] and a broadside-coupled distributed transformer structure [15], as shown in Fig. 11.9b,c, respectively. The gain and relative phase shift of the STPS are shown in Fig. 11.10. The peak gain is 0.2 dB at 267 GHz and the 3 dB bandwidth ranges from 235 to 279 GHz. This relatively high gain of the STPS is obtained at the expense of extra dc power consumption of the active switches,

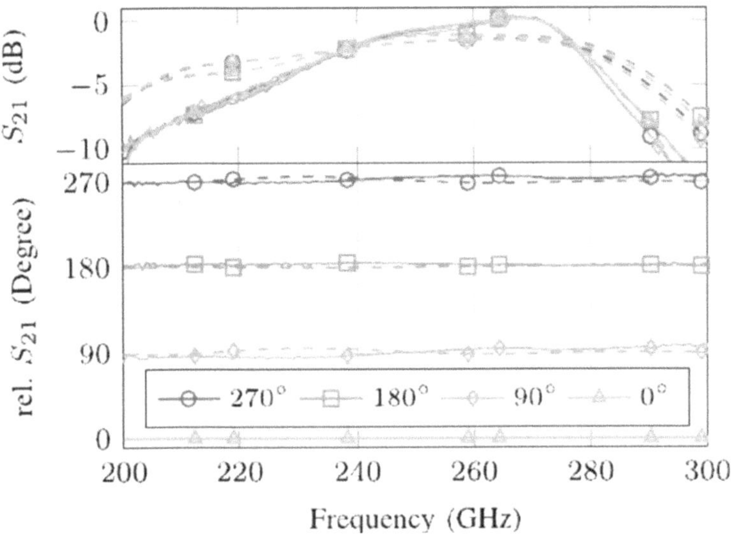

Figure 11.10 Gain and relative phase shift of the WR-3.4-band 2-bit STPS (solid lines: measured, dashed lines: simulated) [12], reproduced by permission of © IEEE.

which is 50.4 mW. A 2-bit phase shift for 0°, 90°, 180°, and 270° is well achieved over the full WR-3.4 band. The RMS phase and amplitude errors are below 0.18 dB and 2.8°, respectively, within the 3 dB bandwidth.

11.2.3 Vector-Sum Phase Shifters (VSPS)

Compared to the previous conventional passive topologies, a VSPS belongs to an active topology that achieves 360° continuous phase shift with relatively low insertion loss (or possibly gain). A basic structure of a VSPS is shown in Fig. 11.11. The amount of phase shift is synthesized by combining two quadrature signal vectors (V_I and V_Q) with different weights (w_I and w_Q) of two individual VGAs. Since both the phase and amplitude of the output signal can be adjusted in an analog manner, the VSPS presents an advantage of larger flexibility than any other types of phase shifters. However, a fine control scheme is required of the VGAs for achieving a desired phase

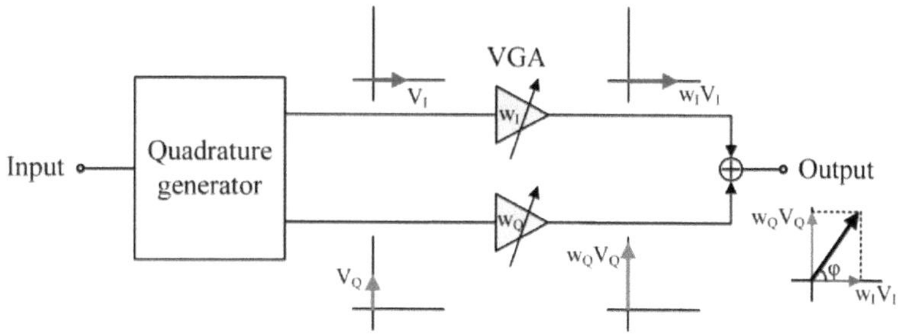

Figure 11.11 Basic structure of a VSPS.

and amplitude synthesis. In addition, the phase-shift performance can be significantly degraded by the mismatch of the quadrature signals.

The design of VGAs and a signal adder becomes challenging as the frequency increases toward THz. Typically, each VGA is implemented using a Gilbert cell containing four transconductance cells. Therefore, totally eight transconductance cells must be combined at the differential output nodes from two VGAs. This obviously lowers the node impedance while increasing the layout complexity and the adverse layout parasitics. Therefore, the bandwidth will be significantly degraded, particularly at the THz band.

To resolve the issue and achieve a wideband operation, a VSPS employing a single Gilbert cell structure (rather than two) was demonstrated at the WR-3.4 band using a 250 nm InP DHBT technology [13]. A schematic and a chip micrograph of the VSPS are shown in Fig. 11.12. It consists of a quadrature generator, a quadrature vector modulator, and an output balun. The quadrature vector modulator employs a single Gilbert cell structure that synthesizes the required phase. The transconductances (G_{mI}^+, G_{mI}^-, G_{mQ}^+, and G_{mQ}^-) of the transistors (Q_3–Q_6) are adjusted by the control voltages (V_I^+, V_I^-, V_Q^+, and V_Q^-). Accordingly, a desired phase is synthesized as depicted in Fig. 11.13. Because a single Gilbert cell is employed for the phase synthesis, only four transconductance cells are combined at the nodes A and B and hence the layout complexity and parasitics are reduced at the nodes. In addition, a lossy matching with resistors of R_7–R_{10} is applied at the

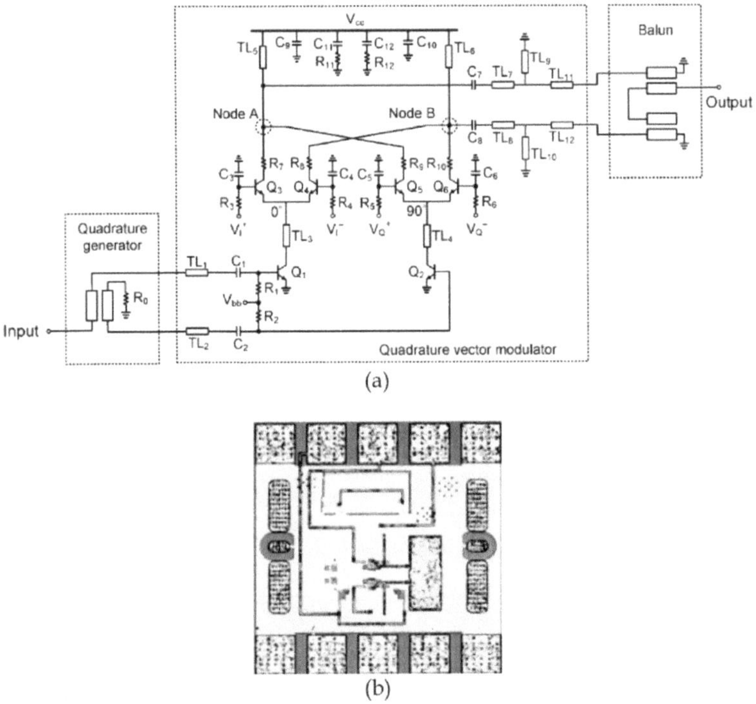

Figure 11.12 Schematic (a) and chip micrograph (size: 0.48 × 0.475 mm^2) (b) of a WR-3.4-band VSPS [13], reproduced by permission of © IEEE.

transistor output matching for extending the operation bandwidth at the expense of gain.

The output balun combines the signals at the nodes A and B while suppressing a common-mode component which intrinsically exists in the single Gilbert cell structure. If the VSPS drives a fully differential load, the output balun may be omitted and the common-mode rejection can be fulfilled by the following differential circuitry such as a typical double-balanced mixer with a tail current source. A quadrature generator of the VSPS is implemented using a broadside-coupled line for compact design.

In Fig. 11.14a, the measured average gain ranges from −15.6 to −11.8 dB over the entire WR-3.4 band. A 4-bit phase-shift performance is shown in Fig. 11.14b. A full 360° phase shift with a resolution of 22.5° is achieved over the entire WR-3.4 band. It is

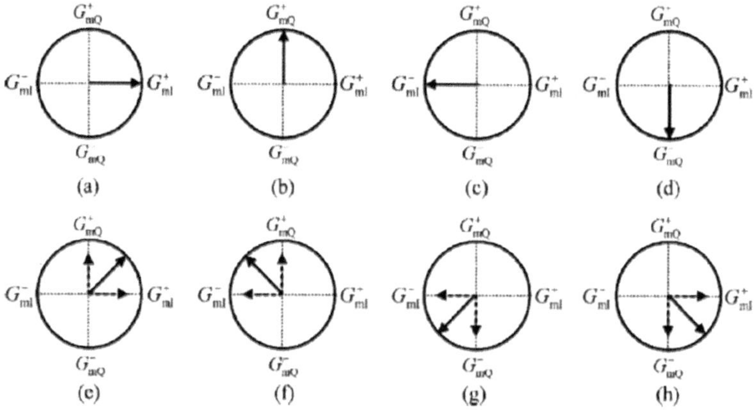

Figure 11.13 Exemplary phase synthesis by adjusting the transconductances [13], reproduced by permission of © IEEE.

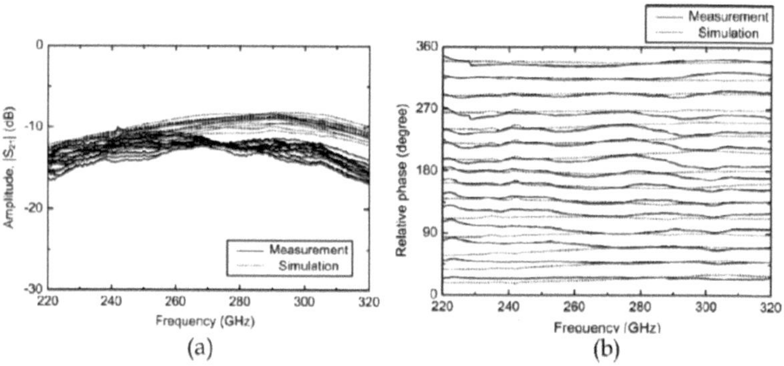

Figure 11.14 Measured performance of the WR-3.4-band VSPS [13]: (a) gain, (b) relative phase shift, reproduced by permission of © IEEE.

noted that a higher resolution than 4 bits is available from the VSPS if the control voltages (V_I^+, V_I^-, V_Q^+, and V_Q^-) are adjusted in a finer way. Practically, the resolution would be limited by the performance of an external DAC that generates the control voltages. The RMS phase and amplitude errors are below 10.2° and 1.2 dB, respectively, and the dc power consumption is 21.8 to 42 mW depending on the phase states.

11.3 Integrated Beamforming THz Transmitters

In this section, several implementation examples of the beamforming THz transmitters are introduced [16–18].

11.3.1 280 GHz CMOS Beamforming Array on Distributed Active Radiators

An early work of an integrated beamforming THz transmitter was reported in [16]. As shown in Fig. 11.15, the beamforming transmitter in a 45 nm SOI CMOS technology combines an array of 4 × 4 distributed active radiators (DARs). Each DAR is a self-oscillating active electromagnetic structure that is composed of two radiating loops which sustain out-of-phase currents at the fundamental frequency, thus being filtered out. On the other hand, the second-harmonic currents keep in-phase to each other, thus being effectively radiated. Therefore, the DAR behaves as a resonator for a traveling-wave oscillation at the fundamental frequency and a selective frequency multiplier and a radiator at the second-harmonic

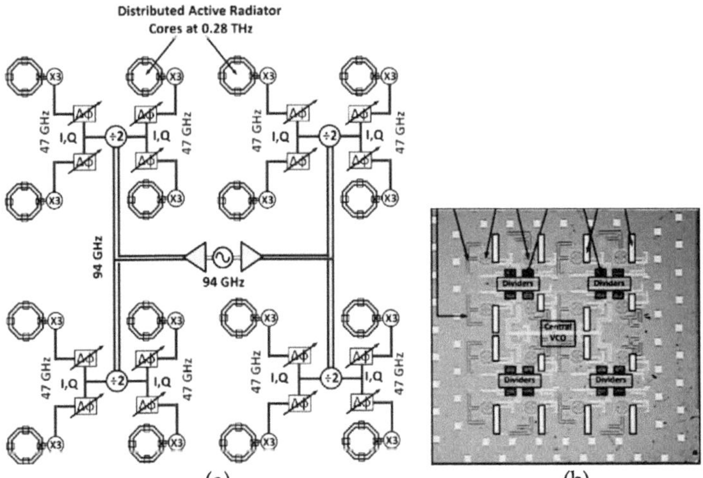

(a) (b)

Figure 11.15 Architecture (a) and chip micrograph (size: 2.7 × 2.7 mm^2) (b) of a 280 GHz CMOS beamforming transmitter [16], reproduced by permission of © IEEE.

frequency. In this way, the radiating frequency can be higher than the f_{max} of the transistor.

A THz signal of the transmitter is generated by an on-chip VCO at around 94 GHz with a 6 GHz tuning bandwidth. Note that the VCO frequency is only one third of the finally radiated signal frequency at 280 GHz. A frequency divider converts the 94 GHz signal into differential quadrature signals, I and Q at 47 GHz. To realize the beamforming and beamsteering functionality, the I and Q signals feed a vector modulator and thus they are added with different weights just like a VSPS described in Section 11.2.3. The VSPS drives a frequency tripler at the output frequency of 140 GHz. Then, the output signal is locked to a fundamental frequency of oscillation of each DAR, thus radiating a 280 GHz wave. Since each DAR is equipped with a VSPS, sixteen independent phases are generated to achieve beamforming at a desired direction.

The frequency of the transmitted signal can be adjusted between 276 and 285 GHz. The peak EIRP was measured to be 9.4 dBm at 281 GHz and varied by less than 0.5 dB over the entire tuning range. The

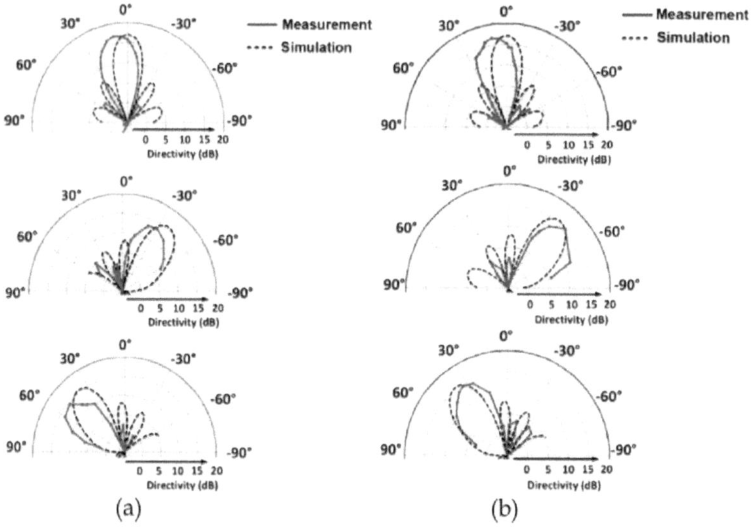

Figure 11.16 Radiation patterns of a 280 GHz CMOS beamforming transmitter [16]: (a) in the azimuthal angles, (b) in the elevation angles, reproduced by permission of © IEEE.

beamsteering performance was measured by applying a progressive phase shift between the adjacent DARs in each of the two orthogonal axes. The measured radiation patterns in the azimuth and elevation angles are shown in Fig. 11.16a,b, respectively. It is observed that approximately 80° of beamsteering range is achieved in each of the two-dimensional axes. The total dc power consumption is 820 mW.

11.3.2 320 GHz BiCMOS Beamforming Transmitter

An integrated 1 × 4 phased-array transmitter was reported using a 0.13 μm SiGe BiCMOS technology [17]. As shown in Fig. 11.17, the transmitter integrates a 20 GHz phased-locked-loop (PLL) frequency synthesizer, an 80 GHz frequency quadrupler, a 1:4 Wilkinson power divider, four-way 80 GHz tunable attenuators, buffer amplifiers, phase shifters, 320 GHz frequency quadruplers, and on-chip antenna arrays.

The frequency synthesizer generates a 20 GHz signal locked to a reference at 75.5 to 80.6 MHz. Multiplied by four, the signal is up-converted to 80 GHz and is distributed to each of four paths. For beamforming, both amplitude and phase of the 80 GHz signal are adjusted by a reflective-type attenuator and an RTPS, respectively. The attenuator presents insertion loss of 2.6 dB and attenuation range of 20 dB. The RTPS achieves a phase-shift range

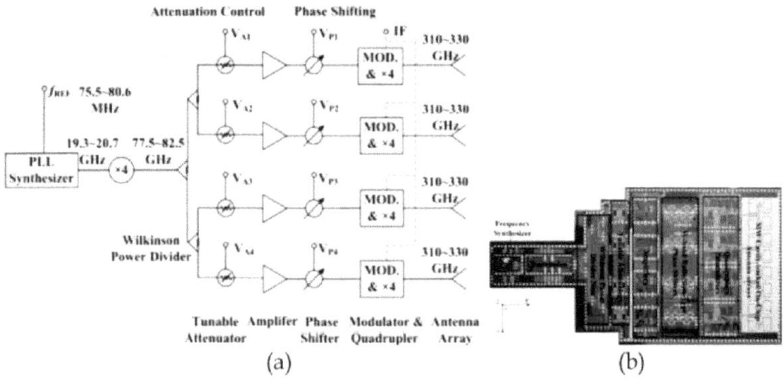

Figure 11.17 Architecture (a) and chip micrograph (size: 8 × 4.3 mm^2) (b) of a 320 GHz BiCMOS beamforming transmitter [17], reproduced by permission of © IEEE.

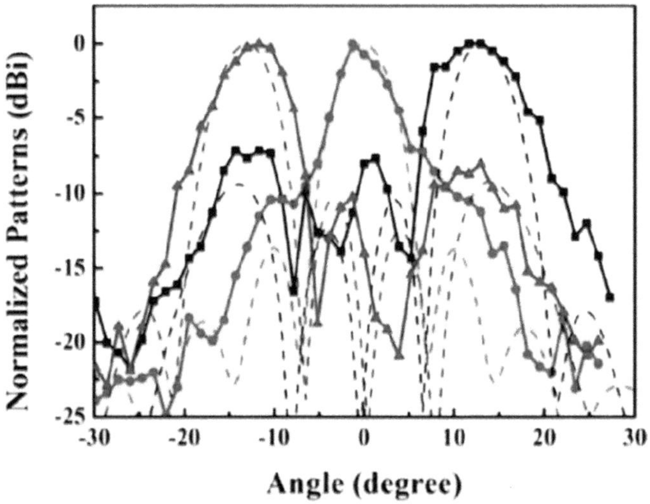

Figure 11.18 Measured (symbol) and simulated (dashed line) radiation patterns of a 320 GHz BiCMOS beamforming transmitter [17], reproduced by permission of © IEEE.

of around 125° at 80 GHz. However, after the subsequent frequency quadrupler, the phase-shift range of the finally radiated signal at 320 GHz becomes more than 360°. The 320 GHz quadrupler behaves as a modulator with the IF signal supplied through a Gilbert cell. Finally, a 1×4 array of substrate-integrated waveguide (SIW) cavity-backed antennas is integrated on a single chip.

The transmitter achieved a peak EIRP of 7.9–10.6 dBm at the frequencies from 310 to 330 GHz. The radiation patterns in the E-plane were measured with each of the four phase shifters adjusted to three conditions, (0°, 0°, 0°, 0°), (0°, 90°, 180°, 270°), and (270°, 180°, 90°, 0°). As shown in Fig. 11.18, a beamsteering range of ± 12° was achieved at 320 GHz. The total dc power consumption is 1000 mW.

11.3.3 370–410 GHz CMOS Beamforming Transmitter

A CMOS eight-element beamforming transmitter was demonstrated at 370–410 GHz in [18]. The architecture and chip micrograph of

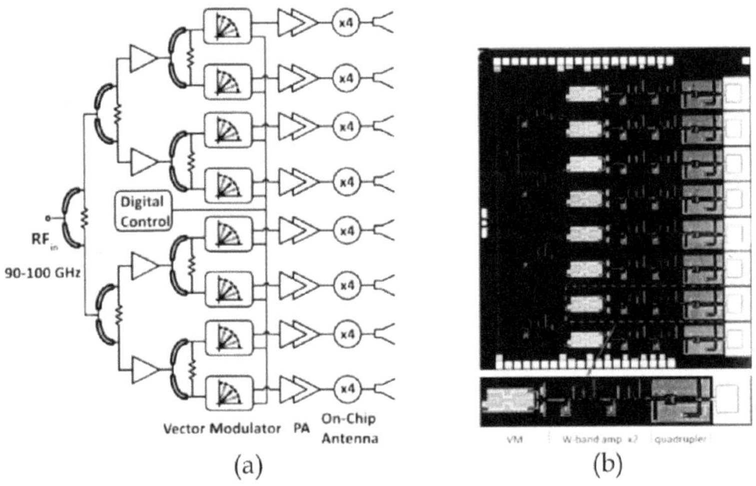

Figure 11.19 Architecture (a) and chip micrograph (size: 3 × 3.5 mm^2) (b) of a 370-410 GHz CMOS beamforming transmitter [18], reproduced by permission of © IEEE.

the transmitter are shown in Fig. 11.19. The input signal at 90 to 105 GHz is distributed into eight paths by seven Wilkinson dividers. A VSPS is followed in each path for phase shift by 90°. Given that a subsequent frequency quadrupler expands the phase shift by a factor of four, it is sufficient to have a phase shift in a single quadrant only from the VSPS. The frequency quadrupler employs a differential common source topology and provides output power of −10 dBm at 400 GHz. A microstrip antenna was integrated on a chip, exhibiting a peak efficiency of 20% and gain of −2 to 0 dBi at 360–400 GHz. To improve the radiation efficiency, a quarter-wave quartz superstrate was assembled on top of the microstrip antenna. According to the simulation, the superstrate increases the peak gain and efficiency by 3.1 and 3.5 dB, respectively.

The measured EIRP of 8–8.5 dBm was achieved at frequencies from 380 to 400 GHz. A 3 dB bandwidth of the EIRP is from 375 to 407 GHz. The radiation patterns in the H-plane at 400 GHz are shown in Fig. 11.20. The beamsteering angle ranges ± 35–40° at 380–400 GHz. The total dc power consumption is 1500 mW.

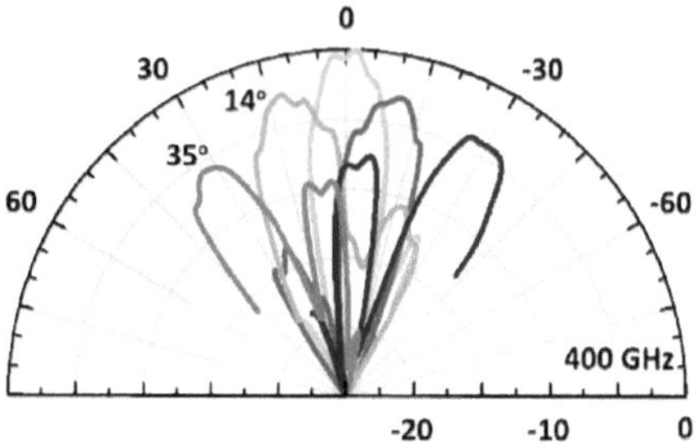

Figure 11.20 Measured radiation patterns of a 370-410 GHz CMOS beamforming transmitter [18], reproduced by permission of © IEEE.

References

1. Siegel, P. H. (2002). Terahertz technology, *IEEE Trans. Microw. Theory Techn.*, **50**, pp. 910–928.
2. Liu, H.-B., Zhong, H., Karpowicz, N., Chen, Y., and Zhang, X.-C. (2007). Terahertz spectroscopy and imaging for defense and security applications, *Proc. IEEE*, **95**, pp. 1514–1527.
3. Mittleman, D. M. (2017). Terahertz science and technology, *J. Appl. Phys.*, **122**, 230901.
4. Schmid, R. L., Ulusoy, A., Zeinolabedinzadeh, S., and Cressler, J. (2015). A comparison of the degradation in RF performance due to device interconnects in advanced SiGe HBT and CMOS technologies, *IEEE Trans. Electron Devices*, **62**, pp. 1803–1810.
5. Urteaga, M., Seo, M., Hacker, J., Grif?th, Z., Young, A., Pierson, R., Rowell, E., Skalare, A., and Rodwell, M. J. W. (2010). InP HBT integrated circuit technology for terahertz frequencies, *Proc. IEEE Compound Semicond. Integr. Circuit Symp.*, pp. 1–4.
6. Kim, J., Jeon, S., Kim M., Urteaga, M., and Jeong, J. (2015). H-band power amplifier integrated circuits using 250 nm InP HBT technology, *IEEE Trans. Terahertz Sci. Tech.*, **5**, pp. 215–222.
7. Griffith, Z., Urteaga, M., and Rowell, P. (2017). 180–265 GHz, 17–24 dBm output power broadband, high-gain power amplifiers in InP HBT, *Proc. IEEE Int. Microwave Sym.*, IEEE, pp. 973–976
8. VDI Inc., https://www.vadiodes.com/.
9. Kallfass, I., Boes, F., Messinger, T., Antes, J., Inam, A., Lewark, U., Tessmann, A., and Henneberger, R. (2015). 64 Gbit/s transmission over

850 m fixed wireless link at 240 GHz carrier frequency, *J. Infrared Milli Terahertz Waves*, **36**, pp. 221–233.

10. Müller, D., Haag, A., Bhutani, A., Tessmann, A., Leuther, A., Zwick, T., and Kallfass, I. (2018). Bandwidth optimization method for reflective-type phase shifters, *IEEE Trans. Microw. Theory Techn.*, **66**, pp. 1754–1763.
11. Müller, D., Beck, A., Massler, H., Tessmann, A., Leuther, A., Zwick, T., and Kallfass, I. (2017). A WR3-band reflective-type phase shifter MMIC with integrated amplifier for error- and loss compensation, *Proc. European Microwave Integrated Circuits Conference*, pp. 1–4.
12. Müller, D., Pahl, A., Tessmann, A., Leuther, A., Zwick, T., and Kallfass, I. (2018). A WR3-band 2-bit phase shifter based on active SPDT switches, *IEEE Microw. Wireless Compon. Lett.*, **28**, pp. 810–812.
13. Kim, Y., Kim, S., Lee, I., Urteaga, M., and Jeon, S. (2015). A 220–320 GHz vector-sum phase shifter using single Gilbert-cell structure with lossy output matching, *IEEE Trans. Microw. Theory Techn.*, **63**, pp. 256–265.
14. Chang, T.-Y., Liao, C.-L., and Chen, C. H. (2003). Coplanar-waveguide tandem couplers with backside conductor, *IEEE Microw. Wireless Compon. Lett.*, **13**, pp. 214–216.
15. Pahl, P., Wagner, S., Massler, H., Diebold, S., Leuther, A., Kallfass, I., and Zwick, T. (2015). A 50 to 146 GHz power amplifier based on magnetic transformers and distributed gain cells, *IEEE Microw. Wireless Compon. Lett.*, **25**, pp. 615–617.
16. Sengupta, K., and Hajimiri, A. (2012). A 0.28 THz power-generation and beam-steering array in CMOS based on distributed active radiators, *IEEE J. Solid-State Circuits*, **47**, pp. 3013–3031.
17. Deng, X.-D., Li, Y., Li, J., Liu, C., Wu, W., and Xiong, Y.-Z. (2015). A 320 GHz 1×4 fully integrated phased array transmitter using 0.13 μm SiGe BiCMOS technology, *IEEE Trans. THz Sci. Technol.*, **5**, pp. 930–940.
18. Yang, Y., Gurbuz, O. D., and Rebeiz, G. M. (2016). An eight-element 370–410 GHz phased-array transmitter in 45 nm CMOS SOI with peak EIRP of 8–8.5 dBm, *IEEE Trans. Microw. Theory Techn.*, **64**, pp. 4241–4249.

Chapter 12

Solid-State THz Power Amplifiers

Ahmed S. H. Ahmed[a] and Munkyo Seo[b]

[a]*Department of Electronics and Electrical Communications Engineering, Cairo University, Egypt*
[b]*Department of Electrical and Computer Engineering, Sungkyunkwan University, South Korea*
asamir@cu.edu.eg, mkseo@skku.edu

In this chapter, we consider the solid-state implementation of terahertz (THz) power amplifiers. As frequency increases, the transistor has a limited gain and power which requires certain technologies and design techniques. The key challenges and design options are discussed. The chapter reviews the amplifier design fundamentals such as loadline matching, unit cell, and different power combining techniques. The pros and cons of III-V and silicon technologies are discussed. We addressed oscillations and thermal heating, which are common practical issues. We discussed the setup of the THz S-parameter and power measurements. Finally, the measurement results of several examples are presented the across 140–600 GHz frequency band.

Advances in Terahertz Source Technologies
Edited by Gun-Sik Park, Masahiko Tani, Jae-Sung Rieh, and Sang Yoon Park

ISBN 978-981-4968-89-8 (Hardcover), 978-1-003-45967-5 (eBook)
www.jennystanford.com

12.1 Introduction

Next-generation communication systems consider the terahertz (THz) bands since those bands support high data rates and short wavelengths allow building compact multi-beam (MIMO) arrays to further increase the transmission capacity. Also, high-resolution imaging, radars, satellite communications, or measurement equipment uses THz bands. It is challenging to get high power at high frequencies since the transistor is working closer to the maximum power gain cut-off frequency (f_{max}) of the technology, which results in low gain (Buckwalter et al., 2021). Also, the interconnect routing and parasitic have a significant impact and add high loss.

The book shows various ways of designing THz sources based on the application and requirements. Here we focus on the solid-state electronics implementation. The main advantages of the solid-state realization are: (1) manufacturers are widely available with mass production capability, which reduces the cost. This is attractive for commercial applications such as cell phones or laptops. (2) It is more convenient to integrate solid stage sources into a system implemented mainly in the same framework. (3) Solid-state usually offers compact and light solutions.

Figure 12.1 shows a survey for the saturated output power vs the frequency using different technologies. The figure shows that frequency multipliers and oscillators could be used as THz sources, yet they typically have low output power. Power amplifiers are better candidates since they have significantly higher output power and efficiency. The application dictates the technology. CMOS works well until D-band with moderate output power and efficiency. However, the power degrades sharply at higher frequencies. SiGe shows higher output power and efficiency. III-V technologies such as GaN, GaAs, and InP have higher breakdown voltage and current density resulting in higher output power. So far, InP shows superior power and efficiency at sub-THz frequencies.

Here, we will briefly cover the power amplifier fundamentals, unit cell design, power combining techniques, and technology considerations then we will present the design details and measurement results of several state-of-the-art amplifiers at 140 GHz

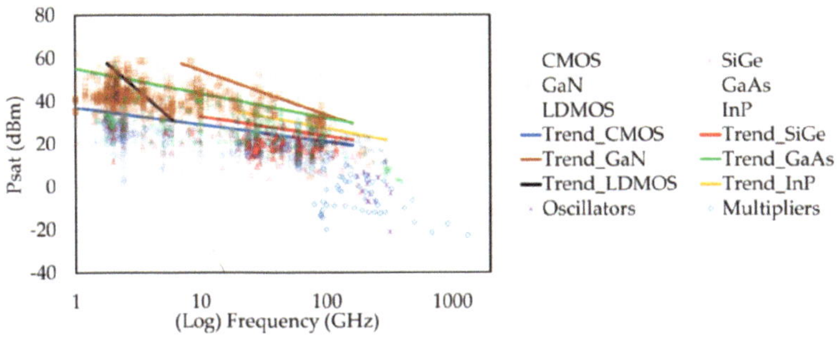

Figure 12.1 Saturated Output Power vs. Frequency (Hua Wang and Smith, 2021).

(Ahmed et al., 2021a, 2020), 210 GHz (Ahmed et al., 2021b), 270 GHz (Ahmed et al., 2021c), and 600 GHz (Seo et al., 2013).

12.2 THz Power Amplifier Fundamentals

The objective of the power amplifier is to deliver the required output power for given input power with the highest efficiency. In power amplifiers, transistors are matched for maximum power transfer (also known as load-pull techniques or loadline matching) or maximum power-added efficiency (PAE) (Cripps, 2006). Technology defines the maximum operating voltage (V_{max}) and maximum safe current (I_{max}). Those limits are known as a safe operating area (SOA). Without proper matching (red contours in Fig. 12.2(b)), the transistor suffers from early saturation and low output power since the voltage or current contours exceed the SOA limits before reaching the maximum available swings. To reach the maximum voltage and current swings, the transistor parasitics must be tuned to have a linear relationship between the voltage and current (blue contours in Fig. 12.2(b)). Also, the slope of the line is defined as $1/R_{loadline}$. The loadline resistance ($R_{loadline}$) is adjusted according to (12.1) to maximize the voltage and current swings. There are many ways of tuning implementations; for example, adding a shunt inductor tunes the transistor reactance, then the transformer adjusts the slope of the line (Fig. 12.2(a)). After proper loadline

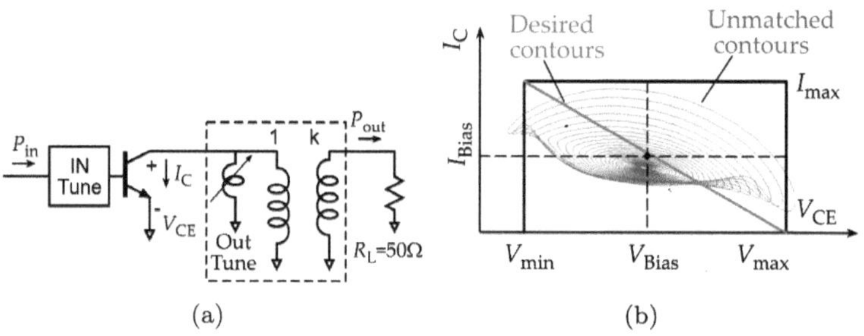

Figure 12.2 (a) Transistor tuned for maximum output power. (b) internal voltages and current contours with loadline matching (blue) and with arbitrarily matching (red).

matching, the maximum output RF power is (12.2), and the DC power is (12.3). In class A, the amplifier is biased according to (12.4) and (12.5). It is worth noting that the loadline matching may differ from the conjugate match for maximum power gain, and that is why many power amplifiers have a poor output reflection coefficient (S_{22}).

$$R_{loadline} = \frac{V_{max} - V_{min}}{I_{max} - I_{min}} \tag{12.1}$$

$$P_{out,\,max} = (V_{max} - V_{min})I_{max}/8. \tag{12.2}$$

$$P_{DC} = I_{Bias}V_{Bias}. \tag{12.3}$$

$$V_{Bias} = (V_{max} + V_{min})/2. \tag{12.4}$$

$$I_{Bias} = I_{max}/2. \tag{12.5}$$

The drain efficiency (η) represents the ratio of the RF output power (P_{out}) to the DC power (12.6). As the amplifier efficiency increases, the heat dissipation decreases, which extends the battery life and requires less complicated heatsinks. The PAE is another definition for efficiency, which counts for the amplifier power gain (12.7).

$$\eta = P_{out}/P_{DC}. \tag{12.6}$$

$$PAE = (P_{\text{out}} - P_{\text{in}})/P_{\text{DC}} \quad (12.7)$$

The efficiency depends on the conduction angle. In class A, the conduction angle is 360°, and the peak drain efficiency is 50%. In class B, the conduction angle is 180° and the peak η is 78.5%. The conduction angle is getting smaller in class C, and the η can reach 100% (but the device has no gain). The gain drops as we decrease the conduction angle, and that is why the class A power amplifier has the highest gain and is most popular at THz frequencies.

In class A, the PAE is linearly proportional to the RF power. The amplifier has low PAE at low input power and reaches the peak PAE at saturation. Improving the efficiency at the backoff efficiency is an active topic, especially at THz (due to low gain), and there are common techniques such as Doherty, out-phasing, or envelope tracking (Cripps, 2006).

12.2.1 Unit Cell Design

The first step in the design is to do back-of-the-envelope calculations to determine the proper number of transistor fingers and the topology. For a certain number of fingers and given SOA, the loadline impedance is computed from (12.1). As the number of fingers increases, the cell delivers more output power. However, the required loadline impedance gets smaller. As the impedance transformation increases, the matching loss increases, and the bandwidth gets narrower. On the other hand, if the cell size is too small, the cell has low output power, and the loadline impedance is too big and difficult to match. So, there is a tradeoff, and typically we prefer to choose the cell's loadline impedance to be close to 50 Ω. There are other considerations for the transistor layout: (1) some foundries offer optimized footprint for a certain number of fingers which becomes more desirable (2) it is recommended to run thermal analysis and see if the cell dissipate the heat properly otherwise, we may reduce the number of fingers or increase the spacing between the fingers. (3) The fingers are usually combined in lumped fashion. This implies that there is un-balancing between the fingers and the un-balance increases with a bigger transistor. This adds another limit on the maximum number of combined fingers.

The topologies under consideration are common base, common base with finite base capacitance (known as a stacked cell) and common emitter in BJT or common gate, common gate with finite gate capacitance, and common source in MOSFET. Common collector and common drain have low power gain and are rarely used at THz. Neutralization techniques could be used with the previous topologies to boost the gain at the expense of lower output power. First, the transistors should be properly biased according to the class of operation and tuned to the maximum output power. Then we use a harmonic balance to perform large-signal simulations between the different topologies. At the desired compression level, we compare the output power, efficiency, and gain. Then select the topology with the highest efficiency. There are more practical considerations that impact the decision such as linearity, backoff efficiency, design sensitivity to a certain circuit parameter, matching loss, RF or bias stability. After accounting for the previous considerations and with experience, the designer reaches the proper cell selection.

12.2.2 Power Combining Techniques

A single unit cell has limited power which might not satisfy the system requirements. There are three main techniques to get more output power: (1) stacking (Ahmed et al., 2018a,b): the maximum voltage swing is limited by the breakdown voltage of the technology. In stacked power amplifiers, transistors are connected in series, so the output voltage swing is much higher than a single transistor (Fig. 12.3a). The crucial issue in the stack is to guarantee that all transistors are clipping simultaneously and have the proper loadline. This happens by proper tuning of the interstage matching along with the base/gate termination to hold the appropriate voltage and current. The conventional design methods rely on the circuit models while the authors proposed a two-port technique approach for design in (Ahmed et al., 2018a) and it is currently extended into a journal. The two-port technique is more convenient at mm-wave frequency since it works with arbitrarily transistor models and interconnects parasitics modeled by EM simulations. (2) Parallel combiners (Ahmed et al., 2021a, 2020, 2021b,c): in this approach, the total output current swing increases by combining

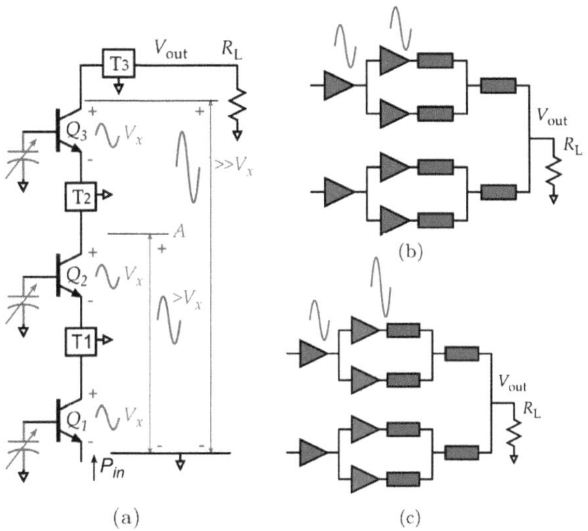

Figure 12.3 (a) Three stacked transistors. (b) Conceptual drawing for 4:1 parallel power combiner. Red indicates non-stacked cell. (c) Conceptual drawing for combining four stacked cells. Purple indicates stacked power cells with higher voltage swings.

many small power cells in parallel (Fig. 12.3b). Hypothetically, we can combine the cells by an ideal short circuit. However, at THz frequency, any short line acts as a transmission line and causes impedance change. Even if we combined the cells with an ideal short circuit, we still need to design an impedance-tuning circuit. And as we combine more cells, the required loadline impedance gets very small, and the matching loss will be unfeasible. The solution is to design a power combiner to provide the necessary loadline impedance for each cell while maintaining a low loss. A clear example is Wilkinson. This is the standard power combiner which provides 50-Ω impedance for all terminals. Other compact power combiners could be used to achieve the same impedance transformation while maintaining lower loss and uses much less area. We will discuss them later in the design example in this chapter. (3) Parallel combining with stacking (Ahmed et al., 2021a, 2020, 2021b,c): this approach is a combination of the previous approaches. Parallel power combiners are used to combine stacked power cells (Fig. 12.3c). We can achieve higher output power, though the design techniques are usually more complicated. All power cells must have the proper loadline and the voltage and current clips simultaneously. This requires extreme care for the interstage matching design. The design examples show this approach without rigorous analysis.

12.2.3 Power Supply Oscillations and Heat Effect

Practical real amplifiers use several stages to get high gain. All stages may share the same DC supply lines. This creates more loops and possible oscillations. Typically, the DC power supply impedances are unknown due to the uncertainty in the DC cable lengths or lack of accurate modeling. This complicates the design since the design must be stable over a wide range of power supply impedances. We should pay extreme care for the biasing network. Usually, we increase the on-chip capacitance as much as possible. The bypass capacitors are usually connected in series with a few ohms to avoid any additional oscillations. More capacitors are added on the packaging side to provide additional isolation between the stages.

There are several concerns regarding the heat dissipation: (1) in BJT, there is a thermal stability phenomenon, which means that some fingers may draw more current than others which causes more heating and draw more current until the transistor burn. The solution is to add the proper amount of ballast resistance and choose the proper number of fingers per cell. (2) the package must have an adequate heatsinking mechanism to dissipate the generated heat from the amplifier to avoid thermal destruction and performance degradation.

12.2.4 Technology Considerations

The key features of any technology are (1) power gain cut-off frequency (f_{max}) which indicates the maximum useable gain, (2) breakdown voltage and maximum current density, (3) wiring stack, and (4) cost and yield. THz amplifiers demand technologies with high f_{max} to provide sufficient power gain. Technologies with high breakdown voltage and current swings deliver higher output power under the stipulation that the matching loss is reasonable. Most technologies offer multiple metals for routing. It is necessary to have the appropriate metal spacing to permit low-loss high-impedance transmission lines with higher current capabilities.

Silicon CMOS works well at D-band with moderate output power but degrades at higher frequencies. Silicon technologies are widely used in mass production which decreases the cost. III-V technologies have significantly higher f_{max} ($\sim$1 THz) which makes THz amplifier design feasible with a reasonable performance. III-V technologies have higher breakdown voltages and higher current density compared to silicon which provides significantly higher power and efficiency. Additionally, the design in III-V is usually easier since the number of layers is typically small and technologies allow solid metals without density rules or maximum line widths constraints. In advanced CMOS nodes, we need to maintain a certain metal density which requires dummy fillings. There are many design rules which we must follow and it consumes more design time. Unfortunately, III-V technologies are usually more expensive, and it is not easy to integrate with silicon technology.

12.3 Design Examples

In this section, we will review several THz amplifiers designed by the authors showing record efficiencies and output power at (140 GHz(Ahmed et al., 2021a, 2020), 210 GHz (Ahmed et al., 2021b), 270 GHz (Ahmed et al., 2021c), and 600 GHz (Seo et al., 2013)). The amplifiers are designed using Teledyne 250 nm InP HBT technology, except for the last 600 GHz design where 130 nm InP HBT process was employed. Technical details of these technologies are summarized in the next subsection. To reduce the redundancy, we will present in detail the 140 GHz power cells then we will show briefly how other designs differ along with the measurement results.

ICs were fabricated in the Teledyne 130 nm/250 nm InP HBT technologies (Urteaga et al., 2017) with 0.3 fF/μm^2 MIM capacitors, 50Ω/square thin-film resistors and three or four Au interconnect layers for 130 nm and 250 nm nodes, respectively. The 250 nm HBTs have a maximum 650 GHz power gain cut-off frequency (f_{max}), a

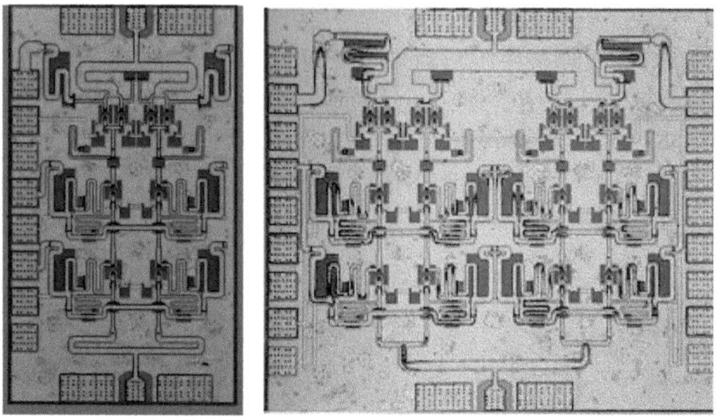

(a) (b)

Figure 12.4 Chip micrograph of 140 GHz power amplifiers (a) Four cells (1.08 mm × 0.63 mm). © [2020] IEEE. Reprinted, with permission, from (Ahmed et al., 2020)(b) Eight cells (1.23 mm × 1.09 mm). © [2021] IEEE. Reprinted, with permission, from (Ahmed et al., 2021a).

maximum 3 mA/μm current density, and 4.5 V BV_{CEO}, while 130 nm HBTs exhibit 1.15 THz peak $f_{\max}$ and 3.5 V BV_{CEO}.

12.3.1 140 GHz Power Amplifier

At 140 GHz, we report two high-efficiency power amplifiers in 250 nm InP HBT Technology. The first design (Ahmed et al., 2021a) has three common-base stages and a low-loss 4:1 transmission-line output power combiner. The amplifier (Fig. 12.4(a)) has 20.5 dBm peak saturated output power with 20.8% PAE and 15 dB associated large-signal gain at 140 GHz. The second design (Ahmed et al., 2020) has three common-base stages and combines eight power cells to double the output power while utilizing similar power and driver cells to the first design. The second amplifier (Fig. 12.4(b)) has 23 dBm peak power with 17.8% power-added efficiency (PAE) and 16.5 dB associated large-signal gain at 131 GHz.

12.3.1.1 Unit cell design

Both designs share the same driver (Fig. 12.5(a)) and power cells (Fig. 12.5(b)). We considered common-emitter (CE) and common-base (CB) designs. Since those amplifiers are intended to be

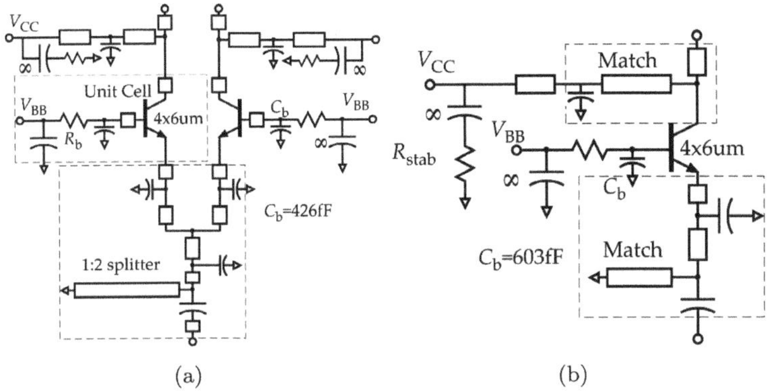

Figure 12.5 Schematic diagram of: (a) Two combined power amplifier cells driven by a single driver cell. (b) Driver stage with input/output matching circuits. © [2020] IEEE. Reprinted, with permission, from (Ahmed et al., 2020).

used in communication systems where linearity is necessary, we determined the cell based on the highest efficiency at OP^{1dB}. At the same bias conditions (1.4 mA/um and $V^{CE} = 2.5$ V) and dimensions (4 × 6 um), large-signal simulations are performed for a CE (Fig. 12.6(a)), grounded common base (Fig. 12.6(b)), and grounded base with finite base capacitance (Fig. 12.6(c)). The latter topology could be viewed as a stacked power cell. The OP^{1dB} for CB with a 600 fF base capacitance is 15.2 dBm with 29.7% PAE compared to 12 dBm with 15.4% PAE in CE and 13.5 dBm with 22.4% PAE for the grounded-base CB stage. We chose the CB with finite base capacitance since it has the highest OP^{1dB} simply because of the contribution of the driver stage power. Additionally, CB with finite base capacitance is biased more efficiently compared to the grounded common base. After determining the proper topology, the output is tuned for maximum PAE using a shunt inductor, and the input is matched for the proper loadline impedance for the drivers. Staggered tuning is used to broaden the bandwidth (BW).

12.3.1.2 Combiner design

Four and eight cells are combined in the first and second designs respectively. Wilkinson power combiners are well-known

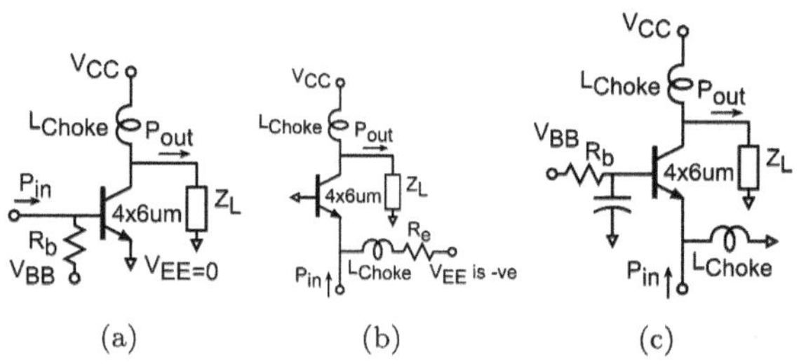

Figure 12.6 (a) Common emitter. (b) Common base with grounded base. (c) Common base with base capacitance. © [2020] IEEE. Reprinted, with permission, from (Ahmed et al., 2020).

techniques to combine 50-Ω power cells. Cascaded Wilkinson combiner can provide 4:1 and 8:1 combining. A 4:1 Wilkinson (Fig. 12.7(a)) requires two cascaded $\lambda/4$ sections with 71-Ω characteristic impedance. As we increase the combining ratio, more $\lambda/4$ sections should be added. Therefore, Wilkinson combiners are bulky and lossy. Also, skinny lines limit the maximum current limits. At the expense of narrower bandwidth, abandoning the Wilkinson design permits the transmission-line combiner to be designed for less loss and a smaller die. In transmission-line combiners the impedance transformation is done using a single $\lambda/4$ section which implies that the combiners are much more compact with lower losses. Also, the required characteristic impedance is low which means that the lines are wide and handle higher current

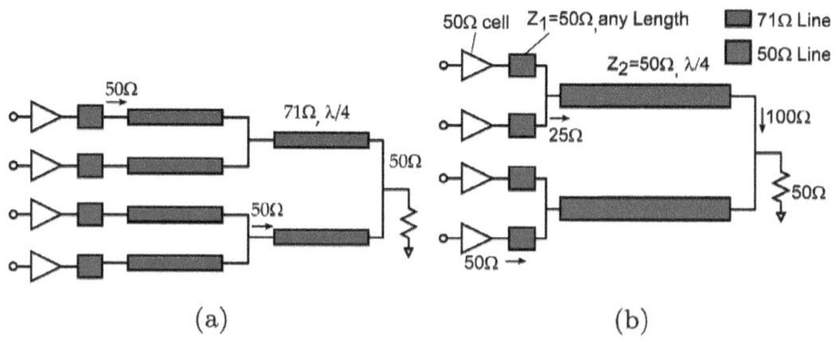

Figure 12.7 (a) 4:1 Wilkinson power combiner (without bridge resistor). (b) proposed 4:1 low-loss transmission-line combiner. Numbers near the arrows show the input impedance. © [2020] IEEE. Reprinted, with permission, from (Ahmed et al., 2020).

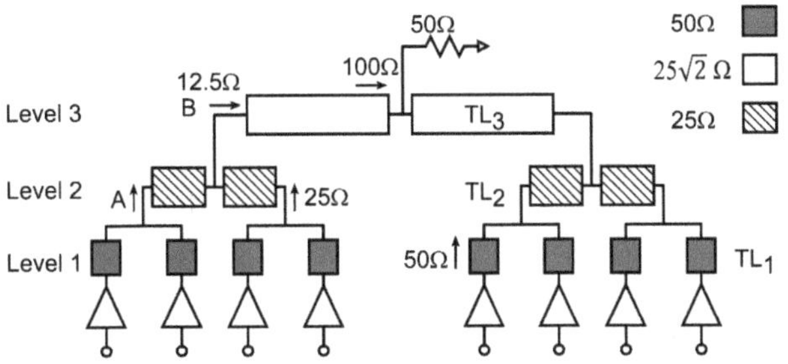

Figure 12.8 Proposed 8:1 transmission-line combiner for 50 Ω load. © [2021] IEEE. Reprinted, with permission, from (Ahmed et al., 2021a).

capability. By tweaking the quarter line's characteristic impedance, the transmission-line combiner works with non-50-Ω power cells. We will explain the design by examples. The first example is the 4:1 power combiner (Fig. 12.7(b)). Two power cells are combined by a 50-Ω line with a minimum length. Note that the length would not affect the impedance transformation but reducing the length decreases the losses. Two combined parallel cells require 25-Ω load impedance. The value of Z_2 is tuned according to the known quarter transformation equation ($Z_2 = \sqrt{100 \times 25}$) to transform the 100 Ω presented at its load to 25 Ω. Another example is the 8:1 transmission-line combiner (Fig. 12.8). We combine the cells in a binary fashion. In the first combining level, we use 50-Ω lines. The required load impedance at node "A" is 25 Ω. In the second level, we use 25-Ω lines. The required impedance at node "B" becomes 12.5 Ω. Finally, we select the proper Z_3 to transform the 100-Ω load impedance to 12.5 Ω. We compared the loss of the 8:1 Wilkinson with the transmission line, and the transmission-line combiner has a loss of 0.63 dB at 140 GHz, compared to 0.96 dB for Wilkinson. The loss remains smaller than that of the Wilkinson over a 47.5 GHz bandwidth. The combiner could be extended to combine more cells using a similar strategy. However, we must watch the stability. In transmission-line combiner, we do not have isolation between the stages. We have not observed any stability issue pertaining to the combiner, but it is a valid concern. For the sake of illustration, we ignored the pad impedance and assumed a 50-Ω load in the previous

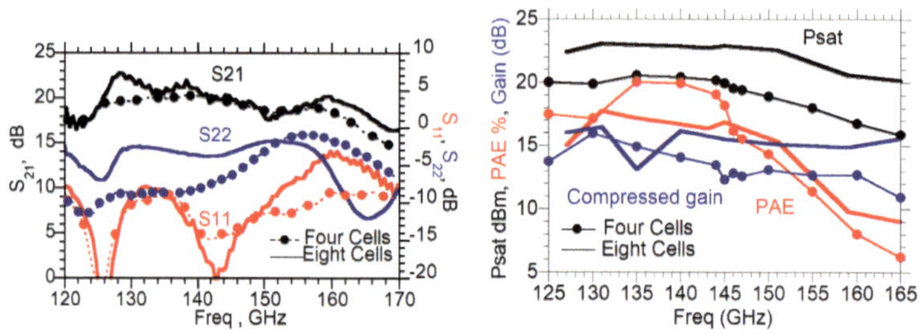

(a) (b)

Figure 12.9 (a) Measured S-parameters for 4-cell (dotted) and 8-cell (solid) amplifiers. (b) measured saturated output power with associated gain and efficiency vs frequency for 4-cell (dotted) and 8-cell (solid) amplifiers. © [2020, 2021] IEEE. Adapted, with permission, from (Ahmed et al., 2020) and (Ahmed et al., 2021a).

examples. We did more optimizations for the lines' characteristic impedances to consider the pad impedance.

12.3.1.3 Measurement results

The two main measurements are small-signal S-parameter measurement and large-signal power measurement. The network analyzer "N5247A" goes only up to 67 GHz, and 110-170 GHz VDI frequency extenders extend the operation up to D-band. DC probes are used to provide the DC power supply, and microwave probes are used for the RF interface. A D-band attenuator is added after the extender to avoid amplifier saturation. The S-parameter reference plane is moved to the probe tips using a SOLT standard on an external calibration substrate. Fig. 12.9(a) shows the measured S-parameters for the 4- and 8-cell amplifiers. For the 4-cell design, the peak measured gain is 20.3 dB at 140 GHz with 25 GHz 3 dB BW while consuming 492 mW DC power and has an area of 1.08 mm × 0.63 mm. For the 8-cell design, the amplifier has a 21.9 dB peak gain at 131 GHz with 20 GHz 3 dB BW while dissipating 1.1 W and has an area of 1.23 mm × 1.09 mm.

For the power measurement, we used the 110–170 GHz VDI frequency extender as the signal source since it delivers up to 10 dBm of output power. The power is swept by a mechanical variable attenuator. Ericson power meter with PM4 head sensor measures the output power. After de-embedding the probe losses, the 4-cell design shows 20.5 dBm with 20.8% PAE and 15 dB gain. Over 125–150 GHz bandwidth, the saturated output power is within 2 dB of its 140 GHz maximum (Fig. 12.9(b)). Over 127–151 GHz, the 8-cell design has 22.3–23 dBm saturated output power with 13–16.5 dB gain and 15–17.8%PAE (Fig. 12.9(b)).

12.3.2 210 GHz Power Amplifier

At 210 GHz, we report four capacitively linearized CB stages. Four power cells are combined by a low-loss 4:1 corporate combiner. At 202 GHz operation, the amplifier (Fig. 12.10) has a compact area of 1.2 mm×0.95 mm and delivers a saturated output power of 18.3 dBm with 7.9% PAE. We followed a similar design approach to Sec. 12.3.1 as follows: (1) capacitively linearized cells show superior performance and are used in the power and driver cells. We used smaller base capacitance since they have less parasitic inductance and hence higher self-resonance frequency. (2) The drivers are scaled to reduce the total DC power consumption and sustain high PAE. (3) We scaled the 4:1 transmission-line combiner presented in Example 1 to 210 GHz. Combiner scaling is fairly easy. The characteristic impedance is a weak function of the frequency so, the line widths do not scale with frequency. We mainly decrease the length of the transformation line section since λ is smaller at 210 GHz.

Measurement results

The S-parameter measurement is quite similar to Section 12.3.1. The PNA can operate with frequency extender modules to push the operation up to high frequency. Here we used the available H-band frequency extender (G-band extender would have been more appropriate). The reference plane is moved to the probe tips using LRRM standards. Circuit measurement does not require tremendous accuracy compared to the device characterization. Therefore, the accuracy of LRRM or SOLT is enough. If higher accuracy is required, we have to build our own calibration kit and move the reference

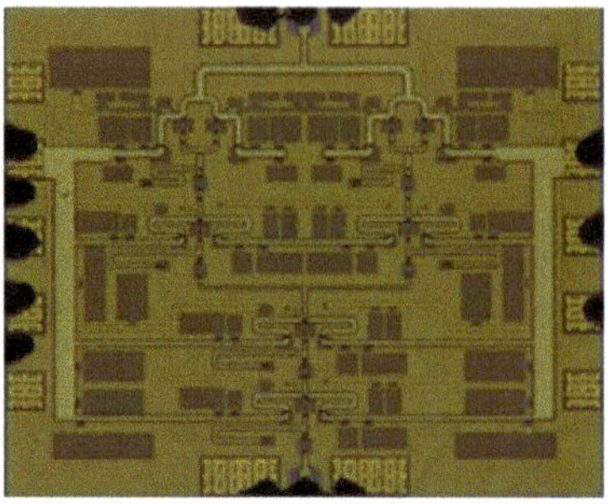

Figure 12.10 Chip micrograph of 4-cell 210 GHz power amplifier (1.2 mm × 0.95 mm). © [2021] IEEE. Reprinted, with permission, from (Ahmed et al., 2021b).

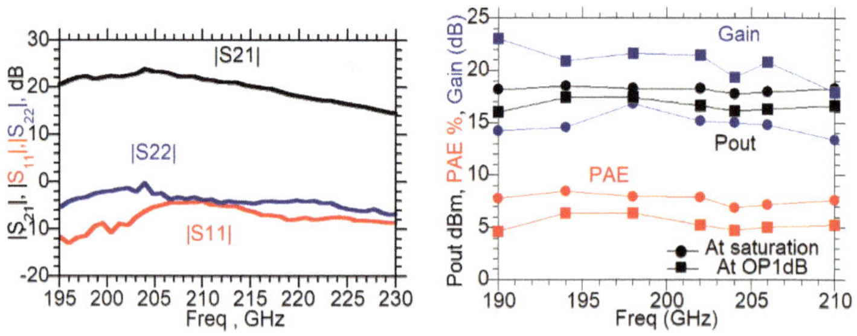

(a) (b)

Figure 12.11 (a) Measured S-parameters. (b) Measured output power with the associated PAE and compressed gain vs. frequency reported at the saturated output power and OP^{1dB}. © [2021] IEEE. Reprinted, with permission, from (Ahmed et al., 2021b).

plane to the device level. The amplifier, dissipating 0.86 W, has a peak small-signal gain of 23.5 dB at 204 GHz with more than 20.5 GHz 3 dB BW (Fig. 12.11(a)). The H-band frequency extender has very low output power and does not provide the necessary driving capability for power measurement. Instead, we used a ×8 frequency multiplier chain. A 20 dB coupler is added after the frequency multiplier to sample the power going to the amplifier. The coupling port goes to another attenuation, then harmonic mixer and spectrum analyzer to monitor the power. The spectrum analyzer readings represent the power by adding the appropriate calibration

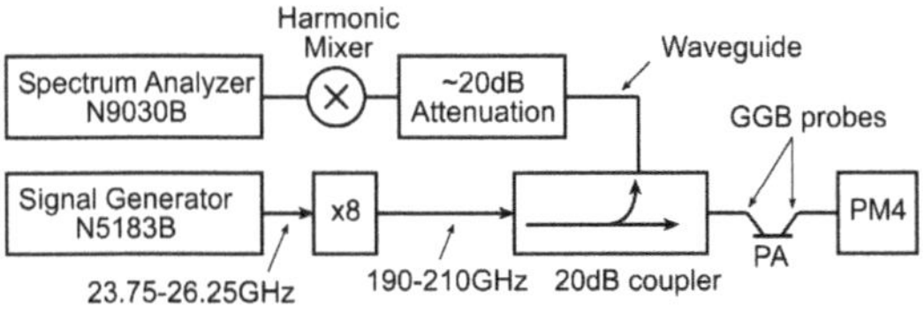

Figure 12.12 Power measurement setup of G-band amplifier. © [2021] IEEE. Reprinted, with permission, from (Ahmed et al., 2021b).

factor. Monitoring the amplifier's input power gives more accurate results since the frequency multiplier output power may change from time to time. PM4 with a power meter is used to measure the output power. Over 190–210 GHz and after calibrating the probe losses (Fig. 12.11(b)), the amplifier shows 17.7–18.5 dBm saturated output power with more than 6.9% PAE and 13.4–16.8 dB associated gain while dissipating 814 mW DC power.

12.3.3 270 GHz Power Amplifier

At 270 GHz, we report a compact H-band power amplifier with high output power. The amplifier (Fig. 12.13(a)) has a compact area of 1.08 m m× 0.77 mm. The amplifier has four stages and combines four power cells by a low-loss transmission-line combiner. The amplifier has 16.8 dBm saturated output power with 4% PAE at 270 GHz. Similar design techniques are used with some adjustments: (1) We are still using linearized CB topologies. However, as the frequency increases, the base inductance modeling error becomes more significant, which may lead to stability problems or huge uncertainty in the gain. We overcame this issue by significantly reducing the base capacitance. This increases the SRF of the capacitor itself and pushes the resonance formed between the base capacitor and the parasitic base inductor to a higher frequency which makes the gain more predictable. The downside is that the gain per stage decreases, which requires more stages for compensation. Additionally, in a lossy network, the efficiency drops. (2) After tuning the transistor parasitics, the required loadline impedance is 18 Ω. This is a relatively small impedance, and clearly, Wilkinson cannot be used since it works with standard 50-Ω cells.

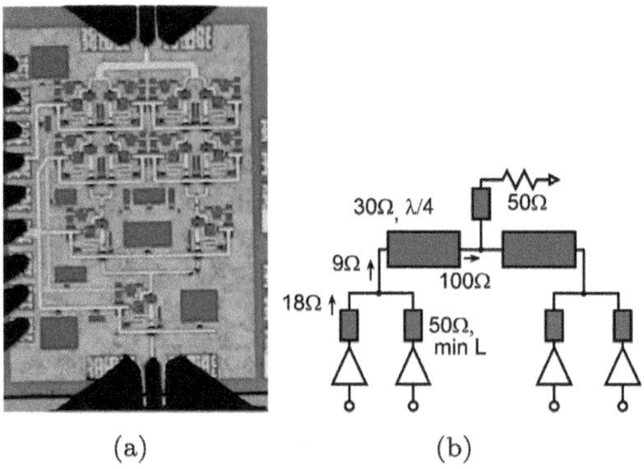

Figure 12.13 (a) Chip micrograph of 8-cell 270 GHz power amplifier (1.08 mm×0.77 mm). (b) 4:1 compact low-loss combiner © [2021] IEEE. Reprinted, with permission, from (Ahmed et al., 2021c).

We used the corporate power combiner (Fig. 12.13(b)), which relies on a single quarter section transformation where we tuned the characteristic impedance according to $Z_o = \sqrt{100 \times 9} = 30$ to transform the 50-Ω load impedance to the load impedance for each power cell. (3) We also scaled the driver to sustain high PAE.

Measurement results

The S-parameters are measured similarly to Section 12.3.2. H-band frequency extenders are used with the PNA. The reference plane is moved to the probe tips using SOLT standards. The amplifier has a peak small-signal gain of 20.5 dB at 264 GHz with 48 GHz 3 dB small-signal BW.

A 110–170 GHz VDI frequency extender drives a 270–290 GHz frequency doubler. An H-band coupler samples the power, and the coupled port goes to a harmonic mixer and spectrum analyzer for monitoring. We are still using PM4 to measure the output power. Interestingly, PM4 operates above ~2 THz with the appropriate adaptors. After calibrating the probe losses, the amplifier shows 14–16.8 dBm saturated output power with 2.2–4% PAE and 9.6–10.9 dB associated gain over 266–285 GHz frequency range while consuming 1.09 W DC power.

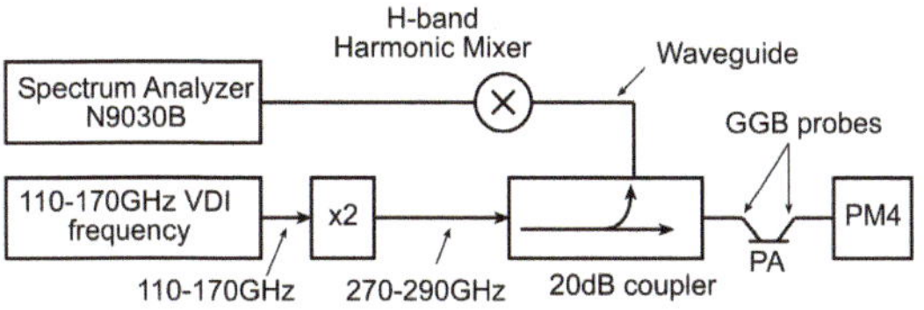

Figure 12.14 Power measurement setup of H-band amplifier © [2021] IEEE. Reprinted, with permission, from (Ahmed et al., 2021c).

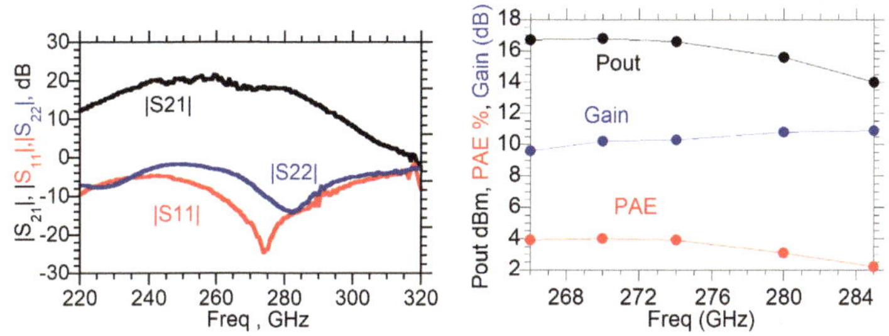

(a) (b)

Figure 12.15 (a) Measured S-parameters. (b) Measured output power with the associated PAE and compressed gain vs. frequency reported at the peak PAE. © [2021] IEEE. Adapted, with permission, from (Ahmed et al., 2021c)

12.3.4 600 GHz Power Amplifier

In this section, we report a 600 GHz amplifier based on 130 nm InP HBT technology. Since the operating frequency is fairly close to the device power gain cut-off frequency, $f_{\text{max}} = 1.1$ THz, it turns out only a few dB of power gain per stage is available after accounting for interstage matching losses. To realize $>$ 25 dB of target gain at 600 GHz, therefore, it was found that more than 10 stages were necessary. Amplifier design with $>$ 10 stages poses several design challenges: large layout size, low power efficiency, and increased sensitivity to various circuit modeling errors including layout parasitics. In this work, three CB stages are stacked, sharing a common collector current, to form a compact gain block. The overall amplifier consists of four-stage cascade of such gain blocks, as shown in Fig. 12.16. To increase the overall output power, each of the differential outputs from the second gain block drives a

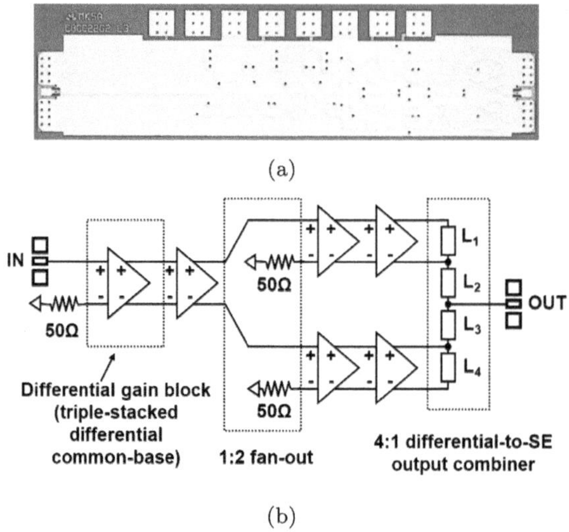

Figure 12.16 600 GHz power amplifier: (a) Chip photograph (chip size: 1.360 mm×0.34 mm). (b) Block diagram. © [2013] IEEE. Reprinted, with permission, from (Seo et al., 2013)

separate amplifier chain, realizing a 1:2 fan-out. The unit gain block is designed for sufficiently low common-mode rejection, so it can be driven single-ended as in Fig. 12.16(b) while keeping the output fully differential. The final amplifier output is generated by a 4:1 differential-to-single-ended power combiner. To make sure the amplifier remains stable under various modeling errors, individual gain stages are stabilized by cross-coupled feedback capacitors with sufficient margins.

12.3.4.1 Unit gain stage

Three types of unit differential gain stages are considered (Fig. 12.17). At 600 GHz, a CB configuration provides 9 dB of MSG/MAG (maximum stable gain: MSG, maximum available gain: MAG), which is significantly higher than 3 dB from a common-emitter (CE) stage. A CB stage is, however, only conditionally stable at 600 GHz, unlike a CE stage. In addition, any underestimated inductance at the base terminal will make the CB more unstable than designed. In this work, individual CB stages were stabilized using cross-coupled feedback

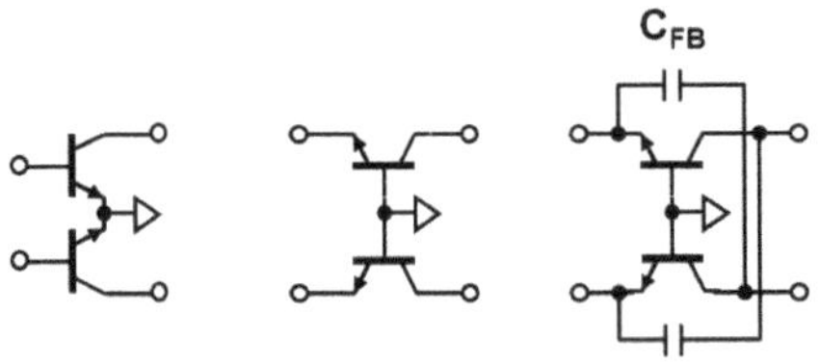

Figure 12.17 Unit gain stages in differential configuration: Common-emitter (CE), CB and CB with cross-coupled feedback capacitors. © [2013] IEEE. Reprinted, with permission, from (Seo et al., 2013)

capacitors, as shown in Fig. 12.17, making use of the differential topology. In a differential CB stage, any low-delay current path between its emitter and collector on the other side provides a negative feedback, since the two node voltages are approximately in-phase. Simulation confirms that a feedback capacitance, as small as $C_{FB} = 0.4$ fF, was sufficient to stabilize a single CB stage with only 0.2 dB reduction in MSG/MAG, even with 50% modeling uncertainty in the base inductance. C_{FB} was implemented by an overlap capacitance between the two lowest metal layers.

Differential configuration brings several benefits to amplifier design: first, the circuit operation is insensitive to the impedance of common-mode bias circuits, and thus more tolerant to associated modeling errors. Second, the virtual ground provided by the differential operation eliminates the degradation in gain and bandwidth resulting from lossy single-ended ac-grounds with finite internal inductance. Last, a single-stage differential amplifier can effectively double as a 1:2 fan-out when only one input is driven with the other input properly terminated. With sufficient common-mode rejection, its output will still be fully differential.

12.3.4.2 Differential gain block

It turns out the estimated layout size of the overall amplifier with > 10 CB stages was beyond the area budget. To reduce the overall chip size, therefore, three CB stages were stacked, as shown in

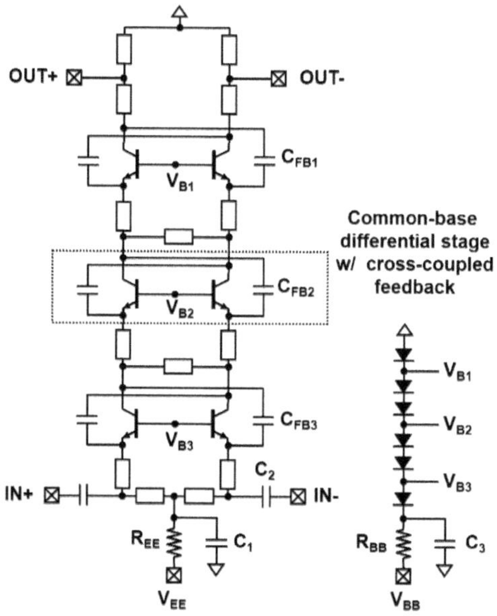

Figure 12.18 Differential gain block with three stacked CB stages. © [2013] IEEE. Reprinted, with permission, from (Seo et al., 2013)

Fig. 12.18. Sharing of collector current enabled compact layout by reducing the number of transmission lines and resistors to provide the collector or emitter current path. The overall 600 GHz amplifier consists of six such differential gain blocks (Fig. 12.16). The entire amplifier circuit is implemented using inverted microstrip lines to utilize a continuous ground plane on the top metal layer.

12.3.4.3 Measurement results

On-wafer testing of the fabricated 600 GHz amplifier IC was performed using VDI WR-1.5 VNA extenders and Dominion WR-1.5 GSG probes. There was reasonable agreement between simulation and measurement as seen in Fig. 12.19. The measured gain was greater than 20 dB up to 620 GHz with > 30 dB peak gain. The amplifier exhibits > 0 dB gain until 655 GHz, consuming 455 mW of total dc power. Large-signal testing of the amplifier was performed using the setup in Fig. 12.20. Measured saturated output power at

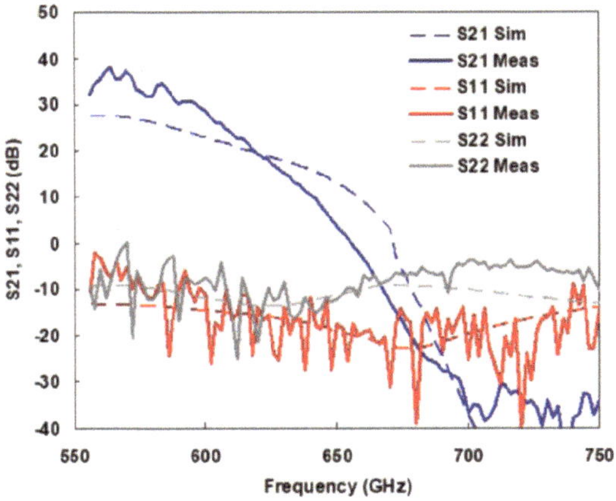

Figure 12.19 S-parameters of the 600 GHz amplifier: Measured (solid lines) and simulated (dotted lines). © [2013] IEEE. Reprinted, with permission, from (Seo et al., 2013)

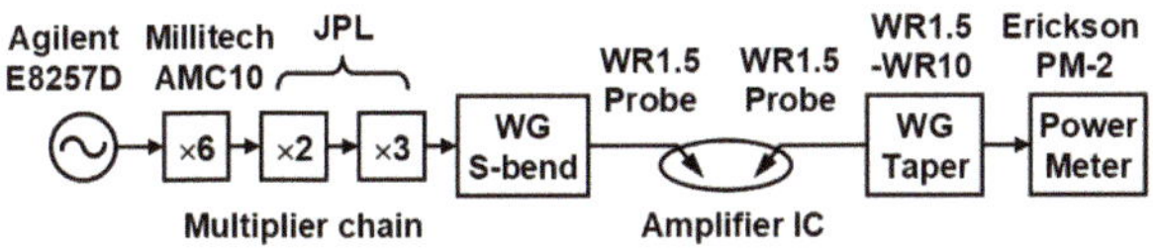

Figure 12.20 Power testing setup (560–585 GHz). © [2013] IEEE. Reprinted, with permission, from (Seo et al., 2013)

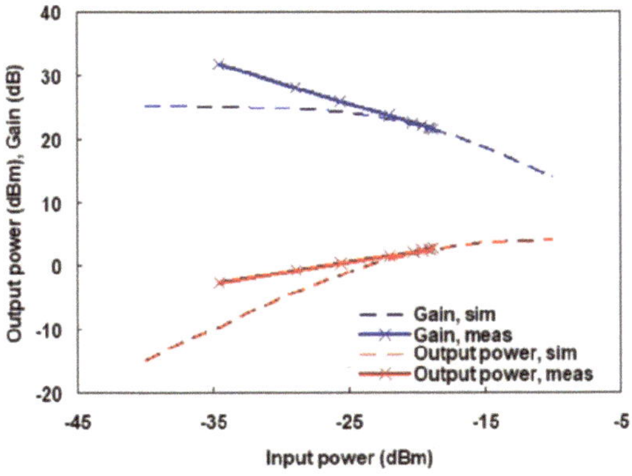

Figure 12.21 Large-signal gain and output power characteristics at 585 GHz: Measured (solid lines) and simulated (dotted lines). All input and output waveguide/probe losses are de-embedded. © [2013] IEEE. Reprinted, with permission, from (Seo et al., 2013)

585 GHz was +2.8 dBm (1.9 mW) after compensating for 7.3 dB of the output probe loss, with 21 dB of compressed gain (Fig. 12.21). Measured saturated output power at 565 GHz was +2.7 dBm.

References

1. Ahmed, A. S., Farid, A. A., Urteaga, M., and Rodwell, M. J. (2018a). 204GHz stacked-power amplifiers designed by a novel two-port technique, in *2018 13th European Microwave Integrated Circuits Conference (EuMIC)*, pp. 29–32, doi:10.23919/EuMIC.2018.8539884.
2. Ahmed, A. S., Seo, M., Farid, A. A., Urteaga, M., Buckwalter, J. F., and Rodwell, M. J. (2021a). A 200mW D-band power amplifier with 17.8% PAE in 250-nm InP HBT technology, in *2020 15th European Microwave Integrated Circuits Conference (EuMIC)*, pp. 1–4, doi:10.1109/EuMIC48047.2021.00012.
3. Ahmed, A. S., Simsek, A., Urteaga, M., and Rodwell, M. J. (2018b). 8.6–13.6 mW series-connected power amplifiers designed at 325 GHz using 130 nm InP HBT technology, in *2018 IEEE BiCMOS and Compound Semiconductor Integrated Circuits and Technology Symposium (BCICTS)*, pp. 164–167, doi:10.1109/BCICTS.2018.8550924.
4. Ahmed, A. S. H., Seo, M., Farid, A. A., Urteaga, M., Buckwalter, J. F., and Rodwell, M. J. W. (2020). A 140GHz power amplifier with 20.5dBm output power and 20.8% PAE in 250-nm InP HBT technology, in *2020 IEEE/MTT-S International Microwave Symposium (IMS)*, pp. 492–495, doi:10.1109/IMS30576.2020.9224012.
5. Ahmed, A. S. H., Soylu, U., Seo, M., Urteaga, M., Buckwalter, J. F., and Rodwell, M. J. W. (2021b). A 190-210GHz power amplifier with 17.7–18.5dBm output power and 6.9–8.5% PAE. in *2021 IEEE/MTT-S International Microwave Symposium (IMS)*.
6. Ahmed, A. S. H., Soylu, U., Seo, M., Urteaga, M., and Rodwell, M. J. W. (2021c). A compact h-band power amplifier with high output power, in *2021 IEEE Radio Frequency Integrated Circuits Symposium (RFIC)*, pp. 123–126, doi:10.1109/RFIC51843.2021.9490426.
7. Buckwalter, J. F., Rodwell, M. J. W., Ning, K., Ahmed, A., Arias-Purdue, A., Chien, J., O'Malley, E., and Lam, E. (2021). Prospects for high efficiency silicon and III-V power amplifiers and transmitters in 100–300 GHz bands, in *2021 IEEE Custom Integrated Circuits Conference (CICC)*, pp. 1–7, doi:10.1109/CICC51472.2021.9431537.

8. S. Cripps, *RF Power Amplifiers for Wireless Communications*. Artech House, 2nd ed. (2006).
9. Wang, H., Huang, T.-Y., Mannem, N. S., Lee, J., Garay, E., Munzer, D., Liu, E., Liu, Y., Lin, B., Eleraky, M., Jalili, H., Park, J., Li, S., Wang, F., Ahmed, A. S., Snyder, C., Lee, S., Nguyen, H. T., and Smith, M. E. D. (2021). Power amplifiers performance survey 2000–2021, [Online]. Available: https://gems.ece.gatech.edu/PA_survey.html.
10. Seo, M., Urteaga, M., Hacker, J., Young, A., Skalare, A., Lin, R., and Rodwell, M. (2013). A 600 GHz InP HBT amplifier using cross-coupled feedback stabilization and dual-differential power combining, in *2013 IEEE MTT-S International Microwave Symposium Digest (MTT)*, pp. 1–3, doi:10.1109/MWSYM.2013.6697692.
11. Urteaga, M., Griffith, Z., Seo, M., Hacker, J., and Rodwell, M. J. W. (2017). InP HBT technologies for THz integrated circuits, *Proceedings of the IEEE* **105**, 6, pp. 1051–1067, doi:10.1109/JPROC.2017.2692178.

Chapter 13

Terahertz Silicon On-Chip Antenna

Jinho Jeong
Department of Electronic Engineering, Sogang University,
35 Baekbeom-ro, Mapo-gu, Seoul 04107, Korea
jjeong@sogang.ac.kr

The antenna is an important component in wireless systems, converting the guided wave in the transceiver circuits into the propagating wave in free space, and vice versa. This chapter provides an overview of various THz antennas integrated in silicon integrated circuits (Si ICs). Firstly, the metal and dielectric stack-up structure of the general Si IC process will be described together with the material properties, followed by the discussion on their impact on the antenna performance in terms of radiation efficiency, gain, and bandwidth. Then, several antenna structures are presented, to overcome the limitations caused by the Si process and to restore the antenna performance. Based on the radiation direction, they are classified in this chapter into the topside radiating antenna with frontside ground, the topside radiating antenna with backside ground, and backside radiating antenna. Each antenna structure is explained including the design examples and measurement results. Finally, the IC design rules that should be taken into account in the design of the on-chip antenna are briefly discussed.

Advances in Terahertz Source Technologies
Edited by Gun-Sik Park, Masahiko Tani, Jae-Sung Rieh, and Sang Yoon Park

ISBN 978-981-4968-89-8 (Hardcover), 978-1-003-45967-5 (eBook)
www.jennystanford.com

13.1 Introduction

Antenna is an essential component of wireless systems for various applications, such as wireless communications, radars, imaging, and sensors. Generally, it is implemented separately from Si ICs, mainly because it is too large to be integrated. Instead, it is designed using sophisticated materials with low dielectric and conductor losses with optimal thickness and size, referred to as off-chip antennas, which provide good radiation performances. Thus, off-chip antennas inherently require interconnections to ICs using wire or flip-chip bonding. In some cases, they also need electromagnetic (EM) mode transition structures depending on the antenna type. They entail parasitic components and package complexity and thus lead to performance degradation such as impedance mismatches and high insertion losses, which are more serious at terahertz (THz) frequencies. Therefore, the off-chip antenna approach results in bulky and high-cost THz systems.

At THz frequencies, the antenna becomes sufficiently small to be integrated into Si IC chips owing to the short wavelength. This on-chip antenna does not require off-chip interconnections and packaging, leading to the full integration of THz Si chips with a compact size, which is beneficial for low-cost mass production. It also allows flexibility in the design of the overall system with optimized performance, because the antenna can be designed together with other circuits in the same Si IC technologies [1, 2].

However, the materials (dielectrics and conductors) and stack-up structure in the Si IC process are not well-suited for antenna design, resulting in low performance, such as low radiation efficiency and low antenna gain with a small bandwidth. The low radiation efficiency and gain imply that a large amount of output power in the transmitter is dissipated in the antenna, which increases the power consumption and reduces the effective isotropic radiated power (EIRP). They also result in a higher noise power and thus lower sensitivity in the receiver. Furthermore, the bandwidth of the on-chip antenna might be too small to limit the operating bandwidth of the overall system. For example, it can overshadow the ultrawideband availability for high-date-rate wireless communications, which is one of the major merits of the

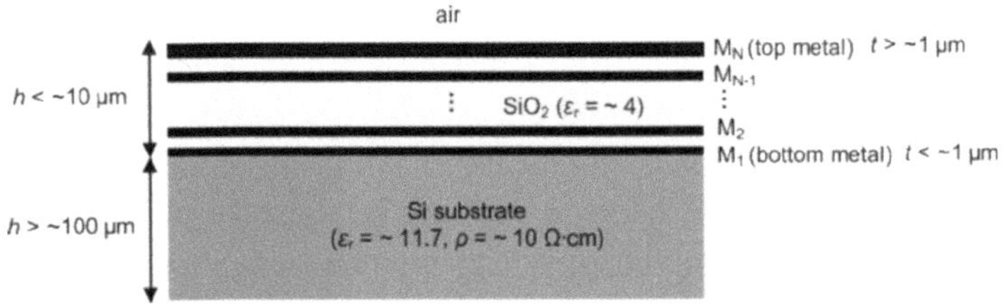

Figure 13.1 Simplified metal/dielectric stack-up structure of the general Si IC process for the on-chip antenna design.

THz frequency band. In this chapter, various antenna structures are presented to improve the performances of THz Si on-chip antennas, based on the understanding of the limitations caused by the Si IC process.

13.2 Si IC Technologies for on-Chip Antenna

An understanding of the metal and dielectric structure of the Si IC process is required for Si on-chip antenna design. Figure 13.1 describes a simplified layer structure of the general Si IC process, consisting of back-end-of-line (BEOL) metal/interdielectric layers on a Si substrate.

The metal layers (M_1–M_N) usually consist of high-conductivity copper or aluminum. Their thicknesses are in the range of a few tenths of micrometers for the bottom and lower metal layers to a few micrometers for the top and upper layers. For the on-chip antenna, the metal layers are utilized as radiators (thick upper layers as patches or dipoles, for example) and ground planes (in thin lower layers). They introduce a conductor loss that reduces the radiation efficiency of the on-chip antenna. Several lower metal layers can be connected using vias to obtain a sufficient thickness to reduce the conductor loss, considering the skin depth.

Dielectric materials (usually SiO_2 with a dielectric constant (ε_r) of approximately 4) fill the space between the metal layers. This dielectric layer can serve as a substrate for a certain type of on-chip antenna. Its thickness can be at most a few tens of micrometers, which is very short relative to the wavelength even at THz frequencies. Thus, it can have a detrimental effect on the antenna performance in terms of radiation efficiency and bandwidth, as discussed later.

The Si substrate has an important role in the design of an on-chip antenna. It has a high dielectric constant ($\varepsilon_r \sim 11.7$) and exhibits a low wave impedance compared to that of air. This implies that the EM waves can be confined and propagate in the Si substrate instead of radiating into the air. This can be considered as power loss in view of the radiation power and is represented by surface wave or substrate mode loss [2, 3].

In the general bulk Si IC process, the Si substrate is N-type-doped to have a low resistivity ($\rho \sim 10\ \Omega\cdot$cm) which introduces dielectric loss into on-chip antennas. The dielectric loss can also be specified by the loss tangent. The height of the Si substrate is approximately a few hundred micrometers, which corresponds to an order of wavelength at THz frequencies. Therefore, the substrate mode can be easily excited and propagated along this electrically thick substrate with a high dielectric constant, resulting in a high substrate mode loss.

In summary, various loss factors and limitations degrade the radiation performance when the antenna is designed for the Si IC process. Various on-chip antennas have been devised to overcome the limitations of the metal/dielectric structures of the Si IC process and thus achieve good radiation performances, such as high radiation efficiency/gain and large bandwidth. Depending on the radiation direction, the on-chip antennas can be classified as topside radiating, backside radiating, and end-fire radiating antennas (based on the plane of the Si substrate). The first two types of antennas are primarily discussed in this chapter, even though the last has some design examples.

13.3 Topside Radiating Antenna with Frontside Ground

13.3.1 Antenna Structure and Design Considerations

The top thick metal layer in the BEOL can be utilized as a radiator of the on-chip antenna. EM waves from the radiator can propagate

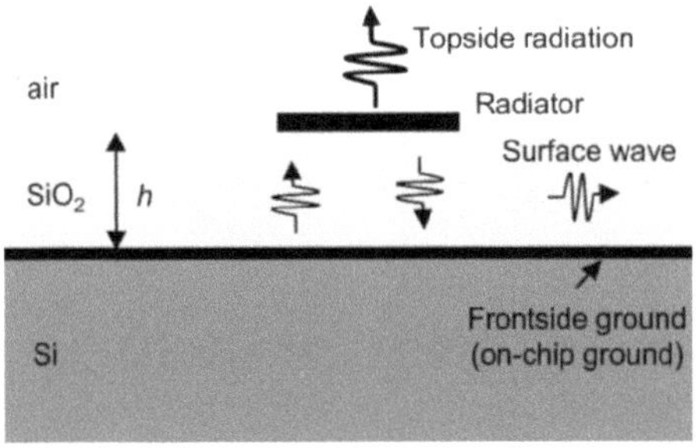

Figure 13.2 Topside radiating antenna with frontside (on-chip) ground plane.

into both the topside and backside, as illustrated in Fig. 13.2. The EM waves in the backside direction can couple into the lossy Si substrate and experience dielectric loss. To minimize this loss, the bottom metal layers can be adopted as the ground plane. This frontside or on-chip ground plane performs the function of an EM-wave reflector. As a result, the EM waves can radiate into the air in the topside direction with the shielded lossy Si substrate.

In this structure, the space between the radiator and reflector is filled with the dielectric (SiO_2), which is electrically very thin compared to the wavelength (λ_0) in free space. The EM waves generated by the radiator are then partly canceled out by those generated by the image currents on the ground plane in close proximity to the radiator. Roughly, this results in a reduction in radiation resistance, limiting the radiation efficiency/gain and bandwidth of the antenna [4].

13.3.2 Design Examples

13.3.2.1 On-chip patch antenna

The rectangular patch antenna can be easily designed and integrated in the Si IC process, even though it exhibits a relatively low gain and small bandwidth. It is a typical topside radiating on-chip antenna with an on-chip ground plane [5, 6]. It can be regarded as a cavity resonator consisting of a top patch/bottom ground (electric walls) and four open side walls (magnetic walls). It radiates EM fields from patch edges in the form of fringing fields [7]. Therefore, a microstrip

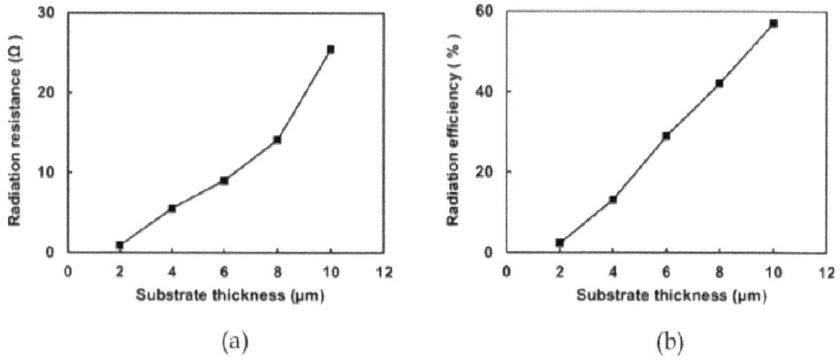

Figure 13.3 Simulated performance of the rectangular patch antenna designed at 300 GHz as a function of the SiO_2 thickness. (a) Radiation resistance. (b) Radiation efficiency.

patch antenna is generally designed in a thick substrate with a low dielectric constant for good radiation performance (high efficiency and large bandwidth) [4, 7].

Note that the substrate of the on-chip patch antenna in this configuration consists of a very thin SiO_2 layer ($\varepsilon_r \sim 4$). Figure 13.3 shows the simulated radiation resistance and efficiency of the rectangular patch designed at 300 GHz as a function of the substrate thickness (h) from 2 to 10 μm, which corresponds to $h/\lambda_0 \leq 0.01$. A thinner substrate leads to lower radiation resistance. This leads to a low radiation efficiency, because the thin substrate decreases the radiation resistance and thus increases the impact of the conductor loss of the metal layers used as the radiator and reflector (ground plane). A peak radiation efficiency is lower than 60% at $h = 10$ μm. If the copper metal is replaced with a perfect electric conductor (PEC), a high radiation efficiency (close to 100%) is achieved regardless of the substrate thickness.

The radiation efficiency can increase as the ratio h/λ_0 increases. That is, higher efficiency can be obtained due to shorter λ_0 as the frequency increases. However, note that there is an optimal h/λ_0 above which the efficiency begins to decrease owing to the surface wave loss [4]. The optimal h/λ_0 is determined by the material properties of the patch antenna such as the dielectric constant and loss tangent of the dielectric and conductivity of the conductors.

The impedance-matching bandwidth (meeting the input reflection coefficient lower than −10 dB) also increases as the substrate thickness increases. However, it reaches only approximately 6% for a substrate thickness of 10 μm. This small bandwidth is also attributed to the low radiation resistance of the thin substrate (or small h/λ_0).

According to the above simulation results, the metal layers in the BEOL for the radiator (patch) and the ground plane should be carefully selected to have a sufficient dielectric thickness. It is also necessary to ensure that the ground metal plane has a sufficient thickness for a low conductor loss, because it is generally formed using very thin bottom metal layers.

For a large bandwidth, the radiator shape can be modified to store less energy and have a lower Q-factor such as an E-shaped patch, slots in the patch, or ring shape. Multi-resonators with slightly different resonant frequencies from that of the radiator can be stacked on top of the radiator to increase the bandwidth, which might result in a reduction in the distance between the radiator and the ground plane. Parasitic strips can be placed along the sides of the patch for a large bandwidth. However, it can cause variations in the radiation pattern and increase the antenna size [4, 7].

Figure 13.4 shows an on-chip E-shaped patch antenna designed at 340 GHz [8]. It is placed on a SIW cavity consisting of BEOL layers (M1 as ground) which shields the EM radiation from the lossy Si substrate and suppresses the surface wave propagation. The patch is utilized as a driver antenna of the DR director, which is placed on top of the patch with a supporter in between. This 3D on-chip antenna exhibits a radiation efficiency as high as 80% and peak gain of 10 dBi with a bandwidth of 12%.

13.3.2.2 Slot antenna

Slot antenna is another type of on-chip antenna that can be easily implemented using the metal layers in the Si IC process. The radiating slot can be formed of various shapes in the top metal plane such as straight lines or circular/elliptical rings. It radiates the EM waves into both topside and backside directions. Then, it is usually

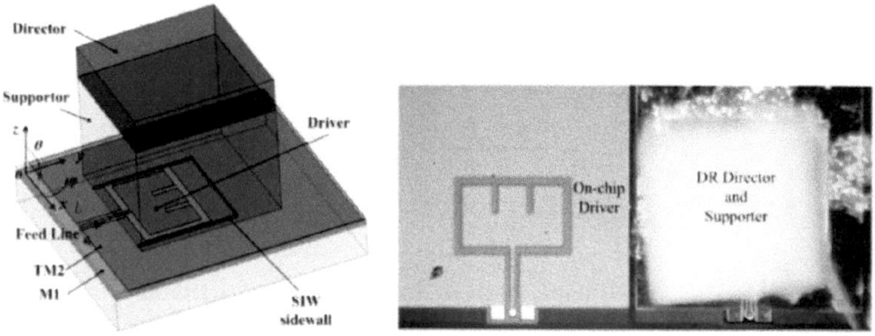

(a) (b) (c)

Figure 13.4 Substrate-integrated waveguide (SIW)-backed E-shape patch antenna with a dielectric resonator (DR) director fabricated with a 0.13 μm SiGe BiCMOS technology (DR: 100 μm-thick Al_2O_3 with $\varepsilon_r \sim 9.8$. Supporter: thickness ~330 μm and $\varepsilon_r = 2.1$. Size: 0.7 mm × 0.7 mm). (a) Three-dimensional (3D) structure. (b) E-shape patch. (c) E-shape patch with a DR director. (© 2015 IEEE [8]).

backed by on-chip ground plane (using bottom metal layers) which reflects the EM waves back to topside.

In this structure, the two metal planes (top and bottom) constitute parallel-plate waveguide in which the surface waves can propagate, lowering the radiation efficiency and degrading the radiation pattern. The SIW structure with sidewalls can be utilized in the slot antenna to alleviate the surface wave generation. It can also perform a function of high-pass filter which can be useful characteristics for the circuit design [9].

Figure 13.5 shows an elliptical slot-ring antenna patterned in the top metal layer backed with the on-chip ground in the bottom metal layer [10]. A quartz superstrate is placed on top of the antenna, to equalize the transverse electromagnetic (TEM) mode in the dielectric between top and bottom grounds, and transverse magnetic (TM_0) mode in the quartz substrate. Then, the radiating slots in the ring antenna are placed at half-wavelength apart for both modes, so that both surface waves are canceled out. It results in the high radiation efficiency of 62% with a gain of 4.5 dB in the simulation at 360 GHz.

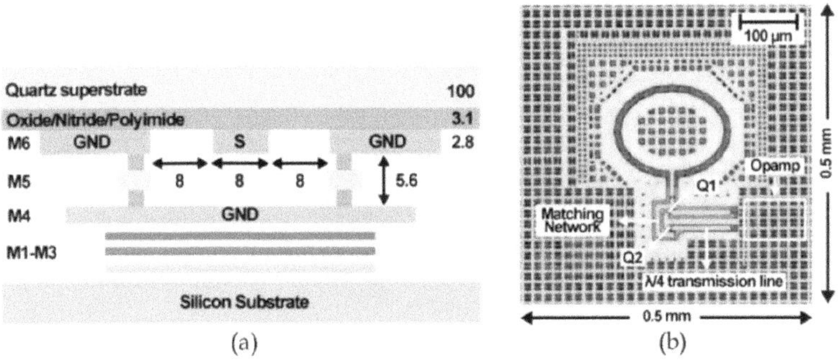

Figure 13.5 360 GHz elliptical slot-ring antenna with a 100 μm-thick quartz superstrate (slot ring in the top metal (M6) and ground in the bottom metal (M1)). (a) Metal stack-up structure in the 0.13 μm SiGe CMOS process. (b) Chip photograph. © 2013 IEEE [10].

13.3.2.3 Antenna with AMC

In the topside radiating antenna with a frontside ground, the radiator and ground are in close proximity to each other. Ideally, the ground plane can be regarded as a PEC with a reflection coefficient (Γ) of -1, so that the reflected waves exhibit a phase shift of 180°, producing a destructive interference. This leads to a low input impedance, which entails a high conductor loss, low radiation efficiency, and small impedance-matching bandwidth.

In contrast, a perfect magnetic conductor (PMC) presents an open impedance with $\Gamma = +1$, resulting in in-phase reflection and constructive interference. Therefore, the PEC ground can be replaced with PMC structure as shown in Fig. 13.6, to increase the radiation resistance and efficiency as well as to prevent the EM waves from coupling to the lossy Si substrate. The PMC can be designed in the periodic structures with various shapes such as dog-bone, snowflake, and cross [11–13]. However, this AMC can be modeled as a parallel-resonant circuit and exhibits band-limited performance. AMC-based on-chip antennas have been reported at millimeter-wave frequencies in the form of microstrip patches and slotted bowties with topside radiation [11, 12]. In [13], end-fire radiation is presented using the AMC Yagi antenna.

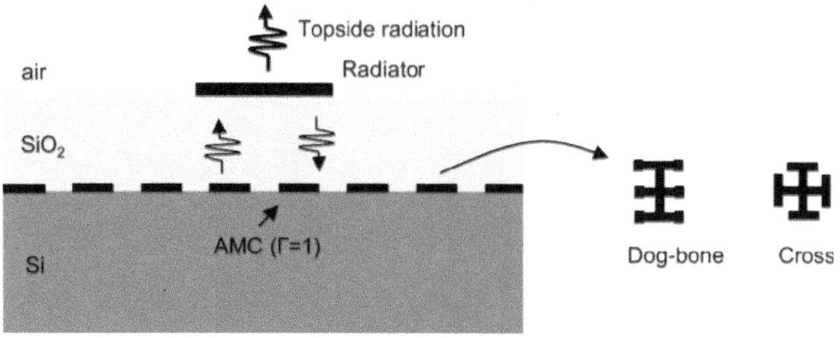

Figure 13.6 On-chip antenna with an artificial magnetic conductor (AMC) in the bottom metal layers [11–13].

13.4 Topside Radiating Antenna with Backside Ground

The ground plane on the frontside of the Si substrate prevents the EM waves radiated by the topside radiator from entering the lossy Si substrate. However, a very thin SiO_2 substrate limits the radiation performance of the antenna. To overcome this problem, the EM fields can be allowed to enter the Si substrate by removing the on-chip ground plane, as shown in Fig. 13.7. The ground plane is moved underneath the Si substrate and reflects the EM waves back to the topside. In this chapter, a topside radiating antenna with a backside ground plane is discussed.

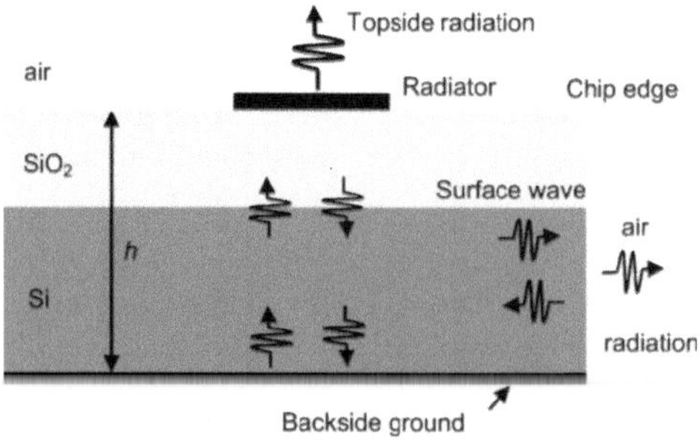

Figure 13.7 Topside radiating antenna with a backside ground plane.

13.4.1 Antenna Structure and Design Considerations

In this configuration, the radiator is, in general, designed in the top thick metal layer for a low conductor loss, similar to the topside radiating antennas with a frontside ground. However, there is no frontside ground at least under the radiators, so that the EM waves can radiate into the Si substrate as well as into the air. Note that on-chip ground planes can exist in the bottom metal layers for other circuits. In addition, the EM power radiated into the Si substrate is considerably higher than that into the air because of the high dielectric constant of the Si substrate and thus low wave impedance [2, 3]. Therefore, it is required to reflect the EM waves at the backside of the Si substrate back to the topside for high radiation efficiency and gain. This is implemented by the backside ground plane underneath the Si substrate as shown in Fig. 13.7.

In the conventional Si IC process, the thickness of the Si substrate is approximately a few hundred micrometers, which can be on the order of the guide wavelength (λ_g) at THz frequencies. Therefore, the reflected waves can cause constructive or destructive interference with the EM fields radiating in the topside direction depending on the substrate thickness and frequency. If the thickness is an odd multiple of $\lambda_g/4$, constructive interference occurs and the gain can peak. The Si substrate has a finite conductivity, so that the dielectric losses increase with the thickness. Therefore, an optimal thickness can exist for the highest radiation efficiency and gain, and thus the substrate can be ground down to this optimal thickness if possible. Alternatively, the lossy Si substrate underneath the radiator can be etched out to remove the dielectric loss.

The EM waves coupled to the Si substrate can propagate along the thick and high-ε_r substrate backed with a ground plane. This surface wave is a dominant loss factor because of high h/λ_0 in this configuration. The surface wave propagation is also dependent on the lateral dimensions of the substrate. At the chip edges, surface waves can be partly reflected from the air or partly transmitted to the air. The latter, the surface wave radiation into the air, distorts the radiation pattern of the antenna. The reflected waves at the chip edges propagate in the Si substrate, which reduces the radiation efficiency owing to the Si loss. It also increases the coupling with other circuits. In general, small chip size is helpful in reducing the substrate mode. It is necessary to optimize the distances of the antenna from the chip edges for high efficiency and gain [4]. The reflected waves from the chip edges can create standing waves inside the finite size of the Si substrate, which can serve as the DR and thus increase the antenna gain if the size is properly determined.

In the antenna array, surface wave propagation can be minimized by properly placing the antenna elements. The elements are placed half-wavelength apart from each other, where the wavelength is for the surface wave mode. Then, the surface waves from each element can be canceled out by 180° out-of-phase interference.

13.4.2 Design Examples

13.4.2.1 Slot-ring antenna

Figure 13.8a illustrates a slot-ring radiator patterned in the top metal layer with a metal reflector at the backside of a 300 μm-thick Si substrate in a 65 nm Si CMOS process [14]. The simulated antenna gain is strongly dependent on the substrate height owing to the constructive and destructive interference of the reflected waves, as shown in Fig. 13.8b, where the optimal height is found to be around 300 μm at 240 GHz. The antenna gain also depends on the lateral size of the Si substrate, as shown in Fig. 13.8c, because a larger substrate leads to strong surface waves. The peak gain is obtained at a chip edge distance of ~150 μm.

13.4.2.2 Dipole antenna

The dipole antenna can be integrated in the Si IC process in a planar form, where the dipole is patterned in the top thick metal layer on the Si substrate. Most of the radiated power is coupled into the Si substrate because of its high dielectric constant. The ground plane beneath the Si substrate reflects the EM power back to the topside, creating a broadside radiation. The Si substrate in [15] is 300 μm-thick, corresponding to $\sim 3\lambda_g/4$ at 210 GHz, which produces constructive reflections and increases the radiation efficiency to 24%. The dipole antennas are arrayed to 2 × 2 to increase the output power of the transmitter through spatial power combining,

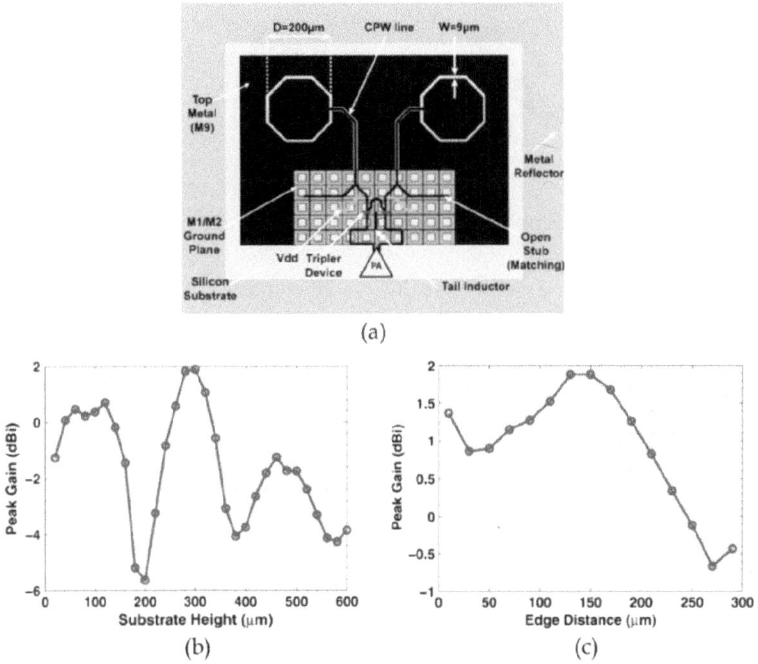

Figure 13.8 240 GHz slot ring antenna (in the top metal) with a Si backside reflector in a 65 nm Si CMOS technology (Si height = 300 μm). (a) Antenna structure. Simulated antenna gain as a function of the (b) substrate height and (c) chip edge distance. © 2015 IEEE [14].

as shown in Fig. 13.9. Ground rings using bottom to top metal layers surround each antenna, reducing the coupling between the antennas and other circuits. The antenna gain is increased from −2.5 dBi for a single element to 4.5 dBi for a 2 × 2 array at 210 GHz. The bandwidth is as wide as 40 GHz.

Four double-folded dipole radiators in the top thick metal layer are arrayed to 1 × 4 with a backside reflector in a 0.13 μm SiGe BiCMOS process, where the bottom metal layer is used as the ground plane of the transmitter IC [16]. The ground plane is open below the dipole antenna so that the EM waves can couple to the Si substrate. In addition, the Si substrate underneath the radiators is removed by the localized backside etch (LBE) process, allowing a high gain of 14 dBi and high efficiency of ∼ 76% at 245 GHz.

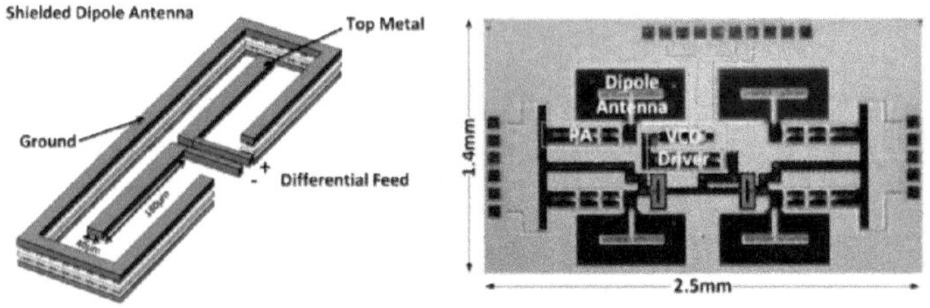

(a) (b)

Figure 13.9 Dipole antenna in a 32 nm silicon-on-insulator (SOI) CMOS process. (a) Layout. (b) Photograph of the transmitter with a 2 × 2 antenna array. © 2014 IEEE [15].

13.4.2.3 Patch antenna with DGS

For the microstrip patch antenna, several slots can be formed in the frontside ground plane to allow the EM fields to enter the Si substrate. This DGS is helpful in increasing the radiation resistance of the antenna for a high efficiency and large bandwidth [17, 18]. Figure 13.10a shows the patch antenna with a DGS designed at 300 GHz. The V-shaped patch is used as a primary radiator on top of which another patch is stacked to generate multiple resonances for a large bandwidth. The antenna gain varies depending on

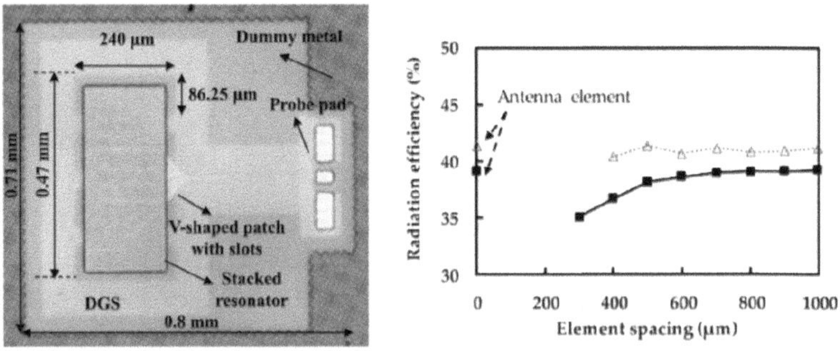

(a) (b)

Figure 13.10 300 GHz patch antenna with a defected ground structure (DGS) in a 65 nm Si CMOS process. (a) Photograph of a V-shaped patch antenna [17]. (b) Simulated radiation efficiency of 1 × 2 array as a function of the element spacing (antenna: rectangular patch, square: without DGS, triangle: with DGS) [18].

the substrate thickness, owing to the leakage waves in the Si substrate. The substrate thickness is 250 μm in this process, which corresponds to $0.86\lambda_g$ (close to $3/4\lambda_g$) at 300 GHz. Thus, it produces an additive interference of reflected waves, allowing a peak gain of 3.4 dB and radiation efficiency of 26.3%. The fractional bandwidth is larger than 28%.

In the antenna array, there can be coupling between adjacent antennas caused by surface waves. This coupling deforms the current distribution in the patches, degrading the radiation performance [19]. The DGS can reduce the coupling between antenna elements, so that it can maintain high efficiency in the close placement of array elements, as shown in Fig. 13.10b [18].

13.4.2.4 Comb-shaped dipole with chip-integrated dielectric resonator

The backside-grounded Si substrate with a finite lateral size can be utilized as a DR, even though it has a dielectric loss, as shown in Fig. 13.11 [20]. The DR is designed to support multiple resonant modes, providing a large bandwidth and high gain. The comb-shaped dipole antenna excites and applies various substrate modes to the Si DR. This topside radiating dipole antenna with CIDR enables

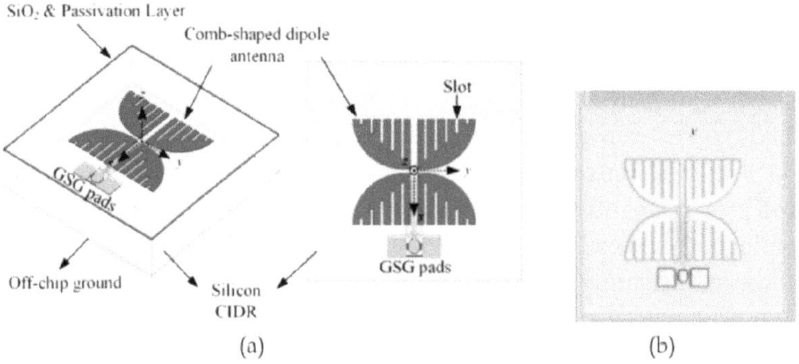

Figure 13.11 Comb-shaped dipole antenna with a chip-integrated dielectric resonator (CIDR) backed by an off-chip ground in a 65 nm Si CMOS process (CIDR size = $0.8 \times 0.8 \times 0.189$ mm^3 ($0.87\lambda_0 \times 0.87\lambda_0 \times 0.2\lambda_0$ at 325 GHz)). (a) Antenna structure. (b) Photograph. © 2021 IEEE [20].

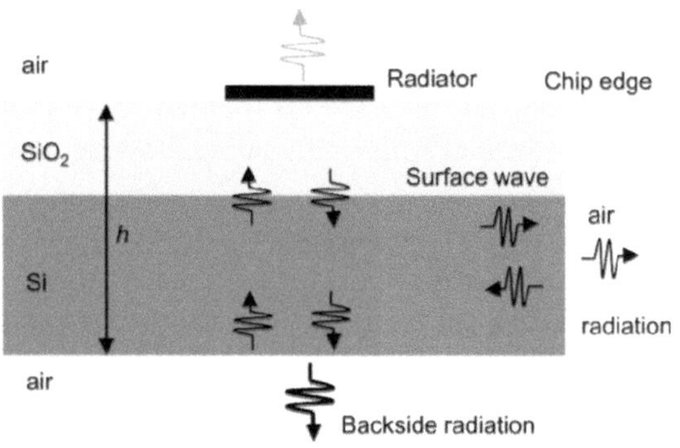

Figure 13.12 Backside radiation without backside ground plane.

a high peak gain of 8.6 dBi and peak efficiency of 44% at 295 GHz in a 65 nm Si CMOS process. The fractional bandwidth is as wide as 50%.

13.5 Backside Radiating on-Chip Antenna

13.5.1 Antenna Structure and Design Considerations

In the previous section, there was a reflector at the backside of the Si substrate so that the EM waves were reflected back to the topside. Without the backside reflector (and on-chip ground plane), the EM fields coupled to the Si substrate can radiate into the air in the backside direction. Figure 13.12 shows this type of backside radiating antenna, where there are two interfaces: air and dielectric (SiO_2) on the top side, and dielectric (Si) and air at the backside.

The EM waves radiated by the radiator on the topside interface can propagate into both air (topside) and substrate (backside). Most of the power is coupled to the high-ε_r Si substrate because it exhibits a lower wave impedance than that of air. For a dipole antenna placed at the interface between air and dielectric with a dielectric constant ε_r, the ratio of the power coupled to air (P_{air}) to the total power (P_{total}) is approximately $P_{\text{air}}/P_{\text{total}} = 1/\varepsilon_r^{1.5}$ [2].

The EM waves coupled to the substrate can propagate and radiate into the air at the backside interface. A part of the EM waves is reflected at this interface because of the mismatch between the wave impedances of the substrate and air. Thus, the power radiated into the air will be strongly dependent on the thickness of the Si substrate. For example, the backside radiated power into the air can be maximized and equal to the frontside radiation power when the substrate thickness is half-wavelength. An impedance-matching layer can also be applied at the backside of the Si substrate to reduce the reflections from the air and increase the radiated power.

A part of the EM fields coupled to the Si substrate propagates along the Si substrate in the lateral direction in the form of surface waves. The substrate mode power can be higher than the radiated power, which can be a dominant loss factor, reducing the radiation efficiency and gain of this type of antenna. Therefore, the substrate mode excitation and propagation should be minimized by modifying the shape of the substrate by, e.g., cutting or thinning the substrate or attaching a dielectric lens. The lens is also useful to focus the radiation beam and increase the directivity and gain of the antenna. An anti-reflection coating (with a quarter-wave thickness) can be applied to the surface of the lens (typically using a high-resistivity Si) for impedance-matching with air.

In general, the Si substrate in the bulk Si CMOS process has a low resistivity which introduces dielectric loss and limits the radiation efficiency of the backside radiating antenna. The substrate might be ground down to 50–100 μm to reduce the dielectric loss. The loss can be further reduced using a high-resistivity Si substrate, which is used in some SOI CMOS processes.

13.5.2 Design Examples

13.5.2.1 Backside radiating antenna with a lens

The dielectric lens can be attached to the backside of the Si substrate to convert the substrate mode power to the power radiated into the air, as illustrated in Fig. 13.13. In [21, 22], a folded dipole antenna was implemented in the top metal layer in a 65 nm Si SOI CMOS process, to illuminate the lens placed at the backside of the Si substrate. The substrate loss is minimized owing to the use of a high-resistivity Si substrate (1 kΩ·cm). The overall antenna exhibits a simulated radiation efficiency over 80%.

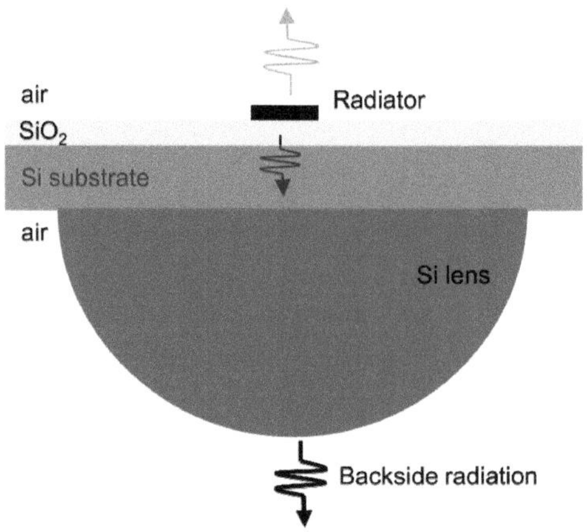

Figure 13.13 Backside radiating antenna using a Si lens.

The dielectric lens effectively eliminates the substrate modes and increases the radiation power. The simulation shows that the radiation efficiency increases from 52% to 80% owing to the lens in a high-resistivity Si substrate. The substrate with a low resistivity (15 Ω·cm) in the bulk CMOS process exhibits a radiation efficiency of 48% in the lens-integrated antenna [23].

At 288 GHz, an antenna consisting of two wire semi-rings on a 185 μm-thick low-resistivity Si substrate is mounted on a hyper-hemispherical lens with a diameter of 4 mm and extension length of 0.55 mm. The overall antenna has a gain of 18.3 dB and radiation efficiency of approximately 54% [21].

13.5.2.2 Backside radiating antenna without lens

Although the lens is helpful in improving the radiation performance of the backside radiating antenna, there are some drawbacks in real applications. Basically, it requires the attachment of the lens at the backside of the Si substrate with a good alignment to the illumination antenna on the top side. This misalignment can cause performance degradation. The glue used for attachment also affects the radiation performance. As a result, the lens attachment increases the packaging complexity, size, and cost.

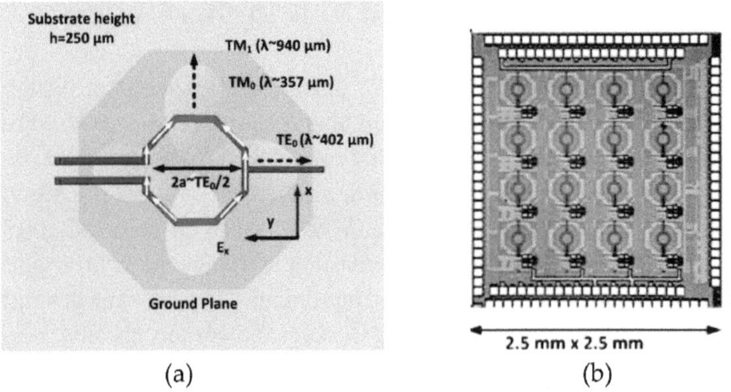

Figure 13.14 Loop antenna with an on-chip ground (open around antenna) in a 0.13 µm SiGe process (Si substrate: thickness = 250 µm, resistivity = 13.5 Ω·cm). (a) Structure of a single loop antenna. (b) Photograph (4 × 4 array). © 2015 IEEE [24].

The substrate modes, a dominant loss factor in this type of antenna, can be suppressed by properly designing the radiators without using an external lens. In [24], a differential loop antenna with an on-chip ground was designed at 280 GHz as shown in Fig. 13.14. The on-chip ground is open around the loop antenna so that the EM fields can be coupled to the Si substrate. It also serves as a reflector that reduces the frontside radiation by the reflected wave from the backside interface. The substrate waves add in phase at the backside interface if the substrate thickness is approximately an odd multiple of the quarter-wavelength. The surface wave modes TE_0 and TM_1 contribute to nearly 83% of the substrate mode power in this loop antenna. The loop diameter is designed to be half of the wavelength of the TE_0 mode in the Si substrate, as illustrated in Fig. 13.14a. Therefore, it effectively suppresses the TE_0 mode excitation by the current elements at half-wavelengths apart in the loop. The TM_1 mode is suppressed by separating the antenna elements by half-wavelength in the 4 × 4 array. In this manner, the substrate modes are effectively suppressed without using a lens or substrate-thinning. A radiation efficiency higher than 40% was achieved for a single antenna at 260 GHz.

13.6 Design Rules Related to Antenna Design

For the high yield and reliability of Si CMOS ICs, there are design rules that should be passed in circuit layouts. Metal density rules are one of the design rules that should be considered in the design of on-chip antennas. They specify the lower and upper limits of the metal occupancy density for a given area. Note that the on-chip radiators such as patches are generally implemented in a thick metal layer and occupy a large area even at THz frequencies. The on-chip ground planes, which can be implemented in several layers in parallel, also have a large metal area. Thus, the upper limit of the metal density can be violated in the patches and on-chip ground planes. Therefore, several slots or holes are inserted in the patch and ground planes to reduce the metal density. The slots are arranged to not disturb the current distribution of the patch so that their effect on the radiation characteristics can be minimized [5, 25, 26].

Generally, no metal patterns are placed below the radiators in the antenna. However, this can result in a violation of the lower limit of the metal densities. In this case, dummy metal fills are carried out by the foundry, which can cause unpredictable performance of the antenna. The dummy metal fills increase the capacitances and losses of the antenna, so that the antenna resonant frequency decreases, and the radiation efficiency and gain can be seriously degraded [10].

In summary, the slots and dummy metals added to pass the design rules can have a detrimental effect on the on-chip antenna performance. Therefore, it is necessary to optimize the dimensions and placements of the slots and dummy metals to avoid degradation of the antenna performance.

References

1. Öjefors, E., Pfeiffer, U. R., Lisauskas, A., and Roskos, H. G. (2009). A 0.65 THz focal-plane array in a quarter-micron CMOS process technology, *IEEE J. Solid-State Circuits*, **44**, pp. 1968–1976.
2. Babakhani, A., Guan, X., Komijani, A., Natarajan, A., and Hajimiri, A. (2006). A 77 GHz phased–array transceiver with on–chip antennas in silicon: receiver and antennas, *IEEE J. Solid–State Circuits*, **41**, pp. 2795–2806.

3. Hajimiri, A. (2007). mm-wave silicon ICs: challenges and opportunities, *2007 IEEE Custom Integrated Circuits Conf.*, pp. 741–747.
4. Jackson, D. R. (2007). Microstrip antennas, in *Antenna Engineering Handbook*, eds. Volakis, J. L., Chapter 7 (Mc Graw Hill, USA), pp. 7-1–7-29.
5. Seok, E., Cao, C., Shim, D. Arenas, D. J., Tanner, D. B., Hung, C. M., and Kenneth, K. O. (2008). A 410 GHz CMOS push-push oscillator with an on-chip patch antenna, *Proc. Int. Solid-State Circuits Conf.*, pp. 472–473.
6. Seok, E., Shuim, E., Mao, C., Han, R., Sankaran, S., Cao, C., Knap, W., and Kenneth, K. O. (2010). Progress and challenges towards terahertz CMOS integrated circuits, *IEEE J. Solid-State Circuits*, **45**, pp. 1554–1564.
7. Pandey, A. (2019) *Practical Microstrip and Printed Antenna Design* (Artech House, USA) pp. 27–50.
8. Deng, K., Li, Y., Liu, C., Wu, W., and Xiong, Y. (2015). 340 GHz on-chip 3-D antenna with 10 dBi gain and 80% radiation efficiency, *IEEE Trans. Terahertz Sci. Technol.*, **5**, pp. 619–627.
9. Hu, S., Wang, L., Xiong, Y. Z., Zhang, B., and Lim, T. G. (2011). A 434 GHz SiGe BiCMOS transmitter with an on-chip SIW slot antenna, *Proc. IEEE Asian Solid-State Circuits Conf.*, pp. 269–272.
10. Uzunkol, M., Gurbuz, O. D., Golcuk, F., and Rebeiz, G. M. (2013). A 0.32 THz SiGe 4 × 4 imaging array using high-efficiency on-chip antennas, *IEEE J. Solid-State Circuits*, **48**, pp. 2056–2066.
11. Chu, H., Guo, Y. X., Lin, F., and Shi, X. Q. (2009). Wideband 60 GHz on-chip antenna with an artificial magnetic conductor, *Proc. IEEE Int. Symp. On Radio-Frequency Integration Technology*, pp. 307–310.
12. Khan, M. S., Tahir, F. A., and Cheema, H. M. (2016). Design of bowtie-slot on-chip antenna backed with E-shaped FSS at 94 GHz, *Proc. 2016 10th European Conference on Antennas and Propagation*, pp. 1–3.
13. Kuo, H., Yue, H., Ou, Y., Lin, C., and Chuang, H. (2013). A 60 GHz CMOS sub-harmonic RF receiver with integrated on-chip artificial-magnetic-conductor Yagi antenna and balun bandpass filter for very-short-range gigabit communications, *IEEE Trans. Microw. Theory Techn.*, **61**, pp. 1681–1691.
14. Kang, S., Thyagarajan, S. V., and Niknejad, A. M. (2015). A 240 GHz fully integrated wideband QPSK transmitter in 65 nm CMOS, *IEEE J. Solid-State Circuits*, **50**, pp. 2256–2267.
15. Wang, Z., Chiang, P., Nazari, P., Wang, C., Chen, Z., and Heydari, P. (2014). A CMOS 210 GHz fundamental transceiver with OOK modulation, *IEEE J. Solid-State Circuits*, **49**, pp. 564–580.

16. Schmaltz, K. (2016). 245 GHz transmitter array in SiGe BiCMOS for gas spectroscopy, *IEEE Trans. Terahertz Sci. Technol.*, **6**, pp. 318–327.
17. Kim, H., Choe, W., and Jeong, J. (2018). A terahertz CMOS V-shaped patch antenna with defected ground structure, *Sensors*, **18**, 2432.
18. Lee, C., and Jeong, J. (2020). THz CMOS on-chip antenna array using defected ground structure, *Electronics*, **9**, 1137.
19. Pozar, D. (1983). Surface wave effects for millimeter wave printed antennas, *Proc. 1983 Antennas and Propagation Society Int. Symp.*, pp. 692–695.
20. Kong, S., and Shum, K. M., Yang C., Gao L., and Chan C. H. (2021). Wide impedance-and gain-bandwidth terahertz on-chip antenna with chip-integrated dielectric resonator, *IEEE Trans. Antennas Propag.*, pp. 1–9.
21. Grzyb, J., Zhao, Y., and Pfeiffer, U. R. (2013). A 288 GHz lens-integrated balanced triple-push source in a 65 nm CMOS technology, *IEEE J. Solid-State Circuits*, **50**, pp. 1751–1761.
22. Öjefors, E., Baktash, N., Zhao, Y., Hadi, R. A., Sherry, H., and Pfeiffer, U. R. (2010). Terahertz imaging detectors in a 65 nm CMOS SOI technology, *Proc. 2010 Eur. Solid-State Circuits Conf.*, pp. 486–489.
23. Hadi, R. A., Sherry H., Crzyb J., Baktash N., Zhao Y., Öjefors, E., Kaiser A., Gathelin A., and Pfeiffer, U. (2011). A broadband 0.6 to 1 THz CMOS imaging detector with an integrated lens, *Proc. 2011 IEEE MTT-S Int. Microwave Symp.*, pp. 1–4.
24. Sengupta, K., Seo, D., Yang, L., and Hajimiri, A. (2015). Silicon integrated 280 GHz imaging chipset with 4 × 4 SiGe receiver array and CMOS source, *IEEE Trans. Terahertz Sci. Technol.*, **5**, pp. 427–437.
25. Shafee, M., Nahar, S., Safwat, A. M. E., El-Hennawy, H., and Hella, M. M. (2014). Stacked resonator patch antenna for wide bandwidth THz detection, *Proc. 2014 IEEE Int. Conf. on Ultra Wideband*, pp. 240–244.
26. Li, C., Ko, C., Kuo, M., and Chang, D. (2016). A 340 GHz heterodyne receiver front end in 40 nm CMOS for THz biomedical imaging applications, *IEEE Trans. Terahertz Sci. Technol.*, **6**, pp. 625–636.

Chapter 14

Package Technologies for THz Devices

Ho-Jin Song
Department of Electrical Engineering,
Pohang University of Science and Technology
77 Cheongam-ro, Namgu, Pohang, Gyeongbuk 36763, Korea
hojin.song@postech.ac.kr

In the last couple of decades, solid-state device technologies, particularly electronic semiconductor devices, have been greatly advanced and investigated for possible adoption in various terahertz (THz) applications, such as imaging, security, and wireless communications. Meanwhile, researchers have been exploring ways to package those THz electronic devices and integrated circuits for practical use. Packages are fundamentally expected to offer physical housing for devices and reliable signal interconnections from the inside to the outside or vice versa. However, as frequency increases, we face several challenges associated with signal loss, dimensions, and fabrication. This chapter provides a broad overview of the recent progress in interconnections and packaging technologies.

Advances in Terahertz Source Technologies
Edited by Gun-Sik Park, Masahiko Tani, Jae-Sung Rieh, and Sang Yoon Park

ISBN 978-981-4968-89-8 (Hardcover), 978-1-003-45967-5 (eBook)
www.jennystanford.com

14.1 Introduction

THz generation and detection have been the most essential challenges for a long time. Plenty of research has been conducted and has finally resulted in quite competitive performance in the signal-to-noise ratio. Accordingly, interest in packaging THz devices has been steadily growing for practical applications.

In general, packaging offers physical support, housing, and protection from mechanical and thermal stress, environmental loads, and electromagnetic interference [8, 9]. In addition, appropriate input and output (I/O) interfaces for signal in/outputs, control, monitoring, and bias are formed on packages and interconnected with the core devices so that they can appropriately operate as designed.

Recently, THz packaging technologies have been investigated in attempts to deal with signal loss, dimension, and fabrication issues for THz electronics. In this chapter, the recent progress in interconnections and packaging technologies will be overviewed, particularly in the frequency range of 0.1–3 THz, where state-of-the-art electronic devices can operate. Emerging concepts based on commercial ceramic technologies for compact and lightweight packaging in practical applications are highlighted, along with metallic split blocks with rectangular waveguides (WGs), which are still considered the most valid and reliable approach.

14.2 Issues in Package at THz Frequencies

14.2.1 Packaging Materials

The package body, including signal pins, covers, and frames, is composed of various materials, such as metals, dielectric, and some composite ones. Core devices in the package are surrounded by some of these materials and signals from or to the devices pass through them. In general, the electrical, thermal, and physical properties of packaging materials predominate the performance of the packages. Therefore, accurate material parameters must be available for designing a desired package. Narrowing the topic to THz packages, accurate measurement of potential materials, particularly for electrical properties, such as dielectric permittivity at those high frequencies, should be precisely carried out. In this chapter, materials highly likely to be used in THz packages are summarized in Table 14.1 at 0.5 and 1 THz for direct comparison. Overall, significant absorption at specific frequencies has rarely been observed for the summarized materials, and loss tangents gradually increase as frequency increases. In addition, there is little deviation between measurements reported by multiple groups. Though it would be hard to generalize with the limited data, the parameters listed in Table 14.1 should give readers a good initial idea about how these dielectrics could be used in THz packages.

14.2.2 Interconnections

In many packaging technologies, bonding wires are being commonly used to interconnect a core integrated circuit (IC) or device to another one or to an additional substrate. As operating frequency increases, the inductances associated with the bond wires cause significant reflection, which degrades the frequency performance of the packaged module. The inductances strongly depend on the geometry of the wires, such as their diameter, loop height, and length [10–12]. Simple circuit models of the wires enable us to compensate for the parasitics with an extra matching circuit in ICs and substrates. However, as the frequency increases further above 100 GHz, the wires start to behave as a distributed component like transmission lines rather than a lumped one. Simply shortening them can effectively reduce the wire inductances even for up to 300 GHz [13–15]. However, this is not a proper solution for practical applications when one considers the need for thermal durability and performance uniformity between modules. Instead, in [17], bonding wires were treated as a high-impedance transmission line with approximately 100 Ω characteristic impedance and used as an impedance transformer between the 50 ohm coplanar waveguide (CPW) to a ridge WG.

Flip-chip bonding is also an attractive technique for interconnecting millimeter and THz signals, owing to the small intrinsic parasitics associated with bump transitions [19]. There have been

Table 14.1 Summary of dielectric constant and tangent loss for various materials at 0.5 and 1 THz reported in the literature

Material	**0.5 THz** ε_r	**0.5 THz** tan δ	**1 THz** ε_r	**1 THz** tan δ	**Ref**	**Note**
InP			12.33	0.009	[1]	
			12.39	0.006	[6]	
	12.7	0.009			Own data	
GaAs	12.9	0.024	12.91	0.013	[7]	
Si (doped)	11.85	0.025	11.85	0.01		Doped wafer (concentration: unclear)
	9.91	0.001	9.91	0.001	[7]	High resistivity (>10 kohm CM)
Sapphire (A-plane)	9.41	0.01	9.43	0.031	[7]	C-axis ⊥ pol
	11.61	0.008	11.66	0.032	[7]	C-axis // pol
Al_2O_3			9.3	0.003	[1]	
			9.28	0.004	[16]	
LTCC #1 (5.53@)	5.43	0.053			Own data	0.1 mm 10 layers
LTCC #2 (7.1)	6.71	0.072			Own data	0.1 mm 10 layers
Quartz			3.84	0.004	[1]	
			3.84	0.0037	[18]	
	4.44	0.002	4.45	0.005	[7]	A-plane, C-axis ⊥ pol
	4.64	0.004	4.64	0.003	[7]	A-plane, C-axis // pol
Pyrex Glass			4.48	0.052	[1]	0.8 THz
			4.45	0.05	[18]	
RO3010	12.5	0.025	14	>0.15	[1]	Graph read

RO3006	8	0.015	9	>0.15	[1]	Graph read
RO3003	3.2	0.01	3.2	0.02	[1]	Graph read
Polytetrafluoroethylene (PTFE)			2.06	0.0004	[1]	
			2.08	0.008	[21]	
Liquid crystal polymer (LCP)	3.5	0.02	3.51	0.05	[1]	Graph read
Polyimide (dry)	3.27	0.021			[1]	
			3.37	0.037	[1]	
Polyimide (air)	3.28	0.098	3.24	0.11	[21]	Graph read
	3.42	0.001			wn data	
Polycarbonate (PC)			2.61	0.027	[1]	
			2.67	0.028	[18]	
Zeonor			2.35	0.001	[1]	
			2.28	0.001	[25]	
			2.36	0.002	[1]	
High-density polyethylene (HDPE)			2.37	0.002	[18]	
	2.36	0.013	2.35	0.009	[21]	
PET	2.98	0.031			[1]	
	2.95	0.054	2.93	0.063	[21]	
Benzocyclobutene (BCB)	2.41–2.53	0.0042-0.039			[29]	Membrane structure
Photopolymer resin (acrylic-based)	2.89–2.56	0.06			[30]	Commercial 3D printing material, FullCUreMaterials®

just a couple of demonstrations of the flip-chip bonding technique at 250 and 300 GHz. In [20], a 10 μm AuSn bump formed by a direct evaporation and lift-off process exhibited insertion loss of less than 1 dB per transition and return loss larger than 10 dB at up to 250 GHz. Kawano et al. reported a flip-chip-mounted InP high electron mobility transistor (HEMT) amplifier THz monolithic integrated circuit (TMIC) on a polyimide motherboard [22]. Suppressing the substrate mode in the motherboard and the resonance of current flowing in ground paths was noted as the most critical point for successful results. Though the bump height of around 40 μm was quite large, no serious degradations of return loss and gain of the TMIC were observed. In spite of these successful demonstrations, a few issues should be considered for THz devices. Chip detuning, dielectric loading at transitions, the influence of underfill materials, and substrate modes coupling at motherboards are common drawbacks of the flip-chip bonding technique [23, 24]. Among them, one dilemma we would face related to the bonding pad structure in TMICs and/or TMIC thickness at THz frequencies will be briefly addressed here. Figure 14.1 shows some examples of bonding pad structures. The structures illustrated in Figs. 14.1b,c are commonly used in TMICs based on silicon complementary metal-oxide-semiconductor (CMOS) technologies and compound semiconductors. However, in those structures, bonding pads are opened to the substrate and would therefore be a source of substrate mode excitation at THz frequencies. To avoid this, the wafers have

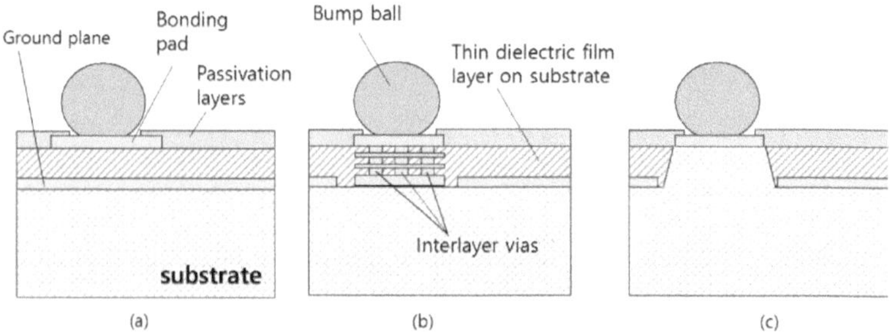

Figure 14.1 Three typical bonding pads in ICs: (a) bond over active structure, (b) bond pad consisting of multi-metal layers connected with inter-layer vias, and (c) bond pad on substrate bump.

been thinned down to 50 μm for 300 GHz [26], 25 μm for 650 GHz [27], and 18 μm for 1 THz [28] amplifiers. As can be easily expected, with the thinned chip, mechanical wafer breakage or cracks would easily occur during the flip-chip bonding process. If a multilayer metallization process is available, like CMOS technologies, a shield ground metal layer can be inserted under the bonding pads, as shown in Fig. 14.1a, to suppress the substrate mode excitation even with no wafer thinning. However, in this case, the ground shield dramatically increases the capacitance of the bonding pad and thereby considerably degrades return loss performance at the transition.

14.2.3 Signal Interfaces

Packages should provide appropriate signal interfaces that can deliver signals from internal TMICs to external circuits or vice versa. To maintain the high-frequency signals, the simple metal wires used at low frequencies should be replaced with transmission line structures that have a given characteristic impedance and are not dispersive in the desired frequency band. It would be highly preferred, in aspect of fabrications, if we could deploy the solder balls commonly incorporated with the reflow soldering process for low-frequency applications to THz devices. However, there is little technical clue about how this might be accomplished.

Incorporating RF connectors generally used for test instruments are an option. In particular, coaxial WGs are attractive due to their broadband operation bandwidth from DC and non-dispersive propagation characteristics in the band. However, as frequency increases, the TE_{11} mode starts to become excited and single-mode operation is no longer guaranteed. Figure 14.2 shows the calculated dimensions of 50 Ω coaxial connectors with respect to the cut-off frequency of the TE_{11} mode. Even with no dielectric between the inner and outer conductor, the diameter of the center pin should be as small as 0.2 mm for the 300 GHz cut-off frequency. Considering a mechanical contact is required for coaxial connectors, such a tiny center pin would not be suitable for the durable and reliable interfaces required for practical applications.

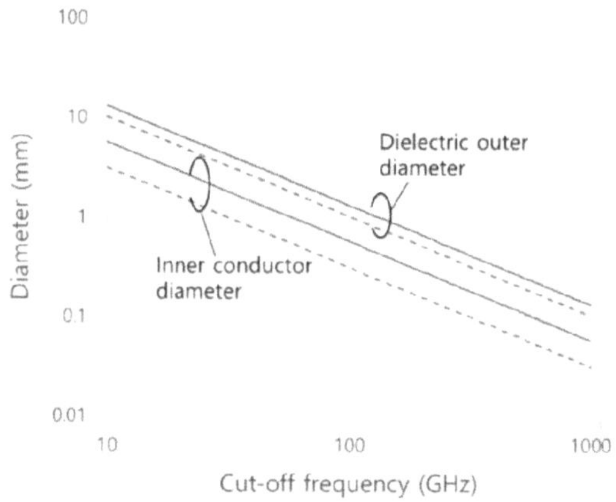

Figure 14.2 Calculated dimensions of 50 W coaxial waveguide with respect to cut-off frequency for TEM single mode operation (Solid line: $\varepsilon_r = 1$. Dashed line: $\varepsilon_r = 2$.)

Even with various drawbacks such as their bulky structure and limited operating bandwidth, rectangular metallic WGs are still considered the most appropriate interfaces for THz packages because of their superior features of low loss, excellent durability, and a reliable and repeatable mating system. Large flange structures with alignment pins and screw taps for mating give us a "bulky" or "mechanical" impression. However, the dimensions of a rectangular metallic WG itself are quite small at THz frequencies. Since WG interfaces are normally deployed in large-scale scientific facilities and military radar systems that require large power- and heat-handling capability [31] and superior electrical performance, WG flanges have so far been developed to improve accuracy and repeatability [32, 33]. In other words, a new concept for a compact mating system for rectangular WGs would enable new sorts of packages at THz frequencies.

14.3 Metallic Waveguide Packages

Rectangular metallic WG split blocks are commonly made one by one by using computer numerical control (CNC) milling processes.

However, recent advanced milling machines provide quite fine manufacturing accuracy of around a few micrometers, which is acceptable for operations up to 3 THz. The waveguide modules should incorporate WG-to-microstrip line (MSL) or CPW transitions between core devices and rectangular metallic WG ports. At millimeter-wave frequencies, transition circuits such as radial E-plane probes have been fabricated on a separate low-dielectric-constant substrate such as Quartz with a bonding wire in the RF path. This approach would work fine up to around 300 GHz [13–15], but, fundamentally, the associated inductance causes issues for bandwidth and module operation at frequencies beyond 200–300 GHz.

The popular approach to implementing a rectangular metallic WG module at THz frequencies is to integrate the transition probes with the core devices or circuits. In those designs, the substrates of the TMICs have been reduced to a few micrometers to enhance RF losses and to minimize dielectric loading of the WGs. These techniques have also been employed for active TMICs, such as amplifiers with radial probe or dipole electromagnetic transitions [34–37]. The active TMICs should also be thinned down to achieve good coupling efficiency at the integrated transitions and suppress substrate modes with dense conductive substrate vias. The cut-off frequencies of the substrate modes are strongly related to substrate thickness. Undesired substrate modes can be suppressed with thinning the substrate, but challenges will arise in fabrication and assembly. According to the literature, substrate thickness should be 50 μm for 300 GHz [26], 25 μm for 650 GHz [27], and 18 μm for 1 THz [28] bands. The thinned TMICs are then mounted on a pedestal in one half of the split blocks so that the transitions can be aligned to the fields on the E-plane of the rectangular metallic WG while they are suspended in the cavity. According to the literature, estimated coupling losses per transition are approximately 1 dB in 340–380 GHz [38], 1.5 dB in 460–500 GHz [35], and 1.5~2 dB in 625–700 GHz bands [27].

One should note that active TMICs require multiple DC pads that are usually located at the edge of the chip beside the RF-core blocks and thus enlarge the chip width. In case of a high-power amplifier, a large number of transistors should be arrayed in parallel to deliver large powers [27, 37], which leads the larger chip widths. In the large-chip-width TMICs, however, the E-plane probes could cause energy leakage from the WG into the chip channel that will finally cause undesired mode excitation in the packages, resulting in potential resonance or oscillation [38, 39]. A simple way to avoid the problem is to remove the corners of the TMICs beside the E-plane probe transitions so that the chip width becomes electrically narrow in the transition region. Such nonrectangular chip dies can be accomplished by chemical etching [27, 40] or a laser dicing singulation process [41–43]. Obviously, these additional processes for nonrectangular chips can be expensive and would cause a yield problem in fabrication, handling, or assembly steps.

Recently, an integrated chip-to-WG transition for large-chip-width TMICs has been reported that provides a large degree of design freedom with no dedicated process [5]. In [5], the dipole probes are located on the corner of the test TMIC, whose width is approximately a couple of times wider than the E-plane width of the WG. To accommodate such a large-chip-width TMIC in the split block modules, one cannot avoid forming a thin slit between blocks. In this structure, it was found that only a certain series of wave modes will be excited and confined in the air gap (see Fig. 14.3). And the parasitic E-fields stand above the TMIC perpendicularly and bypass

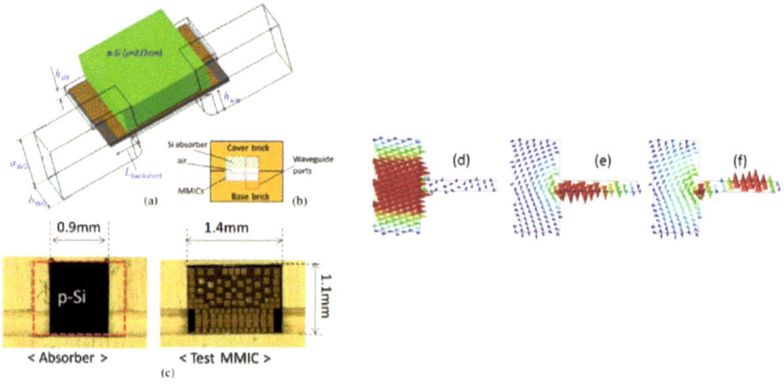

Figure 14.3 (a) Schematic and (b) cross-sectional diagram of WR3.4 package with single slit and p-type Si block as absorber, (c) photographs of mounted silicon absorber and test MMIC, (d)~(f) simulated E-field distribution of first three modes in slit WG. Figure from [5].

the core chip through the chip channel in the module. From the understanding on the field profile of the undesired modes, one can insert a piece of high-permittivity dielectric material such as doped silicon into the chip channel and form a partially loaded space to absorb the parasitic modes with the Si piece. With this technique, 60 dB inter-port isolation was demonstrated with comparable insertion loss per transition probe in the 300 GHz band. Though the technique has not been proved with an active circuit yet, the work described in [5] suggests that a similar in-depth analysis of the undesired modes would give us a chance to effectively suppress them even without expensive and dedicated fabrication processes.

14.4 LTCC Packages at THz Frequencies

Owing to its high integration capability and low cost for high-volume production, low-temperature co-fire ceramic (LTCC) technology has been widely studied for packaging millimeter-wave transceiver ICs, particularly 60 GHz radio front-ends [44, 45]. By stacking of multiple laminated substrates, one can easily implement the substrate-in-waveguides (SIWs) [45–47] and antenna-in-a-package (AIP) [48–50] structures. One example of THz packages with LTCC technology is illustrated in Fig. 14.4. In this concept, a rectangular WG and microstrip-to-WG transition are monolithically integrated into the package body [3] and a TMIC is assumed to be mounted in the package with flip-chip bonding interconnections. As shown in the figure, the rectangular WG can be formed with multiple air-cavity layers and an array of vias along the edges of the air cavity. Since ordinary LTCC technology does not provide a metal coating on the internal sidewalls of the air cavity, vias and metallization on every layer simulate the metallic wall inside the WG. Specifically speaking, it is a sort of corrugated WG with periodic dielectric loading. Therefore, the distance between the air-cavity edges and the interlayer vias, which would be equivalent to the corrugation depth in a corrugated WG, becomes a critical parameter determining the operation bandwidth of the WG [4]. Tajima et al. fabricated the straight WG section with a standard LTCC foundry process and

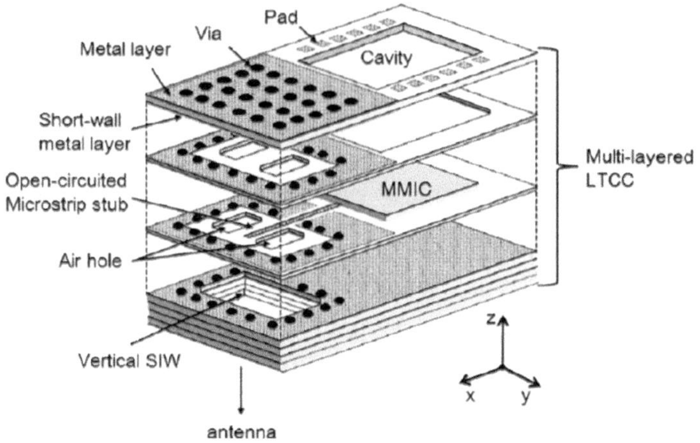

Figure 14.4 Concept of LTCC packages for THz MMICs with integrated microstrip-to-waveguide transition in package frame. Figure from [3].

experimentally demonstrated insertion loss of around 0.6 dB/mm and return loss better than 15 dB in the 240–330 GHz band.

Instead of the WG in Fig. 14.4, a horn antenna can be used as an interface to the air. Based on the corrugated structure, a step-profiled pyramid horn antenna was demonstrated as well in the 300 GHz band [4] as shown in Fig. 14.5. Though the overall structure could not be optimized due to the limitations in the fabrication, the horn antenna yielded promising performance in size and antenna gain; 15 dBi gain in the 250–330 GHz band with its peak at 18 dBi from 2.8 mm length horn fabricated by stacking 27 LTCC layers.

The coupling between the LTCC WG and a TMIC would be achievable with integrated electromagnetic transitions in the package body, as shown in Fig. 14.4. This approach is advantageous for hermetically sealing packages, but with the drawback of transition loss. In [3], a back-to-back microstrip-to-WG transition was experimentally demonstrated. Two WGs were mechanically isolated from each other. In order to enhance the coupling efficiency and bandwidth performance and maintain the hermetic sealing, the dielectric around the electromagnetic couplers was removed and fully dielectric-filled striplines were inserted between the transition probes and TMIC channels as matching circuits. The estimated

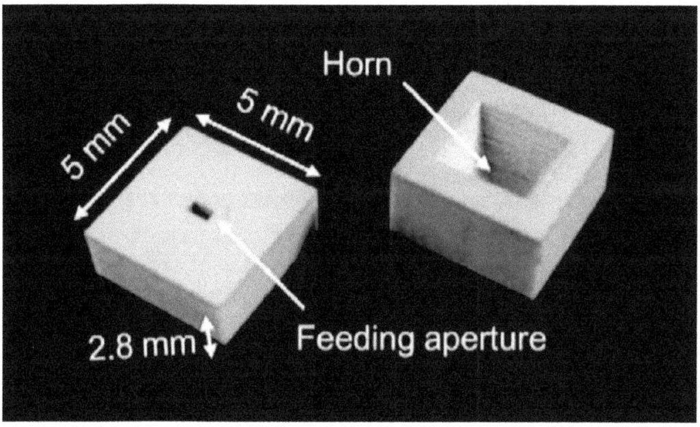

Figure 14.5 Concept of LTCC packages for THz MMICs with integrated microstrip-to-waveguide transition in package frame. Figure from [4].

transition loss per coupler was approximately 1 dB at 300 GHz, which is comparable with that of the integrated transitions on TMICs [5, 38].

We overviewed a few promising results for LTCC technologies at THz frequencies. Nevertheless, obvious limitations that need to be noted are manufacturing tolerance and large signal loss. The waves in the WG will be loaded by the LTCC materials at the corrugations and thus characteristic impedance and propagation loss of the WG will be suffered. In addition, LTCC technologies are generally based on mechanical punching for interlayer vias and thick film patterning for metallization. The stacked base material sheets, commonly called green sheets, are baked at temperatures below 1,000°C, leading to shrinkage of the entire structure. These fabrication processes do not provide tolerances as accurately as semiconductor processes do at this moment. A brief estimation based on the literature indicates that up to 400–500 GHz would be feasible with reasonable performance.

14.5 Concept of Quasi-Optical THz Package

In many THz applications, including spectroscopy, imaging, and communications, devices and components are actually required to

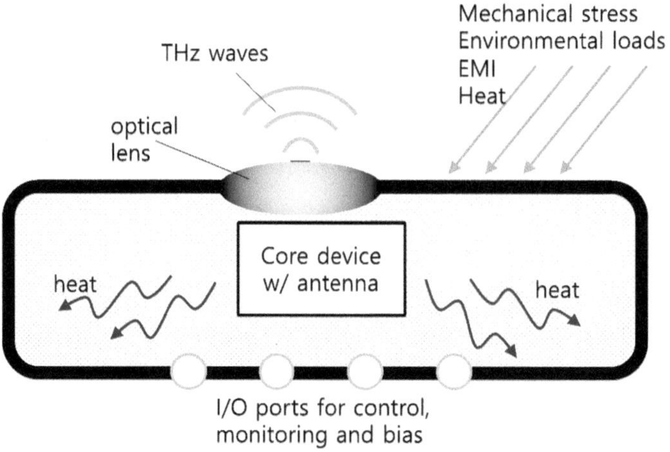

Figure 14.6 Conceptual diagram of quasi-optical THz packages.

radiate THz waves into the air or a sample under test and detect the signals that have passed through a certain medium. Therefore, it is natural that THz packages embed antennas for signal radiation. In the many examples shown in previous sections, horn antennas or radiators were monolithically fabricated with the package body. On the other hand, the antenna or radiators can be integrated into the TMICs or RFICs. In this case, an interconnection for THz signals is not necessary in the packages. Figure 14.6 shows a conceptual diagram of a quasi-optical THz package for core devices with an antenna. Imagining packaged optical components such as discrete light-emitting diodes (LEDs) will help one to have a better understanding of the quasi-optical packages. Assuming a transmitter or emitter, THz waves will be directly radiated into the air from the on-chip antenna. An optical lens would help to improve the directivity of the radiated beam [51–54]. Since there is no interconnection for the high-frequency signal, the assembly will not be so sensitive to manufacturing tolerances or circuit parasitics.

With advanced manufacturing technology, such a lensed package can be miniaturized by following a concept similar to that for optical components such as LEDs. Figure 14.7 shows a compact THz lensed package [2] consisting of an LTCC-based chip carrier, TMIC with an on-chip antenna, and 8 mm-diameter silicon lens.

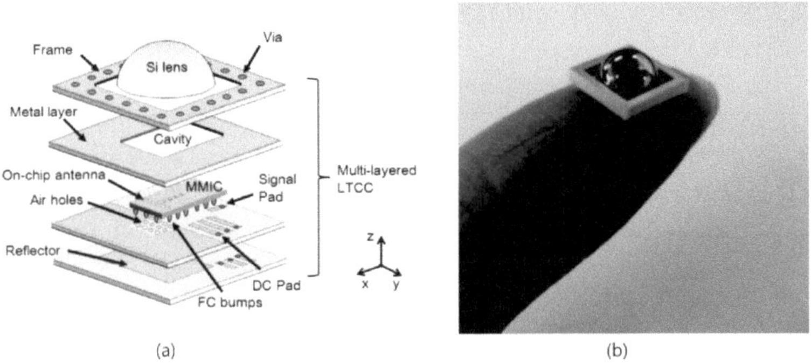

Figure 14.7 (a) Structure of LTCC-based silicon-lens THz package and (b) photograph of implemented module. Figures from [2].

The 300 GHz receiver TMIC for wireless communications [55] was mounted in the LTCC package with a flip-chip bonding technique for DC and data signal read-outs, and the silicon lens was placed on the backside of the TMIC and fixed with a non-conductive adhesive epoxy. The performance of the compact module was evaluated in a data transmission experiment and no degradation was observed in the bit error rate and the Q-factor of the recovered eye-diagram.

References

1. J. A. Hejase, P. R. Paladhi, and P. P. Chahal, Terahertz Characterization of Dielectric Substrates for Component Design and Nondestructive Evaluation of Packages, *IEEE Transactions on Components, Packaging and Manufacturing Technology,* vol. 1, no. 11, pp. 1685–1694, 2011, doi: 10.1109/TCPMT.2011.2163632.
2. T. Tajima, H.-J. Song, and M. Yaita, Compact THz LTCC Receiver Module for 300 GHz Wireless Communications, *IEEE Microwave and Wireless Components Letters,* vol. 26, no. 4, pp. 291–293, 2016.
3. T. Tajima, H.-J. Song, and M. Yaita, Design and Analysis of LTCC-Integrated Planar Microstrip-to-Waveguide Transition at 300 GHz, *IEEE Transactions on Microwave Theory and Techniques,* vol. 64, no. 1, pp. 106–114, 2016.
4. T. Tajima, H.-J. Song, K. Ajito, M. Yaita, and N. Kukutsu, 300 GHz Step-Profiled Corrugated Horn Antennas Integrated in LTCC, *IEEE*

Transactions on Antennas and Propagation, vol. 62, no. 11, pp. 5437–5444, 2014.

5. H.-J. Song, H. Matsuzaki, and M. Yaita, Sub-Millimeter and Terahertz-Wave Packaging for Large Chip-Width MMICs, *IEEE Microwave and Wireless Components Letters,* vol. 26, no. 6, pp. 422–424, 2016.
6. T. D. Dorney, R. G. Baraniuk, and D. M. Mittleman, Material Parameter Estimation with Terahertz Time-Domain Spectroscopy (in English), *Journal of the Optical Society of America a-Optics Image Science and Vision,* vol. 18, no. 7, pp. 1562–1571, Jul 2001. [Online]. Available: <Go to ISI>://WOS:000169453600018.
7. D. Grischkowsky, S. Keiding, M. Van Exter, and C. Fattinger, Far-Infrared Time-Domain Spectroscopy with Terahertz Beams of Dielectrics and Semiconductors, *JOSA B,* vol. 7, no. 10, pp. 2006–2015, 1990.
8. W. Menzel, Packaging and Interconnects for Millimeter Wave Circuits: A Review, *Annales des télécommunications,* vol. 52, no. 3–4: Springer, pp. 145–154, 1997.
9. L. Devlin, The Future of mm-Wave Packaging (in English), *Microwave Journal,* vol. 57, no. 2, p. 24, Feb 2014. [Online]. Available: <Go to ISI>://WOS:000332058700002.
10. J.-Y. Kim, H.-Y. Lee, J.-H. Lee, and D.-P. Chang, Wideband Characterization of Multiple Bondwires for Millimeter-Wave Applications, in *Asia-Pacific Microwave Conference,* IEEE, pp. 1265–1268, 2000.
11. F. Alimenti, P. Mezzanotte, L. Roselli, and R. Sorrentino, Modeling and Characterization of the Bonding-Wire Interconnection, *IEEE Transactions on Microwave Theory and Techniques,* vol. 49, no. 1, pp. 142–150, 2001.
12. A. Sutono, N. G. Cafaro, J. Laskar, and M. M. Tentzeris, Experimental Modeling, Repeatability Investigation and Optimization of Microwave Bond Wire Interconnects, *Advanced Packaging, IEEE Transactions on,* vol. 24, no. 4, pp. 595–603, 2001.
13. P. Huang, R. Lai, R. Grundbacher, and B. Gorospe, A 20-mW G-Band Monolithic Driver Amplifier Using 0.07 μM InP HEMT, in *2006 IEEE MTT-S International Microwave Symposium Digest,* IEEE, pp. 806–809, 2006.
14. A. Tessmann, A. Leuther, V. Hurm, H. Massler, M. Zink, M. Kuri, M. Riessle, R. Losch, M. Schlechtweg, and O. Ambacher, A 300 GHz MHEMT Amplifier Module, in *Indium Phosphide & Related Materials, 2009. IPRM'09. IEEE International Conference on,* IEEE, pp. 196–199, 2009.

15. H. J. Song, K. Ajito, Y. Muramoto, A. Wakatsuki, T. Nagatsuma, and N. Kukutsu, Uni-Travelling-Carrier Photodiode Module Generating 300 GHz Power Greater Than 1 mW (in English), *IEEE Microwave and Wireless Components Letters,* vol. 22, no. 7, pp. 363–365, Jul 2012. [Online]. Available: <Go to ISI>://WOS:000306182400011.
16. P. H. Bolivar, M. Brucherseifer, J. G. Rivas, R. Gonzalo, I. Ederra, A. L. Reynolds, M. Holker, and P. de Maagt, Measurement of the Dielectric Constant and Loss Tangent of High Dielectric-Constant Materials at Terahertz Frequencies, *IEEE Transactions on Microwave Theory and Techniques,* vol. 51, no. 4, pp. 1062–1066, 2003.
17. T. Kosugi, H. Hamada, H. Takahashi, H.-J. Song, A. Hirata, H. Matsuzaki, and H. Nosaka, 250–300 GHz Waveguide Module with Ridge-Coupler and InP-HEMTIC, in *2014 Asia-Pacific Microwave Conference*, IEEE, pp. 1133–1135, 2014.
18. M. Naftaly and R. E. Miles, Terahertz Time-Domain Spectroscopy for Material Characterization, *PROCEEDINGS-IEEE,* vol. 95, no. 8, p. 1658, 2007.
19. W. Heinrich, The Flip-Chip Approach for Millimeter Wave Packaging, *Microwave Magazine, IEEE,* vol. 6, no. 3, pp. 36–45, 2005.
20. S. BuMonayakul, S. Sinha, C.-T. Wang, N. Weimann, F. Schmückle, M. Hrobak, V. Krozer, W. John, L. Weixelbaum, and P. Wolter, Flip-Chip Interconnects for 250 GHz Modules, *IEEE Microwave and Wireless Components Letters,* vol. 25, no. 6, pp. 358–360, 2015.
21. Y.-S. Jin, G.-J. Kim, and S.-G. Jeon, Terahertz Dielectric Properties of Polymers, *Journal of the Korean Physical Society,* vol. 49, no. 2, pp. 513–517, 2006.
22. Y. Kawano, H. Matsumura, S. Shiba, M. Sato, T. Suzuki, Y. Nakasha, T. Takahashi, K. Makiyama, and N. Hara, Flip Chip Assembly for Sub-Millimeter Wave Amplifier MMIC on Polyimide Substrate, in *2014 IEEE MTT-S International Microwave Symposium (IMS2014)*, 1–6 June 2014, pp. 1–4, doi: 10.1109/MWSYM.2014.6848323.
23. A. Jentzsch and W. Heinrich, Theory and Measurements of Flip-Chip Interconnects for Frequencies up to 100 GHz, *Microwave Theory and Techniques, IEEE Transactions on,* vol. 49, no. 5, pp. 871–878, 2001.
24. J. Grzyb, Millimeter-Wave Interconnects, in *Advanced Millimeter-Wave Technologies*, D. Liu, B. Gaucher, U. Pfeiffer, and J. Grzyb Eds. UK: Wiley, Ch. 4, pp. 71–161, 2009.
25. A. Podzorov and G. Gallot, Low-Loss Polymers for Terahertz Applications, *Applied Optics,* vol. 47, no. 18, pp. 3254–3257, 2008.

26. H. Hamada, T. Kosugi, H.-J. Song, M. Yaita, A. El Moutaouakil, H. Matsuzaki, and A. Hirata, 300 GHz Band 20 Gbps Ask Transmitter Module Based on InP-HEMT MMICs, in *2015 IEEE Compound Semiconductor Integrated Circuit Symposium (CSICS)*, IEEE, pp. 1–4, 2015.
27. V. Radisic, K. M. Leong, X. Mei, S. Sarkozy, W. Yoshida, and W. R. Deal, Power Amplification at 0.65 THz Using InP HEMTs, *IEEE Transactions on Microwave Theory and Techniques*, vol. 60, no. 3, pp. 724–729, 2012.
28. X. Mei, W. Yoshida, M. Lange, J. Lee, J. Zhou, P.-H. Liu, K. Leong, A. Zamora, J. Padilla, and S. Sarkozy, First Demonstration of Amplification at 1 THz Using 25 Nm InP High Electron Mobility Transistor Process, *IEEE Electron Device Letters*, vol. 36, no. 4, pp. 327–329, 2015.
29. E. Perret, N. Zerounian, S. David, and F. Aniel, Complex Permittivity Characterization of Benzocyclobutene for Terahertz Applications, *Microelectronic Engineering*, vol. 85, no. 11, pp. 2276–2281, 2008.
30. A. Younus, P. Desbarats, S. Bosio, E. Abraham, J. C. Delagnes, and P. Mounaix, Terahertz Dielectric Characterisation of Photopolymer Resin Used for Fabrication of 3D THz Imaging Phantoms, *Electronics Letters*, vol. 45, no. 13, pp. 702–703, 2009, doi: 10.1049/el.2009.0688.
31. J. H. Booske, Plasma Physics and Related Challenges of Millimeter-Wave-to-Terahertz and High Power Microwave Generationa), *Physics of Plasmas (1994-Present)*, vol. 15, no. 5, p. 055502, 2008.
32. A. Kerr, E. Wollack, and N. Horner, Waveguide Flanges for Alma Instrumentation, *ALMA Memo*, vol. 278, 1999.
33. N. Ridler and R. Ginley, Ieee P 1785: A New Standard for Waveguide above 110 GHz, *Microwave Journal*, vol. 54, no. 3, pp. 20–24, 2011.
34. L. Samoska, W. R. Deal, G. Chattopadhyay, D. Pukala, A. Fung, T. Gaier, M. Soria, V. Radisic, X. Mei, and R. Lai, A Submillimeter-Wave HEMT Amplifier Module with Integrated Waveguide Transitions Operating above 300 GHz, *IEEE Transactions on Microwave Theory and Techniques*, vol. 56, no. 6, pp. 1380–1388, 2008.
35. W. Deal, X. Mei, V. Radisic, K. Leong, S. Sarkozy, B. Gorospe, J. Lee, P. Liu, W. Yoshida, and J. Zhou, Demonstration of a 0.48 THz Amplifier Module Using InP HEMT Transistors, *IEEE Microwave and Wireless Components Letters*, vol. 20, no. 5, pp. 289–291, 2010.
36. V. Radisic, W. R. Deal, K. M. Leong, X. Mei, W. Yoshida, P.-H. Liu, J. Uyeda, A. Fung, L. Samoska, and T. Gaier, A 10-mW Submillimeter-Wave Solid-State Power-Amplifier Module, *IEEE Transactions on Microwave Theory and Techniques*, vol. 58, no. 7, pp. 1903–1909, 2010.

37. V. Radisic, K. M. Leong, S. Sarkozy, X. Mei, W. Yoshida, P.-H. Liu, W. R. Deal, and R. Lai, 220 GHz Solid-State Power Amplifier Modules, *IEEE Journal of Solid-State Circuits,* vol. 47, no. 10, pp. 2291–2297, 2012.
38. K. M. Leong, W. R. Deal, V. Radisic, X. B. Mei, J. Uyeda, L. Samoska, A. Fung, T. Gaier, and R. Lai, A 340–380 GHz Integrated CB-CPW-to-Waveguide Transition for Sub Millimeter-Wave MMIC Packaging, *IEEE Microwave and Wireless Components Letters,* vol. 19, no. 6, pp. 413–415, 2009.
39. K. Leong, X. B. Mei, W. Yoshida, M. Lange, A. Zamora, B. Gorospe, and W. R. Deal, Progress in InP HEMT Submillimeter Wave Circuits and Packaging, in *2015 IEEE Compound Semiconductor Integrated Circuit Symposium (CSICS),* 11–14 Oct. 2015, pp. 1–4, doi: 10.1109/CSICS.2015.7314521.
40. M. Urteaga, M. Seo, J. Hacker, Z. Griffith, A. Young, R. Pierson, P. Rowell, A. Skalare, and M. Rodwell, InP HBT Integrated Circuit Technology for Terahertz Frequencies, in *2010 IEEE Compound Semiconductor Integrated Circuit Symposium (CSICS),* IEEE, pp. 1–4, 2010.
41. A. Tessmann, A. Leuther, V. Hurm, I. Kallfass, H. Massler, M. Kuri, M. Riessle, M. Zink, R. Loesch, and M. Seelmann-Eggebert, Metamorphic HEMT MMICs and Modules Operating between 300 and 500 GHz , *IEEE Journal of Solid-State Circuits,* vol. 46, no. 10, pp. 2193–2202, 2011.
42. A. Tessmann, V. Hurm, A. Leuther, H. Massler, R. Weber, M. Kuri, M. Riessle, H. P. Stulz, M. Zink, M. Schlechtweg, O. Ambacher, and T. Narhi, 243 GHz Low-Noise Amplifier MMICs and Modules Based on Metamorphic HEMT Technology (in English), *International Journal of Microwave and Wireless Technologies,* vol. 6, no. 3–4, pp. 215–223, Jun 2014. [Online]. Available: <Go to ISI>://WOS:000337742800002.
43. V. Hurm, R. Weber, A. Tessmann, H. Massler, A. Leuther, M. Kuri, M. Riessle, H. Stulz, M. Zink, and M. Schlechtweg, A 243 GHz LNA Module Based on MHEMT MMICs with Integrated Waveguide Transitions, *IEEE Microwave and Wireless Components Letters,* vol. 23, no. 9, pp. 486–488, 2013.
44. A. E. Lamminen, J. Saily, and A. R. Vimpari, 60 GHz Patch Antennas and Arrays on LTCC with Embedded-Cavity Substrates, *IEEE Transactions on Antennas and Propagation,* vol. 56, no. 9, pp. 2865–2874, 2008.
45. J. Xu, Z. N. Chen, X. Qing, and W. Hong, Bandwidth Enhancement for a 60 GHz Substrate Integrated Waveguide Fed Cavity Array Antenna on LTCC, *IEEE Transactions on Antennas and Propagation,* vol. 59, no. 3, pp. 826–832, 2011.
46. H. Li, W. Hong, T. J. Cui, K. Wu, Y. L. Zhang, and L. Yan, Propagation Characteristics of Substrate Integrated Waveguide Based on LTCC, in

Microwave Symposium Digest, 2003 IEEE MTT-S International, vol. 3: IEEE, pp. 2045–2048, 2003.

47. H.-Y. Chien, T.-M. Shen, T.-Y. Huang, W.-H. Wang, and R.-B. Wu, Miniaturized Bandpass Filters with Double-Folded Substrate Integrated Waveguide Resonators in LTCC, *IEEE Transactions on Microwave Theory and Techniques*, vol. 57, no. 7, pp. 1774–1782, 2009.
48. Y. Zhang, M. Sun, K. Chua, L. Wai, D. Liu, and B. Gaucher, Antenna-in-Package in LTCC for 60 GHz Radio, in *2007 International Workshop on Antenna Technology: Small and Smart Antennas Metamaterials and Applications*, IEEE, pp. 279–282, 2007.
49. Y. Zhang, M. Sun, K. Chua, L. Wai, and D. Liu, Integration of Slot Antenna in LTCC Package for 60 GHz Radios, *Electronics Letters*, vol. 44, no. 5, p. 1, 2008.
50. W. Hong, A. Goudelev, K.-H. Baek, V. Arkhipenkov, and J. Lee, 24-Element Antenna-in-Package for Stationary 60 GHz Communication Scenarios, *IEEE Antennas and Wireless Propagation Letters*, vol. 10, pp. 738–741, 2011.
51. H. Sherry, J. Grzyb, Y. Zhao, R. Al Hadi, A. Cathelin, A. Kaiser, and U. Pfeiffer, A 1kpixel CMOS Camera Chip for 25 fps Real-Time Terahertz Imaging Applications, in *2012 IEEE International Solid-State Circuits Conference*, IEEE, pp. 252–254, 2012.
52. J. Grzyb, Y. Zhao, and U. R. Pfeiffer, A 288 GHz Lens-Integrated Balanced Triple-Push Source in a 65 Nm CMOS Technology, *IEEE Journal of Solid-State Circuits*, vol. 48, no. 7, pp. 1751–1761, 2013.
53. R. Han and E. Afshari, A 260 GHz Broadband Source with 1.1 mW Continuous-Wave Radiated Power and EIRP of 15.7 dBm in 65 nm CMOS, in *2013 IEEE International Solid-State Circuits Conference Digest of Technical Papers*, IEEE, pp. 138–139, 2013.
54. U. R. Pfeiffer, Y. Zhao, J. Grzyb, R. Al Hadi, N. Sarmah, W. Förster, H. Rücker, and B. Heinemann, A 0.53 THz Reconfigurable Source Array with up to 1 mW Radiated Power for Terahertz Imaging Applications in 0.13 μm SiGe BiCMOS, in *2014 IEEE International Solid-State Circuits Conference Digest of Technical Papers (ISSCC)*, IEEE, pp. 256–257, 2014.
55. H.-J. Song, J.-Y. Kim, K. Ajito, M. Yaita, and N. Kukutsu, Fully Integrated Ask Receiver MMIC for Terahertz Communications at 300 GHz, *IEEE Transactions on Terahertz Science and Technology*, vol. 3, no. 4, pp. 445–452, 2013.

Chapter 15

Semiconductor Technologies for THz Applications

Jae-Sung Rieh
School of Electrical Engineering, Korea University,
145 Anam-Ro, Seongbuk-Gu, Seoul 02841, Korea
jsrieh@korea.ac.kr

Planar solutions for the implementation of THz systems benefit from the small form factor, low power dissipation, and possibly low cost. This aspect applies to THz sources, the topic of this book, as well. The early planar solutions for THz sources were mostly diode-based device-level approaches, such as IMPATT (IMPact ionization Avalanche Transit Time) diode, Gunn diode, and RTD (Resonant Tunneling Diode). However, the advent of modern high-frequency semiconductor transistor technologies, driven by both structural and material innovations, has enabled THz sources based on transistor-based circuit-level solutions. This can be considered as a great milestone for the development of compact THz sources (and other systems) for a couple of good reasons. First, the transistor is a general-purpose device, which would allow the integration of THz sources with other electronic circuitry, both analog and digital. Second, the system optimization can be made on the circuit level rather than the device level, which would provide a much

Advances in Terahertz Source Technologies
Edited by Gun-Sik Park, Masahiko Tani, Jae-Sung Rieh, and Sang Yoon Park

ISBN 978-981-4968-89-8 (Hardcover), 978-1-003-45967-5 (eBook)
www.jennystanford.com

higher level of design flexibility. Considered first as THz-compatible semiconductor solutions were III–V technologies, namely HBT (Heterojunction Bipolar Transistor) and HEMT (High Electron Mobility Transistor) technologies. Now, Si-based technologies, not only SiGe HBT but also Si CMOS (Complementary Metal-Oxide-Semiconductor), are regarded as viable options for the realization of THz systems. In this chapter, the device operation principles and the performance trend will be reviewed with these four major candidate technologies for THz applications.

15.1 Si CMOS Technology

In 1997, a paper entitled "CMOS RF: no longer an oxymoron" appeared at one of the major conferences [1], which was authored by Thomas Lee of Stanford, who later published a very popular textbook on CMOS RF integrated circuits [2]. In the paper, which was intended to dismiss the skepticism about the RF applications of CMOS technology at that time, the author claimed that "Peak CMOS f_T's are now in excess of 30 GHz" and the scaling trend will "enable CMOS IC technology to perform well enough at GHz frequencies". As of today, about 25 years after the appearance of the paper, CMOS technology is not only a mainstream technology for RF applications in GHz frequencies, but also considered as a strong contender for applications in THz frequencies, with peak f_T or f_{max} marking over 300 GHz (a ten-fold increase from 30 GHz) with advanced nodes. In this section, the basic operation of MOSFETs is overviewed and the recent structural innovations for the devices will be introduced. This will be followed by its performance trend and some selected commercially available technology examples.

15.1.1 Device Operation

The field-effect transistor (FET) is a semiconductor device in which the current between two electrodes (drain and source) is controlled by a third electrode (gate) by adjusting the electric field in the current path. While the world's first transistor demonstrated in 1947 was a bipolar transistor [3], the concept of FET is dated back

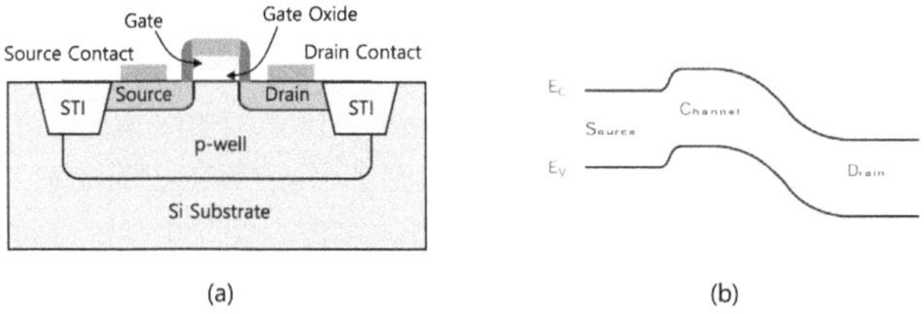

Figure 15.1 (a) Conceptual structure of a standard nMOSFET (not to scale). (b) Energy band diagram along the channel.

to as early as 1926 [4]. While various types of FETs have been suggested since then, MOSFET, first demonstrated by Kahng and Atalla in 1960 [5], has been by far the most successful device type. The cross section of the standard MOSFET is shown in Fig. 15.1 for an *n*-type device (NMOS) as an example. Shown together is the energy band diagram along the channel. The current from the drain to the source is modulated as the density of the inversion electrons at the channel (and thus the channel resistivity) is controlled by the potential level at the gate, which is separated from the channel by a very thin insulating film. As the details on the basic operation principles of MOSFETs can be found in most standard textbooks on semiconductor devices, they will not be repeated here. Instead, a brief description of the RF characteristics of MOSFETs will be provided below.

The operation speed of transistors is typically represented by f_T (cutoff frequency) and f_{max} (maximum oscillation frequency), which are defined as the frequency where the current gain and unilateral power gain U, respectively, become unity. Both can be expressed in terms of the device parameters of MOSFETs. Although there exist more comprehensive expressions, the following can be used for a quick estimation of their values:

$$f_T = \frac{g_m}{2\pi C_{in}} = \frac{g_m}{2\pi \left(C_{GS} + C_{GD}\right)}, \tag{15.1}$$

$$f_{max} = \sqrt{\frac{f_T}{8\pi R_G C_{GD}}}, \tag{15.2}$$

where g_m is the transconductance, C_{GS} and C_{GD} are the G-S (gate-source) and G-D (gate-drain) capacitances, respectively, and R_G is the gate resistance. (15.1) indicates that the most effective way to improve f_T is to increase g_m, for which lateral scaling, mostly the gate length reduction, is desired. The control of the minimum gate length, however, is beyond the capacity of circuit designers. What they can do for g_m boost is to either raise the drain current or increase the gate width. The former does result in f_T enhancement, but at the cost of increased DC power consumption. This leads to the well-known trade-off between speed and power consumption (In reality, there exists a maximum current beyond which f_T begins to roll off). The latter effort will hardly affect f_T since the increased g_m is largely canceled out by the increase in C_{in} that shows up at the denominator of (15.1). On the other hand, circuit designers have a better control over f_{max}, since it is more sensitive to the device layout details such as the number of gate fingers and the interconnect metal routings. For example, R_G, which affects f_{max} along with f_T and C_{GD}, will depend on the unit gate finger width, requiring layout optimization for multi-finger configuration [6]. So, which one is more relevant to RF applications, f_T or f_{max}? There used to be a dispute over this, but there now seems to be a consensus that f_{max} is a better indicator of the available device speed, at least for analog and RF circuit designs.

Noise is another key aspect of RF performance. While there are various parameters to indicate the noise level of a device, high-frequency noise is often represented by the noise factor F, which is defined as the ratio between the input and output signal-to-noise ratio (noise figure NF is given as $20\log F$). The minimum noise factor F_{min}, which is the noise factor when a perfect noise matching is achieved, is given as follows for MOSFETs:

$$F_{min} = 1 + K\frac{f}{f_T}\sqrt{g_m(R_S + R_G)}, \tag{15.3}$$

where R_S is the source resistance and K is a fitting parameter. From (15.3), it is obvious that high operation speed is needed for a low noise level as indicated by its relation to f_T. In addition, process and layout optimization to lower R_S and R_G will also help suppress the noise. It should be noted that a larger g_m will reduce F_{min}, which may appear contrary to the expression given in (15.3). Still, it is true because f_T implicitly includes a linear relation with g_m in itself,

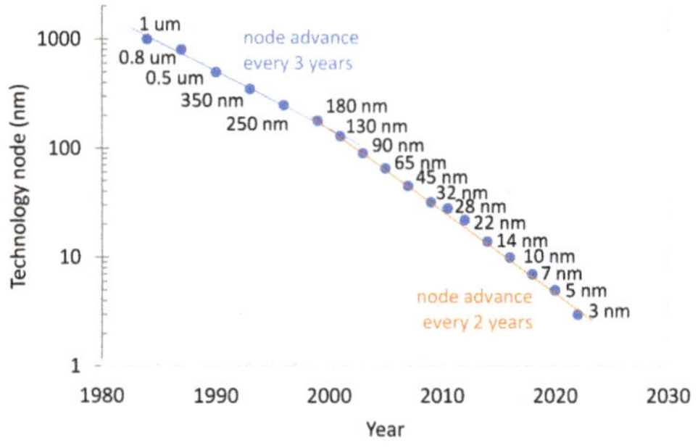

Figure 15.2 The evolution of CMOS technology nodes.

leading to an inverse relation between $F_{\rm min}$ and $g_{\rm m}^{1/2}$. Also, it is clear from (15.3) that the noise will increase with frequency as expected.

15.1.2 Structural Variations

The main driving force for the performance enhancement of MOSFETs since its birth has been the scaling, or the reduction of the device size.

Figure 15.2 shows the trend of the scaling for the past four decades in terms of technology node advance. The technology node, often given by the minimum gate length, typically advances at a rate of $1/\sqrt{2}$, which would result in a two-fold chip size reduction (assuming highly integrated chips). Initially, the technology node advanced every 3 years, but it was accelerated to follow a 2-year cycle since around 2000. It is noted that the transition from 32 to 28 nm was a half node advance, giving a slight disturbance in the trend shown in the plot. While the scaling is still under way continuing its successful ride, some variations in the device structure have also been adopted as a way to complement the scaling. They will be reviewed in this section.

15.1.2.1 SOI MOSFET

SOI MOSFETs (or more typically called SOI CMOS, which tends to imply the technology rather than individual devices) refer to MOSFETs built on a thin insulating layer on top of Si substrate (or

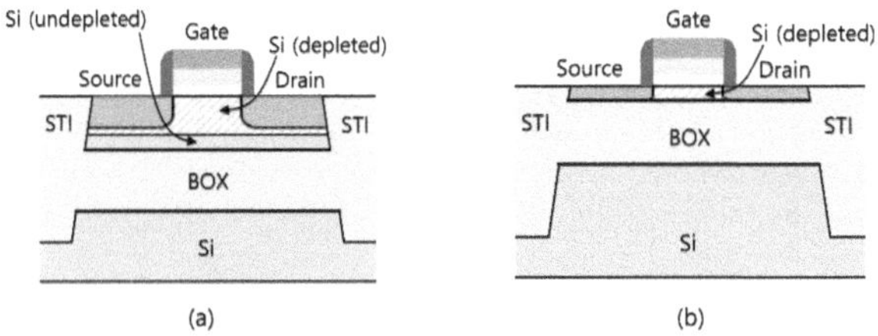

Figure 15.3 Conceptual structures of SOI MOSFETs (not to scale): (a) Partially depleted SOI (PDSOI). (b) Fully depleted SOI (FDSOI).

bulk), as opposed to the standard MOSFETs (sometimes called bulk devices in contrast to SOI devices) which are built directly on top of Si substrate. Conceptual structures of SOI devices are illustrated in Fig. 15.3. As shown in the figure, for SOI devices, a thin Si film is placed on top of the buried oxide (BOX), which is in the order of a few tens of nm (20–25 nm for advanced nodes). Depending on the thickness of the Si film on BOX, two different types of SOI devices can be considered: partially depleted (PD) SOI devices (Fig. 15.3a) and fully depleted (FD) SOI devices (Fig. 15.3b). For the latter case, the Si film is so thin (typically 5–20 nm or ~1/4 of the gate length) that it is fully depleted in normal operation, while for the former case the Si film is thicker (typically 50–100 nm) and only partially depleted as the depleted region is thinner than the film thickness. It was PDSOI technology that was first adopted for commercial applications as it is less challenging to realize. However, FDSOI technology has recently emerged as a mainstream SOI technology from around 22 nm node, owing to the continuing advances in Si process technology. The discussion on SOI provided below will assume FDSOI technology unless indicated otherwise.

There are obvious advantages of FDSOI devices over conventional bulk devices. One immediate observation from Fig. 15.3 is that FDSOI devices provide better isolation as the active Si region is fully isolated from the substrate. This can be compared to bulk devices, where the isolation is only partially provided by shallow-trench isolation (STI), leaving the device bottom open to the substrate. The FDSOI structure will improve AC-wise isolation as well since the oxide layer will significantly reduce the capacitance to the substrate, leading to device performance improvement. On the other hand, the

BOX layer is not thick enough (about 10 times of the gate oxide layer) so that the potential change at the substrate will affect the electrons in the channel region. However, this back gate effect can be in fact exploited as a means to control the threshold voltage (V_T), as is the case for the bulk devices with the body effect.

Other advantages of FDSOI devices include the following. The thin depletion region will lower the required gate potential to induce the inversion charge. This will reduce V_T and thus help enhance the device operation speed and reduce power consumption. Also, FDSOI devices do not need channel and/or halo dopings in the active region, which were dominantly employed for the bulk devices to suppress the short-channel effects and V_T control. This allows the channel region of FDSOI devices to remain undoped, which would reduce the random dopant fluctuation (RDF) and thus suppress the device variability. Considered as other advantages are the elimination of latch-up that often troubles the bulk devices, lowered drain-induced barrier lowering (DIBL) that leads to a more ideal sub-threshold swing slope, and a compact layout that allows higher device density. FDSOI devices have been successfully adopted for RF CMOS technologies, leading to f_{max} exceeding 300 GHz at 22 nm node [7].

15.1.2.2 FinFET and GAA FET

For planar MOSFETs, both bulk and SOI types, one apparent limit in the device operation is that the control of the electron flow in the channel comes almost exclusively from the top-side gate only. As a result, the bottom of the channel remains largely loose and the electric field beneath the channel is not under tight control, causing various problems in device operation, especially for short-channel devices. For more extensive control of the channel, an additional gate control from the opposite side will be very helpful. However, forming a bottom-side gate will be extremely challenging from the fabrication process viewpoint. However, the recent advances in Si technology have finally made such a dual-gate structure possible. The solution was to turn the thin channel layer into a vertical position and form the gates from two lateral sides. Such a device is

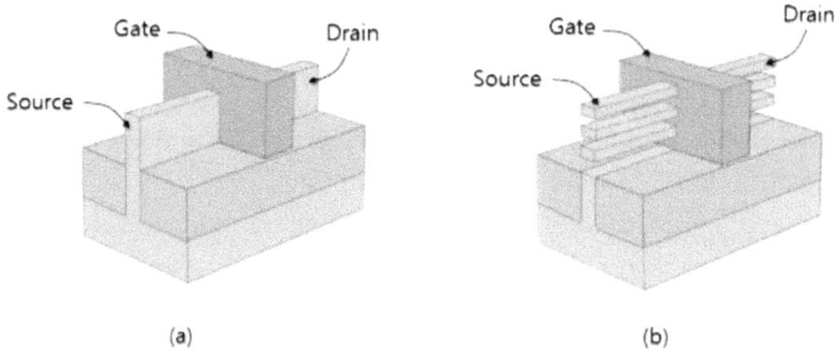

Figure 15.4 Structure of (a) FinFET and (b) GAA transistor.

now called FinFET, after the shape of the standing channel that looks like the fin of fish as depicted in Fig. 15.4a.

The advantages of FinFETs are rather straightforward. One of the main issues with a very short channel is DIBL, with which the drain begins to directly interact with the source through the channel region, causing a serious leakage current problem. This will be mitigated if the electrostatics in the channel layer is controlled from both sides, which is readily achieved with the given FinFET structure. In this structure, the channel is fully depleted and the troublesome parasitic bulk region is completely removed from the active part of the device. Owing to the obvious advantage, FinFET has been adopted as the standard device type for the bulk CMOS technology from 22 nm (or 14 nm depending on foundries) and following nodes. In principle, FinFETs can be applied to both bulk and SOI technologies. However, they have been so far applied to only bulk technologies, one possible reason being that SOI devices address similar issues (such as DIBL) as FinFETs, apparently obviating the need for an additional complicated fabrication process.

If FinFETs control the channel region from two sides and improve the device performance, why don't we control the channel from four sides or from all directions? The answer is the GAA (Gate All Around) FETs. As depicted in Fig. 15.4b, the cylindrically shaped channel region is wrapped around with a gate region, separated by a thin circular layer of the gate dielectric. The channel region is completely depleted and the channel control is highly symmetric. Of course,

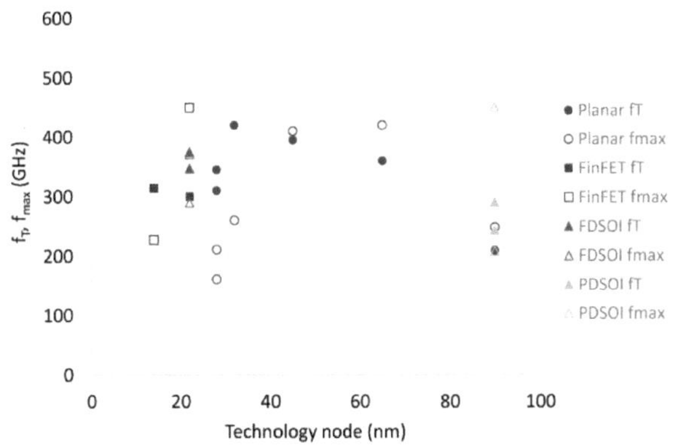

Figure 15.5 Recently reported RF performance of Si CMOS technologies.

fabrication steps will be more challenging than the FinFET case. Nonetheless, the foundries are pushing for the adoption of GAA for bulk CMOS technologies as a replacement for FinFETs from 3 nm or smaller node (again depending on foundries).

15.1.3 Performance Trend

Figure 15.5 shows the RF performance of CMOS technologies recently reported from various major foundries. It includes f_T and f_{max} values of NMOS (PMOS typically shows slightly lower numbers) for various structures, including bulk planar, PDSOI, FDSOI, and FinFET, as a function of the technology node. According to the plot, the maximum f_T and f_{max} have reached as high as 485 and 450 GHz, respectively. On the other hand, there is no obvious increasing trend for both f_T and f_{max} with the technology node advances, which is rather against our expectation as the technology node is supposed to represent the gate length that drives the performance. In fact, the absence of increasing tendency with technology node implies that the trend of f_T and f_{max} of the devices with altered structures such as SOI and FinFET do not follow that of the standard bulk planar devices anymore. With FinFETs, for example, it is expected that the enhancement of f_{max} with scaling will be slowed down or saturated as R_B is significantly affected by the fin structures. Also responsible is the uncertainty in the reported f_T and f_{max} values, since some of them are for intrinsic devices (the effect of metal wiring is subtracted out), while others are for devices including some part of the metal wirings. Due to the effect of metal wiring on extracted values, it is likely that f_T and f_{max} available for practical circuit designs are lower than what is shown in Fig. 15.5, as the metal routings up to top metal levels may be needed for many circuit applications.

The RF performances of some selected technologies of advanced nodes are introduced below along with structure highlights. The RF performance of a 22 nm FDSOI from GlobalFoundries was reported in 2018 [7]. It employs fully depleted SOI with high-k metal gate, which are built upon a triple well structure (comprising nWell, deep nWell, and pWell). NMOS and PMOS exhibit f_T/f_{max} of 347/371 and 242/288 GHz, respectively. It is interesting to note that the thick-oxide device, which handles a larger voltage swing for I/O stages, shows much smaller f_T (60/36 GHz for NMOS/PMOS) but only moderately degraded f_{max} (249/142 GHz for NMOS/PMOS). One unique feature adopted for the RF variant of CMOS technologies is the ultra-thick metal (UTM) formed at the top of the BEOL metal layers, which offers low-loss solutions for various passive devices such as microstrip lines, inductors, transformers, etc., with a thickness reaching up to a few μm. This 22 nm FDSOI technology offers 2-layer and 3-layer options for UTM, providing flexibility in designing high-frequency circuits.

RF performances of FinFETs have also been reported. A 22 nm FinFET technology from Intel exhibits NMOS f_T and f_{max} of 300 and 450 GHz, respectively [8]. For PMOS, those values are dropped down to slightly below 300 GHz. One distinct aspect of FinFET technologies is that the gate access resistance has the vertical component as well as the horizontal component, the latter being the only component for the standard planar devices. Hence, the gate access resistance does not show a linear relation to the gate width anymore, indicating that there exists an optimal number of fins for minimum R_G, which was around 4 for the given technology. It is also noted that peak f_T and f_{max} can be achieved at a lower bias condition for FinFETs compared to the planar devices. This provides an opportunity for reduced power dissipation to attain a similar level of RF performance.

15.2 SiGe HBT Technology

Not widely known to many is the fact that the very first transistor in the world, invented in Bell Labs in 1947, was a germanium transistor [3]. Until Si began to rapidly emerge as a mainstream semiconductor in the early 1960s, it was Ge that was considered as a material of choice for building transistors. However, the strong temperature dependence of the main material properties of Ge, which arises from its relatively small energy bandgap of 0.67 eV at room temperature (vs. 1.12 eV for Si), as well as the lack of stable oxide film, prompted the fall of Ge and the rise of Si. On the other hand, there has been continued interest in the alloy made of Si and Ge, or SiGe. Although the early efforts on SiGe were mainly from academic interests, they eventually ended up with a successful industrial adoption. The first major practical application of SiGe was Si-based bipolar junction transistors (BJTs), which employ SiGe alloy in the base region of the transistors for performance enhancement. Such a device is called the SiGe HBT, first demonstrated in 1987 [9]. In this section, a brief overview of the operation of SiGe HBTs will be provided, followed by their reported performance trend along with an introduction to some notable SiGe BiCMOS technologies commercially available.

15.2.1 Device Operation

The device structure and the energy band diagram of a typical *npn* SiGe HBT operating under the forward active mode are presented in Fig. 15.6. When compared to the conventional BJTs, there are a

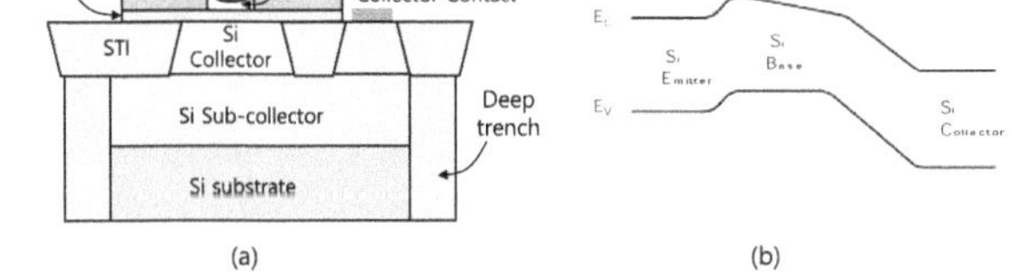

Figure 15.6 (a) Conceptual structure of a typical SiGe HBT (not to scale). (b) Energy band diagram along the vertical direction beneath the emitter contact.

couple of distinct features unique for SiGe HBTs that can be found in the band diagram in Fig. 15.6b: a smaller bandgap in the base compared to emitter/collector and the slope in the conduction band edge along the base. They are due to the following facts. The bandgap of SiGe is smaller than Si, a natural consequence of the smaller bandgap of Ge than that of Si, approximately showing ~80 meV reduction for every 10% Ge added to Si. As the Ge composition is graded across the base region in most cases, increasing toward the collector side, the bandgap is gradually tapered down generating the observed slope. As far as DC characteristics are concerned, it is mainly the addition of Ge in the base region rather than its grading that determines the unique static properties of SiGe HBTs. On the other hand, the grading will significantly affect the RF performance of the devices. In the following, DC characteristics will be first discussed, followed by the RF aspect of the device operation.

SiGe HBTs are a minority carrier device as all the bipolar devices are. Hence, the main operation mechanism is governed by the minority carrier behavior. In the case of the collector current I_C, which is equal to the electron current component of the emitter current I_E, its magnitude is dictated by the minority electron concentration in the *p*-type base region. Since the base is composed of SiGe, which has a narrower bandgap than Si, the base minority electron concentration of a SiGe HBT is much larger than that of a conventional Si BJT when the same base doping (acceptor) concentration is assumed. This leads to I_C being much larger in SiGe HBTs than in Si BJTs. On the other hand, the base current is determined by the minority hole concentration in the *n*-type emitter region. Since the emitter is in Si for both SiGe HBTs and Si BJTs, there is little difference in the I_B levels between the two device types assuming the same emitter doping concentration. The consequence is a much larger current gain for SiGe HBTs than Si BJTs, reaching several hundred or even a few thousand. Now a question follows: Is this large current gain always desired? Not necessarily. In fact, excessive current gains may harm the breakdown voltage (specifically BV_{CEO}). Then why are SiGe HBTs so popular and have largely replaced Si BJTs in most RF applications? The truth is that the large current gain can be traded for improved RF performance, as discussed below.

For bipolar transistors, including SiGe HBTs, f_T is expressed as follows in terms of emitter-to-collector delay τ_{EC}:

$$f_T = \frac{1}{2\pi\,\tau_{EC}}, \tag{15.4}$$

where

$$\begin{aligned}\tau_{EC} &= \tau_E + \tau_C + \tau_B + \tau_{CSCL}\\ &= \frac{C_{EB}}{g_m} + \left(\frac{1}{g_m} + R_C + R_E\right) C_{CB} + \frac{W_B^2}{\gamma\, D_{nB}} + \frac{W_{CSCL}}{2v_{sat}}.\end{aligned} \tag{15.5}$$

Here, k is the Boltzmann constant, C_{EB} and C_{CB} are the E-B and C-B capacitances, and R_C and R_E are the collector and emitter resistances, respectively. Also, W_B and W_{CSCL} represent the neutral base width and the B-C space-charge region (SCR) width, respectively, while γ is for the field factor, D_{nB} for the electron diffusion constant at the base, and v_{sat} for the saturation velocity. The advantages of SiGe HBTs over conventional Si BJTs for improved f_T mostly come from the possibility of reducing the base transit time (τ_B) for the following reasons. Firstly, a smaller W_B can be obtained with SiGe HBTs owing to higher base doping available. For Si BJTs, where the base doping needs to be maintained at a moderate level to guarantee a sufficient current gain, too thin base layer may cause the pinch-off, or the merge of the E-B and B-C depletion regions, limiting the base scaling. On the contrary, for SiGe HBTs, the base doping can be raised as needed since the current gain is already large enough, which facilitates an aggressive base scaling without risking the base pinch-off. That is, the large current gain of SiGe HBTs can be traded for higher speed with reduced τ_B and improved f_T. Secondly, the field factor γ, which is the measure of the slope in the conduction band edge and thus the induced quasi-electric field, can be larger than unity for SiGe HBTs. It will help accelerate the electrons across the base region and thus reduce τ_B, also contributing to the enhanced f_T for SiGe HBTs.

The expression for f_{max} is basically identical to that of MOSFETs, the only difference being R_G and C_{GD} replaced by R_B and C_{CB}, respectively:

$$f_{max} = \sqrt{\frac{f_T}{8\pi\, R_B C_{CB}}}. \tag{15.6}$$

One apparent approach to raise f_{max} is to reduce R_{B}, for which increased base doping concentration will be helpful as it lowers the lateral access resistance along the thin base layer. Again, the availability of the high base doping for SiGe HBTs will enable lowered R_{B}, resulting in f_{max} improvement. Another way to reduce the lateral access resistance and thus R_{B} is to shorten the lateral current path in the base layer, for which the emitter finger needs to be narrow. For this end, emitter finger width is often defined with the minimum feature size with the given technology. Equation (15.6) indicates that the reduction of C_{CB} will also help boost f_{max}, which is obtained with lowered collector doping as it increases the B-C depletion width. However, the extended B-C depletion region will increase τ_{CSCL} and the lowered collector doping will limit the maximum allowed current level by the Kirk effect, both of which will degrade the peak f_{T}. Hence, there exists an apparent trade-off between f_{T} and f_{max} in terms of collector doping. As for high-frequency noise, F_{min} will be approximated as follows assuming high-frequency operation [10].

$$F_{\mathrm{min}} \simeq 1 + \frac{f}{f_{\mathrm{T}}}\sqrt{2g_{\mathrm{m}}R_{\mathrm{B}}}. \tag{15.7}$$

This relation is almost identical to the one for MOSFET given in (15.3). Hence, the approaches to improve F_{min} in SiGe HBTs will be similar to those for MOSFETs discussed earlier.

15.2.2 Performance Trend

It is interesting to note that the RF performance of SiGe HBTs (and Si BJTs as well) is not very sensitive to the technology node that typically determines the emitter finger width. It is true that a smaller emitter finger width will help improve f_{max} as mentioned above. As for f_{T}, a narrower emitter finger will reduce C_{EB} (and C_{CB} for certain fabrication processes) and tend to improve f_T, while it will increase R_{E} at the same time, counteracting the effect of the reduced capacitances. Further, the fringing component of the emitter current will be more pronounced with a narrow emitter finger, which will impose a negative effect on f_{T} as it will increase the average traveling path of electrons across the base region and thus the base transit time. As such, the emitter finger scaling is not a dominating factor

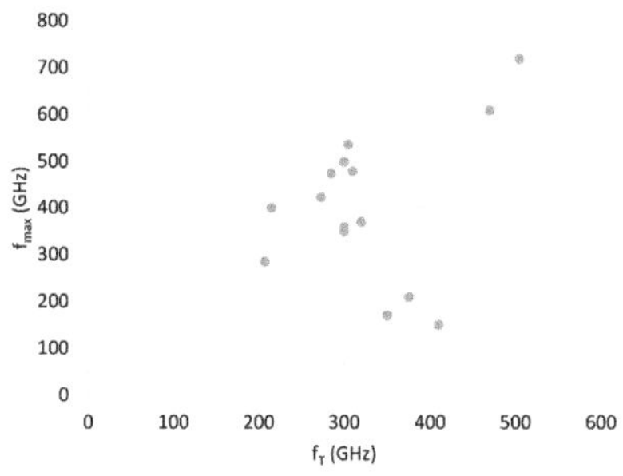

Figure 15.7 Recently reported RF performance of SiGe HBTs.

for f_T improvement. For this reason, the scaling of the emitter finger has been stalled around 50 nm [11], in sharp contrast to MOSFETs whose performance is dictated by the gate length and thus strongly relies on continuing scaling efforts. In a sense, these contrasting behaviors are a natural consequence of the fact that SiGe HBTs are a vertical device (vertical current path) while MOSFETs are a lateral device (lateral current path). In fact, this is an opportunity for SiGe HBTs, since the performance enhancement can be achieved with the vertical scaling, which is far less costly than the lateral scaling, as it does not require a very expensive lithography upgrade for technology node advance. Probably this is the main reason why a ~100 nm SiGe technology exhibits a better performance than a ~10 nm CMOS technology.

From most foundries of today, SiGe HBTs are available as part of BiCMOS technologies, which offer both bipolar and CMOS devices. The performance trend of SiGe HBTs in BiCMOS technologies is shown in terms of f_T and f_{max} in Fig. 15.7. The best f_{max} reported so far is 720 GHz with an associated f_T of 505 GHz, which was attained with a 130 nm BiCMOS technology from IHP [12]. The SiGe HBT, which exhibits breakdown voltages of $BV_{CEO}/BV_{CES} = 1.6/3.2$ V and a peak current gain slightly lower than 1000, is composed of 8 emitter fingers, each with a dimension of $0.105 \times 1\ \mu m^2$. The peak f_T/f_{max} was obtained at the collector current density of 34 mA/μm^2. The gate delay, which is often employed to represent the overall performance of a given technology, was only 1.34 ps as measured with a 31-stage ring oscillator based on the SiGe HBTs.

15.3 III–V HEMT Technology

Although compound semiconductor technologies are not a mainstream technology of today in the semiconductor industry, there is no doubt they outperform Si-based technologies in many performance measures. As such, they are still considered as highly favored options for certain special applications that require extreme performance, including THz sources. There are a number of attractive features in compound semiconductors. First, many compound semiconductors, including GaAs, InP, and alloys based on them, exhibit electron mobility far exceeding that of Si, which greatly benefits high-frequency applications. Second, a wide bandgap is readily available from compound semiconductors, which facilitates high-power applications. Third, the commonly used GaAs and InP wafers provide extremely high substrate resistivity (thus called semi-insulating or SI substrate), which is desirable for realizing low-loss passive devices. Lastly, the availability of various ternary or quaternary alloys based on binary compound semiconductors (typically made of column III and V elements, thus often called III–V semiconductors) offers chances for bandgap engineering that enables bandgaps to be tailored for improved performance as needed, with which various heterojunctions can be readily formed. There are challenges, however, for transforming these great material characteristics into great devices by applying proper fabrication processes. To mention a few, the crystal quality of compound semiconductors is not as good as that of Si, causing a large leakage current. This, combined with their poor thermal conductivity, also causes lots of reliability issues. The lack of stable oxide films also limits the device structure as well as fabrication steps.

HEMTs and III–V HBTs are the most successful electrical devices based on compound semiconductors as of today. Both have been successfully applied to high-frequency applications, while conventional wisdom is that HEMTs are for low-noise applications and HBTs are for high-power applications. HEMTs will be first described in this section for device operation and performance examples, and III–V HBTs will be covered in the next section.

15.3.1 Device Operation

To understand the operation of the HEMT, which was first demonstrated in 1980 [13, 14], the concept of modulation doping needs to be understood first. The readers are reminded that there are two major scattering mechanisms inside semiconductors that dictate carrier mobility: lattice scattering and ionized impurity scattering. The former is inherent and cannot be avoided as semiconductors are built with a lattice of atoms, while the latter is absent if the semiconductor is undoped. However, because carriers are supplied with dopants, you need doping when you need carriers. Is there any way to retain enough carriers inside an undoped semiconductor? The answer is yes, and modulation doping serves that purpose. Consider two stacked layers of semiconductors with different bandgaps, in which the wide-bandgap layer is doped while the narrow-bandgap layer is undoped. In this modulation-doped structure, the carriers in the wide-bandgap layer tend to be transferred to the narrow-bandgap layer due to the potential difference. As a result, the narrow-bandgap layer will now contain carriers although it remains undoped. Further, the carriers in this layer will show great mobility owing to the absence of ionized impurity scattering. HEMTs employ an *n*-type modulation-doped structure for enhanced electron mobility along the channel, for which reason they are also called MODFETs (MOdulation Doped FETs).

Figure 15.8 illustrates a typical cross section of a HEMT explicitly showing the layer structure. Also shown is the band diagram along the vertical direction across the stacked layers. In this exemplary

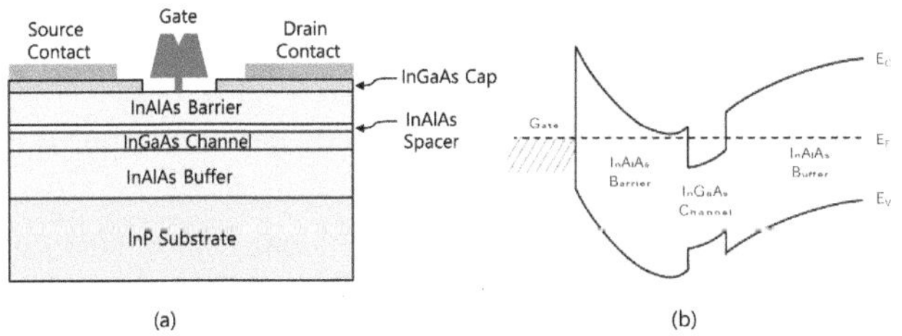

Figure 15.8 (a) Conceptual structure of a typical HEMT (not to scale). (b) Energy band diagram along the vertical direction beneath the gate.

structure, the InAlAs barrier layer (wide bandgap) is *n*-type doped while the InGaAs layer underneath (narrow bandgap), which serves as the channel layer, is undoped. With this modulation-doped structure given, a high mobility is attained for the channel electrons which are supplied from the barrier layer, resulting in high-speed operation of the device. Inside the wide-bandgap barrier layer, the bottom part close to the boundary with the channel is often kept undoped (called the spacer layer) to allow a finite space between the doped region and the channel to prevent the ionized dopants from remotely affecting the channel mobility. Alternatively, a very highly doped 2-D sheet layer can be formed slightly away from the boundary, which is called δ-doping. Often an additional highly doped layer (called cap layer), made of a narrow-bandgap material such as InGaAs as is the case for the structure given in Fig. 15.8, is placed upon the barrier layer, which will help reduce the contact resistance for source and drain. These epitaxial layers are often prepared with MBE (molecular-beam epitaxy) or MOCVD (metal organic chemical vapor deposition), which is typical for most III–V devices including III–V HBTs to be described in the next section. On top of the epitaxial HEMT structure, source/drain and gate contacts are formed. For source/drain, the contacts should be ohmic, which is helped by the highly doped cap layer as mentioned. The gate should be a Schottky contact, for which the cap layer needs to be locally removed by a recess etching.

As in the case of MOSFET, a small gate length is the key to performance enhancement with HEMTs. For this, electron-beam (e-beam) lithography has been widely employed for HEMTs as well as other III–V devices. Ironically, the e-beam lithography, which was developed to provide much narrower lines (thus gate length) than the optical lithography, is now outperformed by modern optical lithography, as indicated by the gate length of the most advanced commercial HEMT technology (about 25 nm) being larger than that of the latest CMOS technology (below 10 nm). Massive investments in Si process technology development have made this possible. Another important aspect of the gate for improved performance is the gate resistance. In contrast to MOSFETs that usually employ multi-finger gates with a unit finger width of 1–2 μm for RF circuits, HEMTs typically adopt single- or dual-finger gate (except for high-power devices), in which the unit finger width ranges easily up to several tens of μm. To minimize the gate resistance with this long gate current path, a T-shaped gate, as indicated by the gate metal profile in Fig. 15.8, has been routinely adopted to maximize the cross-sectional area while maintaining a narrow footprint.

While HEMTs require carefully tailored epitaxial layers as described above, their DC and RF operation principles are largely similar to that of MOSFETs described in Section 15.1.1, which will not be repeated here. Instead, a few unique points for HEMT operation will be listed below. Due to the heterojunctions between the narrow-bandgap channel layer and the adjacent wide-bandgap layers, a quantum well is formed along the channel region. Accordingly, electron energy levels for the vertical dimension are quantized, resulting in two-dimension electron gas (2-DEG) in the channel. This will lead to additional electron mobility enhancement owing to the reduced scattering. It is also worthwhile to note that the mobility enhancement will be more pronounced at low-temperature operation. This is because scatterings at a low temperature are dominated by the ionized impurity scattering, while this scattering mode is effectively suppressed by the modulation doping. As a result, the total scattering rate is greatly reduced at low temperature, leading to mobility and performance boost in cryogenic operation.

15.3.2 Performance Trend

Before discussing the performance trend of HEMTs, it would be useful to review different types of HEMT devices. Some are called pseudomorphic HEMT or pHEMT (It is never meant to indicate *p*-type HEMT. III–V FETs are almost exclusively *n*-type since the hole mobility is much lower than electron mobility in most III–V materials employed). The name is related to the lattice constant matching between different epitaxial layers in the devices. For HEMTs based on InP wafer, InGaAs is widely employed as the high-mobility channel layer. For the $In_xGa_{1-x}As$ channel to match the lattice constant of the InP substrate, the composition of In (x) needs to be 53%, or $In_{0.53}Ga_{0.47}As$. On the other hand, it is known that a higher In composition would enhance the device performance owing to the increased electron mobility and improved electron confinement in the channel layer. In principle, the lattice mismatch is supposed to generate crystal dislocations in the layers, causing lots of problems in device operation. Fortunately, however, a slight mismatch will be allowed if the layer is thin enough, in which case the lattice will be deformed slightly but maintain the overall crystal structure (thus 'pseudo-morphic'). For this reason, a very thin InGaAs layer with In composition larger than 53% can be employed for the channel layer grown on top of the InP substrate. This type of HEMT is pHEMT and shows a better performance than lattice-matched devices owing to the improved electron transport in the pseudomorphic channel layer. For many pHEMTs, In composition is brought up to as high as 70–80% or even 100% (or InAs).

There is another type of HEMT, called metamorphic HEMT or mHEMT. It employs a thick ($\sim$1 μm) buffer layer in the substrate that converts the lattice constant from the substrate value to another by gradually grading the composition (and thus the lattice constant) of ternary or quaternary alloys. It enhances the freedom in choosing the channel material since the lattice constant can be selected as desired by controlling the buffer structure, relaxing the constraint needed for lattice matching. Typically, mHEMTs are built on a GaAs wafer with an InAlGaAs layer employed as the buffer. With this metamorphic structure, enhanced performance can be achieved by employing the material systems lattice-matched to InP, while more cost-effective GaAs wafers can be still used for the substrate. Or, with a better manipulated buffer layer, the lattice constant can be tuned for the InGaAs layer with even higher In composition for further improved performance. Fraunhofer has been successfully developing mHEMT technologies, which have been applied to various mm-wave and THz circuits. A recently reported 35 nm mHEMT technology from Fraunhofer based on 4-inch GaAs wafers exhibited f_T of 515 GHz and f_{max} exceeding 1 THz, which was employed for a 600 GHz amplifier with a gain of 20 dB [15].

The recently reported RF performance of HEMTs is presented in Fig. 15.9a in terms of f_T and f_{max} as a function of the gate length. Overall, both f_T and f_{max} show increasing trends with the gate length scaled down, which is as expected from the well-known correlation between the gate length and the FET performance. Nonetheless, the data points do not tightly follow the inverse relation, exhibiting a

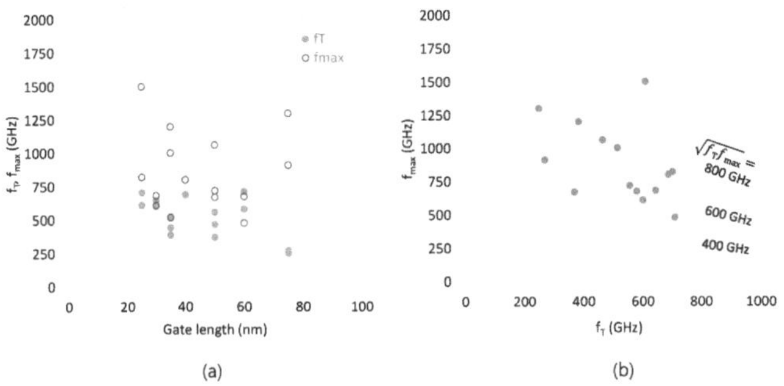

Figure 15.9 Recently reported RF performance of HEMTs: (a) f_T and f_{max} vs. gate length. (b) f_T vs. f_{max}. The lines are contours for constant f_T–f_{max} products.

significant fluctuation, which may stem from the various material systems and structure details employed by different foundries. It is also interesting to note that there is only a weak correlation between f_T and f_{max} of the devices reported. In fact, it appears as if there is a slight negative correlation as shown in Fig. 15.9b (more apparent if one outlier, which happens to be the one for the best f_{max}, is omitted). This indicates there exists a trade-off between f_T and f_{max} in device design, affected by the parameters strongly influencing f_{max} such as R_G and C_{GD}. The highest f_{max} achieved so far is 1500 GHz, with an associated f_T of 610 GHz, which was obtained with a 25 nm gate pHEMT from NGC [16]. Built on a 3-inch InP wafer, this device employs an InAs layer as thin as 9.5 nm for the channel layer. The electron mobility in the channel was estimated to be 13000 cm^2/Vs, which apparently helped boost the performance, together with the scaled gate length. Based on this device, the world's first amplifier operating up to beyond 1 THz was realized [16].

15.4 III–V HBT Technology

The concept of incorporating heterojunctions to bipolar transistors for performance enhancement can be traced back to as early as the 1950s. Shockley, one of the inventors of the transistor, proposed a wide-bandgap emitter for bipolar transistors to improve the performance in his patent issued in 1951 [17]. The idea was further elaborated in a pioneering paper by Kroemer in 1957 [18], in which systematic analyses of HBT were provided. Together with SiGe HBTs, III–V HBTs are regarded as highly promising candidates for various THz applications including THz sources. Moreover, III–V HBTs are the main device for power amplifiers in various wireless devices of today. In this section, an overview of III–V HBTs are presented.

15.4.1 Device Operation

A typical structure and corresponding band diagram of III–V HBTs are shown in Fig. 15.10. Similar to SiGe HBTs described earlier, the base is composed of a narrow-bandgap material, sandwiched between emitter and collector layers with a wider bandgap. The narrow-bandgap cap layer on top of the emitter layer is for a low-loss emitter ohmic contact, a similar function served by the cap layer for HEMTs. The basic operation principles and the advantages expected from employing heterojunctions are similar to the SiGe HBT case discussed in Section 15.2, the highlights of which are recaptured as follows. Owing to the heterojunction between the emitter and the base, a very high level of current gain can be achieved. The sufficient current gain can be traded for increased base doping, which will allow aggressive base scaling, leading to reduced base transit time and thus increased f_T. The highly doped base doping will also lower R_B, resulting in improved f_{max} and F_{min}. There are differences as well, of course, which will be briefly discussed below in terms of material, DC performance, and RF performance.

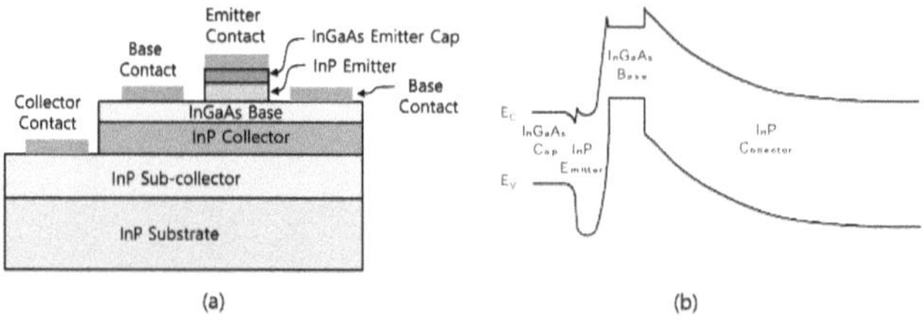

Figure 15.10 (a) Conceptual structure of a typical III–V HBT (not to scale). (b) Energy band diagram along the vertical direction beneath the emitter contact.

It is fair to say that there is only one material choice for SiGe HBTs: Si for wide bandgap and SiGe for narrow bandgap. On the contrary, there are almost unlimited number of options for material selection with III–V HBTs, because of a variety of binary, ternary, and quaternary alloys available with GaAs and InP systems. This flexibility greatly promotes the degree of freedom in device design, allowing detailed tailoring of bandgaps in various regions inside the devices. Widely employed for III–V HBTs are AlGaAs and InAlAs for wide bandgap, InGaAs and GaAsSb for narrow-bandgap materials. InP and GaAs, which are the two main standard substrate (and wafer) options, can be considered as mid-bandgap materials, while InP is often adopted as a wider bandgap material for emitter and/or collector in recent devices.

As for DC characteristics, one notable difference is the much lower current gain for III–V HBTs compared to SiGe HBTs, around a few tens versus a few hundred. Mainly responsible for the low current gain is the low-quality interface/surface inside III–V HBTs, which leads to a significant surface recombination current that becomes a part of the base current. In fact, the poor control of the interface and surface inside the device is one of the major drawbacks of not only III–V HBTs, but III–V devices in general. In contrast, in Si-based devices, high-quality oxides as well as the sophisticated fabrication processes effectively suppress unwanted leakage currents and also contribute to excellent device reliability. On the positive side, III–V HBTs typically exhibit much larger breakdown voltages. For instance, today's most advanced III–V HBT shows BV_{CEO} of around 3.5 V, compared to $\sim$1.5 V for most advanced SiGe HBTs. It is due to the wider bandgap employed for the collector in III–V HBTs.

RF performance of III–V HBTs is obviously superior to that of SiGe HBTs. It brings up one fundamental question: why do III–V devices outperform Si-based devices in general? This question applies not only to III–V HBTs vs. SiGe HBTs but also to III–V FETs vs. Si MOSFETs, thus requiring a more general discussion. Certainly, the higher electron mobility is one reason as mentioned earlier. However, as devices scale down, the importance of carrier mobility diminishes, since mobility is a parameter defined at the linear regime of velocity–electrical field relation, or the region of weak electric field. Think of a FET, for example, with a channel length of 100 nm, across which 1 V is applied. Although the field profile along the channel is not uniform, the average electric field across the channel would be around 100 kV/cm. This level of electric field is far beyond the linear region where the mobility is properly defined, and the electrons under such a high field would travel at the saturation velocity v_{sat}. As far as v_{sat} is concerned, however, the advantage of III–V materials vanishes, as the mainstream III–V semiconductors such as GaAs and InP exhibit v_{sat} similar to or slightly lower than that of Si ($\sim 0.7 \times 10^7$ cm/sec for GaAs and InP vs $\sim 1 \times 10^7$ cm/sec for Si). Then, what else can be responsible for the difference in the RF performance? In fact, it is mainly the velocity overshoot or the ballistic behavior of electrons. When a traveling electron (or hole) experiences an abrupt increase in the electric field in the forward direction, its velocity temporarily soars beyond v_{sat} until it is stabilized down to v_{sat} after a number of scatterings. The ballistic velocity exhibited by electrons in such a situation may reach up to 2–3 times of v_{sat} in III–V compound semiconductors, while it is not so much pronounced for Si, only about 50% increase over v_{sat}. This difference plays a great role in III–V devices exhibiting greater performance. For the case of HBTs, the more pronounced velocity overshoot at the B-C junction for electrons in III–V HBTs is one of the main reasons for its superior RF performance over SiGe HBTs.

15.4.2 Performance Trend

In the device depicted in Fig. 15.10, which is assumed to be based on the InP substrate, the emitter and collector layers are in InP while the base is in InGaAs, either lattice-matched or pseudomorphic. The devices based on this structure is called double heterojunction bipolar transistor (DHBT) as there are heterojunctions at both E-B and B-C junction. One issue confronted with such a structure is the spike in the conduction band near the B-C junction, as indicated in the energy band diagram depicted in Fig. 15.10b. This local energy barrier tends to block the electron injection from the base into the collector, deteriorating the high-speed performance of the device. The effect can be partially suppressed with the grading of the bandgap near the junction. Alternatively, one can introduce a different material for the base. For example, in the devices with a base formed with GaAsSb, the spike will be eliminated [19]. This is due to the unique band offset at the junction between GaAsSb and InP layers, in which the conduction band edge of GaAaSb will be located above that of InP despite the narrower bandgap of GaAaSb (called Type II band offset). Such a band alignment at the B-C junction will not only remove the spike, but will also help accelerate the electrons entering the collector as they will be 'launched' from a higher energy position. One downside of the GaAsSb base is the larger R_B, which originates from the lower hole mobility of GaAaSb than that of InGaAs.

Another approach to avoid the issues related to the parasitic energy barrier at the B-C junction is to have the collector region composed of the same material as the base, typically the InGaAs layer. In this case, there will be a continuum in the energy band across the base and collector boundary, allowing a smooth transition into the collector region. Devices based on such a structure is called single heterojunction bipolar transistor (SHBT) [20]. Another expected advantage of SHBT is the improved speed, as electron transport characteristics are superior in InGaAs compared to InP. An apparent drawback, however, is the reduced breakdown voltage, as expected from the narrow energy bandgap of the InGaAs collector. SHBTs were widely employed for high-frequency operation in the early 2000s, but the advanced HBTs of today are mostly DHBTs.

The RF performance of recently reported III–V HBTs are plotted in Fig. 15.11 in terms of f_T and f_{max} for three different types of devices: SHBT with InGaAs base/collector, DHBT with InGaAs base, and DHBT with GaAsSb base. As of today, the best performance achieved for III–V HBTs in terms of f_{max} is 1150 GHz, which was from an InGaAs DHBT developed by Teledyne [21]. A detailed look into the plot reveals that InGaAs DHBTs tend to show higher f_{max} than GaAsSb DHBTs, which may arise from the larger R_B of GaAsSb devices due to the low hole mobility as mentioned above. Another apparent trend in the plot is the relatively low performance of InGaAs SHBTs. For this, it should be pointed out that the data points for InGaAs SHBTs in the plot are rather old (all before 2005), suggesting it may not necessarily indicate the inherent limit of SHBTs, but possibly be affected by the process technology at that time.

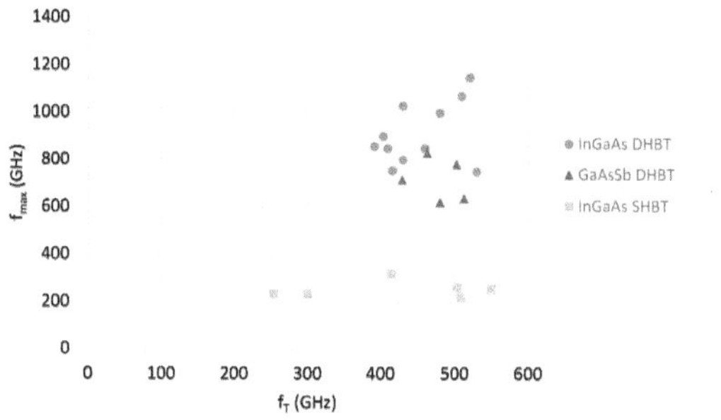

Figure 15.11 Recently reported RF performance of III–V HBTs.

As mentioned earlier with SiGe HBTs, the technology node for bipolar transistors is not as critical as the case for FETs, since they are vertical devices and performance is dictated by the vertical dimension rather than the horizontal dimension. For III–V HBTs, it is the control of the epitaxial layer thickness and doping profile along the depth that affects the performance. The technology node, which defines the emitter finger width, for most advanced III–V HBT technologies is around 130 nm [21], more relaxed than SiGe BiCMOS technologies that still require a decent level of lateral scaling as MOSFETs are offered together with HBTs. It would be worthwhile to mention that the devices in Fig. 15.11 are recently reported state-of-the-art III–V HBTs based on InP wafers, which are surely great candidates for THz applications. However, from the practical point of view, the majority of the III–V HBT production volume as of today is for power amplifier (PA) modules embedded in cellular phones, which need high power at a rather moderate frequency of 1–2 GHz. For such an application, the HBTs are typically based on GaAs wafers for cost-effective mass production. It is very likely that your phone transmits your voice and data through a PA module based on III–V HBTs.

References

1. T. H. Lee, CMOS RF: No longer an oxymoron, in *GaAs IC Symposium. IEEE Gallium Arsenide Integrated Circuit Symposium. 19th Annual Technical Digest 1997*, 1997: IEEE, pp. 244–247.

2. T. H. Lee, *The Design of CMOS Radio-Frequency Integrated Circuits*. Cambridge university press, 2003.
3. J. Bardeen and W. H. Brattain, The transistor, a semi-conductor triode, *Physical Review*, vol. 74, no. 2, P. 230, 1948.
4. J. E. Lilienfeld, Method and apparatus for controlling electric currents, US Patent 1,745,175, 1930.
5. D. Kahng and M. Atalla, Silicon-silicon dioxide field induced devices, in *J. Solid-State Device Research Conference, Pittsburg*, 1960.
6. A. F. Tong, W. M. Lim, C. B. Sia, K. S. Yeo, Z. L. Teng, and P. F. Ng, RFCMOS unit width optimization technique, *IEEE Transactions on Microwave Theory and Techniques*, vol. 55, no. 9, pp. 1844–1853, 2007.
7. S. Ong et al., A 22 nm FDSOI technology optimized for RF/mmWave applications, in *2018 IEEE Radio Frequency Integrated Circuits Symposium (RFIC)*, 2018: IEEE, pp. 72–75.
8. H.-J. Lee et al., Intel 22 nm FinFET (22FFL) process technology for RF and mm wave applications and circuit design optimization for FinFET technology, in *2018 IEEE International Electron Devices Meeting (IEDM)*, 2018: IEEE, pp. 14.1. 1–14.1. 4.
9. S. S. Iyer, G. L. Patton, S. S. Delage, S. Tiwari, and J. M. C. Stork, Silicon-Germanium base heterojunction bipolar transistors by molecular beam epitaxy, in *Techcinal Digest of International Electron Device Meeting*, 1987, pp. 874–876.
10. J.-S. Rieh, D. Greenberg, A. Stricker, and G. Freeman, Scaling of SiGe heterojunction bipolar transistors, *Proceedings of the IEEE*, vol. 93, no. 9, pp. 1522–1538, 2005.
11. P. Chevalier et al., A 55 nm triple gate oxide 9 metal layers SiGe BiCMOS technology featuring 320 GHz f T/370 GHz f MAX HBT and high-Q millimeter-wave passives, in *2014 IEEE International Electron Devices Meeting*, 2014: IEEE, pp. 3.9. 1–3.9. 3.
12. B. Heinemann et al., SiGe HBT with f_T/f_{max} of 505 GHz/720 GHz, in *2016 IEEE International Electron Devices Meeting (IEDM)*, 2016, pp. 3.1.1–3.1.4, doi: 10.1109/IEDM.2016.7838335.
13. T. Mimura, S. Hiyamizu, T. Fujii, and K. Nanbu, A new field-effect transistor with selectively doped GaAs/n-$Al_xGa_{1-x}As$ heterojunctions, *Japanese Journal of Applied Physics*, vol. 19, no. 5, p. L225, 1980.
14. D. Delagebeaudeuf, P. Delescluse, P. Etienne, M. Laviron, J. Chaplart, and N. T. Linh, Two-dimensional electron gas MESFET structure, *Electronics Letters*, vol. 16, no. 17, pp. 667–668, 1980.

15. A. Leuther, A. Tessmann, M. Dammann, H. Massler, M. Schlechtweg, and O. Ambacher, 35 nm mHEMT technology for THz and ultra low noise applications, in *2013 International Conference on Indium Phosphide and Related Materials (IPRM)*, 19–23 May 2013, pp. 1–2, doi: 10.1109/ICIPRM.2013.6562647.
16. X. Mei et al., First demonstration of amplification at 1 THz using 25 nm InP high electron mobility transistor process, *IEEE Electron Device Letters,* vol. 36, no. 4, pp. 327–329, 2015, doi: 10.1109/LED.2015.2407193.
17. W. Shockley, Circuit element utilizing semiconductive material, US Patent 2,569, 347, 1951.
18. H. Kroemer, Theory of a wide-gap emitter for transistors, *Proceedings of the IRE,* vol. 45, no. 11, pp. 1535–1537, 1957.
19. C. Bolognesi, M. Dvorak, P. Yeo, X. Xu, and S. Watkins, InP/GaAsSb/InP double HBTs: a new alternative for InP-based DHBTs, *IEEE Transactions on Electron Devices,* vol. 48, no. 11, pp. 2631–2639, 2001.
20. W. Hafez, J.-W. Lai, and M. Feng, InP/InGaAs SHBTs with 75 nm collector and $f_T > 500$ GHz, *Electronics Letters,* vol. 39, no. 20, pp. 1475–1476, 2003.
21. M. Urteaga, R. Pierson, P. Rowell, V. Jain, E. Lobisser, and M. J. W. Rodwell, 130 nm InP DHBTs with $f_t > 0.52$ THz and $f_{max} > 1.1$ THz, in *Device Research Conference*, 20–22 June 2011, pp. 281–282.

Part III

THz Vacuum Electronic Sources

Chapter 16

Development and Applications of THz Gyrotrons

Svilen Sabchevski,[a] Teruo Saito,[b] and Mikhail Glyavin[c]

[a] *Institute of Electronics of the Bulgarian Academy of Sciences, Sofia, Bulgaria*
[b] *University of Fukui, Fukui, Japan*
[c] *Institute of Applied Physics, Russian Academy of Science, Russia*
sabch@ie.bas.bg

16.1 Introduction

The remarkable history of gyrotrons begins in the distant 1964 [1] when the breakthrough ideas in relativistic electronics [2] and the visionary research of outstanding scientists in Gorky, USSR (now Nizhny Novgorod, Russia) [3], had led to the invention, construction, and experimental demonstration of the first gyrotron. Soon afterward, they gave rise to a new family of fast-wave vacuum tubes that includes amplifiers and oscillators such as Gyro-BWO, CARM (Cyclotron Autoresonance Maser), Gyro-TWT, Gyro-Klystron, etc. An excellent introduction to the physics of gyrotrons can be found in the monographs [4–6]. Their operation is based on the mechanism known as electron cyclotron maser instability which

Advances in Terahertz Source Technologies
Edited by Gun-Sik Park, Masahiko Tani, Jae-Sung Rieh, and Sang Yoon Park

ISBN 978-981-4968-89-8 (Hardcover), 978-1-003-45967-5 (eBook)
www.jennystanford.com

takes place when a synchronization condition for a resonance interaction between a hollow helical electron beam and a high-frequency field of a resonant cavity is fulfilled. The electrons of the beam gyrate in a strong magnetic field B with a cyclotron frequency

$$\Omega_c = \frac{eB}{\gamma m_0}, \tag{16.1}$$

e and m_0 being the charge and rest mass of an electron, γ is the relativistic Lorentz factor given by $\gamma = \left(1 - \frac{v_\perp^2 + v_z^2}{c^2}\right)^{-1/2}$, where $v_\perp$, v_z, are the transverse and axial velocities of an electron, respectively, and c is the speed of light in vacuum. Due to the relativistic dependence of the cyclotron frequency, an initially uniform electron beam undergoes azimuthal bunching, since during the beam-wave interaction the accelerated electrons (with increased γ) rotate slowly and vice versa, the decelerated (with decreased γ) rotate faster. At a proper beam-wave synchronization, given by the following equations, the formed bunches slip to the deaccelerating phase of the high-frequency electromagnetic field where they radiate (transfer their energy associated with the transverse motion) through bremsstrahlung:

$$\omega = s\Omega_c + v_z k_z, \tag{16.2}$$

$$\omega^2 = c^2 \left(k_\perp^2 + k_z^2\right), \tag{16.3}$$

where ω is the circular frequency of the electromagnetic wave, s is the harmonic number of the cyclotron resonance, k_z and $k_\perp = \frac{\upsilon_{mp}}{R_{cav}}$ are the axial and the transverse wavenumber, respectively. Here R_{cav} is the cavity radius and ν_{mp} is the eigenvalue of the operating TE_{mp} mode (p-th zero of the equation $J'_m(\nu) = 0$, $J_m(x)$ being a Bessel function of the first kind of order m). Therefore, the gyrotron operates at the intersection point of the beam line (16.2) and the dispersion curve (16.3) of the working mode. The gyrotron operation is characterized by a small Doppler shift (second term of (16.2)) and thus the frequency of radiation is close to the cyclotron frequency or its harmonics ($\omega \approx s\Omega_c$). As already mentioned, since only the energy of the electrons associated with their transverse motion is available for the beam-wave interaction, the velocity ratio (pitch factor) $g = v_\perp/v_z$ must be greater than unity (typically 1.2–1.5). Such beams are being formed initially by a magnetron injection gun (MIG) and then in the rest of the electron-optical system (EOS) are "pumped" increasing the pitch factor in an adiabatically increasing magnetic field that reaches its maximum in the flat top region inside the cavity resonator.

Since in the gyrotrons the electrons interact with fast waves (i.e., such as having a phase velocity greater than the speed of light) in oversized cavities (with characteristic dimensions much larger than the wavelength of the radiation λ), they, in contrast to the classical slow-wave devices, do not need tiny structures (with sizes of the order of λ) situated close to the electron beam) that can easily be destroyed. On the contrary, in the gyrotron cavities much more powerful electron beam can be used and thus higher output powers can be achieved.

The structure and main components of a gyrotron ate shown schematically in Fig. 16.1.

Each gyrotron has an EOS based on either diode or triode type MIG which forms a hollow helical electron beam with the required

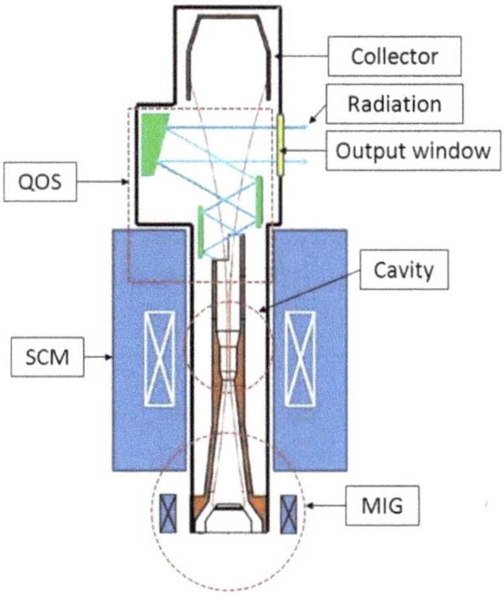

Figure 16.1 General structure and main components of a gyrotron: MIG (Magnetron Injection Gun) with additional coils and beam tunnel; SCM (Superconducting magnet); QOS (Quasi-optical system) of an internal mode converter that includes a wave beam launcher and a system of mirrors; collector of the spent electron beam; and an output window.

parameters (current, voltage, pitch factor, appropriate injection radius in the cavity, etc.) and high-quality (small velocity spread, uniformity, absence of reflected electrons and losses of current on the electrodes and beam tunnel). Besides the superconducting magnet (SCM) the EOS includes additional coils at the gun region for fine-tuning of the initial beam formation. The excitation of the desired operating mode and generation of radiation takes place in the electrodynamical system which is a simple resonant structure – an open gyrotron cavity (usually consisting of a regular cylindrical waveguide section delimited at both ends with up- and down-tapers). Since the generated wave beam has a complex polarization that is not suitable for low-loss transmission and for many practical applications, nowadays most of the gyrotrons have an internal mode converter that forms a well-collimated Gaussian-like beam. The used quasi-optical systems (QOS) utilize either Vlasov- or Denisov-type launchers (antennas) and a set of mirrors. The radiation is coupled to the outside world by an output window made of material with appropriate dielectric properties (sufficient transparency, mechanical strength, and thermal resistance). Among them are the most expensive diamond output windows used in the megawatt levels gyrotrons for fusion. One of the bulkier parts of the gyrotron (alongside the SCM) is the collector, where the energy of the spent electron beam has to be dissipated. Very often this requires the utilization of intensive water cooling and additional magnetic coils for sweeping of the beam in order to decrease the thermal loading of the device. In the most advanced designs, various depressed collectors (e.g., single-stage, multi-stage, sectioned) are used for the recuperation of the electron beam energy.

Nowadays, the gyrotrons are among the most powerful sources of coherent radiation in a broad frequency range that has been extended to the THz frequency. The symbolic threshold of 1 THz has been crossed in pioneering works. Over 1 THz second harmonic oscillation has been demonstrated for the first time with a 20 T pulsed magnet in FIR UF [7]. The pulsed gyrotron developed at IAP-RAS has demonstrated an operation at 1.3 THz (with output power of 1 kW in pulses of 20 μs and energy of 30 mJ) exciting the $TE_{24,4}$ mode (fundamental operation) at magnetic field intensity of 51.5 T [8]. Almost at the same time, at FIR UF, the gyrotron FU CW III with a

20 T superconducting magnet has achieved a continuous wave (CW) second harmonic operation on the mode $TE_{4,12}$ with a frequency of 1.08 THz [9].

The current status of the gyrotron development is well presented in the annually updated report by Professor Manfred Thumm [10]. In recent years, the remarkable progress in this field has opened a broad avenue to many applications in wide scientific and technological fields that are overviewed in many recent review papers (see, for example [11–19]). In this book chapter, we focus on the new trends in both the development and application of THz gyrotrons.

16.2 Development of THz Gyrotrons

Great efforts for higher frequency oscillation have been made in the eighties and nineties of the last century [20, 21]. The highest frequency oscillation realized in this period was 889 GHz. Then, the symbolic threshold of 1 THz has been crossed at two institutions as already mentioned above. Following the success in over 1 THz oscillation, many CW gyrotrons in the sub-THz range were developed.

Gyrotrons, in principle, oscillate at discrete frequencies defined by Eq. (16.3), where k_z is nearly equal to zero. Moreover, high-order waveguide modes are used for high-frequency oscillations, which leads to a very complicated spatial structure of the wave radiated from a traditional gyrotron. For application to various studies, multi-frequency oscillation and/or continuous frequency tunability are highly desirable. Gaussian beam radiation makes gyrotrons very valuable THz radiation sources. Continuous frequency tunability was demonstrated by an experiment carried out in FIR UF by utilizing the finite k_z effect in Eq. (16.3) (Gyro-backward-wave oscillation, Gyro-BWO) [22]. A wide-band frequency variation of 6 from 134 to 140 GHz was obtained. Later experiments also verified the effectiveness of the concept of Gyro-BWO in the sub-THz frequency range up to 460 GHz at the fundamental harmonic and the second harmonic oscillations [23, 24]. The 400 GHz range gyrotron developed in FIR UF was frequency tunable by about 4 GHz and delivered more than 50 W over the tunability band. It was installed in a 15 T SC magnet and was operated at the fundamental harmonic resonance [23]. While second harmonic oscillation delivers rather moderate power of the order of several watts to tens of watts, it works at the reduced magnetic field strength (half of that for the fundamental harmonic oscillation). Furthermore, efforts for continuous frequency tunability in the higher frequency range are currently in progress [25]. High-frequency stability is also, on the other hand, very important for some applications. Gyrotrons (263, 395, and 527 GHz) with very high-frequency stability developed in CPI have been utilized as radiation sources of DNP-NMR devices [26].

Gaussian beam radiation was first realized in fusion-oriented gyrotrons by using a built-in mode converter [see, e.g., 27]. Non-fusion gyrotrons equipped with an internal mode converter have been developed for various applications. Some examples are cited below. A 300 GHz gyrotron FU CWI developed in IAP-RAS [28] was very carefully tested in FIR UF with a 12 T SC magnet [29]. A maximum CW power of 2.3 kW was eventually attained. This gyrotron was used for ceramic sintering (see Section 16.4). Gyrotrons developed in MIT and CPI also deliver a Gaussian beam with high TEM00 mode purity [13, 26]. FU CW G series gyrotrons have been developed in FIR UF also [16, 17]. Among FU CW G series gyrotrons, FU CW GV is a very versatile gyrotron [30]. It delivers Gaussian beams at ten frequencies from 162 to 265 GHz with an interval of about 10 GHz. It is installed in a 10 T SC magnet and works at the fundamental harmonic resonances. A double disk window of two single crystal Sapphire disks optimizes the transmission efficiency for different frequencies. CW powers higher than 1 kW are radiated at each oscillation frequency. The installed internal mode converter was designed to work for different modes in the wide frequency range [31]. Oscillation modes were also carefully selected to meet the design condition of the mode converter. See Section 16.4 for applications of this gyrotron. FU CW GVI and GVIA radiating 460 GHz beam at the second harmonic resonance have been used for 700 MHz DNP-NMR experiments at Osaka University [32]. Another multi-frequency second harmonic gyrotron FU CW GVII was developed. This gyrotron radiates Gaussian beams at nine frequencies from 297 to 420 GHz [33]. It is also equipped with a double disk window of the same type as that of FU CW GV. Because of the second harmonic resonances, radiation power ranges from 0.2 W to tens of watts.

Usually, heavy and bulky high-field magnets are used for designing high-frequency gyrotrons. An alternative is to utilize compact cryo-free SC magnets that are already available. Following such a concept, two compact gyrotrons were developed with an 8 T compact SC magnet. Its height, outer diameter, and inner room-temperature bore diameter are 338, 350, and 52 mm, respectively. One gyrotron, FU CW CI, was developed for use in 600 MHz DNP-NMR spectroscopy [34]. The height of the gyrotron was about 1 m. Its oscillation frequency was about 395 GHz at the second harmonic resonance. The output power was around 10 W. The other gyrotron, FU CW CII was a fundamental harmonic gyrotron equipped with an internal mode converter [35]. It oscillated at 203 GHz with a power level of about 0.8 kW. The oscillation frequency was the same as that necessary in the positronium experiment described in Section 16.4.

High-power gyrotrons in the sub-THz frequency band have been developed for application to collective Thomson scattering diagnostics of high-temperature fusion plasmas. A second harmonic gyrotron attained 83 kW at 389 GHz [36]. The maximum power of this gyrotron was limited due to mode competition with the fundamental harmonic modes [37]. A much higher power of 320 kW at 303 GHz was achieved with fundamental harmonic oscillation [38]. Almost the same power level of 330 kW was also realized at 250 GHz [39].

The performance of any THz gyrotron strongly depends on the quality of the annular electron beam. A magnetron injection gun was designed to generate high-quality electron beams [40]. This electron gun generates an electron beam with a very low velocity spread.

16.3 THz Gyrotrons: New Concepts, Challenges, and Trends in Their Development

Although the THz gyrotrons are structurally identical to their lower-frequency counterparts (e.g., millimeter-wave tubes), they are characterized by some distinguishing features and specific problems that are encountered during their development and operation.

Due to the limited maximum field intensity of the currently available superconducting magnets, the THz gyrotrons have to operate inevitably at higher harmonics of the cyclotron frequency. This causes several severe complications. First of all, in this frequency range, the mode spectrum is much denser, making both the mode selection and preventing the competition of the desired operating mode with the parasitic ones much more difficult. Moreover, at higher harmonics, the efficiency of the operation is significantly lower, and correspondingly, the output power also is decreased. Besides, the zone of stable oscillation is narrower which makes the range of frequency tunability constricted as well. Apart from the mode competition other types of mode interactions (including nonlinear) such as, for example, automodulation, mode switching, mode cooperation, etc., could take place and complicate the operation of the tube.

The well-known classical approaches, namely electronic and electrodynamic mode selection used to alleviate the problems related to the mode competition in the THz gyrotrons are often insufficient, and more sophisticated solutions have to be implemented. One such advanced concept is the multi-beam (usually twin-beam aka double-beam) system that provides additional means for mode selection. There are two varieties of *double-beam (DB) gyrotrons* [41–45], particularly with two generating beams as well as a combination one generating and one absorbing (nonoscillating) electron beam (Fig. 16.2). In such devices, an improved mode selection is being achieved by adjusting properly the injection radii R_1 and R_2 of both electron beams inside the resonant cavity (see Fig. 16.3) with respect to the maxima of the coupling factors of the operating $G_{op}(R_1)$, $G_{op}(R_2)$ and the competing (parasitic) $G_{par}(R_1)$, $G_{par}(R_2)$ modes given by

$$G(R) = J^2_{m\mp s}\left(\frac{\nu_{m,p}R}{R_{cav}}\right)/(\nu^2_{m,p} - m^2)J^2_m(\nu_{m,p}), \qquad (16.4)$$

The indices $m-s$ and $m+s$ correspond to both possible polarizations of the wave, i.e., to co-rotating and counter-rotating (with respect to the gyrating electrons) modes, respectively. In the DB gyrotron

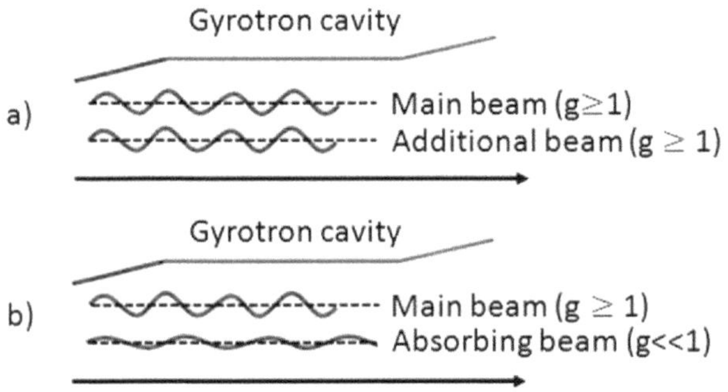

Figure 16.2 Schematic of a double-beam gyrotron with two generating electron beams (a), and one generating and one absorbing beam (b).

with two generating electron beams, their pitch factors are almost equal and $g_1 \approx g_2 > 1$, while in the alternative type, the generating beam has $g_1 > 1$, whereas the absorbing one has $g_2 \ll 1$. The second (absorbing) beam is used as an absorber, which suppresses the excitation of the neighboring parasitic mode. As illustrated in Fig. 16.3, in the former case the injection radii correspond to the maxima of $G(R_1)$ and $G(R_2)$ of the operating mode while for the parasitic mode these values are close to the corresponding minima. In the latter scheme, the absorbing beam (with $g \gg 1$) is targeting the maximum of the coupling factor of the competing mode and the generating beam aims for the maximum of the operating mode.

The concept of a double-beam gyrotron (DBG) with two generating beams has been realized recently in a novel gyrotron operating at a frequency of 0.8 THz operating at the second harmonic of the cyclotron frequency (mode $TE_{8,5}$) [46–48]. Its general view is shown in Fig. 16.4. It has been conceived as a radiation source for the envisaged next-generation ultrahigh field 1.2 GHz DNP-NMR spectrometers utilizing magnets with a field intensity of about 28.2 T. The tube has been built using a cryogen-free superconducting magnet (JASTEC, Ltd.) with a maximum field intensity of 15 T. The EOS of the DBG is based on a triode MIG with two emitting rings on the thermionic cathode (consisting of two conical sections) and forms two helical electron beams with equal pitch factors ($g \approx 1.2 - 1.3$) and small velocity spread. The nominal total current of 1.0 A is distributed between the main and the additional beams in a

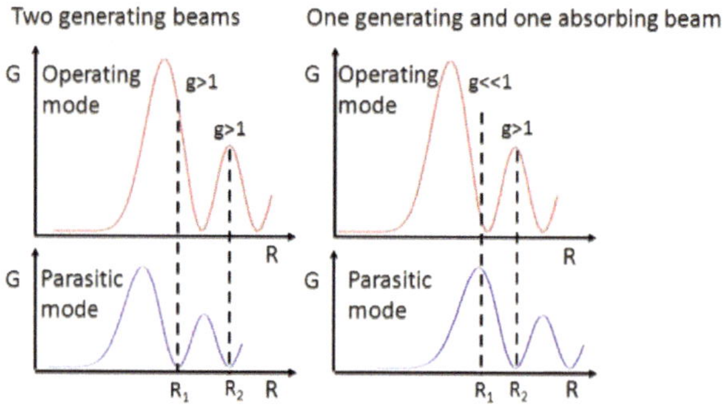

Figure 16.3 Selection of the injection radii of the main electron beam (R_2) and the additional beam (R_1) with respect to the maxima of the coupling factors of the operating and parasitic modes in a DBG.

Figure 16.4 General view of the 0.8 THz double-beam gyrotron (photo taken at FIR UF).

proportion 2:1. Besides the design mode ($TE_{8,5}$), a series of other fundamental and second harmonic modes have been excited in a broad frequency range.

Another proven concept that provides prominent mode selectivity at high-harmonic operation ($s \gg 1$) is the so-called *Large Orbit Gyrotron (LOG)* [49–53] which, in contrast to the conventional

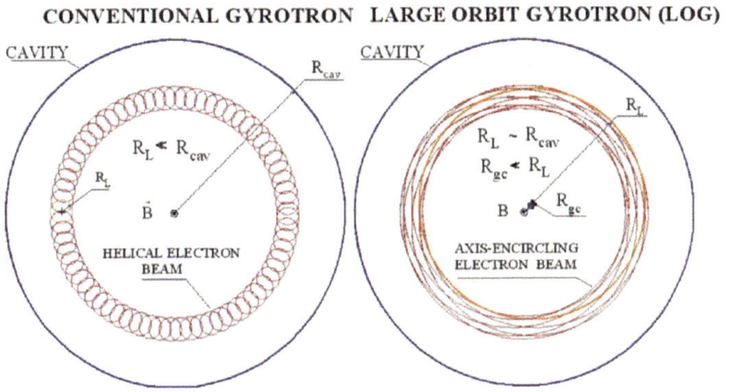

Figure 16.5 Electron orbits in helical beam of a conventional gyrotron and in an axis-encircling (uniaxial) beam of LOG (R_L, Larmor radius; R_{gc}, radius of the guiding center).

gyrotrons, utilizes an axis-encircling (uniaxial) electron beam as illustrated in Fig. 16.5. Such beam can excite only the co-rotating modes with the azimuthal index equal to the number of resonant harmonic (the case $m - s = 0$ and $R = 0$ in Eq. (16.4)).

The formation of high-quality uniaxial beams, however, is much more difficult than the generation of helical electron beams by MIG. The EOSs used for this are of two types. The first one utilizes an abrupt (cusp) or appropriately tailored magnetic field reversal while the second one uses kickers The magnetic system of the kicker forms two regions with magnetic fields directed perpendicularly to the axis of symmetry and separated by a distance equal to half cyclotron wave. The first magnetic field abruptly "kicks" the thin electron beam with rectilinear trajectories (produced by a quasi-Pierce type gun) apart from the axis and the electrons start to follow circular orbits with centers shifted with respect to the symmetry axis. The second transverse magnetic field imparts an azimuthal momentum to the electrons and, in such a way, hollow axis-encircling electron beam rotating around the symmetry axis with cyclotron frequency is formed. In both varieties, the main challenge is to minimize the beam ripple (inflicted by the spread of the guiding centers) since it decreases significantly the efficiency of interaction.

A good example of LOG operating in the THz frequency range is presented in [53]. It has an EOS with a cusp electron gun that forms an 80 keV/0.7 A beam of gyrating electrons in a wide range of voltages and magnetic fields. This gyrotron has demonstrated stable single-mode generation with a power of 0.3–1.8 kW in microsecond pulses at four frequencies in the range 0.55–1.00 THz at resonant magnetic fields 10.5–14 T.

New versions of LOG developed at IAP-RAS on the basis of two (80 keV pulsed and 30 keV CW) experimental setups are overviewed in [54, 55], where selective operation at the second (0.267 THz) and at the third (0.394 THz) cyclotron harmonics has been reported. In order to improve the operation at the 3rd harmonic as well as to achieve fourth harmonic operation at frequencies up to 0.65 THz special cavities have been developed. They have periodic phase correctors, where a far-from-cutoff axial mode with a decreased diffraction Q-factor is excited in a gyrotron-like regime.

A design of a 1 THz 4th harmonic LOG operating in the CW regime has been proposed in [56]. It includes a 10T. superconducting magnet, and is driven by an 80 kV, 0.7 A electron beam. As operating one, the mode $TE_{4,8}$ has been selected. The output power evaluated by numerical simulations is 1.15 kW can be achieved.

A promising variety of LOG is the one that uses permanent magnets (PM) [32, 52, 57]. Their development is stimulated by the progress in the permanent magnet technology and is motivated by such advantages as compactness and instantaneous startup of the operation in a CW regime. For example, the LOG with a 1 T PM has demonstrated successful operation at the third, fourth, and fifth harmonics with frequencies of 89.3, 112.7, and 138 GHz, respectively, exciting the modes TE_{311}, TE_{411}, and TE_{511} [52]. The PM system consists of many magnet elements made of NbFeB and additional coils for controlling the field intensities in the electron gun and cavity regions. The output powers at the third and the fourth harmonics are 1.7 and 0.5 kW, respectively. The gyrotron oscillates in a pulsed regime (pulse length of 1 ms and repetition frequency of 1 Hz). The nominal electron beam parameters are energy of 40 kV and current of 1.2–1.3 A.

An effective means in mode selection is provided by the concept of *coaxial-cavity gyrotron* [58–60]. It allows one to rarefy the eigenfrequency spectrum as well as to lower the total quality factor of the competing modes. There are different varieties of coaxial gyrotrons. For example, in a cavity with a tapered insert (towards the collector end of the tube), a radial mode selection can be implemented. The presence of such an insert lowers the diffractive *Q*-factor of the modes with large radial index and in such a way suppresses their excitation. The azimuthal mode selection can be realized in cavities with perturbed azimuthal symmetry. For instance, in the split cavity gyrotrons, the two halves can be considered as two mirrors similar to that in the quasi-optical gyrotron. In such a configuration, the modes with one azimuthal variation at the reflector surface have the highest diffraction *Q*. Furthermore, the profiling of the insert together with an appropriate corrugation of the cavity wall allows improving further the mode selectivity in both the radial and azimuthal indices. The presence of an insert eliminates the problem of beam voltage depression (due to the space charge) and thus allows using volume operating modes that have very low Ohmic losses.

In recent years, an advanced coaxial gyrotron for the International Thermonuclear Experimental Reactor (ITER) is being developed by the European Gyrotron Consortium (EGYC). The initial experiments obtained with the 170 GHz 2 MW short-pulse coaxial-cavity pre-prototype at a pulse length of a few milliseconds (ms) have shown the potential of the coaxial-cavity concept in the multi-MW operation regime [61]. Besides EGYC, several other organizations are pursuing projects on the development and investigation of coaxial-cavity gyrotrons (see, for example [62, 63]).

Another promising variety of the so-called "non-canonical gyrotrons" is the *gyrotron with echelette-type resonator* [64, 65]. In them, the minimum value of the diffraction *Q*-factor is not limited so rigidly by their length as in the conventional resonators. Additionally, the power flux density to the wall here can be reduced several folds. As shown in [40], it is possible to realize smooth frequency tuning within the band of about 1% in the gyrotrons with two-mirror echelette resonators operating on the first to third harmonics of the gyrofrequency. Despite the advantages of such resonators their development still encounters serious technological problems. Even more severe hurdle stem from the fact that the output radiation has a complex structure that requires new variants of mode converters.

We end up this section by only mentioning several other advanced concepts that are considered very promising for the further development of high-performance THz gyrotrons. Among them is the CW LOG with sectioned Klystron-Type cavity and cavities with phase correctors [66, 67]. In such resonant structures, an extended length of the electron-wave interaction region can be combined with a relatively low diffraction Q-factor of the system. In the first experiment on a gyrotron with a sectioned cavity, selective excitation of higher (second and third) cyclotron harmonics has been observed in the terahertz frequency range (0.55 and 0.74 THz). It is reported that the Ohmic losses in such cavities are reduced to 20–25% of the generated microwave power which is significantly lower in comparison to the losses of about 85% in the conventional resonator. It has been shown that the utilization of quasi-regular cavities, i.e., resonators with short irregularities, which provide correction of the wave phase, in low-power gyrotrons operated at higher cyclotron harmonics is beneficial for solving two problems, namely, increasing the selectivity of excitation of a higher cyclotron harmonic and decreasing the diffraction Q-factor of the gyrotron wave excited in an extended cavity [68].

Another promising, attractive, albeit challenging, novel concept is that of a *planar gyrotron* that uses a sheet (ribbon) helical electron beam that interacts with the electromagnetic field in a planar waveguide [69, 70]. The proposed scheme is with a transverse diffractive output of the radiation. An advantage of this scheme in comparison with conventional cylindrical geometry is the possibility to ensure effective mode selection over the open transverse coordinate in combination with radiation outcoupling that leads to a substantial reduction of Ohmic losses. Mode selection along the axis collinear with the radiation direction is due to the difference of the diffractive losses of the modes with a different number of variations of the field, while the mode selection along the normal of the waveguide is accomplished by adjusting the separation between the plates. Additionally, in the direction collinear with the electron beam an electronic mode selection takes place. In a recent study [71], a method for the production of high-power, multi-THz radiation (1–3 THz) based on frequency multiplication in a planar gyrotron has been proposed. It provides a much higher efficiency of nonlinear conversion when compared with conventional cylindrical gyrotron schemes. In the numerical studies of a 1 THz planar gyrotron, it has been demonstrated that it is possible to generate kilowatt-level radiation at the second cyclotron harmonic and power of several watts at the third one.

There is also active research on the *gyromultiplyers* in various configurations [72–74]. They fall into two varieties, namely single-cavity and multiple cavity arrangements. Here we mention only a few of the most recent studies. For example, in [47] a new scheme of a two-resonator self-exciting frequency gyromultiplier based on the use of two coaxial hollow electron beams has been proposed and theoretically investigated. The effect of frequency multiplication in a fundamental harmonic 0.263 THz kW-level gyrotron, in which a certain fraction of its radiation was observed experimentally at the doubled operating frequency, has been analyzed in [73]. The maximum power ratio of the second harmonic to the fundamental one is about 10^{-4} (an absolute power of 10–15 mW at a frequency of 0.526 THz).

A scheme of fourth-harmonic multiplying gyro-TWAs has been studied numerically in [75]. The simulations suggest the feasibility of fourth-harmonic TE_{41}-mode multiplying gyro-TWAs, where the drive stages operate in the fundamental harmonic TE_{11} mode. It is expected that this gyro-TWA can yield a peak output power of 2.7 kW at 400.6 GHz with a saturated gain of 75 dB and a bandwidth of 0.7 GHz for a 75 kV/2 A electron beam.

The second harmonic generation of sub-THz gyrotron radiation by frequency doubling in InP:Fe and its application for magnetospectroscopy have been studied in [76]. This approach is completely different (*semiconductor-based frequency multiplication*) from the above-mentioned schemes and is a promising route to obtain THz radiation by doubling the frequency of intense gyrotron radiation. It has been estimated that at least 1% conversion efficiency is achievable in InP wafers, corresponding to the second harmonic power of 60–100 W/cm^2 in the 0.5–1.2 THz frequency range. This is an order of magnitude higher than the intensity of the second harmonic available from the gyrotron without frequency multiplication.

Mode selectivity strongly affects the success in the operation of frequency-tunable THz gyrotrons in which many modes are distributed very densely. In particular, this is the case in a continuously frequency-tunable second harmonic gyrotron, because mode competition with nearby fundamental harmonic modes is a very serious problem for its stable operation. It has recently been shown that a complex cavity resonator is promising for improvement of the mode selectivity [77]. The complex cavity resonator consists of two longitudinally connected cavities with different radii R_1 and R_2 (they should not be confused with R_1 and R_2 in Fig. 16.3). The cavity of the radius R_1 ($R_1 < R_2$) is located at the electron gun side. When the following condition holds,

$$\frac{\nu_{m,\,p}}{R_1} = \frac{\nu_{m,\,p+n}}{R_2} \tag{16.5}$$

the coupling coefficients for the electron beam (see Eq. (16.4)) in both cavities can be set simultaneously at the maximum. Then, two modes, $\mathrm{TE}_{m,\,p}$ and $\mathrm{TE}_{m,\,p+n}$, oscillate with the same frequency. It is expected that two modes strengthen each other and overcome competing modes. Experiments that follow this concept are now under way in FIR UF.

The two cavities work independently except for one particular combination of m, p, and n that satisfies Eq. (16.5). By changing the magnetic field strength in the cavity and adjusting the radius of the annular electron beam, several modes consecutively oscillate so that the following condition is satisfied.

$$c\frac{\nu_{k,l}}{R_{cav}} \approx s\Omega_c. \tag{16.6}$$

Since a complex cavity resonator has two cavities with different radii, more modes have a chance to be excited compared to a single-cavity resonator. Recently, an experiment verified this concept [78]. Twenty-two modes altogether were identified in the two cavities with frequencies ranging from 120 to 220 GHz at fundamental harmonic resonances. Some modes are continuously frequency tunable and 27% of the frequency band from 120 to 220 GHz is covered. Efforts for super multi-frequency oscillations in a higher frequency band at second harmonic resonances are now under way.

16.4 Some of the Most Prominent Applications of THz Gyrotrons

The gyrotrons have numerous applications in diverse scientific and technological fields and their list is growing continuously. Some of them have been overviewed in several recent papers [10–19, 79–82]. They are devoted mainly to the gyrotrons for fusion research [10, 79–81], materials processing [19], sensing and imaging [17], advanced spectroscopic studies [12–16], etc. For completeness, we mention them only briefly (due to the limited space here) and refer the reader for more details to the most recent comprehensive review [82] and the references therein. Strictly speaking, most of these applications (e.g., thermal treatment) are based on millimeter-wave gyrotrons rather than on their sub-millimeter and THz counterparts. This is one more reason to allocate less space to the former and more to the latter group of radiation sources.

16.4.1 Controlled Thermonuclear Fusion

Nowadays, the most powerful (megawatt level) CW gyrotrons are considered indispensable sources of strong microwaves for additional heating of magnetically confined plasma in various reactors for controlled thermonuclear fusion (e.g., tokamaks and stellarators). Besides the electron cyclotron resonance heating (ECRH) and current drive (ECRCD) they are used also for the initial ignition (startup), ramp down, plasma control, and diagnostics. Compared with other (less powerful) tubes their design is characterized by the most sophisticated implementation of three of their key components, namely an efficient internal mode converter, depressed collector, and diamond output window. The development of these gyrotrons is being carried out in coordination with other parts (for example low-loss transmission lines, matching optics units (MOU), launchers) of the electron cyclotron heating (ECH) system. Many innovative design solutions that have initially been demonstrated by the fusion gyrotrons nowadays are proliferating to other similar tubes. The current state-of-the-art in this field is well represented in [10, 79–81].

Although the gyrotrons are used for ECH in many fusion facilities around the world, currently the biggest efforts are focused on the development of gyrotrons for the ITER project and its sequel DEMO. ITER needs 24 gyrotrons operating at a frequency of 170 GHz with an output power of 1 MW each in the CW regime with a duration of 1000 s. This demanding task is being pursued by several parties involved in ITER. Among them are the EGYC, Japan (QST), and the Russian Federation (GYCOM). Their gyrotrons have already passed successfully the Factory Acceptance Tests demonstrating full compliance with the stringent requirements and high standards of ITER. The same organizations are working on a new generation of tubes operating at higher frequencies (up to 250 GHz) for DEMO [61, 81, 83].

16.4.2 Materials Treatment

The microwave thermal treatment of various materials is a matured industrial-grade technology that has many advantages compared with conventional heating. It allows achieving thermal cycles and heating rates that are not possible when using other methods. Due to the volumetric nature of microwave energy absorption, the temperature inside the heated objects is higher than that on the surface. Besides this specific feature (reverse temperature gradient that is not attainable by other means), microwave heating is characterized by higher efficiency, reduced processing time, and precise control of the overall process. In industrial microwave technologies (e.g., ceramic sintering, melting, drying, joining, etc.), more often as radiation sources are used magnetrons operating at a frequency of 0.9 and 2.45 GHz. The gyrotrons, however, operate at significantly higher frequencies and, respectively, at shorter wavelengths which provide additional benefits such as more uniform heating.

Both the advantages and the current status of the development of gyrotron-based technological systems have been discussed in the recent review [19]. That is why here we confine ourselves only to some of the most representative examples. For instance, the equipment developed at IAP-RAS and manufactured by GYCOM utilizes gyrotrons generating at 24 GHz (3–5 kW CW) [84], 30 GHz

(10–15 kW CW) [85], 83 GHz (15 kW CW) [85], 24–84 GHz [86], and 300 GHz (3 kW CW) [28, 29]. A great number of different materials have been treated using these technological systems. They include B4C (boron carbide) which is used for the manufacture of control rods of nuclear power plant reactors [87]; zirconia (ZrO_2) ceramic with outstanding mechanical properties; sintering of different ceramics based on silica xerogel (SX) derived from sago waste ash; mullite; just to mention a few. More importantly, not only the number but also the type of processed materials is increasing rapidly. For example, a system for the production of nanopowders has been developed recently at IAP-RAS. As a radiation source, it uses a sub-THz gyrotron with an output frequency of 263 GHz and an output power of 1 kW [88]. Since the gyrotron radiation can be delivered easily (using simple quasi-optical transmission systems and reflectors) into various systems for materials treatment, different hybrid technological schemes are possible. One such example is the millimeter-wave reactor for microwave plasma-assisted chemical vapor deposition (MPACVD) for the enhanced growth of diamond films [89].

Advanced commercially available gyrotron-based technologies for the thermal treatment of various materials include also ultra-rapid polymer curing, semiconductor annealing, ceramic-to-ceramic, and metal-to-ceramic joining, deposition of coatings, drying, and food decontamination [90].

16.4.3 Advanced Spectroscopic Techniques

16.4.3.1 DNP-NMR spectroscopy

The classical spectroscopy based on nuclear magnetic resonance (NMR) is a powerful method for studying the physical, chemical, and biological properties of matter but is characterized by a low signal-to-noise ratio (S/N) and long spectrum acquisition time. This problem has been resolved radically by a method pioneered at MIT [91]. The novel technique that increases significantly S/N and the overall sensitivity is called DNP-NMR spectroscopy since it involves a signal enhancement through Dynamic Nuclear Polarization (DNP). DNP-NMR requires doping the sample with an appropriate polarizing agent the electron spins of which can be polarized to a large degree. Then, their polarization is transferred to the neighboring nuclear spins of the sample, irradiating it by gyrotron radiation with a frequency that is equal to or close to the electron spins resonance ESR. Since the signal enhancement is proportional to $\gamma_e/\gamma_p \approx 657$, where γ_e and γ_p are the electron and the proton gyro-magnetic ratios, theoretically S/N can be increased more than two orders of magnitude. In practice, however, the enhancement depends on many factors and most notably on the quality of the used instrumentation in which the operational performance of the gyrotron is of paramount importance. Some of the challenging requirements with regard to the gyrotrons for DNP-NMR are: (i) output power of 15–20 W at the output window and at least several watts at the irradiated sample; (ii) stable output power (fluctuations less than $\pm$ 0.5%); (iii) frequency stability $\leq$ 2 MHz; (iv) frequency tunability in a wide frequency range ($\geq$1 GHz); (v) high mode purity (>90%), and additionally, (vi) all these output parameters and spectral characteristics must be maintained in a CW mode of operation during long spectrum acquisition times (> 24 h). The spectacular progress in the development of magnets for spectroscopy reflects in the fact that in each new generation of such instrumentation, the field intensity increases steadily. Therefore, the high-field DNP-NMR requires, respectively, gyrotrons operating at elevated frequencies. For modern 400–1200 MHz spectrometers, they are in the range 260–800 GHz.

Some of the gyrotrons developed specially for high-field DNP-NMR spectroscopy are presented in Table 16.1 (data from [13–16, 26, 92]).

16.4.3.2 ESR spectroscopy

The spectroscopy based on the electron spin resonance (ESR), also known as electron paramagnetic resonance (EPR), is a versatile analytical method that is used for studies on the electron state of magnetic materials and semiconductors, investigation of the structure of glasses, and other amorphous materials, tracking of catalytic reactions, etc. It is being applied also for the analysis of radicals of polymerization processes and active oxygen radicals

Table 16.1 Gyrotrons for DNP-NMR

Institution/Gyrotron	Frequency, GHz	Spectrometer
MIT	140	210 MHz (the 1st in the world)
MIT	250	380 MHz
CPI	263	400 MHz
MIT	330	500 MHz
CPI	395	600 MHz
MIT	460	700 MHz
MIT, CPI	527	800 MHz
FIR FU/ FU CW IV	131	200 MHz at FIR UF
FIR FU/ FU CW VII	187	300 MHz at Warwick Univ.
FIR FU/ FU CW II, FU CW VI	394	600 MHz at Osaka Univ.
FIR FU/ FU CW GVI FU CW GVIA	460	700 MHz
IAP-RAS	260	400 MHz at Goethe University, Germany
EPFL, Switzerland	265–530	400–800 MHz

related to aging and diseases of living organisms as well as oxidative degeneration of lipids in foods, petroleum products, and so on. Moreover, the ESR instruments provide the only means of selectively measuring free radicals non-destructively and in any sample phase (gas, liquid or solid). Analogously to DNP-NMR, this technique also benefits from gyrotrons as radiation sources that provide higher frequencies than the other microwave sources and allows high-field ESR to be realized. The latter is characterized by increased sensitivity, spectral resolution, and precision.

An important milestone in the realization of the high-field ESR spectroscopy was the development of a system that uses a magnet with a field intensity of up to 40 T (in pulses of several milliseconds and the gyrotron FU IVA as a radiation source) [93]. An alternative approach for increasing the sensitivity and reducing the spectrum acquisition time is offered by a pulsed ESR (also known as Fourier transform ESR) that uses coherent pulses instead of CW radiation. This concept has been demonstrated successfully in a system that uses as a radiation source the gyrotron FU CW VIIB (154 GHz, 150 W) [94]. The necessary short millimeter-wave pulses with a precisely adjustable time delay are generated by a light-controlled semiconductor shutter. This system allows implementing the method of electron spin-echo envelope modulation (ESEEM) in the measurements, which allows measuring the relaxation times of the electron spins. The developed instrumentation is equipped with a quadrature detection system that increases the resolution of the FT-ESR spectra obtained from the FID (Free Induction Decay) signal [95]. Quite recently, a new system for high-frequency ESR studies based on the measurement of force-detected electron spin resonance (FDESR) has been presented in [96]. It uses as a radiation source FU CW VIIB operating at a frequency of 154 GHz with sufficiently high output power that permits using a strong transverse magnetic field (exceeding 10^{-4} T), which is enough to saturate the ESR signal. The obtained FDESR spectrum is characterized by high spin sensitivity of the order of 1012 spins/G at a temperature of 280 K.

16.4.3.3 XDMR spectroscopy

X-ray detected magnetic resonance (XDMR) is a promising advanced spectroscopic technique in which X-ray magnetic circular dichroism (XMCD) is used to probe the resonant precession of spin or orbital magnetization components pumped by the magnetic field of strong sub-terahertz gyrotron radiation in a plane perpendicular to the static bias magnetic field. This method is element- and edge-selective and is expected to become a unique tool to investigate the role of precession dynamics of orbital magnetization components. The first feasibility tests have been carried out at the European synchrotron radiation facility (ESRF) in Grenoble (France) using the gyrotron FU II built at FIR UF [97]. The same approach can be used also for the investigation of various new X-ray electro-optical and magneto-electric effects.

16.4.3.4 Measuring the energy levels of positronium

Positronium (Ps) is an exotic Hydrogen-like atom (a bound state of an electron and a positron), which can exist in two states, namely a triplet (ortho-positronium, o-Ps) and a singlet (parapositronium, p-Ps). The hyperfine energy splitting (HFS) between them corresponds to a frequency close to 203.4 GHz. The previously used indirect methods are prone to systematic errors and have indicated a significant discrepancy between the measured HFS values and the theoretical prediction of the quantum electrodynamics. In contrast to them, the new approach is based on a stimulated transition from o-Ps to p-Ps induced using irradiation with a strong electromagnetic wave produced by a gyrotron operating with a frequency of about 203 GHz produced by the gyrotron FU CW V [98]. The sophisticated experimental setup includes also a high-finesse Fabry-Pérot resonator in which the positronium is being formed using a ^{22}Na source of positrons, a transmission line that delivers and couples the wave beam to the cavity, as well as a set of gamma detectors and an electronic control system. When irradiated by an electromagnetic wave with a frequency of about 203 GHz, some of the o-Ps (decaying into three γ rays) transit into much shorter life time p-Ps (decaying into two γ rays with the energy of 511 keV), and consequently, the ratio of two γ ray events increases. This process is monitored by the photon detectors ($LaBr_3$(Ce) scintillators) that are located around the cavity and count co-incident events.

In order to observe the Breit–Wigner function of the transition, a Gaussian beam gyrotrons of demountable type FU CW GI was developed in FIR UF [99]. The gyrotron frequency has been varied in a very wide range from 201 to 205 GHz by changing successively several gyrotron cavities of different radii. The hyperfine transition has been observed with a significance of 5.4 standard deviations. The Einstein transition probability that has been measured directly for the first time is found to be $A = 3.69 \pm^{0.48}_{0.29} \times 10^{-8}$ s^{-1}, which is in good agreement with the theoretical value of 3.37×10^{-8} s^{-1} [100].

16.4.3.5 Radioacoustic spectroscopy using gyrotron radiation

Gas molecular spectroscopy is a powerful method used in many fundamental and applied studies (e.g., qualitative and quantitative gas analysis, atmospheric remote sensing, and so on). The current levels of the achieved sensitivity in the conventional schemes of mm-wave spectrometers, however, have already reached the physical limits. A promising approach for further increase of the sensitivity is based on the optoacoustic (aka photo- or radioacoustic) detection of absorption. Recently, high-resolution molecular spectroscopy in the sub-THz range using a gyrotron as a source of high-power, continuous, and frequency-tunable CW radiation has been presented in [101]. The obtained results indicate that the sensitivity increases in direct proportion to the radiation power until the molecular line saturation is reached. For unsaturated lines corresponding to rotational transitions of methane, a sensitivity of the order of 10^{-11} cm^{-1} is achieved for a power level of about 20 W. The analysis shows that there are good prospects for a further increase in the sensitivity of the method by increasing the radiation power by one or two orders of magnitude. Additionally, the recorded spectral lines permit one to examine the spectral content of gyrotron radiation. For the first time, the presence of high harmonics without any external frequency multiplier is revealed and evaluated.

16.4.4 Plasma Physics and Localized Gas Discharges

An active field of research is related to various applications of gas discharges in air sustained by gyrotron radiation. Such discharges can be controlled precisely and maintained in free space as well as near the surface of different materials and are characterized by high plasma density. Using simple quasi-optical components gyrotron radiation can easily be transmitted and focused on the region where the discharge is initiated. In contrast to the plasma torches (plasmatrons) that produce equilibrium plasma, these discharges produce non-equilibrium plasma. Recently, it has been demonstrated that they can be used effectively for the generation of extreme ultraviolet (UVL) light with a potential application to high-resolution projection lithography for microelectronic chip production. The first extreme UV radiation has been obtained at IAP-RAS using a 250 GHz gyrotron with an output power of 250 kW in 50 μs pulses [102]. In these experiments, a peak electron density of up to 3×10^{17} electrons/cm^3 and a size of 150 μm has been reached, and the detected emission power ranges from ~20 W at 18–50 nm for Ar and Xe to 0.3 W at 13–17 nm for Xe.

In another study, the gas breakdown thresholds in a focused beam of CW THz radiation have been investigated theoretically and experimentally using a gyrotron operating at a frequency of 0.263 THz with an output power of 1 kW [103]. The wave beam has been focused by quasi-optical reflectors into a spot with a diameter of less than 3 mm yielding a power density of 15 kW/cm^2 in argon. The achieved electrical field intensity is sufficient for initiating the breakdown of Ar gas with pressures ranging from 10 to 300 Torr. The above-mentioned experimental arrangements for the generation of non-equilibrium localized plasma can be used also in different plasma-chemical processes. Recently, a discharge experiment with a high-power 303 GHz gyrotron [38] was carried out. A Gaussian beam radiated from the gyrotron was focused with a parabolic mirror and a plasma was ignited in the air [104]. An array parallel to the wave electric field appeared with a spacing of $\lambda/4$ at the focal point. Later, a new comb-shaped array was created by a standing wave structure in the electric field plane under the subcritical condition. The produced filaments are parallel to the incident Gaussian beam and bulk plasma with a sharp vertex has been formed in the magnetic field plane.

In a proof-of-principle experiment, focused continuous radiation of the gyrotron at a frequency of 24 GHz and power up to 5 kW has been used for the decomposition of CO_2 at atmospheric pressure. The highest conversion of 31.3% was observed for a mixture of CO_2 and Ar with a ratio of 1:5 at the optimum total gas flow rate of 30 l/min, while the energy efficiency was about 9.5% at SEI (specific energy input) of 4.7 eV/molecule. The results of optical spectroscopy indicate that non-equilibrium plasma of microwave discharge with electron temperature in the range 4000–8000 K (0.3–0.7 eV) and gas temperature at the level of 2000–3500 K, is well suited for the effective decomposition of CO_2 [105].

The remote detection of concealed radioactive materials is of paramount importance for national and collective security. The concept that uses gyrotron radiation for this purpose involves focusing a high-power wave beam in a small spot where the electromagnetic field intensity exceeds the breakdown threshold [106]. Provided there are some seed electrons, this focused radiation initiates an RF breakdown. At THz frequencies, the breakdown-prone volume is small. More importantly, the breakdown rate in the case of a natural ambient electron density will be low. On the contrary, a high breakdown rate will indicate that in the vicinity of a focused wave beam, there are some additional sources of air ionization (presumably radioactive material). The envisaged scheme for the detection of a distant breakdown is based on observing the THz signal reflected from the discharge. A dedicated gyrotron for such a system has been designed [107]. It operates at a frequency of 0.67 THz and provides a maximum output power of 300 kW (3 Joules in a pulse of 10 μs) using a pulsed solenoid (cooled by liquid nitrogen) with a maximum magnetic field intensity of 28 T.

16.4.5 Electron Cyclotron Resonance Ion Sources

Beams of multiply charged ions accelerated to high energies in particle accelerators (cyclotrons, linacs, and synchrotrons) are in great demand for many applications in atomic physics. Currently, the most advanced sources of highly charged ions are the 4th generation of electron cyclotron resonance ion sources (ECRIS) operating at frequencies of more than 40 GHz. They benefit from the higher frequency because the plasma density scales with the square of operating frequency. Nowadays gyrotron-based ECRIS are under development worldwide. A good example that illustrates the current state-of-the-art in this field is the ECRIS which uses as a source of radiation a gyrotron operating at a frequency of 45 GHz (in both pulsed and CW regimes) with an automated smoothly controlled output power in the 0.1–20 kW range (up to 26 kW in manual mode) with an efficiency of up to 50% (using a depressed collector) [108]. The overall setup of this system has been developed and built specially for the advanced fourth-generation superconducting ECRIS called FECRAL (the first fourth-generation ECR ion source with Advanced design in Lanzhou, China) [109].

A novel concept, namely gas-dynamic ECRIS, has been pioneered at IAP-RAS recently [110]. It utilizes a confinement mechanism in a magnetic trap that is different from Geller's classical ECRIS confinement [111]. In it, the used quasi gas-dynamic mechanism is similar to that in fusion mirror traps. The experimental studies have been performed using gyrotron radiation with frequencies of 37.5 and 75 GHz and output power up to 100 kW. The irradiation by strong microwaves at high frequency allows creating and sustaining plasma with significant density (up to $8 \times \cdot 10^{13}$ cm^{-3}).

The gas-dynamic ECRIS are considered especially appropriate for the formation of low emittance hydrogen and deuterium beams. High current H+ and D+ beams extracted from the plasma of ECR discharge sustained by 75 GHz / 200 kW gyrotron radiation in an open magnetic trap are reported in [112]. At such elevated microwave power and frequency significantly higher density of hydrogen plasma (up to 7×10^{13} cm^{-3}), in comparison to the conventional sources, has been produced. An additional advantageous feature is the low ion temperature (only a few eV) which allows the formation of high-brightness hydrogen ion beams with a low emittance (in these experiments the measured rms normalized emittance is 0.03 $\pi \cdot$ mm $\cdot$ mrad) and current up to 500 mA.

16.4.6 Applications in Bioscience and Material Science Areas

Applications in bioscience and material science areas have already been reviewed above [11–15]. Some additional new results are presented here. Hydrothermally grown bulk crystals with high conductivity have been irradiated at room temperature with up to 60 W output of a sub-terahertz gyrotron wave source (FU CW IIB) [113]. During the gyrotron irradiation, the high-conductivity crystals exhibit intense yellow emissions. The sample temperature has been raised to above 1000 K. Heating up to 1250 K using a heater, no visible emission has been observed. The emission peak intensity can be enhanced by the gyrotron-induced non-equilibrium states. As the spatial distribution of the yellow emission reflects the gyrotron beam pattern, the bulk ZnO single-crystals can then be utilized for the quick diagnosis of the gyrotron beam patterns and positions.

Techniques of manipulating actin polymerization, including actin-binding chemicals, have been developed for understanding and regulating multiple biological functions. When the actin polymerization reaction is performed under irradiation with 460 GHz waves generated by a Gyrotron FU CW GVIB manufactured with the same design as that of FU CW GVIA, actin polymerization has been activated [114]. This phenomenon has been observed by monitoring the fluorescence of pyrene actin fluorophores. The number of actin filaments has been increased by 3.5-fold after 460 GHz wave irradiation for 20 min. These results suggest that the THz waves could be applied for manipulating biomolecules and cells.

A reliable database for future safety regulation of THz wave exposure is very important. Gyrotron FU CW GV has been used in such a study. An ocular-damaged rabbit model for exposure of 162 GHz wave was developed [115]. One day after 162 GHz wave exposure revealed a round area of opacity, characterized by fluorescein staining indicating damaged epithelial cells in the central pupillary zone. Corneal edema, indicative of corneal stromal damage, peaked 2–3 days after the exposure, with thickness gradually subsiding to normal by 9 days after exposure. Power densities of the 162 GHz wave causing the observed ocular damage with probabilities of 10, 50, and 90% were 173, 252, and 368 mW/cm^2, respectively.

Application studies with THz gyrotrons are currently under way. See, for example, [116, 117], etc.

16.5 Conclusions and Outlook

The current status of the development and application of THz gyrotrons presented in this chapter demonstrates the remarkable potential of these devices for bridging the THz power gap ("the last frontier") in this region of the electromagnetic spectrum where other radiation sources (both electronic and photonic) provide significantly lower (orders of magnitude) output power. We are witnessing a growing demand for such devices that will continue acting as a strong motivation and a driving force for further advancement toward this frequency range through designing high-performance gyrotrons. This will involve the utilization of novel approaches, advanced concepts, adequate physical models, and numerical tools for computer-aided design (CAD).

Besides the high power, the gyrotron radiation is characterized by such advantageous features as precise control of the output parameters, step-wise and continuous frequency tunability, possibility to modulate both the power and frequency, excellent spectral characteristics, multiple-frequency Gaussian beam radiation, availability of quasi-optical methods and components for transmission and manipulation (e.g., steering, scanning) of the wave beam, etc.

In a number of feasibility studies and proof-of-principle experiments, many of the advanced concepts mentioned above (for example, multi-beam systems, LOG, gyrotrons with complex cavities, planar gyrotrons, tubes with frequency multiplication, super multiple-frequency oscillation combined with continuous frequency tunability, wide frequency band with gyro-TWA, and so on) have already demonstrated remarkable potential for implementation in the THz gyrotrons but are still in an initial stage in comparison with the most matured conventional gyrotrons. It is expected, however, that the active research which is being carried out on such promising concepts will eventually lead to further progress in their realization.

Another well-established tendency, dictated by many applications, is the constant improvement and optimization of the gyrotron operation using a computer-controlled system for measuring, diagnostic, and stabilization of the output parameters based on different techniques (such as phase locking, stabilization of gyrotron frequency by appropriate reflection from a remote load or by PID feedback control) as well as for automation of the overall work of the device.

Taking into account all these advantageous features of gyrotrons as high-power sources of coherent radiation, we anticipate that they will continue to open new directions to many new scientific and technological fields and will lead to the creation of synergy links between them as well as to the conversion of different THz techniques.

References

1. Nusinovich G. S., Thumm M. K., Petelin M. I., The gyrotron at 50: historical overview, *J. Infrared Millim. Terahertz Waves*, vol. 35, no. 4 (2014) 325–381. DOI: 10.1007/s10762-014-0050-7.
2. Petelin M. I., The gyrotron: physical genealogy, *Terahertz Sci. Technol.*, vol. 8, no. 4 (2015) 157-169. DOI: 10.11906/TST.157-169.2015.12.15.

3. Pliński E. F., Gorky's gyrotron heroes, *Bull. Pol. Acad. Sci. Tech. Sci.*, vol. 68, no. 6 (2020). DOI: 10.24425/bpasts.2020.135392.
4. Kartikeyan M. V., Borie E., Thumm M., Gyrotrons: high-power microwave and millimeter wave technology, in *Advanced Texts in Physics*, Springer (2003).
5. Nusinovich G. S., *Introduction to the Physics of Gyrotrons*, The Johns Hopkins University Press, Baltimore-London (2004).
6. Chu K. R., The electron cyclotron maser, *Rev. Mod. Phys.*, vol. 76 (2004) 439. DOI: 10.1103/RevModPhys.76.489.
7. Idehara T., et al., The first experiment of a THz gyrotron with a pulse magnet, *Int. J. Infrared Milli. Waves*, vol. 27 (2006) 319. DOI: 10.1007/s10762-006-9084-9.
8. Glyavin M. Y., Luchinin A. G., Golubiatnikov G. Y., Generation of 1.5 kW, 1 THz coherent radiation from a gyrotron with a pulsed magnetic field, *Phys. Rev. Lett.*, vol 100 (2008) 015101. DOI: 10.1103/PhysRevLett.100.015101.
9. Idehara T., Ogawa I., Mori H., Kobayashi S., Mitsudo S., Saito T., A THz gyrotron FU CW III with a 20T superconducting magnet, *J. Plasma Fusion Res. Series*, vol. 8 (2009) 1508–1501.
10. Thumm M., State-of-the-art of high power gyro-devices and free electron masers, *J. Infrared Millim. Terahertz Waves*, vol. 41 (2020) 1–140. DOI: 10.1007/s10762-019-00631-y.
11. Bratman V., et al., Review of subterahertz and terahertz gyrodevices at IAP RAS and FIR FU, *IEEE Trans. Plasma Sci.*, vol. 37 (2009) 36. DOI: 10.1109/TPS.2008.2004787.
12. Idehara T., Sabchevski S. P., Development and applications of high-frequency gyrotrons in FIR FU covering the sub-THz to THz range, *J. Infrared Millim. Terahertz Waves*, vol. 33 (2012) 667. DOI: 10.1007/s10762-011-9862-x.
13. Temkin R. J., Development of terahertz gyrotrons for spectroscopy at MIT, *Terahertz Sci. Technol.*, vol. 7, no. 1 (2014) 1–9. DOI: 10.11906/TST.001-009.2014.03.01.
14. Glyavin M. Y., Idehara T., Sabchevski S. P., Development of THz gyrotrons at IAP RAS and FIR UF and their applications in physical research and high-power THz technologies, *IEEE Trans. Terahertz Sci. Technol.*, vol. 5, no. 5 (2015) 788–797. DOI: 10.1109/TTHZ.2015.2442836.
15. Idehara T., Sabchevski S., Gyrotrons for high-power terahertz science and technology at FIR UF, *J. Infrared Millim. Terahertz Waves*, vol. 38 (2017) 62. DOI: 10.1007/s10762-016-0314-5.

16. Idehara T., Sabchevski S. P., Development and application of gyrotrons at FIR UF, *IEEE Trans. Plasma Sci.*, vol. 46, no. 7 (2018) 2452–2459. DOI: 10.1109/TPS.2017.2775678.

17. Idehara T., Sabchevski S., Glyavin M., Mitsudo S. The gyrotrons as promising radiation sources for THz sensing and imaging, *Appl. Sci.*, vol. 10, no. 3 (2020) 980. DOI: 10.3390/app10030980.

18. Kumar N., Singh U., Singh T. P., Sinha A. K., A review on the applications of high power, high frequency microwave source: gyrotron, *J. Fusion Energy* vol. 30, no. 4 (2011) 257–276. DOI: 10.1007/s10894-010-9373-0.

19. Glyavin M., Sabchevski S., Idehara T., Mitsudo S., Gyrotron-based technological systems for material processing—current status and prospects, *J Infrared, Millimeter, and Terahertz Waves*, vol. 41 (2020) 1022–1037. DOI: 10.1007/s10762-020-00727-w.

20. Brand G. F., et al., Operation of a tunable gyrotron at the second harmonic of electron cyclotron frequency, *Int. J. Electronics*, vol. 68 (1990) 1099. DOI: 10.1080/00207219008921249.

21. Idehara T., et al., Development of frequency tunable, medium power gyrotrons (gyrotron FU series) as submillimeter wave radiation sources, *IEEE Tr. Plasma Sci.*, vol. 27 (1999) 340. DOI: 10.1109/27.772260.

22. Chang T. H., Idehara T., Ogawa I., Agusu L., Kobayashi S., Frequency tunable gyrotron using backward-wave components. *J. Appl. Phys.*, vol. 105 (2009) 063304. DOI: 10.1063/1.3097334.

23. Ikeda R., et al., Broadband continuously frequency tunable gyrotron for 600 MHz DNP-NMR spectroscopy, *Plasma Fusion Res.*, vol. 9 (2014) 1206058. DOI: 10.1585/pfr.9.1206058.

24. Torrezan A. C., et al., Continuous-wave operation of a frequency-tunable 460 GHz second-harmonic gyrotron for enhanced nuclear magnetic resonance, *IEEE Tr. Plasma Sci.*, vol. 38 (2010) 1150. DOI: 10.1109/TPS.2010.2046617.

25. Song T., et al., Experimental investigations on a 500 GHz continuously frequency-tunable gyrotron, *IEEE Electron Dev. Lett.*, vol. 42 (2021) 1232. DOI: 10.1109/LED.2021.3092379.

26. Blank M., Development of high-frequency cw gyrotrons for DNPNMR applications, *Terahertz Sci. Tech.*, vol. 9 (2016) 177. DOI: 10.11906/TST.177-186.2016.12.17.

27. Denisov G. G., et al., 110 GHz gyrotron with a built-in high-efficiency converter, *Int. J. Electronics*, vol. 72 (1992) 1079. DOI: 10.1080/00207219208925634.

28. Zapevalov V. E., et al., High-power oscillator of continuous electromagnetic radiation with a frequency of 300 GHz, *Radio Phys. Quantum Electronics*, vol. 50 (2007) 420. DOI: 10.1007/s11141-007-0039-3.
29. Saito T., et al., Performance test of CW 300 GHz gyrotron FU CW, *Int. J. Infrared Mill. Waves*, vol. 28 (2007) 1063. DOI: 10.1007/s10762-007-9291-z.
30. Tatematsu Y., et al., Development of the multifrequency gyrotron FU CW GV with gaussian beam output, *J. Infrared Mill. Terahz Waves*, vol. 36 (2015) 697. DOI: 10.1007/s10762-015-0173-5.
31. Tatematsu Y., et al., Characteristics of the mode converter of gyrotron FU CW GII radiating Gaussian beams in both the fundamental and second harmonic frequency bands, *J. Infrared Mill. Terahz Waves*, vol. 35 (2014) 517. DOI: 10.1007/s10762-014-0072-1.
32. Idehara T., et al., The Development of 460 GHz gyrotrons for 700 MHz DNP-NMR spectroscopy, *J. Infrared Mill. Terahz Waves*, vol. 36 (2015) 613. DOI: 10.1007/s10762-015-0150-z.
33. Tatematsu Y., et al., Development of gyrotron FU CW GVII a second harmonic, multifrequency gyrotron that radiates Gaussian beams, *J. Infrared Mill. Terahz Waves*, vol. 41 (2020) 576. DOI: 10.1007/s10762-020-00681-7.
34. Idehara T., et al., Development of a Compact sub-THz gyrotron FU CW CI for application to high power THz technologies, *J. Infrared Mill. Terahz Waves*, vol. 33 (2012) 724. DOI: 10.1007/s10762-012-9911-0.
35. Ikeda, et al., Development of a table-top 200 GHz gyrotron FU CW CII with an internal mode convertor, *37th International Conference on Infrared, Millimeter and Terahertz Waves (IRMMW-THz)*, Sep 30–Oct 05, 2012, Wollongong, Australia, Mon-C-3-4. DOI: 10.1109/IRMMW-THz.2012.6380361.
36. Saito T., et al., Generation of high power sub-terahertz radiation from a gyrotron with second harmonic oscillation, *Phys. Plasmas*, vol. 19 (2012) 063106. DOI: 0.1063/1.4729316.
37. Saito T., et al., Observation of dynamic interactions between fundamental and second-harmonic modes in a high-power sub-terahertz gyrotron operating in regimes of soft and hard self-excitation, *Phys. Rev. Lett.*, vol. 109 (2021) 155001. DOI: 0.1103/PhysRevLett.109.155001.
38. Saito T., et al., Oscillation characteristics of a high power 300 GHz band pulsed gyrotron for use in collective thomson scattering diagnostics, *Plasma Fusion Res.*, vol. 14 (2019) 1406104. DOI: 10.1585/pfr.14.1406104.

39. Denisov G. G., et al., First experimental tests of powerful 250 GHz gyrotron for future fusion research and collective Thomson scattering diagnostics, *Rev. Sci. Instrum.*, vol. 89 (2018) 084702. DOI: 10.1063/1.5040242.
40. Yamaguchi Y., et al., Formation of a laminar electron flow for 300 GHz high-power pulsed gyrotron, *Phys. Plasmas*, vol. 19 (2012) 113113. DOI: 10.1063/1.4768959.
41. Zapevalov V. E., Tsimring Sh. E., Multibeam gyrotrons, *Radiophys. Quantum Electron.,* vol. 33, no. 11, pp. 594–960, 1990. DOI: 10.1007/BF01039240.
42. Zapevalov V. E., Manuilov V. N., Tsimring Sh. E., Double-beam gyrotron electron-optical systems, *Radiophys. Quantum Electron.*, vol. 34, no. 2 (1991) 174–179. DOI: 10.1007/BF01045526.
43. Zapevalov V. E., Manuilov V. N., Malygin O. V., Tsimring S. E., High-power twin-beam gyrotrons operating at the second gyrofrequency harmonic, *Radiophys. Quantum Electron.*, vol. 37, no. 3 (1994) 237–240. DOI: 10.1007/BF01054034.
44. Liu D., Yuan X., Liu Sh., Coupled-mode theory of coaxial gyrotron with two electron beams, *Fusion Eng. Des.*, vol. 83 (2008) 606–612.
45. Fu W., Yan Y., Yuan X., Liu Sh., Two-beam magnetron injection guns for coaxial gyrotron with two electron beams, *Phys. Plasmas*, vol. 16 (2009) 023103.
46. Manuilov V. N., Glyavin M. Yu., Sedov A. S., Zaslavsky V. Yu., Idehara T., Design of a second harmonic double-beam continuous wave gyrotron with operating frequency of 0.79 THz, *J. Infrared Millim. Terahertz Waves*, vol. 36, no. 12 (2015) 1164–1175.
47. Idehara T., Glyavin M., Kuleshov A., Sabchevski S., Manuilov V., Zaslavsky V., Zotova I., Sedov A., A novel THz-band double-beam gyrotron for high-field DNP-NMR spectroscopy, *Rev. Sci. Instrum.*, vol. 88, no. 9 (2017) 094708. DOI: 10.1063/1.4997994.
48. Mitsudo S., Glyavin M., Khutoryan E., Bandurkin I., Saito T., Ishikawa Y., Manuilov V., Zotova I., Fedotov A., Kuleshov A., Sabchevski S., Tatematsu Y., Zaslavsky V., Idehara T., An experimental investigation of a 0.8 THz double-beam gyrotron, *J. Infrared Millim. Terahertz Waves*, vol. 40 (2019) no. 11–12, 1114–1128. DOI: 10.1007/s10762-019-00629-6.
49. McDermott D. B., Luhmann Jr, N. C., Kupiszewski A., Jory H. R., Small-signal theory of a large-orbit electron-cyclotron harmonic maser, *Phys. Fluids*, vol. 26, no. 7 (1983) 1936–1941. DOI: 10.1063/1.864341.

50. Lawson W., Destler W. W., Striffler C. D., High-power microwave generation from a large-orbit gyrotron in vane and hole-and-slot conducting wall geometries, *IEEE Trans. Plasma Sci.* vol. 13 (1985) 444. DOI: 10.1109/TPS.1985.4316458.
51. Bratman, V. L., Fedotov A. E., Kalynov Y. K., Manuilov V. N., Ofitserov M. M., Samsonov S. V., Savilov A. V., Moderately relativistic high-harmonic gyrotrons for millimeter/submillimeter wavelength band, *IEEE Trans. Plasma Sci.*, vol. 27, no. 2 (1999) 456–461. DOIi: 10.1109/27.772273.
52. Idehara T., Ogawa I., Mitsudo S., Iwata Y., Watanabe S., Itakura Y., Itakura Y., Ohashi K., Kobayashi H., Yokoyama T., Zapevalov V. E., Glyavin M. Yu., Kuftin A. N., Malygin O. V., Sabchevski S. P., A high harmonic gyrotron with an axis-encircling electron beam and a permanent magnet, *IEEE Trans. Plasma Sci.*, vol. 32, no. 3 (2004) 903–909. DOI: 10.1109/TPS.2004.827614.
53. Bratman V. L., Kalynov Yu. K., Manuilov V. N., Large-orbit gyrotron operation in the terahertz frequency range, *Phys. Rev. Lett.* vol. 102 (2009) 245101. DOI: 10.1103/PhysRevLett.102.245101.
54. Bandurkin I., Bratman V. L., Kalynov Y. K., Osharin I. V., Savilov A. V., Terahertz large-orbit high-harmonic gyrotrons at IAP RAS: Recent experiments and new designs, *IEEE Trans. Electron Devices*, vol. 65, no. 6 (2018) 2287–2293.
55. Kalynov Y. K., Manuilov V. N., Osharin I. V., Savilov A. V., Fiks A. S. Universal subterahertz large-orbit gyrotron: operation at the second and third cyclotron harmonics, *Radiophys. Quantum Electron.*, vol. 63, no. 5 (2020) 32. DOI: 10.1007/s11141-021-10057-z.
56. Li F.-H., et al., Development of a 1 THz 4th-harmonic gyrotron, *2020 13th UK-Europe-China Workshop on Millimetre-Waves and Terahertz Technologies (UCMMT)*, 2020, pp. 1–2. DOI: 10.1109/UCMMT49983.2020.9296084.
57. Xu, Shou Xi, et al. Design of a high-harmonic gyrotron with a permanent magnet system. *IEEE Trans. Electron Devices*, vol. 60, no. 10 (2013) 3570–3575. DOI: 10.1109/TED.2013.2276858.
58. Nusinovich G. S., Read M. E., Dumbrajs O., Kreischer, K. E., Theory of gyrotrons with coaxial resonators, *IEEE Trans. Electron Devices*, vol. 41, no. 3 (1994) 433–438. DOI: 10.1109/16.275231.
59. Iatrou C. T., Mode selective properties of coaxial gyrotron resonators, *IEEE Trans. Plasma Sci.*, vol. 24, no. 3 (1996) 596–605. DOI: 10.1109/27.532942.

60. Dumbrajs O., Nusinovich G. S., Coaxial gyrotrons: past, present, and future, *IEEE Trans. Plasma Sci.*, vol. 32, no. 3 (2004) 934–946. DOI: 10.1109/TPS.2004.829976.

61. Ruess S., et al. KIT coaxial gyrotron development: from ITER toward DEMO, *Int. J. Microwave Wireless Technol.*, vol. 10, no. 5–6 (2018). 547–555. DOI: 10.1017/S1759078718000144.

62. Singh S., Yuvaraj S., Baghel G. S., Kartikeyan M. V., Design studies of a RF interaction cavity for a 4 MW, 170 GHz triangular corrugated coaxial cavity gyrotron, *2018 IEEE International Vacuum Electronics Conference (IVEC)* (2018) pp. 35–36 Doi: 10.1109/IVEC.2018.8391539.

63. Glyavin M. Yu., Tsvetkov A. I., Pagonakis I. Gr., Jelonnek J., Avramidis K. A., Mitsudo S., The project of a 2nd harmonic 0.78 THz gyrotron with a coaxial cavity, *Proc. SPIE 11582, Fourth International Conference on Terahertz and Microwave Radiation: Generation, Detection, and Applications*, 1158212 (17 November 2020). DOI: 10.1117/12.2580564.

64. Vlasov S. N., Koposova E. V., Pavel'ev A. B., Khizhnyak V. I. Gyrotrons with echelette resonators, *Radiophys. Quantum Electron.*, vol. 39, no. 6 (1996) 458–462. DOI: 10.1007/BF02122392.

65. Zapevalov V. E., Vlasov S. N., Koposova E. V., Kuftin A. N., Paveliev A. B., Zavolsky N. A., Various types of echelette resonators for gyrotrons, *EPJ Web of Conferences EDP Sciences*, vol. 195 (2018) 01022. DOI: 10.1051/epjconf/201819501022.

66. Bandurkin I. V., Kalynov Y. K., Savilov A. V. Experimental study of a gyrotron with a sectioned klystron type cavity operated at higher cyclotron harmonics, *Radiophys. Quantum Electronics*, vol. 58, no. 9 (2016). DOI: 10.1007/s11141-016-9641-6.

67. Bandurkin I. V., Bratman V. L., Kalynov Y. K., Osharin I. V., Savilov A. V., Terahertz large-orbit high-harmonic gyrotrons at IAP RAS: recent experiments and new designs, *IEEE Trans. Electron Devices*, vol. 65, no. 6 (2018) 2287–2293. DOI: 10.1109/TED.2018.2797311.

68. Bandurkin I. V., Glyavin M. Y., Zavolsky N. A., Kalynov Y. K., Osharin I. V., Savilov A. V., Use of quasiregular resonator cavities with short phase correctors in gyrotrons operated at higher cyclotron harmonics, *Radiophys. Quantum Electron.*, vol. 59, no. 8–9 (2017) 655–666. DOI: 10.1007/s11141-017-9732-z.

69. Ginzburg N. S., Zotova I. V., Sergeev A. S., Zaslavsky V. Y., Zheleznov I. V., High-power terahertz-range planar gyrotrons with transverse energy extraction, *Phys. Rev. Lett.*, vol. 108, no. 10 (2012) 105101. DOI: 10.1103/PhysRevLett.108.105101.

70. Zotova I. V., Ginzburg N. S., Malkin A. M., et al., Terahertz-range high-order cyclotron harmonic planar gyrotrons with transverse energy extraction, *J. Infrared Milli., Terahertz Waves*, vol. 41 (2020) 152–163. DOI: 10.1007/s10762-019-00661-6.
71. Zaslavsky V. Y., Zheleznov I., Ginzburg N., Zotova I., Malkin A., Sergeev A., Frequency multiplication in planar gyrotrons as a method for production of high-power multi-THz radiation, *IEEE Trans. Electron Devices*, vol. 68, no. 3 (2021) 1267–1270. DOI: 10.1109/TED.2020.3049108.
72. Bandurkin I. V., Glyavin M. Y., Idehara T., Savilov A. V., Double-beam gyrotron with frequency multiplication, *IEEE Trans. Electron Devices*, vol. 66, no. 5 (2019) 2396–2400. DOI: 10.1109/TED.2019.2905047.
73. Glyavin M., Zotova I., Rozental R., et al., Investigation of the frequency double-multiplication effect in a sub-THz gyrotron, *J. Infrared Milli., Terahertz Waves*, vol. 41 (2020) 1245–1251. DOI: 10.1007/s10762-020-00726-x.
74. Constable D. A., Phelps A. D., Whyte C. G., He W., Cross A. W., Ronald K., Numerical and experimental validation of the passive performance of a co-harmonic gyro-multiplier interaction region, *IEEE Trans. Microwave Theory Tech.* (2021) 1–9. DOI: 10.1109/TMTT.2021.3094219.
75. Yeh Y. S., Luo C. C., Huang B. H., Wang C. C., Hung S. X., Comparative analysis of terahertz fourth-harmonic gyro-TWAs and multiplying gyro-TWAs with axis-encircling electron beams, *AIP Adv.*, vol. 10 (2020) 045325. DOI: 10.1063/1.5129728.
76. Rumyantsev V. V., Maremyanin K. V., Fokin A. P., et al., Second-harmonic generation of subterahertz gyrotron radiation by frequency doubling in InP:Fe and its application for magnetospectroscopy of semiconductor structures, *Semiconductors*, vol. 53 (2019) 1217–1221. DOI: 10.1134/S1063782619090185.
77. Melnikova M. M., et al., Electromagnetic modeling of a complex-cavity resonator for the 0.4 THz second-harmonic frequency-tunable gyrotron, *IEEE Tra. Electron Devices*, vol. 64 (2017) 5141. DOI: 10.1109/TED.2017.2764874.
78. Yamaguchi Y., et al., Super multi frequency oscillations at fundamentarl harmonics with a complex cavity gyrotron, *IEEE Electron Device Lett.*, vol. 41 (2020) 1241. DOI: 10.1109/LED.2020.3000640.
79. Thumm M. K. A., Denisov G. G., Sakamoto K., Tran M. Q., High power gyrotrons for electron cyclotron heating and current drive, *Nucl.*

Fusion, vol. 59, no. 7 (2019) Art. no. 073001. DOI: 10.1088/1741-4326/ab2005.

80. Litvak A. G., Denisov G. G., Glyavin M. Y., Russian gyrotrons: achievements and trends, *IEEE J. Microwaves*, vol. 1, no. 1 (2021) 260–268. DOI: 10.1109/JMW.2020.3030917.
81. Oda Y., Ikeda R., Kajiwara K., Kobayashi T., Hayashi K., Takahashi K., Moriyama S., Sakamoto K., Eguchi T., Kawakami Y., Development of the first ITER gyrotron in QST, *Nuclear Fusion*, vol. 59, no. 8 (2019) 086014. DOI: 10.1088/1741- 4326/ab22c2.
82. Sabchevski S., Glyavin M., Mitsudo S., Tatematsu Y., Idehara T., Novel and emerging applications of the gyrotrons worldwide: current status and prospects. *J. Infrared Milli., Terahertz Waves* (2021). DOI: 10.1007/s10762-021-00804-8. (Invited Review Article, Published: 29 July 2021).
83. Yuvaraj S., Kartikeyan M. V., Thumm M. K., Design studies of a 3 MW, multifrequency (170/204/236 GHz) DEMO class triangular corrugated coaxial cavity gyrotron, *IEEE Trans. Electron Devices*, vol. 66, no. 1 (2018) 702–708. DOI: 10.1109/TED.2018.2876870.
84. Bykov Y., Eremeev A., Glyavin M., Holoptsev V. V., Plotnikov I. V., Pavlov 3.5 kW 24 GHz compact gyrotron system for microwave processing of materials. In: Willert-Porada M. (eds.) *Advances in Microwave and Radio Frequency Processing*. Springer, Berlin, Heidelberg. (2006) DOI: 10.1007/978-3-540-32944-2_3.
85. Sano S., Makino Y., Miyake S., Bykov Yu. V., Eremeev A. G., Egorov S. V., 30 and 83 GHz millimeter wave sintering of alumina, *J. Mater. Sci. Lett.*, vol. 19, no. 24 (2000) 2247–2250. DOI: 10.1023/A:1006733125930.
86. Bykov Yu., Eremeev A., Glyavin M., Kholoptsev V., Luchinin A., Plotnikov I., Denisov G., Bogdashev A., Kalynova G., Semenov V., Zharova N., 24–84 GHz gyrotron systems for technological microwave applications, *IEEE Trans. Plasma Sci.*, vol. 32, no. 1 (2004) 67–72. DOI: 10.1109/TPS.2004.823904.
87. Hoshizuki H., High temperature thermal insulation system for millimeter wave sintering of B4C, *Int. J. Infrared Mill. Waves*, 26 (2007) 1531. DOI: 10.1007/s10762-005-0030-z.
88. Vodopyanov A. V., Samokhin A. V., Alexeev N. V., Sinayskiy M. A., Tsvetkov A. I., Glyavin M. Y., Fokin A. P., Malygin V. I., Application of the 263 GHz/1 kW gyrotron setup to produce a metal oxide nanopowder by the evaporation-condensation technique, *Vacuum*, vol. 145 (2017) 340–346. DOI: 10.1016/j.vacuum.2017.09.018.

89. Chernov V. V., Gorbachev A. M., Vikharev A. L., Lobaev M. A., Criterion for comparison of MPACVD reactors working at different microwave frequencies and diamond growth conditions, *Phys. Status Solidi A*, vol. 213 (2016) 2564–2569. DOI: 10.1002/pssa.201600193.
90. Visit: https://www.gyrotrontech.com/.
91. Becerra L. R., Gerfen, G. J., Temkin R. J., Singel D. J., Griffin R., Dynamic nuclear polarization with a cyclotron resonance maser at 5 T, *Phys. Rev. Lett.*, vol. 71 (1993) 3561–3564. DOI: 10.1103/PhysRevLett.71.3561.
92. Alberti S., Ansermet J.-P., Avramides K. A., Fasel D., Hogge J.-P., Kern S., Lievin C., Liu Y., Macor A., Pagonakis I., Design of a frequency-tunable gyrotron for DNP-enhanced NMR spectroscopy, *2009 34th International Conference on Infrared, Millimeter, and Terahertz Waves*, 2009, pp. 1–2. DOI: 10.1109/ICIMW.2009.5324906.
93. Mitsudo S., Higuchi T., Kanazawa K., Idehara T., Ogawa I., Chiba M., High field ESR measurements using gyrotron FU series as radiation sources, *J. Phys. Soc. Jpn.*, vol. 72, Suppl. B (2003) 172–176. DOI: 10.1143/JPSJS.72SB.172.
94. Mitsudo S., Umegaki C., Hiiragi K., Narioka M., Fujii Y., Tatematsu Y., Development of a millimeter wave pulsed ESR system by using a gyrotron as a light source, *Proceedings of the 2016 41st International Conference on Infrared, Millimeter, and Terahertz waves (IRMMW-THz)*. DOI: 10.1109/IRMMW-THz.2016.7758476.
95. Mitsudo S., Dono K., Hayashi K., Ishikawa Y., Fujii Y., Improvement in sensitivity of FT- ESR measurements by using a gyrotron as high-power millimeter wave source, *2020 45th International Conference on Infrared, Millimeter, and Terahertz Waves (IRMMW-THz)*, Buffalo, NY, USA (2020) pp. 1-1. DOI: 10.1109/IRMMW-THz46771.2020.9370426.
96. Takahashi H., Ishikawa Y., Okamoto T., Hachiya D., Dono K., Hayashi K., Asano T., Mitsudo S., Ohmichi E., Ohta H., Force detection of high-frequency electron spin resonance near room temperature using high-power millimeter-wave source gyrotron, *Appl. Phys. Lett.*, vol. 118, no. 2 (2021) 022407. DOI: 10.1063/5.0036800.
97. Rogalev A., Goulon J., Goujon G., Wilhelm F., Ogawa I., Idehara T., X-ray detected magnetic resonance at Sub-THz frequencies using a high power gyrotron source, *J. Infrared, Millimeter, and. Terahertz Waves*, vol. 33 (2012) 777–793. DOI 10.1007/s10762-011-9855-9.
98. Yamazaki T., Miyazaki A., Suehara T., Namba T., Asai S., Kobayashi T., Saito H., Ogawa I., Idehara T., Sabchevski S., Direct observation of the

hyperfine transition of ground-state positronium, *Phys. Rev. Lett.*, vol. 108 (2012) 253401. DOI: 10.1103/PhysRevLett.108.253401.

99. Tatematsu Y., et al., Development of a kW level-200 GHz gyrotron FU CW GI with an internal quasi-optical mode convertor, *J. Infrared Mill. Terahz Waves*, vol. 33 (2012) 292. DOI: 10.1007/s10762-012-9881-2.
100. Miyazaki A., Yamazaki T., Suehara T., Namba T., Asai S., Kobayashi T., Saito H., Tatematsu Y., Ogawa I., Idehara T., First millimeter-wave spectroscopy of ground-state positronium, *Prog. Theor. Exp. Phys.* (2015) 011C01. DOI: 10.1093/ptep/ptu181.
101. Golubiatnikov G. Y., Koshelev M. A., Tsvetkov A. I., Fokin A. P., Glyavin M. Y., Tretyakov M. Y., Sub-terahertz high-sensitivity high-resolution molecular spectroscopy with a gyrotron, *IEEE Trans. Terahertz Sci. Technol.*, vol. 10, no. 5 (2020) 502–512. DOI: 10.1109/TTHZ.2020.2984459.
102. Shalashov A. G., Vodopyanov A. V., Abramov I. S., Sidorov A. V., Gospodchikov E. D., Razin S. V., Chkhalo N. I., Salashchenko N. N., Glyavin M. Yu., Golubev S. V., Observation of extreme ultraviolet light emission from an expanding plasma jet with multiply charged argon or xenon ions, *Appl. Phys. Lett.*, vol. 113 (2018) 153502. DOI: 10.1063/1.5049126.
103. Sidorov A. V., Razin S. V., Tsvetkov A. I., Fokin A. P., Veselov A. P., Golubev S. V., Vodopyanov A. V., Glyavin M.Yu., Gas breakdown by a focused beam of CW THz radiation, *2017 Progress in Electromagnetics Research Symposium-Spring (PIERS)* (2017) pp. 2600–2602. DOI: 10.1109/PIERS.2017.8262191.
104. Fukunari M., et al., Observation of a comb-shaped filamentary plasma array under subcritical condition in 303 GHz millimetre-wave air discharge, *Sci. Rep.*, 9 (2019) 17972. DOI: 10.1038/s41598-019-54333-5.
105. Mansfeld D., Sintsov S., Chekmarev N., Vodopyanov A., Conversion of carbon dioxide in microwave plasma torch sustained by gyrotron radiation at frequency of 24 GHz at atmospheric pressure, *J. CO_2 Utilization*, vol. 40 (2020) 101197. DOI: 10.1016/j.jcou.2020.101197.
106. Nusinovich G. S., Pu R., Antonsen T. M., Sinitsyn O. V., Rodgers J., Mohamed A., Silverman J., Al-Sheikhly, M., Dimant Y. S., Milikh G. M., Glyavin M. Yu., Luchinin A. G., Kopelovich E. A., Granatstein V. L. Development of THz-range gyrotrons for detection of concealed

radioactive materials, *J. Infrared Millim. Terahertz Waves*, vol. 32, no. 3 (2011) 380–402. DOI: 10.1007/s10762-010-9708-y.

107. Glyavin M. Yu., Luchinin A. G., Bogdashov A. A., Manuilov V. N., Morozkin M. V., Rodin Yu. V., Denisov G. G., Kashin D., Rogers G., Romero-Talamas C. A., Pu R., Shkvarunetz A. G., Nusinovich G. S., Experimental study of the pulsed terahertz gyrotron with record-breaking power and efficiency parameters, *Radiophys. Quantum Electron.*, vol. 56, no. 8–9 (2014) 497–507. DOI: 10.1007/s11141-014-9454-4.
108. Denisov G., Glyavin M., Tsvetkov A., Eremeev A., Kholoptsev V., Plotnikov I., Bykov Yu., Orlov V. C., Morozkin M., Shmelev M., Kopelovich E., Troitsky M., Kuznetsov M., Zhurin K., Novikov A., Bakulin M., Sobolev D., Tai E., Soluyanova E., Sokolov E., A 45 GHz/20 kW gyrotron-based microwave setup for the fourth-generation ECR ion sources, *IEEE Trans. Electron Devices*, vol. 69 (2018). DOI: 10.1109/TED.2018.2859274.
109. Zhao H. W., Sun L. T., Guo J. W., Zhang W. H., Lu W., Wu W., Wu B. M., Sabbi G., Juchno M., Hafalia A., Ravaioli E., Xie D. Z., Superconducting ECR ion source: From 24–28 GHz SECRAL to 45 GHz fourth generation ECR, *Rev. Sci. Instrum.*, vol. 89 (2018) 052301. DOI: 10.1063/1.5017479.
110. Skalyga V. A., Golubev S. V., Izotov I. V., Kazakov M. Yu., Lapin R. L., Razin S. V., Sidorov A. V., Shaposhnikov R. A., Bokhanov A. F., Tarvainen O., Status of new developments in the field of high-current gasdynamic ECR ion sources at the IAP RAS, *AIP Conference Proc.*, vol. 2011 (2018) 020018. DOI: 10.1063/1.5053260.
111. Geller R., *Electron Cyclotron Resonance Ion Sources and ECR Plasmas*, Institute of Physics Publishing, London (1999) ISBN 0-7503-0107-0.
112. Skalyga V., Izotov I., Golubev S., Razin S., Sidorov A., H+ and D+ high current ion beams formation from ECR discharge sustained by 75 GHz gyrotron radiation, *AIP Conference Proc.*, vol. 1771 (2016) 070012. DOI: 10.1063/1.4964236.
113. Kato K., et al., Strong yellow emission of high-conductivity bulk ZnO single crystals irradiated with high-power gyrotron beam, *Appl. Phys. Lett.*, 111 (2017) 031108. DOI: 10.1063/1.4994316.
114. Yamazaki S., et al., Actin polymerization is activated by terahertz irradiation, *Sci. Rep.*, vol. 8 (2018) 9990. DOI: 10.1038/s41598-018-28245-9.
115. Kojima M., et al., Clinical course of high-frequency millimeter-wave (162 GHz) induced ocular injuries and investigation of damage

thresholds, *J. Infrared Mill. Terahz Waves*, vol. 41 (2012) 834. DOI: 10.1007/s10762-020-00714-1.

116. Kawasaki T., et al., Irradiation effect of a submillimeter wave from 420 GHz gyrotron on amyloid peptides in vitro, *Biomed. Opt. Exp.*, vol. 11 (2020) 5341. DOI: 10.1364/BOE.395218.

117. Mizuno M., et al., Dielectric property measurements of corneal tissues for computational dosimetry of the eye in terahertz band in vivo and in vitro, *Biomed. Opt. Exp.*, vol. 12 (2021) 1295. DOI: 10.1364/BOE.412769.

Chapter 17

Extended-Interaction Klystrons

Khanh T. Nguyen and John Pasour
Beam-Wave Research, Inc.
US Naval Research Laboratory, USA
knguyen@beamwave.com

Extended-interaction klystrons (EIKs) are resonant cavity-based circuits. First proposed by Chodorow and Wessel-Berg [1], the EIK has evolved into a compact device of choice for the THz regime. Similar to klystrons, EIKs are comprised of an input cavity, which imparts a velocity modulation on the beam to start the beam bunching process, one or more idler cavities to improve beam bunching (gain) and/or bandwidth, and an output cavity to extract *RF* power from the optimally bunched beam. The key difference between a standard klystron and an EIK is that EIK cavities have several interaction gaps versus one for a standard klystron, which raises the cavity interaction impedance. The higher cavity interaction impedance is a key factor in compensating for the unavoidable lower beam current at higher operating frequencies. This is because beam current tends to scale as $\sim 1/f^2$ for any given beam-forming approach.

But more than just the number of gaps per cavity, these gaps are also coupled together within the EIK cavity via coupling slots in a manner similar to that in coupled-cavity TWTs. In this context,

Advances in Terahertz Source Technologies
Edited by Gun-Sik Park, Masahiko Tani, Jae-Sung Rieh, and Sang Yoon Park

ISBN 978-981-4968-89-8 (Hardcover), 978-1-003-45967-5 (eBook)
www.jennystanford.com

EIK cavities can be thought of as a shorted section of a slow-wave structure. The shorting introduces a feedback mechanism, further enhancing the strength of beam-wave interaction, resulting in substantially higher gain per unit length compared to that of coupled-cavity TWTs. This is an important consideration for the millimeter and submillimeter wave regimes, as the shorter circuit length reduces the potential for beam interception due to alignment issues in the progressively smaller diameter beam tunnel. For this reason, commercially available continuous-wave (cw) EIKs in the submillimeter wave regime exist today in an extremely compact form factor, as exemplified by the 9 W cw, 220 GHz EIK (VKY2444) produced by Communications and Power Industries (CPI) Canada [2].

Due to the advantages described above and the strong interest in high-power THz *RF* sources for various applications, the EIKs have been a focused research area in the millimeter and submillimeter wave regimes. CPI-Canada has been a pioneer since the 1970s and continues to be the leading developer of round-beam EIKs, with peak power as high as 60 W in G-band [3, 4]. In the United States, the Naval Research Laboratory (NRL) has demonstrated a sheet-beam EIK of more than 7.5 kW at 94 GHz with a solenoidally focused 20 kV, 3.5 A sheet-beam [5]. A design for a 453 W at 220 GHz EIK amplifier with similar circuit topology as the W-band device driven by a 16.5 kV, 0.52 A sheet-beam has also been performed [6]. More recently, there has been an intensive effort in the People's Republic of China on EIK designs at 220 and 340 GHz. [9–13].

17.1 EIK Cavity

The basic building block in an EIK is the multi-gap cavity. In the millimeter and submillimeter wave regimes, high impedance gaps with reentrant nose to maximize beam-wave interaction are no longer practical from the fabrication standpoint. Thus, the basic building blocks are pillbox-type interaction gaps. The box height is d, which is also the gap width. A beam of current I_b, voltage V_b, and radius R_b is assumed to pass through a beam tunnel of radius R_T normal to the center of the cross-section. The inter-gap

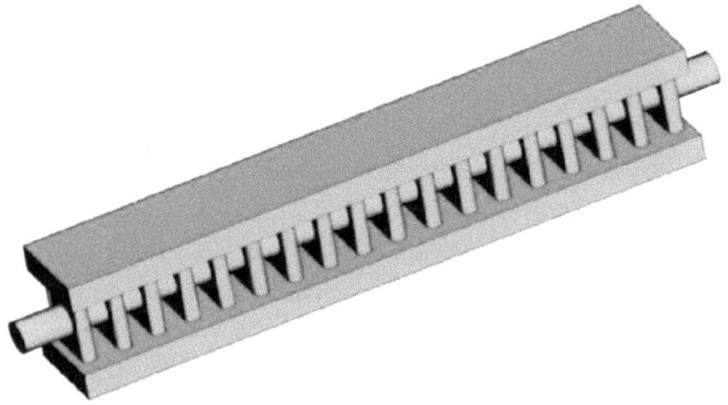

Figure 17.1 Circuit topology of an 18-gap EIK cavity for 2π-mode interaction. For π mode, the ladder width generally staggers from gap to gap.

distance (gap-to-gap axial spacing) is defined as p. These pillbox gaps are coupled together by top and bottom coupling slots. This cavity topology describes the traditional ladder circuit. Figure 17.1 illustrates one such cavity with 18 gaps.

In addition, the number of axial resonance modes in the EIK cavity is proportional to the number of gaps coupled together. These discrete modes are characterized by the phase shift from one gap to the next. Most EIKs operate at either π or 2π modes. However, for broadband applications, it is also possible to broaden the bandwidth by interacting with more than one axial mode. In such a case, each cavity must be comprised of many gaps to produce mode proximity and must be heavily loaded for merging the modes and to ensure stability. The topology as shown in Fig. 17.1 is more suitable for 2π mode of operation. For π-mode, the ladder width tends to be staggered from gap to gap [9].

EIKs that employ this ladder circuit generally are designed to operate in either the π or 2π modes, where the gap characteristic impedance is the highest. [staggered ladder for π-mode] For a given beam voltage, the gap-to-gap distance p must be approximately equal to v_z/f for 2π-mode operation and half of that for π mode operation to ensure beam-wave synchronism. Here, v_z is the beam axial velocity corresponding to the beam voltage, and f is the operating frequency. From this synchronism relation, it is obvious that the π mode operation is a more desirable choice since one can pack twice as many gaps per unit length, which results in a more compact device due to the higher gain per unit length. As the π mode circuit is shorter, it also helps with beam alignment resulting in lower beam interception; hence, higher operating duty. However, the π mode may not be practical at higher frequencies due to geometric considerations, since p reduces inversely with the operating frequency. For example, for beam voltage at approximately 20 kV, π-mode operation can be the appropriate approach for frequency of approximately 300 GHz or below. Above this frequency, 2π-mode operation is the more appropriate approach. Conversely, the transition frequency between π and 2π modes of operation can be shifted up to higher frequency with higher beam voltage. Of course, one must be mindful that such higher voltage may substantially increase the size, weight, and cost of the power modulator.

17.2 Beam-Wave Interaction in EIK

In a standard klystron, the cavity gap voltage induced by a bunched electron beam can be shown to be approximately:

$$V_g \approx Q_L(R/Q)M(f_b I_b),$$

where f_b is the beam fundamental bunching factor, Q_L is the total cavity quality factor (external and ohmic), M is the beam-wave gap coupling coefficient, and R/Q is the interaction gap impedance. The stored energy per cell (gap and inter-cavity coupling slots) is given by:

$$W_c = \frac{1}{2\omega}\frac{V_g^2}{R/Q}.$$

Since RF power $P_{RF} = \omega/QW_c$, it is obvious from the above equations that the quantity, $(R/Q)M^2$, which is essentially the effective gap impedance K_c, must be maximized for maximum RF excitation in the cavity by the beam RF current $(f_b I_b)$.

The gap impedance is different for different cavity resonant modes. Obviously, for a given unit of cavity RF stored energy, it is highly desirable that the gap voltage V_g be as high as possible to impart the maximum voltage modulation on the electron beam. This is the key reason that the operating mode for a pillbox cavity is invariably the TM_{11} mode (or TM_{01} mode for a cylindrical cavity) when feasible. In addition to higher gap voltage, the other important factor in beam-wave interaction is the gap coupling coefficient M, which is less than unity as a result of the time-varying sinusoidal nature of the electric field as the beam crossing the interaction gap. For a single gap, M can be approximated by:

$$M \equiv \frac{\sin(\theta_g)}{\theta_g},$$

where $\theta_g \equiv 2\pi f d_e / v_z$ is the effective gap transit angle, and d_e is the effective gap width, which can be approximated by

$$d_e \approx d + 0.7R_t,$$

where d is the physical gap width, and the R_T term accounts for fringing RF electric fields due to beam tunnel size. Since $R/Q \sim f d_e \sim \theta_g$, it can then be shown that

$$K_c = \frac{R}{Q}M^2 \sim \theta_g \times \left[\frac{\sin\theta_g}{\theta_g}\right]^2.$$

Figure 17.2 illustrates the approximated behavior of the normalized gap impedance as a function of the effective gap transit angle. The optimal operating point is when the gap transit angle is approximately 1.15 radians.

Beyond this point, EIKs operate very differently from standard klystrons. This is due to the fact that each EIK cavity is comprised of multiple single resonance gaps coupled together. For example, in addition to selecting the beam voltage and gap width to maximize beam-wave coupling at each interaction gap, the gap-to-gap distance p must be chosen to ensure beam-wave synchronism. Therefore, based on Fig. 17.1, in an ideal world, one would select the effective gap width, d_e, and beam voltage so that the gap transit angle is approximately 1.15 radians to maximize interaction impedance. Then for synchronism, the ratio of p/d_e would be approximately 2.73 for π-mode and 5.46 for 2π-mode. However, in the millimeter

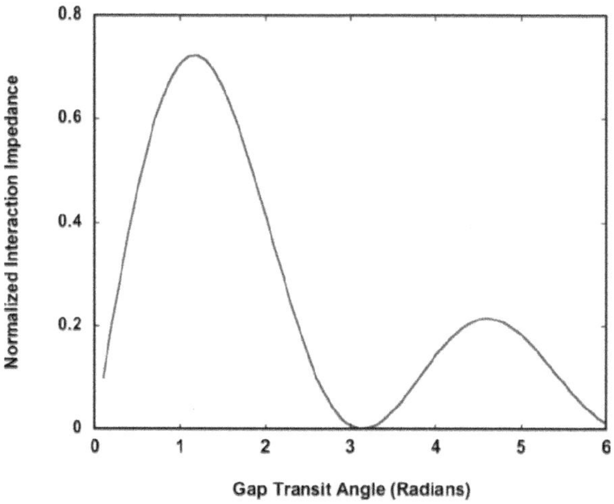

Figure 17.2 Normalized interaction impedance as a function gap transit angle.

and submillimeter regime, other considerations, including high ohmic loss due to surface roughness, relatively large beam tunnel size from beam optics thermal effects, beam voltage impacts on modulator size and weight, etc., render the design of EIKs to be an extremely complex endeavor with multiple trade-offs that must be carefully considered.

Another major consideration in EIK design is oscillations. In fact, a single EIK cavity can also be and often is used as an oscillation circuit, in which case it is known as an extended-interaction oscillator (EIO). A pulsed EIO with frequency as high as 280 GHz and pulsed power of 5 W has been demonstrated and is commercially available. For amplifiers, the number of gaps in each cavity has to be carefully selected to avoid potential oscillation. This is, however, also the EIK's key advantage, that by operating just below the start oscillation threshold, it can provide extremely high gain in a very short distance. Consequently, it is very compact, which mitigates the potential for beam interception, and it operates at much higher current density and magnetic field than other devices, as high magnetic field intensity is much easier to achieve for small volume.

For an EIK cavity of N gaps, it can be shown that the fundamental bunching factor, f_b, at gap number j is given by:

$$f_b^j = M\theta_p \frac{V_g}{2V_b} \frac{j(j-1)}{2},$$

where V_g is gap voltage, θ_p is inter-gap transit angle (which is π or 2π), and M and V_b are again gap coupling coefficient and beam voltage, respectively. Thus, the average bunching factor in the cavity is:

$$f_b^{Ave.} = \pi \frac{MV_g}{4V_b} \frac{1}{N} \sum\nolimits_{j=1}^{N} j(j-1)$$

The cavity can freely oscillate if the beam generates more RF than can be dissipated. This condition is met when:

$$\alpha_N = \frac{1}{N} \sum_{j=1}^{N} \frac{j(j-1)}{2} \geq \frac{2}{\pi} \frac{V_b/I_b}{Q_L K_c} = \alpha_B$$

The quantity (α_N) on the left-hand side of the above equation is plotted in Fig. 17.3 for reference. From the above equation, the maximum number of gaps, N, that can be employed in an EIK cavity for a given beam and circuit parameters, can then be determined.

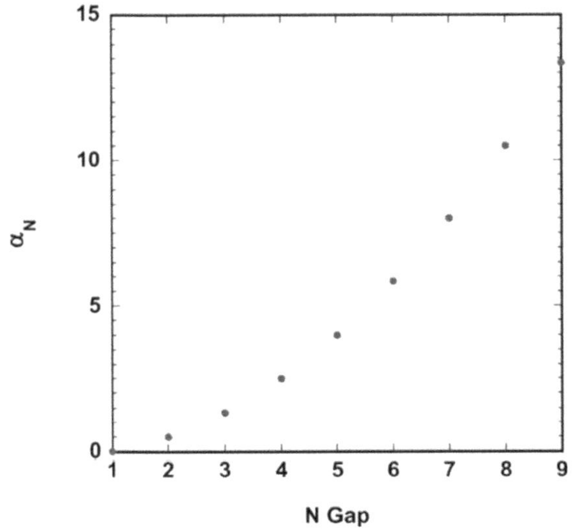

Figure 17.3 Average gap impedance enhancement as a function of gap number in an EIK cavity.

17.3 Gain

We can now calculate the gain per cavity or stage. Let us assume the beam enters the cavity first gap with a bunching factor, f_b^0. Then the induced cavity gap voltage is:

$$V_g \approx MQ_L R/Q \frac{f_b^0 I_b}{1 - \alpha_N/\alpha_B}$$

The key difference between the gap voltage of an EIK cavity and a single gap klystron cavity is the factor, $1 - \alpha_N/\alpha_B$, in the denominator, which greatly enhances the gap voltage in the EIK case. With this gap voltage, the bunching factor at the cavity last gap $j = N$ is approximately:

$$f_B^N = f_B^0 \left[1 + \frac{N(N-1)}{\alpha_B - \alpha_N}\right]$$

Further ballistic bunching will occur in the drift section between circuit cavities. However, for EIKs, drift sections are generally kept to a minimum as interaction gaps are much more effective in inducing *RF* bunching on the electron beam. Again, this is different from klystrons as ballistic bunching plays a major role in klystron gain.

The gain per stage (cavity plus drift length) can now be expressed in a very simple form:

$$G_s\,(\mathrm{dB}) = 20\mathrm{Log}\left(\frac{f_B^f}{f_B^0}\right) = 20\mathrm{Log}\left[1 + \frac{N(N-1) + \frac{NL_d}{p}}{\alpha_B - \alpha_N}\right]$$

Here, f_B^f is the beam bunching factor at the start of the next cavity and includes further bunching in the inter-cavity drift length, L_d. In comparing various amplifiers, it is not the total gain but the gain per unit length that is the key metric since it is used in trading off between amplification strength and size. Here, the total length per stage is $L_s = N^* p + L_d$. Thus, the gain per unit length is simply G_s/L_s. Higher gain per unit length reduces circuit length, which in turn reduces the likelihood of beam interception. This is an important consideration given the tiny size of the beam tunnel as we approach the true THz regime.

We end the discussion on gain by noting that in the high-frequency limit, where the highest current density achievable is already employed to optimize power and efficiency, current will scale as $1/f^2$. It may be tempting to increase the beam tunnel radius to accommodate more beam current. However, at relatively constant beam voltage, the gap-to-gap distance p will scale as $1/f$. Thus, at a given frequency, increasing the beam tunnel radius R_T will also increase the effective gap width d_e due to fringing *RF* fields as noted above. The fringing fields in conjunction with the $1/f$ scaling of p will result in gap-to-gap field overlapping. Such overlapping will cause field cancellation in the $\theta_p = \pi$ mode of operation or a baseline uniform electric field for the 2π mode of operation. In either case, the net effect is a reduction in gap coupling coefficient. Of course, the lower gap coupling coefficient will reduce beam-wave interaction impedance. However, in EIKs, this can be compensated for by adding more gaps.

17.4 *RF* Power

At the output cavity, the beam is already tightly bunched and no further enhancement in beam bunching can be expected, as is the case in the input or buncher cavities. In fact, beam bunching is reduced as *RF* energy is being extracted from it. A general rule of thumb is that the average fundamental current bunching factor in the output cavity at saturation is approximately 70% of its peak value, which for a well-optimized narrow band circuit is 1.6. Thus, for this discussion, we will assume the saturated bunching factor in the output cavity is 1.12.

The maximum beam power converted to *RF* power is then:

$$P_T = \frac{1}{2} Q_T K_c N (1.12 I_b)^2$$

Here, the number of gaps is similarly determined as in the previous section on gain. It is important to note that the power value given above is power converted into *RF*, which is different from *RF* output power. Extracted output *RF* power is related to the total converted power as:

$$P_{RF} = \frac{Q_T}{Q_E}(1 - \rho_W) P_T,$$

where Q_E is the external coupling quality factor. The difference between total converted power and extracted power is the power loss due to ohmic loss in the output cavity and output waveguide. The ohmic quality factor due to the cavity alone can be calculated using the approach discussed earlier. The term ρ_w represents the loss in the output waveguide, which is difficult to calculate since it depends on numerous factors, such as guide width and length. However, it is reasonable to assume that output waveguide loss will have the same dependence on frequency ($f^{1/2}$) as in the cavity ohmic loss and that the waveguide length is an integral value of the axial wavelength, then, $\rho_w \sim f^{1/2}$. Thus, we can relate all the output cavity quality factors together as:

$$\frac{1}{Q_T} = \frac{1-\rho_W}{Q_E} + \frac{1}{Q_0}$$

It is obvious from the above equation that maximum output power is extracted, if bandwidth is not an issue, is given by:

$$Q_E = \frac{Q_0}{1-\rho_W}$$

Then, since $I_b \sim 1/f^2$ and $Q_0 \sim 1/f^{1/2}$, the maximum peak *RF* output power achievable will scale as a function of frequency as:

$$P_{RF}^{\max} \propto S(f) = \frac{1}{f^{4.5}} \frac{1-\rho_W (f/f_0)^{1/2}}{1-\rho_W}$$

This is a key result, and it shows how once a basic design has been demonstrated at a lower frequency, it can be expected to perform if scaled to a higher frequency. The key, of course, is a known low-frequency anchor point.

In Fig. 17.4, we use the achieved CPI-Canada's W-band EIK peak power as the anchor point and extrapolated to a higher frequency using the scaling law described above. Also in this plot are actual peak powers for higher frequencies EIKs as tabulated by Table 17.1 in [2]. Strong correlations can be observed in this plot indicating the validity of the scaling laws. This plot only gives the peak power and not the average power (*peak-power x duty cycle*), which is much lower than peak power. A key reason for this decrease is the beam interception in the very small beam tunnel, especially toward the output cavity, and also power absorbed by ohmic loss in the output

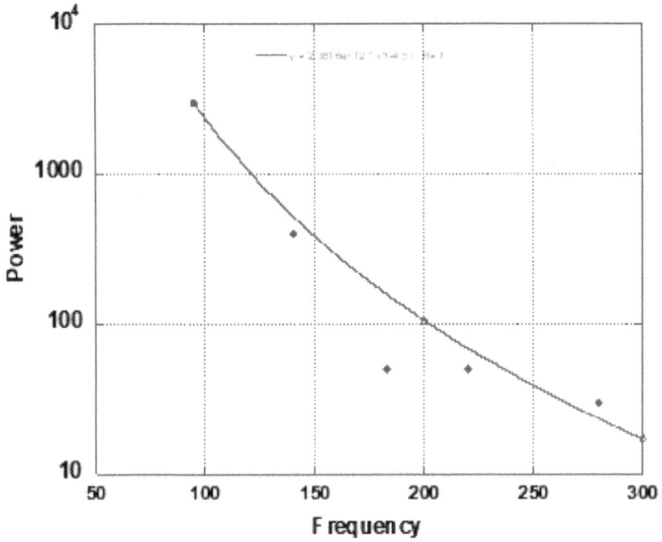

Figure 17.4 Peak power for EIKs as a function of frequency. Solid curve is the theoretical projection based on the *S*(*f*) scaling law using 94 GHz EIK power as anchor point. Dots are actual demonstrated amplifier data from CPI-Canada as denoted in Table 1 of Reference [2].

cavity. These power sources contribute to the heat flux on the circuit (primarily at the output). Consequently, the thermal issue is one of two key factors in limiting an EIK average power. The other limiting key factor is the beam power available from the electron gun. This is because, for the same electron gun, the highest beam voltage achievable is also a function of pulse length with long pulse and cw operation has lower operating voltages than that of short pulse operation. The correlation between achieved performance by the same group and predicted trend is quite remarkable given the complexity of implementation as the frequency increases.

17.5 Sheet-Beam EIKs

To further enhance the performance of this promising technology, one would need a higher power anchor point while maintaining the relatively low beam voltage [5]. Such an anchor point can be found in the new breakthrough in sheet-beam technology developed by the U.S. NRL, which permits the formation and transport of an intense sheet electron beam with a solenoidal magnetic field at a beam voltage of approximately 20 kV [14]. Prior to this breakthrough, it was thought that a solenoidally focused intense electron beam was not feasible due to issues associated with ExB drift. Thus, a periodic-cusp magnet was the only solution to defeat the ExB issues in sheet-beam transport [15], which, of course, would limit the beam current density at a lower threshold than the higher strength and simpler solenoidal magnetic field. The key to this breakthrough is through the beam formation and beam matching approach to the solenoidal magnetic field [16]. This beam-forming and matching concept was demonstrated via a beam stick with the successful transport of a 19.5 kV, 3.5 A sheet-beam in 0.4 mm high × 5 mm wide × 2 cm-long beam tunnel focused by an 8.5 kG solenoidal magnetic field [17]. This successful sheet-beam transport demonstration establishes the key enabling technology not only for the next generation of high-power EIKs but also for other devices such as coupled-cavity TWTs as well [18].

The success of sheet-beam formation and transport introduces another challenge. The wide rectangular beam tunnel for such sheet-beams can allow the propagation of the TE_{10} beam tunnel mode at the operating frequency, which by itself is not an issue if everything is symmetrically aligned [19–20]. However, even a small asymmetry in the transverse direction along the beam narrow dimension, such as caused by fabrication errors or beam misalignment, can provide a feedback loop from the output cavity back to the input cavity, which is just the recipe for potential oscillations. In the case of transverse beam misalignment, the space charge of the bunched beam will excite the TE_{10} beam tunnel mode at the operating frequency, even if the circuit is perfectly fabricated. Another potential source for TE_{10} excitation is circuit asymmetry as illustrated by Fig. 17.5a, where the beam tunnel is transversely misaligned with the cavity. In such a case, the misalignment will cause a conversion of the cavity operating mode to the beam tunnel mode, which provides another feedback loop mechanism during amplification. As illustrated by Fig. 17.5b, HFSS [21] analysis shows that a 50 μm displacement of the beam tunnel relative to the cavity structure (in the narrow beam

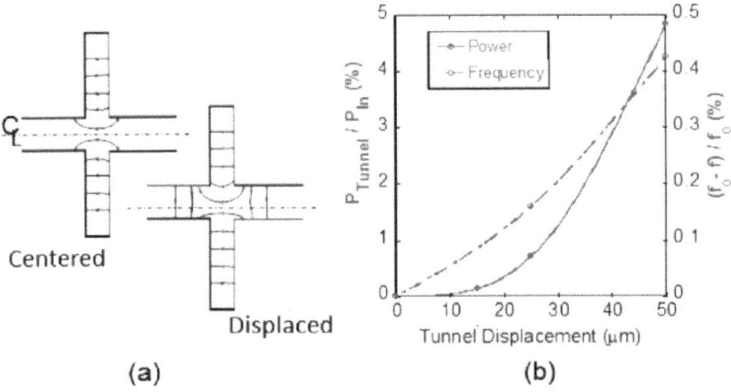

Figure 17.5 (a) Illustration of TE_{10} mode excitation in displaced beam tunnel by asymmetric gap field. (b) Power flow into beam tunnel and resonant frequency shift as a function of beam tunnel displacement.

direction) causes almost 5% of the cavity power to escape into the beam tunnel. The resonant frequency is also decreased by ~0.4%. If left unmitigated, such fabrication errors coupled together with beam transverse misalignment would lead to disastrous oscillations based on Neptune particle-in-cell (PIC) simulations [22].

The simple and obvious solutions to mitigate such potential oscillation are either to reduce the beam tunnel width to below the cutoff for the operating frequency, which of course limits beam current, or to substantially reduce the gain of the operating mode. Both of these are not very satisfactory solutions, as they limit the promise of sheet-beam EIKs. The breakthrough solution resides in the use of $\lambda/4$ chokes to suppress beam tunnel coupling between cavities due to these misalignment or fabrication errors [19, 22]. Both the depth of each choke and its spacing from the adjacent outer cavity gap are $\lambda_g/4$, where λ_g is the wavelength in the beam tunnel. Various configurations were tested via simulation, including placing chokes on both sides of the beam tunnel or placing chokes only on each side of the buncher cavity. The final configuration, as shown in Fig. 17.6, was determined based on Neptune simulations and strongly suppressed cavity-to-cavity feedback loop in most of the cases tested.

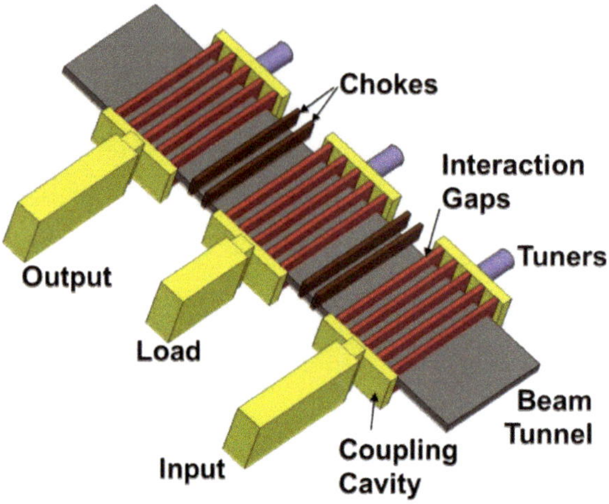

Figure 17.6 W-band sheet-beam EIK circuit topology. Total length is 2.2 cm.

With these breakthroughs, a W-band amplifier employing three identical cavities of five gaps each as shown in Fig. 17.6 was fabricated and tested. The amplifier operates in the 2π-mode with a demonstrated 7.7 kW peak power at the output flange. This corresponds to approximately 10 kW of power being generated in the output cavity. The EIK is driven by a 19–22 kV, 3.5–4.2 A sheet-beam in a permanent magnet solenoid, with 99% of the beam current transmitted through a 0.4 × 5 mm × 2.2 cm-long beam tunnel. Considering the power level, this amplifier is quite compact as built. However, a new magnet design topology was implemented on the Ka-band traveling-wave amplifier that uses the same sheet-beam gun [18]. A permanent magnet solenoid based on that design would be ~50% smaller and lighter for the same field and interaction length and could be employed in future builds.

Under tests, the amplifier is very stable at all beam voltages up to the maximum of 22 kV. This is remarkable considering the highly over-moded nature of sheet-beam amplifiers. As built, each of the cavities incorporates a tuner and can be tuned individually. Tuning the buncher cavity allows the gain to be traded for peak power. Shown in Fig. 17.7 are the transfer curves showing the impact of buncher cavity tuning relative to its initial value (Δf)

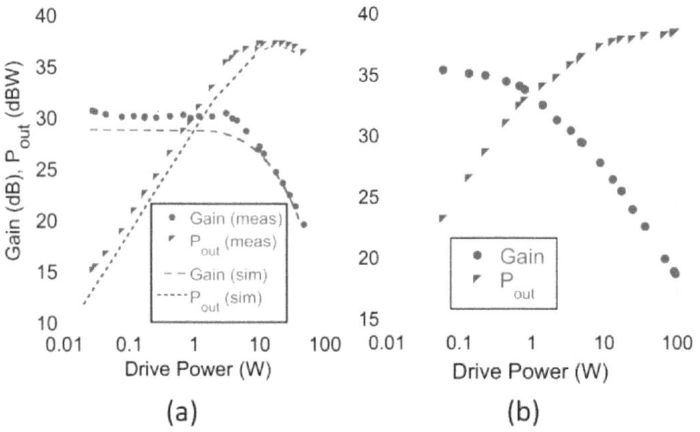

Figure 17.7 Drive curves for two operating conditions. (a) $\Delta f = 150$ MHz, 20.4 kV, 3.9 A. Results from Neptune are also shown. (b) $\Delta f = 200$ MHz, 21.3 kV, 4.2 A.

The corresponding electronic efficiency is 8.6%, and the net efficiency is 17.2% with a relatively simple single-stage collector depressed to -11 kV. A multi-stage collector can further improve the overall efficiency.

Also shown in Fig. 17.7a is a comparison with the Neptune PIC prediction, which shows the good agreement between simulation and measured data. This agreement, together with the absolute stability of the amplifier as predicted, provides a very important validation point for future EIK designs. In addition, the high fidelity of the MICHELLE gun code [23] in designing the beam-forming system and transport in this complex sheet-beam amplifier is another critical element. In conjunction with electromagnetic and magnet design codes, such as HFSS [20], Maxwell-3D [24], and Analyst [25], these codes allow the designers to use their insights to explore and test various technical ideas and concepts prior to full engineering design and implementation. This approach can result in a substantial reduction in the need for multiple builds, resulting in time and cost savings, minimizing the risks and maximizing the potential returns. The technological breakthrough embodied in this W-band sheet-beam EIK in its successful first build is an example of the advantage of such an approach.

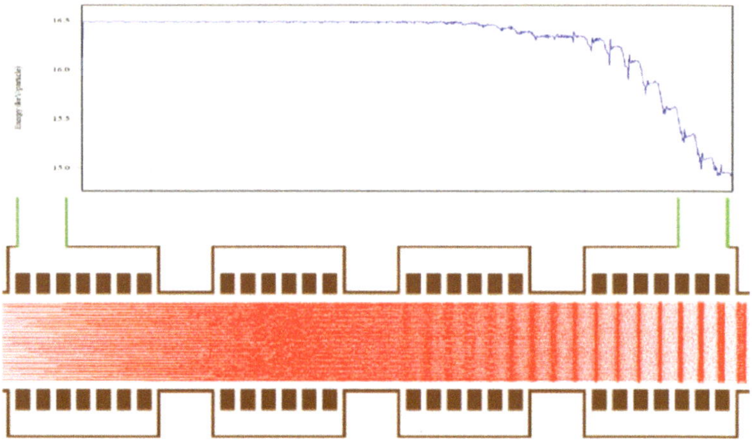

Figure 17.8 Particle trajectories across the beam width as the beam propagates through the 1.16 cm four-cavity circuit (bottom), and beam averaged kinetic energy as a function of axial distance (top) from a MAGIC-3D simulation with 25 mW of input power at 219.4 GHz.

The circuit topology employed in the 94 GHz sheet-beam EIK can also be scaled up to a 220 GHz EIK [6], as illustrated by Fig. 17.8, based on simulations performed with the MAGIC-3D PIC code [26]. It is worth noting that MAGIC-3D was also employed in the preliminary exploration that led to the successful development of the W-band EIK. As shown, the circuit is driven by a 16.5 kV, 0.52 A sheet-beam. The synchronously tuned circuit is comprised of four cavities with a circuit length of 1.2 cm. The input and output cavities have eight gaps each, and both gain cavities have seven gaps. The peak power at the output waveguide is 453 W with 0.25 W of drive power, corresponding to a net electronic gain of 42.6 dB. The high gain in such a short distance is a result of efficient beam bunching. This can be noted by the tight and uniform bunching of the beam across its width at the output cavity, where peak *RF* beam current reaches approximately 800 mA. This corresponds to a fundamental bunching factor $f_b = 1.54$, which is an impressive bunching factor at this frequency for any device and, particularly, for a sheet-beam

Also shown in Fig. 17.8 is the beam kinetic energy averaged over one *RF* period. The plot illustrates the beam power being converted into *RF* power as reflected by the drop in beam kinetic

Figure 17.9 Photograph of five-gap G-band test cavity top view under a high-power microscope. The interaction gap width is only ~75 microns.

energy. In total, 780 W of beam power is converted into *RF* power, or 9.1% conversion efficiency. However, due to the high ohmic loss in this frequency regime, approximately 80 W is dissipated in the idler cavities, and an additional 243 W is dissipated in the output circuit. This level of ohmic loss emphasizes the important fact that careful attention must be paid to the fabrication process of high-frequency amplifiers, not only on fabrication tolerancing but also on the surface finish, as surface roughness will lower conductivity. Lower conductivity will result in more ohmic loss in the circuit and will also degrade the amplifier's overall performance.

At G-band, circuit fabrication within required tolerancing is quite a challenge, but it is not insurmountable. This statement is supported by the fact that EIKs up to 280 GHz have been fully demonstrated by CPI-Canada. For the sheet-beam EIK shown above, test cavities have also been fabricated. A sample photograph of such a cavity is shown in Fig. 17.9 for illustrative purposes. The results have been very promising, with a reasonably smooth surface finish and dimensional uniformity. This is particularly impressive in

Figure 17.10 MAGIC-3D simulation of a 670 GHz EIK (not 1:1 aspect ratio.) Colors correspond to axial electric field strength showing strong amplification.

consideration of the fact that the groove width of the interaction gaps is only ~75 microns.

17.6 THz EIKs

As we progress toward the true THz frequency, two factors are key to the successful implementation of such EIKs. These factors are: a higher current beam-forming system at a relatively low voltage to break free from the $1/f^2$ scaling limit for round beams, and a high precision fabrication approach with low surface roughness to ensure consistent and strong amplification with less ohmic loss. Shown in Fig. 17.10 is a simulation by MAGIC-3D for a synchronous-tuned six-cavity 670 GHz EIK. The amplifier is driven by a 100 mA, 25 kV electron beam in a 125 μm diameter beam tunnel. The circuit has a ladder topology operating in 2π-mode. The predicted output power is > 6000 mW with a drive power of 15 mW [8].

The circuit length is approximately 1.7 cm, which corresponds to an extremely high aspect ratio of approximately 270. As the formation and transport of such an intense electron beam with a current density of ~800 A/cm^2 in a tiny beam tunnel with a very large aspect ratio has never before been attempted, a beam stick with this beam tunnel dimension was built [8] and carefully tested [27]. The best beam transmission at 25 kV was 71% with an emitted beam current of 94 mA, which is quite remarkable for a first build with such challenging requirements.

The observed beam-stick performance over a wide range of parameters was methodically analyzed with the MICHELLE gun code, and two key factors were found to be responsible for the 29% beam interception. One was that the cathode was slightly forward of its designed position in the build, which had a minor impact on beam transmission. The other factor was a very slight 0.1° misalignment between the magnetic axis and beam tunnel. Given the extreme tunnel aspect ratio of 270, this very small tilt was sufficient to account for most of the beam interception. Once these factors were built into the MICHELLE model, agreement between observed data and code prediction was excellent over a wide range of operation parameters [28]. This type of analysis is critical for future improvements of the beam-forming and transport system, if feasible. Nevertheless, even as built and demonstrated, this beam-forming and transport system is eminently suitable for any EIKs that require slightly larger beam tunnel diameters, such as those at 500 GHz or below.

Another critical factor is the fabrication tolerance of the circuit. Irrespective of the circuit type, the critical dimensions for the fundamental mode of operation scale essentially as $\lambda/2$. Consequently, it is obvious that fabrication errors must also be reduced by $1/f$ as frequency increases. The importance of fabrication tolerance is illustrated by Fig. 17.12, which shows the behavior of peak power and bandwidth of the EIK as a function of fabrication errors with everything else being held constant. In this figure, the critical dimensions for each gap of the ladder circuit are randomly selected within the error limits specified. Peak power will drop sharply with fabrication errors by almost an order of magnitude for each half percentage point increase in fabrication errors with a concomitant increase in bandwidth. It should be noted that the results shown in Fig. 17.11 are representative of random errors only. The actual behavior will depend on the actual distribution of the errors within the circuit. In general, the circuit would perform better if the error distribution by chance was lower in the initial portion of the circuit, because in such a case the *RF* current will have a chance to form which results in better performance. Of course, depending on chance is not a design mode. However, continual advances in fabrication approaches, as detailed elsewhere in this volume [29], will at a minimum reduce the design risks due to fabrication errors associated with THz EIKs in implementation.

The THz regime also introduces another challenge besides fabrication errors. Based on careful measurements, Kirley and

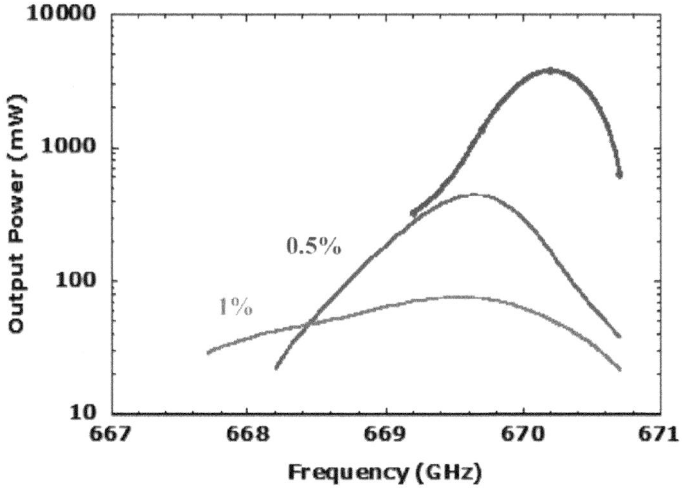

Figure 17.11 Impact of random errors of interaction-gap transverse dimensions on power and bandwidth.

Booske have found that THz dissipation loss in surfaces is extremely sensitive to the roughness, grain size, and purity of the material [30]. Consequently, in addition to fabrication errors, new fabrication approaches must also be mindful of the effective surface conductivity, as it also has a substantial impact on amplifier performance. This is as illustrated by Fig. 17.12(top), which plots the peak power as a function of effective conductivity with everything else held constant. The peak power scales approximately as effective surface conductivity to the 3.5 power, which demonstrates the strong impact of surface conductivity on THz EIKs. Figure 17.12(bottom) shows the behavior of peak power and bandwidth of the EIK at two surface conductivity values: 3.8 and 1.0 $\times$ 10^7 S/m. It is interesting to note that the impact of a factor 3.8 change in conductivity is very similar to that shown in Fig. 17.12 for the case of no fabrication error and that of 1% random errors.

Over the long run, the fabrication challenges that have been the key obstacle in the march toward a true THz EIK demonstration will very likely be resolved with continual progress in modern fabrication approaches. It should be emphasized that these fabrication challenges are not particular to only EIKs but also to

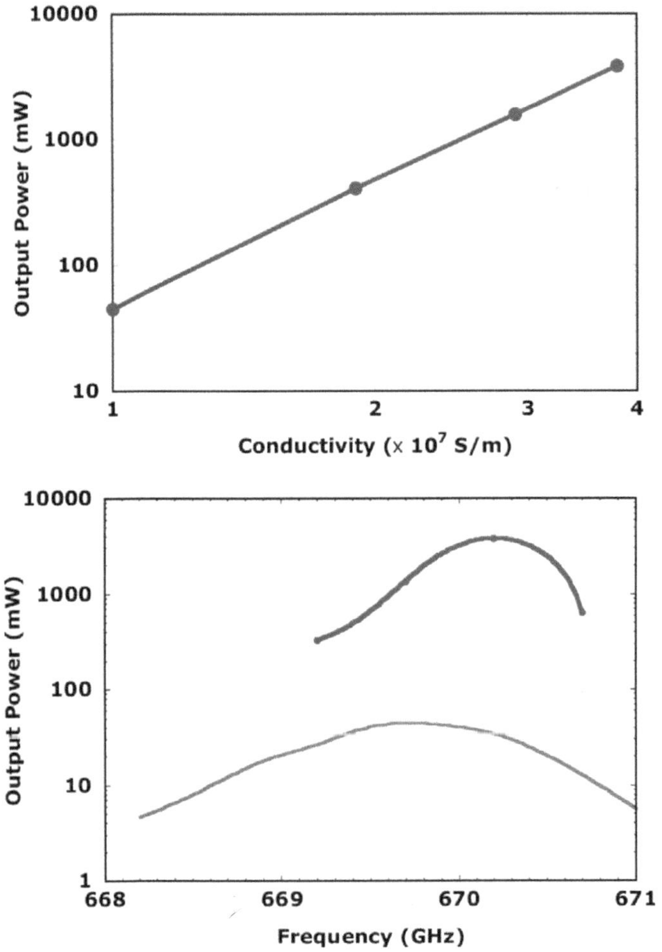

Figure 17.12 Impact of surface conductivities on peak power (*top*) and on power and bandwidth (*bottom*).

other amplifiers, where beam-wave interaction depends critically on the cutoff frequency of critical dimensions, such as traveling-wave tubes (TWTs). However, for the intermediate term, with the beam-forming approach demonstrated as discussed earlier, higher-order-mode EIKs can provide a potential solution. For round-beam ladder circuits, the TE_{33} mode appears to be the ideal higher-order mode step up from the fundamental mode TE_{11}. We note parenthetically that the mode number is always odd to accommodate the electron beam at the center. Thus, by employing the third-order mode, fabrication tolerance can be relaxed by approximately a factor of three. We note with interest that fundamental mode EIKs up to 264 GHz have been demonstrated by CPI and are commercially available from CPI. Thus, it appears that much higher frequency EIKs with acceptable fabrication tolerance can be achieved with existing fabrication approaches. A key trade-off in higher-order mode operation is the lower gap R/Q, which must be carefully compensated for with a larger number of gaps in the design, taking into account potential beam interception. Another solution to compensate for low R/Q is the use of a higher perveance sheet-beam. In the sheet-beam case, the operating frequency is controlled by the narrow cavity dimension, which is essentially half wavelength. The higher mode number of 3 in that direction will also enable high-frequency sheet-beam EIKs with existing fabrication techniques.

17.7 Conclusions

In this chapter, we have presented an overview of the current status of EIK with a particular emphasis on demonstrated amplifiers and existing designs. This overview is not meant to be exhaustive, but to serve as a brief survey of this exciting and unique type of amplifiers, particularly, in the challenging quest to sub-mm and THz frequency.

References

1. M. Chodorow and T. Wessel-Berg, A high-efficiency klystron with distributed interaction, *IRE Transactions on Electron Devices*, vol. 8, issue 1, page 44, Jan. 1961.
2. https://www.cpii.com/product.cfm/4/40.
3. D. Steer, A. Roitman, P. Horoyski, M. Hyttinen, R. Dobbs, and D. Berry, Advantages of extended interaction klystron technology at millimeter and sub-millimeter frequencies, *2007 16th IEEE International Pulsed Power Conference*, June 2007.

4. D. Berry, H. Deng, R. Dobbs, P. Horoyski, M. Hyttinen, A. Kingsmill, R. MacHattie, A. Roitman, E. Sokol and B. Steer, Practical aspects of EIK technology, *IEEE Transactions on Electron Devices*, vol. 61, issue 6, page 1830, June 2014.
5. J. Pasour, E. Wright, K. Nguyen, A. Balkcum, F. Wood, R. Myers, and B. Levush, demonstration of a multikilowatt, solenoidally focused sheet beam amplifier at 94 GHz, *IEEE Transactions on Electron Devices*, vol. 61, issue 6, page 1630, June 2014.
6. K. Nguyen, J. Pasour, E. Wright1, D. Pershing, and B. Levush, Design of a G-band sheet-beam extended-interaction klystron, *Proceedings of the IEEE International Vacuum Electronics Conference (10th) (IVEC 2009)*, page xxx, 2009.
7. K. Nguyen, et al., Design of terahertz extended interaction klystrons, *Proceedings of the IEEE International Vacuum Electronics Conference (11th) (IVEC 2010)*, page xxx, 2010.
8. D. Chernin, et al., Development of a 670 GHz extended interaction klystron amplifier, *IEEE International Conference on Plasma Science*, June 2011.
9. R. Li, C. Ruan, A. Kosar Fahad, C. Zhang, and S. Li, Broadband and high-power terahertz radiation source based on extended interaction klystron, published online: Scientific Reports https://doi.org/10.1038/s41598-019-41087, 14 March 2019.
10. N. Guo, Q. Xue, Z. Qu, K. Liu, W. Song, X. Zhang; D. Zhao, and H. Ding, Study of a 0.34 THz ladder-type extended interaction klystron with narrow coupling cavities, *IEEE Transactions on Electron Devices*, vol. 68, issue 11, page 5851, Nov. 2021.
11. S. Li, J. Wang, H. Xi, D. Wang, B. Wang, G. Wang, and Y. Teng, Optimum design and measurement analysis of 0.34 THz extended interaction klystron, published online: *AIP Advances*, 8, 025101 (2018); https://doi.org/10.1063/1.5020703.
12. S. Li, F. Zhang, C. Ruan, Y. Su, and P. Wang, A G-band high output power and wide bandwidth sheet beam extended interaction klystron design operating at TM31 with 2π mode, published online: *Electronics*, 10, 1948. https://doi.org/10.3390/electronics10161948, 12 August 2021.
13. K. Zhang, K. Chen, Q. Xu, W. Xu, N. Xiong, X. Chen, and D. Liu, Primary design of extended interaction klystron with multi-gap cavity at 225 GHz, published online: *Particles* 2018, 1, 260–266; https://doi:10.3390/particles1010020, 11 Nov. 2018.

14. K. Nguyen, J. Pasour; T. Antonsen, Paul B. Larsen, J. Petillo, and Baruch Levush, Intense sheet electron beam transport in a uniform solenoidal magnetic field, *IEEE Transactions on Electron Devices*, vol. 56, issue 5, page 744, May 2009.
15. J. Booske, M. Basten, A. Kumbasar, T. Antonsen, Jr, S. Bidwell, Y. Carmel, W. Destler, Granatstein, D. Radack, Periodic magnetic focusing of sheet electron beams, *Physics of Plasmas*, vol. 1, page 1714, 1994.
16. K. Nguyen, J. Pasour, E. Wright, J. Petillo, and B. Levush, High-perveance W-band sheet-beam electron gun design, *Proceedings of the IEEE International Vacuum Electronics Conference (9th) (IVEC 2008)*, page 179, 2008.
17. J. Pasour, K. Nguyen, E. Wright, A. Balkcum, J. Atkinson, M. Cusick, and B. Levush, Demonstration of a 100 kW solenoidally focused sheet electron beam for millimeter-wave amplifiers, *IEEE Transactions on Electron Devices*, vol. 58, issue 6, page 1792, June 2011.
18. D. Pershing, K. Nguyen D. Abe, E. Wright, P. Larsen, J. Pasour, S. Cooke, A. Balkcum, F. Wood, R. Myers, and B. Levush, Demonstration of a wideband 10 kW Ka-band sheet beam TWT amplifier, *IEEE Transactions on Electron Devices*, vol. 61, issue 6, page 1637, June 2014.
19. J. Pasour, K. Nguyen, E. Wright, and B. Levush, Sheet beam EIK sensitivity to multimoding and circuit imperfections, *Proceedings of the IEEE International Vacuum Electronics Conference (11th) (IVEC 2010)*, page 45, 2010.
20. D. Yu and P. Wilson, Sheet-beam klystron *RF* cavities, *Proceedings Particle Accelerator Conference*, page 2681, May 1993.
21. ANSYS Inc., Canonsburg, PA 15317, USA [Online]. Available: http://www.ansys.com.
22. S. Cooke, GPU-accelerated 3D large-signal device simulation using the particle-in-cell code 'Neptune', *Proc. IEEE 13th IVEC*, p. 21, April 2012.
23. J. Petillo, K. Eppley, D. Panagos, P. Blanchard, E. Nelson, N. Dionne, et al., The MICHELLE three-dimensional electron gun and collector modeling tool: theory and design, *IEEE Transactions on Plasma Science*, vol. 30, issue 3, page 1238, Jun. 2002.
24. ANSYS Inc., Canonsburg, PA 15317, USA [Online]. Available: http://www.ansys.com.
25. Analyst Reference.
26. B. Goplen, L. Ludeking, D. Smith, and G. Warren, User-configurable MAGIC for electromagnetic PIC calculations, *Computer Physics Communications*, vol. 87, issue 1–2, page 54, May 1995.

27. J. Calame, B. Levush, R. Dobbs, Dave Berry, K. Nguyen, E. Wright, and D. Chernin, Experimental testing of an electron gun and beam transport system for a 670 GHz extended interaction klystron, *Proceedings of the IEEE International Vacuum Electronics Conference (13th) (IVEC 2012)*, page xxx, 2012.
28. K. Nguyen, E. Wright, J. Calame, B. Levush, J. Pasour, R. Dobbs, D. Berry, D. Chernin, and J. Petillo, Terahertz beamstick performance diagnostics using MICHELLE, *Proceedings of the IEEE International Vacuum Electronics Conference (13th) (IVEC 2012)*, page xxx, 2012.
29. C. Joye, A. Cook, et al., Fabrication chapter, in these proceedings.
30. M. Kirley and J. Booske, Terahertz conductivity of copper surfaces, *IEEE Transactions on Terahertz Science and Technology*, vol. 5, issue 6, page 1012, November 2015.

Chapter 18

THz Oscillators Based on Cherenkov, Smith–Purcell and Hybrid Radiation Effects

Konstantin Lukin,[a] Eduard Khutoryan,[a] Alexei Kuleshov,[a] Sergey Ponomarenko,[a] Matlab Sattorov,[b] and Gun-Sik Park[b]

[a] *O. Ya. Usikov Institute for Radiophysics and Electronics of NAS of Ukraine, Kharkiv, Ukraine*
[b] *Seoul National University, Seoul, Republic of Korea*
lukin.konstantin@gmail.com

18.1 Introduction

Considered in this chapter, classical vacuum electron devices based on "electron beam–slow-wave" synchronism, such as backward wave oscillator (BWO) [1], Carcinotron [2], clinotron [3], orotron [4], and diffraction radiation oscillator (DRO) [5, 6], have been widely used in microwaves due to outstanding performances combining high levels of output power and wide frequency tuning range. Therefore, at present many research and development activities are focused on filling the THz gap with the help of mentioned above

Advances in Terahertz Source Technologies
Edited by Gun-Sik Park, Masahiko Tani, Jae-Sung Rieh, and Sang Yoon Park

ISBN 978-981-4968-89-8 (Hardcover), 978-1-003-45967-5 (eBook)
www.jennystanford.com

devices. However, there are several reasons for a drastic output power drop in those devices with the wavelength shortening that results in small efficiency of those tubes in the THz frequency range and, therefore, it limits their practical applications. Among these reasons, the most special ones are as follows: (1) technological constraints in the manufacturing of small-scale slow-wave circuits and other components of the tubes; (2) requirements in generation and transportation of the intense electron beams including the problems of fabrication of magnetic system; (3) increase in electromagnetic wave attenuation caused by ohmic losses due to metal surface roughness and skin layer effects; (4) problems of electromagnetic energy extraction from the small-scaled slow-wave circuits, as well as mode competitions, etc.

In this chapter, the consideration and discussions of physical principles and the state of the art in the design of these devices, as well as some earlier and current activities for mastering the THz gap using Cherenkov devices, are presented. Special attention is paid to the theory of Cherenkov and Smith–Purcell radiation (SPR) effects (Section 18.2). The review of the design and performance of the THz BWOs and clinotrons is presented in Section 18.3. Mode transformation in the oversized cavities and its effect on the output performances of the Cherenkov devices are considered in Section 18.4. The characteristics of the devices based on Smith–Purcell/Diffraction radiation are discussed in Section 18.5. A new concept of Cherenkov device based on the hybrid modes of a cavity containing a biperiodic structure that combines the features of both the surface waves and the bulk waves with unique performance is presented in Section 18.6.

18.2 Theory of Cherenkov and Smith–Purcell/Diffraction Radiation Effects

The motion of electron beams along and close to a periodic structure (grating) may cause radiation effects of different types. Consider a simple case of the electron beam with current density modulated according to a harmonic law:

$$\vec{j}(zt) = \vec{z}_0 \rho_{e0} v_e \delta(y - a) \exp\left[i(kz/\beta - \omega t)\right], \tag{18.1}$$

where ρ_e and v_e are the density and speed of the electron beam, respectively; $\beta = v_e/c$; $k = \omega/c$, c is the speed of light; ω is modulation frequency; a is the beam-grating distance; and $\delta(y)$ is the Dirac delta function.

If we may ignore the back influence of the radiated electromagnetic field onto the electron's motion, then a simplified analysis of the radiation effects may be reduced to the analysis of the scattering of the eigen field of the electron beam in open space by a periodic structure [6].

In this case, from Maxwell's equations and Fouquet theorem it follows that the electric component of the scattered field may be represented as a set of spatial harmonics [6]:

$$E_z(yz) = \frac{1}{ik} \sum_s A_s g_s \exp\left\{i\left[k_{ys}(y + a) + k_{zs} z\right]\right\}, \tag{18.2}$$

where A_s is the amplitude of the s-th spatial harmonic; l is the basic grating period as shown in Fig. 18.1, and

$$k_{zs}^2 = k/\beta + 2\pi s/l;\ k_{ys}^2 = k^2 - k_{zs}^2 \tag{18.3}$$

are the wave numbers of spatial harmonics corresponding to the space coordinates along and across electrons' motion.

From Eqs. (18.2) and (18.3), it follows that in the case of $k^2 < k_{zs}^2$ for any n, the scattered field is a set of *inhomogeneous plane waves*, with the amplitudes exponentially decaying when the distance from the periodic structure surface is growing. In other words, electromagnetic energy propagates along the grating. In

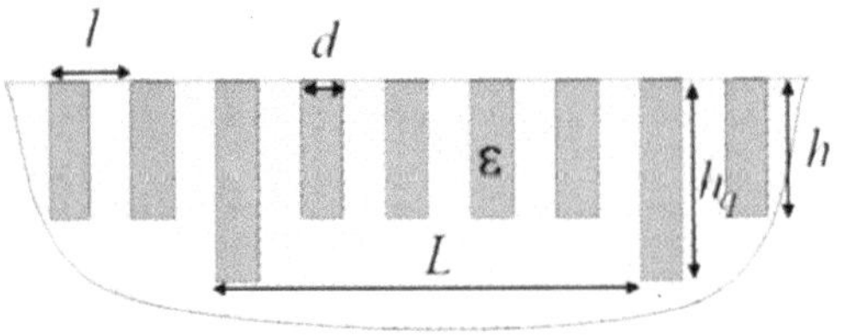

Figure 18.1 Example of biperiodic grating with period $L = 5\,l$ ($N = 5$).

microwave vacuum electronics, devices based on such radiation effect are related to the class of *Cherenkov devices* [1–3, 7–9].

In long-wavelength approximation ($l \ll \lambda$, where $\lambda = 2\pi/k$ is the wavelength) that holds for nonrelativistic vacuum electron devices, the eigenmodes of a uniform slow-wave periodic structure (usually, a *reflective grating*) are the surface waves propagating along the grating axis [3, 6, 7–9].

In a nonuniform, or biperiodic, grating (in which every N^{th} groove is of the modified depth) having second period $L = Nl$, among its eigenmodes may exist a leaky wave mode [3, 10–16]. The set of equations to find the fields which could be excited by the bunched beam having current density (18.1) with $a = 0$ in long-wavelength approximation ($l \ll \lambda$) was obtained in [10–12] and is given below:

$$\sum_{p_0=1}^{N} D_{p_0}\left[\delta_p^{p_0}\cos\left(\sqrt{\varepsilon}kh_p\right) + i\frac{kd}{L}\sin(\sqrt{\varepsilon}kh_{p_0})\sum_{s=-\infty}^{\infty}\left(\frac{\sin(k_{zs}d/2)}{k_{zs}d/2}\right)^2\frac{e^{ik_{zs}(p-p_0)l}}{k_{ys}}\right] = j_0 e^{ipk_0 l}, \tag{18.4}$$

where $k_{zs} = k_0 + 2\pi s/L$; $L = Nl$; l is the period of a basic grating; d is groove width; h_q is q^{th} groove height, $q = 1, \ldots N$. A dielectric with complex permittivity $\varepsilon = 1 + i\varepsilon''$ was placed into the grooves to study the effect of dielectric and ohmic loss, D_q are the mode amplitudes in q^{th} groove.

From the analysis of the dispersion equation (Eq. (18.4) with zero right-hand side) and RF field pattern (Fig. 18.2), it follows that in leaky mode, along with the surface wave, propagating along the grating surface, appears a radiation outgoing into free space. This

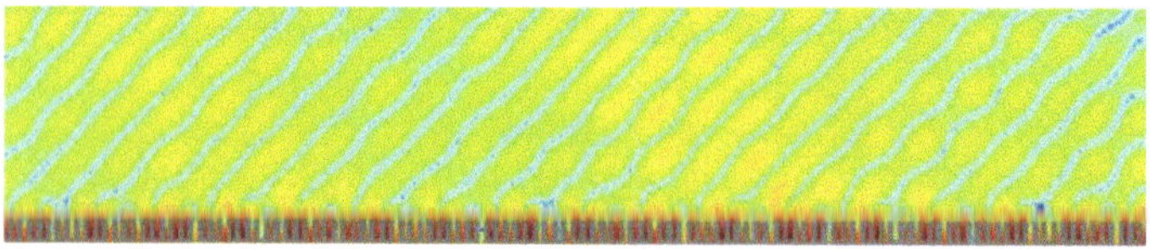

Figure 18.2 RF field excited by a bunched electron beam moving above nonuniform grating.

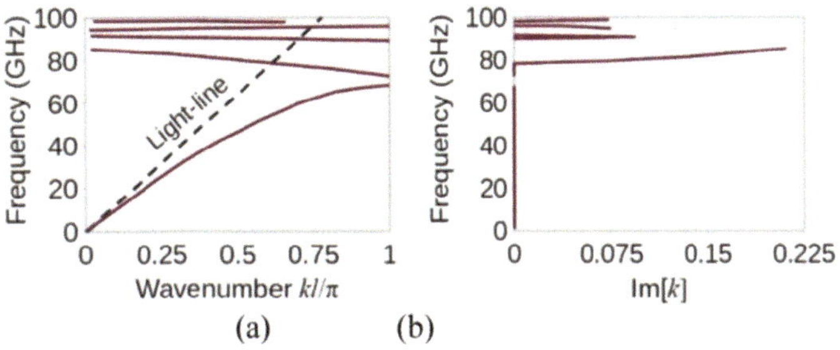

Figure 18.3 Example of dispersion of nonuniform grating with $N = 5$ designed for 100 GHz clinotron [12]. (a) Real part of wavenumber; (b) imaginary part of wavenumber when its real part is above the light line.

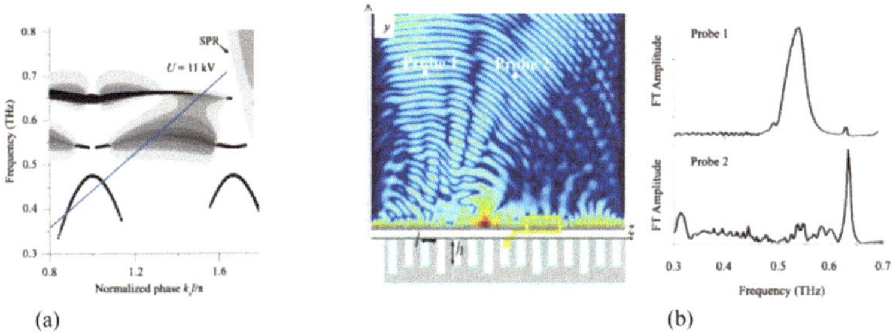

Figure 18.4 (a) $\omega - k$diagram of open three-stage grating (every third groove is of modified groove depth $h_3 \neq h_1$)$l = 0.07$ mm, $h_1 = 0.09$ mm, $h_3/h_1 = 1.3$; (b) MACIC2D simulation of radiation excited by a single bunch (RF field pattern and spectra of probes signal).

corresponds to appearing of the imaginary part of the wavenumber when its real part is above the light line (Fig. 18.3).

When the electron bunch moves above such grating with velocity coinciding with some harmonic of leaky wave mode, the radiation is excited at the frequency of the eigenmode (Fig. 18.4) providing enough strength of beam-wave coupling. The radiation angle is defined in the same way as for the diffraction radiation (SPR) case. Theoretical and PIC simulations have shown that at high enough grating material conductivity, the intensity of leaky wave mode radiation into free space is much higher than the pure diffraction radiation (SPR) intensity as shown in Fig. 18.5.

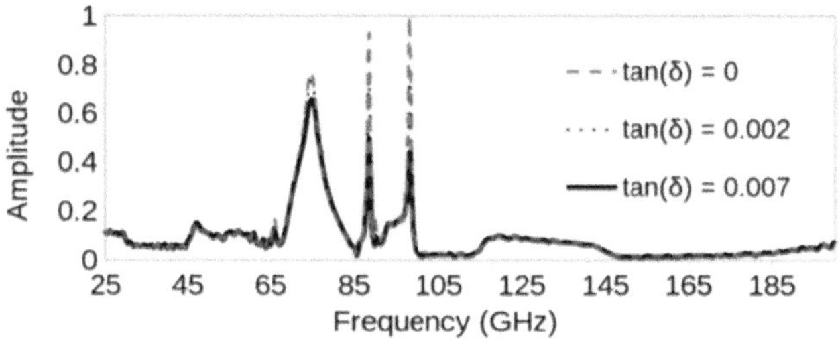

Figure 18.5 Simulations of the leaky wave intensity versus beam modulation frequency for $\beta = 0.12$ and for different dielectric losses.

Unlike Cherenkov devices, the DRO uses a special case of individual radiation of charged particles—diffraction radiation (Smith–Purcell radiation) of electrons moving over a periodic structure as a single-velocity beam [7, 17]. If for some s (or several s) the following condition is satisfied:

$$k^2 > k_{zs}^2, \tag{18.5}$$

then a plane uniform wave (or several waves) appears in the spectrum of the scattered field, leaving the structure without damping at the angle $\theta_s = \arccos(k_{zs}/k)$ to the direction of electrons' motion. Such a regime is called the diffraction radiation regime [6, 17], and condition (18.5) is called the radiation condition. Introducing the notation, $æl/\lambda = kl/2\pi$, we rewrite this condition in the form:

$$|\cos\theta| = \left|\frac{æ/\beta + s}{æ}\right| < 1.$$

Since $\beta < 1$ always, it follows from (18.5) that the field of diffraction radiation is described only by harmonics with negative numbers $s = -1, -2, \ldots$ Condition (18.5) is usually satisfied in some ranges of variation of the parameters æ, β, and s. These areas are called radiation zones and are clearly displayed in the Brillouin diagram constructed on the basis of condition (18.5) [18].

When æ $> 1/2$, several spatial harmonics can be radiated simultaneously at one frequency, each propagates at its own angle θ_s . Such modes can be useful for generating powerful microwave oscillations in relativistic and weakly relativistic DROs.

To create an oscillator based on the effect of diffraction radiation, it is necessary to organize feedback on the radiated field. For this purpose, a metal surface is placed above the diffraction grating, which reflects the emitted wave onto the electron beam. At certain ratios between the wavelength and the distance from the grating to the reflector, positive feedback is realized in the system.

It can be seen [6, 7] that the DRO is built according to the scheme of an optical laser. The difference is that "electron beam + periodic structure" is used as an active medium in the DRO. This means that the energy of free (and not bound) electrons is converted into the energy of the electromagnetic field. Therefore, DRO is sometimes related to the class of electron devices called free electron lasers (FEL).

18.3 Principles of THz BWO Design and Challenges for Efficient Generation in the THz Range. The Clinotron Effect

BWOs belong to Cherenkov devices, i.e., vacuum electron devices based on slow electromagnetic waves. To slow down the phase velocity of electromagnetic waves in the interaction area of the tube, there are many configurations of slow-wave circuits applied in BWOs such as helix, folded waveguide, corrugated waveguide, coupled resonators, planar grating, multistage grating, double-side grating, staggered grating, meander, interdigital lines, etc. The choice of the slow-wave circuit is conditioned by the required frequency tuning range, levels of output power, technological limits in manufacturing, matching of the slow-wave circuit with the output waveguide or transmission line, configuration of electron beams, beam focusing systems, etc.

In THz BWOs produced by "Istok," Russia, the application of either interdigital lines or gratings along with the sheet electron beams allowed to cover the frequency range from 0.18 to 1.5 THz with the help of nine tubes [19]. Typically, the accelerating voltage range of BWOs produced by Istok is from 1 to 6 kV, while the electron beam current is less than 50 mA, which offers several mW of output power in the mentioned frequency ranges. It should be noticed that such BWOs offer the widest range of electronic frequency tuning determined by the dispersion characteristics of the slow-wave circuit.

In BWOs, *the carcinotrons*, manufactured by Thomson-CSF, the applied vane type circuit with the slot for the transportation of the pencil-like electron beam provided higher levels of output power in comparison to the Istok tubes, however, the frequency tuning ranges of carcinotrons are much narrower [20]. The watt level of the output power of both BWOs and carcinotrons in the frequency range lower than 100 GHz was achieved in the accelerating voltage range from 6 to 10 kV [21].

An attempt to develop 0.65 THz BWO with a helical slow-wave circuit with a predicted start current as low as 3.5 mA at 7 kV was done in order to increase the output power up to 60 mW in the wide frequency band [22]. The helix was fabricated lithographically from the gold wire grown onto a diamond sheet and mounted within a copper block [23]. Pencil-like electron beam was transported above the helix in a magnetic focusing field, however, in experiments the misalignments of the cathode position with the helix resulted in no detection of RF power [22].

Two different topologies of slow-wave circuits such as double-staggered grating and double-corrugated waveguide were proposed and manufactured for 0.346 THz BWO [24]. The double-staggered circuit was developed to operate with an elliptical electron beam at 17 kV, while a double-corrugated waveguide was designed for the operation with a cylindrical electron beam at voltages less than 13 kV. Fabricated slow-wave circuits were matched with the output waveguides and the cold test results agreed well with the simulation results obtained with the help of CST MWS [25]. Currently, the fabrication of the tubes is in progress and the results of the BWOs performance will be reported later.

In order to improve the coupling of the excited modes in the BWO interaction area with the modes of the output waveguide, the photonic crystal corrugated waveguide with a photonic crystal coupler was proposed and simulated for BWOs with wide sheet electron beams [26]. The results of 3-D particle-in-cell simulations show the output power higher than 70 mW for the 0.65 THz BWO with the 120 periods of slow-wave circuits and 11 kV, 6 mA electron beam [26].

Several reports were dedicated to the development of powerful BWOs with folded waveguides [27, 28]. Application of a cylindrical electron beam at 10 kV, 50 mA in BWO based on folded waveguide with the period 332 um predicted the watt level of output power at 220 GHz [27], however, no experimental results on the performance of the BWOs are available.

18.3.1 Principle of the Clinotron

Application of thick sheet electron beams moving along the surface of the slow-wave circuit with the small inclination angle was proposed in order to essentially increase the levels of the output power of BWOs, DROs and orotrons based on the single-side grating in millimeter and THz ranges [3, 7, 29]. The powerful type of BWO based on the grating with the inclined electron beam, *the clinotron*, was found in IRE NASU in 1954. Modern modifications of these tubes are attractive for a big number of practical applications in millimeter and THz ranges [3, 30–33]. In modern constructions of clinotron tubes great attention is paid to the conditions of reflections of surface wave from both ends of the grating [31], thus, additional mechanical frequency tuning in the tubes of millimeter range is realized with the help of the plunger in the waveguide at the collector area as it is shown in Fig. 18.6.

Inclination of the electron beam to the surface of the slow-wave circuit provides the condition that every layer of thick electron beam interacts with the surface wave, hence, the output power of the clinotron can be considered as a sum of every layer of electron beam [3]:

$$P_\Sigma = P_i\,(L_i)\,n\,(\varphi)$$

The optimal inclination angle of sheet electron beam varies from several degrees at frequencies about 30 GHz down to several minutes at frequencies higher than 200 GHz. The typical cross section of sheet electron beam is 2.5 × 0.15 mm^2 for the tubes in THz range, while the beam current up to 200 mA should be provided in the accelerating voltage range from 4 to 6 kV. The diode electron

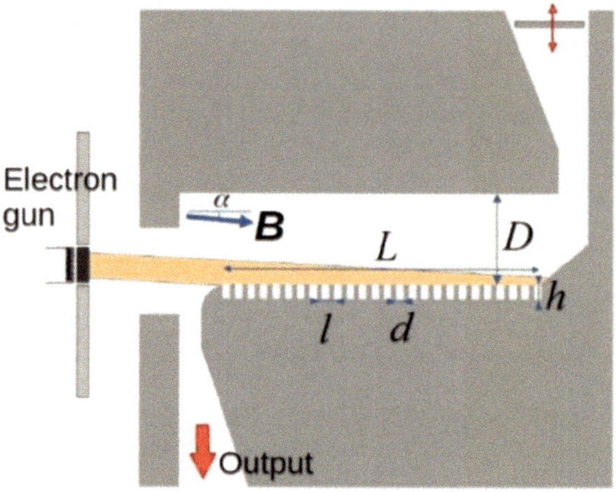

Figure 18.6 Sketch of the clinotron.

(a)

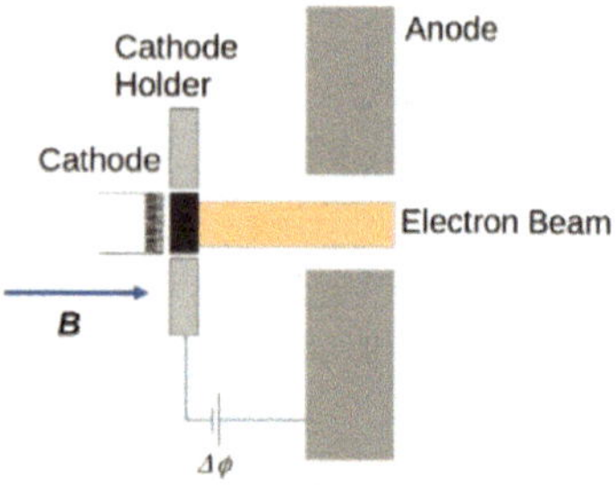

(b)

Figure 18.7 Photo of the thermionic cathode mounted on the ceramic insulator (a) and the scheme of diode electron gun operation (b).

gun with a thermionic cathode generates the intense sheet electron beam in the presence of focusing magnetic field (Fig. 18.7).

The series of compact magnetic system was developed to focus the intense sheet electron beam in the clinotron tubes of different frequency ranges [31, 35]. The clinotrons operating in the frequency range from 20 to 150 GHz require the magnetic field of 0.5 T in the gap between the poles of 32 mm, while THz tubes require the magnetic field higher than 1 T. The sketch of the magnetic system and the distribution of the magnetic longitudinal component along

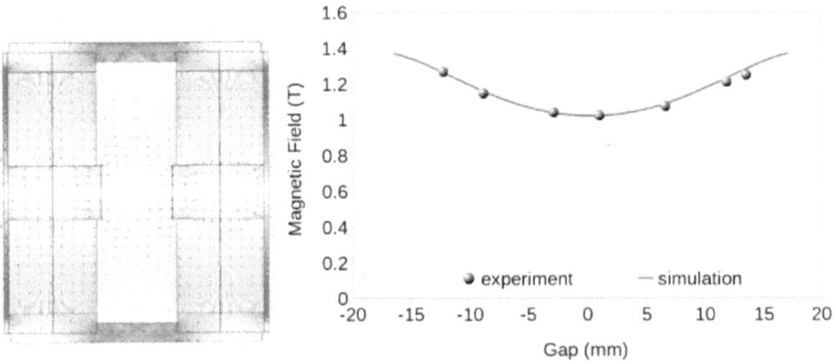

Figure 18.8 The simulation and experimental test of 1 T magnetic focusing system.

the gap between the poles for the 1 T magnetic system is shown in Fig. 18.8.

The application of the special nonsymmetrical electron optical system in the THz clinotrons provide the increase of the beam-wave interaction efficiency [3, 36]. In early works of the founders of the clinotron, the idea of an application of nonsymmetrical optics was to reject the pulsing boundary layers of electron beam close to the grating surface [3]. However, in recent works [33, 36] the nonsymmetrical electron gun was applied to provide the electron velocity distribution of the sheet beam when the electron longitudinal velocities decrease with the distance from the grating surface.

According to the dependence of RF field amplitude along the grating in the THz clinotrons (Fig. 18.9a) and taking into account the dependence of the optimal mismatch between electron and wave velocity on the RF field amplitude (Fig. 18.9b), the application of the thick electron beam with the profiled distribution of electron longitudinal velocities provides the interaction of faster electrons with the surface wave in the area of maximal RF field amplitude, while the slower electrons interact with the RF field of lower amplitude in the case of the beam inclination to the grating surface. The optimal profiled distribution of electron velocities of sheet beam provides the substantial increase of the beam-wave interaction efficiency in the THz clinotrons [36].

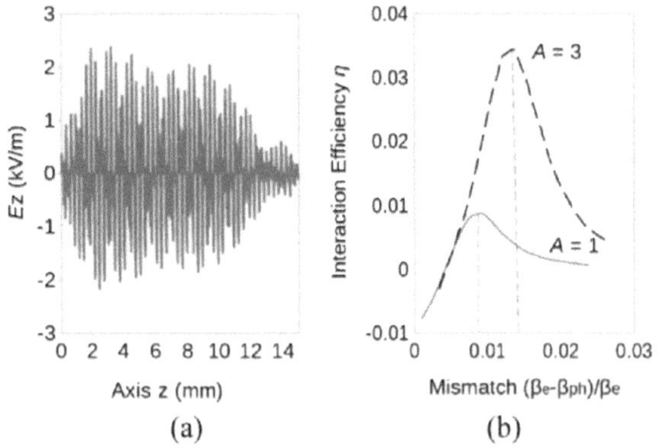

Figure 18.9 Distribution of RF field amplitude of a surface wave along the slow-wave circuit in the THz clinotron (a) and the dependence of the optimal mismatch between electron and wave velocity for different RF field amplitudes (b).

Manufacturing tolerances in the fabrication of the slow-wave circuit of the grating type for the THz vacuum electron devices should provide minimal attenuation of the propagating surface wave. Electrical discharge machining offers metal surface roughness of microns level after the chemical polishing and the surface wave attenuation can be estimated with the help of Hammerstad or Huray models [37]. In clinotrons, the inclined electron beam heats the grating, therefore, both surface roughness and grating heating effects should be considered in simulations of ohmic losses and output power levels [38]. Such effects decrease the efficiency of the clinotron tubes from 10% in the Ka band down to about 1% in the W band. It should be mentioned that typically the power of a sheet electron beam heating the grating of the clinotron is in the range from 300 W (Ka band) to 700 W (W band) and the water-cooling system removes the thermal energy from the grating.

In the frequency ranges from 100 to 200 GHz the clinotron interaction cavity becomes oversized since the tubes utilize the grating with a width of 2.5 mm and the sheet electron beam with the current density less than 100 A/cm^2. Oversized waveguide with plunger becomes less efficient in the mechanical tuning of the operating frequency in comparison with the case of the surface

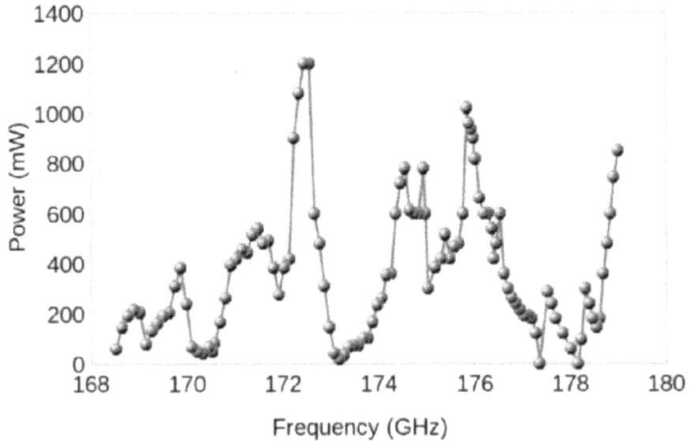

Figure 18.10 Dependence of the output power on operating frequency for the 175 GHz CW clinotron tube operating in the accelerating voltage range from 3.8 to 5 kV with the sheet electron beam current of 150 mA.

wave resonator at lower frequencies [39]. The output power of the clinotron tubes drops from several Watts at 100 GHz down to about one watt at 175 GHz (Fig. 18.10).

The investigation of radiation spectra of THz clinotron has been conducted in order to comply with the requirements of spectroscopic applications such as DNP NMR, ESR, etc. In the case of the Cherenkov synchronism condition the operating frequency strongly depends on accelerating voltage. Thus, the developed high-voltage power supply with the levels of voltage ripples less than 5 ppm [40] has provided the stable long-term operation of the CW clinotron tubes with the radiation linewidth less than 1 MHz in the frequency range 300–400 GHz (Fig. 18.11). The application of both PLL and PID algorithms can offer the narrowing of the radiation linewidth to comply with the specific requirements of practical applications [33, 41].

18.4 Mode Transformation in Oversized Circuits in the THz Range

Dimensions of the THz VEDs are usually much longer than the operating wavelength, especially when wide sheet beams are em-

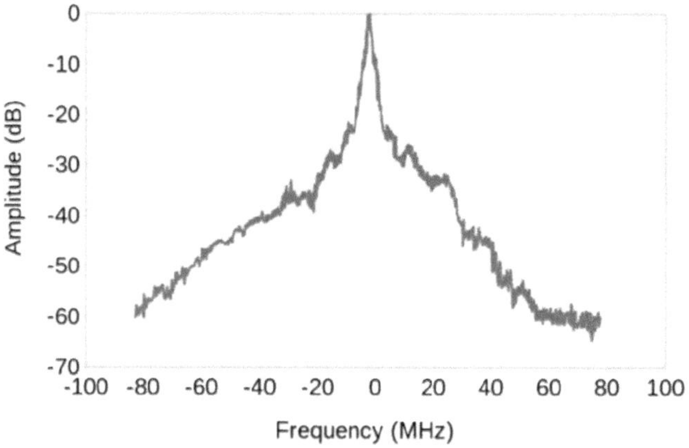

Figure 18.11 The 350 GHz CW clinotron spectrum in the case of 5 kV, 135 mA electron beam, the output power is 100 mW.

ployed [42–51]. Therefore, cavities of the THz BWOs are oversized ones and inhomogeneities located in cavity ends cause not only reflection of the surface mode which interacts with an electron beam [43] but also the transformation of this surface mode into fast modes and vice versa [34, 42]. This results in a redistribution of the power radiated into output waveguide, gun area and back into interaction space. Fast modes reflected back into interaction space might strongly affect the resonant property of a BWO since they are much less sensitive to the ohmic loss than the surface mode. Let's notice that experimentally the effect of the fast mode on a clinotron output parameters was discussed in [3] when the output power of a surface wave clinotron strongly depended on the grating-roof distance D, when it was more than several wavelengths. Obviously, the consideration of the mode transformation would not only provide a more accurate calculation of the THz VED, but also assist the VED design with improved performances due to (1) the optimization of the radiation propagating into the output waveguide; (2) the decrease of the radiation heating a cathode; (3) the decrease of the starting current due to increase of the Q-factor and hence, the feedback by the fast mode, etc.

Let's consider the theoretical model accounting the mode transformation and the example of the simulation of the THz

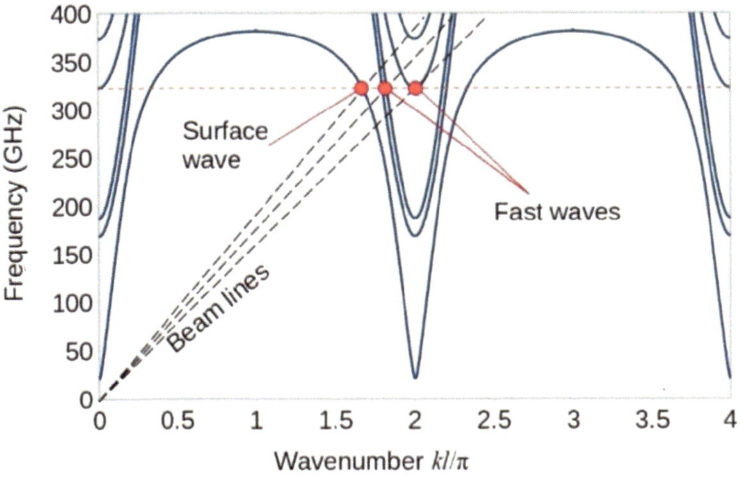

Figure 18.12 2D SWS dispersion of the 0.3 THz CW clinotron.

clinotron [32–34]. For simplicity, Fig. 18.12 shows 2D dispersion of the closed waveguide containing the grating with parameters: period l = 100 um, groove height h = 170 um, D = 800 um. One can see that even in the 2D case, one surface and three fast modes exist at the frequency 0.34 THz. The clinotron design consists of the closed waveguide with a grating, output waveguide and anode hole forming T-junction at the one end of a grating. In the collector grating end a plunger is located for the mechanical frequency tuning [3, 34] (Fig. 18.13). The problem of finding the scattered at the grating ends EM fields and their decomposition by eigenmodes of corresponding waveguide was solved in [34, 46, 47]. The found matrix elements $S_{km}(i, j)$ show j^{th} mode transformation of the mth waveguide (port) into i^{th} mode of the kth waveguide (port). Therefore, coefficients $S_{22}(i, j)$ indicate reflection of EM wave back into the interaction space, $S_{02}(i, j)$ indicate propagation to the output and $S_{12}(i, j)$ indicate the radiation propagating through anode slot and showing the power heating a cathode.

The electric field is represented as the sum of forward and backward waves of all modes propagating in the interaction waveguide. Every mode in its turn can be represented as Floquet

Figure 18.13 Sketch of a clinotron electromagnetic system.

series:

$$\vec{E} = e^{-i\omega t} \sum_m \sum_n \left[C_m^+ \left(z, t\vec{E}_{mn}(x, y, k)\, e^{ik_{mn}z} \right) + \right.$$
$$\left. + C_m^- \left(z, t\vec{E}_{-mn}(x, y, k)\, e^{-ik_{mn}z} \right) \right], \quad (18.6)$$

where E_{mn}, k_{mn} are the electric field transverse distribution and the wavenumber of the nth harmonic of the mth mode.

Boundary conditions at the SWS ends ($z = 0$; $z = L$) are set using the found S-matrix as transformations of the backward wave into the forward wave at the T-junction and vice versa at the collector end as follows:

$$C_m{}^+(0, t) = \sum_j S_{22}(m, j, f_0) C_j{}^-(0, t) + \sum_j \frac{dS_{22}(m, j, f_0)}{df} \frac{\partial C_j^-(0, t)}{\partial t}$$
$$C_m{}^-(L, t) = \exp(i2k_m L)$$
$$\times \sum_j \left(\begin{array}{l} R_{11}(m, j, f_0) C_j{}^+(L, t) + \frac{dR_{11}(m, j, f_0)}{df} \frac{\partial C_j{}^+(L, t)}{\partial t} + \\ + R_{10}(m, j, f_0) \sum_i R_{01}(i, j, f_0) C_j{}^+(L, t - t_{pl}) \exp(i2kL_{pl}) \end{array} \right)$$
$$(18.7)$$

Here, derivatives on frequency and on time arise because the process is not purely harmonic one, and nonstationary equations are used and because of the electron frequency shift. L_{pl} is the length of the plunger and t_{pl} is the time delay of the wave propagating to the plunger short and back to the grating.

In the case of BWO and clinotron, only the 1st harmonic of a surface wave interacts with an electron beam. In the 3D case when grating width equals to several wavelengths $w > \lambda$, transverse (with variations m_0 across lamella, $m_0 \approx 2\ w/\lambda$) surface mode competition may occur. However, bulk modes are not synchronous with electrons as one can see from the dispersion in Fig. 18.12. Therefore, the propagation of backward and forward waves is governed by nonstationary equations [34]

$$v_{gri}^{-1}\frac{\partial C_i^-}{\partial t} - \frac{\partial C_i^-}{\partial z} + \gamma_i C_i^- = \frac{2}{N_{-i}}\int_S \vec{j}(t)\vec{E}_{-i}^{*} e^{i\omega t} dS,\ i < m_0$$

$$v_{grm}^{-1}\frac{\partial C_m^-}{\partial t} - \frac{\partial C_m^-}{\partial z} + \gamma_m C_m^- = 0,\ m > m_0 \tag{18.8}$$

$$v_{grm}^{-1}\frac{\partial C_m^+}{\partial t} + \frac{\partial C_m^+}{\partial z} + \gamma_m C_m^+ = 0,\ m = 0,\ 1,\ 2, \ldots$$

Beam current $\vec{j}(t)$ is found from the solution of three-dimensional motion equations for N macro particles with the right-hand part consisting of the electric RF synchronous with an EB and the static focusing magnetic fields. Since the attenuation constant γ_i for surface mode is high, it is strongly attenuated during traveling in the forward direction after being reflected back into the interaction space. For the fast bulk wave, the attenuation constant γ_i is much lower that results in resonant feedback by the fast mode.

After determining all amplitudes, the output power is found as the sum of power of modes propagating into the output waveguide:

$$P_{out} = \sum_m \left| \sum_j S_{02}(m,\ j,\ f_0) C_j^-(0,\ t) + \sum_j \frac{dS_{02}(m,\ j,\ f_0)}{df}\frac{\partial C_j^-(0,\ t)}{\partial t} \right|^2 \tag{18.9}$$

18.4.1 Simulation and Experimental Results

Motion equations together with excitation Eq. (18.8) have been solved numerically at given beam velocity till steady state. Then the velocity was increased the same procedure was repeated. The dependence of the output power on frequency and of frequency on the beam voltage calculated in such way is shown in Fig. 18.14 for the clinotron with parameters shown in Table 18.1. Experimental results of the 300 GHz clinotron packaged into the 0.8 T permanent magnet [32–34] are shown in Fig. 18.14 by dots. One can see that

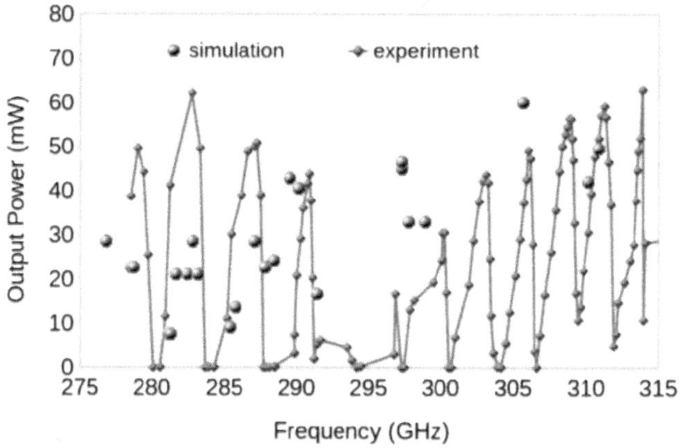

Figure 18.14 Output power as a function of frequency, simulated for 300 GHz clinotron and compared with experimental data.

Table 18.1 Grating and beam parameters

Electron beam voltage U, kV	2–5
EB Current I, mA	150
Beam thickness a, mm	0.14
Grating period l, mm	0.1
Groove width d	$l/2$
Grating length L_{sws}, mm	15.2
Groove depth h, mm	0.17
Grating length L, mm	15.5
Magnetic field B, T	0.8

both the simulation and the experimental output power dependence behavior is resonant one, despite to the surface mode attenuation of 10–20 dB during propagation through interaction space due to ohmic loss [38]. Therefore, the resonant behavior is due to the slow mode transformation into fast ones. Consideration of the plunger length effect has showed that not only increase of interaction power is possible with a proper length, but also it may effect on decrease of radiation striking a cathode. Analysis made in [34] demonstrated at least three resonances reasons: (1) resonance by slow wave causing electronic efficiency change; (2) resonance by fast wave in the cavity that causes change of the radiated power without change of electron efficiency and (3) redistribution of power propagating into anode slot and output waveguide.

The up going activity using the developed theory includes synthesis of the T-junction and the plunge section to optimize the geometry of the output and plunger section to (1) minimize power portion into the cathode area; (2) minimize reflection coefficient of the surface mode back into the interaction space; (3) increase the feedback by fast modes [34].

18.5 Principles for Design of the THz Diffraction Radiation Oscillator

The principle of DRO and orotron operation is based on the Smith–Purcell radiation when electron beam moves above a grating [4, 6, 7, 52–69]. The upper mirror, which reflects SPR back towards the beam area and form an open resonator (OR), that provides feedback in the oscillator. Initially, in orotron, a grating occupied the entire bottom mirror (Fig. 18.15). In DRO and in modern orotrons, grating occupies only small part of the bottom mirror. Another version of DRO with double comb was proposed in [7] and it resembles some designs of EIOs.

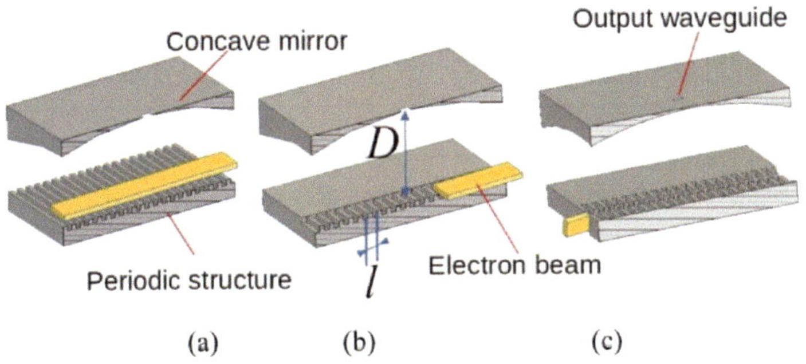

Figure 18.15 The schemes of orotron and DRO: (a) grating occupies the whole mirror providing TEM_{00q} operating mode; (b) grating occupies a small part of the mirror providing TEM_{20q} operating mode; (c) double grating DRO.

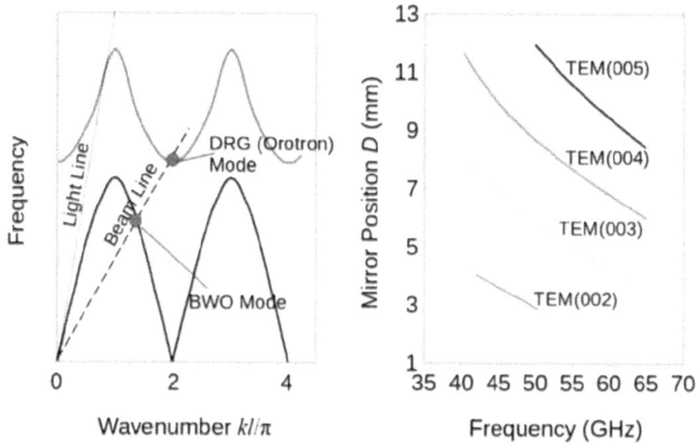

Figure 18.16 (a) Dispersion of BWO and DRO (orotron). (b) Performance of the electron-mechanical frequency tuning in DRO.

The synchronism condition here is the same as in all O-type Cherenkov devices: $v_e = v_{phn}$ and the operating point on dispersion is shown in Fig. 18.16a. One can see from the dispersion that the 0th harmonic is fast one and all other harmonics are slow, and hence they exponentially decay from the grating surface that provides the same challenges for THz generators as in other Cherenkov devices (high beam current density, etc.). However, there are several advantages of DRO compared with BWOs operating on surface waves. One of them is the absence of severe mode competition as in cavities that enables the application of wide electron beams. Also, fixed RF field pattern (uniform, Gaussian, etc.), radiation output by the bulk wave, and high-frequency stability due to high Q-factor are the advantages of DRO.

For strong feedback, the radiation angle is close to the grating normal that corresponds to the 2π point on dispersion in Fig. 18.16a and to Gaussian field distribution with one spot with a spherical mirror [6, 7]. Also, it was shown that excitation of eigenmodes with several spots along the electron beam is possible [58, 59], and even their cooperation with fundamental mode was shown in [65], unlike the usual mode competition regime for single spot OR modes [66]. From Eqs. (18.2) and (18.3) it follows that for 2π point $\beta \approx l/\lambda$ and the effective beam thickness $\Delta_{eff} \cong l/2\pi \cong \beta\lambda/2\pi$. Due to the high

Q-factor, the frequency tuning merely by voltage is quite narrow. However, combined electron-mechanical frequency tuning both by the beam voltage ($\lambda \cong l/\beta$) and by the OR mirrors separation varying (roughly $\lambda \sim 2\ D/n$, n is transverse mode number) can be up to 15% [6, 7, 55, 57] (Fig. 18.16b).

The self-consistent theory of DRO includes a set of differential equations for both electron motion and excitation of the OR field. The simplest motion equation considers the interaction of a sheet electron beam infinitely thin beam (assumption of infinite magnetic focusing field) with a longitudinal component of the OR field and space-charge electric field inside the electron beam itself [60–62]. More advanced theoretical models take into account finite nonuniform magnetic field and both longitudinal and transverse components of the RF and space-charge electric fields [67–69]. Excitation equations, in the general case, are a set of multimode nonstationary (slow-changing amplitude) equations for resonant modes [70]. More rigorous OR excitation theory includes so-called nonresonant terms [70], which may account for the loss of electrons energy for SPR excitation due to the discrepancy of Smith–Purcell and open resonator radiation patterns [71], etc.

Since the effective beam thickness Δ_{eff} in the THz range is quite small (10–50 μm), very similar methods as considered above for BWO are used to increase the interaction impedance, namely: double grating (Fig. 18.15), double-row periodic structure [72, 74], multipin gratings [56], clinotron effect (inclined beam) [3, 7], local magnetic inhomogeneity [69, 73]. Obviously, the application of the double grating scheme allows a double increase in the effective beam thickness. The mentioned above modification of DRO with double grating was very promising in the mm wavelength range, however, in the THz range, employment of a sheet beam is limited. The use of multirow and multipin gratings is usually possible only in the pulse regime. In principle, in the clinotron regime and with local magnetic inhomogeneity, the effective beam thickness may be several times bigger than in the case of a sheet beam [3, 29, 30, 73]. It is quite a promising way for power increase in the pulse regime, however, in the CW mode, the beam current density should be limited to not overheat a grating and to not severely increase the ohmic loss [38].

Another challenge for the THz DRO and orotron is an increase in the interaction length for the starting current decrease. One way for this is the implementation of multifocusing mirrors that provides forming of quite uniform RF field longitudinal distribution over long grating that provided 300 mW operation at 300 GHz [74].

Another way to increase interaction length is to reflect a beam back to the interaction space, similar to this procedure in the reflective klystron. In the mm range, in the reflective DRO, the multitransit of a beam has provided essential decrease in the starting current down to less than 1 mA [7].

Recently, the mechanism for the enhancement of spontaneous Smith–Purcell radiation has been considered promising for the generation of the THz radiation. The superradiant SPR is excited as a result of a beam bunching due to the self-excitation of a surface eigenwave of a grating and holding of the SPR condition for the temporal harmonic of the surface wave [75–85]. Due to an excitation of relatively low frequency BWO, the starting current is relatively low. An increase in the power of superradiant SPR is possible due to stimulated radiation when placing the upper mirror which forms an open resonator as in orotron. Such a regime is known as oromultiplier [54, 80]. An efficient oromultiplier regime occurs when four conditions are met [76, 85]: (SPR), $h = lam/4$, $f_{OR} = nf_{BWO}$,

18.6 Excitation of THz Self-Oscillations in Resonant Systems Supporting Hybrid Bulk-Surface Modes: Cavity with Bieriodic Grating and Electromagnetic Mode Interaction

As shown in Section 18.2, the leaky modes excited by an electron beam moving above a nonuniform grating comprise surface and radiating waves that make them a kind of mix of the Cherenkov and Smith–Purcell radiation. Placing the upper wall above a grating, which, as in a DRO provides feedback by the bulk (fast) wave, causes appearing of hybrid bulk-surface modes with simultaneous propagation of both bulk and surface waves [10, 12]. The dispersion of a closed waveguide with nonuniform grating contains branches, which correspond to both pure surface waves of an open grating and pure bulk waves of a closed waveguide. The area of the branches' intersection corresponds to the resonance coupling of the radiating wave with the surface plasmon polariton [3, 13, 86]. In this area, the so-called "mode intertype interaction" occurs [87]. The EM excited by a bunched beam with current density $j = j_0 \delta(y) e^{i\left(\frac{\omega}{v_e} z - \omega t\right)}$ (v_e is electron velocity) can be found from the solution of the following set of equations [12]:

$$\sum_{p_0=1}^{N} D_{p_0} \left[\begin{array}{l} \delta_p^{p_0} \cos\left(\sqrt{\varepsilon} k h_p\right) \\ +\frac{kd}{L} \sin\left(\sqrt{\varepsilon} k h_{p_0}\right) \sum\limits_{r=-\infty}^{\infty} \left(\frac{\sin(k_{zs} d/2)}{k_{zs} d/2}\right)^2 \frac{e^{i k_{zs}(p-p_0) l}}{\tan(k_{ys} D) k_{ys}} \end{array} \right] = j_0 e^{i p k_0 l} \tag{18.10}$$

When the right-hand side of Eq. (18.10) is zero (no current), it reduces to the dispersion equation. The solution of (18.10) for the parameters $D = 2$ mm, $l = 70$ μm, $h_1 = 90$ μm, $h_3 = 118$ μm, $N = 3$ is shown in the dispersion diagram in Fig. 18.17, and as one can see, such parameters correspond to the 0.5–0.7 THz oscillator.

Several hybrid modes are marked in the dispersions by A-D. As one can see, they differ by mutual directions of propagation

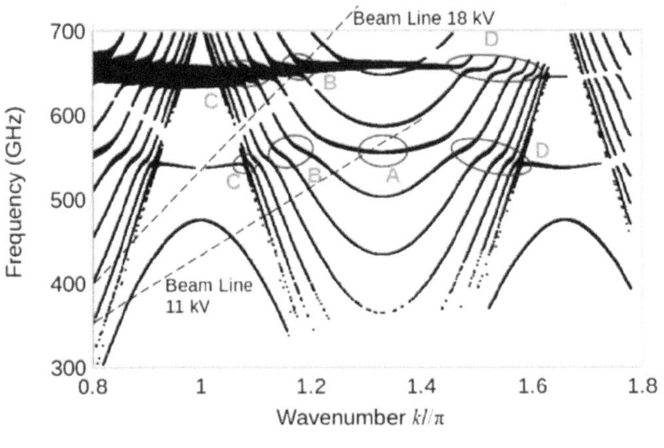

Figure 18.17 Dispersions of waveguide containing grating with period $l =$ 0.07 mm, regular groove depth $h_1 = 0.09$ mm, $D = 2$ mm, $h_3 = 1.3h_1$.

of surface and bulk wave, by the group velocity, by the radiation angle, etc. Correspondingly, oscillators based on those hybrid waves differ by the feedback nature, the coupling with a load, frequency tuning range, etc. Some simulation and experimental results of these regimes are presented below.

18.6.1 Feedback by the Backward Radiating Harmonic

This regime corresponds to the region on the dispersion diagram in Fig. 18.17 marked as B with the radiation angle within 120–150°. In [10] it was shown that self-oscillation is excited in the case when the feedback length L_{fb} (see Fig. 18.18c) is shorter than the grating length. In its turn, as one can see from the rays sketch in Fig. 18.18c, L_{fb} depends on the radiation angle and waveguide height D. RF field pattern of the simulated by MAGIC2D stationary oscillation in Figs. 18.18a,b demonstrates that at $D = 2$ mm, multiple reflections from the upper wall occur, whereas at $D = 8$ mm there is only a single reflection. At $D > 10$ mm at $L_g = 17$ mm and $U = 19$ kV, the feedback length becomes longer of the grating one: $L_{fb} > L_g$ and oscillations drop. This feedback nature provides nonsensitivity to the shortening of several grooves in the grating [10] as well as the possibility of the increase of the radiation output width until holding the condition

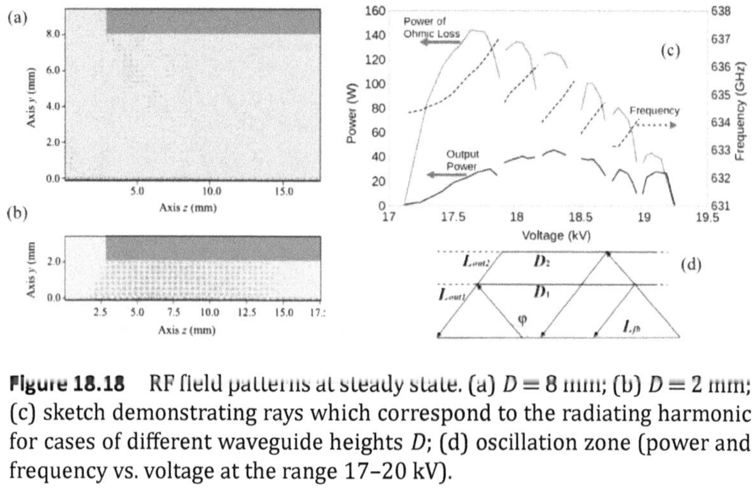

Figure 18.18 RF field patterns at steady state. (a) $D = 8$ mm; (b) $D = 2$ mm; (c) sketch demonstrating rays which correspond to the radiating harmonic for cases of different waveguide heights D; (d) oscillation zone (power and frequency vs. voltage at the range 17–20 kV).

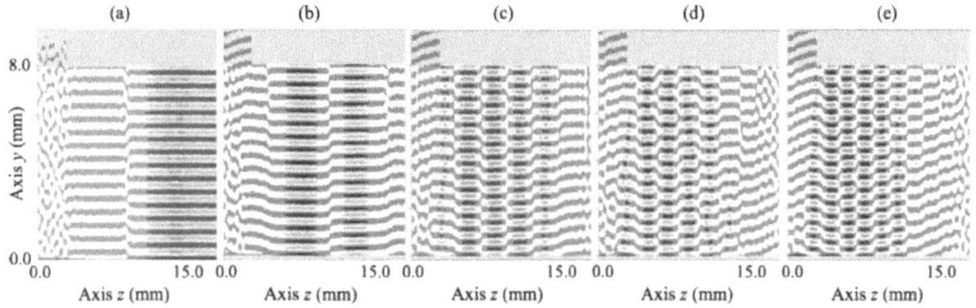

Figure 18.19 RF field patterns (y and z scales are different) at an increase of beam voltage which results in excitation of HOAMs. (a) $U = 9.9$ kV; (b) $U = 10$ kV; (c) $U = 10.3$ kV; (d) $U = 10.4$ kV; (e) $U = 10.6$ kV.

$L_{out} = D \tan \alpha$ which, for the considered waveguide parameters, corresponds to about 5 mm and provides output efficiency about 1% at 0.65 THz (Fig. 18.18d).

18.6.2 Radiation Angle Is Normal to the Grating

In this regime (marked as A in Fig. 18.17) the RF field pattern in the volume is close to this in devices based on Smith–Purcell radiation such as DRO and orotron [54, 88, 89]. However, due to the surface wave, the RF field near the grating is quite strong that provides such distinction of operation of the considered regime from DRO like wide continuous electron frequency tuning (see Figs. 18.19 and 18.20), unnecessity of complicated multifocusing mirrors, wide radiation output [90].

18.6.3 Regime of Grazing Radiation Angle

At the radiation angle close to 180°, there is almost no role of the upper reflector for the feedback by the backward radiating harmonic. Thus, the interaction power is almost independent from D, which was proved by hot simulations. This regime corresponds to the dispersion area C in Fig. 18.17. The simulations have demonstrated the excitation of an oscillation with an interaction efficiency of about 4% at 0.625 THz and with a field pattern shown in Fig. 18.21a. The feedback here is also due to backward radiating harmonic, which has relatively low loss and provides quite uniform RF field distribution along grating, and, hence, all

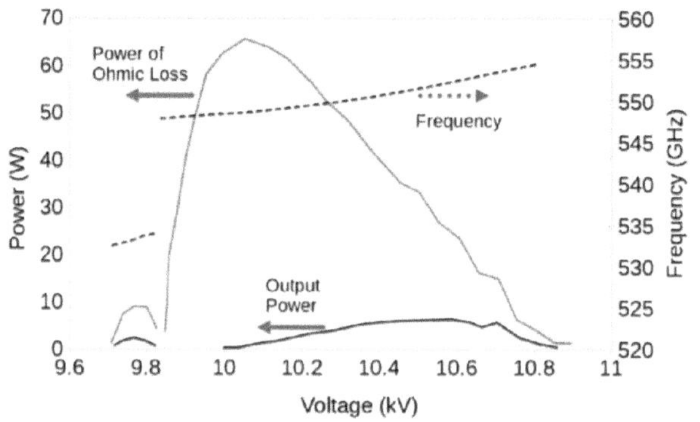

Figure 18.20 Frequency and the power of loss and output radiation versus beam voltage at $D = 8$ mm.

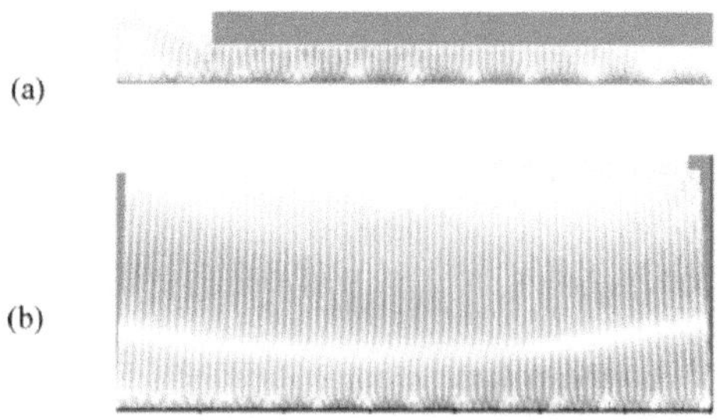

Figure 18.21 Steady-state RF field pattern at $U = 21.9$ kV (a). Steady-state RF field patterns at open resonator formed by reflectors with angle 5° at ends: (b) $U = 21.9$ kV.

electron layers effectively interact with the wave. In this regime, additional resonance feedback is possible when placing reflectors at grating ends, which creates an open (without upper wall) or closed resonator (Fig. 18.21b). Obviously, the radiation output considered above for such a regime is not effective and should be different from this, for example, output from the sidewall (however, the magnet system is an obstacle for this), or upward by means of some reflector.

The frequency tuning range of every considered regime was about 5 GHz (at 0.65, 0.62, and 0.56 THz) and the total continuous tuning range can be extended up to 20 GHz by proper choice of the grating profile [10].

Table 18.2 Parameters of designed clinotrons with nonuniform grating

Operating frequency f, GHz	100	130
Grating period l, mm	0.28	0.21
Groove width d	$l/2$	$l/2$
Groove depth h_1, mm	0.6	0.43
Modified groove depth h_1/h_q	1.3	1.1
Grating length L_{SWS}, mm	13.72	16.27
Waveguide height D, mm	0.8–2.5	0.8

18.6.4 Experimental Results

The experiments for the study of the hybrid bulk-surface wave oscillations have been carried out for the 0.1 and 0.13 THz oscillators with parameters presented in Table 18.2 [11–13]. There were three radiation outputs corresponding to different directions (forward, FW; backward, BW) of the surface and bulk radiation (Fig. 18.22).

Oscillations in the 0.1 THz oscillator were obtained during voltage sweep as shown on oscillograms in Fig. 18.23a–c. The measured frequency versus wavenumber is plotted on the dispersion diagram together with 2D cold calculation in Fig. 18.23d.

As follows from the dispersion, the radiating harmonic angle of the hybrid mode at $D = 0.8$ mm is 0–15° that in good agreement with the enhanced signal at the output in the upper wall and in the output corresponding to the forward wave (Fig. 18.23d).

Moreover, the axial modes competition due to the resonances provided by reflections from side walls, as shown in the previous

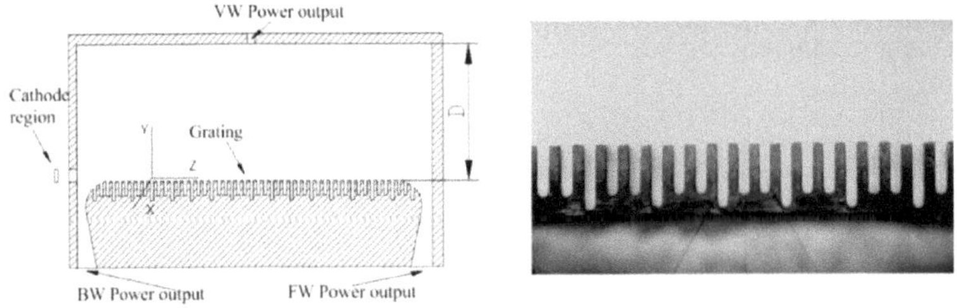

Figure 18.22 Sketch of the experimental prototype and photo of the fabricated grating.

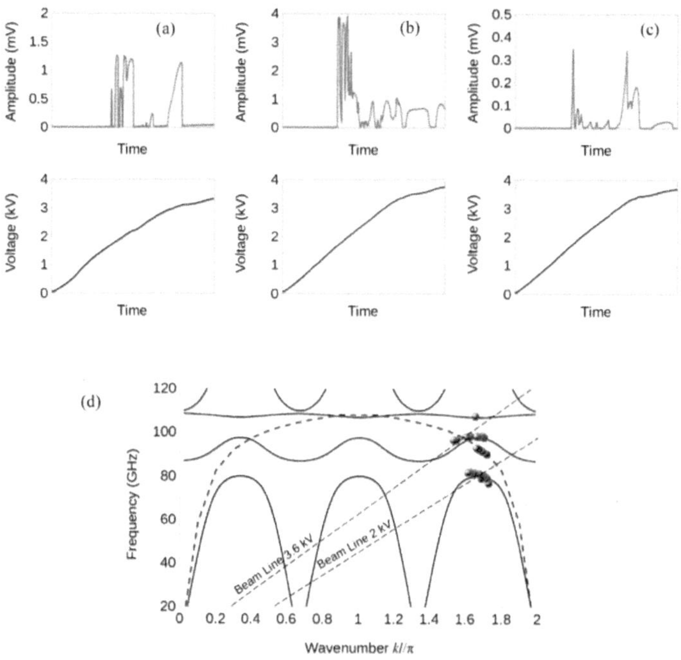

Figure 18.23 Detected signal during voltage sweep at (a) FW output; (b) BW output; (c) VW output. (d) the measured frequency versus wavenumber.

subsection (see Figs. 18.18 and 18.21), causes voltage hysteresis shown in Figs. 18.23 and 18.24. In the absence of reflections, such a regime is also very attractive for a traveling-wave amplifier [11].

Preliminary results of both simulations and experiments proved that a hybrid bulk-surface wave regime is quite promising for a watt-level oscillator at the range up to 1 THz that encourages further study activities.

18.7 Conclusion

Based on the presented survey, the existing nonrelativistic Cherenkov devices provide levels of ouput power below 1 watt at frequencies higher than 200 GHz. Such radiation power levels are sufficient for many practical applications requiring compact-

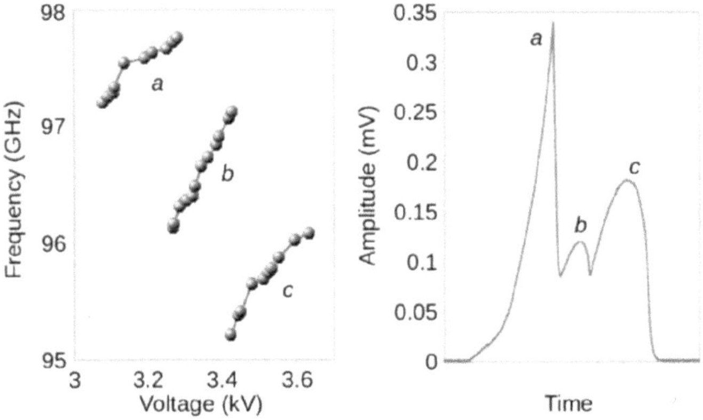

Figure 18.24 Frequency (a) and detected signal (b) vs. voltage demonstrating voltage hysteresis [12].

size frequency-tunable THz sources. As it was shown and discussed in the chapter, the current research activity is dedicated to the development of more powerful sources with a watt level of output power that would essentially increase the number of their practical applications and would help to fill the THz gap from the sub-THz part. Modern technologies allow the fabrication of slow-wave circuits with micron-scale periods and nanoscale tolerances together with nanoscale surface roughnesses that would provide an essential decrease of electromagnetic wave losses at higher frequencies, hence improving the performance of existing Cherenkov devices, and also help to extend the device operating frequency into the THz range. Application of both modern technologies and new physical concepts discussed in the chapter is promising for achieving the breakthrough of 1 THz, watt-level compact frequency-tunable sources.

References

1. J. H. Booske, J. Dobbs, C. D. Joye, C. L. Kory, G. R. Neil, G. S. Park, J. Park, R. J. Temkin, Vacuum electronic high power terahertz sources, *IEEE Trans. Terahertz Sci. Technol.*, vol. 1, no. 1, pp. 0–75, doi: 10.1109/tthz.2011.2151610, 2011.

2. G. Kantorowicz, P. Palluel, Backward Wave Oscillators, in *Infrared and Millimeter Waves*, K. Button, ed., vol. 1, Chapter 4, Academic Press, 1979.
3. G. Ya. Levin, A. I. Borodkin, A. Ya. Kirichenko, A. Ya. Usikov, ed., Churilova, S. A. *Clinotron*. Kyiv: Naukova Dumka Publ. (in Russian), 1992.
4. F. S. Rusin, G. D. Bogomolov, Orotron as the generator of the millimeter range, *Powerful Electronics*, no. 5, pp. 45–51, 1968.
5. V. G. Kurin, B. K. Skrynnik, V. P. Shestopalov, Experimental studies of generators of diffraction radiation, *J. Commun. Technol. Electronics*, vol. 42, no. 11, pp. 1309–1314, 1997.
6. V. P. Shestopalov, *Diffraction Electronics*. (Kharkov: Vyscha shkola), 1976 [in Russian].
7. V. P. Schestopalov, A. A. Vertiy, G. P. Yermak, B. K. Skrynnik, G. I. Khlopov, A. I. Tsvyk, *Difraction Radiation Oscillators*. Kyiv: Naukova Dumka Publishing, 1991.
8. E. Khutoryan, M. Sattorov, K. Lukin, O. Kwon, S. Min, R. Bhattacharya, I. Baek, S. Kim, M. Yi, J. So, G.-S. Park, Theory of multimode resonant backward-wave oscillator with an inclined electron beam, *IEEE Trans. Electron Devices*, vol. 62, no. 5, pp. 1628–1634, doi: 10.1109/TED.2015.2411680, May 2015.
9. M. Sattorov, E. Khutoryan, K. Lukin, O. Kwon, G. Park, Improved efficiency of backward-wave oscillator with an inclined electron beam, *IEEE Trans. Electron Devices*, vol. 60, no. 1, pp. 458–463, doi: 10.1109/TED.2012.2225837, Jan. 2013.
10. E. Khutoryan, A. Kuleshov, S. Ponomarenko, K. Lukin, Y. Tatematsu, M. Tani efficient excitation of hybrid modes in a THz clinotron, *J. Infrared Millim. Terahertz Waves*, doi: 10.1007/s10762-021-00800-y, June 2021.
11. S. S. Ponomarenko, A. A. Likhachev, S. A. Vlasenko, Yu. S. Kovshov, V. V. Stoyanova, S. A. Kishko, E. M. Khutoryan, A. N. Kuleshov Traveling-wave amplification in a circuit with nonuniform grating, *IEEE Trans. Electron Devices*, pp. 1–6, doi: 10.1109/TED.2021.3105951, August 2021.
12. E. M. Khutoryan, Yu. S. Kovshov, A. S. Likhachev, S. S. Ponomarenko, S. A. Kishko, K. A. Lukin, V. V. Zavertanniy, T. V. Kudinova, S. A. Vlasenko, A. N. Kuleshov, T. Idehara, Excitation of hybrid space-surface waves in clinotrons with non-uniform grating, *J. Infrared, Millim. Terahertz Waves*, March 2018, vol. 39, no. 3, pp. 236–249.
13. S. S. Ponomarenko, S. A. Kishko, E. M. Khutoryan, A. N. Kuleshov, B. P. Yefimov, Development of 94 GHz BWO-Klynotron with 3-stage grating, *Telecommun. Radio Eng.*, vol. 73, pp. 271–281, 2014.

14. G.-S. Park, S. Kim, I. Baek, R. Bhattacharya, D. Hong, M. Sattorov, A. Bera, J. So, D. Kim, High-Q metallic fano metamaterial for highly efficient Cerenkov lasing, *Adv. Opt. Mater.*, vol. 6, no. 12, doi: 10.1002/adom.201800041, June 2018.
15. A. Bera, R. Barik, M. Sattorov, O. Kwon, S. Min, I. Baek, S. Kim, J. So, G.-S. Park, Surface-coupling of Cerenkov radiation from a modified metallic metamaterial slab via Brillouin-band folding, *Optics Express*, vol. 22, no. 3, pp. 3039–3044, https://doi.org/10.1364/OE.22.003039, 2014.
16. B. Luk'yanchuk, N. I. Zheludev, S. A. Maier, N. J. Halas, P. Nordlander, H. Giessen, C. T. Chong, The Fano resonance in plasmonic nanostructures and metamaterials, *Nat. Mater.*, 9, 707–715, 2010.
17. S. J. Smith, E. M. Purcell, Visible light from localized surface charges moving across a grating, *Phys. Rev.*, vol. 92, pp. 1069–1069, 1953.
18. V. L. Bratman, N. S. Ginzburg, M. I. Petelin, Common properties of free electron lasers, *Opt. Commun.*, vol. 30, no. 3, pp. 409–412, https://doi.org/10.1016/0030-4018(79)90382-1, 1979.
19. G. A. Komandin, S. V. Chuchupal, S. P. Lebedev, Y. G. Goncharov, A. F. Korolev, O. E. Porodinkov, I. E. Spektor, A. A. Volkov, BWO generators for terahertz dielectric measurements, *IEEE Trans. Terahertz Sci. Technol.*, vol. 3, no. 4, pp. 440–444, 2013.
20. C. Paoloni, D. Gamzina, R. Letizia, Y. Zheng, N. C. Luhmann Jr., Millimeter wave traveling wave tubes for the 21st Century, *J. Infrared Millim. Terahertz Waves*, vol. 35, no. 5, pp. 567–603. https://doi.org/10.1080/09205071.2020.1848643, 2021.
21. D. C. Forster, High power millimeter wave sources, *Adv. Microwaves*, vol. 3, pp. 301–346, 1968.
22. J. A. Dayton, C. L. Kory, G. T. Mearini, D. Malta, M. Lueck, B. Vancil, Fabrication and testing of the 0.650 THz helical BWO, *IEEE International Vacuum Electronics Conference (IVEC) 2012*, Monterey, USA, pp. 33–34, 2012.
23. J. A. Dayton, Jr. C. L. Kory, G. T. Mearini, D. Malta, M. Lueck, K. Gilchrist, *Applying Microfabrication to Helical Vacuum Electron Devices for THz Applications*, Tenth IVEC, Rome, Italy, 2009.
24. C. Paoloni, D. Gamzina, L. Himes, B. Popovic, R. Barchfeld, L. Yue, Y. Zheng, X. Tang, Ye Tang, P. Pan, H. Li, R. Letizia, M. Mineo, J. Feng, N. C. Luhmann, THz backward-wave oscillators for plasma diagnostic in nuclear fusion, *IEEE Trans. Plasma Sci.*, vol. 44, no. 4, pp. 369–376, April 2016.

25. CST AG, Darmstadt, Germany. CST Studio Suite. [Online]. Available: http://www.cst.com.
26. R. Letizia, M. Mineo, C. Paoloni, Photonic crystal-structures for THz vacuum electron devices, *IEEE Trans. Electron Devices*, vol. 62, no. 1, pp. 178–183, Jan. 2015.
27. A. Srivastava, S. Penmetsa, L. Christie, K. N. Bhat, Electromagnetic design of a 220 GHz BWO with experimental study of micro-fabricated folded waveguide structure, *J. Electromagnetic Waves Appl.*, pp. 1–14, doi: 10.1080/09205071.2019.1645744, 2019.
28. J. Cai, L. Hu, H. Chen, X. Jin, G. Ma, H. Chen, Study on the increased threshold current in the development of 220 GHz folded waveguide backward-wave oscillator, *IEEE Trans. Microwave Theory Techniques*, vol. 64, no. 11, pp. 3678–3685, 2016.
29. A. E. Fedotov, A theory of the orotron with an inclined electron beam, *Int. J. Infrared Millim. Waves*, vol. 29, no. 11, pp. 997–1003, 2008.
30. K. Schunemann, D. M. Vavriv, Theory of the clinotron: A grating backward-wave oscillator with inclined electron beam, *IEEE Trans. Electron Devices*, vol. 46, no. 11, pp. 2245–2252, Nov 1999.
31. S. S. Ponomarenko, et al., 400 GHz continuous-wave clinotron oscillator, *IEEE Trans. Plasma Sci.*, vol. 41, no. 1, pp. 82–86, Jan. 2013.
32. Y. S. Kovshov, S. S. Ponomarenko, S. A. Kishko, E. M. Khutoryan, A. N. Kuleshov, Numerical simulation and experimental study of Sub-THz and THz CW clinotron oscillators, *IEEE Trans. Electron Devices*, vol. 65, no. 6, pp. 2177–2182, doi: 10.1109/TED.2018.2792258, June 2018.
33. A. Likhachev, A. Danik, S. Kishko, S. Ponomarenko, O. Martseniak, E. Khutoryan, I. Ogawa, T. Idehara, A. Kuleshov, Compact radiation module for THz spectroscopy using 300 GHz continuous-wave clinotron, *Rev. Sci. Instrum.*, vol. 90, no. 3, p. 034703, doi: 10.1063/1.5064796, 2019.
34. Yu. S. Kovshov, S. S. Ponomarenko, S. A. Kishko, A. Likhachev, A. Danik, L. Mospan, S. Steshenko, E. M. Khutoryan, A. N. Kuleshov, Effect of mode transformation in THz clinotron, *J. Infrared Millim. Terahertz Waves*, vol. 39, no. 11, pp. 1055–1064, doi: https://doi.org/10.1007/s10762-018-0534-y, 2018.
35. V. V. Zavertanniy, S. A. Kishko, S. S. Ponomarenko, B. P. Yefimov, A. F. Zabrodskiy, L. A. Kirichenko, T. V. Kudinova, A. N. Kuleshov, Magnetic focusing system for intense electron beams of submillimeter clinotrons. Prikladnaya nelineynaya dinamika. *Commun. Syst.*, vol. 20, no. 5, pp. 112–120 (in Russian), doi: https://doi.org/10.18500/0869-6632-2012-20-5-112-120, 2012.

36. A. A. Likhachev, A. A. Danik, Yu. S. Kovshov, S. A. Kishko, S. S. Ponomarenko, V. N. Zheltov, E. M. Khutoryan, A. N. Kuleshov, Effect of electron beam velocity spread in a clinotron, *IEEE Trans. Electron Devices*, vol. 66, no. 3, pp. 1540–1544, doi: 10.1109/TED.2019.2891547, 2019.

37. P. G. Huray, *The Foundations of Signal Integrity*, John Wiley & Sons, Inc., N. J. Hoboken, USA, 2009.

38. Yu. S. Kovshov, S. S. Ponomarenko, S. A. Kishko, A. A. Likhachev, S. A. Vlasenko, V. V. Zavertanniy, E. M. Khutoryan, A. N. Kuleshov, High frequency ohmic losses in terahertz frequency range CW klynotrons, *Telecommun. Radio Eng.*, vol. 76, no. 10, pp. 929–940, 2017.

39. B. P. Yefimov, M. V. Milcho, O-type millimeter-wave band generator of amplitude-modulated oscillations with resonance interaction space, *Int. J. Infrared Millim. Waves*, vol. 18, pp. 1681–1687, https://doi.org/10.1007/BF02678279, 1997.

40. S. S. Ponomarenko, A. A. Likhachev,V. V. Stoyanova, Y. S. Kovshov, S. A. Vlasenko, S. A. Kishko, E. M. Khutoryan, A. N. Kuleshov, Spectral characteristics of THz CW clinotrons, *IEEE Trans. Electron Devices*, vol. 67, no. 12, pp. 5766–5770, doi: 10.1109/TED.2020.3032940, 2020.

41. Y. Matsuki, T. Fujiwara, Advances in high-field DNP methods, in *Experimental Approaches of NMR Spectroscopy: Methodology and Application to Life Science and Materials Science*, ed., The Nuclear Magnetic Resonance Society of Japan, Springer, Chapter 4, pp. 91–134, 2017.

42. Vlasov, A. N., et al., Overmoded GW-class surface-wave microwave oscillator, *IEEE Trans. Plasma Sci.*, vol. 28, no. 3, pp. 550–560, 2000.

43. B. Levush, T. M. Antonsen, A. Bromborsky, W.-R. Lou, Y. Carmel, Theory of relativistic backward-wave oscillators with end reflectors, *IEEE Trans. Plasma Sci.*, vol. 20, no. 3, pp. 263–280, doi: 10.1109/27.142828, Jun 1992.

44. I. Chernyavskiy, T. Antonsen, J. Rodgers, A. Vlasov, D. Chernin, B. Levush, Modeling vacuum electronic devices using generalized impedance matrices, *IEEE Trans. Electron Devices*, vol. 64, no. 2, pp. 536–542, doi: 10.1109/TED.2016.2640205, Feb. 2017.

45. V. Jabotinski, D. Chernin, T. Antonsen, A. Vlasov, J. Rodgers, I. Chernyavskiy B. Levush, Efficient calculation of impedance matrices for vacuum electronic device circuit structures, *IEEE Trans. Electron Devices*, vol. 65, no. 6, pp. 2264–2271, doi: 10.1109/TED.2018.2809638, June 2018.

46. S. O. Steshenko, Generalization of the mode-matching technique to the problems of scattering by semi-infinite slow-wave structures, *Radiofiz. Electron.*, vol. 25, no. 4, pp. 10–17, doi: https://doi.org/10.15407/rej2020.04.010, 2020.
47. A. A. Kirilenko, S. L. Senkevich, S. O. Steshenko, Application of the generalized scattering matrix technique for the dispersion analysis of 3D slow-wave structures, *Telecommun. Radio Eng.*, vol. 74, no. 17, pp. 1497–1511, doi: 10.1615/TelecomRadEng.v74.i17.10, 2015.
48. L. Shuang, J. G. Wang, G.-Q. Wang, P. Zeng, D.-Y. Wang, Mode analysis and design of 0.3 THz Clinotron, *Chinese Phys. B*, vol. 25, no. 10, p. 108401, October 2016.
49. A. V. Maksimenko, V. I. Shcherbinin, A. V. Hlushchenko, V. I. Tkachenko, K. A. Avramidis, J. Jelonnek, Starting currents for eigenmodes of a gyrotron cavity with mode conversion, *IEEE Trans. Electron Devices*, vol. 66, no. 3, pp. 1552–1558, doi: 10.1109/TED.2019.2893888, March 2019.
50. D. Wagner, M. Thumm, G. Gantenbein, W. Kasperek, T. Idehara, Analysis of a complete gyrotrons oscillator using the scattering matrix description, *Int. J. Infrared Millim. Waves*, vol. 19, no. 2, 185–194, 1998.
51. M. Thumm, Modes and mode conversion in microwave devices, in *Generation and Application of High-Power Microwaves*, R. A. Cairns, A. D. R. Phelps, eds., IOP, Bristol, UK, pp. 121–171, 1997.
52. K. Mizuno, S. Ono, Y. Shibata, Two different mode interaction in electron tube with a Fabry-Perot resonator - The ledatron, *IEEE Trans. Electron Devices*, vol. 20, no. 8, pp. 749–752, doi: https://doi.org/10.1109/T-ED.1973.17737, 1991.
53. L. A. R. Leavitt, D. Wortman, H. Dropkin, Millimeter-wave orotron oscillation—Part I: Theory, *IEEE J. Quantum Electron.*, vol. 17, no. 8, pp. 1333–1340, Aug. 1981.
54. V. L. Bratman, B. S. Dumesh, A. E. Fedotov, P. B. Makhalov, B. Z. Movshevich, F. S. Rusin, Terahertz orotrons and oromultipliers, *IEEE Trans. Plasma Sci.*, vol. 38, no. 6, June 2010.
55. F. S. Rusin, G. D. Bogomolov, Orotron—an electronic oscillator with open resonator and reflecting grating, *Proc. IEEE*, vol. 57, no. 4, pp. 720–722, Apr. 1969.
56. V. L. Bratman, B. S. Dumesh, A. E. Fedotov, Yu. A. Grishin, F. S. Rusin, Broadband orotron operation at millimeter and submillimeter waves, *Int. J. Infrared Millim. Waves*, vol. 23, no. 11, November 2002.
57. V. L. Bratman, A. E. Fedotov, P. B. Makhalov, Theory of Smith–Purcell frequency multiplier with transverse diffraction, *Phys. Rev. ST Accel. Beams*, 2010.

58. Yu. Evdokimenko, K. A. Lukin, I. D. Revin, B. K. Skrynnik, V. P. Shestopalov, A new mechanism of excitation of a diffraction-radiation generator (free-electron laser), *Sov. Phys. Doklady*, vol. 27, p. 554, 1982.

59. Yu. I. Evdokimenko, K. A. Lukin, I. D. Revin, B. K. Skrynnik, V. P. Shestopalov, On one feature of the energy transformation in the diffraction radiation generators – free-electron lasers, *Dokl. Akad. Nauk SSSR*, vol. 268, no. 4, pp. 853–856, 1983.

60. K. A. Lukin, V. P. Shestopalov, Nonlinear theory of diffraction-radiation oscillators. Akademiia Nauk Ukrains'koi RSR, Dopovidi, Seriia A - Fiziko-Matematichni ta Tekhnichni Nauki, Apr., pp. 335–338 (in Ukrainian), 1978.

61. A. S. Bakai, K. A. Lukin, V. P. Shestopalov, Nonlinear nonstationary theory of generators of diffraction radiation, *Radiofizika*, vol. 22, #9, pp. 1117–1123. *Translation: Radiophys. Quantum Electron.*, vol. 22, pp. 774–778, https://doi.org/10.1007/BF01035286, 1979.

62. K. A. Lukin, Theoretical analysis of transient and steady-state conditions in a diffraction-radiation oscillator, *Radiofizika*, vol. 22, no. 12, 1979, pp. 1514–1522. *Translation: Radiophys. Quantum Electron.*, vol. 22, no. 12, pp. 1056–1062, June 1980.

63. S. J. Papadakis, J. A. Hoffmann, A. H. Monica, D. M. Deglau, J. Yu, G. S. Nusinovich, T. M. Antonsen, R. Osiander Orotron-based sub-millimeter-wave source, in *Passive Millimeter-Wave Imaging Technology XIII*, D. A. Wikner, A. R. Luukanen, ed., *Proc. SPIE*, vol. 7670, 767000, 2010.

64. E. J. Price, J. E. Walsh, M. F. Kimmitt, Short submillimeter operation of the planar orotron, *Nucl. Instrum. Methods Phys. Res. A Accel. Spectrom. Detect. Assoc. Equip.*, vol. 304, no. 1–3, pp. 133–136, Jul. 1991.

65. Ermak, G. P. Lukin, K. A. Shestopalov, V. P., Mode interaction in a diffraction radiation generator, *Radiofizika* (ISSN 0021-3462), vol. 29, no. 5, 1986, pp. 610–617 (in Russian).

66. Lukin K. A. Khutoryan E. M. Tsvyk A. I., Mode self-synchronization in diffraction radiation oscillator, *IEEE 2008 IEEE International Vacuum Electronics Conference (IVEC) - Monterey*, CA, USA, pp. 62–63, doi: 10.1109/ivelec.2008.4556416, 2008.

67. Yu. I. Evdokimenko, K. A. Lukin, B. K. Skrynnik, Theory of diffraction-radiation generator on basis of discrete interaction mode, *Radiophys. Quantum Electronics*, vol. 27, pp. 1013–1027, 1984.

68. Yu. I. Evdokimenko, K. A. Lukin, V. P. Shestopalov. On 2D nonlinear nonstationary theory of diffraction radiation oscillator, *Electronnaya Technika. Electronika SVCh*, vol. 10, no. 334, pp. 35–40 (in Russian), 1981.

69. E. Odarenko, A. Shmat'ko, Nonlinear theory of O-type microwave oscillators with nonuniform magnetostatic field (two-dimensional model), *J. Commun. Technol. Electron.*, vol. 39, no. 9, pp. 1–8, September 1994.
70. K. A. Lukin, A. E. Poedinchuk, V. P. Shestopalov, A theory for the excitation of open resonators by nonlinear currents, *Akademiia Nauk SSSR Doklady*, vol. 286, no. 3, pp. 625–629, 1986.
71. A. I. Tsvyk, E. M. Khutoryan, Influence of the Smith-Purcell radiation on electron beam in DRO, *Telecommun. Radio Eng.*, vol. 68, pp. 217–229, 2009.
72. Ye. A. Myasin, V. V. Evdokimov, A. Yu. Il'yin, 1 mm wavelength orotron with DRPS and broad electron beam, *22nd International Crimean Conference Microwave and Telecommunication Technology*, Sevastopol, Ukraine, 10–14 Sept. 2012.
73. A. I. Tsvyk, G. S. Vorob'yov, A. V. Nesterenko,V. N. Zheltov, Influence of the local inhomogeneity of a magnetostatic field on the start and output characteristics of a diffraction radiation generator, *Radiophys. Quantum Electron.*, vol. 43, no. 2, pp. 132–141, doi: 10.1007/BF02677146, February 2000.
74. E. A. Myasin, V. V. Evdokimov, A. Yu. Il'yn, "300–350 GHz frequency range orotron with two electron beams, *J. Radio Electron.*, no. 7, doi: 10.30898/1684-1719.2019.7.7, 2019.
75. C. S. Liu, V. K. Tripathi, Stimulated coherent Smith–Purcell radiation from a metallic grating, *IEEE J. Quantum Electron.*, vol. 35, no. 10, pp. 1386–1989, Oct. 1999.
76. B. K. Skrynnik, V. K. Korneenkov, M. Y. Demchenko, On the feedback in the Smith–Purcell experiments, *Proc. 4th Int. Kharkov Symp. Phys. Eng. Mm Submm Waves*, vol. 2, pp. 550–552, 2001.
77. H. L. Andrews, C. A. Brau, Gain in the Smith–Purcell free-electron laser, *Phys. Rev. ST Accel. Beams*, vol. 7, no. 7, p. 070701, Jul. 2004.
78. J. T. Donahue, J. T. Gardelle, Simulation of Smith–Purcell radiation using a particle-in-cell code, *Phys. Rev. ST Accel. Beams*, vol. 8, no. 6, p. 060702, Jun. 2005.
79. V. Kumar, K.-J. Kim, Analysis of Smith–Purcell free-electron lasers, *Phys. Rev. E Stat. Phys. Plasmas Fluids Relat. Interdiscip. Top.*, vol. 73, no. 2, p. 026501, Feb. 2006.
80. V. L. Bratman, A. E. Fedotov, I. M. Khaimovich, P. B. Makhalov, Excitation of orotron oscillations at the double frequency of the surface wave, *Radiophys. Quantum Electron.*, vol. 50, no. 10/11, pp. 780–785, Oct. 2007.

81. J. Urata, M. Goldstein, M. F. Kimmitt, A. Naumov, C. Platt, J. E. Walsh, Superradiant Smith–Purcell emission, *Phys. Rev. Lett.*, vol. 80, no. 3, pp. 516–519, Jan. 1998.
82. J. Gardelle, P. Modin, H. Bluem, R. Jackson, J. Jarvis, A. Todd, J. Donohue, A Compact THz source: 100/200 GHz operation of a cylindrical Smith–Purcell free-electron laser, *IEEE Trans. Terahertz Sci. Technol.*, vol. 6, no. 3, pp. 497–502, doi: 10.1109/TTHZ.2016.2543603, May 2016.
83. Y. Yin, H. Chen, Enhanced superradiant Smith–Purcell radiation in a three-mirror quasi-optical cavity, *IEEE Trans. Plasma Sci.*, vol. 37, no. 6, pp. 1057–1061, Jun. 2009.
84. D. Li, K. Imasaki, X. Gao, Z. Yang, G.-S. Park, Reduce the start current of Smith–Purcell backward wave oscillator by sidewall grating, *Appl. Phys. Lett.*, vol. 91, no. 22, p. 221506, Nov. 2007.
85. K. Lukin, E. Khutoryan, G.-S. Park, Interaction of evanescent wave and Smith-Purcell radiation modes in resonant BWO-DRO device, *2009 IEEE International Vacuum Electronics Conference*, 28–30, doi: 10.1109/IVELEC.2009.5193529, April 2009.
86. O. Pishko, V. Chumak, S. Churilova, *Radiofizika and Elektronika*, 12, 2007 (in Russian).
87. V. P. Shestopalov, Yu. K. Sirenko, *Dynamic Theory of Gratings.* Kiev: Naukova Dumka, 1989 (in Russian).
88. K. Lukin, et al., State of the art and future development of vacuum electronics in the IRE NAS of Ukraine, *2006 IEEE International Vacuum Electronics Conference Held Jointly with 2006 IEEE International Vacuum Electron Sources*, pp. 457–458, doi: 10.1109/IVELEC.2006.1666380, 2006.
89. E. A. Myasin, V. V. Evdokimov, A. Y. Il'in. A terahertz orotron with double-row periodic structure and broad electron beam, *Tech. Phys. Lett.*, vol. 38, pp. 497–498, https://doi.org/10.1134/S1063785012060119, 2012.
90. E. A. Myasin, A. N. Solov'ev, Specific features of an open resonator of orotron with a two-row periodic structure. *J. Commun. Technol. Electron.*, vol. 66, pp. 844–852, 2021.

Chapter 19

Folded Waveguide Traveling Wave Tube

Shengpeng Yang, Duo Xu, Ningjie Shi, and Yubin Gong
National Key Lab on Vacuum Electronics,
School of Electrical Science and Engineering,
University of Electronic Science and Technology of China,
Chengdu 610054, China
ybgong@uestc.edu.cn

19.1 Introduction

The traveling wave tube (TWT), which has been widely used as an amplifier in radar, satellite transponders, electronic warfare, and self-protection systems, is an important member of vacuum electron devices. Nowadays, people can enjoy broadband, phone, and TV signals almost anywhere on the globe, thanks to satellite-based TWT.

The first TWT was invented by Rudolf Kompfner in 1943. It consists of three core components: a helix to slow down the electromagnetic wave, a precision electron gun to provide a high-energy electron beam, and a collector to collect the electron beam at the end. Since then, the helix became the best choice for ultra-wide-band TWTs, the bandwidth of which could reach several octaves. However, the power capacity of the helix TWT was limited

Advances in Terahertz Source Technologies
Edited by Gun-Sik Park, Masahiko Tani, Jae-Sung Rieh, and Sang Yoon Park

ISBN 978-981-4968-89-8 (Hardcover), 978-1-003-45967-5 (eBook)
www.jennystanford.com

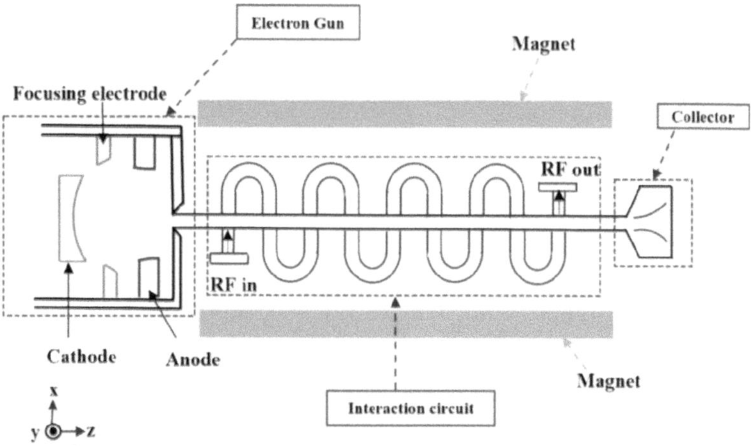

Figure 19.1 Schematic of FW-TWT.

by the thin wire or strip. The coupled-cavity slow-wave structure (SWS) could overcome the power capacity issue but will reduce the bandwidth notably. During the past decades, helix and coupled-cavity SWSs always have played an important role in TWT SWSs.

When the frequency goes up to the terahertz band, the helix and coupled-cavity have to face manufacturing problems. In this case, the folded waveguide (FW) became the most popular SWS for TWTs operating at the terahertz band. Figure 19.1 shows a schematic of FW-TWT. It mainly consists of an electron gun, magnets, SWS (interaction circuit), and a collector. The electron gun is used to emit, accelerate, and focus electrons into a beam. After the electron beam is formed, it enters the SWS through a beam tunnel. The magnets surrounding the SWS focus the electron beam to avoid beam diverging due to the space-charge force. In the SWS, the input RF fields become slow waves so that they can effectively exchange energy with the electron beam under synchronization conditions. Finally, the beam will deliver its energy to the slow waves, and the remaining beam energy will be transferred into thermal energy as the beam hit on the collector.

The core part of FW-TWT is the FW SWS. As shown by the interaction circuit in Fig. 19.1, FW SWS is usually obtained by folding the broad edge of a rectangular waveguide in a periodic serpentine path

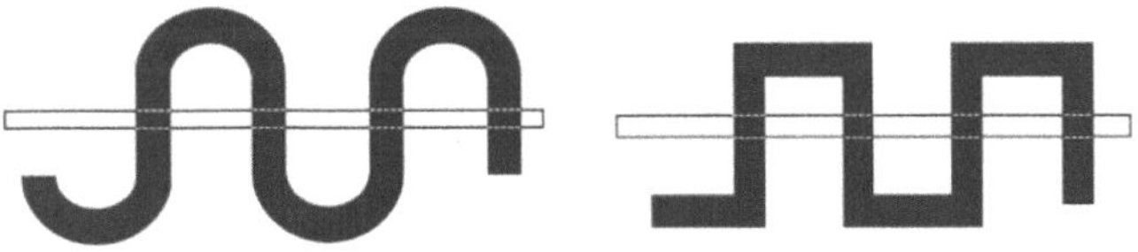

Figure 19.2 Planar structure diagrams of (a) SW and (b) FW.

along the broad edge. This periodically loaded waveguide can reduce the phase velocity of waves to the electron beam velocity. Then, a cylindrical or rectangular tunnel is drilled along the centerline to let the electron beam pass through the FW SWS and interact with the slow waves. Figure 19.2 shows two typical structures of FW SWSs. As shown in Fig. 19.2a, the FW with a smooth S-shaped path is usually called a serpentine waveguide (SW), which is essentially the same kind of SWS as the FW with right-angle bends shown in Fig. 19.2b. To facilitate the discussions in this chapter, the two types of structures shown in Fig. 19.2 are collectively referred to as FW.

An FW-TWT amplifier is suitable for higher power with reasonably wide bandwidth in comparison with helix TWT, and it has the advantages of simpler coupling structures and robust structure over the conventional helix TWT. Especially in high-frequency TWT operating at millimeter waveband and above, FW SWSs have great advantages and competitiveness over helix and coupled-cavity SWSs. On the one hand, in terms of working performance, FW SWS can reach a good compromise between bandwidth and power capacity. On the other hand, its unique planar structure is suitable for advanced processing methods, which provides not only higher precision but also lower cost.

The research on FW SWSs originated in the 1980s. G. Dohler first reported the hot measurement experiment of FW-TWT in 1987 and achieved a maximum output power of 300 W and a gain of 14 dB in the frequency range of 40–54 GHz [1]. Over the next 40 years, the theoretical research on FW SWSs has been gradually deepened, new fabrication methods have emerged one after another, and then the operating performance of TWT has been continuously improved.

In order to clearly describe the development history and level of FW-TWT, the main content of this chapter is organized as following:

Section 19.2: Theory and algorithm; Section 19.3: Improvement of high-frequency structure; Section 19.4: Electron optical system; Section 19.5: Fabrication technology; Section 19.6: Performance of FW-TWT.

19.2 Theory and Algorithm

In 1987, G. Dohler gave a simple derivation of the dispersion characteristics and coupling impedance of FW SWSs according to Waveguide Handbook written by Marcuvitz and established a preliminary model of FW SWSs. In recent years, the theoretical research of FW-TWTs has been carried out by many institutes, including the U.S. Naval Research Laboratory (NRL), Science Applications International Corporation (SAIC), the University of Wisconsin-Madison, France Thales (Thales), India Bharat Electronics (BE), Indian Institute of Science (IIS), Seoul National University (SNU), the Institute of Electronics of the Chinese Academy of Sciences (IECAS) and University of Electronic Science and Technology of China (UESTC), etc.

19.2.1 High-Frequency Characteristics

The early theoretical research of the high-frequency characteristics of FWG TWT was mainly completed by Professor G. S. Park of Seoul National University in South Korea and Professor Shunkang Liu of Southeast University in China. In 1995, Shunkang Liu gave a more detailed derivation process and supplemented the expression of the coupling impedance of high-order space harmonics [2]. In 1998, G. S. Park et al. derived the dispersion relation of FW SWS in their linear interaction theory [3]. In 2000, Shunkang Liu analyzed the influence of the waveguide's narrow side and the diameter of the electron beam tunnel on the transmission bandgap of FW SWSs according to the equivalent circuit model proposed by G. S. Park [4]. Between 2008 and 2011, M. Sumathy of BE published several papers on the dispersion characteristics and coupling impedance of FW SWSs, using the equivalent circuit method [5, 6] and conformal transformation method [7, 8], respectively. In their series studies, a simple equivalent circuit model for the analysis of dispersion and interaction impedance characteristics of FW SWS was developed by considering the straight and curved portions of the structure supporting the dominant TE_{10} mode of the rectangular waveguide. Expressions for the lumped capacitance and inductance per period of the slow-wave structure were derived in terms of the physical dimensions of the structure, incorporating the effects of the beam-hole in the lumped parameters. The lumped parameters were subsequently interpreted for obtaining the dispersion and interaction impedance characteristics of the structure. The analysis was simple yet accurate in predicting the dispersion and interaction impedance behavior at millimeter-wave frequencies. The analysis was benchmarked against measurement as well as with 3D electromagnetic modeling using MAFIA for two typical SWSs [5–8].

Here, we briefly introduce the dispersion relation and the coupling impedance based on the model of reference [3], as shown in Fig. 19.3.

Let us consider that the operation mode of electromagnetic wave is TE_{10} mode. In the waveguide, the TE_{10} mode propagates along the folded path z' shown as a dashed line. The electron beam travels along the z-axis through the holes in the waveguide walls. The axial phase velocity decreases because the wave propagates in the

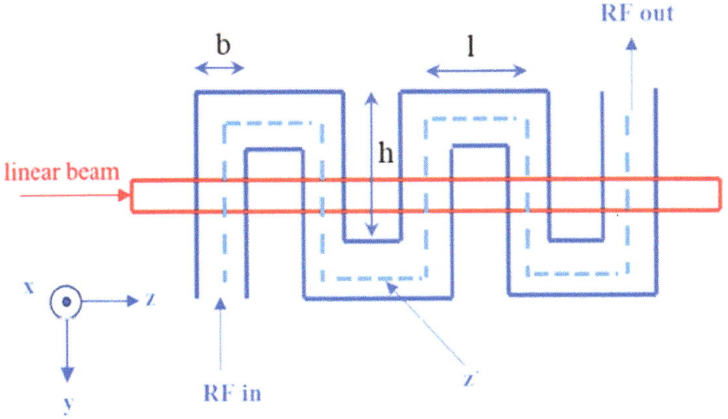

Figure 19.3 The schematic diagram of interaction in a folded waveguide.

waveguide along a circuitous path. The phase velocity of the wave along the z' path is written as

$$v_{ph} = \frac{\omega}{k_{wg}}, \tag{19.1}$$

where $\omega^2 = \omega_c^2 + c^2k_{wg}^2$, ω_c is the cutoff frequency, c is the light velocity in vacuum, k_{wg} is the wave number along the waveguide. The effective phase velocity v'_{ph} along the path z and the related axial wave vector k' are given by

$$v'_{ph} = \frac{l}{l+h}v_{ph}, \quad k' = \frac{l+h}{l}k_{wg}. \tag{19.2}$$

The axial and temporal variation of the traveling wave which interacts with the electron beam could be represented by $e^{i\left(\omega t - \frac{l}{l+h}k'z\right)}$ and the unperturbed dispersion relation for the fundamental space harmonic ($m = 0$) is given by

$$\omega = \sqrt{\omega_c^2 + \left(\frac{l}{l+h}\right)^2 (ck')^2}. \tag{19.3}$$

From the equations above, the frequency frequencies of higher space harmonics can be represented by

$$\omega = \sqrt{\omega_c^2 + \left(\frac{l}{l+h}\right)^2 c^2 \left(k_m - \frac{(2m+1)\pi}{l}\right)^2}, \tag{19.4}$$

where $k_m = k' + \frac{(2m+1)\pi}{l}$ is the phase constant of the m-th space harmonic. The result of Eq. (19.4) is shown in Fig. 19.4.

Moreover, the dominant contribution to the radiation field components of the TE_{10} mode in a waveguide is given by

$$E_y = \left(jE_0 \sin\frac{\pi}{a}x\right) e^{j\left(\omega t - \frac{l}{l+h}k'z\right)} \tag{19.5}$$

$$H_x = -\sqrt{\frac{\varepsilon_0}{\mu_0}}\frac{l}{l+h}\frac{ck'}{\omega}E_y \tag{19.6}$$

$$H_z - \sqrt{\frac{\varepsilon_0}{\mu_0}}\frac{jc}{\omega}\frac{\partial E_y}{\partial x}, \tag{19.7}$$

where ε_0 is the permittivity of vacuum, μ_0 is the permeability of vacuum, c is the light velocity in vacuum, and a is the width of the FW

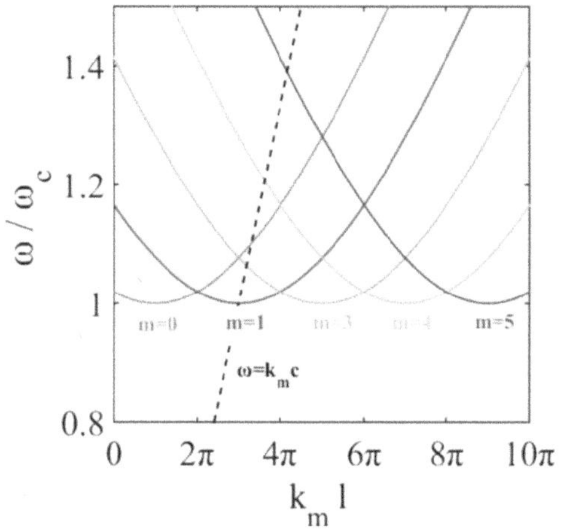

Figure 19.4 Dispersion relation of FW SWS.

in the x-direction. The power P in the z-direction can be obtained from the time-averaged Poynting flux

$$P = \frac{1}{4}\sqrt{\frac{\varepsilon_0}{\mu_0}}E_0^2 ab\frac{ck'l}{\omega(l+h)}. \tag{19.8}$$

The coupling impedance can be written as

$$K = \frac{E_0^2}{2\beta^2 P} = 2\sqrt{\frac{\mu_0}{\varepsilon_0}}\left(\frac{l+h}{lk'}\right)^3 \frac{\omega}{abc}, \tag{19.9}$$

where $\beta = \frac{l}{l+h}k'$.

19.2.2 Theory of Beam-Wave Interaction

The theory of beam-wave interaction is a quantitative explanation for the power transfer between the beam and wave. The aim of the beam-wave interaction is to predict the performance of the TWT accurately, by solving a set of equations. A classical way to establish these equations considers three steps or sections, as following: (1) The effect of the wave on the beam, which is the electronics equation; (2) The effect of the beam on the wave, which is the circuit equation; (3) Combine the two equations and solve the desired physical parameters. According to the ratio of the perturbed (AC) part to the unperturbed (DC) part, the beam-wave interaction theory is divided into two categories: small-signal theory (linear) and large-signal theory (nonlinear).

A. Small-signal model

Small-signal condition is that the AC part is much less than the DC part. So, the second or higher order of the AC part is negligible in the equations. That is, the equations can be linearized and therefore can be solved analytically. One can predict the small-signal gain of FW-TWTs as long as an accurate value of the coupling impedance has been obtained.

In 1998, G. S. Park et al. proposed a linear interaction theory of FW SWSs based on the field distribution in FW SWSs, and predicted the small-signal gain of FW-TWT [3]. In 2003, G. S. Park et al. extended their original theory and proposed the design idea and method of FW SWSs, and an FW-TWT in Ka band was designed using this method [9]. The predicted dispersion characteristics are in good agreement with the simulation results of HFSS. The results of gain are in good agreement with the particle-in-cell (PIC) simulation results of MAGIC 3D in a certain frequency range. These comparisons showed reasonable agreement, which verified the validity of the theory. As a result, this method can be directly applied to the designs for robust high-power millimeter-wave circuits with reasonably broad bandwidth and high gain, such as folded waveguide or interdigital line.

At the same time, an equivalent circuit model of FW SWSs [3] is also given, as shown in Fig. 19.5. The dispersion characteristic curve calculated by this model is in good agreement with the simulation results of HFSS.

The kinetic equation of the electron beam is

$$\frac{d\vec{v}}{dt} = -\eta(\vec{E} + \vec{v} \times \vec{B}), \tag{19.10}$$

where $\vec{v}$ is the velocity of the electrons; $\eta = e/m$ is the specific charge of electrons.

Then, linearize and simplify the kinetic equation by the assumed conditions of non relativistic, 1 D, and a time dependence of $e^{-i\omega t}$. The linearized simplified kinetic equation is

$$j\omega v_1 + v_0\frac{\partial v_1}{\partial z} = -\eta E_1, \tag{19.11}$$

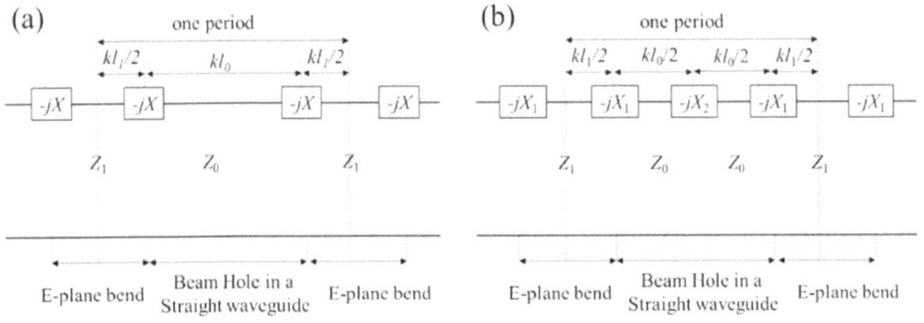

Figure 19.5 Equivalent circuit model of FW SWSs proposed by G. S. Park et al. [3]: (a) Without an electron beam tunnel; (b) Includes an electron beam tunnel.

where the subscripts 1 represents the AC parts, 0 represents the DC parts; E_1 consists of two parts, the RF field E_c and the space-charge field E_s. Introducing the 1-D continuity equation of current and Poisson's equation, and substituting E_1 and v_1, (19.11) is transformed to be

$$\frac{\partial^2 J_1}{\partial z^2} + 2j\beta_e \frac{\partial J_1}{\partial z} - (\beta_e^2 - \beta_q^2) J_1 = \frac{j\beta_e J_0}{2V_0} E_c, \tag{19.12}$$

where J is the current density of the beam; $\beta_e = \omega/v_0$ is the wave number of the electrons; $\beta_q = \omega_p/v_0$ is the plasma wave number of the electrons, $\omega_p = (\eta\rho_0/\varepsilon_0)^{1/2}$ is the plasma frequency of the electrons; ρ_0 is the charge density of the electrons. That is the electronics equation for TWTs. In 1D assumption, (19.12) could be written as

$$\frac{\partial^2 i_1}{\partial z^2} + 2j\beta_e \frac{\partial i_1}{\partial z} - (\beta_e^2 - \beta_q^2) i_1 = \frac{j\beta_e i_0}{2V_0} E_c, \tag{19.13}$$

where i is the current of the beam. The circuit equation is derived from Pierce's equivalent circuit model as shown in Fig. 19.6.

This model assumes a voltage V and a current I in a circuit with distributed inductor L and capacitor C for representing the propagation of the RF field in the SWS. The effect of the beam on the circuit is considered to be an external excitation current J_l. So, the relation between J_l and i_1 is

$$J_l = -\frac{\partial i_1}{\partial z}. \tag{19.14}$$

There are three preconditions which make sure that the model is equivalent to the SWS:

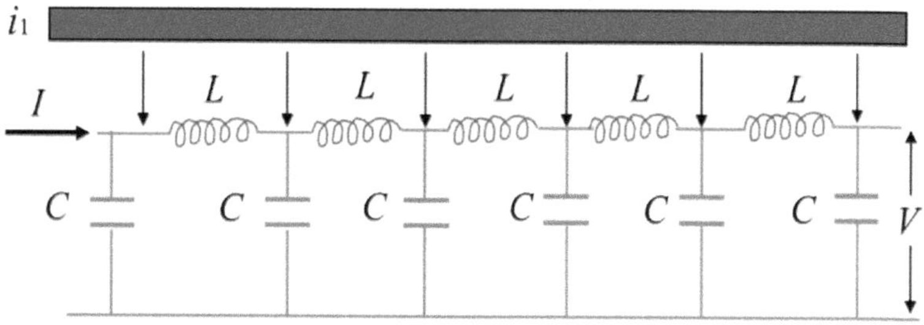

Figure 19.6 Pierce's equivalent circuit model for TWT.

1. The propagation constant of the wave in the model should be equal to that in the SWS. That is

$$\Gamma_0 = j\omega\sqrt{LC} = j\beta, \tag{19.15}$$

   where Γ_0 represents the complex propagation constant of the wave in the circuit and in the SWS simultaneously. β is the wave number of the RF field in the SWS.

2. The RF field E_c and the voltage V should have a relation of

$$E_c = -\frac{\partial V}{\partial z}. \tag{19.16}$$

3. The characteristic impedance of the circuit should be equal to the coupling impedance of the SWS. That is

$$Z_c = \sqrt{\frac{L}{C}} = K_c, \tag{19.17}$$

   where Z_c is the characteristic impedance of the circuit; K_c is the coupling impedance of the SWS.

Substituting (19.14–19.17) into the telegraph's equation [10] one obtains

$$\frac{\partial^2 E_c}{\partial z^2} - \Gamma_0^2 E_c = -\Gamma_0 K_c \frac{\partial^2 i_1}{\partial z^2}. \tag{19.18}$$

That is the circuit equation for TWTs.

Combining the electronics Eq. (19.13) and the circuit Eq. (19.18) by substituting E_c and re-arranging terms, one gets characteristic equation of TWTs

$$(\Gamma^2 - \Gamma_0^2)[(\Gamma - j\beta_e)^2 + \beta_q^2] + 2j\beta_e C^3 \Gamma_0 \Gamma^2 = 0, \qquad (19.19)$$

where $C = [K_c I_0/(4V_0)]^{1/3}$ is called the gain parameter of TWTs or Pierce's gain parameter. C reflects the strength of the coupling between the beam and wave. Solving (19.19) with some ingenious mathematic techniques one then obtains the small-signal gain of TWTs

$$G = -9.54 + 47.26CN, \qquad (19.20)$$

where $N = \beta ez/2\pi$ is the number of electron wavelength; z is the length of the interaction area. It should be noted that Eq. (19.20) is obtained under three assumed conditions: without space-charge field, perfect synchronization of the beam and wave, and free of loss.

The tedious derivation processes of (19.19) and (19.20) are omitted in this section as this section is merely a brief introduction of the core principle of the classical small-signal gain to the readers. One can find the detailed derivation process in [11].

B. Large-signal model and algorithm

When the AC part is comparable to or even larger than the DC part, a large-signal model is required. As mentioned above, there are many assumed conditions for simplifying in the derivation of the small-signal gain. While in the large-signal situation, most of these conditions are not satisfied. Especially, the linearization of the electronics equation would be impractical. Therefore, there is usually not an analytical solution for the large-signal characteristic equation of FW-TWTs. In general, the solution of the large-signal gain has to take advantage of computer numerical computation with various equivalent physical models, mostly active equivalent circuit models.

In 2004, Professor J. H. Booske et al. from the University of Wisconsin-Madison proposed a simplified analytical model and an equivalent circuit model for FW SWSs [12]. The predictions of output power or gain of FW-TWTs established by them agree with 1-D parametric TWT models if phase velocity and interaction impedance functions are sufficiently well characterized. The most sensitive performance parameter was observed to be the small-signal gain. It was observed in experimental measurements on a 40 $\sim$ 55 GHz

FW-TWT that the accuracy of the small-signal gain was confirmed to be within 4 dB over the entire operating band. In addition, the saturated output powers between CHRISTINE-1D predications and experimental measurements are basically in agreement.

In 2004, Shunkang Liu proposed the large-signal operating interaction equation of FW-TWT, and discussed the relationship between design parameters and interaction efficiency of TWT. A fast engineering design method of enhancing efficiency for FW-TWT by velocity re-synchronization was described. The computation was shown that the efficiency of Ka-band folded waveguide TWT with double taper can be enhanced over twice as original tube [13].

In 2013, T. M. Antonsen et al. with SAIC improved the equivalent circuit model parameters of FW based on the calculation results in the electromagnetic simulation software, and obtained a more accurate model. The dispersion characteristic error in the operating frequency band is as low as 0.1%, and the error of coupling impedance is as low as 1% [14].

The related studies have also been conducted by UESTC and IECAS. UESTC has studied a variety of large-signal-interaction models for conventional FW-TWTs, including a large-signal-interaction model based on the velocity modulation theory [15], a 1D large-signal-interaction model based on the time-varying RF field equation [16] and a three-port network-based 1D/3D interaction model named BWISFW-1-D/3-D [17–19]. A steady-state 3-D large-signal model of the beam-wave interaction in FW-TWTs was developed. A time-dependent nonlinear theory including the generalized time-dependent radio frequency field equations was presented to simulate the beam-wave interaction of FW-TWTs. The analytical RF fields in FW-TWTs were replaced by digitized RF field profiles obtained from electromagnetic simulations. Moreover, a 1D large-signal-interaction model based on active transmission matrix has been proposed in UESTC [20]. IECAS has focused on the nonlinear beam-wave interactions in over-mode double-beam FWs [21] and sheet-beam FWs [22].

In terms of the algorithm of the beam-wave interaction, related research in the United States started earlier, mainly including two series, CHRISTINE and TESLA, which were jointly developed by NRL and SAIC.

CHRISTINE [23], a 1D beam-wave interaction algorithm for TWTs, was developed in 1997. In 2000, NRL and SAIC extended the algorithm to 3D and named it CHRISTINE 3D [24]. A C-band helix TWT was used to validate the calculation results of CHRISTINE 3D. Good agreement of the saturated output power between the predictions and experimental measurements was obtained. In addition, the calculation results of the beam transparency also agree with the experimental results. In the following 10 years, CHRISTINE experienced a series of updates and verifications [25–29]. A 1D parallel-loaded transmission line model and algorithm for FW-TWTs were proposed by D. Chernin et al. in 2014 [26]. Its predictions agree well with the results of MAGIC and TESLA-FW.

TESLA, one of the design tools currently being used by NRL, is a beam-wave interaction algorithm for pencil-beam amplifiers [29–37]. TESLA was used in klystrons [36] at first and then has been developed into a series, including TESLA-CC for coupled-cavity TWTs and TESLA-FW for FW-TWTs. Reference [26] compares the CHRISTINE and Tesla-FW in calculating small-signal gain and large-signal gain of FW-TWTs. The calculation results of TESLA are also compared with the experimental results for various pencil-beam amplifiers such as multibeam klystron (MBK), extended interaction klystron (EIK), FW traveling wave tube (FW-TWT), and multibeam cascaded FW-TWT (MBC-FW-TWT) [31, 35, 37, 38]. These comparisons demonstrated the excellent accuracy of TESLA of high efficiency.

Thales also conducted research on the beam-wave interaction theory of FW-TWT. It has successively developed KlysTOP [38] and DIMOHA algorithms [39, 40] in cooperation with the Hamburg University of Technology in Germany and the French National Space Research Center, respectively.

KlysTOP is a frequency domain algorithm. Its idea is to separate the transmission of electromagnetic waves in SWSs from the movement and radiation of electrons in the tunnel. The transmission of waves in SWSs adopts the method described by the transmission line model [14] proposed by T. M. Antonsen. The movement and radiation of electrons in the electron beam tunnel are calculated using the PIC algorithm, and they are coupled using gap voltage and induced currents as interface data. Finally, the quasi-Newton iteration is used to perform iterative operations on the entire model until the convergence requirements are met.

DIMOHA is a relatively novel time-domain algorithm. It is a 1D time-domain Hamiltonian discrete model method for large-signal interaction based on the theory of multibody physics [41], and the complexity of the model is reduced by using the periodicity of SWSs. Compared with the frequency domain algorithm, its main advantage is that it can simulate complex input signal conditions such as multicarrier signals and physical phenomena including harmonic excitation and oscillation. A comparison of the DIMOHA simulation results against the experimental measurements is shown in reference [40].

19.3 Improvement of High-Frequency Structure

To promote the performance of FW-TWTs, many improved FW SWSs have been proposed in recent years. Some studies focused on reaching higher gain, higher output power, broader bandwidth, and so on, to improve the corresponding performances of the devices or systems where the TWTs serve. Other researchers paid more attention to the accessibility of higher-frequency TWTs. These improved structures can be divided into the following six categories.

19.3.1 Ridge/Groove-Loaded FW SWS

By loading ridges or grooves in conventional FW SWS, the high-frequency characteristics of FW-TWT can be adjusted, so that the gain or operating bandwidth can be increased. Based on this idea, many improved structures have been proposed [42–59]. The common methods of ridge/groove loading are shown in Figs. 19.7 and 19.8. Figure 19.7 shows four ridge-loading methods proposed by (a) M. Sumathy et al. [42], (b) J. He et al. [43], (c) Y. Hou et al. [44], and (d) S. Meyne et al. [45]. M. Sumathy et al. analyzed the dispersion and interaction impedance of the FW SWS with ridge loading on one of its broadsides and found this loading method has the ability to increase the bandwidth of an FW-TWT. J. He et al. compared the FW-TWT with horizontal ridge loading on the E-plane against

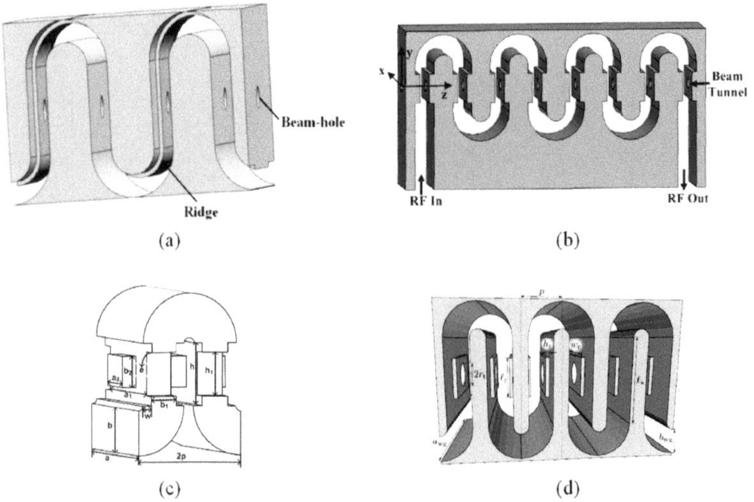

Figure 19.7 Several common ridge-loading methods in FW SWS. (a) Loading on one of the broadsides [42]. (b) Horizontal loading on the E-plane [43]. (c) Ridge-vane loading [44]. (d) Nose-cone loading [45].

a conventional FW-TWT by PIC simulation and found the former can produce a higher saturated output power and a higher gain. Figure 19.8 are two groove-loading methods proposed by J. He et al. [46]. and M. Liao et al. [47], respectively. J. He et al. investigated the high-frequency characteristics of a novel rectangular groove-loaded FW SWS using the field-matching method and numerical calculation and proved that this loading method would increase the interaction impedance while making the SWS more dispersive.

19.3.2 Metamaterial Structure Loaded FW SWS

Metamaterials, like photonic crystal and left-hand material (LHM), are artificial materials designed around unique patterns or structures to have abnormal physical properties that natural materials do not have. They have become a rather hot research topic in this century. Some researchers had investigated the application of metamaterials in FW-TWTs and found that the introducing of metamaterial can provide some unique improvements in performance.

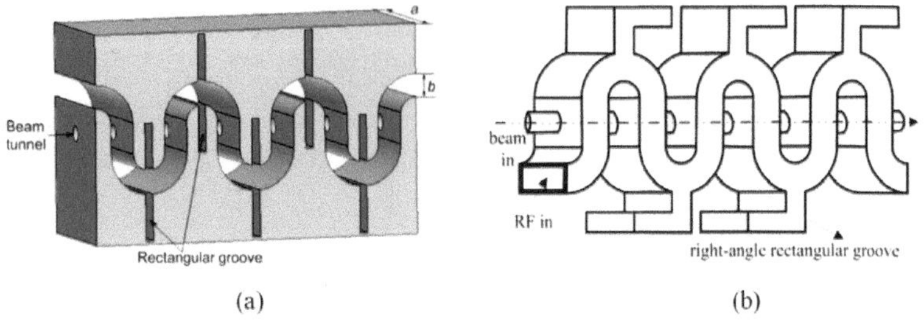

Figure 19.8 Several common groove-loading methods in FW SWS [46]. (a) Rectangular groove loading. (b) Right-angle rectangular groove loading [47].

In 2010, UESTC proposed a photonic-crystal FW SWS [60], in which the metallic wall of the waveguide was replaced by 2D photonic crystals (Fig. 19.9a). The high-order modes and clutter in the SWS, which may cause oscillations, can be eliminated owing to the photonic bandgap of the photonic crystal, while the simulated key performances, such as output power, gain, and efficiency, of the photonic-crystal FW SWS were similar to that of the conventional FW SWS.

In 2014, University of Wisconsin-Madison reported a preliminary investigation on epsilon negative (ENG) metamaterial-loaded FW-TWT [61] and found that the FW SWS loaded with ENG metamaterial slabs (Fig. 19.9b) had a remarkably enhanced coupling impedance than that without loading. PIC simulation results showed that the output power of the ENG metamaterial-loaded FW-TWT is approximately 50 times as much as that of the unloaded FW-TWT. However, there was a real sacrifice of operating voltage and bandwidth.

In 2016, A novel gap-groove FW SWS was proposed by Shahed University in Iran [62–64]. The gap-groove consisted of a bed of nails (a kind of 2D photonic crystals), a folded groove, and a sheet-beam tunnel between them, as shown in Fig. 19.9c. The bed of nails created a high-impedance surface over a certain frequency range and prevented the fields from leaking transversely. Compared to the conventional FW SWS, the gap-groove FW SWS had a higher coupling impedance and the ability to suppress oscillations.

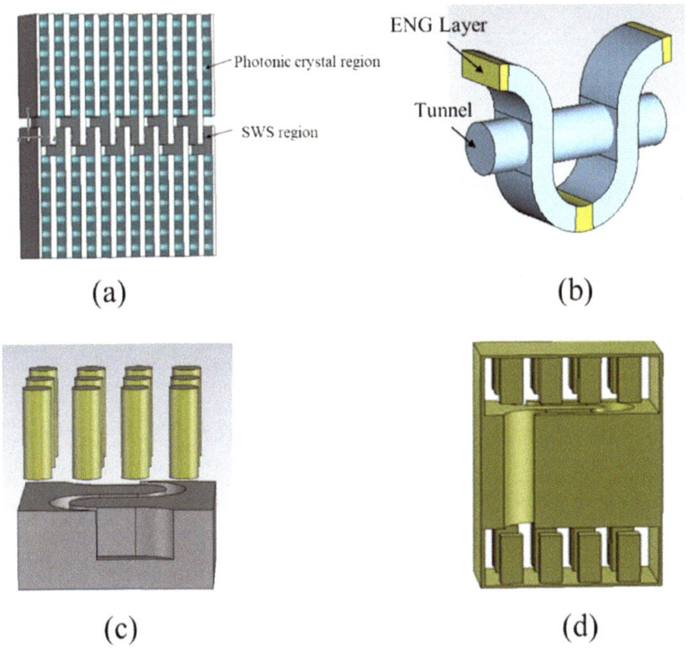

Figure 19.9 Examples of metamaterial structure loaded FW SWS. (a) Photonic-crystal FW SWS proposed by UESTC in 2010 [60]. (b) FW SWS loaded with ENG metamaterial proposed by the University of Wisconsin-Madison in 2014 [61]. (c) Gap-groove FW SWS proposed by Shahed University in 2016 [62–64]. (d) Dual-beam photonic-crystal FW SWS proposed by UESTC in 2019 [65].

In 2020, Southeast University reported a study that used lumped resistance metamaterial absorber (LR-MMA) as an alternative to the conventional concentrated attenuator in FW-TWTs [66]. LR-MMA with impedance matching can effectively reduce the wave reflection at the cutting position in FW-TWT, which can be used to suppress the reflection oscillation. Moreover, the absorption of LR-MMA is easier to adjust than that of conventional medium attenuators.

19.3.3 Nonuniform-Unit FW SWS

Nonuniform-unit design is a conventional performance enhancement method for TWTs, such as the phase velocity tapering

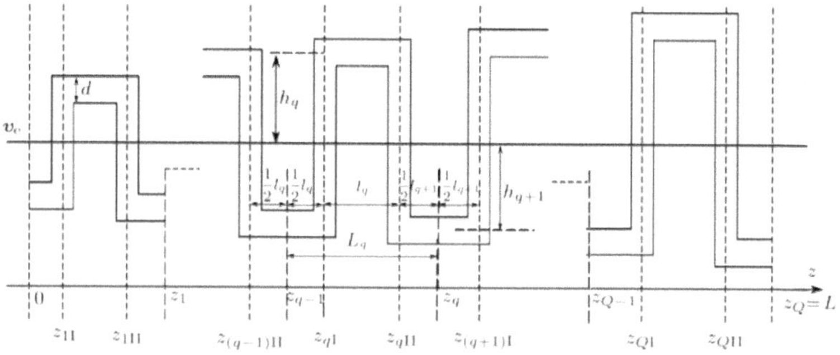

Figure 19.10 A design of nonuniform-unit FW SWS [67].

technology, which is used for improving the output power and efficiency of TWTs by decreasing the phase velocity of waves at the end of the SWSs. However, nonuniform-unit introduces more improvements to FW-TWTs.

Figure 19.10 shows a schematic of a nonuniform-unit FW SWS proposed by A. Xu et al. The corresponding investigation of synchronization [67] and space harmonic selectivity effect [68] was reported in 2009 and proved two advantages of the nonuniform-unit FW SWS. The one was the re-synchronization of the beam and wave, as same as that of a phase velocity tapered one. The other was that the nonuniform-unit FW SWS had a selectivity of the space harmonics, which meant a proper design of the dimensions could improve the coupling impedance of selected space harmonics and increase the performances of the FW-TWT.

The unit of the modified angular log-periodic FW SWS [69] is shown in Fig. 19.11. Its most remarkable advantage was the ability to suppress oscillation owing to the unique characteristics of the quasi-synchronized area. The designed modified angular log-periodic FW-TWT was predicted to have a stable ultra-high gain exceeding 40 dB, without the demand of attenuator.

19.3.4 Resonant Cavity Loaded FW-TWT

As previously mentioned, the most attractive advantage of FW-TWT is their wide-band amplification with a power level which is much higher than that of the available solid-state devices. But the

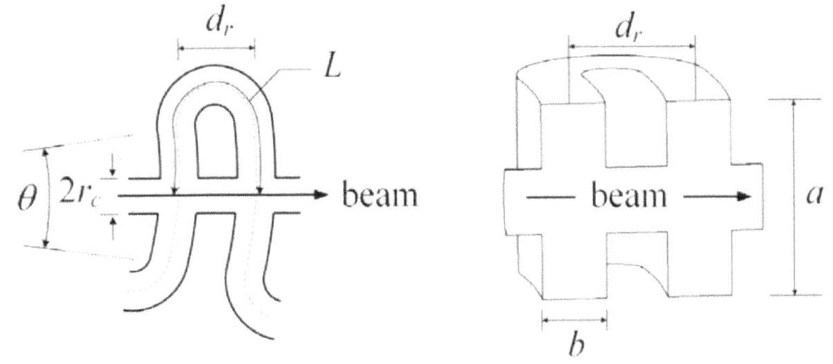

Figure 19.11 Unit of modified angular log-periodic FW SWS [69].

gain and efficiency of the terahertz TWTs are relatively low due to inherently high ohmic loss at terahertz frequencies as well as the low interaction impedance of the SWS. In order to obtain a high gain (>30 dB), the interaction circuit is inevitably long. As a result, the electron beam transmission is typically only 70–80%. On the other hand, klystrons involve very short length interaction circuits and provide high gain and power but with limited bandwidth. The EIK uses multigap resonant cavities as the interaction circuit [70–72]. Recently, a G-band EIK has been reported [73]; at 237.5 GHz, a considerably high power of 125 W has been indicated, with electronic efficiency and gain of about 2.5% and 33 dB, respectively. The multigap resonant cavity has a high characteristic impedance (R/Q), enabling an EIK to achieve high gain in a very short circuit length. This is a crucial advantage which can be of great significance for the terahertz VEDs.

Considering the disadvantages of the conventional FW-TWTs, UESTC has proposed a novel high gain FW SWS combing the extended interaction cavity. To shorten the length of the interaction circuit and to enhance the performance of the THz TWT, this novel scheme is inspired by the advantages of the EIK [74]. As shown in Fig. 19.12, the scheme involves interposing one or more multigap resonant cavities between two sections of a SWS. The two-stage TWT layout is popular in practice since it helps to achieve high gain and stable operation. The first slow-wave section initiates modulation and bunching of the beam. The inserted cavities serve

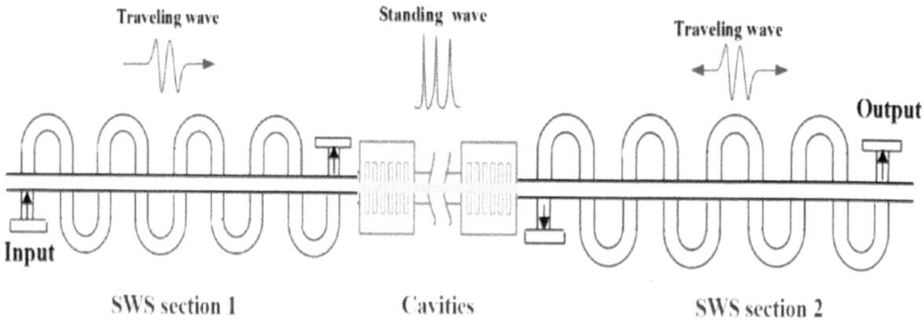

Figure 19.12 The scheme of the novel FW SWS combing with the extended interaction cavities.

as the intermediate or penultimate cavities. The cavities provide additional modulation to the electron beam, which enhances the RF current component in the electron beam before it enters the second slow-wave section. The second slow-wave section, besides providing further modulation to the beam, can be regarded as an output cavity for the final energy conversion and power extraction; as a result, more power can be extracted from the beam, along with higher efficiency, high gain, and shorter circuit length. The efficiency and bandwidth of the second slow-wave section determine the overall performance of the device directly.

A remarkable performance enhancement has been shown due to the insertion of the cavities. The design has been verified by the PIC simulations, using the CST Particle Studio. A round beam with a radius of 0.6 mm, propagating in a cylinder tunnel of radius 0.1 mm, is used in the simulations. The operating voltage is 21.3 kV, the beam current is 60 mA, and the input power is 15 mW. With cavities inserted between the two sections of the SWS, the maximum gain is predicted to be 34.45 dB, corresponding to the output power of 41.87 W and an efficiency of 3.27%. The 3 dB bandwidth is 4.5 GHz, from 215.5 GHz to 220 GHz, within a circuit length of 46 mm.

19.3.5 High-Order Harmonic Amplifier FW SWS

The high-order harmonic amplifier FW-TWT [75–77] was proposed by H. Gong et al. Unlike conventional FW-TWTs, whose output signal frequency is the same as the input signal frequency, the high-order harmonic amplifier FW-TWT can generate a high-order harmonic

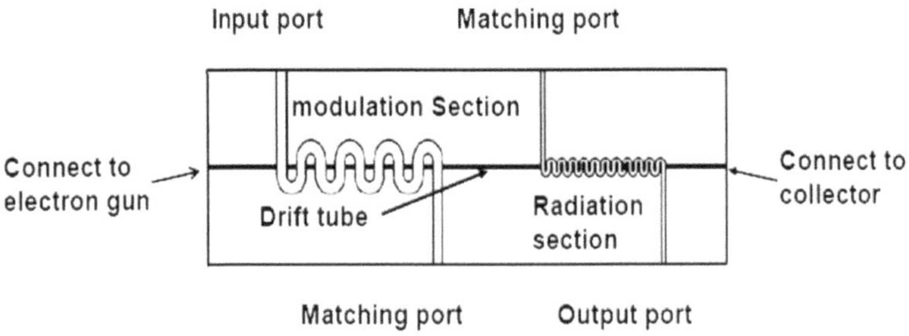

Figure 19.13 Schematic diagram of high-order harmonic amplifier FW SWS [75].

output signal, whose frequency is several times the input signal. It consisted of two sections, the modulation section and the radiation section, and had the ability to amplify the high-order harmonics of the input signals, as shown in Fig. 19.13.

19.3.6 Multibeam/Sheet-Beam FW SWS

By applying multibeam or sheet-beam SWS, the cross section of the electron beam tunnel can be increased, and so that the current and power of FW-TWT can be effectively enhanced [78–90]. Some of the structures of this kind are designed to operate at high-order modes. Figure 19.14 demonstrates some interesting concepts of multibeam FW-TWTs, including fundamental-mode and overmoded designs. Figure 19.15 shows groove FW SWSs, which is a special kind of sheet-beam FW SWSs with ultrawide beam tunnel, with different shapes of meander paths and cross sections.

19.4 Electron Optical System

Compared with FW SWS, the electron optical system of FW-TWT has much fewer innovations in theory and structure. The dominant electron optical system for FW-TWTs remains the traditional combination of a round beam and a periodic permanent magnet (PPM) focusing system. Although a sheet beam may improve the performance, the practical focusing methods for high-frequency

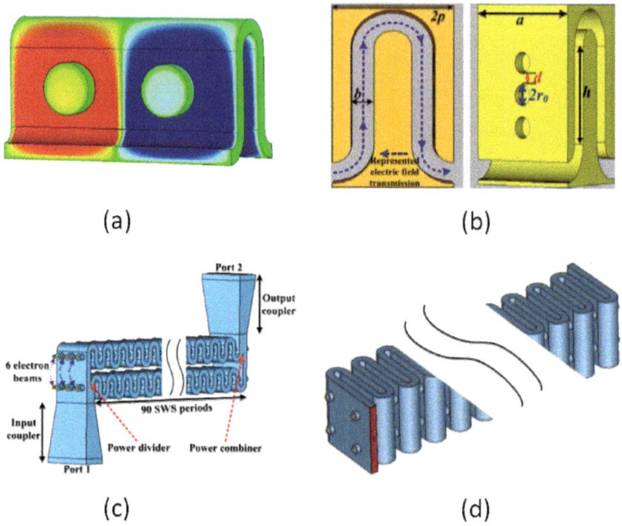

Figure 19.14 Multibeam FW SWSs. (a) Overmoded broadside dual-beam FW SWS [78] (b) fundamental-mode triple-beam FW SWS [79] (c) fundamental-mode broadside triple-beam FW SWS with power combining [80] (d) Overmoded array quadruple-beam FW SWS [81].

sheet-beam TWTs are under developing. Hence, this section merely provides a brief introduction of the round-beam electron optical system with a PPM focusing system.

An electron gun is the source of electrons in which the electrons are emitted from the surface of the cathode and accelerated to a specific velocity. A typical structure of an electron gun consists of a cathode, a control electrode, and an anode. Figure 19.16 shows a schematic diagram of the traditional Pierce gun for generating a round beam.

After the electrons leave the electron gun, a focusing system should be applied to ensure the stable transmission of the beam. The most conventional focusing system for the FW-TWTs is the PPM focusing system of light weight. There are two critical formulas for the design of PPM. The one is the Brillouin magnetic field

$$B_b = \frac{0.833 \times I^{\frac{1}{2}}}{rU^{\frac{1}{4}}} \left(\frac{2}{1+\gamma}\right)^{\frac{1}{4}}, \tag{19.21}$$

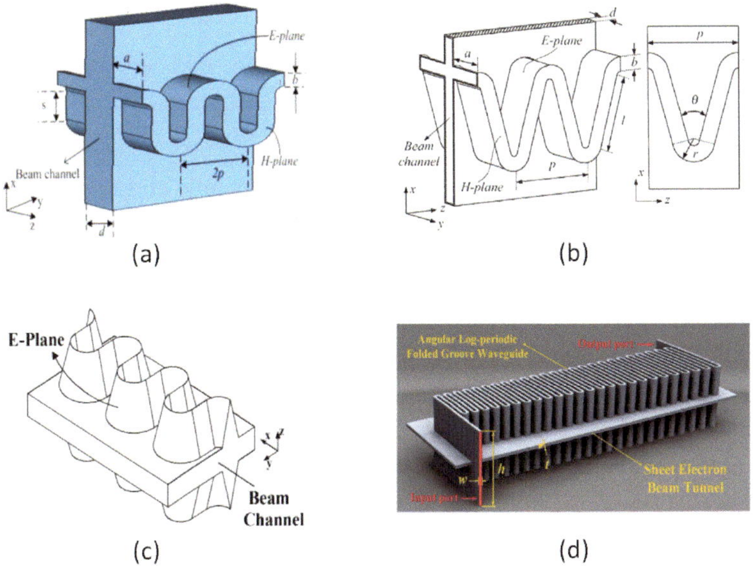

Figure 19.15 Sheet-beam groove FW SWSs: (a) U type [82]; (b) V type [83]; (c) V-cross section type [84]; (d) Angular log-periodic type [85].

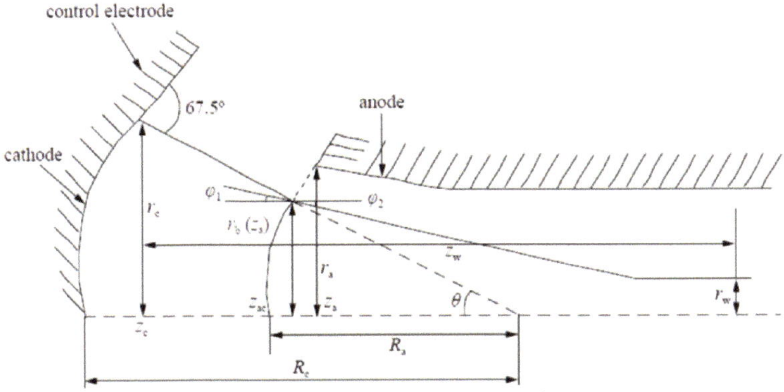

Figure 19.16 Schematic diagram of the traditional Pierce gun.

where U and I are the operating voltage (V) and current (A) of the electron gun, respectively; r is the waist radius of the beam with the unit of mm; γ is the relativistic factor. The unit of B_b is T.

Another formula is the plasma wavelength of the electron beam

$$\lambda_p = 3.592 \times 10^{-2}\frac{rU^{\frac{3}{4}}}{I^{\frac{1}{2}}}. \quad (19.22)$$

The unit of λ_p is mm.

The axial magnitude of the on-axis magnetic field and the period length of the PPM should be approximately 2 times of B_b and a third of λ_p, respectively. One may refer to [91–93] for the detailed theoretical base and design flow.

19.5 Fabrication Technology

Owing to the planarity of FWs, they are more suitable for new planar processing methods with higher precision and easier to be fabricated than traditional helix SWSs and coupled-cavity SWSs. Common processing methods for FW SWSs include electrical discharge machining (EDM), computer numerical control machining (CNC), lithography, electroforming, and molding (LIGA), and deep reactive ion etching (DRIE).

19.5.1 EDM

The principle of EDM is to etch and cut metal workpieces by pulse-spark discharge of a continuously moving fine metal wire electrode. This technology can be applied to processing electron tunnels and cavities of FWs. A feature of this technology is that the entire workpiece has to be penetrated. Therefore, the FW SWS should be divided into at least three parts, which are the cavity in the middle and the two metal cover plates on both sides.

The main process method used in the FW-TWT reported by G. Dohler in 1987 was EDM [1]. In this work, EDM was used to form the main feature of the SWS, and then the whole SWS was brazed with the aid of thin silver plates. At 21.6 kV operating voltage and 0.2 A current, the FW-TWT achieves a maximum small-signal gain of 14 dB and a maximum saturated output power of 250 W in the frequency range of 40–50 GHz, and the maximum saturated output power can be increased to 300 W by adjusting the operating voltage. In 2006, Calabazas Creek Research Inc. and the University of Wisconsin-Madison jointly developed a W-band FW SWS using EDM process [94]. In the frequency range of 81–86 GHz, the measured S_{11} of the SWS without attenuator is about −4 dB, and the measured insertion loss of the SWS with attenuator is about −17 dB. In 2013, Bharat Electronics reported the cold test results of their W-band FW SWS [95], which is also manufactured by EDM process. The results showed that the S_{11} was less than −15 dB in the range of 90–100 GHz, and S_{21} was higher than −7 dB in the range of 97–98 GHz.

19.5.2 CNC

CNC is a processing technology for cutting workpieces using computer-numerical-controlled tools. It has the advantages of high precision and low surface roughness. Due to the limitation of the aspect ratio of the tools, an SWS usually needs to be split into two symmetrical halves for the processing of the electron tunnel. The electron tunnel can also be processed by combining CNC with EDM technology.

In 1999, Seoul National University reported the cold test experiment of a Ku-band FW-TWT fabricated by CNC [96]. The SWS was designed to operate at 14.0–14.5 GHz, and its S_{21} in this frequency band was about -1 dB. In 2020, Thales reported the test of a W-band FW-TWT [97]. The SWS of this TWT was divided into two symmetrical halves and processed by CNC. Although the hot test results of the TWT were not ideal, the cold test experimental results were good in the frequency range of 78–110 GHz, where the S_{11} of the SWS is lower than −18 dB. In 2021, Bharat Electronics reported a 220 GHz FW SWS machined by CNC provided by Kern Co., Ltd. [98]. The measurement results showed that the linear tolerance of the narrow side width in the depth of the waveguide is 30 μm; the linear tolerances of the other major dimensions were about 3–5 μm; the roughness of each inner surface of the waveguide was about 45 nm. In 2016, the University of California at Davis, Lancaster University, and Ulsan National Institute of Science and Technology jointly published their research progress of using CNC to manufacture various THz vacuum electronic devices [99], demonstrating the applications of CNC in klystron, staggered-double-vane TWT, FW-TWT, sheet-beam BWO, and some other devices.

19.5.3 DRIE

DRIE is a dry etching process utilizing accelerated ions to etch the wafer and replicate the planar pattern of a mask on the wafer. It is widely used in the manufacture of integrated circuits. The DRIE process for FW SWSs can be summarized into the next three steps. First, serpentine trenches need to be etched symmetrically on the surfaces of two wafers. Then, the surfaces of the wafers are coated by spraying metal. Finally, the two symmetrical halves are brazed together to form a complete SWS. Figure 19.17 shows the process of etching a trench on a silicon substrate with DRIE. The main advantage of DRIE lies in its high precision and good repeatability, while its disadvantage is that the sectional shape of the waveguide is not easy to control, especially for the perpendicularity and roughness of the side surfaces. Another problem of the DRIE process is that it can only form rectangular electron beam tunnels rather than the more commonly used round electron beam tunnels.

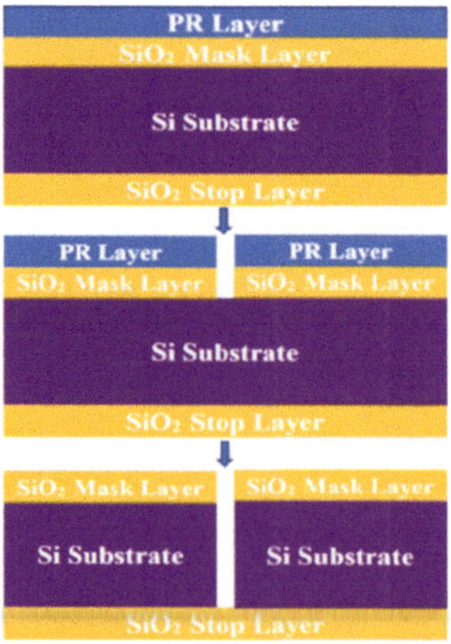

Figure 19.17 Schematic diagram of etching a trench on a silicon substrate with DRIE process.

In 2006, the Georgia Institute of Technology published a study on the influence of re-entrant sidewall and vertical striation generated by DRIE processing of a straight waveguide and an FW SWS on the electromagnetic transmission characteristics of the structures [100]. The simulation results showed that both of them have little influence on the transmission characteristics of the straight waveguide, but the experimental results showed that the insertion loss of the W-band straight waveguide manufactured by the DRIE process was 0.23–0.42 dB/cm, which was still far from the commercial straight waveguide products. The insertion loss of the W-band simplified FW SWS was 0.41–0.69 dB/cm.

19.5.4 LIGA

LIGA is a planar micro-machining process based on lithography technology, which is the abbreviation of three German words: lithographie (lithography), galvanoformung (electroforming), and abformung (molding). It is the most widely used machining technology for FW SWSs operating at millimeter-wave band or even higher frequencies. The process of LIGA is shown in Fig. 19.18. LIGA usually uses X-ray with extremely high collimation and energy intensity to ensure the high precision of micro or nano level.

Such X-ray can be realized in synchrotron radiation facilities, which is high-quality but extremely expensive. To reduce the cost and the technical difficulty, some lower-cost quasi-LIGA processes have emerged, such as UV-LIGA which uses ultraviolet light instead of X-ray for exposure. Despite these, the processing of electron channels is still a difficulty for LIGA or UV-LIGA. Generally, there are three methods to solve this problem: (1) use a removable inlay to form an electron channel during the forming process; (2) use EDM after the forming process; (3) use a two-step LIGA process [101] to produce a rectangular electron channel.

In 2002, Seoul National University reported a Ka-band FW-TWT fabricated by LIGA process [102]. The study found that LIGA with high precision can eliminate bandgaps of wave transmission caused by the asymmetry of the SWS fabricated by low-precision processes.

In 2004, the University of Wisconsin-Madison reported a study of FWs without electron channels fabricated by three different

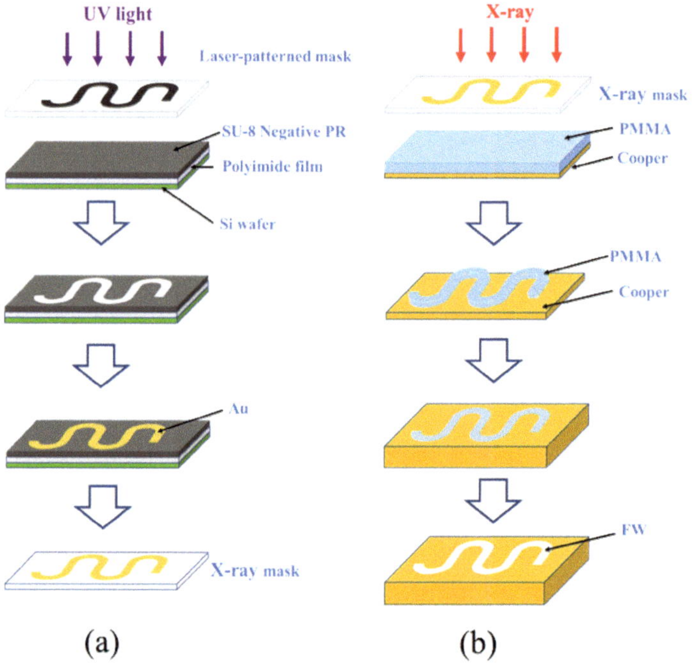

Figure 19.18 Fabrication procedures of FW SWS by LIGA process: (a) fabrication of mask; (b) fabrication of waveguide cavity.

LIGA processes and one DRIE process (Fig. 19.18) [103]. The three LIGA processes include (1) using X-rays together with polymethylmethacrylate (PMMA) photoresist to fabricate masks and molds; (2) using UV and SU-8 photoresist to make masks, and then using X-ray and PMMA photoresist to make molds; (3) using UV and SU-8 photoresist to fabricate masks and molds. The experimental results showed that (1) in the first LIGA process, using graphite as the substrate instead of copper can improve the adhesion between the PMMA photoresist and the substrate, but it needs additional procedures; (2) the surface roughness of the SWS manufactured by the first LIGA process was much less than 1 μm; (3) the processing cycle can be shortened by using UV instead of X-ray, and in particular, the third pure UV-LIGA process has accelerated the processing speed greatly; (4) in the DRIE process, the tolerance of the SWS is about 1 μm and the surface roughness is about 0.1 μm. In the same year,

Seoul National University also reported a preliminary experiment on the processing of FW SWS using LIGA [104].

In 2013, the U.S. Office of Naval Research (ONR) reported a new UV-LIGA process for FW-TWTs [105]. Different from the common processes, it used an embedded polymer monofilament to occupy the space of the electron beam channel, so that the SWS and the electron beam tunnel were formed as a unity. Such a process had several advantages. On the one hand, it can be used to fabricate small-size tunnels, which is unachievable by EDM, and it provided a higher precision. On the other hand, it can avoid the misalignment of two separate halves. Meanwhile, ONR also pointed out a new problem in this process: It was not easy to maintain the electron beam tunnel on the symmetrical plane of the structure. In their prototypes, the deviation from the center of the electron beam tunnel to the symmetrical plane of the structure was about 100 μm.

Since 2011, Beijing Vacuum Electronics Research Institute (BVERI) has conducted a series of studies on FW-TWTs fabricated by UV-LIGA combining with DRIE or EDM [106–110]. The experiments of manufacturing FW SWSs by UV-LIGA and DRIE were reported from 2011 to 2012. In 2014, their hot test on a W-band FW-TWT was also carried out. In 2016, BVERI carried out a hot test of a G-band FW-TWT fabricated by UV-LIGA and EDM. In 2019, the hot test experiment of a W-band ultra-wide-band FW-TWT processed by UV-LIGA and EDM was also reported.

Shizuoka University in Japan and Kwangwoon University in Korea also carried out experimental researches on FW SWS with X-ray LIGA processes, and the relevant results were published in 2016 and 2019, respectively [111, 112].

In 2020, IECAS manufactured a W-band FW-TWT using X-ray LIGA process [113]. Compared with the W-band TWT using UV-LIGA process reported by BVERI in 2019, it obtained a higher processing precision, a broader operating bandwidth, and a higher electronic efficiency.

19.5.5 Other Technologies

In addition to the four processes mentioned above, some researchers explored some new processes in manufacturing FW SWSs.

For example, NRL manufactured W-band and D-band FW SWSs using 3D printing technology in 2019 [114]. The W-band FW has an excellent transmission performance, and the D-band FW also has a good transmission performance despite a frequency shift between the test result and the simulation result. In 2020, IIS demonstrated a new fabrication method using photochemical machining (PCM) [115]. In the first step, copper sheets of 200 μm thickness were fabricated by photochemical machining. Next, multilayers of these copper sheets were aligned together to fabricate the FW SWS. PCM had the advantage of high machining efficiency and low cost, but this technology still cannot ensure good flatness of the FW inner walls. The problem will cause the measured S-parameters to be worse than expected, even after pickling and gold-plating on the copper sheets.

19.6 Performance of FW-TWT

Table 19.1 summarizes some reports of SW/FW-TWTs mainly from institutions including NRL, BVERI, Northrop Grumman Corporation (NGC), UESTC, and so on, mostly distributed in the U.S. or China.

Most experimental data listed in Table 19.1 were from saturated TWTs. Some particular cases and special TWTs were noted in the "Notes" column. The "Type of Bandwidth" was defined by the writers for clarifying the "Bandwidth" as many authors of the references used various criteria to evaluate the "Bandwidth". The "Tiny" in the "RF Efficiency" column means the maximum RF efficiency was lower than 1%.

As shown in Table 19.1, the most concerned operating frequency bands among these SW/FW-TWTs are W-band and G-band.

19.7 Conclusion

As an important type of TWT, FW-TWT is attracting more and more attention due to its good performance in bandwidth, efficiency, and power capacity. In particular, it has aroused considerable research enthusiasm in recent years because of its potential of balancing wide-bandwidth and high-power capacity in high frequencies such

Table 19.1 Major operating parameters of some reported SW/FW-TWTs

Institution	Year	Voltage (kV)	Current (mA)	Max. power	Gain (dB)	RF efficiency (%)	Duty cycle	Type of bandwidth	Operating frequency (GHz)	Notes	Ref.
BVERI	2014	22	180	~ 170 W	~ 38–52	2.53–4.3	20%	over 100 W	91.2–101.3		[116]
BVERI	2014	16	80	~ 40 W	~ 30–45	2.34–3.13	CW	over 30 W	91–101		[117, 118]
BVERI	2014	21.74	171	~ 130 W	35–52	3–3.5	20%	over 100 W	90–95		[108]
BVERI	2014	21.88	186	~ 230 W	32–38	2.4–5.6	20%	over 100 W	89–95.7		[108]
BVERI	2015	21.5	150	~ 500 mW	10–38	Tiny	20%	over 100 mW	171.4–182.8	2nd harmonic amplifier	[119]
BVERI	2017	22.5	189	424 W	32–45	5.88–9.97	1%	over 250 W	89.6–97.6		[120]
BVERI	2019	21.95	170	143 W	30–42	2.86–4.08	1%	over 100 W	94–101.8	partly un-saturated	[110]
BVERI	2021	15.66	145	213 W	37.6–38.5	7.8–9.38	30%	over 180 W	91.3–97		[121]
BVERI	2021	24.25	59	56.7 W	35–39	3.5–4	5%	Over 50 W	215.4–219		[122]
NGC	2007	9.2–10	~ 3	~ 16 mW	~ 13	Tiny	1%	/	~ 650		[123]
NGC	2012	9.72	4.8	108 mW	/	Tiny	0.5%	over 25 mW	644–661	combined with a SSPA	[124]
NGC	2013	~ 19	~ 260	30–55 W	≤ 28.5	≤ 3.4	/	/	~ 215	5-beam 5 cascaded TWTs	[125]
NGC	2014	11.4	2.8	39 mW	≤ 23.3	Tiny	1%	over 19 mW	848–857		[126]

(*Continued*)

Table 19.1 *(Continued)*

Institution	Year	Voltage (kV)	Current (mA)	Max. power	Gain (dB)	RF efficiency (%)	Duty cycle	Type of bandwidth	Operating frequency (GHz)	Notes	Ref.
NGC	2016	20.95	114	79 W	21–24	1.7–3.3	50%	over 50 W	232.4–234.8		[127]
NGC	2016	12.1	2.3	29 mW	≤ 20	Tiny	0.3%	/	~ 1030		[128]
NRL	2014	~ 11.7	~ 104	~ 60 W	≤ 15	~ 5	/	over 10 dB	208–225		[129]
NRL	2021	20/20.8	144/139	215 W/285 W	≤ 22	3.6–10.2	0.1%	over 100 W	88–98/87–95		[130]
UESTC	2011	24.5	446	1100 W	30–40	7–10	10%	over 700 W	32.2–34.6		[131]
UESTC	2014	25.2	436	1200 W	43–60	6.3–11.2	10%	over 1000 W	32–34		[132]
UESTC	2017	17.45	110	/	18–25	Tiny	/	over 18 dB	137–144	unsaturated	[133]
UESTC	2018	18.6	76	280 mW	/	Tiny	/	/	~ 133	3rd harmonic amplifier	[76]
L-3	2010	20/21	295/315	~ 230 W/~ 360 W	30–36/25–29	≤ 5.4	/	Over 100 W	91.2–91.6/90.4–90.9		[134]
L-3	2013	~ 20.5	~ 220	176 W	34–40	2.2–4	30%	over 100 W	91–95.5		[135]
Kwang-woon Univ.	2017	15	50	/	10–17	/	/	/	82–96	unsaturated	[136]
CAEP	2018	16.9	16	134 mW	3–19.6	Tiny	10%	/	~ 320	unsaturated	[137]
CAEP	2021	20.5	52.4	~ 30 W	22–31	1–2.79	CW	over 16 W	213–220		[138]
AIRCAS	2021	17	71	~ 60 W	22–35	3.3–5	20%	over 40 W	210–216	2-beam 2 cascaded TWTs	[139]
IECAS	2020	19.4	90	123 W	28.7–38.4	4.35–7.04	10%	over 87 W	93–100		[113]

as millimeter-wave band and terahertz band. In this chapter, we have reviewed the basic concept, development, and technologies of FW-TWT, including the theoretical model, improved high-frequency structure, electronic optical system, fabrication technology, and device performances. We hope that the content could be helpful for graduate students and researchers to study or conduct further research.

References

1. G. Dohler, D. Gagne, D. Gallagher, and R. Moats, Serpentine waveguide TWT, in *1987 International Electron Devices Meeting*, IEEE, pp. 485–488, 1987.
2. S. Liu, Folded waveguide circuit for broadband MM wave TWTs, *International Journal of Infrared and Millimeter Waves*, vol. 16, no. 4, pp. 809–815, 1995.
3. H.-J. Ha, S.-S. Jung, and G.-S. Park, Theoretical study for folded waveguide traveling wave tube, *International Journal of Infrared and Millimeter Waves*, vol. 19, no. 9, pp. 1229–1245, 1998.
4. S. Liu, Study of propagating characteristics for folded waveguide TWT in millimeter wave, *International Journal of Infrared and Millimeter Waves*, vol. 21, no. 4, pp. 655–660, 2000.
5. M. Sumathy, K. Vinoy, and S. Datta, Equivalent circuit analysis of serpentine folded-waveguide SWSs for millimeter-wave traveling-wave tubes, *Journal of Infrared, Millimeter, and Terahertz Waves*, vol. 30, no. 2, pp. 151–158, 2009.
6. M. Sumathy, K. Vinoy, and S. K. Datta, A simple equivalent circuit analysis of rectangular folded-waveguide SWS, *AEU-International Journal of Electronics and Communications*, vol. 64, no. 12, pp. 1192–1195, 2010.
7. M. Sumathy, K. Vinoy, and S. Datta, Analysis of rectangular folded-waveguide millimeter-wave SWSs using conformal transformations, *Journal of Infrared, Millimeter, and Terahertz Waves*, vol. 30, no. 3, pp. 294–301, 2009.
8. M. Sumathy, K. Vinoy, and S. K. Datta, Analysis of serpentine folded-waveguide SWS using elliptical conformal transformation, *AEU-International Journal of Electronics and Communications*, vol. 65, no. 2, pp. 161–164, 2011.

9. S. T. Han, J. I. Kim, and G. S. Park, Design of a folded waveguide traveling-wave tube, *Microwave and Optical Technology Letters*, vol. 38, no. 2, pp. 161–165, 2003.
10. D. M. Pozar, *Microwave Engineering*. John Wiley & Sons, 2011.
11. B. N. Basu, *Electromagnetic Theory and Applications in Beam-Wave Electronics*. World Scientific, 1996.
12. J. H. Booske, et al., Accurate parametric modeling of folded waveguide circuits for millimeter-wave traveling wave tubes, *IEEE Transactions on Electron Devices*, vol. 52, no. 5, pp. 685–694, 2005.
13. S. Liu, Engineering computation and analysis of enhancing efficiency for folded waveguide TWT in MMW, *International Journal of Infrared and Millimeter Waves*, vol. 25, no. 3, pp. 471–480, 2004.
14. T. M. Antonsen, A. N. Vlasov, D. P. Chernin, I. A. Chernyavskiy, and B. Levush, Transmission line model for folded waveguide circuits, *IEEE Transactions on Electron Devices*, vol. 60, no. 9, pp. 2906–2911, 2013.
15. Y. Hai-Rong, X. Jin, Y. Ling-Na, G. Yu-Bing, and W. Yan-Yu, A wave-beam interaction theory for folded-waveguide traveling wave tubes, *Acta Physica Sinica*, vol. 61, no. 24, 2012.
16. W. F. Peng, et al., Time-dependent nonlinear theory and numerical simulation of folded waveguide traveling wave tubes, *Journal of the Korean Physical Society*, vol. 62, no. 6, pp. 949–953, 2013.
17. Y. Wei-Zhong, H. Yu-Lu, L. Jian-Qing, Y. Zhong-Hai, T. Yun-Xian, and L. Bin, Research on the beam-wave interaction theory of folded waveguide traveling wave tubes based on three-port network model, *Acta Physica Sinica*, vol. 63, no. 23, 2014.
18. W. Yan, et al., A one-dimensional large signal simulation of folded waveguide TWTs, in *IEEE International Vacuum Electronics Conference*, IEEE, pp. 481–482, 2014.
19. W.-Z. Yan, Y.-L. Hu, Y.-X. Tian, J.-Q. Li, and B. Li, A 3-D large-signal model of folded-waveguide TWTs, *IEEE Transactions on Electron Devices*, vol. 63, no. 2, pp. 819–826, 2016.
20. R. Zhang, et al., An active transmission matrix-based nonlinear analysis for folded waveguide TWT, *IEEE Transactions on Electron Devices*, vol. 67, no. 3, pp. 1205–1210, 2020.
21. W. Liu, K. Li, P. Gao, C. Zhao, X. Guo, and Z. Zhang, Nonlinear theory for beam-wave interactions of two electron beams with higher order TE20 mode in serpentine waveguide traveling wave amplifier, *Physics of Plasmas*, vol. 25, no. 12, p. 123106, 2018.

22. F. Lu, M. Grieser, C. Zhang, and Y. Wang, 3-D nonlinear theory for sheet-beam folded-waveguide traveling-wave tubes, *IEEE Transactions on Electron Devices,* vol. 65, no. 11, pp. 5103–5110, 2018.
23. T. M. Antonsen Jr and B. Levush, CHRISTINE: a multifrequency parametric simulation code for traveling wave tube amplifiers, Naval Research Lab Washington DC Vacuum Electronics Branch, 1997.
24. D. Chernin, T. Antonsen, B. Levush, and D. Whaley, CHRISTINE 3D: a 3D multi-frequency large signal simulation code for helix traveling wave tubes, in *Abstracts. International Vacuum Electronics Conference 2000 (Cat. No. 00EX392)*, IEEE, p. 2, 2000.
25. R. Begum and J. Ramirez-Aldana, Christine-CC evaluation of an As-built Ka-band CCTWT to predict drive induced oscillation, in *IEEE International Vacuum Electronics Conference*, IEEE, pp. 77–78, 2014.
26. D. Chernin, T. M. Antonsen, A. N. Vlasov, I. A. Chernyavskiy, K. T. Nguyen, and B. Levush, 1-D large signal model of folded-waveguide traveling wave tubes, *IEEE Transactions on Electron Devices,* vol. 61, no. 6, pp. 1699–1706, 2014.
27. D. Chernin, D. Dialetis, T. Antonsen, J. Legarra, J. Qiu, and B. Levush, Validation studies for CHRISTINE-CC using a Ka-band coupled-cavity TWT, in *2006 IEEE International Vacuum Electronics Conference held Jointly with 2006 IEEE International Vacuum Electron Sources*, IEEE, pp. 399–400, 2006.
28. P. N. Safier, D. K. Abe, T. M. Antonsen, B. G. Danly, and B. Levush, Simulation of noise-power ratio with the large-signal code CHRISTINE, *IEEE Transactions on Electron Devices,* vol. 48, no. 1, pp. 32–37, 2001.
29. A. N. Vlasov, et al., Modeling of the NRL G-band TWT amplifier using the CHRISTINE and TESLA simulation codes, in *2013 IEEE 14th International Vacuum Electronics Conference (IVEC)*, IEEE, pp. 1–2, 2013.
30. I. A. Chernyavskiy, et al., Current status of the large-signal code TESLA: recent development and new applications, in *IVEC 2012*, IEEE, pp. 533–534, 2012.
31. I. A. Chernyavskiy, T. M. Antonsen, A. N. Vlasov, D. Chernin, K. T. Nguyen, and B. Levush, Large-signal 2-D modeling of folded-waveguide traveling wave tubes, *IEEE Transactions on Electron Devices,* vol. 63, no. 6, pp. 2531–2537, 2016.
32. I. A. Chernyavskiy, A. N. Vlasov, T. Anderson, S. J. Cooke, B. Levush, and K. T. Nguyen, Large-signal code TESLA: improvements in the implementation and in the model, in *2006 IEEE International Vacuum*

Electronics Conference Held Jointly with 2006 IEEE International Vacuum Electron Sources, IEEE, pp. 441–442, 2006.

33. I. A. Chernyavskiy, T. M. Antonsen, J. C. Rodgers, A. N. Vlasov, D. Chernin, and B. Levush, Modeling vacuum electronic devices using generalized impedance matrices, *IEEE Transactions on Electron Devices*, vol. 64, no. 2, pp. 536–542, 2017.
34. I. A. Chernyavskiy, A. N. Vlasov, B. Levush, T. M. Antonsen, and K. T. Nguyen, Parallel 2D large-signal modeling of cascaded TWT amplifiers, in *2013 IEEE 14th International Vacuum Electronics Conference (IVEC)*, IEEE, pp. 1–2, 2013.
35. I. A. Chernyavskiy, et al., TESLA modeling of the linear-beam amplifiers, in *2009 IEEE International Vacuum Electronics Conference*, IEEE, pp. 121–122, 2009.
36. A. N. Vlasov, T. M. Antonsen, D. Chemin, B. Levush, K. T. Nguyen, and S. Cooke, TESLA: large signal simulation code for klystrons, in *Fifth IEEE International Vacuum Electronics Conference (IEEE Cat. No. 04EX786)*, IEEE, pp. 314–315, 2004.
37. I. A. Chernyavskiy, A. N. Vlasov, B. Levush, T. M. Antonsen, D. P. Chernin, and J. R. Legarra, Validation study for the large-signal code TESLA-CC based on experimental Ka-band Coupled-Cavity TWT, in *IVEC 2012*, IEEE, pp. 535–536, 2012.
38. S. Meyne, P. Bernadi, P. Birtel, J.-F. David, and A. F. Jacob, Large-signal 2.5-D steady-state beam-wave interaction simulation of folded-waveguide traveling-wave tubes, *IEEE Transactions on Electron Devices*, vol. 63, no. 12, pp. 4961–4967, 2016.
39. D. F. Mincnna, et al., DIMOHA: traveling-wave tube simulations including band edge and multiple-carriers operations, in *2019 International Vacuum Electronics Conference (IVEC)*, IEEE, pp. 1–2, 2019.
40. D. F. Mincnna, Y. Elskens, F. André, A. Poyé, J. Puech, and F. Doveil, DIMOHA: a time-domain algorithm for traveling-wave tube simulations, *IEEE Transactions on Electron Devices*, vol. 66, no. 9, pp. 4042–4047, 2019.
41. D. F. Minenna, Y. Elskens, and F. André, Electron-wave momentum exchange and time domain simulations applied to traveling wave tubes, in *2017 Eighteenth International Vacuum Electronics Conference (IVEC)*, IEEE, pp. 1–2, 2017.
42. M. Sumathy, K. Vinoy, and S. Datta, Analysis of ridge-loaded folded-waveguide SWSs for broadband traveling-wave tubes, *IEEE Transactions on Electron Devices*, vol. 57, no. 6, pp. 1440–1446, 2010.

43. J. He, Y. Wei, Y. Gong, W. Wang, and G.-S. Park, Investigation on a W band ridge-loaded folded waveguide TWT, *IEEE Transactions on Plasma Science,* vol. 39, no. 8, pp. 1660–1664, 2011.
44. Y. Hou, J. Xu, S. Wang, Z.-G. Lu, Y.-Y. Wei, and Y.-B. Gong, Study of high efficiency novel folded waveguide traveling-wave tube with sheet electron beam, *Progress In Electromagnetics Research,* vol. 141, pp. 431–441, 2013.
45. S. Meyne, P. Birtel, and A. F. Jacob, Nose-cone loaded folded-waveguide delay lines for Q-band traveling-wave tubes, in *2016 German Microwave Conference (GeMiC),* IEEE, pp. 373–376, 2016.
46. J. He, Y. Wei, and G. Park, Nonlinear investigation of beam-wave interaction in double-groove loaded folded-waveguide traveling-wave tube, *Science China Physics, Mechanics and Astronomy,* vol. 56, no. 7, pp. 1366–1372, 2013.
47. M. Liao, et al., A novel folded waveguide for V-band TWT, *IEEE Transactions on Plasma Science,* vol. 43, no. 12, pp. 4088–4091, 2015.
48. J. Cai, J. Feng, and X. Wu, Folded waveguide SWS with modified circular bends, *IEEE Transactions on Electron Devices,* vol. 61, no. 10, pp. 3534–3538, 2014.
49. G. Guo, Y. Wei, Y. Gong, Z. Duan, and G.-S. Park, Analysis of the three types of folded waveguide SWS for 140 GHz traveling wave tube, in *IEEE International Vacuum Electronics Conference,* IEEE, pp. 307–308, 2014.
50. G. Guo, et al., A tapered ridge-loaded folded waveguide SWS for millimeter-wave traveling-wave tube, *Journal of Infrared, Millimeter, and Terahertz Waves,* vol. 33, no. 2, pp. 131–140, 2012.
51. J. He and Y. Wei, Study of the slow-wave properties of a rectangular groove-loaded folded waveguide for millimeter traveling-wave tubes, *IEEE Transactions on Plasma Science,* vol. 42, no. 1, pp. 55–61, 2014.
52. J. He, Y. Wei, Z. Lu, Y. Gong, and W. Wang, Investigation of a ridge-loaded folded-waveguide slow-wave system for the millimeter-wave traveling-wave tube, *IEEE Transactions on Plasma Science,* vol. 38, no. 7, pp. 1556–1562, 2010.
53. Y. Hou, et al., A novel ridge-vane-loaded folded-waveguide SWS for 0.22 THz traveling-wave tube, *IEEE Transactions on Electron Devices,* vol. 60, no. 3, pp. 1228–1235, 2013.
54. Z. Lu, Z. Su, and Y. Wei, Design and cold test of period-tapered double-ridge-loaded folded waveguide SWS for Ka band TWTs, *AIP Advances,* vol. 8, no. 5, p. 055105, 2018.

55. L. Ming-Liang, W. Yan-Yu, H. Jun, G. Yu-Bin, W. Wen-Xiang, and G.-S. Park, Analysis of a novel Ka-band folded waveguide amplifier for traveling-wave tubes, *Chinese Physics Letters,* vol. 26, no. 11, p. 114207, 2009.
56. L. Su-Ye, Z. Chang-Qing, and W. Yong, The high-frequency characteristics of an open ridged loaded folded waveguide slow-wave circuit, *Journal of Infrared and Millimeter Waves,* vol. 36, no. 2, pp. 246–251, 2017.
57. Y. Wei, et al., Novel W-band ridge-loaded folded waveguide traveling wave tube, *IEEE Electron Device Letters,* vol. 35, no. 10, pp. 1058–1060, 2014.
58. F. Xie, G. Ding, X. Zhao, and P. Cheng, Design, fabrication and measurement of a novel 140 GHz folded waveguide based on SU-8 UV-LIGA technology, *Journal of Micromechanics and Microengineering,* vol. 25, no. 8, p. 085010, 2015.
59. X. Zhang, J. Feng, J. Cai, Y. Du, R. Dong, and X. Wu, Power Enhancement in W-Band Pulsed Folded Waveguide TWT, *IEEE Transactions on Electron Devices,* vol. 68, no. 5, pp. 2504–2508, 2021.
60. Y. Gong, et al., Study of traveling wave tube with folded-waveguide circuit shielded by photonic crystals, *IEEE Transactions on Electron Devices,* vol. 57, no. 5, pp. 1137–1145, 2010.
61. A. Rashidi and N. Behdad, Metamaterial-enhanced traveling wave tubes, in *IEEE International Vacuum Electronics Conference*, IEEE, pp. 199–200, 2014.
62. E. Tahanian and G. Dadashzadeh, A novel gap-groove folded-waveguide SWS for G-band traveling-wave tube, *IEEE Transactions on Electron Devices,* vol. 63, no. 7, pp. 2912–2918, 2016.
63. E. Tahanian and G. Dadashzadeh, Two novel kinds of the G-band travelling-wave tubes with multiple gap-groove folded-waveguides, *IEEE Transactions on Plasma Science,* vol. 45, no. 2, pp. 223–228, 2017.
64. E. Tahanian and G. Dadashzadeh, Gain analysis of the gap-groove folded-waveguide travelling-wave tube, *Journal of Electromagnetic Waves and Applications,* vol. 31, no. 4, pp. 363–374, 2017.
65. N. Shi, et al., Study of 220 GHz dual-beam overmoded photonic crystal-loaded folded waveguide TWT, *IEEE Transactions on Plasma Science,* vol. 47, no. 6, pp. 2971–2978, 2019.
66. N. Bai, W. Xiang, J. Shen, C. Shen, and X. Sun, A Ka-band folded waveguide traveling wave tube with lumped resistance metamaterial absorber, *IEEE Transactions on Electron Devices,* vol. 67, no. 3, pp. 1248–1253, 2020.

67. A. Xu, et al., The study of synchronization in the periodic nonuniform folded waveguide, *Chinese Physics B*, vol. 18, no. 2, pp. 810–814, 2009.
68. A. Xu, et al., Analysis of space harmonic selectivity of period-varying folded waveguide, *Chinese Physics B*, vol. 18, no. 3, pp. 1270–1275, 2009.
69. D. Xu, et al., Theory and experiment of high-gain modified angular log-periodic folded waveguide SWS, *IEEE Electron Device Letters*, vol. 41, no. 8, pp. 1237–1240, 2020.
70. M. Chodorow, and T. Wessel-Berg, A high-efficiency klystron with distributed interaction, *IRE Transactions on Electron Devices*, vol. 8, no. 1, pp. 44–55, Jan. 1961.
71. M. Chodorow, and B. Kulke, An extended-interaction klystron: efficiency and bandwidth, *IEEE Transactions on Electron Devices*, vol. 13, no. 4, pp. 439–447, Apr. 1966.
72. J. Pasour, E. Wright, K. T. Nguyen, A. Balkcum, F. N. Wood, R. E. Myers, and B. Levush, Demonstration of a multikilowatt, solenoidally focused sheet beam amplifier at 94 GHz, *IEEE Transactions on Electron Devices*, vol. 61, no. 6, pp. 1630–1636, Jun. 2014.
73. L. Hu, Z. Zeng, H. Chen, G. Ma, and F. Meng, Design and simulation of the high-frequency structure for a G-band extended interaction klystron, *The Journal of Engineering*, vol. 2018, no. 14, pp. 689–691, 2018.
74. N. Shi, C. Zhang, et al. A novel scheme for gain and power enhancement of THz TWTs by extended interaction cavities, *IEEE Transactions on Electron Devices*, vol. 67, no. 2, pp. 667–672, 2020.
75. H. Gong, G. Travish, J. Xu, Y. Wei, J. Feng, and Y. Gong, High-power tunable terahertz radiation by high-order harmonic generation, *IEEE Transactions on Electron Devices*, vol. 60, no. 1, pp. 482–486, 2013.
76. H. Gong, et al., Third-harmonic traveling-wave tube multiplier-amplifier, *IEEE Transactions on Electron Devices*, vol. 65, no. 6, pp. 2189–2194, 2018.
77. R. Zhang, et al., Novel dual beam cascaded schemes for 346 GHz harmonic-enhanced TWTs, *Electronics*, vol. 10, no. 2, p. 195, 2021.
78. Y. Gong, et al., A 140-GHz two-beam overmoded folded-waveguide traveling-wave tube, *IEEE Transactions on Plasma Science*, vol. 39, no. 3, pp. 847–851, 2011.
79. S. Yan, W. Su, Y. Wang, and A. Xu, Design and theoretical analysis of multibeam folded waveguide traveling-wave tube for subterahertz radiation, *IEEE Transactions on Plasma Science*, vol. 43, no. 1, pp. 414–421, 2015.

80. S. Yan, W. Su, A. Xu, and Y. Wang, Analysis of multi-beam folded waveguide traveling-wave tube for terahertz radiation, *Journal of Electromagnetic Waves and Applications*, vol. 29, no. 4, pp. 436–447, 2015.
81. D. Xu, et al., Investigation on 0.1 THz array beams folded waveguide traveling wave tube, in *2019 Photonics & Electromagnetics Research Symposium-Fall (PIERS-Fall)*, IEEE, pp. 2737–2740, 2019.
82. Y. Tian, et al., A novel SWS—Folded rectangular groove waveguide for millimeter-wave TWT, *IEEE Transactions on Electron Devices*, vol. 59, no. 2, pp. 510–515, 2012.
83. Y. Liu, L. Yue, Y. Tian, J. Xu, and W. Wang, V-shape folded rectangular groove waveguide for millimeter-wave traveling-wave tube, *IEEE Transactions on Plasma Science*, vol. 40, no. 4, pp. 1027–1031, 2012.
84. Y. Tian, L. Yue, Q. Zhou, Y. Wei, Y. Wei, and Y. Gong, Investigation on sheet beam folded V-shape groove waveguide for millimeter-wave TWT, *IEEE Transactions on Plasma Science*, vol. 44, no. 8, pp. 1363–1368, 2016.
85. H. Wang, et al., Investigation of angular log-periodic folded groove waveguide SWS for low voltage Ka-band TWT, *AIP Advances*, vol. 10, no. 3, p. 035030, 2020.
86. F. Lu, Numerical simulation and theoretical analysis of terahertz sheet beam TWTs based on a ridge-loaded folded waveguide SWS, *IEEE Transactions on Plasma Science*, vol. 49, no. 4, pp. 1333–1339, 2021.
87. F. Lu, C. Zhang, M. Grieser, Y. Wang, S. Lü, and G. Zhao, Study of rectangular beam folded waveguide traveling-wave tube for terahertz radiation, *Physics of Plasmas*, vol. 24, no. 10, p. 103132, 2017.
88. Y. Sheng-Mei, et al., Theoretical and simulation study of 0.14 THz fundamental mode multi-beam folded waveguide traveling wave tube, *Acta Physica Sinica*, vol. 63, no. 23, 2014.
89. Y. Tian, et al., Investigation of ridge-loaded folded rectangular groove waveguide SWS for high-power terahertz TWT, *IEEE Transactions on Electron Devices*, vol. 65, no. 6, pp. 2170–2176, 2018.
90. A. N. Vlasov, et al., Design of a low voltage folded waveguide four beam Mini-TWT, in *2018 IEEE International Vacuum Electronics Conference (IVEC)*, IEEE, pp. 47–48, 2018.
91. B. Du and J. Wang, *Electron Optics (in Chinese)*, Tsinghua University Press, 2002.
92. A. S. Gilmour, Klystrons, traveling wave tubes, magnetrons, crossed-field amplifiers, and gyrotrons. Artech House, 2011.

93. B. N. Basu, *Electromagnetic Theory and Applications in Beam-Wave Electronics*. World Scientific, 1996.
94. C. Kory, et al., Overview of W-band traveling wave tube programs, in *2006 IEEE International Vacuum Electronics Conference held Jointly with 2006 IEEE International Vacuum Electron Sources*, IEEE, pp. 447–448, 2006.
95. M. Sumathy, D. Augustin, S. K. Datta, L. Christie, and L. Kumar, Design and RF characterization of W-band meander-line and folded-waveguide SWSs for TWTs, *IEEE Transactions on Electron Devices,* vol. 60, no. 5, pp. 1769–1775, 2013.
96. H.-J. Ha, et al., Experimental investigation of Ku-band folded waveguide TWT, *International Journal of Infrared and Millimeter Waves,* vol. 20, no. 12, pp. 2099–2106, 1999.
97. F. André, et al., Technology, assembly, and test of a W-band traveling wave tube for new 5G high-capacity networks, *IEEE Transactions on Electron Devices,* vol. 67, no. 7, pp. 2919–2924, 2020.
98. R. K. Bhardwaj, H. Sudhamani, V. Dutta, and N. Bhatnagar, Micromachining and characterisation of folded waveguide structure at 0.22 THz, *Journal of Infrared, Millimeter, and Terahertz Waves,* vol. 42, no. 3, pp. 229–238, 2021.
99. D. Gamzina, et al., Nano-CNC machining of sub-THz vacuum electron devices, *IEEE Transactions on Electron Devices,* vol. 63, no. 10, pp. 4067–4073, 2016.
100. Y. Li, P. L. Kirby, and J. Papapolymerou, Silicon micromachined W-band folded and straight waveguides using DRIE technique, in *2006 IEEE MTT-S International Microwave Symposium Digest,* IEEE, pp. 1915–1918, 2006.
101. G. Park, et al., Feasibility study of two-step LIGA-fabricated circuits applicable to millimeter/submillimeter wave sources, in *AIP Conference Proceedings,* 2006, vol. 807, no. 1: American Institute of Physics, pp. 299–308.
102. S.-T. Han, J.-I. Kim, B. Jia, G.-S. Park, and S.-S. Chang, Ka-band folded waveguide traveling-wave tube experiment, in *2002 3rd International Conference on Microwave and Millimeter Wave Technology, 2002. Proceedings. ICMMT 2002.*, IEEE, pp. 190–193, 2002.
103. S. Bhattacharjee, et al., Folded waveguide traveling-wave tube sources for terahertz radiation, *IEEE Transactions on Plasma Science,* vol. 32, no. 3, pp. 1002–1014, 2004.

104. S.-T. Han, et al., Low-voltage operation of Ka-band folded waveguide traveling-wave tube, *IEEE Transactions on Plasma Science,* vol. 32, no. 1, pp. 60–66, 2004.
105. C. D. Joye, et al., Microfabrication and cold testing of copper circuits for a 50 watt 220 GHz traveling wave tube, in *Terahertz, RF, Millimeter, and Submillimeter-Wave Technology and Applications VI,* 2013, vol. 8624: International Society for Optics and Photonics, p. 862406.
106. H. Li, J. Feng, and G. Bai, Microfabrication of W-band folded waveguide SWS using DRIE and UV-LIGA technology, in *2011 IEEE International Vacuum Electronics Conference (IVEC),* IEEE, pp. 379–380, 2011.
107. H. Li and J. Feng, Microfabrication of W band folded waveguide SWS using two-step UV-LIGA technology, in *IVEC 2012,* IEEE, pp. 387–388, 2012.
108. J. Feng, et al., Development of W-band folded waveguide pulsed TWTs, *IEEE Transactions on Electron Devices,* vol. 61, no. 6, pp. 1721–1725, 2014.
109. P. Pan, et al., Development of G band folded waveguide TWTs, in *2016 IEEE International Vacuum Electronics Conference (IVEC),* IEEE, pp. 1–2, 2016.
110. Y. Du, et al., Experimental investigation of an ultrawide bandwidth W-band pulsed traveling-wave tube with microfabricated folded-waveguide circuits, *IEEE Transactions on Plasma Science,* vol. 47, no. 1, pp. 219–225, 2019.
111. K. Tsutaki, Y. Neo, H. Mimura, N. Masuda, and M. Yoshida, Design of a 300 GHz band TWT with a folded waveguide fabricated by microelectromechanical systems, *Journal of Infrared, Millimeter, and Terahertz Waves,* vol. 37, no. 12, pp. 1166–1172, 2016.
112. K. H. Jang, J. J. Choi, and J. H. Kim, X-ray LIGA Microfabricated circuits for a sub-THz wave folded waveguide traveling-wave-tube amplifier, *Journal of the Korean Physical Society,* vol. 75, no. 9, pp. 716–723, 2019.
113. L. Fei, et al., Development of W-band folded waveguide TWT with lowered operating voltage and improved gain flatness, *IEEE Transactions on Plasma Science,* vol. 48, no. 8, pp. 2939–2947, 2020.
114. A. M. Cook, C. D. Joye, and J. P. Calame, W-band and D-band traveling-wave tube circuits fabricated by 3D printing, *IEEE Access,* vol. 7, pp. 72561–72566, 2019.
115. R. Panigrahi, M. J. Thomas, and K. Vinoy, A new fabrication method for serpentine-folded waveguide SWS at W-band, *IEEE Transactions on Electron Devices,* vol. 67, no. 3, pp. 1198–1204, 2020.

116. J. Cai, J. Feng, Y. Hu, X. Wu, Y. Du, and J. Liu, 10 GHz bandwidth 100 watt W-band folded waveguide pulsed TWTs, *IEEE Microwave and Wireless Components Letters,* vol. 24, no. 9, pp. 620–621, 2014.
117. Y. Hu, et al., Design and experimental study of a widebandwidth W-band folded waveguide continuous-wave TWT, *IEEE Transactions on Plasma Science,* vol. 42, no. 10, pp. 3380–3386, 2014.
118. Y. Hu, et al., Progress of wide bandwidth W-band 20 W CW TWT, in *IEEE International Vacuum Electronics Conference,* IEEE, pp. 179–180, 2014.
119. J. Cai, X. Wu, and J. Feng, Traveling-wave tube harmonic amplifier in terahertz and experimental demonstration, *IEEE Transactions on Electron Devices,* vol. 62, no. 2, pp. 648–651, 2015.
120. X. Zhang, et al., Design and experimental study of 250 W W-band pulsed TWT with 8 GHz bandwidth, *IEEE Transactions on Electron Devices,* vol. 64, no. 12, pp. 5151–5156, 2017.
121. Y. Du, J. Cai, X. Zhang, R. Dong, X. Wu, and J. Feng, A substantially lowered-voltage high power W-band pulsed traveling wave tube, *IEEE Microwave and Wireless Components Letters,* DOI:10.1109/LMWC.2021.3070637, 2021.
122. X. Bian, et al., Demonstration of a pulsed G-band 50 W traveling wave tube, *IEEE Electron Device Letters,* vol. 42, no. 2, pp. 248–251, 2020.
123. J. Tucek, K. Kreischer, D. Gallagher, R. Vogel, and R. Mihailovich, Development and operation of a 650 GHz folded waveguide source, in *2007 IEEE International Vacuum Electronics Conference,* IEEE, pp. 1–2, 2007.
124. J. C. Tucek, et al., A 100 mW, 0.670 THz power module, in *IVEC 2012,* IEEE, pp. 31–32, 2012.
125. K. E. Kreischer, J. C. Tucek, M. A. Basten, and D. A. Gallagher, 220 GHz power amplifier testing at Northrop Grumman, in *2013 IEEE 14th International Vacuum Electronics Conference (IVEC),* IEEE, pp. 1–2, 2013.
126. J. C. Tucek, et al., 0.850 THz vacuum electronic power amplifier *15th IEEE International Vacuum Electronics Conference (IVEC),* APR 22–24, 2014.
127. M. Basten, J. Tucek, D. Gallagher, and K. Kreischer, 233 GHz high power amplifier development at Northrop Grumman, in *2016 IEEE International Vacuum Electronics Conference (IVEC),* IEEE, pp. 1–2, 2016.
128. J. C. Tucek, et al., Operation of a compact 1.03 THz power amplifier, *17th IEEE International Vacuum Electronics Conference (IVEC),* APR 19–21, 2016.

129. C. D. Joye, et al., Demonstration of a high power, wideband 220 GHz traveling wave amplifier fabricated by UV-LIGA, *IEEE Transactions on Electron Devices,* vol. 61, no. 6, pp. 1672–1678, 2014.

130. A. M. Cook, et al., Demonstration of a W-band traveling-wave tube power amplifier with 10 GHz bandwidth, *IEEE Transactions on Electron Devices,* vol. 68, no. 5, pp. 2492–2498, 2021.

131. H. Gong, Y. Gong, T. Tang, J. Xu, and W.-X. Wang, Experimental investigation of a high-power Ka-band folded waveguide traveling-wave tube, *IEEE Transactions on Electron Devices,* vol. 58, no. 7, pp. 2159–2163, 2011.

132. H. Gong, et al., A 1 kW 32–34 GHz folded waveguide traveling wave tube, *IEEE Transactions on Plasma Science,* vol. 42, no. 1, pp. 8–12, 2014.

133. Z. Wang, et al., Development of a 140 GHz folded-waveguide traveling-wave tube in a relatively larger circular electron beam tunnel, *Journal of Electromagnetic Waves and Applications,* vol. 31, no. 17, pp. 1914–1923, 2017.

134. A. J. Theiss, C. J. Meadows, R. Freeman, R. B. True, J. M. Martin, and K. L. Montgomery, High-average-power W-band TWT development, *IEEE Transactions on Plasma Science,* vol. 38, no. 6, pp. 1239–1243, 2010.

135. R. Kowalczyk, et al., A 100 watt W-band MPM TWT, in *2013 IEEE 14th International Vacuum Electronics Conference (IVEC),* IEEE, pp. 1–2, 2013.

136. K. H. Jang, J. J. Choi, and J. H. Kim, Experiments on sub-terahertz folded waveguide traveling wave tubes, in *2017 Eighteenth International Vacuum Electronics Conference (IVEC),* IEEE, pp. 1–2, 2017.

137. P. Hu, et al., Development of a 0.32 THz folded waveguide traveling wave tube, *IEEE Transactions on Electron Devices,* vol. 65, no. 6, pp. 2164–2169, 2018.

138. Y. Jiang, et al., Demonstration of a 220 GHz continuous wave traveling wave tube, *IEEE Transactions on Electron Devices,* vol. 68, no. 6, pp. 3051–3055, 2021.

139. W. Liu, et al., Demonstration of a high-power and wide-bandwidth G-band traveling wave tube with cascade amplification, *IEEE Electron Device Letters,* vol. 42, no. 4, pp. 593–596, 2021.

Chapter 20

Vacuum Nanoelectronics and Electron Emission Physics

Kevin L. Jensen
Code 6362, MSTD Naval Research Laboratory Washington, DC 20375 USA
kljensen@ieee.org

"... we must take the current when it serves,
Or lose our ventures."
William Shakespeare (*Julius Caesar*, IV.4)

Abstract: The theory and simulation of vacuum nanoelectronics (VNE) often uses standard models, in particular the canonical equations of field emission, thermal emission, and photoemission, but these equations are increasingly challenged by processes and configurations characteristic of VNE, in particular, size, shape, and temperature. Simple models are used to explore the physics behind the approximations in the Canonical equations. Next, the effects of processes such as heating, as attends current flow through nanowires and nanofibers, is considered. Time factors associated with emission, in particular, tunneling time and the Hartman effect,

Advances in Terahertz Source Technologies
Edited by Gun-Sik Park, Masahiko Tani, Jae-Sung Rieh, and Sang Yoon Park

ISBN 978-981-4968-89-8 (Hardcover), 978-1-003-45967-5 (eBook)
www.jennystanford.com

and transit time across an anode-cathode (AK) gap, are examined. Lastly, the emission barrier is affected by a non-planar image charge, field variation near highly curved surfaces, depletion barriers of semiconductors, and other mechanisms. The modifications lead to a general quadratic barrier model examined in terms of the Gamow tunneling factor and a shape factor method.

Vacuum nanoelectronics has evolved considerably from the early days when the promise of field emission from tungsten wires for microwave devices [2, 9, 14, 20] was envisioned to be harnessed by microfabricated field emitters created by thin-film deposition and micromachining [27, 90–93, 98]. But harnessed it was through the development of emitters capable of rapid emission modulation and operation in inductive output amplifiers, traveling wave tubes, and high power microwave devices [35, 70, 72, 75, 88, 89, 96, 101, 102] using a variety of sources such as Spindt-type field emitter arrays, semiconductor field emitters, carbon fibers, and others. Progress in integrating photoemission processes [34, 76, 95] addresses the call for very short bunches of electrons for particle accelerators, rf guns, and free electron lasers [32, 68, 69]. The engineering behind these applications is profound, but fortunately, a basic understanding of the physics governing how electrons are coaxed into vacuum is understandable using simple models and the effects of that physics on the beams produced. Particular attention is directed to tunneling behavior, thermal-field effects on current density, photoemission effects, space charge, transit time, and beam properties.

20.1 Background

The study of electron emission spans over a century, making the evolution in the units used to describe emission unavoidable. It is common to encounter equations in centimeter-gram-second (cgs) units, meter-kilogram-second-amp (MKSA) units, and combinations of mixed units to reflect the needs of experiment or of a particular community. Odd hybrid units such as current in amperes but area in square centimeters emerge, causing current density and field described by units of [A/cm^2] and [volts/cm] while the image charge contribution to the emission barrier were represented as $V_i(x) = q^2/4x$ (a cgs convention), or temperatures reported in centigrade when the theory required Kelvin (see, *e.g.* Refs. [20, 33, 74]). This can create problems for those adhering to modern conventions based on the *International System of Units* [97], or SI units. Therefore, in what follows, the following conventions shall be exclusively used. The Schottky–Nordheim barrier from which the canonical equations of electron emission [39] are based is given in SI units by

$$q\varphi(x) \equiv \mu + \Phi - q|\mathcal{E}|x - \frac{q^2}{16\pi\varepsilon_0 x}, \tag{20.1}$$

where φ is the potential, $\mathcal{E}$ is the electric field, $-q$ is the charge on the electron, ε_0 is the permittivity of free space, μ is the chemical potential (identified with the Fermi level E_F at zero temperature), Φ is the work function of the metal, x is position such that the surface is at $x = 0$, the bulk material is to the left ($x < 0$) and vacuum to the right ($x > 0$). This equation is transformed by conjoining $q\varphi(x) \to V(x)$, $q|\mathcal{E}| \to F$, and invoking the fine structure constant $\alpha_{fs} \equiv q^2/4\pi\varepsilon_0\hbar c \approx 1/137$ to define $Q \equiv \alpha_{fs}\hbar c/4 = 0.36$ eV nm; it becomes

$$V(x) \equiv \mu + \Phi - Fx - \frac{Q}{x} \tag{20.2}$$

The convention to be used herein follows Ref. [39] regarding units and is summarized as follows. Energy is given in electron-volts [eV], distance in nanometers [nm], time in femtoseconds [fs], temperature in Kelvin [K], and the unit charge term q is absorbed into volts and 10^9 volts/meter to form V in units of [eV] and F in units of [eV/nm] (that is, $q = 1$q, or treated as unity) in the equations below and in Tables 20.1 and 20.2. As an example, the Schottky factor $\sqrt{4QF}$ for a "field" of $F = 1$ eV/nm (equivalent to $\mathcal{E} = 1$ GV/m) is then $\sqrt{4(0.36 \text{ eV nm})(1\text{eV/nm})} = 1.2$ eV.

Because of the image charge term (Q/x), the maximum of the barrier is no longer $\mu+\Phi$. Setting $\partial_x V(x_o) = 0$ identifies the location of the maximum to be $x_o = \sqrt{Q/F}$, and so $V(x_o) = \mu + \Phi - \sqrt{4QF}$. Because the Fermi energy (treated as equivalent to the chemical potential when the temperature dependence of the latter is ignored) retains a privileged position in emission theory, it is common to talk of an *effective* work function $\phi \equiv \Phi - \sqrt{4QF}$ instead. This convention

Table 20.1 The fundamental constants associated with emission in the units of [eV, fs, nm, K] and the charge on the electron of $(-q) = -\text{eV/volt}$

Symbol	Definition	Value	Unit
$-q$	Electron charge	-1	q
c	Speed of light	299.792	nm/fs
k_B	Boltzmann's constant	1/11604.5	eV/K
$\hbar$	Planck's constant/2π	0.658212	eV fs
ε_0	Electric constant	0.0552635	eV/(nm V^2)
α_{fs}	Fine structure constant	$q^2/4\pi\varepsilon_0\hbar c$	1/137.02
m	Electron rest mass	5.68563	eV/c^2
a_0	Bohr radius $\hbar/m\alpha_{fs}c$	0.052918	nm
Q	$\alpha_{fs}\hbar c/4$	0.359991	eV nm

Table 20.2 The parameters of electron emission

Symbol	Definition	Formula	Unit
T	Temperature	-	K
$V(x)$	Energy barrier	-	eV
μ	Chemical potential	$\hbar^2 k_F^2/2m$	eV
Φ	Work function	-	eV
$\hbar k_F$	Fermi momentum	$\sqrt{2m\mu}$	1/nm
F	Force	$q\mathcal{E}$	eV/nm
ϕ	Schottky-lowered Φ	$\Phi - \sqrt{4QF}$	eV
$\hbar\omega$	Photon energy	-	eV
θ	Gamow factor	$2\sigma\kappa L$	-
σ	Shape factor	Eq. 20.14	-
β_T	Temperature slope factor	$1/k_B T$	1/eV
β_F	Field slope factor	$-\partial_E\theta(E)$	1/eV
J	Current density	Eq. (20.3)	A/cm^2

of the lower case ϕ denoting the *reduced* or Schottky-lowered work function is used throughout.

Current density J is charge flow past the barrier described by Eq. (20.2). From basic considerations in statistical mechanics [79] leading to for example the Boltzmann Transport Equation, a current of particles past a boundary situated at $x = 0$ is a product of charge q and a velocity $v_x = \hbar k_x/m$ averaged over a distribution $f(x, k)$ on the downstream side of the boundary. The distribution may be approximated as being position-independent and given by the product of the Fermi–Dirac distribution for electrons with a transmission probability $D(k_x)$, resulting in

$$J(F,T) = \frac{2q}{(2\pi)^3}\int \frac{\hbar k_x}{m} D(k_x)dk_x \int 2\pi k_\perp f_{FD}(E_k)dk_\perp \tag{20.3}$$

where a factor of 2 in the coefficient accounts for electron spin, and all transverse momentum components are integrated over. For an electron with total energy $E(\mathbf{k}) = \hbar^2(k_x^2 + k_\perp^2)/2m$, the Fermi–Dirac distribution is given by

$$f_{FD}(E) \equiv \frac{1}{1+\exp[\beta_T(E-\mu)]} \tag{20.4}$$

where $\beta_T \equiv 1/k_B T$ is the temperature slope factor, k_B is Boltzmann's constant and T is temperature such that $k_B T = 0.025$ eV for $T = 290.1126$ K. The integration over $k_\perp$ is analytical and results in the supply function $f(k_x)$ defined by

$$\begin{aligned} f(k_x) &= \frac{2}{(2\pi)^2}\int_0^\infty \frac{2\pi k_\perp dk_\perp}{1+e^{\beta_T[\hbar^2(k_x^2+k_\perp^2)/2m-\mu]}} \\ &= \frac{m}{\pi\beta\hbar^2}\ln\left[1+e^{\beta_T(\mu-E_x)}\right] \end{aligned} \tag{20.5}$$

where $E_x \equiv \hbar^2 k_x^2/2m$. As a result, the equation for current density is generically given by

$$J(F,T) \equiv \frac{q}{2\pi}\int_0^\infty \frac{\hbar k_x}{m} D(k_x) f(k_x) dk_x \tag{20.6}$$

Switching the integration from k_x to E_x is often done and results in [39, 71]

$$J(F,T) \equiv \frac{q}{2\pi\hbar}\int_0^\infty D(E_x) f(E_x) dE_x \tag{20.7}$$

where $D(k_x) = D(\sqrt{2mE_x}/\hbar) \rightarrow D(E_x)$ and similarly with $f(E_x)$. With the introduction of the supply function, though, the problem has become one-dimensional, and so retaining the x-subscript is an inconvenience. Below, therefore, E_x is understood to be the "normal" (into the barrier) energy unless otherwise explicitly stated, and the x-subscript discontinued.

20.2 Emission Equations

20.2.1 Thermal Emission

For thermal emission, the transmission probability is often such that $D(E) = 0$ if the energy of the electron is below the barrier maximum $E < \mu+\phi$. Letting $D(E) = \Theta(\mu+\phi-E)$, where $\Theta(s)$ is the Heaviside step function (0 for negative argument, 1 for positive argument) and noting that $f(E) \approx (m\pi/\beta_T \hbar^2)\exp(-\beta_T E)$ for $E > \mu + \phi$, then the Richardson–Laue–Dushman (RLD) equation for thermal emission results and is

$$\begin{aligned} J_{RLD}(T) &= \frac{q}{2\pi\hbar}\left(\frac{m}{\pi\beta_T\hbar^2}\right)\int_{\mu+\phi}^{\infty} e^{-\beta_T(E-\mu)}dE \\ &= A_{RLD}T^2\exp(-\beta_T\phi) \end{aligned} \tag{20.8}$$

Numerically, $A_{RLD} = mqk_B^2/2\pi^2\hbar^3 = 1.2017 \times 10^6$ A/K^2 m^2. A plot of $\ln[J_{RLD}/T^2]$ *vs.* $1/T$ is linear on a so-called Richardson plot. A characteristic current density is for $\phi = 1.8$ eV and $T = 1173$ K, for which $J_{RLD}(T) = 30530$ A/m^2 (or 3.053 A/cm^2 in a notation common to the literature).

20.2.2 Photoemission

In the case of photoemission, the transmission probability is still treated as a step function, but the total electron energy is augmented by the photon energy $\hbar\omega$. A return to Eq. (20.3) in a low-temperature limit in order to do the transverse integrations correctly is required [40], but if (as DuBridge assumed) to a good approximation $\hbar\omega$ can be simply added to the normal energy E, then the probability of emission for an electron excited by a photon is $P(\hbar\omega)$ and should resemble

$$P(\hbar\omega) = \frac{\int D(E)f(E)dE}{\int f(E)dE} \rightarrow \frac{\int_{\Delta}^{\infty} D(E)f(E)dE}{\int_0^{\infty} f(E)dE} \tag{20.9}$$

where $\Delta \equiv \mu + \phi - \hbar\omega$ and comes from treating $D(E)$ as a step function. In the low-temperature limit, $f(E < \mu) \propto \beta_T(\mu - E)$ for $E < \mu$ and $f(E > \mu) \propto e^{-\beta_T(E-\mu)} \ll 1$, the latter of which is negligible, and so

$$P(\hbar\omega) \approx \frac{\int_{\Delta}^{\mu}(\mu - E)dE}{\int_{0}^{\mu}(\mu - E)dE} = \frac{(\hbar\omega - \phi)^2}{\mu^2} \tag{20.10}$$

Insofar as quantum efficiency QE, or the number of electrons out compared to the number of absorbed photons, is proportional to $P(\hbar\omega)$, then $QE \propto (\hbar\omega - \phi)^2$, which is the leading order term of the Fowler–DuBridge (FD) equation of photoemission [17, 39]. Good quantum efficiency photocathodes are characterized by QE ranging from several percent to an order of magnitude larger, depending on material and surface coatings; such photocathodes are generally semiconductors, as metal photocathodes have QE's at least two orders of magnitude smaller.

20.2.3 Tunneling Emission

Field emission requires the consideration of tunneling through a barrier, meaning that a full quantum mechanical treatment of the transmission probability is required. Fortunately, for a rectangular barrier of the form $V(x) = V_o\Theta(x)\Theta(L - x)$, the transmission probability is well known from elementary quantum mechanics [7, 39] to be

$$D(E) = \left\{1 + \frac{V_o^2}{4E(V_o - E)}\sinh^2[\kappa(E)L]\right\}^{-1} \tag{20.11}$$

where $\kappa(E) \equiv \sqrt{2m(V_o - E)}/\hbar$ (see also Eq. (20.49) below). Thus, $D(E)$ exponentially increases with E, but $f(E) \propto \mu - E$ decreases, and the maximum of their product occurs near $E = \mu$. Insofar as $V_o = \mu + \Phi$ and μ and Φ are comparable (*e.g.*, for copper, they are 7 eV and 4.5 eV, respectively) then $\sinh^2(\kappa L) \approx e^{2\kappa L}/4$. Sweep all terms other than the exponential into a coefficient, evaluate it at $E = \mu$, and call it $C(\mu)$. Thus, $D(E) \approx C(\mu)e^{-2\kappa L}$ with $C(\mu) = 16\mu\Phi/(\mu + \Phi)^2$ (3.8109 for copper). The zero temperature supply function and the exponential transmission probability then result in

$$J = \frac{mq}{2\pi^2\hbar^3}C(\mu)\int_0^{\mu}(\mu - E)e^{-2\kappa(E)L}dE \tag{20.12}$$

where $\kappa(E) = \sqrt{2m(\Phi + \mu - E)}/\hbar$. Let $(\mu - E)$ be taken as small: emission tends to come from around the Fermi level, and so $\kappa(E) \approx$

$\kappa(\mu)[1+(\mu-E)/2\Phi]$ resulting in an integral that is easily evaluated. It is found

$$J = \frac{q\Phi}{4\pi^2\hbar L^2} C(\mu) e^{-2\kappa(\mu)L} \tag{20.13}$$

20.2.4 Gamow and Shape Factors

The current for a rectangular barrier can be generalized. Let $\theta(E) = 2\kappa(E)L$ in Eq. (20.12), so that $D(E) = e^{-\theta(E)}$. The factor $\theta(E)$ is known as the Gamow tunneling factor, and governs how fast the wave function is attenuated inside the barrier. For the rectangular barrier, it is the product of the width of the barrier and the maximum (or "height") of $\kappa(E)$, which is constant for a rectangular barrier, and a factor of 2. For a more general barrier which varies with position, the Gamow factor can be generalized to $\theta(E) \to 2\sigma\kappa(E)L(E)$, where $L(E)$ is the width of the barrier to a tunneling electron of energy E, and σ is the shape factor, or

$$\sigma \equiv \int_{x_-}^{x_+} \left[\frac{V(x)-E}{V(x_o)-E}\right]^{1/2} \frac{dx}{L} \tag{20.14}$$

where $V(x_o)$ is the maximum of the barrier, $V(x_\pm) - E = 0$, and $L(E) = x_+ - x_-$. For general barriers, σ will change with field, but a subset of barriers have constant shape factors. Barriers of the form [44]

$$V_n(x) = V_o\left[1 - \left|\frac{x}{L_o}\right|^{2/n}\right] \tag{20.15}$$

result in

$$\sigma_n = \frac{\Gamma[(n/2)+1]\Gamma(3/2)}{\Gamma[(n+3)/2]} \tag{20.16}$$

for which special cases are rectangular ($n = 0$, $\sigma_\square = 1$), isosceles triangular ($n = 2$, $\sigma_\triangle = 2/3$), and parabolic ($n = 1$, $\sigma_\cap = \pi/4$). Additional analytic cases exist. A trapezoidal barrier described by $V_\diamond(x) = (V_o - Fx)\Theta(x)\Theta(L_o - x)$ for which

$$\sigma_\diamond(E) = \frac{2}{3\lambda}\left[1 - (1-\lambda)^{3/2}\right]; \; \lambda(E) = \frac{L_o F}{V_o - E} \tag{20.17}$$

models metal-insulator-metal (MIM) or metal-oxide-semiconductor (MOS) barriers when no image charge contribution is present, but as that configuration is outside the quadratic barrier considered below, its treatment is left to the literature [56]. Lastly, a barrier of the form $V_{\cup}(x) = V_o[1 - (x/L_o)]^2\Theta(x)\Theta(L_o - x)$ for which

$$\sigma_{\cup}(E) = \frac{1}{2(1-\eta)}\left\{1 - \frac{2\eta^2}{\sqrt{1-\eta^2}}\ln\left[\frac{\sqrt{1+\eta}+\sqrt{1-\eta}}{\sqrt{2\eta}}\right]\right\} \tag{20.18}$$

where $\eta \equiv \sqrt{E/V_o}$, models depletion barriers when no image charge contribution is present. For the special case $\eta = 0$, then $\sigma_{\cup}(0) = 1/2$.

It is seen that the $V_n(x)$ barriers result in a constant shape factor independent of energy, but MIM and depletion barriers have an energy-dependent $\sigma(E)$.

20.2.5 Field Emission

The potential barrier of Fowler and Nordheim [26] is a right triangular barrier of the form $V(x) = (\mu + \Phi - Fx)\Theta(x)$ and is therefore related to the isosceles triangular case, but to show the shape factor explicitly [44]

$$\sigma_{fn} = \frac{1}{L(E)}\int_0^{L(E)}\left[\frac{V_o - Fx - E}{V_o - E}\right]^{1/2} dx = \frac{2}{3} \tag{20.19}$$

because $L(E) = (V_o - E)/F$. This is seen to be $\sigma_\diamond$ for $\lambda = 1$. Consequently, $\theta(\mu) = 2\sigma\kappa(\mu)L(\mu) = (4/3\hbar)\sqrt{2m\Phi}(\Phi/F)$ because $V_o = \mu + \Phi$ and $L(\mu) = \Phi/F$, and now

$$\theta(E) \approx \theta(\mu)\left[1 + \frac{3(\mu - E)}{2\Phi}\right] \equiv \theta(\mu) - \beta_F(\mu)(\mu - E) \tag{20.20}$$

or $\beta_F(\mu) = 3\theta(\mu)/2\Phi$ and so

$$J(F) = \frac{mq}{2\pi^2\hbar^3\beta_F(\mu)^2}C(\mu)e^{-\theta(\mu)} \tag{20.21}$$

Including $C_{FN}(\mu) = 4\sqrt{\mu\Phi}/(\mu+\Phi)$ (1.952 for copper) that requires a solution of Schrödinger's Equation to determine [25, 26, 39], $J(F) \to J_{FN}(F)$ is the triangular barrier current density of Fowler and Nordheim (FN) given by

$$J_{FN}(F) = \frac{q}{4\pi^2\hbar\Phi}\frac{\sqrt{\mu}}{\sqrt{\Phi}(\mu+\Phi)}F^2 e^{-4\sqrt{2m\Phi^3}/3\hbar F} \tag{20.22}$$

using $F = q\mathcal{E}$. For more complex barriers, and in particular, for the Schottky–Nordheim barrier, σ depends on F and is more complicated, but the present result is enough to show that $\ln[J_{FN}/F^2]$ *vs* $1/F$ is linear on a Fowler–Nordheim plot. A characteristic current density for $F = 5$ GV/m and $\Phi = 4.5$ eV is $J_{FN}(F) = 3.794 \times 10^9$ A/m^2.

20.2.6 Beyond the Simple Models

Such are the simplest of theoretical models for the thermal (RLD), photo (FD), and field emission (FN), which when modified to include Schottky lowering, an image charge barrier that introduces the Schottky–Nordheim functions $v(y)$ and $t(y)$ to modify $\theta(\mu)$ and $\beta_F(\mu)$, respectively, and temperature effects, become the canonical equations:

- Richardson–Laue–Dushman [33]

$$J_{RLD}(T) = A_{RLD}T^2 \exp\left[-\frac{1}{k_BT}\left(\Phi - \sqrt{4QF}\right)\right] \quad (20.23)$$

- Fowler–Nordheim (Murphy and Good formulation) [74]

$$J_{MG}(F) = \frac{qF^2}{16\pi^2\hbar\Phi\, t(y)^2} \exp\left(-v(y)\frac{4\sqrt{2m\Phi^3}}{3\hbar F}\right) \quad (20.24)$$

- Fowler–DuBridge [17] (where I_ω is light intensity)

$$QE \equiv \frac{\hbar\omega}{q}\left(\frac{J}{I_\omega}\right) \propto (\hbar\omega - \phi)^2 + \frac{\pi^2}{3}(k_BT)^2 \quad (20.25)$$

where, recall, $F = q\mathcal{E}$, and the analytical approximations of Forbes and Deane [25] are the recommended versions of $v(y)$ and $t(y)$. With $y = \sqrt{4QF}/\Phi$, they are given by

$$\begin{aligned} v(y) &\approx 1 - \frac{y^2}{3}[3 - \ln(y)] \\ t(y) &\approx 1 + \frac{y^2}{9}[1 - \ln(y)] \end{aligned} \quad (20.26)$$

in terms of which Eq. (20.24) is recast as [37] (compare Eq. (6.3) of Ref. [25])

$$J_{MG}(F) \equiv \frac{A_o}{\Phi t_o^2}\left(\frac{\Phi^2 e^6}{4Q}\right)^{\nu} F^{2-\nu} \exp\left(-B_o \frac{\Phi^{3/2}}{F}\right) \quad (20.27)$$

where $A_o \equiv q/16\pi^2\hbar$, $B_o \equiv (4/3\hbar)\sqrt{2m}$, $\nu = 2B_o Q/3\sqrt{\Phi}$, $t_o = 1 + (1/6e) = 1.06131$, which neglects the weak variation of $1 < t(y) < 11/10$ with field by approximating it as a constant. In the literature, it is common to call Eq. (20.24) the "Fowler–Nordheim Equation" but it is emphasized that the FN equation is Eq. (20.22), the FN equation with image charge modifications due to Murphy and Good is Eq. (20.24), and the Murphy and Good version of the FN equation with the approximations of Forbes and Deane is Eq. (20.27).

The usage of these emission equations in simulations that launch electrons under steady-state conditions from macroscale and mesoscale cathodes is generally acceptable because at length scales comparable to $L(\mu)$ or the radius of curvature of surface roughness, the cathode appears essentially flat. For a nanoscale emitter, or a mesoscale emitter with microscale surface features, that is no longer true. Consider the emission pattern photographs of Dyke, et al. [19] for tungsten needles or Schwoebel, et al. [85, 86] for molybdenum cones (Spindt-type field emitters) that show clear regions of work function variation, field enhancement variation, or both. A schematic of the work function variation at the apex of a tungsten needle based on Figure 6 of Ref. [21] is shown in Figure 20.1(a). The consequences of increased temperature of the needle based on data from Ref. [18] showing total current from a tungsten needle with multiple contributing work function regions as a function of temperature, compared to the same current at room temperature, is shown in Figure 20.1(b): an increase $10^{1.2} = 16$ represents a substantial change. In the upper right corner of that figure, the gray region shows entry into a "thermal-field" (TF) regime. Cross-over between thermal and field regimes are likewise shown in the thermal dispenser cathode data because of roughness effects on the surface, as shown in Figure 20.2 for data extracted from Figure (2) of Ref. [28].

A complication to the usage of these simple equations, however, is the presence of emission statistics, in which a variation in a key parameter exists over the emission sites from which current is

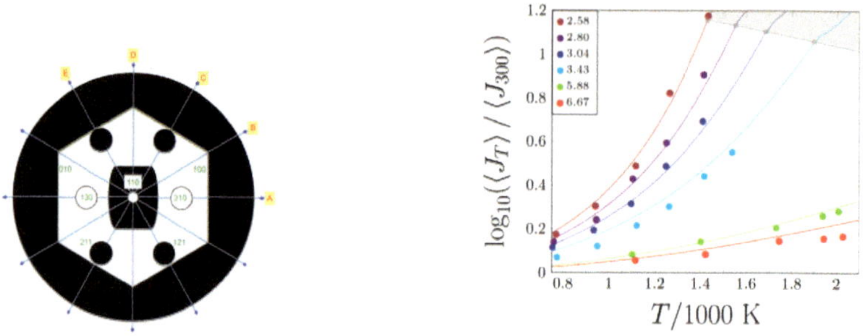

Figure 20.1 (*Left*) Schematic of work function regions on the apex of a tungsten needle emitter, modeled after photographs of Ref. [21]. (*Right*) Current as a function of temperature for data digitally extracted from Figure (4) of Ref. [18].

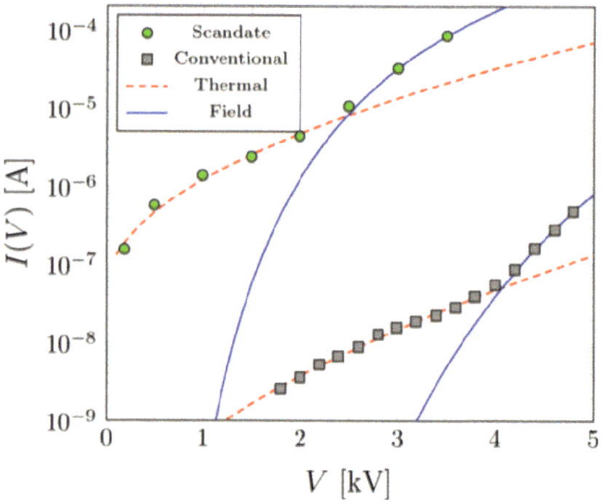

Figure 20.2 Onset of field emission contributions to the emitted thermal current from dispenser cathodes for both scandate and conventional sources held at a fixed temperature as a function of applied anode voltage. Data digitally extracted from Figure (2) of Geittner, et al. [28]. Lines are based on RLD (Eq. (20.23)) or FN (Eq. (20.24)) parameterizations.

drawn. Consider the usage of Eq. (20.27) for field emission, for which the field at the emission site scales as $1/a$, where a is the radius of the emission site (see discussion surrounding Eq. (20.95) below). Suppose that the emitters are approximately log-normal distributed [29, 38, 49], as in Figure 20.3(a). A log-normal distribution in radius

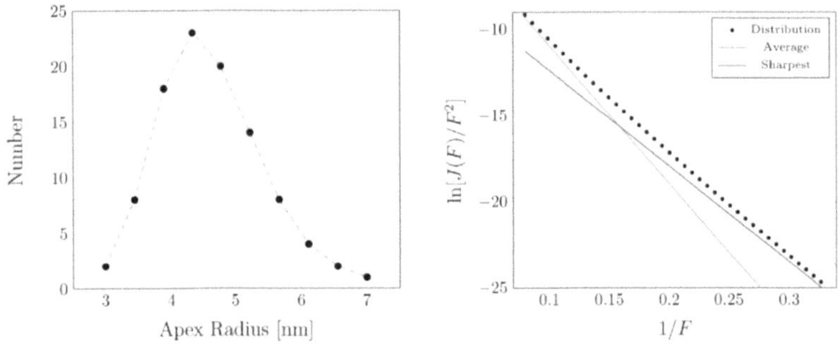

Figure 20.3 (*Left*) A distribution in radius of 100 emitters, based on the log-normal distribution for which the mean radius $\langle a \rangle$ is 9 nm and the spread factor is $\sigma = 0.2$. (*Right*) The resulting array current on a Fowler–Nordheim plot. Symbols are a numerical evaluation of Eq. (20.28); red line is only the sharpest ($j = 1$) emitters contribute, and the blue line is assuming all emitters are identical to the average emitter.

a behaves like a Gaussian $e^{-(x-x_o)^2/2\sigma^2}$ but for $x \rightarrow \ln(a)$. Let the current for an individual emitter of radius a_j be given by the product of the area of the emission site $2\pi a_j^2$, a notional area factor $g(F_j)$ governing the fraction of the apex that is emitting [40], and the current density given by $J_{MG}(F)$ of Eq. (20.27) on the assumption that $\Phi = 4$ eV. That is,

$$I_{tot} = \sum_{j=1}^{N} n_j \, 2\pi a_j^2 \, g(F_j) \, J_{MG}(F_j) \tag{20.28}$$

where n_j is the number of emitters characterized by a_j, and for simplicity approximate $g(F)$ by $g(F) = 1/[(B_o \Phi^{3/2}/F) + 1 - \nu]$ (ellipsoidal emitter [38]). Assume that $F_j = F a_1/a_j$, where F refers to the field on the sharpest emitter ($j = 1$). The total current may then be compared on a so-called Fowler–Nordheim plot, in which $\ln[I/F^2]$ versus $1/F$ is shown, and compared to two limiting cases: the first case will be where *only* the sharpest emitters characterized by a_1 emit; the second case will be where all the emitters are taken as *identical to the average emitter* characterized by $\langle a \rangle$, and so the total current will be the number of emitters times the average emitter. The comparison is shown in Figure 20.3(b). At very low fields, only the sharpest emitters contribute, and so the FN data is best represented by their characteristics (red line). On the other hand, when the field is strong enough that all emitters contribute, then the FN data approximates the product of the number of emitters with the average emitter. It is clear, therefore, that convexity is introduced on an FN plot due to statistics, and second, that a simple slope-intercept characterization of emission data on an FN plot may describe the data, but it does not describe properties of the emission sites. The effect is very similar to the problems associated with thermal-field emission characterization using the RLD or FN equations as well [50] and for analogous reasons.

When cathodes are driven hard, emission is no longer characterized as thermal *or* field *or* photo, but rather thermal-field, field-assisted photoemission, and other combinatoric designations, and how the resulting beams spread - that is, their *emittance* - often becomes of paramount concern. High current is also associated with increased charge between the cathode surface and the collection anode (the AK gap) or near the surface of the emitter, which can reduce the field at the surface substantially and affects all emission mechanisms because of the dependence of emitted current on ϕ. Lastly, at very high frequency and very small AK gap, the fields can be switched sufficiently rapidly that a question of time effects, such as delayed photoemission or tunneling time, arises. The remainder of the present work will account for such complications using simple but representative models.

20.3 Heating Effects in Field Emission

20.3.1 Simple Model

A description of the general thermal-field-photoemission equation (GTFP) [36, 39] followed by its implementation [42, 51] and refinement [41, 56] takes into account barriers modified by image charge contributions ($Q > 0$). However, to understand the mechanics behind the methodology, the simpler Fowler–Nordheim triangular barrier model ($Q = 0$) for $\theta_\triangle(E)$ reveals the physics with a minimum of complication. In particular, the energy dependence of the shape factor σ, which is generally small but which nevertheless matters in exponentials, does not occur because $\sigma_\triangle(E) = 2/3$. For the triangular barrier $V_{\triangle}(x) = \mu + \Phi - Fx$, therefore, the Gamow $\theta_{\triangle}(E)$ and field slope factor $\beta_{\triangle}(E)$ are used exclusively, but the inconvenient triangle subscript is temporarily suppressed. Thus,

$$\theta(E) \equiv \frac{4}{3}\kappa(E)L(E) = \frac{4\sqrt{2m}}{3\hbar}(V_o - E)^{3/2} \tag{20.29}$$

$$\beta_F(E) \equiv -\frac{d\theta}{dE} = \frac{2\sqrt{2m}}{\hbar F}(V_o - E)^{1/2} \tag{20.30}$$

where $V_o = \mu + \Phi$. The Kemble approximation to the transmission probability is used for all of the GTF approaches so as to transition from thermal to field emission smoothly; it is given by

$$D(E) \equiv \frac{1}{1 + e^{\theta(E)}} \tag{20.31}$$

with the approximation that $\theta(V_o + \Delta E) = -\theta(V_o - \Delta E)$ as the energy passes the barrier maximum [56]. As a result, the general equation for current density becomes

$$J(F,T) = \frac{qm}{2\pi^2 \beta_T \hbar^2} \int_0^{\infty} \frac{\ln\{1 + \exp[\beta_T(\mu - E)]\}}{1 + \exp[\theta(E)]} dE \tag{20.32}$$

while the extension to the Fowler–DuBridge model would replace $\theta(E) \to \theta(E + \hbar\omega)$. A representation of the integrand dJ for different fields at a temperature of $T = 1000$ K is shown in Figure 20.4. Replace $\theta(E)$ by its linear approximation $\theta(E) \to \theta(E_m) + \beta_F(E_m)(E_m - E)$ where E_m is the energy corresponding to the maximum of the integrand: the approximation is shown by the lines in the same figure, and bound shaded areas corresponding to the integrand of Eq. (20.32) when using the linear $\theta(E)$ approximation. The integrand is seen to be reasonably sharply peaked, especially at low and high fields, where the area under the exact $\theta(E)$ curve and that under the linear $\theta(E)$ curve are close for low and high field. At the intermediate field, while the linear $\theta(E)$ overestimates the extent of the integrand, and therefore overestimates the integral by a fraction, it does track the behavior of $J(F,T)$ in the thermal-field regime; the small discrepancy is corrected by the reformulated GTF approach [41]. Evaluating $dJ(E)/dJ(E_m)$ for a range of fields F and presenting the resulting surface as a contour plot reveals what is happening: as the value of n moves from below unity to

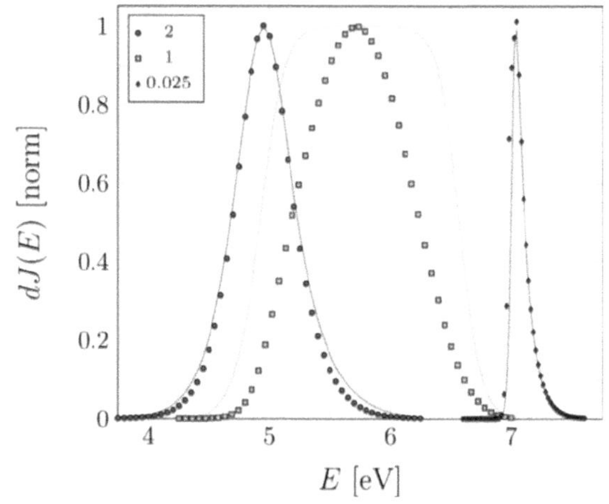

Figure 20.4 Integrand of Eq. (20.32) (normalized) for $\mu = 5$ eV, $\Phi = 2$ eV, and $T = 1000$ K for the fields $F = 2$ eV/nm (circle, blue), $F = 1$ eV/nm (square, green) and $F = 0.025$ (diamond, red). Lines correspond to $\theta(E) \to \theta(E_m) - \beta_F(E_m)(E - E_m)$ and bound the shaded color areas denoting the integrand of Eq. (20.32) when using the linear $\theta(E)$ approximation.

above (namely, as F increases), the peak of $dJ(E)/dJ(E_m)$ shifts from near the barrier maximum $\mu + \Phi$ to near the Fermi level $E_m \approx \mu$. This behavior is shown in Figure 20.4. The range $(0.95 < n < 1.05)$ regime is identified as the "Thermal-Field" regime. In the same scheme, $n > 1$ corresponds to the "Field" regime, and $n < 1$ corresponds to the "Thermal" regime.

The justification for replacing $\theta(E)$ by its linear approximation $\theta(E) \approx \theta(E_m) - \beta_F(E_m)(E - E_m)$ is because doing so leads to an analytic general thermal-field equation [36, 41] whereby

$$
\begin{aligned}
J(F, T) &\approx A_{RLD} T^2 N(n, s) \\
n(F, T) &\equiv \frac{\beta_T}{\beta_F(E_m)} \\
s(F, T) &\equiv \theta(E_m) + \beta_F(E_m)(E_m - E)
\end{aligned}
\tag{20.33}
$$

where

$$
\begin{aligned}
N(n, s) &\approx e^{-s} n^2 \Sigma\left(\frac{1}{n}\right) + e^{-ns} \Sigma(n) \\
\Sigma(x) &\approx \frac{1 + x^2}{1 - x^2} - 0.36\, x^2 - 0.106\, x^4
\end{aligned}
\tag{20.34}
$$

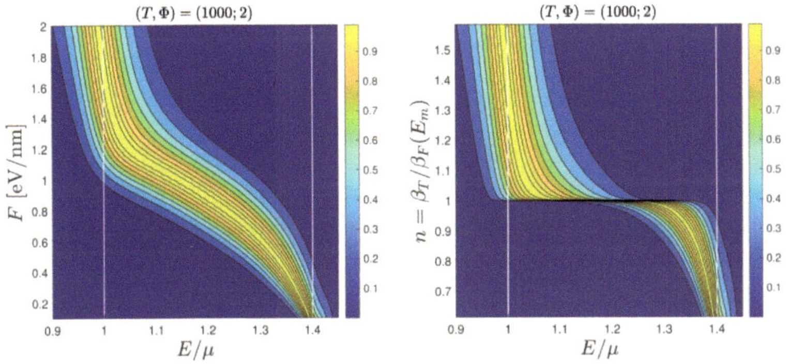

Figure 20.5 Contour plots of $dJ(E)/dJ(E_m)$ for the same parameters as Figure 20.4. (*Left*) As a function of field F and E. (*Right*) As a function of $n = \beta_T/\beta_F(E_m)$ and E. Thin white vertical lines at 1 and 1.4 mark the locations of μ and $(\mu + \Phi)$. The temperature is $T = 1000$ K. Were the same contours shown when using the linear θ approximation, the full widths at half maximum (FWHM) would be larger than here.

Although this equation is approximate, it can be shown that for $n = 1$, $N(1, s) = (s + 1)e^{-s}$ is exact [12]. Importantly, even if more complex barriers than the triangular barrier are considered, the methodology continues to work as $\theta(E)$ can be replaced by a more general $2\sigma(E)\kappa(E)L(E)$ so long as the linear approximation is evaluated at E_m: in fact, using the image charge barrier, then Eq. (20.34) enables the recovery of the FN and RLD equations in the ($n \gg 1$) and ($n \ll 1$) limits, respectively, so long as $E_m \to \mu$ in the former and $E_m \to \mu + \phi$ in the latter. Even for barriers modified by curvature of the emitting surface, the method can be adapted [66]. Much has been hidden in this short reconstruction: the evaluation of E_m for general barriers is non-trivial, particularly in the thermal-field regime, although in the original GTF formulation, it was approximated by a Fowler–Nordheim parameterization involving $v(y)$ and $t(y)$ [40] that reformulated methods discontinue [41]. To extend to photoemission, observe that for a positive s,

$$N(n, s) + N(n, -s) = \frac{1}{2}(ns)^2 + \zeta(2)\left(n^2 + 1\right) \tag{20.35}$$

where $\zeta(2) = \pi^2/6$ is the Riemann zeta-function. Under photoemission conditions, s in Eq. (20.33) becomes negative when $E \to E + \hbar\omega$, and so $N(n, -|s|)$ dominates. The right hand side of

Eq. (20.33) is then proportional to the Fowler–DuBridge equation (compare Eq. (20.25)) [39].

20.3.2 Heating of Wires and Nanotubes

Passing current through a non-planar electron source is generally accompanied by heating due to resistive losses along the emitter body, and called Joule heating. A second source of heating, called Nottingham heating [8, 10, 94] due to the emission process itself, which generally occurs close to the emission site and for field emission, relies on physics very similar to the GTF equation [46, 99], but is outside the present scope, even though the temperature rises can be substantial [65]. Here, consider instead the heating that attends transport along nanotubes and fibers, for which a simple model is available [43].

Temperature as a function of position along the length of a cylindrical conductor of length L and radius r, for which the base is held at a temperature T_o [78], is governed by

$$\frac{d}{dx}\left(\kappa\frac{dT}{dx}\right)-\frac{2\sigma_B}{r}\left(T^4-T_o^4\right)=-\frac{I_o^2}{\pi^2r^4\sigma(T)} \tag{20.36}$$

Here, several terms have changed their meaning: $\kappa(T)$ is the thermal conductivity, $\sigma_B T^4$ is radiative heat loss, $\sigma(T)$ is electrical conductivity, and I_o is the total current passing through the cylinder. Consider the following simplifications. First, the radiative heat loss makes the differential equation non-linear in T, so it can be treated as a perturbation to whatever analytical equation is derived here. Second, express the electrical conductivity using the Drude model [39] as

$$\sigma(T)=\frac{q^2}{m}\rho\tau(T) \tag{20.37}$$

where $\tau(T)$ is the electron relaxation time. At high temperatures compared to the Debye temperature T_D that characterizes electron-phonon scattering rates, the relaxation time is dominated by acoustic scattering $\tau\rightarrow\tau_{ep}$, which can be approximated by

$$\tau_{ep}(T)=\frac{\hbar}{2\pi\lambda k_B T} \tag{20.38}$$

where the dimensionless parameter λ can be taken as 1/2, but is better treated here as an adjustable parameter to obtain the conductivity of a material. Eq. (20.37) with Eq. (20.38) implies a resistivity linear in temperature. Assuming thermal and electrical relaxation times are the same, then κ and σ are related by $\kappa = L_o T \sigma(T)$, where $L_o = (\pi^2/3)(k_B/q)^2$ is the Lorentz number [39]. These approximations make $\kappa(T)$ temperature-independent and given by

$$\kappa = \frac{\pi \hbar k_B \rho}{6\lambda m} \tag{20.39}$$

As a result

$$\frac{d}{dx}\left(\kappa \frac{dT}{dx}\right) \to \kappa \frac{d^2 T}{dx^2} \tag{20.40}$$

It is convenient to reduce matters to a dimensionless equation, as theorists often do. Let $x = sL$ and $T = (f+1)T_o$, where T_o is the temperature of the base contact, *e.g.*, 300 K. Introduce

$$\omega_o \equiv \frac{2\sqrt{3}\lambda I_o L m}{\pi q \hbar r^2 \rho} \tag{20.41}$$

Then Eq. (20.36) becomes (after neglecting radiative heat loss factors, or $\sigma_R \to 0$)

$$\frac{d^2}{ds^2} f = -\omega_o^2 (1+f) \tag{20.42}$$

This equation is immediately recognized as a slight modification of the harmonic oscillator equation: as a consequence, it has solutions based on trigonometric functions fixed by the boundary conditions $f(0) = 0$ and $\partial_s f(1) = 0$. Note that the second boundary condition is commonly made but is speculative. The solution is

$$f(s) = \frac{\cos[\omega_o(1-s)]}{\cos \omega_o} - 1 \tag{20.43}$$

When $\omega_o = \pi/2$, the equation diverges, entailing that the maximum current I_{max} is governed by when $\omega_o = \pi/2$, or

$$I_{max} = \frac{\sqrt{3}\pi q r^2}{2k_B L}\kappa = \frac{\sqrt{3}\pi^2 \hbar q \rho r^2}{12\lambda m L} \tag{20.44}$$

This equation implies more current may be drawn from fibers that exhibit a higher thermal conductivity. Consider some values related to experiment [43]: carbon fibers with a length of 1.5 mm and diameter of 35 μm can carry 25 mA without failing. Eq. (20.39) for the typical values of $\kappa = 15$ W/m K $= 9.362 \times 10^{-5}$ [eV/K nm fs] implies $\rho = 8.962$ nm^{-3}. Eq. (20.44) then entails that $I_{\text{max}} = 96.69$ mA which is compatible with observations.

Consider two representative cases drawn from the study of Ref. [43]:

- Multi-walled carbon nanotube (MWNT) parameters that replicate performance in the literature [3, 107] suggest $\kappa = 2$ W/m K, chosen low to emphasize self-heating effects during field emission. The length is $L = 4$ μm, radius is $r = 14$ nm, current $I_o = 25$ μA, $I_{max} = 31$ μA, $\lambda = 1/2$, and boundary conditions of $T(0) = 300$ K. The relaxation time is $\tau_{ep} = 8.104$ fs. It is found $\omega_o = 1.2692$. The temperature at the end is then $T = T_o/\cos\omega_o = 3.3665T_o$ or 1010 K.
- Carbon fiber parameters where different than those used for the MWNT, are $L = 1.5$ mm, radius is $r = 17.5$ μm, current $I_o = 9$ mA, $I_{\text{max}} = 12.892$ mA. It is found $\omega_o = 1.0966$. The temperature at the end is then $T = T_o/\cos\omega_o = 2.19T_o$ or 657 K. However, in the carbon fiber case, neglected factors can increase that estimate to 1200 K at the apex: the analytical model performs well over half the length of the emitter, but beyond that, the temperature rise can be dramatic.

The analytical model gives an indication of the heating that can be encountered, showing that the longer the wire or nanotube is, the higher the temperature gets along the length, to the point where the general thermal-field equation is needed to account for total emitted current. Moreover, studies have shown that temperature excursions of emitters can be large enough to cause loss of material and failure, in keeping with these expectations.

20.4 Time Factors

The passage of time matters. How fast emission occurs or how long is required before emitted current is beyond the influence of the local conditions of the cathode affects the utility of an electron source: fundamental limits on the switching speed of thermal emitters (related to the transconductance given by the change in current with change in extraction voltage [75]), for example, has led to the development of laser-triggered photocathodes [15, 32, 52, 69] for free electron lasers and field emitter arrays [68, 101, 102] for accelerators and microwave amplifiers. Such considerations also affect how beams are launched in particle-in-cell simulations [77]. Although delayed photoemission effects [55] are beyond the present scope, consideration of tunneling time and transit time are well-described by simple models.

20.4.1 Tunneling Time

The dimensions of some nanogap devices, in particular, the anode-cathode (AK) spacing, have become so small [30, 63, 105] that a finite tunneling time of emission may have an impact if the tunneling barrier and AK gap are comparable. Therefore, a simple model of tunneling time is given. The most often discussed model of tunneling time is the barrier width-dependent semiclassical time of Büttiker and Landauer [6]: for a rectangular barrier $V(x) = (\mu + \Phi)\Theta(x)\Theta(L - x)$ leading to the tunneling probability of Eq. (20.11) and a tunneling electron at $E = \mu$, it is defined by

$$\tau_{sc} \equiv \frac{L}{|v|} = \frac{L}{\sqrt{2\Phi/m}} \tag{20.45}$$

where the electron "velocity" under the barrier is taken to satisfy $(1/2)mv^2 = \mu + \Phi - E$. For comparisons, if $L = 1$ nm and $\Phi = 1$ eV is taken as a baseline case, then $\tau_{sc} = 1.6861$ fs. Compare this to the time scale of atomic physics units, or $\tau_o = \hbar a_o/2Q = 0.048378$ fs.

Hartman [31] studied electron wave packets incident on rectangular barriers modeled after metal-insulator-metal thin films. He concluded that a group delay tunneling time asymptotically approached

$$\tau_g = \frac{\hbar}{\sqrt{\mu\Phi}} \tag{20.46}$$

in the limit of large L: this independence of τ_g on L, which seems paradoxical when compared to τ_{sc}, is known as the "Hartman effect."

Winful [103] demonstrated that $\tau_g(k) = \tau_d(k) + \tau_i(k)$ is the sum of a dwell time τ_d and a self-interference time τ_i, both of which are given by

$$\tau_d(k) \equiv \left(\frac{m}{\hbar k}\right) \int_0^L |\psi_k(k)|^2 \, dx \tag{20.47}$$

$$\tau_i(k) \equiv -\frac{\hbar}{k} \Im[r(k)] \left(\frac{dk}{dE}\right) \tag{20.48}$$

where the simple wave function $\psi_k(x < 0) = e^{ikx} + r(k)e^{-ikx}$ and $\psi_k(x > L) = t(k)e^{ikx}$ (familiar from elementary quantum mechanics), $\Im[r(k)]$ is the imaginary part of the reflection coefficient $r(k)$, and $dk/dE = m/\hbar^2 k$. The transmission and reflection coefficients are well known for a rectangular barrier, and are [7, 39]

$$\begin{aligned} t(k) &= \frac{2k\kappa e^{-ikL}}{2k\kappa \cos(L\kappa) - i(k^2 + \kappa^2)\sin(L\kappa)} \\ r(k) &= \frac{-i(k^2 - \kappa^2)\sin(L\kappa)}{2k\kappa \cos(L\kappa) - i(k^2 + \kappa^2)\sin(L\kappa)} \end{aligned} \tag{20.49}$$

where $\kappa(k) = \sqrt{k_v^2 - k^2}$ and $k_v = \sqrt{2m(\mu + \Phi)}/\hbar$. Note that $D[E(k)] = |t(k)|^2$ recovers Eq. (20.11). In terms of the Gamow factor $\theta(k) \equiv 2\kappa(k)L$, then for $k < k_v$

$$\Im[r(k)] = -\frac{2k\kappa k_v^2 \sinh\theta}{k_v^4[\cosh\theta - 1] + 2(2k\kappa)^2} \tag{20.50}$$

and therefore

$$\frac{\tau_i(k)}{\tau_o} = \left(\frac{\kappa k_v^4}{k}\right) \frac{\sinh\theta}{k_v^4(\cosh\theta - 1) + 2(2k\kappa)^2} \tag{20.51}$$

To find τ_d, the wave function inside the barrier $\psi_k(x) = a(k)e^{-\kappa x} + b(k)e^{\kappa x}$ is required. Matching of wave function and first derivative at $x = 0$ shows that the coefficients satisfy

$$\begin{pmatrix} a \\ b \end{pmatrix} = \frac{1}{2\kappa} \begin{bmatrix} \kappa - ik & \kappa + ik \\ \kappa + ik & \kappa - ik \end{bmatrix} \begin{pmatrix} 1 \\ r \end{pmatrix} \tag{20.52}$$

The integration in τ_d can then be performed analytically, and gives

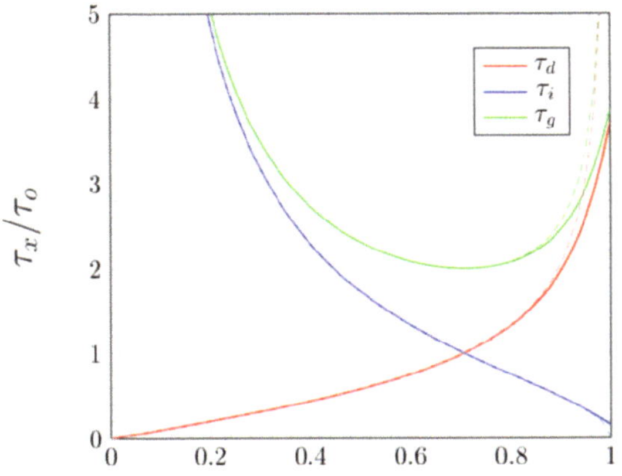

Figure 20.6 A comparison of the asymptotic tunneling times of Eq. (20.54) (lines) to Eqs. (20.51) and (20.53) for the parameters $\mu + \Phi = 5$ eV and $L = 0.5$ nm.

$$\frac{\tau_d(k)}{\tau_o} = \left(\frac{kk_v^2}{\kappa}\right) \frac{k_v^2 \sinh\theta - (k^2 - \kappa^2)\theta}{k_v^4(\cosh\theta - 1) + 2(2k\kappa)^2} \tag{20.53}$$

The wide barrier limit of Hartman is then tantamount to the large θ limit, and results in [48]

$$\begin{aligned} \tau_d(k) &= \tau_o \frac{k}{\kappa} \tanh\theta \\ \tau_i(k) &= \tau_o \frac{\kappa}{k} \tanh\theta \end{aligned} \tag{20.54}$$

The Hartman effect is then seen to correspond to when $\tanh\theta \approx 1$, which occurs for $\theta \gtrsim 4$. A comparison of Eq. (20.54) to the exact evaluation is shown in Figure 20.6 for generic values. Although a relatively narrow barrier was considered to keep parameters similar to the prior analyses, the form of $\theta = 2\kappa L$ insures that the curves have the same behavior for large L and small $\mu + \Phi$ except for close to $k \approx k_v$.

When generalized to the triangular barrier of Fowler and Nordheim and examined more carefully [57], however, it can be shown that as $k \to k_v$, then in fact τ_d does become barrier width dependent, and a different asymptotic relation holds. Nevertheless, it is seen that for conditions typical of field emission, τ_g is quite short and differs in fundamental ways from the semiclassical time.

20.4.2 Transit Time

Transit time τ refers to how long an electron takes to cross the AK gap D under a gap field of $F_o = q\mathcal{E}_a$: since tunneling is not under consideration at the moment, neither D nor τ should be confused with transmission probability or tunneling time from Section 20.4.1. Simple physics suggests that $D = v(0)\tau + (F_o/2m)\tau^2 = (v(\tau) + v(0))\tau/2$, but all electron sources emit with a distribution in velocity as a consequence of a thermal distribution and/or the tunneling probability changing with energy. Therefore, the simple physics viewpoint must be amended. Consider thermal emission first, and take the zero point in energy to now be the vacuum level (such a view ignores the Sommerfeld model of electrons inside the bulk material behind the canonical equations). For a thermal distribution, and letting for simplicity $\beta \equiv 1/k_B T$, then at time $t = 0$ [47]

$$\langle v(0) \rangle \approx \frac{\int_0^\infty (\hbar k/m)\, e^{-\beta E} dk}{\int_0^\infty e^{-\beta E} dk} = \sqrt{\frac{2k_B T}{\pi m}} \tag{20.55}$$

where $E = (\hbar k)^2/2m$, or about 0.1076 nm/fs for $T = 1000$ K, for a kinetic energy (KE) of about 27.43 meV. When tunneling is involved the complexity increases, but neglect that for a moment. At a short distance away from the cathode surface in the presence of a strong field, the electron energy can increase: for anodes held at tens of volts to kilovolts, it is clear that the initial KE for thermally emitted electrons is negligible, and so neglect it. Define the ballistic τ_o by the simple ballistic relation

$$\tau_o = \sqrt{\frac{2mD}{F_o}} \tag{20.56}$$

In the presence of space charge, the definition is altered to

$$\tau \equiv \int_0^D \frac{dx}{\langle v \rangle} = \frac{q\sigma(D)}{J} \tag{20.57}$$

where $q\sigma(x)$ is a charge per unit area, and is defined by

$$\sigma(x) = \int_0^x \rho\left(x'\right) dx' \tag{20.58}$$

where $q\rho(x)$ is the charge density: by keeping q as a coefficient, σ and ρ are explicitly number densities, not charge densities. The boundary conditions for $V(x)$ are $V(0) = 0$, $V(D) = V_a$. For electrons, $\partial_x V(0) = F$. By treating the charge q as a unit, these formulae have mild differences compared to their usual electrostatic representations. The anode potential energy is (remember that $V(x)$ is an energy barrier)

$$V_a = FD + 16\pi Q \int_0^D \sigma(x)dx \tag{20.59}$$

$$\equiv FD + 16\pi QDN_\tau \sigma(D) \tag{20.60}$$

where, in conventional SI units, $16\pi Q = q^2/\varepsilon_0$. The dimensionless factor N_τ is of order unity and corresponds

$$N_\tau \equiv \frac{\int_0^D \sigma(x)dx}{D\sigma(D)} \tag{20.61}$$

To evaluate N_τ, rewrite Poisson's equation as $d[F(x)^2] = (2qJ/\varepsilon_0 v(x))dV(x)$ but which in the units favored here becomes

$$d[F^2] = \frac{32\pi Q}{qv} \tag{20.62}$$

where, remember, $E = \hbar^2 k^2/2m = mv^2/2$. The velocity is $v(x) = \sqrt{2V(x)/m}$ if the initial kinetic energy is negligible. Integration is simple and gives

$$F(x) = \left[F^2 + 32\pi Q \left(\frac{J}{q}\right)\sqrt{2mV(x)}\right]^{1/2} \tag{20.63}$$

The units are worth examining: F (the field at the surface) has units of [eV/nm]. Q has units of [eV nm]. J/qc has units of $1/\text{nm}^3$. Lastly, $\sqrt{mV}$ has units of eV/c. Therefore, the second term has units of $[\text{eV/nm}]^2$ as it should [47, 83]. As a result[1],

$$\tau = \frac{\sqrt{8mV_a}}{F}\left[1 + \left(1 + 16\pi Q \frac{J}{qF^2}\sqrt{8mV_a}\right)^{1/2}\right]^{-1} \tag{20.64}$$

Then, from Eqs. 20.57–20.59,

$$\frac{V_a}{D} = F_o = F + \frac{16\pi Q}{q} N_\tau J\tau \tag{20.65}$$

$N_\tau J\tau$ is the average charge per unit area within the gap. Eq. 20.65 is the defining relation between transit time and current density. N_τ

[1]This equation corrects Eq. (12) of Ref. [47] which neglected the power of 2 on F.

can be computed, but two simple models here suffice. First, space charge limited current is defined by $J \rightarrow J_{CL}$ or Child's Law

$$J_{CL} \equiv \frac{4\varepsilon_0}{9D^2}\sqrt{\frac{2q}{m}}\varphi_a^{3/2} = \frac{q}{18\pi Q\sqrt{2m}}\left(\frac{V_a^{3/2}}{D^2}\right) \tag{20.66}$$

in SI units and then present units. For space charge limited flow using J_{CL} and letting $\tau = \tau_o$, then $N_\tau = 9/8$, as first found in Ref. [100]. This approximation underestimates the influence of space charge. Second, for τ from Eq. 20.64, then $N_\tau = 3/4$ when $F \rightarrow 0$. The $F = 0$ limit of Eq. 20.64 overestimates the influence of space charge. Nevertheless, both are reasonably close parametrically.

The model so far assumed charge was distributed throughout the gap. Both photoemission and field emission can inject so many electrons so quickly that the charge is seemingly confined to a single sheet: the field produced by a sheet of charge is independent of distance, as simple electrostatic arguments show [22]. The location of the sheet of charge, designated by $\bar{x}$, is the solution of

$$V_a = F_+(D - \bar{x}) + F_-\bar{x} \tag{20.67}$$

where F_- is the field between the cathode surface and the sheet, and F_+ between the sheet and anode. For a sheet charge density σ

$$F_+ - F_- = \frac{q^2}{\varepsilon_0}\sigma = 16\pi Q\sigma \tag{20.68}$$

where the first form is in SI units and the second in present units. The equations can be solved simultaneously to find

$$F_+(\bar{x}(t)) = F_o + 16\pi Q\sigma\left(\frac{\bar{x}(t)}{D}\right) \rightarrow F_o\left(1 + \frac{\bar{x}(t)}{D}\right) \tag{20.69}$$

$$F_-(\bar{x}(t)) = F_o - 16\pi Q\sigma\left(1 - \frac{\bar{x}(t)}{D}\right) \rightarrow F_o\frac{\bar{x}(t)}{D} \tag{20.70}$$

where the arrows indicate the replacement $\sigma \rightarrow \sigma_o$. The force under which $\bar{x}$ evolves is the average of F_+ and F_- (see Ref. [47] for a discussion) and so

$$\frac{d^2}{dt^2}\bar{x}(t) = \frac{F_o}{2mD}\left[D + 2\bar{x}(t)\right] \tag{20.71}$$

the solution of which is

$$\bar{x}(t) = \frac{D}{2}\left(\cosh(at) - 1\right) + \frac{v_o}{a}\sinh(at) \tag{20.72}$$

using the notation $v_o = \sqrt{2E_o/m}$ as the initial velocity, $a \equiv \sqrt{F_o/mD} = \sqrt{2}/\tau_o$, and recall $\tau_o = \sqrt{2mD/F_o}$. When the initial velocity is negligible, then $v_o = 0$ and the transit time τ becomes

$$\tau = \frac{1}{a}\cosh^{-1}(3) = 1.2465\tau_o \tag{20.73}$$

This is an intriguing result: it says that the transit time of the electron in the AK gap when the maximum amount of charge is present to shut down the field at the cathode is only some 25% larger than the transit time when there is no other charge in the gap. It suggests that in understanding the effects of space charge on transit, particularly for emission mechanisms that are very dependent on surface field (all are, but field emission is extremely sensitive, followed by photoemission, then thermal emission), the physics depends a great deal on how J varies and not so much *by comparison* on how τ changes. Therefore, looking at the variations in J is the priority.

20.5 Quadratic Barriers

For both the canonical emission equations and the GTFP equation, the barrier was presumed to be a Schottky–Nordheim barrier (*aka* image charge barrier) described by Eq. (20.2). Three processes are known to separately change the form of that barrier by effectively adding on another term of γx^2 to give rise to "quadratic barriers" defined by

$$V_q(x) = V_o - Fx - \frac{Q}{x} + \gamma x^2 \tag{20.74}$$

The first cause to be considered is due to space charge in the AK gap or near the nanoemitter [104, 106]. The second cause is due to the image charge departing from its planar approximation [38, 62]. The third is due to non-planar emitters causing curved field lines, which causes electron trajectories near the surface to become curved as well [11, 64]. Lastly, the depletion layer of a semiconductor [13, 58] can introduce a quadratic term.

20.6 Space Charge

A rapid method to determine how the nature of the variation of J with field affects how much current can be pushed across an AK gap makes use of a dimensionless analysis by Forbes [23] and separately by Rokhlenko, et al. [83]. In one dimension, Poisson's equation is $d^2V/dx^2 = 16\pi Q\rho$, and the current density is $J = qv\rho$ with $mv^2/2 = V(x)$. If the current density is divided by J_o where $J_o = 9J_{CL}/4$ to define j, the barrier function $V(x)$ divided by the anode value V_a to define φ, and the position x by the AK gap separation D to define y, then combining all these equations after eliminating ρ gives

$$\frac{d^2}{dy^2}\varphi(y) = \frac{j}{\sqrt{\varphi}} \tag{20.75}$$

Introduce $F = fV_a/D$ to define a dimensionless field term f. Then a first integration yields

$$\frac{d\varphi}{dy} = \left[4j\sqrt{\varphi} + f^2\right]^{1/2} \tag{20.76}$$

and a second integration gives

$$\sqrt{4j\sqrt{\varphi} + f^2}\,\left(2j\sqrt{\varphi} - f^2\right) + f^3 = 6j^2y \tag{20.77}$$

The boundary conditions are that at $y = 1$, then $\varphi = 1$ and so

$$\left(4j + f^2\right)^{1/2}\left(2j - f^2\right) + f^3 = 6j^2 \tag{20.78}$$

Carefully isolating j so as not to introduce unphysical solutions,

$$j = \frac{1}{9}\left[2 + (2 - 3f)\sqrt{1 + 3f}\right] \tag{20.79}$$

which can be manipulated into the far more suggestive form

$$3f^2(1 - f) = j(4 - 9j) \tag{20.80}$$

but this form allows unphysical roots that retention of the radical in Eq. (20.78) does not (*e.g.* $f = 1$ and $j = 4/9$ are not physical solutions but satisfy the equation). Of the physical roots, a vanishing field on the cathode ($f \to 0$) is associated with the current density approaching the CL limit ($j \to 4/9$).

The form of Eq. (20.80) allows for an investigation of space charge and emission mechanism. Eq. (20.79) is one equation with two unknowns. A second equation is the dependence of j on f due to emission. Where these curves intersect corresponds to a solution of both equations. Consider two representative cases.

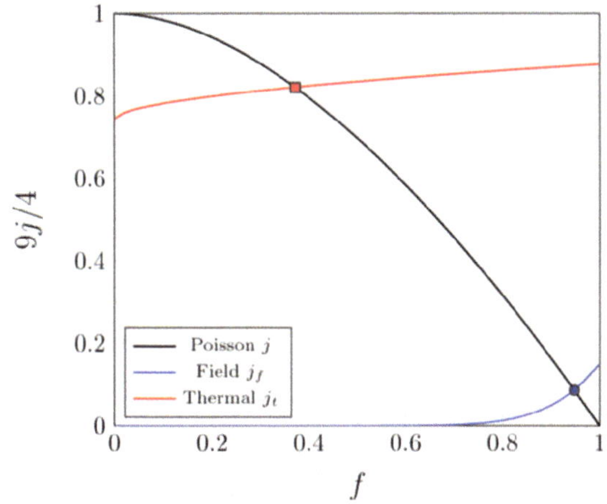

Figure 20.7 Curves associated with Poisson (Eq. (20.79)), field emission (Eq. (20.82)) and thermal emission (Eq. (20.81)). Red square is at (0.4021, 0.7947); blue circle is at (0.9481, 0.0862).

- Thermal: Eq. (20.8) with $T = 1173$ K, $\Phi = 1.95$ eV, $D = 500$ μm, and $V_a = 100$ eV is

$$j_t = 0.32955\, e^{0.16789\sqrt{f}} \tag{20.81}$$

- Field: Eq. (20.27) with $\nu = 0.87628$, $\Phi = 3.5$ eV, $D = 100$ nm, and $V_a = 500$ eV is

$$j_f = 509.36 f^{1.1237}\, e^{-8.9456/f} \tag{20.82}$$

As expected, the greater sensitivity of field emission on field at the surface compared to thermal emission is reflected in an intersection between j and j_f for large f / small j at the blue circle than shown by the intersection between j and j_t for small f / large j at the red circle.

The presence of charge in the AK gap entails that the field across the gap is no longer constant, but changes with position. For one-dimensional problems, the impact on the tunneling barrier is akin to the quadratic barrier model. When emission is from a single emitter or an array of emitters [45, 53], space charge near the emitter has the opportunity to spread and thereby affect the emittance of the beam [54, 61], rendering the problem exceedingly nontrivial and generally requiring the intervention of particle-in-cell codes to model accurately. For that reason, it is not considered further here, apart from noting that if the quadratic factor γ can be estimated by analytical models or simulation, then the impact on current density can be estimated by shape factor methods below.

20.6.1 Non-Planar Image Charge

It is noted that curvature also affects the image charge term Q/x, but the effects of changes to it as a consequence of surface curvature are easily handled. The image charge model of a point charge near a conducting sphere is a standard model in electrostatics [22], and can be extended to hyperbolic emitters [59]. In the simpler conducting sphere case, the electrostatic potential $\varphi(r, \theta)$ in polar coordinates is

$$q\varphi(r, \theta) = 4Q\left(\frac{1}{r_2} - \frac{a}{dr_1}\right) \tag{20.83}$$

where r_2 is the distance from the external charge to the observation point (r, θ) and r_1 is the distance from the image charge inside the sphere to the observation point, a is the radius of the emitter, and d is the distance from the center of the sphere to the external charge. They are

$$\begin{aligned} r_2^2 &= r^2 + d^2 - 2rd\cos\theta \\ r_1^2 &= \left(\frac{rd}{a}\right)^2 - a^2 - 2rd\cos\theta \end{aligned} \tag{20.84}$$

Taking $\theta \to 0$ and $r \to d$ gives the image charge barrier contribution

$$V_{image}(r) = -\frac{2Qa}{d^2 - a^2} \to -\frac{2aQ}{x(x + 2a)} \tag{20.85}$$

where the second form is after making the replacement $d = x + a$ to reference distances x from the surface. A Taylor expansion gives

$$V_{image}(x) \approx \frac{Q}{2a} - \frac{Q}{x} - \frac{Qx}{4a^2} + \frac{Qx^2}{8a^3} \tag{20.86}$$

It is seen, therefore, surface curvature modifies the one-dimensional barrier with the replacements $V_o \to V_o + (Q/2a)$, $F \to F + (Q/4a^2)$ and $\gamma = Q/(8a^3)$, that is, the barrier height and field terms are modified, and a parabolic factor γx^2 is introduced, as a consequence of emission from a curved sphere. Except for very sharp nanoscale emitters, the quadratic contribution to the image charge is generally (but not always) negligible.

20.6.2 Conical Emitters

Prolate spheroidal coordinates are useful for describing conical emitters in a diode configuration (no gate), and thereby reveal additional physics associated with multi-dimensionality. The advantage to prolate spheroidal coordinates is that field and potential lines (and therefore boundaries of conductors) are exactly solvable in Poisson's equation in the absence of charge within the AK gap. This allows curvature of the emitter surface, field enhancement, and electron trajectories (if electrons are assumed to follow the field lines) to be analytically found. Prolate spheroidal coordinates for a hyperbolic geometry are [39, 60]

$$\rho = L_h \sinh(\eta) \sin(\upsilon) \tag{20.87}$$

$$z = L_h \cosh(\eta) \cos(\upsilon) \tag{20.88}$$

where L_h is a length scale. The origin is therefore on the anode. A hyperbolic emitter is defined by constant υ; an ellipsoidal emitter is described by constant η. The hyperbolic case shall be considered, but the methodology applies to both. Laplace's equation $\nabla^2\varphi = 0$ is separable [1], or $q\varphi[\rho(\eta, \upsilon), z(\eta, \upsilon)] \rightarrow V(\eta)U(\upsilon)$, for which the equation for $V(\eta)$ is

$$\left(\sin\upsilon\, \partial_\upsilon^2 + \cos\upsilon\, \partial_\upsilon\right) V(\upsilon) = 0 \tag{20.89}$$

The anode corresponds to $\upsilon = \pi/2$. The surface of the emitter corresponds to υ_o. The equipotential lines are given by

$$V_h(\upsilon) = V_a\left(1 - \frac{Q_0(\cos\upsilon)}{Q_0(\cos\upsilon_o)}\right) \tag{20.90}$$

$$Q_0(s) \equiv \frac{1}{2}\ln\left(\frac{1+s}{1-s}\right) \tag{20.91}$$

where $Q_0(s)$ is a Legendre Polynomial of the Second Kind [1] for a dummy argument s (and not to be confused with an image charge). A three-dimensional cone is being treated, but a two-dimensional wedge can be had by replacing ρ with y and reconsidering the gradient (see Ref. [39]).

A representative hyperbolic emitter configuration is shown in Figure 20.8. The field $F = q\mathcal{E}$ is given by the gradient of V_h according to

$$\hat{\upsilon} \cdot \vec{F} = F(\eta, \upsilon) = -\frac{\partial_\upsilon V_h(\upsilon)}{L_h\left(\sinh^2\eta + \sin^2\upsilon\right)} \tag{20.92}$$

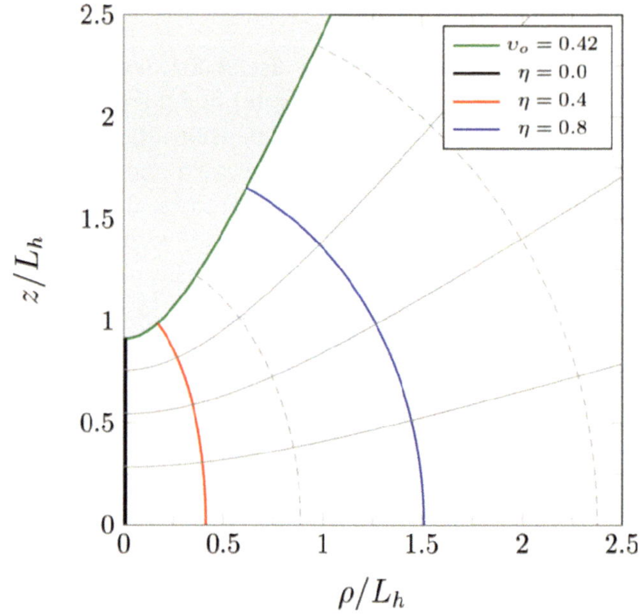

Figure 20.8 Hyperbolic coordinates of Eq. (20.87). Emitter is gray shaded region, anode is x-axis. Field lines are dashed gray along with red and blue; equipotential lines are solid gray. Parameters chosen such that $a_{tip} = 20$ and $z_o = 100$.

Along the surface ($\upsilon = \upsilon_o$) in particular, then, it is

$$F(\eta, \upsilon_o) = \frac{\sin \upsilon_o}{\sqrt{\sinh^2 \eta + \sin^2 \upsilon_o}} F_{tip} \tag{20.93}$$

$$F_{tip} = \frac{V_a}{L_h Q_0 (\cos \upsilon_o) \sin^2 \upsilon_o} \tag{20.94}$$

where F_{tip} is the value at the apex of the emitter ($\rho = 0$, $z = z_o$). The field falls off as $F(\eta, \upsilon_o)$: the barrier must therefore be evaluated along the trajectory line, which is approximated by the field line here. The angle υ_o is specified by the anode-cathode (AK) separation z_o. The radius of the apex a_{tip} is found by demanding that $z_s(\rho) \approx z_o + \rho^2/2a_{tip}$, that is, the apex is approximated as a hemisphere close to the apex, and so

$$\cos \upsilon_o \equiv \sqrt{\frac{z_o}{z_o + a_{tip}}}; \quad L_h = \sqrt{z_o(z_o + a_{tip})} \tag{20.95}$$

relates the physical parameters z_o and a_{tip} to the hyperbolic parameterization governed by υ_o and L_h. Because of the presence of $\sin^2 \upsilon_o = a_{tip}/(z_o + a_{tip})$ in the denominator of Eq. (20.94) for the apex field, it is seen that the field enhancement factor $\beta = F_{tip} z_o / V_a$ (sometimes referred to as the "beta-factor") scales as $1/a_{tip}$. This feature makes the sharpest emitter sites in field emission stand out. Such behavior is general: in the very different hemisphere on a cylindrical post geometry [24], for example, the field enhancement factor is $\beta \approx 0.7h/r$, where h is the height of the cylinder and r is the radius of its cap which functions as a_{tip}.

It is seen that the trajectory paths are curved. The curved hyperbolic trajectory modifies the barrier equation by [60]

$$V[x(\eta, \upsilon)] \equiv \mu + \Phi + V_h(\upsilon) - \frac{Q}{x(\eta, \upsilon)}$$
$$x(\eta, \upsilon) = L_h \int_{\upsilon}^{\upsilon_o} \sqrt{\sinh^2 \eta + \sin^2 u}\, du \qquad (20.96)$$

and is seen to be an arc length along the field/trajectory lines. Observe that the image charge for a sphere is not the same as the planar image charge, but for field emission where the width of the barrier is smaller than the radius of curvature, an expansion of the spherical image charge contribution can be approximated as modifying both the work function and field terms [38, 39], to a good approximation, and so are not considered further. To enable evaluation of the transmission probability, accounting for curvature is thus approximated by including a quadratic term on the barrier equation to find

$$V_q[x(\eta, \upsilon)] \equiv \mu + \Phi - F_\eta\, x + \gamma_\eta\, x^2 - \frac{Q}{x} \qquad (20.97)$$

for each η corresponding to a trajectory, where a different F_η and γ_η is found for each launch site, and the subscript q on V_q denotes *quadratic*. For the trajectories identified in Figure 20.8, the corresponding barriers are shown in Figure 20.9, showing that the quadratic approximation is good, where the parameters F_η and γ_η are determined by

$$\begin{pmatrix} -x_a & x_a^2 \\ -x_b & x_b^2 \end{pmatrix} \begin{pmatrix} F_\eta \\ \gamma_\eta \end{pmatrix} = \begin{pmatrix} V_h(\upsilon_a) \\ V_h(\upsilon_b) \end{pmatrix} \qquad (20.98)$$

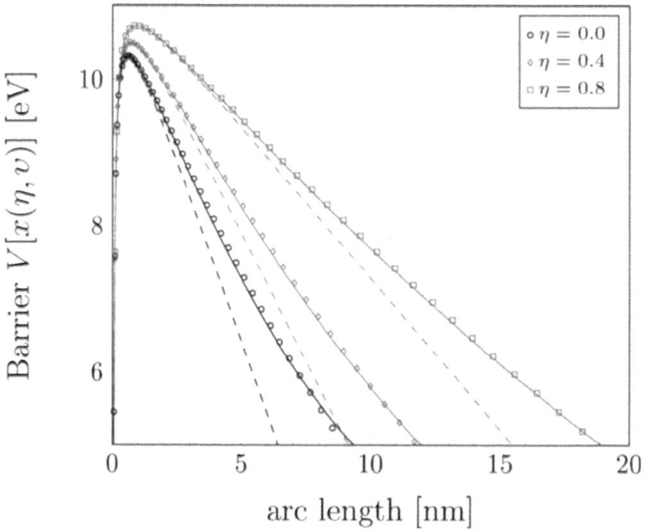

Figure 20.9 Energy barriers for the black ($\eta = 0$), red ($\eta = 0.4$) and blue ($\eta = 0.8$) trajectory lines of Figure 20.8. The dashed lines use $F_\eta = F_o$ and $\gamma = 0$ in Eq. (20.97) (standard image barrier).

The matrix equation is easily inverted to find F_η and γ_η given x_a and x_b, for which suitable choices are near the maximum and larger zero of $V_i(x)$, or

$$x_a = \sqrt{\frac{Q}{F_o}}; \quad x_b = \frac{\mu + \Phi}{F_o} \tag{20.99}$$

where $F_o = F(\eta, \upsilon_o)$ is the actual field factor at the surface where the trajectory originates. The performance of the procedure is shown in Figure 20.9. The approximation improves nearer the apex, as the fields are higher and the width of the barrier is less there. The estimate of F_η is less than $F(\eta, \upsilon_o)$ evaluated from Eq. (20.93) with Eq. (20.94), the departure increasing as γ_η increases.

20.6.3 Depletion Barrier

Depletion barriers are encountered when modeling internal field emission effects from metals into insulators such as diamond or other wide band-gap semiconductors [58, 67, 73, 81, 82]. The barrier is of the form [39]

$$V_{dep}(x) = V_o\left(1 - \frac{x}{L_o}\right)^2 - \frac{K_s - 1}{K_s + 1}\left(\frac{Q}{x}\right) \qquad (20.100)$$

for which the associated shape factor is $\sigma_{\text{U}}(E)$ of Eq. (20.18) if the dielectric constant $K_s = 1$. Expanding, it is seen that the field term is $F = 2V_o/L_o$ and the quadratic term is $\gamma = V_o/L_o^2$. The factor L_o corresponds to the depletion width, and is given by $L_o = K_s F_o/16\pi Q N_a$, where N_a is the number density of the impurity concentration that ionizes and causes the shape of the barrier, and F_o is the field across the insulating layer in the absence of charged impurities. Consequently, $V_{dep}(x)$ is already seen to be in the form of a quadratic barrier.

As a final note, the depletion barrier is a consequence of space charge that is immobile, that is, the charge density is constant. This is in contrast to the 1D space charge analysis of Section 20.6 which was such that the current density instead was constant.

20.6.4 Shape Factors Including Image Charge

The Schottky–Nordheim barrier of Eq. (20.2) has been examined extensively in the literature. An alternate way of treating it is now given based on a root method. The method can then be simply generalized to consider the quadratic barrier.

Observe that the barrier of Eq. (20.2) can be rewritten as

$$V(x) - E \equiv H = \frac{F(x - x_1)(x_2 - x)}{x} \qquad (20.101)$$

where $V_o = \mu + \Phi$ for compactness, $H = V_o - E$ is introduced for greater compactness, and the roots $x_j(E)$ satisfy $V(x_j) - E = 0$. Expanding the equation and comparing to Eq. (20.2) shows that the roots are related to the usual V_o and Q parameters by

$$Q = F x_1 x_2; \;\; H = F(x_1 + x_2) \qquad (20.102)$$

and the maximum height of the barrier, dictated by where $dV(x)/dx = 0$ is

$$x_o = \sqrt{\frac{Q}{F}} \qquad (20.103)$$

which entails that $V(x_o) = V_o - \sqrt{4QF}$ as known from standard accounts. It is trivial to show using the quadratic equation that

$$x_{\pm}(E) = \frac{H}{2F}\left[1 \pm \sqrt{1-y^2}\right] \tag{20.104}$$

where $x_1 \leftrightarrow x_-$ and $x_2 \leftrightarrow x_+$, $L(E) = (H/2)\sqrt{1-y^2}$, and the $y(E)$ factor is

$$y(E) \equiv \frac{V_o - V(x_o)}{V_o - E} = \frac{\sqrt{4QF}}{H(E)} \tag{20.105}$$

which generalizes that introduced by Murphy and Good in their Eq. (20.24), such they use $y(\mu) = \sqrt{4QF}/\Phi$. In terms of the roots x_j, the shape factor is seen to be

$$\begin{aligned}\sigma_n(E) &\equiv \int_{x_1}^{x_2}\left[\frac{V_n(x) - E}{V_n(x_o) - E}\right]^{1/2}\frac{dx}{L(E)} \\ &= \int_{x_1}^{x_1+L}\left[\frac{x_o}{x}\prod_{j=1}^{n}\left(\frac{x - x_j}{x_o - x_j}\right)\right]^{1/2}\frac{dx}{x_2 - x_1}\end{aligned} \tag{20.106}$$

It is seen that *the shape factor depends only on the roots* x_j. As a result, the shape factor can be evaluated rapidly by numerical means. Its weak variation with $y(E)$ for field emission conditions entails that a polynomial approximation may often be adequate for simulation [41].

With the recasting of a familiar model of the Schottky–Nordheim barrier, the quadratic barrier represents a straightforward generalization. Using the same methods,

$$\begin{aligned} Q &= \gamma x_1 x_2 x_3 \\ F &= \gamma(x_1 + x_2 + x_3) \\ H &= \gamma(x_1x_2 + x_1x_3 + x_2x_3)\end{aligned} \tag{20.107}$$

and the right hand side of Eq. (20.106) continues to hold, but now $n = 3$. For field emission, there are two observations: first, it often happens that x_3 is sufficiently large that the integrand of Eq. (20.106) can be Taylor expanded to first order in x/x_3 to find the effects on the Fowler–Nordheim equation [60] using the Murphy and Good form; second, when x_3 is sufficiently larger than x_2, then to a good approximation, Eq. (20.106) is almost independent of x_3, allowing for rapid approximation using the analytical results

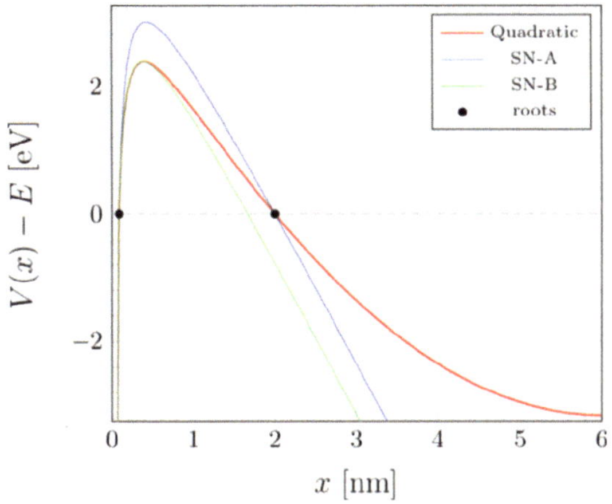

Figure 20.10 Comparison of the quadratic barrier (red) to the SN barrier having the same roots x_j (blue), and to the SN barrier having the same H and F (green). The roots were chosen to be $x_1 = 0.09$ nm, $x_2 = 2$ nm, and $x_3 = 10$ nm. γ is set to $\gamma = Q/x_1x_2x_3$.

for the Schottky–Nordheim barrier: these savings can substantially simplify usage of the Gamow factor $\theta(E)$ because although $\kappa(E)$ and $L(E)$ will still depend on all three roots, they are analytic or readily approximated. An indication of the ease in constructing the barrier is shown in Figure 20.10 where $V(x)$ from Eq. (20.74) and using Eq. (20.107) (red) is compared to the Schottky–Nordheim barrier of Eq. (20.2) having the same roots x_2 and x_1 (blue), and then to the SN barrier having the same H and F (green). A ratio of the red and blue shape factors σ show that the ratio of the quadratic σ to the SN sigma is greater than 0.98, showing that the shape factor for the quadratic barrier is close in value to that of the SN barrier for the same roots x_1 and x_2, as conjectured.

20.7 Concluding Remarks

The equations of electron emission have been examined with regard to their modeling and simulation of emitter configurations characteristic of Vacuum Nanoelectronics (VNE), for which anode-cathode gaps nanoscale. Several complicating mechanisms that alter the canonical emission equations have been considered: heating effects in thermal-field and photoemission, time factors associated with tunneling and transit times, and changes to the emission barrier. Special consideration was given to the quadratic barrier, as it flexibly models modifications to the Schottky-Nordheim barrier due to space charge, curvature of the emitter, depletion barriers associated with semiconductors, and non-planar image charges. The shape factor method, when used in conjunction with the Gamow factor method behind the thermal-field equations, provides an elegant means to include various effects into the standard emission equations of VNE.

In modeling electron sources for VNE applications, the properties of the emitters matter greatly, far more than simply the total current they produce. The tendency of a beam to spread, or emittance (ε) joins with the current from cathode I to define the brightness of a source $B \propto I/\varepsilon^2$. Emittance is affected by the energy spread of emission, the shape of the emission site, and the surface roughness of the cathode, all of which are affected by material, geometric, and operational factors. It, and space charge, govern the ability to use a beam produced by the various sources described herein. Therefore, to close the discussion, a brief account of how beams are affected by emittance and space charge is considered (see also [4, 5, 16, 39, 84, 102]).

An electron emitted from a surface of a cathode immersed in a magnetic field will rotate about the magnetic field lines because the magnetic force in it is perpendicular to both the velocity $\vec{v}$ and the magnetic field $\vec{B}$, or $\vec{F} = q\vec{v} \times \vec{B} = qv_\theta B$ [22]. The frequency with which the electron rotates about the field lines is governed by $\omega_o = qB/m$ as shown from equating the centrifugal force mv_θ^2/r to the magnetic force. Importantly, ω_o is independent of the transverse velocity v_θ of the electron, and so all emitted electrons rotate with the same frequency, although the radius of those orbits will be distributed to reflect the distribution in $v_\theta = r\omega_o$. If the size of the emission area is comparable to the size of the orbits, then the beam appears to have a larger size than the cathode area itself, an effect schematically illustrated in Figure 20.11. A moment's thought convinces one that a measure of the radius of the beam will oscillate with a frequency governed by ω_o. This can be seen

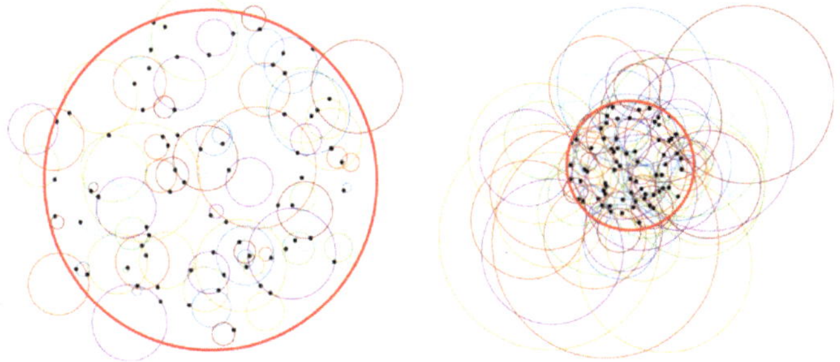

Figure 20.11 Cathode boundary indicated by red circle; individual electron orbits (72 in number shown) are individually color coded. Magnetic field is normal to the plane of the cathode. The initial velocities are Maxwell-Boltzmann distributed and randomly placed. Black dots indicate position at time $t = 0$ and all lie within the red circle. (*Left*) Emission from a cathode for which the cathode area is large compared to the orbit radii. (*Right*) Emission from a cathode for which the cathode area is comparable to the orbit radii.

directly by considering the root mean square (rms) average of the radial coordinates $r_j(t)$: if (x, y) are the cartesian coordinates of the electron in the $x - y$ plane, and the beam is propagating along the $\hat{z}$-axis, then consider

$$R(t) = \left\{ \frac{1}{N} \sum_{j=1}^{N} r_j(t)^2 \right\}^{1/2} = \left\{ \frac{1}{N} \sum_{j=1}^{N} \left[x_j(t)^2 + y_j(t)^2 \right] \right\}^{1/2} \tag{20.108}$$

Doing so results in Figure 20.12. It is seen straight away that $R(t)$ executes harmonic oscillations for $z(t) = v_z t$. For beams accelerated to a fraction of the speed of light c, the total velocity is dominated by v_z, therefore d/dt in the harmonic equation can be transitioned to $vd/dz = \beta c d/dz$. Putting in a factor to account for space charge expansion of the beam and a factor to account for emittance ε then gives the Beam Envelope equation for the $R(z)$ [80, 87]

$$\frac{d^2}{dz^2} R + \left(\frac{qB}{2\beta\gamma mc} \right)^2 R - \frac{2I_a}{(\gamma\beta)^3 I_o} \frac{1}{R} - \frac{\varepsilon^2}{R^3} = 0 \tag{20.109}$$

where $\beta = v/c, \gamma = 1/\sqrt{1-\beta^2}$, I_a is beam current, I_o is the Alfvén-Lawson current $I_o = (qc/\alpha^2 a_o) = 17.045$ kA. The last two terms account for space charge and emittance, respectively. The factors in

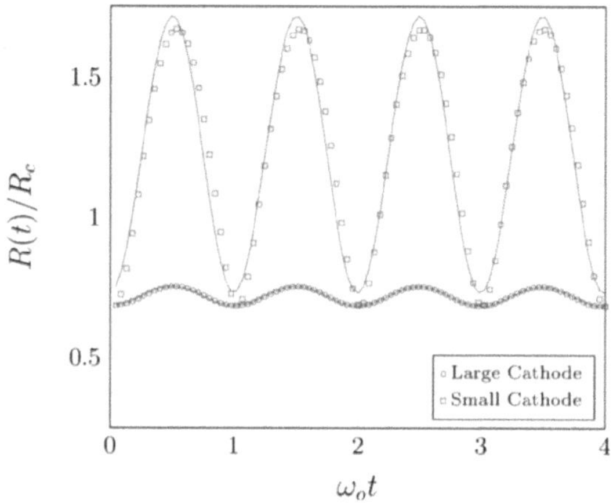

Figure 20.12 Symbols: $R(t)$ defined by Eq. (20.108) for the orbits of Figure 20.11. Lines: *ad hoc* fit to an approximation given by $R_a(t) = R_o - A\cos(6.3\omega_o t)$, where $R_o = (\max[R(t)] + \min[R(t)])/2$ and $A = (\max[R(t)] - \min[R(t)])/2$.

the above equation are called the magnetic term, the space charge term, and the emittance term, respectively. Brillouin flow occurs when the sum of the last three terms cancels. The relativistic factors can be replaced by the beam kinetic energy K_b via $(\beta\gamma)^2 = 2K_b/mc^2$ to leading order. Dividing the beam current I_a by the area of the beam is then

$$\frac{I_a}{\pi R^2} = \frac{I_o}{8\pi}\left(\frac{2K_b}{mc^2}\right)^{1/2}\left(\frac{qB}{mc}\right)^2\left\{1 - \frac{8mK_b\varepsilon^2}{q^2B^2R^4}\right\} \tag{20.110}$$

Calling the left hand side $J_{beam}(\varepsilon)$ then shows

$$J_{beam}(\varepsilon) = J_{beam}(0)\{1 - \delta(\varepsilon)\} \tag{20.111}$$

where $\delta(\varepsilon) = 8mK_b\varepsilon^2/\left(qBR^2\right)^2$ is the fractional amount that the beam current density is decreased because of emittance. Cathode emittance is therefore seen to be a limitation on brightness. Because emittance is becoming a limiting factor for electron sources, its reduction is increasingly important, for once it is generated at the cathode (*aka* intrinsic emittance) [16], it cannot be corrected for. As a consequence, emission physics is an important line of research in the future of vacuum nanoelectronics, particle accelerators, applications requiring high-brightness beams, and other technologies reliant on electron emission.

References

1. Arfken, G. B. (1970). *Mathematical Methods for Physicists* (Academic Press, New York).
2. Barbour, J. P., Dolan, W., Trolan, J., Martin, E. and Dyke, W. (1953). Space-charge effects in field emission, *Phys. Rev.* **92**, 1, pp. 45–51, doi:http://dx.doi.org/10.1103/PhysRev.92.45.
3. Bonard, J., Salvetat, J., Stockli, T., Forro, L. and Chatelain, A. (1999). Field emission from carbon nanotubes: Perspectives for applications and clues to the emission mechanism, *Appl. Phys. A. - Mater.* **69**, 3, pp. 245–254, doi:https://doi.org/10.1007/s003390050998.
4. Booske, J. H., Dobbs, R. J., Joye, C. D., Kory, C. L., Neil, G. R., Park, G.-S., Park, J. and Temkin, R. J. (2011). Vacuum electronic high power terahertz sources, *IEEE Transactions on Terahertz Science and Technology* **1**, 1, pp. 54–75, doi:http://dx.doi.org/10.1109/TTHZ.2011.2151610.
5. Brau, C. (2004). What brightness means, in J. Rosenzweig, G. Travish and L. Serafini (eds.), *Physics and Applications of High Brightness Electron Beams: Proceedings of the ICFA Workshop Chia Laguna, Sardinia, Italy 1-6 July 2002* (World Scientific Publishing Company), pp. 20–27.
6. Büttiker, M. and Landauer, R. (1982). Traversal time for tunneling, *Phys. Rev. Lett.* **49**, 23, pp. 1739–1742, doi:http://dx.doi.org/10.1103/physrevlett.49.1739.
7. Cahay, M. and Bandyopadhyay, S. (2017). *Problem Solving in Quantum Mechanics* (John Wiley & Sons, Inc., Hoboken, New Jersey).
8. Cahay, M., Zhu, W., Fairchild, S., Murray, P., Back, T. and Gruen, G. (2016). Multiscale model of heat dissipation mechanisms during field emission from carbon nanotube fibers, *Appl. Phys. Lett.* **108**, 3, pp. 033110 / 1–5, doi:http://scitation.aip.org/content/aip/journal/apl/108/3/10.1063/1.4940390.
9. Charbonnier, F. M., Dyke, W. P., Garrett, L. and Barbour, J. P. (1963). Basic and applied studies of field emission at microwave frequencies, *Proceedings of the IEEE* **51**, 7, pp. 991–1004.
10. Charbonnier, F. M., Strayer, R. W., Swanson, L. W. and Martin, E. E. (1964). Nottingham effect in field and t-f emission: Heating and cooling domains, and inversion temperature, *Phys. Rev. Lett.* **13**, 13, pp. 397–401, doi:http://link.aps.org/abstract/PRL/v13/p397.

11. Chatziafratis, A., Fikioris, G. and Xanthakis, J. P. (2018). Exact eigenstates of a nanometric paraboloidal emitter and field emission quantities, *Proc. R. Soc. A* **474**, 2214, p. 20170692, doi:http://dx.doi.org/10.1098/rspa.2017.0692.
12. Chernin, D. (2019). Exact evaluation of K. Jensen's general thermal-field-photoemission integral, *IEEE Trans. Plas. Sci.* **47**, 1, pp. 336–338, doi:https://doi.org/10.1109/TPS.2018.2871331.
13. Cutler, P., Miskovsky, N., Lerner, P. and Chung, M. (1999). The use of internal field emission to inject electronic charge carriers into the conduction band of diamond films: A review, *Appl. Surf. Sci.* **146**, 1-4, pp. 126–133.
14. Dolan, W. W. and Dyke, W. P. (1954). Temperature-and-field emission of electrons from metals, *Phys. Rev.* **95**, 2, pp. 327–332, doi:http://link.aps.org/abstract/PR/v95/p327.
15. Dowell, D., Bazarov, I., Dunham, B., Harkay, K., Hernandez-Garcia, C., Legg, R., Padmore, H., Rao, T., Smedley, J. and Wan, W. (2010). Cathode randd for future light sources, *Nuclear Instruments and Methods in Physics Research Section A* **622**, pp. 685–697, doi:http://dx.doi.org/10.1016/j.nima.2010.03.104.
16. Dowell, D., Padmore, H., Vecchione, T., Schmerge, J. and Wan, W. (2012). Models for cathode roughness, *Photocathode Physics for Photoinjectors Workshop* .
17. DuBridge, L. A. (1933). Theory of the energy distribution of photoelectrons, *Phys. Rev.* **43**, 9, pp. 0727–0741, doi:https://dx.doi.org/10.1103/PhysRev.43.727.
18. Dyke, W. P., Barbour, J. P., Martin, E. E. and Trolan, J. K. (1955). T-FEmission: Experimental measurement of the average electron current density from tungsten, *Phys. Rev.* **99**, 4, pp. 1192–1195, doi: http://dx.doi.org/10.1103/physrev.99.1192.
19. Dyke, W. P., Trolan, J., Martin, E. and Barbour, J. P. (1953). The field emission initiated vacuum arc.1. experiments on arc initiation, *Phys. Rev.* **91**, 5, pp. 1043–1054.
20. Dyke, W. P., Trolan, J. K., Dolan, W. W. and Barnes, G. (1953). The field emitter: Fabrication, electron microscopy, and electric field calculations, *J. Appl. Phys.* **24**, 5, pp. 570–576, doi:http://link.aip.org/link/?JAP/24/570/1.
21. Dyke, W. P., Trolan, J. K., Dolan, W. W. and Grundhauser, F. J. (1954). Field emission current-density distribution, *J. Appl. Phys.* **25**, 1, pp. 106–112, doi:http://dx.doi.org/10.1063/1.1721491.

22. Eyges, L. (1972). *The Classical Electromagnetic Field* (Addison-Wesley Pub. Co, Reading, Mass).
23. Forbes, R. G. (2008). Exact analysis of surface field reduction due to field-emitted vacuum space charge, in parallel-plane geometry, using simple dimensionless equations, *J. Appl. Phys.* **104**, 8, p. 084303, doi: http://dx.doi.org/10.1063/1.2996005.
24. Forbes, R. G. (2016). Physical electrostatics of small field emitter arrays/clusters, *J. Appl. Phys.* **120**, 5, p. 054302, doi:http://dx.doi.org/10.1063/1.4959150.
25. Forbes, R. G. and Deane, J. H. B. (2007). Reformulation of the standard theory of fowler-nordheim tunnelling and cold field electron emission, *Proc. R. Soc. A* **463**, 2087, pp. 2907–2927, doi:https://doi.org/10.1098/rspa.2007.0030.
26. Fowler, R. H. and Nordheim, L. (1928). Electron emission in intense electric fields, *Proc. R. Soc. A* **119**, 781, pp. 173–181, doi:http://dx.doi.org/10.1098/rspa.1928.0091.
27. Fursey, G. (2005). *Field Emission in Vacuum Microelectronics* (Kluwer Academic / Plenum Publishers, New York), doi:http://www.loc.gov/catdir/enhancements/fy0663/2002034120-d.html.
28. Geittner, P., Gartner, G. and Raasch, D. (2000). Low temperature properties of Ba-dispenser cathodes, *J. Vac. Sci. Technol. B* **18**, 2, pp. 997–999, doi:https://doi.org/10.1116/1.591315.
29. Guerrera, S., Velasquez-Garcia, L. and Akinwande, A. (2012). Scaling of high-aspect-ratio current limiters for the individual ballasting of large arrays of field emitters, *IEEE Trans. Electron Devices* **59**, 9, pp. 2524–2530, doi:https://doi.org/10.1109/TED.2012.2204262.
30. Han, J.-W., Moon, D.-I. and Meyyappan, M. (2017). Nanoscale vacuum channel transistor, *Nano letters* **17**, 4, pp. 2146–2151, doi:https://pubs.acs.org/doi/abs/10.1021/acs.nanolett.6b04363.
31. Hartman, T. E. (1962). Tunneling of a wave packet, *J. Appl. Phys.* **33**, 12, pp. 3427–3433, doi:https://aip.scitation.org/doi/pdf/10.1063/1.1702424.
32. Hernandez-Garcia, C., O'Shea, P. G. and Stutzman, M. L. (2008). Electron sources for accelerators, *Physics Today* **61**, 2, pp. 44–49, doi:https://doi.org/10.1063/1.2883909.
33. Herring, C. and Nichols, M. (1949). Thermionic emission, *Rev. Mod. Phys.* **21**, 2, pp. 185–270, doi:https://doi.org/10.1103/RevModPhys.21.185.

34. Hommelhoff, P., Sortais, Y., Aghajani-Talesh, A. and Kasevich, M. A. (2006). Field emission tip as a nanometer source of free electron femtosecond pulses, *Phys. Rev. Lett.* **96**, 7, p. 077401, doi:http://link.aps.org/doi/10.1103/PhysRevLett.96.077401.

35. Jensen, K. L. (1999). Field emitter arrays for plasma and microwave source applications, *Phys. Plasmas* **6**, 5, pp. 2241–2253.

36. Jensen, K. L. (2007). General formulation of thermal, field, and photoinduced electron emission, *J. Appl. Phys.* **102**, 2, pp. 024911–024911, doi:http://link.aip.org/link/?JAP/102/024911/1.

37. Jensen, K. L. (2012). A quantum dipole–modified work function for a simplified electron emission barrier, *J. Appl. Phys.* **111**, pp. 054916 / 1–10, doi:https://dx.doi.org/10.1063/1.3692571.

38. Jensen, K. L. (2014). Field emission: Fundamental theory to usage, in J. G. Webster (ed.), *Wiley Encyclopedia of Electrical and Electronics Engineering* (John Wiley & Sons, Inc., New York), pp. 1–29, URL http://dx.doi.org/10.1002/047134608X.W3129.pub2.

39. Jensen, K. L. (2017). *Introduction to the Physics of Electron Emission* (John Wiley & Sons, Inc., Hoboken, New Jersey), doi:https://doi.org/10.1002/9781119051794.

40. Jensen, K. L. (2018). A tutorial on electron sources, *IEEE Trans. Plas. Sci.* **46**, 6, pp. 1881 – 1899, doi:https://doi.org/10.1109/TPS.2017.2782485.

41. Jensen, K. L. (2019). A reformulated general thermal-field emission equation, *J. Appl. Phys.* **126**, 6, pp. 065302 / 1–13, doi:http://dx.doi.org/10.1063/1.5109676.

42. Jensen, K. L. (2020). Chapter 8: A thermal-field-photoemission model and its application, in G. Gaertner, W. Knapp and R. Forbes (eds.), *Modern Developments in Vacuum Electron Sources* (Springer Nature, Cham, Switzerland), pp. 345–386.

43. Jensen, K. L., Connelly, J. M., Petillo, J. J., Harris, J. R., Ovtchinnikov, S., Jensen, A. J., Burke, J., Cahay, M., Ludwick, J., Tripathi, G., Sanchez-Roddy, J. and Puentes, D. (2021). Semi-analytic model of a carbon fiber thermal-field emitter, *J. Appl. Phys.* **129**, 9, p. 095107, doi:http://dx.doi.org/10.1063/5.0044800.

44. Jensen, K. L., Finkenstadt, D., Shiffler, D. A., Shabaev, A., Lambrakos, S. G., Moody, N. A. and Petillo, J. J. (2018). Analytical models of transmission probabilities for electron sources, *J. Appl. Phys.* **123**, 6, p. 065301, doi:https://doi.org/10.1063/1.5018602.

45. Jensen, K. L., Kodis, M., Murphy, R. and Zaidman, E. G. (1997). Space charge effects on the current-voltage characteristics of gated field emitter arrays, *J. Appl. Phys.* **82**, 2, pp. 845–854.

46. Jensen, K. L., Lau, Y. Y., Feldman, D. W. and O'Shea, P. G. (2008). Electron emission contributions to dark current and its relation to microscopic field enhancement and heating in accelerator structures, *Phys. Rev. ST Accel. Beams* **11**, 8, pp. 081001 / 1–17, doi:http://link.aps.org/abstract/PRSTAB/v11/e081001.

47. Jensen, K. L., Lebowitz, J., Lau, Y. Y. and Luginsland, J. (2012). Space charge and quantum effects on electron emission, *J. Appl. Phys.* **111**, pp. 054917–054919, doi:https://dx.doi.org/10.1063/1.3692577.

48. Jensen, K. L., Lebowitz, J. L., Riga, J. M., Shiffler, D. A. and Seviour, R. (2021). Wigner wave packets: Transmission, reflection, and tunneling, *Phys. Rev. B* **103**, p. 155427, doi:https://doi.org/10.1103/PhysRevB.103.155427.

49. Jensen, K. L. and Marrese-Reading, C. (2003). Emission statistics and the characterization of array current, *J. Vac. Sci. Technol. B* **21**, 1, pp. 412–417, doi:https://doi.org/10.1116/1.1524142.

50. Jensen, K. L., McDonald, M., Chubenko, O., Harris, J. R., Shiffler, D. A., Moody, N. A., Petillo, J. J. and Jensen, A. J. (2019). Thermal-field and photoemission from meso- and micro-scale features: Effects of screening and roughness on characterization and simulation, *J. Appl. Phys.* **125**, 23, pp. 234303 / 1–25, doi:https://doi.org/10.1063/1.5097149.

51. Jensen, K. L., McDonald, M., Harris, J. R., Shiffler, D. A., Cahay, M. and Petillo, J. J. (2019). Analytic model of a compound thermal-field emitter and its performance, *J. Appl. Phys.* **126**, 24, p. 245301, doi:http://dx.doi.org/10.1063/1.5132561.

52. Jensen, K. L., Moody, N. A., Feldman, D. W., Montgomery, E. J. and O'Shea, P. G. (2007). Photoemission from metals and cesiated surfaces, *J. Appl. Phys.* **102**, 7, pp. 074902–074919, doi:https://dx.doi.org/10.1063/1.2786028.

53. Jensen, K. L., Mukhopadhyay-Phillips, P., Zaidman, E. G., Nguyen, K., Kodis, M., Malsawma, L. and Hor, C. (1997). Electron emission from a single Spindt-type field emitter: Comparison of theory with experiment, *Appl. Surf. Sci.* **111**, pp. 204–212, doi:https://dx.doi.org/10.1016/S0169-4332(96)00726-X.

54. Jensen, K. L., O'Shea, P. G., Feldman, D. W. and Shaw, J. L. (2010). Emittance of a field emission electron source, *J. Appl. Phys.* **107**, 1, pp. 014903–014914, doi:http://link.aip.org/link/?JAP/107/014903/1.
55. Jensen, K. L., Petillo, J. J., Ovtchinnikov, S., Panagos, N., Dimitrios, Moody, N. A. and Lambrakos, S. G. (2017). Modeling emission lag after photoexcitation, *J. Appl. Phys.* **122**, 16, p. 164501, doi:https://doi.org/10.1063/1.5008366.
56. Jensen, K. L., Shabaev, A., Lambrakos, S. G., Finkenstadt, D., Moody, N. A., Neukirch, A. J., Tretiak, S., Shiffler, D. A. and Petillo, J. J. (2020). Analytic model of electron transport through and over non-linear barriers, *J. Appl. Phys.* **127**, 23, p. 235301, doi:http://dx.doi.org/10.1063/5.0009759.
57. Jensen, K. L., Shabaev, A., Riga, J. M., Shiffler, D. A., Lebowitz, J. L. and Seviour, R. (2021). Reevaluating the Hartman effect for field emission, *(submitted)*.
58. Jensen, K. L., Shaw, J. L., Yater, J. E. and Pate, B. B. (2015). Enhancing secondary yield of a diamond amplifier using a nitrogen layer, *J. Appl. Phys.* **117**, 21, p. 214501, doi:http://scitation.aip.org/content/aip/journal/jap/117/21/10.1063/1.4921804.
59. Jensen, K. L., Shiffler, D. A., Harris, J. R., Rittersdorf, I. M. and Petillo, J. J. (2017). 2D/3D image charge for modeling field emission, *J. Vac. Sci. Technol. B* **35**, 2, p. 02C101, doi:https://doi.org/10.1116/1.4968007.
60. Jensen, K. L., Shiffler, D. A., Peckerar, M., Harris, J. R. and Petillo, J. J. (2017). Current from a nano-gap hyperbolic diode using shape-factors: Theory, *J. Appl. Phys.* **122**, 6, p. 064501, doi:https://doi.org/10.1063/1.4997457.
61. Jensen, K. L., Shiffler, D. A., Petillo, J. J., Pan, Z. and Luginsland, J. W. (2014). Emittance, surface structure, and electron emission, *Phys. Rev. ST Accel. Beams* **17**, 4, pp. 043402–043419, doi:https://dx.doi.org/10.1103/PhysRevSTAB.17.043402.
62. Jensen, K. L. and Zaidman, E. G. (1993). Field-emission from an elliptic boss - exact versus approximate treatments, *Appl. Phys. Lett.* **63**, 5, pp. 702–704.
63. Kr´uger, M., Schenk, M., F´orster, M. and Hommelhoff, P. (2012). Attosecond physics in photoemission from a metal nanotip, *J. Phys. B Atomic Mol. Opt. Phys.* **45**, 7, p. 074006, doi:https://doi.org/10.1088\%2F0953-4075\%2F45\%2F7\%2F074006.
64. Kyritsakis, A., Kokkorakis, G., Xanthakis, J., Kirk, T. and Pescia, D. (2010). Self focusing of field emitted electrons at an ellipsoidal tip,

Appl. Phys. Lett. **97**, 2, pp. 023104 / 1–3, doi:https://doi.org/10.1063/1.3462934.

65. Kyritsakis, A., Veske, M., Eimre, K., Zadin, V., Djurabekova and F. (2018). Thermal runaway of metal nano-tips during intense electron emission, *J. Phys. D Appl. Phys.* **51**, 22.
66. Kyritsakis, A. and Xanthakis, J. P. (2016). Extension of the general thermal field equation for nanosized emitters, *J. Appl. Phys.* **119**, 4, pp. 045303 / 1–6, doi:https://doi.org/10.1063/1.4940721.
67. Lerner, P., Cutler, P. H. and Miskovsky, N. (1997). Theoretical analysis of field emission from a metal diamond cold cathode emitter, *J. Vac. Sci. Technol. B* **15**, 2, pp. 337–342, doi:https://doi.org/10.1116/1.589317.
68. Lewellen, J. W. and Noonan, J. (2005). Field-emission cathode gating for RF electron guns, *Phys. Rev. ST Accel. Beams* **8**, 3, p. 033502, doi: http://link.aps.org/abstract/PRSTAB/v8/e033502.
69. Liu, W., Poelker, M., Smedley, J. and Ganter, R. (2020). Photocathodes modern developments in vacuum electron sources, in G. Gaertner, , W. Knapp, and R. G. Forbes (eds.), *Modern Developments in Vacuum Electron Sources* (Springer International Publishing, Cham), pp. 293–344.
70. Lockwood, N., Cartwright, K., Gensheimer, P., Shiffler, D., d'Aubigny, C., Walker, C., Young, A., Fairchild, S. B. and Maruyama, B. (2010). 1.5: Development of field emission cathodes, electron gun and a slow wave structure for a terahertz traveling wave tube, *IEEE International Vacuum Electronics Conference (IVEC)* , pp. 25–26.
71. Lundstrom, M. (2000). *Fundamentals of Carrier Transport* (Cambridge University Press, Cambridge, U.K.; New York), doi:http://www.loc.gov/catdir/description/cam0210/99049980.htmlhttp://www.loc.gov/catdir/toc/cam025/99049980.html.
72. Makishima, H., Miyano, S., Imura, H., Matsuoka, J., Takemura, H. and Okamoto, a. (1999). Design and performance of traveling-wave tubes using field emitter array cathodes, *Appl. Surf. Sci.* **146**, 1-4, pp. 230–233.
73. Mönch, W. (1995). *Semiconductor Surfaces and Interfaces*, Vol. Springer series in surface sciences; 26 (Springer-Verlag, New York).
74. Murphy, E. L. and Good, R. H. (1956). Thermionic emission, field emission, and the transition region, *Phys. Rev.* **102**, 6, pp. 1464–1473, doi:https://doi.org/10.1103/PhysRev.102.1464.

75. Murphy, R. and Kodis, M. (2001). Chpt. 8: Cold cathode microwave devices, in W. Zhu (ed.), *Vacuum Microelectronics* (Wiley, New York), p. 349.
76. Patterson, B. D., Abela, R., Braun, H.-H., Flechsig, U., Ganter, R., Kim, Y., Kirk, E., Oppelt, A., Pedrozzi, M., Reiche, S., Rivkin, L., Schmidt, T., Schmitt, B., Strocov, V. N., Tsujino, S. and Wrulich, A. F. (2010). Coherent science at the SwissFEL x-ray laser, *New J. Phys.* **12**.
77. Petillo, J. J., Nelson, E., Deford, J., Dionne, N. and Levush, B. (2005). Recent developments to the MICHELLE 2-D/3-D electron gun and collector modeling code, *IEEE Trans. Electron Devices* **52**, 5, pp. 742–748, doi:https://doi.org/10.1109/TED.2005.845800.
78. Purcell, S. T., Vincent, P., Journet, C. and Binh, V. T. (2002). Hot nanotubes: Stable heating of individual multiwall carbon nanotubes to 2000 k induced by the field-emission current, *Phys. Rev. Lett.* **88**, 10, p. 105502, doi:http://link.aps.org/abstract/PRL/v88/e105502.
79. Reichl, L. E. (1980). *A Modern Course in Statistical Physics* (University of Texas Press, Austin).
80. Reiser, M. (1994). *Theory and Design of Charged Particle Beams*, Wiley series in beam physics and accelerator technology (Wiley, New York), doi:http://www.loc.gov/catdir/description/wiley032/93033110.htmlhttp://www.loc.gov/catdir/toc/onix02/93033110.htmlhttp://www.loc.gov/catdir/enhancements/fy0607/93033110-b.html.
81. Rhoderick, E. (1970). Physics of Schottky barriers, *J. Phys. D Appl. Phys.* **3**, 8, pp. 1153–and, doi:https://doi.org/10.1088/0022-3727/3/8/203.
82. Robertson, J. (1999). Mechanisms of electron field emission from diamond, diamond-like carbon, and nanostructured carbon, *J. Vac. Sci. Technol. B* **17**, 2, pp. 659–665, doi:https://doi.org/10.1116/1.590613.
83. Rokhlenko, A., Jensen, K. L. and Lebowitz, J. L. (2010). Space charge effects in field emission: One dimensional theory, *J. Appl. Phys.* **107**, 1, pp. 014904–014910, doi:http://link.aip.org/link/?JAP/107/014904/1.
84. Rosenzweig, J., Cahill, A., Carlsten, B., Castorina, G., Croia, M., Emma, C., Fukusawa, A., Spataro, B., Alesini, D., Dolgashev, V., Ferrario, M., Lawler, G., Li, R., Limborg, C., Maxson, J., Musumeci, P., Pompili, R., Tantawi, S. and Williams, O. (2018). Ultra-high brightness electron beams from very-high field cryogenic radiofrequency photocathode sources, *Nucl. Instrum. Methods Phys. Res. Section A Accelerators Spectrometers Detectors Assoc. Equipment* **909**, pp. 224–228, doi:http://dx.doi.org/10.1016/j.nima.2018.01.061.

85. Schwoebel, P. R. and Brodie, I. (1995). Surface-science aspects of vacuum microelectronics, *J. Vac. Sci. Technol. B* **13**, 4, pp. 1391–1410.

86. Schwoebel, P. R., Spindt, C. A. and Brodie, I. (1995). Electron-emission enhancement by overcoating molybdenum field-emitter arrays with titanium, zirconium, and hafnium, *J. Vac. Sci. Technol. B* **13**, 2, pp. 338–343.

87. Serafini, L. and Rosenzweig, J. B. (1997). Envelope analysis of intense relativistic quasilaminar beams in rf photoinjectors: A theory of emittance compensation, *Phys. Rev. E* **55**, 6, p. 7565, doi:http://link.aps.org/abstract/PRE/v55/p7565.

88. Shiffler, D., Fairchild, S., Tang, W., Maruyama, B., Golby, K., LaCour, M., Pasquali, M. and Lockwood, N. (2012). Demonstration of an acid-spun single-walled nanotube fiber cathode, *IEEE Trans. Plas. Sci.* **40**, 7, pp. 1871–1877, doi:https://doi.org/10.1109/TPS.2012.2195328.

89. Shiffler, D. A., Luginsland, J., Ruebush, M., Lacour, M., Golby, K., Cartwright, K., Haworth, M. and Spencer, T. (2004). Emission uniformity and shot-to-shot variation in cold field emission cathodes, *IEEE Trans. Plas. Sci.* **32**, 3, pp. 1262–1266, doi:https://doi.org/10.1109/TPS.2004.827608.

90. Spindt, C., Holland, C., Brodie, I., Mooney, J. and Westerberg, E. (1989). Field-emitter arrays applied to vacuum fluorescent display, *IEEE Trans. Electron Devices* **36**, 1, pp. 225–228, doi:https://doi.org/10.1051/jphyscol:1988625.

91. Spindt, C. A. (1968). A thin-film field-emission cathode, *J. Appl. Phys.* **39**, 7, pp. 3504–3505, doi:http://dx.doi.org/10.1063/1.1656810.

92. Spindt, C. A., Brodie, I., Holland, C. E. and Schwoebel, P. R. (2001). Chapter 4: Spindt field emitter arrays, in W. Zhu (ed.), *Vacuum Microelectronics* (Wiley, New York), p. 105.

93. Spindt, C. A., Brodie, I., Humphrey, L. and Westerberg, E. R. (1976). Physical properties of thin-film field emission cathodes with molybdenum cones, *J. Appl. Phys.* **47**, 12, pp. 5248–5263, doi:https://dx.doi.org/10.1063/1.322600.

94. Swanson, L., Crouser, L. and Charbonnier, F. (1966). Energy exchanges attending field electron emission, *Phys. Rev.* **151**, 1, pp. 327–and.

95. Tafel, A., Meier, S., Ristein, J. and Hommelhoff, P. (2019). Femtosecond laser-induced electron emission from nanodiamond-coated tungsten needle tips, *Phys. Rev. Lett.* **123**, 14, p. 146802, doi:http://dx.doi.org/10.1103/physrevlett.123.146802.

96. Takemura, H., Tomihari, Y., Furutake, N., Matsuno, F., Yoshiki, M., Takada, N., Okamoto, A. and Miyano, S. (1997). A novel vertical current limiter fabricated with a deep trench forming technology for highly reliable field emitter arrays, *International Electron Devices Meeting. IEDM Technical Digest* , pp. 709–712doi:https://doi.org/10.1109/IEDM.1997.650481.
97. Taylor, B. N. (1995). *Guide for the use of the International System of Units (SI)*, Vol. NIST special publication; 811 (U.S. Dept. of Commerce, Technology Administration, National Institute of Standards and Technology, Gaithersburg, MD : Washington, D.C.), doi:https://www.nist.gov/pml/special-publication-811.
98. Temple, D. (1999). Recent progress in field emitter array development for high performance applications, *Mater. Sci. Eng. R Rep.* **24**, 5, pp. 185–239.
99. Tripathi, G., Ludwick, J., Cahay, M. and Jensen, K. L. (2020). Spatial dependence of the temperature profile along a carbon nanotube during thermal-field emission, *J. Appl. Phys.* **128**, 2, p. 025107, doi: http://dx.doi.org/10.1063/5.0010990.
100. Umstattd, R., Carr, C., Frenzen, C., Luginsland, J. and Lau, Y. Y. (2005). A simple physical derivation of Child-Langmuir space-charge-limited emission using vacuum capacitance, *Am. J. Phys.* **73**, 2, pp. 160–163.
101. Whaley, D., Duggal, R., Armstrong, C., Bellew, C., Holland, C. and Spindt, C. (2009). 100 W operation of a cold cathode TWT, *IEEE Trans. Electron Devices* **56**, 5, pp. 896–905, doi:https://doi.org/10.1109/TED.2009.2015614.
102. Whaley, D. R. (2014). Practical design of emittance dominated linear beams for RF amplifiers, *IEEE Trans. Electron Devices* **61**, 6, pp. 1726–1734.
103. Winful, H. G. (2003). Delay time and the Hartman effect in quantum tunneling. *Phys. Rev. Lett.* **91**, 26 Pt 1, p. 260401, doi:https://doi.org/10.1103/PhysRevLett.91.260401.
104. Zhang, P., Ang, Y. S., Garner, A. L., Valfells, Á., Luginsland, J. W. and Ang, L. K. (2021). Space–charge limited current in nanodiodes: Ballistic, collisional, and dynamical effects, *J. Appl. Phys.* **129**, 10, p. 100902, doi: http://dx.doi.org/10.1063/5.0042355.
105. Zhang, P. and Lau, Y. Y. (2016). Ultrafast and nanoscale diodes, *J. Plasma Phys.* **82**, 5, p. 595820505, doi:https://doi.org/10.1017/S002237781600091X.

106. Zhang, P., Valfells, A., Ang, L. K., Luginsland, J. W. and Lau, Y. Y. (2017). 100 years of the physics of diodes, *Appl. Phys. Rev.* **4**, 1, p. 011304, doi: https://doi.org/10.1063/1.4978231.

107. Zhu, W., Cahay, M., Ludwick, J., Jensen, K., Forbes, R., Fairchild, S., Back, T., Murray, P., Harris, J. and Shiffler, D. (2019). Multiscale modeling of field emission properties of carbon-nanotube-based fibers, in *Nanotube Superfiber Materials* (Elsevier), pp. 541–572.

Chapter 21

Terahertz Free-Electron Laser

Andrea Doria, Gian Piero Gallerano, and Emilio Giovenale
ENEA-Frascati Research Center, 00044 Frascati (RM), Italy
andrea.doria@enea.it, gpgallerano@gmail.com, emilio.giovenale@enea.it

21.1 Historical Introduction

When the free-electron laser (FEL) appeared in the scenario of lasers, it was immediately clear that a new family of coherent sources of radiation was born, having unique and peculiar characteristics [1]. It must be pointed out that the story of the free-electron-based sources is older than that of conventional lasers [2] and starts from the technological development of electronic tubes (ET), that goes back to the '30s of the last century [3]. The FEL is capable to exploit all the positive aspects of the ET, such as the possibility of continuously tuning the output frequency, the wide range of output power operation, and a temporal structure of the emitted radiation that can be chosen during the design phase.

Moreover, FELs have exploited some of the features of the conventional lasers as, *in primis*, the use of an optical cavity that, together with the possibility of manipulating the transverse phase

Advances in Terahertz Source Technologies
Edited by Gun-Sik Park, Masahiko Tani, Jae-Sung Rieh, and Sang Yoon Park

ISBN 978-981-4968-89-8 (Hardcover), 978-1-003-45967-5 (eBook)
www.jennystanford.com

front, introduces mechanisms for dynamical processes leading to saturation, similar to that of conventional lasers [4].

The most intriguing feature that distinguishes an FEL is that it is capable to overcome some of the limitations of both ET and Laser sources. The frequencies it can generate can span from the X-ray (even beyond the water window) down to the sub-THz spectral region. This may happen due to the linkage between the emitted radiation wavelength and the electron energy that, nowadays, can span from keV up to several GeV, making use of different accelerators. This fact makes FEL to be preferable with respect to ET, which frequencies are limited to the smallest cavities dimensions that can be realized, but also with respect to the lasers, which emissions are related to specific and limited quantum transitions in different materials. The second important feature to be mentioned is that an FEL can generate electromagnetic radiation with, virtually, no power or energy limitation except the one given by the electron beam power available; this detail is a direct consequence of the fact that the electron beam does not interact with a medium or with a resonant cavity, as it happens for lasers and ETs, respectively, which thermal or electrical physical properties establish an actual constraint to the sustainable energy or power.

All these features, which will be deeply analyzed in the following, allowed the realization of a number of FEL devices [5–8] right after the first lasing operation [1]. The first experiment was performed at Stanford University in 1977 by John Madey's group; the operation of the device, as an oscillator, was at a wavelength of $\lambda = 3.4$ μm, right well into the infrared region. A second experiment, carried out at the University of Santa Barbara, CA [5], extended the operation toward the far infrared (FIR or THz as commonly known today) around $\lambda = 400$ μm. Other experiments in the early eighties at the ACO storage ring in France [6], at the ENEA-Frascati Research Centre in Italy [7] and at Novosibirsk in Russia [8], all operated in spectral regions ranging from the visible to the THz frequencies. Furthermore, in the 1990s, two user facilities were realized (CLIO in Orsay, France [9], and FELIX in Nieuwegein, The Netherlands [10]) to offer a reliable source of electromagnetic radiation to users from wide different scientific disciplines. Both CLIO and FELIX operated in the mid-infrared and THz region performing tunable emission at relatively high peak power levels. All these historical examples reveal that the first FEL devices were built at frequencies "easy" to be obtained with conventional electron beam sources, with energies ranging from few MeV up to tens of MeV, and with established magnetic undulators, like the ones used as insertion devices in the synchrotron radiation sources. As will be clear in the forthcoming sections, the electron beam parameters and the magnetic undulator parameters available at the beginning of the FEL history allowed the generation of radiation mainly in the visible or at longer wavelengths.

However, in the same years, conventional lasers were capable to cover a similar spectral range with devices much easier to use and undoubtedly less expensive. Therefore, during the 1990s, there was great excitement about the possibility to realize FEL devices in spectral regions not accessible to conventional lasers. To this purpose two main regions of the spectrum were identified by the FEL community: the first region addresses the very high frequencies [11], in order to go beyond the limitations of the storage ring synchrotron sources, regarding power and pulse duration. FELs operating in the XUV and X-Ray regions of the spectrum are very huge devices, which utilize very high energy electron beams (above 1 GeV) and exploit sophisticated emission mechanisms to overcome the lack of good reflectors at such high frequencies, but, as a counterpart, require very long undulator sections in order to get saturation. On the other side, the second region was identified, extending from the far-Infrared to sub-THz and THz frequencies, so as to "bridge" a region not covered by ET sources and where the only laser available is the gas laser, with individual discrete lines and limited power. It will be clearer, in the following, that it's much simpler to design, construct, use, and maintain compact-FEL devices in this longer wavelength range.

21.2 Theory

21.2.1 The Basis

It is time now to deal with the mechanism underlying the physical process in an FEL; as the name expresses well, this device is a laser, which uses free electrons, that means, not bound to atomic, molecular, or reticular structures, or to particular states of matter, such as plasmas. It is known that an electron and a photon cannot interact in vacuum, because the principles of conservation of energy and momentum, which are exchanged in the processes of emission and absorption, would be violated. The basic mechanism can be related to a contextual emission and absorption mechanism that requires a passage through a virtual state, allowed in the context of the so-called Fermi–Weiszacker–Williams approximation. According to this theory, relativistic electrons "carry" a cloud of virtual photons with them, since the overall electric and magnetic fields are transverse because $\vec{E} \approx \vec{B}$ (in Gaussian units) is valid [12]. This method explains the mechanism that is known as Compton Scattering, which is the basis of conventional FELs based on magnetic undulator, as we will see shortly. Compton scattering is the quantum vision of a mechanism which in classical mechanics is known as synchrotron emission. In circular accelerators and storage rings, this process was found to be responsible for the energy losses, which the electrons undergo during their circular motion, associated to a magnetic field orthogonal to the orbit plane. Such losses need to be restored to prevent the electrons themselves from coming out of the delicate synchronism that the dynamics of those accelerators require. The phenomenon of radiation emission by a charged particle, following a curvilinear trajectory, due to the centripetal acceleration, follows the 19th-century Larmor theory for charged particles in accelerated motion [13]. The more general theory, for the calculation of the radiation emission by a charged particle in motion, involves the so-called Lienard–Wiechert retarded potentials [13, 14]. Through the Poynting vector theorem, one ends up with the general expression for the energy emission $\mathcal{E}$:

$$\frac{d\mathcal{E}}{d\omega d\Omega} = \frac{q^2\omega^2}{4\pi^2 c}\left|\int \left[\vec{n} \wedge \left(\vec{n} \wedge \vec{\beta}\right)\right] e^{i\omega\left(\tau - \vec{n}\cdot\frac{\vec{r}_0(\tau)}{c}\right)} d\tau\right|^2, \tag{21.1}$$

where q is the particle charge ω is the radiation frequency $\vec{n}$ is the versor linking the particle and the observer, $\vec{\beta}$ is the normalized particle velocity and $\vec{r}_0$ its position. Applying the general Eq. (21.1)

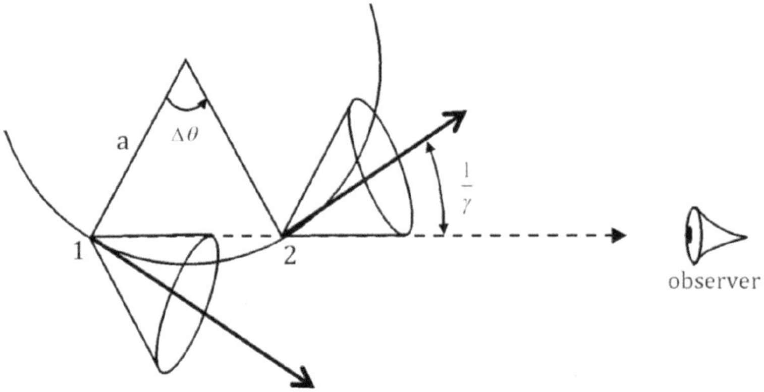

Figure 21.1 Emission cones of a moving charge along a circular trajectory [14].

to a circular geometry, as the one reported in Fig. 21.1, we get [14]:

$$\frac{d\mathcal{E}}{d\omega d\Omega} = \frac{q^2\omega^2}{3\pi^2 c}\left\{\theta^2\left[\frac{a^2\left(1+\gamma^2\theta^2\right)}{\gamma^2 c^2}\right]K_{1/3}^2(\eta) + \left[\frac{a\left(1+\gamma^2\theta^2\right)}{\gamma^2 c}\right]^2 K_{2/3}^2(\eta)\right\}, \tag{21.2}$$

where a is the radius of curvature γ is the relativistic factor θ is the angle between the vectors $\vec{n}$ and $\vec{\beta}$, and $K_{1/2}(\eta)$, $K_{2/3}(\eta)$ are the modified Bessel functions of the argument $\eta \equiv \left(\omega a/3\gamma^3\beta c\right)\left(1+\gamma^2\theta^2\right)^{3/2}$.

The emission of the radiation occurs in a cone with angle $\theta \approx 1/\gamma$ as reported in Fig. 21.1. An observer can record a radiation pulse whose duration is linked both to the transit time of the arc of circumference (from point 1 to point 2 in Fig. 21.1) and to the angle aperture of the aforementioned emission cone. Since we are dealing with relativistic particles, the Lorentz contraction will give a very short pulse duration $\delta t \approx 1/\left(\gamma^3\omega_B\right)$, being ω_B the Cyclotron frequency. In the classical context can be introduced a Critical Frequency $\omega_C = 3\gamma^3\beta c/(2a)$ that is proportional to the cube of the energy of the particles and inversely proportional to the osculating radius of the orbit. In so doing, the argument η became $\eta = (\omega/2\omega_C)\left(1+\gamma^2\theta^2\right)^{3/2}$ showing, for Eq. (21.2) a behavior

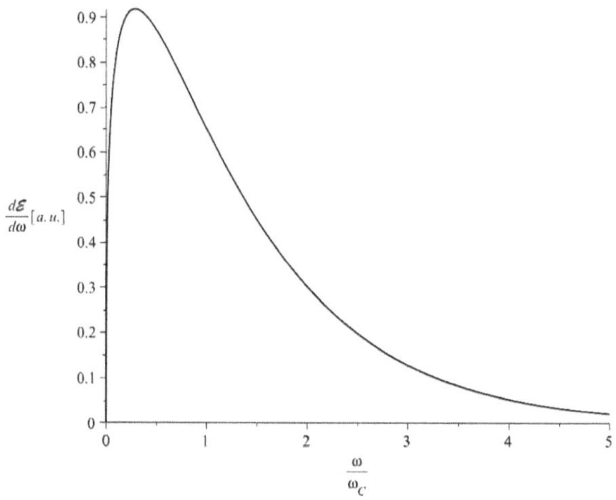

Figure 21.2 Synchrotron radiation spectrum [14].

related to the ratio between the radiation frequency and the critical frequency as illustrated in Fig. 21.2, that reports the plot of the spectrum of the total energy emitted.

Due to the properties of the integral in equation

$$\frac{d\mathcal{E}}{d\omega} = 2\pi \int_{-\pi/2}^{\pi/2} \frac{d\mathcal{E}}{d\omega d\Omega} \cos\theta d\theta \approx \sqrt{3}\frac{q^2}{c}\gamma\frac{\omega}{\omega_C}\int_{\omega/\omega_C}^{\infty} K_{5/3}(x)\,dx,$$

the behavior of the spectrum in the limit $\omega << \omega_C$ reduces to $\omega^{1/3}$. This gives a very broad flat spectrum at frequencies below ω_C. For the high-frequency limit $\omega >> \omega_C$, we have a trend given by $(\omega/\omega_C)^{1/2}\exp(-\omega/\omega_C)$ corresponding to an exponential damping [14].

What we have seen up to now is that the synchrotron radiation spectrum, associated to a charged particle in circular motion, is very broad. Many applications require bright sources and narrower spectra. To satisfy these requests, a specific device, called magnetic undulator [15], was realized as an insertion device in storage ring accelerators for synchrotron radiation users (see Fig. 21.3a).

The magnetic field associated with an undulator is such that a charged particle undergoes an oscillating trajectory in the plane orthogonal to the magnetic field. This fact can be interpreted as an oscillating dipole in motion which will then emit radiation like

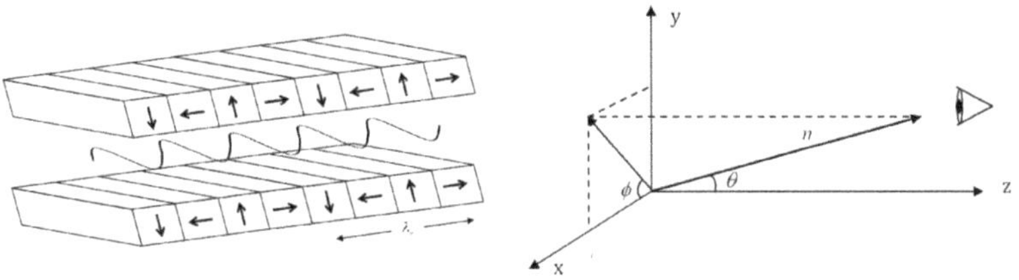

Figure 21.3 (a) Schematic view of a planar magnetic undulator and the associated sinusoidal charged particle trajectory (b) Radiation directions and angles in the laboratory reference frame [14].

any antenna. It is easy to understand that if the particle angular excursion, along the undulating trajectory, is smaller than the emission cone width (as shown in Fig. 21.1), the observer will see an emission duration longer with respect to that previously described. As a result, a much narrower spectrum is expected from this longer pulse.

The complete power distribution, in terms of frequency and angular directions, can be calculated by means of the evaluation of the electron trajectories in the undulator magnetic field that are expressed as:

$$\begin{cases} x(t) = \dfrac{K}{\beta\gamma k_u}\sin(\omega_u t) \\ z(t) = \bar{v}_z t - \dfrac{K^2}{8\gamma^2 k_u}\sin(2\omega_u t) \end{cases}, \tag{21.3}$$

where

$$\omega_u = c\bar{\beta}k_u;\ K = \frac{qB_0}{mc^2 k_u};\ k_u = \frac{2\pi}{\lambda_u}$$

By inserting Eq. (21.3) into (21.1) and omitting the long algebraic calculations, we finally obtain [14]:

$$\frac{d\mathcal{E}}{d\omega d\Omega} = \frac{8(q\gamma N)^2}{c}\sum_{m=1}^{\infty}\left(\frac{m\xi}{K}\frac{\sin{}^{\nu_m}/_2}{{}^{\nu_m}/_2}\right)^2 \times\left[\frac{1}{2}\left(\frac{\gamma\theta}{K}\right)^2 A_{0,m}^2 + \sqrt{2}\left(\frac{\gamma\theta}{K}\right)\cos\phi A_{0,m}A_{1,m} + A_{1,m}^2\right], \tag{21.4}$$

where

$$A_{\alpha,m} = (-1)^{m+\alpha} \sum_{m'=-\infty}^{\infty} (-1)^{m'} J_{m'}(m\xi)$$
$$\times \left[J_{m-2m'-\alpha}\left(m\sqrt{2}\chi\cos\phi\right) + J_{m-2m'+\alpha}\left(m\sqrt{2}\chi\cos\phi\right)\right]$$
$$\alpha = 0, 1 \quad ; \quad \xi = \frac{1}{2}\frac{K^2}{1+K^2+\gamma^2\theta^2} \quad ; \quad \chi = \frac{2K\gamma\theta}{1+K^2+\gamma^2\theta^2};$$
$$\nu_m = 2\pi N\left(\frac{\nu - m\nu_0}{\nu_0}\right).$$

N is the number of periods of the undulator, ν_0 is the central frequency emitted (we will shortly express it) and m refers to the harmonic number. The geometry explaining the angles involved in Eq. (21.4) is reported in Fig. 21.3b). Equation (21.4) is not easy to read except for a couple of facts. The first is the presence of sinc-function concerning the frequency behavior; this is strictly related to the lengthening of the pulse duration and the consequent restriction of the spectrum (see Fig. 21.4). The specific sinc-like behavior is related to the Fourier transformation, over a finite undulator length $L = N\lambda_u$, of the oscillating field. The second fact is related to the presence of the harmonics m. All the other characteristics are very difficult to extract at first glance [14].

The central emission frequency can be determined simply by imposing the so-called synchronism condition. In order for the electron to emit radiation in phase, during each oscillation, it is sufficient to request that after an undulator period, the field oscillates with a phase difference equal to 2π.

$$(\omega t - kz) = 2\pi \quad \text{at} \quad \begin{cases} z = \lambda_u \\ t = \dfrac{\lambda_u}{c\beta_z} \end{cases} \Rightarrow \quad \frac{\omega}{c}\frac{1}{\beta_z} - k - k_u = 0, \quad (21.5)$$

where $k_u = 2\pi/\lambda_u$. The expression reported in Eq. (21.5) is referred to as the beam line equation and is completely general and can be applied to very different environments as long as to different dispersion relations that link the frequency ω to the moment k. In the most emblematic case of a vacuum space, frequency and moment are equal, except for the velocity $c(\omega/c = k)$, and the beam line takes the following form: $\omega/c = \beta_z(1+\beta_z)\gamma_z^2 k_u$, where γ_z is linked to β_z by the known relativistic relationship. With a little algebra,

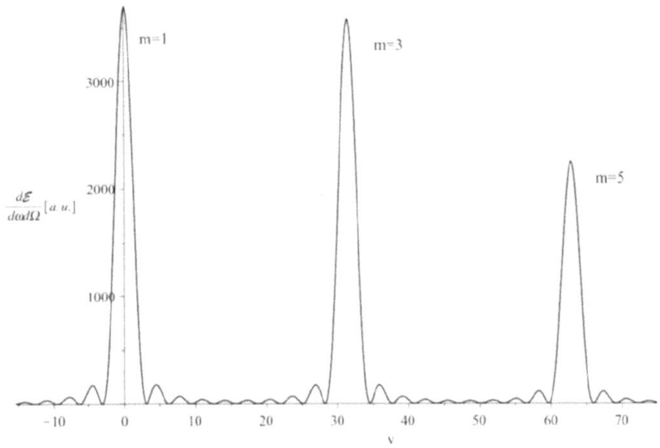

Figure 21.4 On-axis ($\theta = 0$) electromagnetic spectrum emitted by an electron passing through a magnetic undulator with $N = 5, \phi = 1$ and $K = 1$ [14].

we also obtain the relation between the total and longitudinal relativistic factor γ_z and the Lorentz factor: $\gamma_z^2 = \gamma^2/(1 + K^2)$ which, as easily understood, scales through the undulator parameter K. Combining all the previous relations and specializing them for relativistic electrons ($\beta \sim 1$) we get finally the most relevant and known cinematic equation for FEL:

$$\lambda = \frac{\lambda_u}{2\gamma^2}(1 + K^2) \tag{21.6}$$

This equation is very easy to read and expresses all the peculiarities of the FEL sources: the wavelength of the emitted radiation is proportional to the undulator period λ_u and inversely proportional to the square of the electron energy γ^2; this first term can be considered as related to the relativistic double Doppler shift. Finally the emitted wavelength is weakly proportional to the magnetic field B_0 expressed by means of the undulator parameter K (see Eq. (21.3)). This equation is extremely explanatory regarding the peculiarities of a FEL. The emitted radiation can be continuously tuned in frequency by varying parameters such as the energy of the electrons, the magnetic field of the undulator and, in principle, the undulator period itself. No conventional laser can do this.

FEL devices are, in any case, real laser sources because they obey to dynamical processes associated to stimulated emission that is ruled by a gain mechanism. Such a mechanism establishes a relative energy increase of the electromagnetic radiation field, during one passage inside the magnetic undulator, with respect to the pre-existing field. The gain parameter can be defined as:

$$G \equiv \frac{\Delta W_L}{W_L^0} = -\frac{m_0 c^2 \langle \Delta \gamma \rangle}{W_L^0},$$

where W_L^0 represents such an existing field [4].

To be more explicit, with an electromagnetic wave-field stored inside the undulator volume, where also the static magnetic field is present, it is possible to experience an increase of the wave amplitude if a charge particle interact, having an appropriate phase, with the electric field associated to the wave and, at the same time, with the magnetic field of the undulator. This condition is achievable with several configurations, but the first, and most common one, is the one using an optical cavity, as for conventional lasers, capable to store the spontaneously emitted radiation. The exchange of energy, therefore, takes place between the current vector and the electric field vector, as indicated in the following Eq. (21.7):

$$\Delta W_L = -\int_0^{T_L} \int_V \vec{J}(\vec{x}, t) \cdot \vec{E}(\vec{x}, t) d\vec{x} \tag{21.7}$$

The above energy exchange, which allows us to calculate the gain of the process, occurs along the transverse x-direction (see Fig. 21.3a,b) that is the electron motion component induced by the undulator magnetic field. The radiation electric field can be expressed, in a general way, as:

$$E_L = E_0 \cos(\omega t - kz + \varphi_L);$$

as can be seen, the electric field contains the phase term of the propagation of a wave $(\omega t - kz)$ plus a generic phase term φ_L. The electron beam current is slightly more complicated to calculate because it is necessary to evaluate the electrons' trajectories inside the magnetic undulator that, for particles out of the symmetry axis, is not an easy task [14, 15]. Generally speaking, the transverse electron velocity, induced by the undulator, can be expressed as:

$v_T = -cK\sqrt{2}/\gamma \cdot \sin(2\pi z/\lambda_u)$. By a proper substitution of such expressions into Eq. (21.7) and after some algebra, we can finally end up to:

$$\dot{\gamma} = \frac{eE_0K}{m_0c\gamma\sqrt{2}}\sin(\psi) \quad \text{where} \quad \psi = \omega t - \left(k + \frac{2\pi}{\lambda_u}\right)z + \varphi_L \quad (21.8)$$

Equation (21.8) tells us that the energy exchange is a periodic function of the phase term ψ that regulates the electron-radiation interaction. A complete presentation can be found elsewhere [4, 14], but is important to explain the dynamical behavior of the phase ψ. The proper equation can be found from the analysis of the electron trajectories and the energy conservation that lead us to the expression of the electron longitudinal acceleration:

$$\ddot{z} = c\frac{\dot{\gamma}}{\gamma^3}\frac{(1+K^2)}{\beta_z} \quad (21.9)$$

The acceleration depends on the energy variation (21.8), that can be inserted in the right-hand side of Eq. (21.9); on the other hand, from (21.8) we also have: $\ddot{z} \approx \ddot{\psi}$. Therefore, after some algebra we get:

$$\ddot{\psi} = -\Omega^2 \sin(\psi) \quad (21.10)$$

where

$$\Omega^2 = \left(k + \frac{2\pi}{\lambda_u}\right)\frac{eE_0K(1+K^2)}{m_0\gamma^4\beta_z\sqrt{2}}$$

The result reported in Eq. (21.10) is a Pendulum Equation that contains a frequency Ω that can be associated to the Rabi frequency, because, in it, we find a direct proportionality with the electric field amplitude E_0, associated with the electromagnetic wave, which acts as a forcing term. This problem is therefore similar to that of the Bloch equations usually used for two-level systems problems [16] and for electrons in a lattice problem; in this last case the Mathieu or Hill equation is often used [17], as well as a linearized form of Eq. (21.10). Understandably, Eq. (21.10) cannot completely describe the dynamics of an FEL which must include also the evolution of the electric field and, consequently, the evolution of the electromagnetic intensity if we want to follow the complete dynamical process from the quite-start, up to the saturation.

Due to the direct link expressed by $\ddot{z} \approx \ddot{\psi}$, we get, from the first relation of Eq. (21.8) that $\ddot{\psi} \approx \dot{\gamma}$ and, therefore, $\Delta\gamma(t) \approx \Delta\dot{\psi}(t)$

that reveals the deep linkage between the energy exchange, between particles and field, and the phase derivative variation. From the above gain definition, therefore, we finally obtain $G \approx \Delta\gamma \approx \Delta\dot{\psi}$, that, with the complete calculation writes:

$$G = \frac{\gamma^3 \beta_z \left(m_0 c^2\right)}{c\left(k + \frac{2\pi}{\lambda_u}\right)\left(1 + K^2\right) W_L^0} \left\langle \Delta\dot{\psi} \right\rangle \tag{21.11}$$

The <...> operator in Eq. (21.11) indicates an average over the particle phases in order to express the gain coefficient of a real charged beam. Equation (21.11) tells us that, to evaluate the gain process in an FEL, we need to express such an average; if we restrict our analysis to the "small signal" regime we just need to solve the pendulum Eq. (21.10), neglecting the field evolution. Moreover we see that Eq. (21.10) cannot be solved analytically and requires a perturbative approach by virtue of the "small signal" hypothesis; this is allowed in the so-called "small gain" regime. The perturbative analysis involves the Rabi frequency $\Omega^2 \equiv \varepsilon$ as expansion parameter allowing us to write: $\psi = \psi_0 + \varepsilon\psi_1 + \varepsilon^2\psi_2 + \ldots$. With the proper substitutions and after some algebra [4, 14] we get, at the end, for the gain coefficient:

$$\left\langle \Delta\dot{\psi} \right\rangle = \left(\frac{\Omega L}{c}\right)^4 \frac{c}{4L\beta_z^3} \underbrace{\frac{d}{d\nu}\left[\frac{\sin(\nu/2)}{\nu/2}\right]^2}_{g(\nu)}$$

$$\Rightarrow \qquad G = \pi \frac{\left(1 + K^2\right) K^2 \lambda_u^3 N^3}{\beta_z^5 \gamma^5} \frac{I}{\Sigma_L I_0} \left(k + \frac{2\pi}{\lambda_u}\right) g(\nu), \tag{21.12}$$

where the parameter ν, expresses the deviation from the central frequency of spontaneous emission, as described in Eq. (21.4). We observe that the term $g(\nu)$ shows that the frequency gain curve is none other than the derivative, in frequency, of the spontaneous emission curve, which, as we have seen, is a "sinc" function (see Eq. (21.4)). The spontaneous emission and gain spectra are illustrated in Fig. 21.5. This fact is extremely important and not found in conventional lasers; it highlights that the process can be both of emission, if the gain is positive, and of absorption, which results in an acceleration of the electrons if the gain is negative. The importance of this phenomenon lies in the fact that a FEL can be designed to emit or absorb electrons without the need for a third level

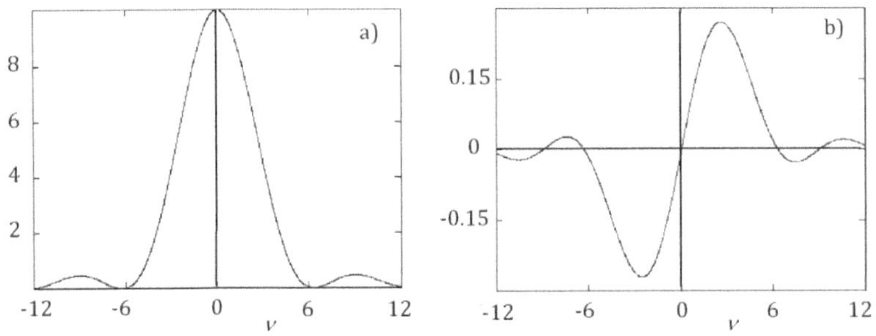

Figure 21.5 Spectral behavior of the spontaneous emission and gain. (a) Spontaneous emission vs. ν behaves like a "sinc" function (see Eq. (21.4)). (b) Gain coefficient vs. ν performs like the spontaneous emission derivative (see Eq. (21.12)) [14].

which in conventional laser systems is necessary precisely to avoid the continuous reabsorption and emission which are equiprobable processes. The FEL is a two-level laser system.

21.2.1.1 Considerations about efficiency

The FEL gain described in (21.12) depends on many parameters. From the analysis of this expression it is easy to notice that it is much more difficult to build a short wavelength FEL than a long wavelength FEL. The shorter the wavelength, the higher must be the electron energy, which, apart from technological difficulties, causes a strong decrease of the gain, that goes as $1/\gamma^3$. The maximum value of K is limited by the maximum achievable B values in the undulator. Gain is also proportional to the electron current and to N^3, it is possible to increase the gain using a longer undulator. As a counterpart, the efficiency of the system decreases when increasing N. If we define the efficiency η as the ratio between the energy extracted from the beam, converted into radiation, and the total energy beam, we can write: $\eta = \frac{\Delta E}{E}$

If in the FEL interaction the electron transfers energy to the radiation beam, it's energy will decrease. This is represented by the blue dot moving toward the left in Fig. 21.6, where the small signal gain is reported as a function of the so-called detuning parameter $\theta = 2\pi N\frac{\omega-\omega_0}{\omega_0}$.

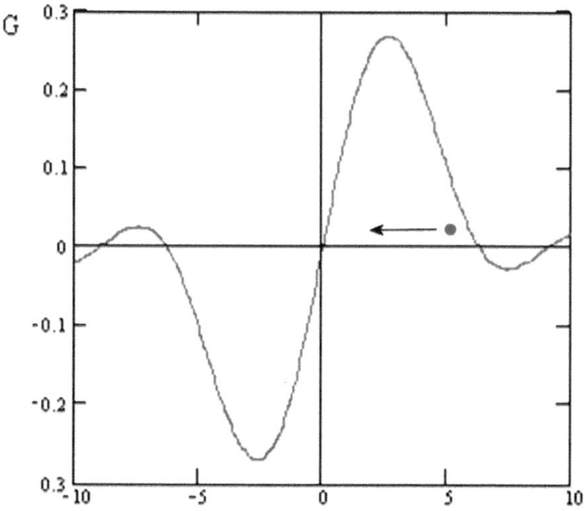

Figure 21.6 Small signal Gain as a function of the detuning parameter θ; the blue dot represents an electron losing energy due to the interaction [14].

If the electron loses enough energy it finds itself in the region where the gain is negative, and the FEL emission stops. This happens when $\Delta\theta > 2\pi$. Thus the maximum allowed value of $\Delta\theta$ for the FEL emission to occur is: $\Delta\theta \leq 2\pi$. This corresponds to a maximum value for the energy variation of the electron ΔE. In the FEL the emission frequency is proportional to the square of the electron energy, so that $\omega_0 \propto E^2 \rightarrow \Delta\omega_0 \propto 2E\Delta E$. Reminding the expression of the detuning parameter we have thus a limit for the efficiency, the so-called Renieri limit [18]:

$$\eta = \frac{\Delta E}{E} = \frac{1}{2}\frac{\Delta\omega_0}{\omega_0} = \frac{\Delta\theta}{4\pi N} \leq \frac{1}{2N}$$

This means that increasing the undulator length does affect gain, but the dynamic of the process limits the efficiency at the same time, so the number of periods of the undulator must be carefully chosen as a right compromise between gain and efficiency.

21.2.2 Waveguide Operation and Dispersion Relations

Following the analysis of the emission process reported in [19], in a first approximation, the volume integral (21.7) representing the energy exchange between an electron beam and the radiation field, can be factorized in two terms, a form factor F given by the surface integral of the normalized electron density distribution over the cross section of the mode, and a coupling integral over the interaction length L, which describes the beating between the electron motion and the wave:

$$C = \int_0^L \left\{ \exp\left[i \left(\frac{\omega}{c\beta_z} - k_e - k \right) z \right] \right\} dz = L e^{i\theta/2} \frac{\sin(\theta/2)}{\theta/2}, \tag{21.13}$$

where ω and k are the frequency and wave vector of the wave, respectively, and k_e is the wave vector of the electron oscillations. $\theta = \left(\frac{\omega}{c\beta_z} - k_e - k \right) L$ is the so-called phase shift parameter, which is a function of frequency and plays a crucial role in all free-electron devices.

In free space, if there is no transverse motion of the electrons, $k_e = 0$ and the velocity of the wave $\omega/\kappa = c$ will always be greater of the electron drift velocity $c\beta_z$. The integrand in (21.13) will be rapidly oscillating giving rise to a negligible amount of coupling. In a guided structure, synchronism can be achieved in the absence of transverse motion if the phase velocity of the wave equals the electron velocity. However, synchronism can also be achieved in free space if the transverse motion provides a wave vector $k_e > 0$.

Synchronism and therefore the resonant frequency of the emission process are defined by the condition $\theta(\omega) = 0$, once the dispersion relation of the electromagnetic waves in the medium is known. The radiated power is proportional to the square module of (21.13) and, its dependence on frequency will show a narrow linewidth with a line-shape function $\frac{sin^2(\theta/2)}{(\theta/2)^2}$.

Some insight into the various emission mechanisms of free-electron devices can be gained by a graphical representation of the synchronism condition. In the plane $(k, \omega/c)$ the condition $\theta = 0$ is represented by a straight line called "beam line"; the resonance of the emission process is given by the intersection of this line with the dispersion relation $\omega = \omega(k)$ of e.m. waves in the medium in which the interaction with electrons occurs. To get a deeper physical insight into this process, we can discuss the FEL resonance condition in a rectangular waveguide. In this case, the dispersion relations of

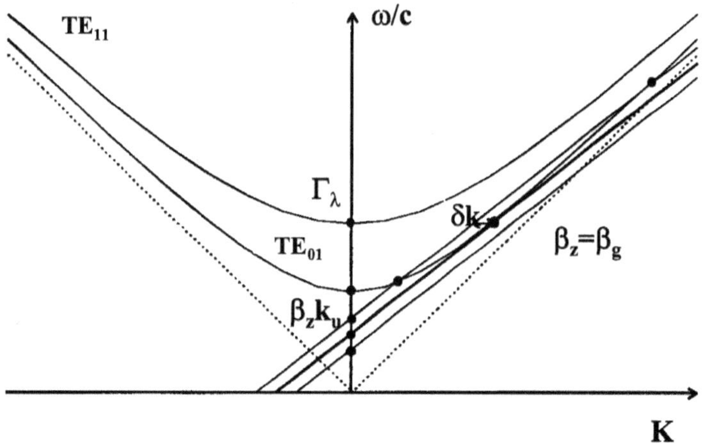

Figure 21.7 Graphical representation of the waveguide dispersion relation and the FEL resonance condition (see text for details).

the $TE_{0,1}$ and $TE_{1,1}$ modes in a rectangular waveguide are plotted in Fig. 21.7 together with the beam line and the light line of unity slope. The dispersion relation intercepts the ω/c axis at the cut-off frequency $\omega/c_{co} = \Gamma_{0,n}$. Due to the transverse oscillations of the electrons $k_e = k_u$, the beam line also intercepts the ω/c axis at βk_u. Depending on the value of the parameter k_u and of $\Gamma_{0,n}$, one can distinguish between the following three cases [20]:

1. $k_u < \Gamma_{0,n}/\beta_z$ - There are two distinct solutions corresponding to the two resonant frequencies

$$\frac{\omega_\pm}{c} = \beta_z \gamma_z^2 k_u \left\{1 \pm \beta_z \Delta\right\}$$

$$\text{with} \quad \Delta = \left[1 - \left(\frac{\Gamma_{0,n}}{\beta_z \gamma_z k_u}\right)^2\right]^{1/2}, 0 \leq \Delta \leq 1 \quad (21.14)$$

The parameter Δ characterizes the downshift of the resonant frequencies with respect to the free space resonance $\frac{\omega_v}{c} = (1 + \beta_z)\, \gamma_z^2 k_u$

2. $k_u = \Gamma_{0,n}/\beta_z$ - There is a degenerate solution ($\Delta = 0$) occurring when the beam line is tangent to the dispersion relation curve. This means that the group velocity of a propagating wave is equal to the electron drift

velocity:

$$\beta_g \equiv \frac{\partial \omega / c}{\partial k} = \beta_z \qquad (21.15)$$

This is the so-called "zero slippage" condition, which allows a wave packet to travel with optimum overlap with an electron bunch through the whole undulator length. In this case, the resonant frequency for relativistic electrons approaches one-half the free space value.

3. $k_u < \Gamma_{0,n}/\beta_z\gamma_z$ - There are no real solutions of the resonance conditions. Electrons cannot radiate energy. This can be understood considering that the e.m. wave associated with the undulator is below the cut-off in the electron rest frame and cannot propagate through the waveguide.

21.3 The Source Survey

Historically, the first free-electron laser to provide THz radiation to users for a variety of applications has been the UCSB-FEL [5], which is driven by an electrostatic accelerator and operates in a quasi-CW mode. It currently provides tunable terahertz radiation in the region from 120 GHz to 4.8 THz (2.5 mm–60 μm) with an output power in the range from 500 W to 5 kW and a pulse duration of 1–20 μs at 1 Hz repetition rate.

Since then, several FEL facilities operated and some of them are still operating in the THz spectral region. Besides the UCSB-FEL, we recall here the Stanford FEL (15–80 μm) [21]; The Vanderbilt FEL MK-III, Nashville (TN) operating in the range 2.1–9.8 μm [22]; The ENEA Compact-FEL [23] and the ENEA FEL-CATS experiment [24], operating in the sub-mm spectral region; the Jefferson Lab. facility—Newport News (VA), which reached record power THz radiation [25]; the compact THz FEL at KAERI, which uses a hybrid electromagnetic undulator [26]; the FELs at ILE-Osaka [27] and Kyoto University [28] in Japan, both operating at about 40 μm wavelength; the Beijing FEL [29], FELiChem at Hefei [30], the CAEP THz FEL facility [31] in China; the NovoFEL at Novosibirsk, Russia, based on a CW energy recirculated LINAC (ERL) and providing about 100 W in the spectral range 120–180 μm [32]; the Israeli EAFEL (100 GHz), based on a 6 MeV EN-Tandem Van der Graaf accelerator [33].

A complete list and related table of FELs that operated or are operating worldwide is reported in (http://sbfel3.ucsb.edu/www/vl_fel.html).

A picture of the current FEL development in Europe is given by the initiative FELS of Europe (https://www.fels-of-europe.eu/), a collaboration FEL facilities in Europe, with the goal to provide unique research infrastructures that enable exploiting the full scientific potential of these accelerator based short-pulse light sources. The collaboration includes 14 facilities in 10 countries; a few of them are briefly recalled in the following:

- CLIO at LURE-Orsay, a pulsed free-electron laser continuously tunable between 3 and 120 μm. It has a double pulsed structure: "macropulses" 10 μs long (at 1 to 25 Hz) contain "micro-pulses" of length adjustable from 0.3 to 5 ps (with Fourier transform spectral width).
  The micropulse interval is 16 ns. CLIO can lase simultaneously at two different, independently tunable frequencies, allowing 2-color pump-probe experiments [9, 34] (https://www.fels-of-europe.eu/fel_facilities/ir_facilities/clio/);
- the FELIX (Free Electron Lasers for Infrared eXperiments) facility in Nijmegen [10, 35], an international user facility providing the scientific community with tunable radiation of high brightness in the mid- and far-infrared as well as the THz regime. The facility houses two independent accelerators that together drive four FELs: FELIX-1, FELIX-2, FELICE, and FLARE. The total spectral range covered extends from 3 to 1500 μm. (https://www.fels-of-europe.eu/fel_facilities/ir_facilities/felix/);
- FELBE at Dresden, Germany, operated by the Helmholtz-Zentrum Dresden-Rossendorf (HZDR), provides picosecond infrared pulses. Two free-electron lasers [36] cover the mid- and far-infrared spectral range from 4–250 μm. The electron bunches for the FELBE are provided by the superconducting

radio-frequency linear accelerator ELBE. FELBE can operate in three different modes: continuous pulsing with a repetition rate of 13 MHz, pulsing with 1 kHz by applying a pulse picker, and macrobunch operation with bunch length > 100 μs and macrobunch repetition rates ≤ 25 Hz. Pulse duration and pulse energy vary with wavelength and lie in the range from 1–25 ps and 100 nJ to a few μJ, respectively (https://www.fels-of-europe.eu/fel_facilities/ir_facilities/felbe/);

- TeraFERMI at Trieste, Italy, the THz beam line [37] of the FERMI FEL facility. TeraFERMI collects the THz radiation (0.3–10 THz) naturally emitted by the spent FEL electron beam, as it passes through a Coherent Transition radiator in the beam dump section, and guides it into a dedicated THz Laboratory.

21.3.1 Undulator Based FELs

One of the critical issues of a long wavelength FEL, when driven by an RF accelerator, is the effect of "slippage," i.e., the lack of overlapping between electron bunches and wave packets as they travel through the undulator with different velocities. In free space the slippage length over the undulator length is $\delta = N\lambda$; at millimeter wavelengths, it can be quite large causing a gain reduction and an increase of the threshold for oscillation. In a waveguide, the slippage can be strongly reduced by controlling the group velocity of the excited mode, i.e., by varying the waveguide boundary conditions.

At the "zero-slippage" condition stated in (21.15), which can be expressed as $\lambda_u \approx 2\frac{\gamma}{\sqrt{1+K^2}}b$ for the fundamental $TE_{0,1}$ mode in a planar or rectangular waveguide, additional features appear which make the waveguide operation particularly attractive:

- The group velocity of wave packets gets close to the e-beam velocity and allows operation with short electron bunches.
- The merging of the two FEL resonances results in a broadband gain curve with a half-width of the order of $1/\sqrt{N}$ (instead of $1/N$) [20].
- The intersection of the $TE_{0,1}$ mode at grazing incidence results in the operation on a single transverse mode even in an overmoded waveguide.

21.3.1.1 The ENEA THz Compact-FEL

The Compact-FEL built at ENEA-Frascati employs a permanent magnet undulator with eight periods of 2.5 cm and shows all the above features [23, 38]. Zero slippage is reached at a wavelength of 2.6 mm (120 GHz) using 2.3 MeV electron energy and a waveguide gap of 4.32 mm. The resonator is composed of a 30 cm long WR42 rectangular waveguide placed within the undulator gap and terminated at both ends by wire grid mirrors used as electron transparent mirrors (ETM). The upstream mirror has 80 μm wire spacing with a reflectivity > 99% at wavelengths longer than 2 mm. It can be translated along the resonator axis to allow tuning of the resonator length. The downstream mirror is used as an output coupler and has therefore a larger wire spacing. The maximum output power of about 1.5 kW in 4 μs pulses was obtained with a 14% output coupler.

The wide gain curve obtained at zero slippage is shown in Fig. 21.8. It can be exploited to tune the emission wavelength. Indeed the round-trip time of a wave packet in the resonator has to be tightly synchronized to the time distance between electron bunches entering the undulator. Since the group velocity in a waveguide is a

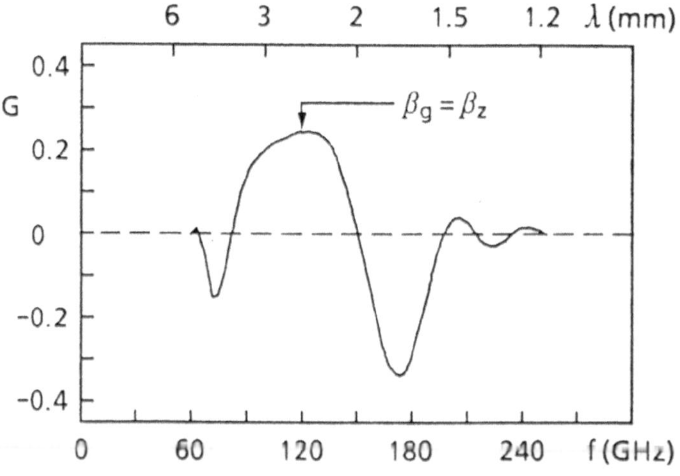

Figure 21.8 Calculated gain curve of the Compact-FEL as a function of operation frequency [23].

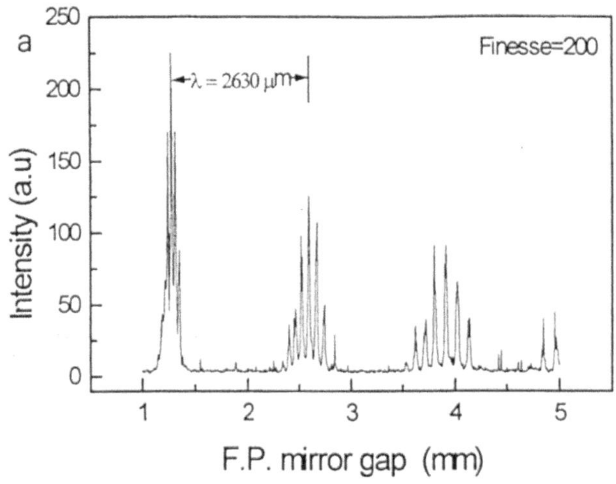

Figure 21.9 Fabry–Perot interferogram of the Compact-FEL lasing emission [23].

function of frequency, this allows tuning the frequency by varying the resonator length [11].

A measurement of the spectrum of the laser output at maximum power is shown in Fig. 21.9, where a high finesse ($F = 200$) Fabry–Perot (FP) interferogram is reported. The line structure is extremely clear with a measured relative bandwidth of the FEL emission of about 7%.

21.3.1.2 Coherent spontaneous emission and energy-phase correlation

Let us now consider the emission of radiation from a single rectangular electron bunch of duration (length Δx) and charge Q. The total radiated power will be proportional to the square of the sum of the fields, with their relative phase, generated by individual electrons in the bunch.

If the bunch length is much longer than the wavelength of the radiation (typically: $\tau = 10$ ps implies 100 optical cycles at $\lambda = 30$ μm) the average over the phase distribution will in general approach zero, apart from a contribution from a small fraction of the total charge, $\Delta Q = fQ$, which is assumed to be distributed over a length scale of half a wavelength or less. The total radiated power will then be:

$$P \propto N + (fN)^2, \tag{21.16}$$

where N is the number of electrons in the bunch.

The first term in (21.16) represents the incoherent spontaneous emission, while the second term represents the coherent one. The coherent emission will dominate if

$$f^2 N \gg 1 \tag{21.17}$$

For a typical bunch of duration $\tau = 10$ ps and peak current $I_p = 1.6$ A, the total charge is $Q = 1.6 \times 10^{-11} C$ and the number of electrons is $N = 10^8$. The condition (21.17) will then require $f \gg 10^{-4}$.

Such a small "excess" charge will be distributed over a fraction of the bunch length according to: $f\Delta x \leq \lambda/2$, where the relational operator $<$ holds true for a bunch shape with smoother edges than a rectangular one. This implies that coherent radiation will in general overcome incoherent radiation already at wavelengths in the medium infrared.

If we now consider a beam composed of an infinite train of electron bunches spaced at the RF period T_{RF}, utilizing the physical model described in [19], it is possible to calculate the field amplitude of the total coherent emission. Since the electron current density is periodic in time, it can be expanded in series of harmonics of the fundamental $\omega_{RF} = 2\pi/T_{RF}$. As a consequence, radiation will be also emitted only at discrete frequencies which are harmonics of ω_{RF} and fall within the FEL resonance curve. The total radiated power is then calculated as the time average of the flux of the Poynting Vector:

$$P_{l,0,n} = \frac{\beta_{gl}}{2Z_0} |A_{l,0,n}|^2, \tag{21.18}$$

where

$$A_{l,0,n} = -\frac{Z_0}{\beta_{gl}} I_p \frac{C_l}{2} \frac{KL}{\sqrt{ab}} F \frac{1}{\beta\gamma} \frac{sin(\theta/2)}{\theta/2} i e^{i\frac{\theta}{2}},$$

where $Z_0 = 377\ \Omega$ is the free space impedance, β_{gl} is the normalized group velocity of the waveguide mode at the frequency $\omega_l = 2\pi l/T_{RF}$ with wave vector $k_{0,n}$, C_l is the Fourier coefficient of the electron current density at the l-th harmonic, F is the form factor describing the overlapping between the e-beam transverse distribution and the waveguide mode, and a and b are the transverse waveguide dimensions. The factor $e^{i\theta/2}$ in (21.18) shows the dependence of the phase of the radiated field on the electron drift velocity, which is implicit in θ.

Expression (21.18) for the coherent radiated power from a modulated electron beam predicts a quadratic dependence on the electron current. This has been verified on the two sources built at ENEA-Frascati, the Compact-FEL and the FEL-CATS, which are described in greater detail in this section. Spectral measurements of the coherent spontaneous emission from the Compact-FEL [38] show the striking feature of emission at discrete frequencies that are harmonics of the fundamental RF. A Fabry–Perot interferogram of the coherent spontaneous emission, taken with a resolution of 1 GHz, is shown in Fig. 21.10. Two bands, typical of the FEL emission in a waveguide close to the zero-slippage condition, can be observed at the first interferometric order for values of the mirror gap between 1 and 2 mm. A line structure clearly appears within these bands and is well resolved at the second interferometric order,

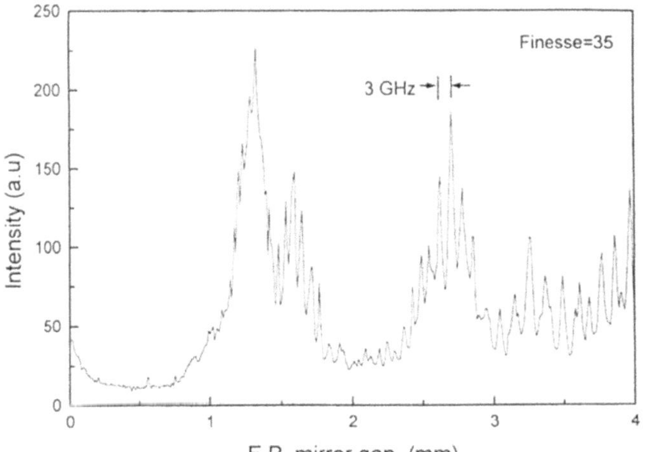

Figure 21.10 Fabry–Perot interferogram of the Compact-FEL coherent spontaneous emission. Source [23].

showing a separation of 3 GHz between adjacent lines equal to the fundamental frequency driving the accelerator.

The width of the individual lines will be related to the number of correlated bunches. In practice, if correlation between bunches is assumed to occur over the whole macropulse duration T_M, typically 1–10 μs, a relative linewidth $\Delta\nu/\nu \sim T_{RF}/T_M \sim 3 \times 10^{-4} - 3 \times 10^{-5}$ is expected. On the other hand, current amplitude fluctuations during the macropulse over a time interval of about 100 ns would increase the linewidth to: $\Delta\nu/\nu \sim 3 \times 10^{-3}$.

Another interesting feature is the possibility of enhancing the coherent spontaneous emission from an RF-modulated electron beam by a proper manipulation of the electron distribution in the longitudinal phase space.

Indeed the phase factor present in the expression of the radiated field (21.18) shows that the electron current and the radiated field are no longer in phase when the electron drift velocity does not match the resonance condition $\theta = 0$. If each electron bunch is treated as a collection of particles each with its energy γ and position or phase ψ along z, the total radiated power is maximum when the single electron contributions in the expansion coefficient $A_{l,0,n}$ interfere constructively with each other. This happens when the electrons are distributed in the longitudinal phase space (ψ, γ) as close as possible to the "phase-matching" curve [39]:

$$\psi = -\pi \frac{L}{cT_{RF}} \left(\frac{1}{\beta_z(\gamma)} - \frac{1}{\beta_{z0}} \right) \tag{21.19}$$

Calculations show that an increase of about one order of magnitude in the radiated power is expected for a proper correlation between the energy and phase of the electrons in the bunch.

The method of enhancing the coherent emission by energy-phase correlation has been successfully tested on the compact advanced terahertz source (FEL-CATS) [24] at the ENEA laboratories in Frascati. The layout of the experimental setup occupies a space of about 0.5 m × 1 m × 2 m, comparable to that of a standard optical table.

The electron beam source is a 2.998 GHz RF linear accelerator (LINAC) capable of generating an electron current of 250 mA in 5 to 10 μs macropulses at a kinetic energy of about 2.5 MeV. The electron macropulse is composed of a train of 15 ps bunches spaced at the RF period of 330 ps. The electron beam is generated by a pulsed triode gun, equipped with a 7.7 mm diameter osmium-treated dispenser thermionic cathode and is accelerated to the 13 kV anode potential before entering the LINAC structure through a magnetic lens assembly. The electrons, after having been accelerated in the LINAC, enter into a second RF section, called phase-matching device (PMD), where the correlation in the longitudinal phase space takes place. The RF system has been described in detail in [40]. The distribution of the bunched electrons in the phase space at the PMD output can be modified by varying the phase and the amplitude of the RF field driving the PMD with respect to the LINAC. Two sets of steering coils and a triplet of quadrupoles transport the "energy-phase correlated" electron bunch to the undulator entrance. The undulator is a 40 cm long permanent magnet linear device. It is realized with *NdFeB* magnets in the Halbach configuration and it consists of 16 periods of 2.5 cm each. The gap is variable and can be remotely controlled to vary the undulator parameter K between 0.5 and 1.4.

THz radiation is generated in a rectangular waveguide with cross-sectional dimensions $a \times b = 24.67 \times 6.32\ \text{mm}^2$ placed inside the undulator. Immediately after the undulator, a copper horn and a 45° copper-mesh reflector are used to extract the THz radiation. The spent electron beam passes through the mesh reflector and is sent into a beam dump. At the end of the light-pipe, the radiation is analyzed by means of an FP interferometer equipped with mesh reflectors and a pyroelectric detector (Molecron P4-35). To analyze the coherence of the emitted radiation the signal of the pyroelectric detector was recorded as a function of the e-beam current measured at the entrance of the undulator and at the beam dump. The result is shown in Fig. 21.11 in a double logarithmic scale; the output power clearly increases as the square of the e-beam current, confirming the occurrence of coherent spontaneous emission.

A maximum emitted power of about 1.5 kW in a 5 μs pulse duration was measured at the peak of the phase-tuning curve when the RF field in the *PMD* (E_{PMD}) was set to about 0.5 the field in the *LINAC* (E_{linac}). This power level was obtained after a single pass of the electron through the undulator without any optical cavity. The

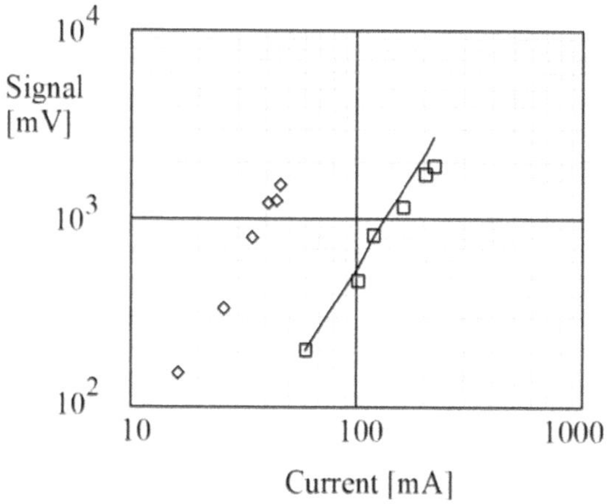

Figure 21.11 P4-35 signal as a function of the electron current collected by the upstream (boxes) and downstream (diamonds) targets, respectively. The straight line shows the expected quadratic dependence [24].

central wavelength of the emission in these operating conditions is 760 μm (0.4 THz).

Fabry–Perot interferograms of the output radiation showed a spectrum with a relative bandwidth of about 10%. The finesse of the instrument was calculated to be $F = 22$ at $\lambda = 700$ μm and was enough to resolve the structure within the output bandwidth due to the emission at discrete frequencies, which are integer harmonics of the 3 GHz RF [24]. Easy and reproducible wide-band tunability of FEL-CATS was achieved by varying the phase in the PMD demonstrating operation between 600 and 800 μm (0.4–0.5 THz) as it is shown in Fig. 21.12. As the phase is varied, the requirement on the energy-phase correlation is gradually released and the mean kinetic energy of the e-beam is either increased or decreased. This results in the emission at a different frequency and, in general, at a lower power level.

21.3.2 Cerenkov, Smith–Purcell and Other Devices

As it has been explained in sect. 2, in free space, if there is no transverse motion of the electrons, the coupling integral (21.13) will be rapidly oscillating giving rise to a negligible amount of

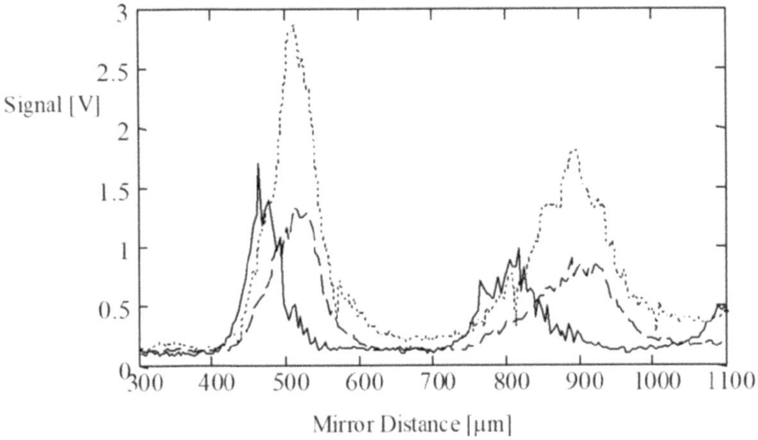

Figure 21.12 FP interferograms of the emission at different values of the PMD phase $\Delta\theta$: dotted line $\Delta\theta = 0$ (zero-crossing); dashed line $\Delta\theta = +9°$; solid line $\Delta\theta = -9°$ [24].

energy exchange. In a guided structure, however, synchronism can be achieved in the absence of transverse motion if the phase velocity of the wave equals the electron velocity. This is the case of the Cerenkov FELs (C-FEL) [41] and orotrons, or grating FELs (G-FEL) [42]. In such structures, TM modes can be excited, which have a longitudinal component of the electric field and can directly couple to the longitudinal component of the current density due to the drift motion of the electrons.

In a Cerenkov FEL an electron beam passes close to the surface of a dielectric film waveguide, thus producing polarization in the medium, which in turn excites TM mode in the waveguide. The evanescent field above the slab surface drives the beam and produces electron bunching, which in turn produces stimulated Cerenkov emission, see Fig. 21.13. Synchronism can be achieved if the electron velocity v_e approaches the phase velocity of the light in the dielectric waveguide $v_{ph} = \omega/k$.

The emission wavelength is proportional to the energy of the electron beam according to the formula:

$$\lambda = 2\pi d\gamma \frac{\varepsilon - 1}{\varepsilon} \tag{21.20}$$

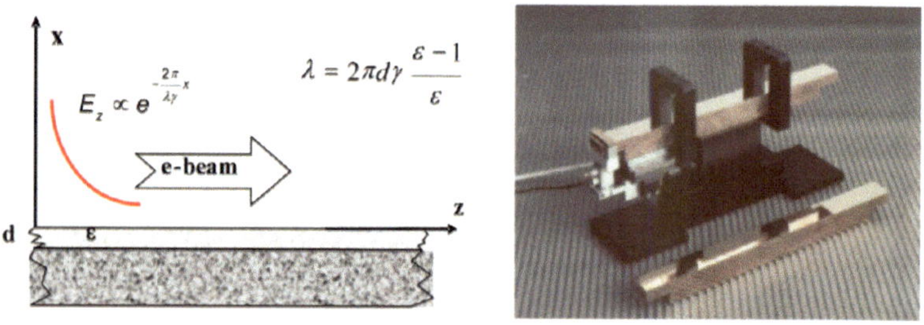

Figure 21.13 Scheme of the FEL-Cerenkov interaction and a photo of the interaction structure utilized at the ENEA center of Frascati [41].

The Smith–Purcell FEL is based on the interaction of an electron beam with the periodic structure of a metallic grating. This interaction produces radiation whose wavelength depends on the period l of the grating, on the harmonic number n, on the speed of the electron, expressed in terms of $\beta = v/c$ and also on the angle of observation θ relative to the electron's velocity:

$$\lambda = \frac{l}{n}\left(\frac{1}{\beta} - \cos\theta\right) \tag{21.21}$$

We can imagine that when the electron passes close to the surface of the grating, the position of the image charge inside the metal changes with the periodicity of the grating period. This is like having an oscillating dipole, which emits radiation. A more refined analysis can ascribe the origin of the emitted radiation to surface currents induced on the grating by the electron beam, accelerated by the grating periodic profile [43, 44]. This model describes correctly the angular and frequency distribution of the emitted light.

It is interesting to notice that it's easy to select a wavelength simply by adjusting the optics to collect light at different angles, as in the device shown in Fig. 21.14, which uses a rotating mirror to select the proper wavelength to be delivered to the output.

In several cases THz radiation is obtained as a parasitic radiation from bending magnets, transition radiation and edge radiation devices placed on large FEL facilities that are mainly built to operate in other regions of the spectrum. Nevertheless, there are undulator FEL facilities, such as FELBE, that have been specifically designed to operate in a wide range, extending from 1.2 to 60 THz (λ = 5–250 μm). At the same site, a Superradiant THz source (TELBE), has

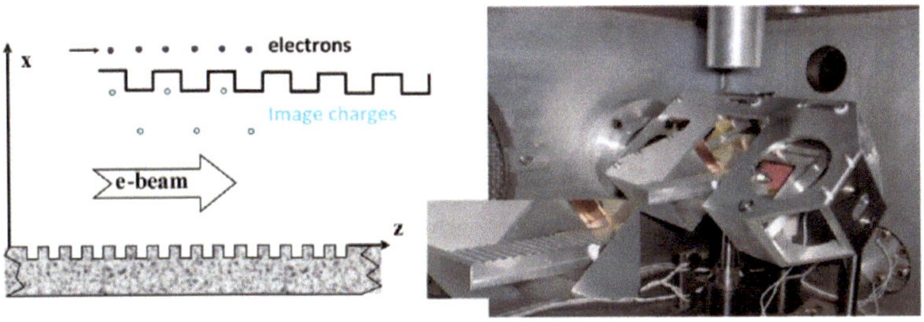

Figure 21.14 Scheme of the FEL-Grating interaction and a photo of the interaction structure utilized at the ENEA center of Frascati [42]. In the inlet a detail of the metal grating.

been developed, which includes both a coherent, broadband THz radiation (transition/diffraction radiation) source and a coherent, narrow-band THz radiation source based on an undulator with eight periods without resonator. These sources, all together, extend the operating range in the low frequency side of the spectrum from 0.1 to 2 THz [45].

21.4 Gimmicks: Novel Schemes

21.4.1 Tailoring THz Radiation Properties

Free-electron laser sources at long wavelength offer a series of features that drive the design project and realization toward small-scale dimensions devices. Equation (21.6) clearly indicates that small energy electron accelerators are suitable for long-wavelength FEL operations; considering real numbers that consist in undulator periods in the centimeter range and magnetic fields in the 5 to 8 kGauss range, we get electron energy values below 10 MeV for FEL radiation emission in the THz region of the spectrum. Moreover, Eq. (21.12) reveals us that FELs with low energy perform high gain coefficient values due to the inverse 5th power scaling; such a result has a direct and important consequence to offer the possibility to reduce the interaction region length provided that a proper balance of the gain coefficient is kept. So far the small electron energy and the possible small size interaction region length go in the joint direction of the dimension reduction of the THz FEL source device. But this is not the only advantage: as has been deeply analyzed in Section 21.2.1, the emission spectral band has a sinc shape, and its width is inversely proportional to the magnetic undulator length; therefore a short magnetic undulator, designed for a THz FEL, shall generate a large bandwidth coherent emission. This feature will be discussed in the next sub-session.

The electron beam accelerator is very relevant in order to determine the final radiation properties. Low-energy electron beams can be generated by many different types of accelerators capable to exhibit different varieties of characteristics, unlike high-energy electron that can be solely generated by LINACs or by Storage Rings. Among the possible low energy accelerators it is worth mentioning: Pulsed Diodes ($E_{beam} < 2$ MeV, $I_{peak} = 1$–10 kA), Cockroft–Walton and Van der Graaff Electrostatic accelerators ($E_{beam} < 10$ MeV, $I_{peak} < 10$ A), RF-Microtron ($E_{beam} < 30$ MeV, $I_{peak} < 10$ A), RF-LINAC ($E_{beam} = 2$–1000 MeV, $I_{peak} = 1$–100 kA). All the above-mentioned accelerators exhibit different temporal structures that are reflected on that of the generated radiation. This means that when designing an FEL source, in the THz region, one should make the first choice on the electron accelerator to be used depending on the radiation time structure needed; short single pulse with high power: pulsed diode, long single pulse with high energy: electrostatic accelerator, train of medium level power pulses at an RF repetition rate and for an indefinitely long time emission: microtron or LINAC (see Table 21.1 for further details).

Table 21.1 Characteristics of electron beam accelerators categories for FELs realization

Accelerator	Energy	Peak current	Pulse duration	Spectral region
Pulsed diode	< 2 MeV	1–10 kA	0.1–1 μs	mm-wave
Electrostatic	< 10 MeV	< 10 A	1–100 μs	FIR, mm-wave
Microtron	< 30 MeV	< 10 A	10–30 ps (1–20 μs)	IR, FIR, mm-wave
Induction LINAC	< 50 MeV	10 kA	100 ns	IR, FIR, mm-wave
RF-LINAC	< 100 MeV	100 A	1–20 ps (1–100 μs)	VIS,IR, (UV)
Storage Ring	0.2–2 GeV	100 A	0.1–1 ns	XUV, UV, VIS

21.4.2 Techniques to Optimize FEL Performance in the THz Range

21.4.2.1 Wide-band emission

In Section 21.2.2, it has been already explained how the waveguide properties contribute to the modification of the gain spectral bandwidth due to the dispersion effects. A generic quadratic dispersion relation, that expresses two distinct solutions for the synchronism, produces a broadening of both the spontaneous emission and gain spectra. A broadband/short-pulse THz emission easily reminds the classic Time-Domain based THz emitters [46], it is therefore evident that an FEL source can be designed to operate as a conventional THz source with some more important features that can distinguish it from the panorama of the far-Infrared devices. In order to maintain the emission bandwidth as broad as possible it is necessary to avoid the use of any optical cavity, which certainly reduces the frequency components. It is therefore compulsory to extract as much power is possible in a single electron passage inside the undulator. This is possible only if we are able to create a high degree of coherence inside the electron bunch. The problem has been deeply analyzed in [14] and [19] and can be analytically and numerically evaluated starting from the classical electromagnetic theory from which we can write the emission from N charged particles, of charge q, into a solid angle $d\Omega$ in the frequency interval $d\omega$ (see also Eq. (21.1)):

$$\frac{dI}{d\omega d\Omega} = \left[N + N\left(N-1\right)f\left(\omega\right)\right]\frac{q^2\omega^2}{4\pi^2 c} \left|\int \left[\vec{n}\wedge\left(\vec{n}\wedge\vec{\beta}\right)\right]e^{i\omega\left(\tau-\vec{n}\cdot\frac{\vec{r}_0(\tau)}{c}\right)}d\tau\right|^2 \tag{21.22}$$

The function $f(\omega)$ in Eq. (21.22) describes the coherence of the emitted radiation; its values range from 0 to 1. The limit $f(\omega) = 0$ indicates an incoherent emission from the bunch that means that the power emitted by N particles is just the sum of N individual contributions. The upper limit $f(\omega) = 1$ represents the total coherence, when the power emitted by N charged particles is given by N^2 times the emission of the single particle. This result is fully equivalent to the emission of a single particle having a charge $Q =$

Nq, because looking at Eq. (21.22), power is proportional to $Q^2 = N^2q^2$.

A second degree of coherence can be exploited considering the relationship among all the electron bunches as established by the RF electromagnetic wave. The consequence of that has been discussed in Section 21.3.1.2 and reported in Eq. (21.18), and proves that the radiation is emitted as a series of narrow-band frequencies that are the harmonics of the RF.

A further and relevant degree of coherence can be considered if we treat the bunch as a collection of N particles distributed in the longitudinal phase space [39]. Each electron carries its own energy γ_j and a proper position or phase ψ_j equal to the RF times the time t_j along the bunch. The result is a rewriting of Eq. (21.18) as follows (see refs. [39] and [47] for further details):

$$A_{l,0,n} = -\frac{Z_0}{\beta_{gl}} I_p \frac{C_l}{2} \frac{KL}{\sqrt{\Sigma}} F \sum_{j=1}^{N_e} \frac{1}{\beta_{zj}\gamma_j} \frac{\sin(\theta_{lj}/2)}{\theta_{lj}/2} ie^{i\left(\frac{\theta_{lj}}{2}+l\psi_j\right)};$$

$$\theta_{lj} = \left(\frac{\omega_l}{c\beta_{zj}} - \frac{2\pi}{\lambda_u} - k_{0,n}\right) \tag{21.23}$$

The previous equation clearly expresses that the phase factor gives rise to an interference mechanism; the contributions to the total power, given by the sum of all electrons, may result in constructive or destructive interference. The maximum power extraction is obtained by minimizing the negative interference, i.e., by keeping constant the phase term in Eq. (21.23) with respect to the contribution of each electron (see refs. [39] and [47]). The result is an ideal distribution of the electrons in the longitudinal phase space that can be obtained by proper manipulation of the particles before their entrance in the field of the magnetic undulator. Generally speaking, this manipulation compresses the electron bunches increasing the energy spread, but maintaining the required correlation; the result is an increase of the current, which increases the gain helping the dynamics of the process and, moreover, a reduction of the bunch duration that corresponds to an enhancement of the coherent contributions in the emission bandwidth.

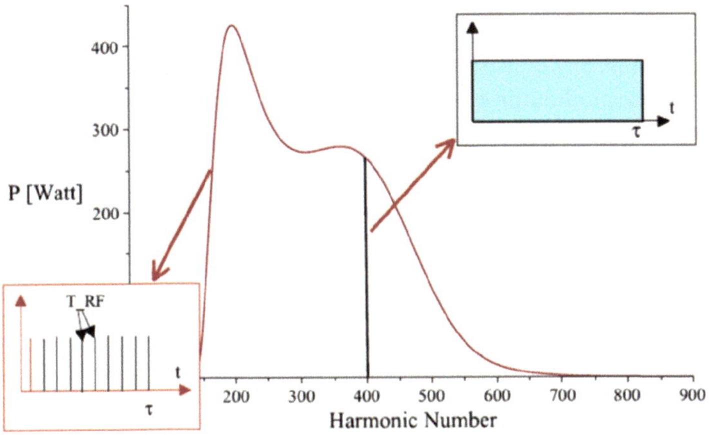

Figure 21.15 Radiation power spectrum as a comb of frequencies related to the harmonics of the fundamental RF ($\nu_{RF} = 3$ GHz) [47].

The result of the above discussion is reported in Fig. 21.15, which was elaborated in a design case studied in [47]. In this case, the spectrum ranges from 0.5 to 1.5 THz with an integrated power over the macropulse duration and over the total spectral bandwidth of about 90 kW. It is very important to stress that such a THz FEL source exploits all the features of a coherent broadband emitter, avoiding the use of any optical cavity or resonator recirculating the radiation. Moreover, due to the RF properties of the accelerator, it is possible to isolate the single harmonic with an interferometer, for instance, still having an average power for the single frequency of the order of hundreds of watts. This is not possible with conventional THz sources. Furthermore, it is worth stressing that the single frequency, being a harmonic of the RF, has a temporal structure equal to that of the RF macropulse. On the other hand, if we look at the whole bandwidth, the temporal structure is the well-known train of microbunches separated by the RF period as discussed in Section 21.3.1.2.

In conclusion, an FEL THz radiator can operate as a natural "frequency-comb" emitter or a single-frequency emitter; therefore, it can be considered a convenient, flexible, and powerful source for the generation of coherent radiation in the THz spectral region.

21.4.2.2 Buncher-emitter scheme

Another novel appealing FEL scheme consists of a first section, called the modulator, emitting at a specific synchronous frequency, followed by a frequency multiplier that exploits the harmonic content in the modulated beam [48, 49]. Such an arrangement may be reproduced several times, creating a kind of cascade [50] until it leads to a final beam degradation. Following what has been discussed in the previous chapters, we can imagine designing a hybrid system consisting in a slow-wave guiding structure acting as a buncher (Cerenkov FEL, Smith–Purcell FEL, etc.), followed by an undulator-based FEL as an emitter. The following discussion directly traces ref. [51].

The hybrid THz FEL system can be assumed to be composed of three main elements: (1) an electron beam accelerator; (2) a Cerenkov slow-wave guiding structure, with an adequate radiation resonator for the saturation regime accomplishment; and lastly, (3) a magnetic undulator acting as a radiator. The accelerated electron beam generates radiation interacting with the Cerenkov resonator. The electric beam associated with the stored radiation modulates the electrons' velocities in order to obtain the most efficient emission in the magnetic undulator radiator (see Fig. 21.16).

A proper design of this kind of device starts from the estimation of the electron energy that shall be common for both the modulator and the radiator. The starting point, for such an evaluation, are the synchronism equations reported in Eq. (21.6) for the undulator, and in Eq. (21.20) for the Cerenkov FELs, respectively: $\lambda_U = \lambda_W/(2\gamma^2)\left(1 + K^2\right)$ and $\lambda_C = 2\pi\gamma d\left[(\varepsilon - 1)/\varepsilon\right]$. An effective modulation is obtained when the radiation wavelength generated from the Cerenkov oscillator FEL is related to that of the undulator radiator by the condition: $\lambda_U = \lambda_C/n$, where $n = 1,2,3.\ldots$ specifies the harmonic number. From the above relations, we get an equation for the electron energy:

$$\gamma^3 - \frac{\lambda_W}{4\pi} n\left(1 + K^2\right)\frac{\varepsilon}{d\left(\varepsilon - 1\right)} \qquad (21.24)$$

Equation (21.24) gives the value of the electron energy as a function of all the parameters involved in both FELs.

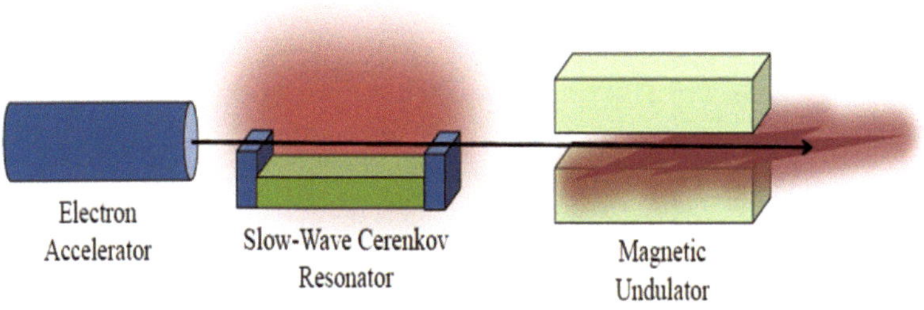

Figure 21.16 Scheme of the hybrid THz FEL [51].

Let us consider some reasonable values for the involved parameters in Eq. (21.24) such as: $\lambda_W = 2.5$ cm, $K = 1$, $d = 5$ μm, $n = 1$, $\varepsilon = 5$ (that corresponds to the dielectric constant of quartz) from which we get, for the electron relativistic parameter, $\gamma \sim 10$ that corresponds to moderate relativistic beam energy. With all the above numbers we eventually get a radiation emission wavelength $\lambda_U \sim \lambda_C \sim 250$ μm that is well within the THz spectral range.

The complete analysis of the dynamical process of the growth of the signal in the Cerenkov FEL oscillator, up to saturation, is reported in [51] and references therein. The relevant parameters for the gain calculations are reported in Table 21.2.

In Fig. 21.17, we report the behavior of the gain coefficient and intracavity peak power as a function of the round-trip number.

Analyzing Fig. 21.17 we can estimate the peak power at saturation that is about $P_{SAT} \sim 4.2 \times 10^7 W$; considering an average transverse radiation dimensions obtained from the data reported in Table 21.2, of about $\Sigma_L = 1$ cm^2, the electric field associated to the intracavity radiation at saturation is: $E_{SAT} = \sqrt{2P_{SAT}/(\varepsilon_0 c \Sigma_L)} \approx 1.782 \cdot 10^7\,[V/m]$. Moreover, saturation occurs after about 100 round-trips that, for the cavity length considered, correspond to about 200 nanoseconds, which is a quite short time if compared to the macrobunch duration of an S-band RF accelerator that is of the order of several microseconds.

When new electron bunches enter the Cerenkov optical cavity, they superimpose with the optical pulses. In such a situation, the electrons of the bunch, generally equally spaced in phase, will experience the oscillating longitudinal electric field with a spatial

Table 21.2 Cerenkov FEL gain parameters @ $\lambda = 250\ \mu m$

Electron bunch current	Angular distribution	Transverse distribution	Peak gain coefficient	Total resonator losses	Intracavity equilibrium power	Resonator length	Resonator width	Double slab height
$I = 20.0$ A	$\sigma' = 5 \times 10^{-4}$ (r.m.s.)	$\sigma_x = 1$ mm (r.m.s.)	$g_{0MAX} = 0.4$	$\Gamma = 0.042$	$I_e = 7.22 \times 10^7$ W	$L = 35$ cm	$W = 1$ cm	$D = 2$ cm

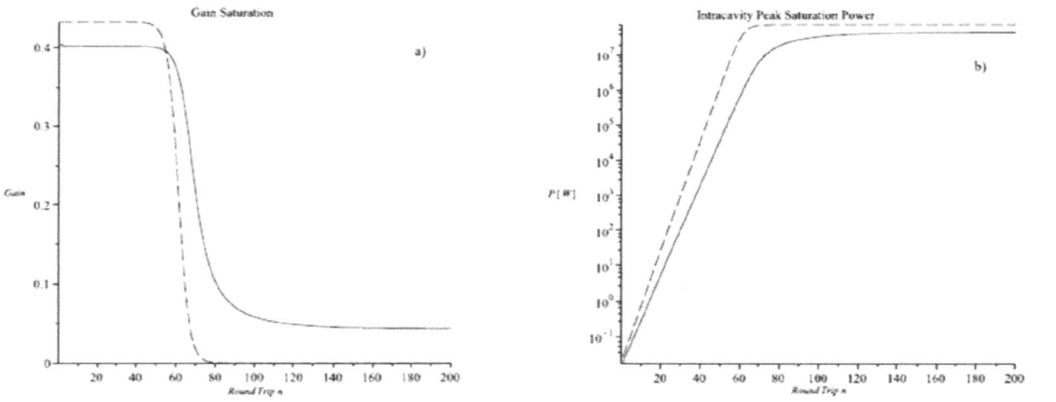

Figure 21.17 Saturation behavior of the Cerenkov THz FEL oscillator: (a) gain as a function of the round-trip number; (b) intracavity power as a function of the round-trip number [51].

periodicity of λ_C. The situation is similar to that of an electron beam accelerated by a series of RF cavities such as those of a standing-wave LINAC. The number of cavities can be estimated by calculating the number of the electric field oscillations "contained" within the electron micro-bunch length ($\delta_z \sim \tau_e \beta_e c$) divided by the wavelength λ_C: $N_c \sim (\delta_z/\lambda_C)$. For S-band accelerators, the typical duration of the micro-bunches is $\tau_e \sim 15$ ps, which corresponds to a bunch length of about $\delta_z \sim 5$ mm. For the parameter values considered, we obtain $N_c \sim 20$. What eventually happens, when electrons and radiation run together, is that the electrons undergo an energy variation (velocity modulation) that is connected to the particle-radiation relative phase. Such an energy variation will be positive for electrons close to the so-called accelerating phase and negative for those with an opposite phase. The result is a bunching of the electrons into N_c "slices". Any of these "slices" is an electron pulse interacting with the radiation field like in an RF cavity.

The energy variation of a single electron occurs along the time needed for the particle, to cross the RF cavity length L. The particle, during this time $T = L/(c\beta_e)$, experiences a quasi-static electric field due to the fact that, even with small differences, its speed is almost synchronous with the electromagnetic pulse that travels with its group velocity and therefore, $\beta_g \sim \beta_e$. The energy variation, for each electron, is consequently (see [51] for details):

$$\Delta\gamma = \int_0^T \frac{e\mathbf{E}(t)}{m_0 c^2} \cdot \mathbf{v}_e dt = \frac{eE_{0z}}{m_0 c^2} L \left[\frac{\sin\left(\frac{\omega}{c}\frac{L}{\beta_e}\right)}{\left(\frac{\omega}{c}\frac{L}{\beta_e}\right)} \cos\left(\phi \frac{\lambda_{RF}}{\lambda_C}\right) + \frac{\cos\left(\frac{\omega}{c}\frac{L}{\beta_e}\right)}{\left(\frac{\omega}{c}\frac{L}{\beta_e}\right)} \sin\left(\phi \frac{\lambda_{RF}}{\lambda_C}\right) \right] \tag{21.25}$$

The result is that the energy modulation expressed by Eq. (21.25) contributes to creating a strong electron bunching that can be exploited inside the magnetic undulator used here as the radiator.

It has been already discussed in Section 21.3.1.2 and in Section 21.4.2.1 how a bunched electron beam can generate a coherent emission up to saturation levels because each electron has the correct energy and phase to generate radiation in phase with all the others in the same bunch (see Eqs. (21.18) and (21.23)). The

Table 21.3 Undulator radiator FEL parameters @$\lambda_U = 250$ μm

Electron energy	Undulator period	Undulator parameter	Number of periods	Waveguide width	Waveguide height
$\gamma = 10$	$\lambda_W = 2.5$ cm	$K = 1$	$N_U = 50$	$W = 0.5$ cm	$H = 0.5$ cm

magnetic undulator FEL considered has the characteristics reported in [51] and summarized in Table 21.3.

The effect of the energy modulation introduced by the electron–field interaction inside the Cerenkov FEL optical resonator can be assessed by comparing the emission from an electron distribution $D(\gamma_j, \psi_j)$ and the same distribution to which an energy modulation is added $D(\gamma_j + \Delta\gamma(\psi_j)_j, \psi_j)$, where the contribution $\Delta\gamma(\psi_j)_j$ is calculated from Eq. (21.25) for each electron. To this aim, an ad hoc computer code has been realized calculating the dynamics of the emission process with the two different electron distributions in the longitudinal phase space.

In Fig. 21.18a, we report the power spectrum corresponding to the aforementioned "simple" electron distribution $D(\gamma_j, \psi_j)$ calculated with the Eqs. (21.23) and (21.18). The total power, integrated over the whole bandwidth, results in $P_{TOT} \sim 55$ W. A similar calculation can be performed for the electron distribution to which the energy modulation $\Delta\gamma(\psi_j)_j$ is added. The amplitude of the electric field in Eq. (21.25) is the one evaluated from the saturated intracavity radiation power of the Cerenkov FEL oscillator $E_{0z} \sim 1.8 \times 10^7$ [V/m]. In Fig. 21.18b, we report the power spectrum, again calculated by means of Eqs. (21.23) and (21.18). As can be clearly seen, the emission is strongly increased by orders of magnitude due to the modulation introduced. It is further worth underlining how the total bandwidth is slightly reduced with respect to the unmodulated case. The total power, integrated over the whole bandwidth, now results to be $P_{TOT} \sim 1.28 \times 10^7$ W.

The example reported indicates how a hybrid scheme for free-electron devices can be effective for the power enhancement of the electromagnetic radiation generated in particular in the THz region of the spectrum where the electron energies, though relativistic, are small.

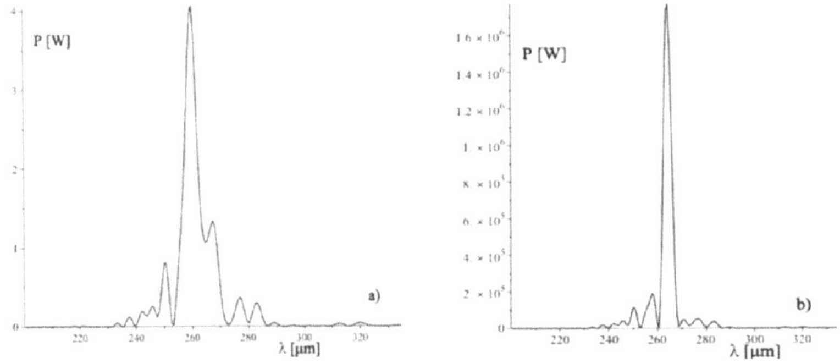

Figure 21.18 Emission by the undulator radiator FEL: (a) not modulated case power spectrum; (b) modulated case power spectrum [51].

This two-element scheme, based on an oscillator, in which intracavity radiation acts as an energy modulator, and a magnetic undulator acts as a radiator, is a versatile arrangement because the oscillator section can be chosen among a series of different devices. Besides the Cerenkov device, in fact, other sources can be taken into account. Among these, the Smith–Purcell is an interesting choice because it offers characteristics that lay between that of a Cerenkov FEL (being a slow-wave device) and an undulator FEL (wavelength scales inversely with the electron energy, as can be seen in Eq. (21.21)).

Moreover the harmonic number n in Eqs. (21.21) and (21.24) is a pivotal parameter for extending the spectral emission of such a coherent source, that, with a proper combination of all the parameters involved, for both the oscillator and the radiator, proves to be a promising and versatile device, with the additional feature of compactness.

References

1. Deacon, L. Elias, J. Madey, G. Ramian, H. Schwettman, T. Smith, First operation of a free-electron laser, *Phys. Rev. Lett.* 38, 16, 892 (1977).
2. M. A. Piestrup, R. N. Fleming, R. H. Pantell, Continuously tunable submillimeter wave source, *Appl. Phys. Lett.* 26, 418 (1975).

3. R. J. Barker, J. H. Booske, N. C. Luhmann, G. S. Nusinovich, *Modern Microwave and Millimeter Wave Power Electronics,* Wiley-New York (2005).
4. G. Dattoli, A. Renieri, Experimental and theoretical aspects of free electron lasers, in *Laser Handbook,* vol. 4, M. L. Stitch, M. Bass, ed., Elsevier Science Publisher E.V. (1985).
5. L. R. Elias, High-power, cw, efficient, tunable (uv through ir) free-electron laser using low-energy electron beams, *Phys. Rev. Lett.* 42, 977 (1979).
6. M. Billardon, P. Elleaume, J. M. Ortega, C. Bazin, M. Bergher, M. Velghe, Y. Petroff, D. A. G. Deacon, K. E. Robinson, J. M. J. Madey, First operation of a storage-ring free-electron laser, *Phys. Rev. Lett.* 51, 1652 (1983).
7. U. Bizzarri, F. Ciocci, G. Dattoli, A. De Angelis, G. P. Gallerano, I. Giabbai, G. Giordano, T. Letardi, G. Messina, A. Mola, L. Picardi, A. Renieri, E. Sabia, A. Vignati, E. Fiorentino, A. Marino, Above threshold operation of the ENEA free electron laser, *Nucl. Instr. Meth. Phys. Res. A.* 250, 1–2, 254 (1986).
8. A. S. Artamonov, N. A. Vinokurov, P. D. Voblyi, E. S. Gluskin, G. A. Kornyukhin, V. A. Kochubei, G. N. Kulipanov, V. N. Litvinenko, N. A. Mezentsev, A. N. Skrinsky, The first experiments with an optical klystron installed on the VEPP-3 storage ring, *Nucl. Instrum. Methods* 177, 247 (1980).
9. F. Glotin, J. M. Berset, R. Chaput, B. Kergosien, G. Hambert, D. Jaroszynski, J. M. Ortega, R. Prazeres, M. Velghe, J. C. Bourdon, M. Bernard, M. Dehamme, T. Garvey, M. Mencik, B. Mouton, M. Omeich, J. Rodier, P. Roudier, First lasing of the CLIO FEL, in *Proceedings of the 3rd European Particle Accelerator Conference,* Frontières, Gif-sur-Yvette, ed., Berlin, 1992, p. 620) (1991).
10. P. W. van Amersfoort, R. J. Bakker, J. B. Bekkers, R. W. B. Best, R. van Buuren, P. F. M. Delmee, B. Faatz, C. A. J. van der Geer, D. A. Jaroszynski, P. Manintveld, W. J. Mastop, B. J. H. Meddens, A. F. G. van der Meer, J. P. Nijman, D. Oepts, J. Pluygers, M. J. van der Wiel, First lasing with FELIX, *Nucl. Instrum. Methods* 318, 42 (1992).
11. R. Bonifacio, C. Pellegrini, L. M. Narducci, Collective instabilities and high-gain regime in a free electron laser, *Opt. Comm.* 50, 373 (1984).
12. E. Fermi, Über die Theorie des Stoßes zwischen Atomen und elektrisch geladenen Teilchen, *Z. Phys.* 29, 315 (1924).
13. J. D. Jackson, *Classical Electrodynamics,* 2nd ed., J. Wiley & Sons, New York (1975).

14. G. Dattoli, A. Doria, E. Sabia, M. Artioli, *Charged Beam Dynamics, Particle Accelerators and Free Electron Lasers,* IOP Publishing, Bristol, UK (2017).
15. H. Motz, Applications of the radiation from fast electron beams, *J. Appl. Phys.* 22, 527 (1951).
16. R. Loudon, *The Quantum Theory of Light,* 3rd ed., Oxford University Press, New York (USA) (2003).
17. H. T. Davies, *Introduction to Nonlinear Differential and Integral Equations,* Dover Pubns (1960).
18. A. Renieri, Storage ring operation of the free-electron laser: the amplifier, *Nuovo Cimento B* 53, 160 (1979).
19. A. Doria, R. Bartolini, J. Feinstein, G. P. Gallerano, R. H. Pantell, Coherent emission and gain from a bunched electron beam, *IEEE J. Quantum Electron.* QE-29, 1428–1436 (1993).
20. A. Doria, G. P. Gallerano, A. Renieri, Kinematic and dynamic properties of a waveguide FEL *Opt. Commun.* 80, 417–424 (1991).
21. K. W. Berryman, E. R. Crosson, K. N. Ricci, T. I. Smith, Coherent spontaneous radiation from highly bunched electron beams, *Nucl. Instr. Meth. A* 375, 526 (1996).
22. C. A. Brau, The Vanderbilt University free-electron laser center, *Nucl. Instr. Meth. A* 318, 38 (1992).
23. F. Ciocci, R. Bartolini, A. Doria, G. P. Gallerano, E. Giovenale, M. F. Kimmitt, G. Messina, A. Renieri, Operation of a compact free-electron laser in the millimeter-wave region with a bunched electron beam, *Phys. Rev. Lett.* 70, 928–931 (1993).
24. A. Doria, G. P. Gallerano, E. Giovenale, G. Messina, I. Spassovsky, Enhanced coherent emission of terahertz radiation by energy-phase correlation in a bunched electron beam, *Phys. Rev. Lett* 93, 264801 (2004).
25. G. L. Carr, M. C. Martin, W. R. McKinney, K. Jordan, G. Neil, G. P. Williams, High-power terahertz radiation from relativistic electrons, *Nature* 420, 153 (2002).
26. S. Miginsky, S. Bae, K. H. Jang, Y. U. Jeong, K. Lee, J. Mun, S.Saitiniyazi, A Compact FEL at KAERI: the project and the status, *Proceedings of 38th International Free Electron Laser Conference FEL2017,* Santa Fe, NM, USA; doi:10.18429/JACoW-FEL2017-MOP048.
27. H. Horiike, et al., Status of the institute of free-electron laser, Osaka University, *Jpn. J. Appl. Phys.* 41, 10 (2002).

28. T. Yamazaki, et al., Present status of Kyoto University free electron laser, *Proceedings of 23rd Int. Free Electron Laser Conference*, pp. II-13-14 (2002).
29. J. Xie et al, First lasing of the Beijing FEL, *Nucl. Instr. Meth. A* 341, 34 (1994).
30. L. Heting, H. Zhigang, J. Qika, Z. Shancai, W. Lin, L. Qing, Status of FELiChem, a new IR-FEL in China, *Proceedings of IPAC2016*, Busan, Korea, MOPOW026 (2016).
31. Dai Wu, First Lasing of the CAEP THz FEL facility Driven by a Superconducting Accelerator, J. of Phys.: Conf. Ser. (2018) 1067 032010 available from: https://www.researchgate.net/publication/328056737_First_Lasing_of_the_CAEP_THz_FEL_facility_Driven_by_a_Superconducting_Accelerator.
32. G. N. Kulipanov, et al., Novosibirsk free electron laser facility: description and recent experiments, *IEEE Trans. THZ Sci. Tech.* 5, (2015).
33. A. Gover, A. Faingersh, A. Eliran, M. Volshonok, H. Kleinman, S. Wolowelsky, Y. Yakover, B. Kapilevich, Y. Lasser, Z. Seidov, M. Kanter, A. Zinigrad, M. Einat, Y. Lurie, A. Abramovich, A. Yahalom, Y. Pinhasi, E. Weisman, J. Shiloh, Radiation measurements in the new tandem accelerator FEL, *Nucl. Instr. Meth. A* 528, 23 (2004).
34. R. Prazeres, F. Glotin, J. M. Ortega, New results of the 'CLIO' infrared FEL, *Nucl. Instr. Meth. A* 528, 83 (2004).
35. A. F. G. van der Meer, FELs, nice toys or efficient tools? *Nucl. Instr. Meth. A* 528, 8 (2004).
36. U. Lehnert, P. Michel, W. Seidel, D. Stehr, J. Teichert, D. Wohlfarth, R. Wünsch, Optical beam properties and performance of the MID-IR FEL at ELBE, *Proceedings of the 27th International Free Electron Laser Conference*, pp. 286–287 (2005).
37. P. Di Pietro, et al., TeraFERMI: A superradiant beamline for THz nonlinear studies at the FERMI free electron laser facility, *Synchrotron Radiation News A* 30, 36–39 (2017).
38. G. P. Gallerano, A. Doria, E. Giovenale, G. Messina, Coherence effects in FEL radiation generated by short electron bunches, *Nucl. Instr. Meth. Phys. Res. A* 358, 78–81 (1995).
39. A. Doria, G. P. Gallerano, E. Giovenale, S. Letardi, G. Messina, C. Ronsivalle, Enhancement of coherent emission by energy-phase correlation in a bunched electron beam, *Phys. Rev. Lett.* 80, 2841–2844 (1998).
40. A. Doria, V. B. Asgekar, D. Esposito, G. P. Gallerano, E. Giovenale, G. Messina, C. Ronsivalle, Long wavelength compact-FEL with controlled

energy-phase correlation, *Nucl. Instr. Meth. Phys. Res. A* 475, 296–302 (2001).

41. F. Ciocci, A. Doria, G. P. Gallerano, I. Giabbai, M. F. Kimmitt, G. Messina, A. Renieri, J. E. Walsh, Observation of coherent millimeter and submillimeter emission from a microtron-driven Cherenkov free-electron laser, *Phys. Rev. Lett.* 66, 699 (1991) and refs. therein.
42. G. Doucas, M. F. Kimmitt, A. Doria, G. P. Gallerano, E. Giovenale, G. Messina, H. L. Andrews, J. H. Brownell, Determination of longitudinal bunch shape by means of coherent Smith-Purcell radiation, *Phys. Rev. ST Acc. and Beams* 5, 072802 (2002), and refs. therein.
43. J. H. Brownell, J. E. Walsh, G. Doucas, Spontaneous Smith-Purcell radiation described through induced surface currents, *Phys. Rev. E* 57, 1075 (1998).
44. S. R. Trotz, J. H. Brownell, J. E. Walsh, G. Doucas, Optimization of Smith-Purcell radiation at very high energies, *Phys. Rev. E* 61, 7057 (2000).
45. https://www.hzdr.de/db/Cms?pOid=34100&pNid=2609.
46. K. Sakai, *Terahertz Optoelectronics*, ed., Springer, Berlin (2005).
47. A. Doria, G. P. Gallerano, E. Giovenale, Novel schemes for compact FELs in the THz region, *Condens. Matter* 4, 90 (2019).
48. R. Bonifacio, L. De Salvo Souza, P. Pierini, E. T. Scharlemann, Generation of XUV light by resonant frequency tripling in a two Wiggler FEL amplifier, *Nucl. Instr. Meth. A* 296, 787–790 (1990).
49. G. Dattoli, A. Doria, L. Giannessi, P. L. Ottaviani, Bunching and exotic undulator configurations in SASE FELs, *Nucl. Instr. Meth.* A507, 388–391 (2003).
50. F. Ciocci, G. Dattoli, A. De Angelis, B. Faatz, F. Garosi, L. Giannessi, P. L. Ottaviani, A. Torre, Design considerations on a high-power VUV FEL, *IEEE J. Quantum Electron.* QE-31, 1242–1252 (1995).
51. A. Doria, Hybrid (oscillator-amplifier) free electron laser and new proposals, *Appl. Sci.* 11, 5948 (2021).

Chapter 22

Cathode Technologies for Terahertz Source

Ranjan Kumar Barik,[a] Matlabjon Sattorov,[b] and Gun-Sik Park[b]

[a]*CSIR-Central Electronics Engineering Research Institute, Pilani, India*
[b]*Seoul National University, Seoul, South Korea*
ranjan.ceeri@gmail.com

22.1 Introduction

There is a requirement of free electrons in any kind of vacuum electron device. The source of free electrons is called the cathode, which generates the required amount of electron emission when external energy in the form of either high field or thermal energy is supplied. The term "cathode" is the Latinized form of the Greek word *kathodos*, which means "way down" or "descent" and refers to the setting sun. The name was suggested by William Whewell in 1834, who had been consulted by English chemist and physicist Michael Faraday while writing his research paper related to electrolysis where free electrons move through liquid. The invention of cathode rays by Julius Plücker and Johann Wilhelm Hittorf and a demonstration by Arthur Schuster in 1890 prove an electron can be

Advances in Terahertz Source Technologies
Edited by Gun-Sik Park, Masahiko Tani, Jae-Sung Rieh, and Sang Yoon Park

ISBN 978-981-4968-89-8 (Hardcover), 978-1-003-45967-5 (eBook)
www.jennystanford.com

moved and can be deflected under vacuum by a proper application of the electric field. The discovery of X-ray tubes in 1895 by the German physicist Wilhelm Conrad Röntgen ignited the research on cathode technology. Later on, the demand for cathode technology was enhanced drastically after the invention of various kinds of high-power high-frequency vacuum electron devices. This chapter is devoted to various cathode technologies aimed toward making vacuum electron sources, particularly for vacuum electron terahertz devices. Before going into the technologies involved, the basic emission physics and the classification of cathodes are presented for a better understanding and to get an overview.

22.2 Emission Physics of Cathode

From elementary physics, it is known that negatively charged electrons are tightly bound by the positively charged protons inside atoms to maintain charge neutrality. However, an electron can freely move from one atom to another inside a metal but cannot emerge from the metal surface. If they try to do so, the resulting positive charges at the metal surface attract the emerged electrons and pull back into the metal. This attractive force binds the electrons to remain inside the metal, or in other words, it creates a barrier which prevents it from escaping from the metal surface called the surface barrier.

The surface barrier can be broken by applying external energy to free the electrons by increasing their kinetic energy and consequently help them to escape from the metal surface. The minimum external energy required to escape a free electron from the surface is called the work function of that particular material. Therefore, the work function is a material-dependent property which varies from material to material.

22.2.1 Fermi Level

Let us understand the concept of Fermi level with a simpler case. In a metal, the outermost electrons are not bound to individual atoms.

Therefore, the electrons are free to roam within the entire crystal due to kinetic energy which is governed by thermal energy. What will happen to the electrons when we decrease the temperature of the solid metal to absolute zero? The electrons would lose all kinetic energy and their total energy would drop to zero. As a result, they would cease to move and would settle into the lowest energy states available which means all the same energy electrons would drop down to the same quantum state which contradicts quantum mechanics theory. It states that each allowed quantum state can only be occupied by one electron. At absolute zero, once the lowest energy state is filled, the next electron has to drop into the second lowest energy state, and so on. The energy of the highest possible quantum state of the electron is known as the Fermi energy of Fermi level. At zero kelvin temperature, there is a lack of sufficient energy, so the Fermi level can be regarded as a sea of fermions or electrons above which no electron subsists. It can be visualized using a simple example. Let us consider there is some water in a bowl at room temperature. Water molecules can move freely inside the bowl from one place to another. Once you cool down the temperature, the movement of the water molecule is gradually reduced and at a freezing temperature each molecule ceases to move. Once you increase the temperature, again the water molecule starts moving and further increasing the temperature the water molecule will start evaporating when the temperature reaches a boiling point.

22.2.2 Vacuum Level

Electrons remain naturally bound inside a solid metal which possesses potential and kinetic energies. The potential energy is due to the electrostatic field of the nuclei. A reference level is required to measure the potential energy. When the potential energy becomes maximum, the kinetic energy reaches zero and vice versa. The energy level for a free electron having zero kinetic energy is referred to as the vacuum level. The standard definition of vacuum level is the energy level of an electron having zero kinetic energy which is practically a "few nanometers" outside the solid surface.

22.2.3 Work Function

In any metal, there are one or two electrons per atom which are free to move from atom to atom but are strongly bound inside the metal. These free electrons are sometimes referred to as a "sea of electrons" or "electron gas." The velocities of these free electrons follow a statistical distribution, rather than being uniform. Sometimes, by the application of some external energy, the velocity of these electrons becomes high enough to exit the metal without being pulled back inside. The minimum external energy required to emit the free electrons from the surface of a metal is called the work function.

Metals with low work function require less external energy to emit free electrons from the metals. Thus, the work function is a characteristic signature of a material and is measured from the energy difference between the vacuum level and the Fermi level which is of the order of several electron-volts. The work function can also be explained in terms of image force. Initially the metal is assumed to be electrically neutral. After the removal of an electron, it can be assumed to acquire a unit positive image charge. The image charge tends to attract the real charge toward the surface. The energy required to pull an electron out to infinity against its image charge is called the work function of the material.

A graphical representation of Fermi level, Vacuum Level and work function for an n-type semiconductor are shown in Fig. 22.1 Valence bands and conduction bands are separated by the band gap

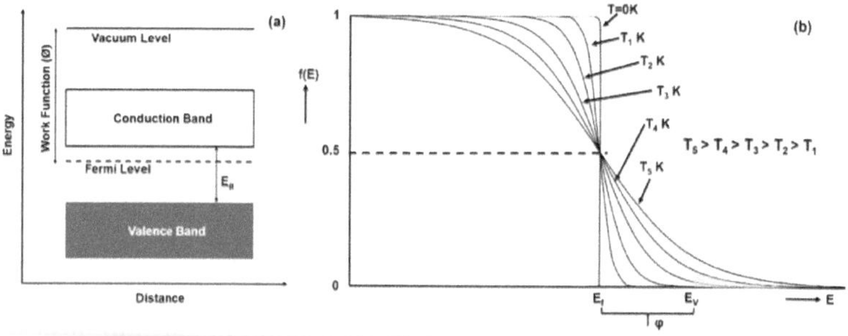

Figure 22.1 (a) Schematic of energy band diagram (b) Fermi–Dirac distribution function explaining emission mechanism. $N(E)$ represents the number of electrons with energy E.

(Eg). The energy difference between the vacuum level and the Fermi level represents the work function.

Work function and electron emission from a metal can be explained from Fermi–Dirac distribution function as shown in Fig. 22.1b. This distribution function $f(E)$ is the probability density function which gives the probability of a particular particle, fermions, (here fermions are the electrons of an atom) can occupy a particular energy level.

Mathematically, the probability that a particular state of energy E at the temperature T is filled with an electron is given by Fermi–Dirac distribution function $f(E)$, as:

$$f\,(E) = \frac{1}{1 + \exp\left(\frac{E - E_F}{k_B T}\right)}, \tag{22.1}$$

where, $K_B = 8.617 \times 10^{-5}$ eV/K is the Boltzmann constant, T the absolute temperature, and E_f the Fermi level or the Fermi energy.

At $T = 0$ K, all the electrons will have low energy and thus occupy lower energy states, i.e., below the Fermi level. The probability of occupancy below Fermi energy is 1 and above Fermi energy is zero.

As the temperature of the material increases from absolute zero to $T_1, T_2, T_3, \ldots,$ the Fermi–Dirac distribution curve starts to deform from a step-like function as shown in Fig. 22.1b. This depicts that more and more electrons occupy higher energy states, i.e., above the Fermi energy with an increase in temperature. At energy $E = E_f$, all these curves pass through a point where $f(E) = 1/2$, which indicates that the Fermi level is the level at which the probability of occupancy of an electron is 50%.

If E_V is the vacuum level, the energy difference between the vacuum level and Fermi level is the work function (φ) of the material. The electrons having probabilities of occupying energy states greater than E_V, as shown by the shaded region in Fig. 22.1b for the temperature T_4, only contribute to the emission current.

22.3 Classifications of Emission Mechanism

Largely the energy of free electrons inside the metal is less than the vacuum level. Therefore, some external energy is required to break the attractive force by nuclei to escape from the metal. The external energy to dislodge the electrons from the surface could be by: (a) supplying thermal energy, (b) applying high electric field, (c) shining photons, and (d) bombarding by primary electrons. Based on the applied external energy, the cathode emission can be classified into the following categories.

- Thermionic emission: the thermal energy is supplied using a filament to produce sufficient heat at the surface to generate the required emission.
- Field emission: a strong electric field is applied at the cathode surface with the use of an anode to lower the surface barrier that generates emission.
- Photoemission: the electron emission takes place by shining photons through an external source.
- Secondary emission: Electron emission from a metallic surface takes place by the bombardment of high-speed electrons or other particles.

22.3.1 Thermionic Emission

The word "thermionic" was coined from "therm" and "ionic," which means generation of ions (electrons) due to thermal energy. Thermionic emission can be defined as the process in which free electrons are generated from a metal surface by supplying heat as external thermal energy. As temperature increases, the thermal energy of electrons is increased resulting in the increment in their kinetic energy. Owen Willans Richardson observed during his experiment that the emission current from a heated wire depends exponentially on the temperature of the wire and published a mathematical form of his experimental results in 1901. Later on, in 1923, Saul Dushman demonstrated the modern form of this law known as Richardson–Dushman equation which states that the emitted current density J (A/m^2) is related to the temperature T (K) by the equation:

$$J = AT^2 e^{-\frac{\psi}{kT}}, \tag{22.2}$$

where T is the temperature of the metal body in kelvin, ϕ is the work function of the metal eV and k is the Boltzmann constant. A is the proportionality constant which is also known as Richardson's constant and is given by

$$A = \frac{4\pi mk^2 e}{h^3} = 1.20173 \times 10^6 \text{ Am}^{-2}\text{K}^{-2}, \qquad (22.3)$$

where m and e are the mass and charge of an electron, and h is Planck's constant.

Richardson–Dushman equation reveals that the thermionic emission depends on mainly two parameters, *viz* temperature of the metal body and work function of the metal.

The metallic solid body which is used to generate thermionic emission is called a thermionic emitter or a thermionic cathode. Now the obvious question is, what will be the criteria while choosing a material as a thermionic cathode for obtaining higher current density? Equation (22.2) depicts that the emission current density is inversely proportional to the exponential of the work function of the material. Hence, for a lower work function material the emission current will be higher compared to higher work function material. Again Eq. (22.2) depicts that the emission current density is directly proportional to the square of the temperature of the material. Thus, the surface temperature of the cathode plays a vital role for thermionic emission. The surface temperature of the cathode is restricted by the melting point and the mechanical strength of the cathode material. Thus, the cathode material should have the following properties:

(i) Low work function: Cathode material with a low work function property can produce electron emission by applying a small amount of heat energy or at a lower operating temperature.

(ii) High melting point: Electron emission from the metal body occurs at very high temperatures (>1500°C). Hence the material used as a cathode should have a high melting point. For example, copper has a low work function, but it cannot be used as a cathode due to its lower melting point. It will be vaporized before it starts emitting electrons.

(iii) High mechanical strength: The emitter material must possess a high mechanical strength even at high temperatures so that the shape of the cathode is not deformed, which affects the electron optical conditions. There is another issue of positive ion bombardment to the negatively biased cathode material. In any type of vacuum tube, there are always some residual gas molecules present. The flowing electrons strike these gas molecules due to which positive ions are generated. As the cathode is kept at a high negative potential its surface is subjected to a considerable bombardment under the influence of a high electric field and may be damaged.

(iv) Vacuum compatibility: The cathode is heated to the required temperature in a vacuum system to emit electrons. If it was heated in open air, it would burn out due to the presence of oxygen in the air. There are some materials which absorb a lot of atmospheric gases and release slowly inside the vacuum environment. Since the cathode works under vacuum, these kinds of materials cannot be used for cathode application.

22.3.2 Evolution of Thermionic Cathode

The development of thermionic cathode was started in early 1900 by simply heating a pure tungsten wire. Subsequently, thorium was added to tungsten to enhance emission which was later on called a thoriated tungsten cathode.

Oxide cathode: In 1904 Wehnelt [1] developed oxide cathode, which produced an emission density of 1 A/cm^2 under CW operation. The cathode was fabricated by spray coating of alkaline earth oxides on a nickel base. Some materials such as Mg, Mn, Si, Al, Ti, and C were added as impurities to the nickel base, which reduced the interface resistance (between the oxide and the nickel base) to supply free barium through a reduction process. This type of cathode offers a low work function of about 1.5 eV. However, it has limited use for the following reasons.

(i) Due to joule heating, there is a limitation of emission current to a maximum of 1 A/cm^2 under CW operation which when exceeded there would be permanent damage to the coating material.

(ii) Under poor vacuum conditions, the material gets poisoned easily due to which there would be emission degradation.

Also, at poor vacuum conditions, this type of cathode cannot withstand continuous ion bombardment.

(iii) Due to brittleness of the spray coating, the cathode may be easily damaged by mechanical shock.

The above limitations are overcome by a new version of cathode which was invented by Lemmens [2] at Philips Research Laboratories, Holland, and is called L-cathode. This L-cathode happened to be the first commercial dispenser cathode. The term "dispenser cathode" can be used to describe any cathode on which a thin film of emitting material is produced at the surface and is continuously replenished at the operating temperature.

L-cathode: It comprises a porous tungsten disc attached to a molybdenum sleeve. Behind the tungsten disk is a reservoir containing a solid solution of barium and strontium carbonates. Upon heating, the carbonates break down to oxides and the gas is pumped off through the pores in the tungsten. The chemical reaction between the oxides and the hot tungsten lead to the production of free barium which diffuses through the pores and reaches the surface to maintain a near monolayer coverage at the surface. In spite of the good performance, there are limitations with the use of these cathodes. The cathode requires a long activation time. Also, it is difficult to manufacture these types of cathode due to barium leakage which degrades vacuum condition.

To make the manufacturing techniques simpler and reduce the processing time, Hughes, Coppola and Rittner [3] modified L-cathode by pressing barium aluminate and tungsten powder which was vastly superior to the L-cathode. Hughes [4] found that further improvements could be made if the cathode was first formed from porous tungsten and then impregnated with barium aluminate. This cathode was found to be superior to the pressed cathode, because this type of cathode could be easily machined to required tolerances and with better dimensional accuracies.

Coated cathode: A coating on an impregnated cathode with a metal of higher work function than that of a base metal improves the emission. This phenomenon was first explained by Zalm and van Stratum [5] using an electrostatic model. According to this model, the effective work function reduces with the higher work function of the coated metal. This model stimulated the study of metal coatings and alloy coatings [6]. However, this model does not agree with certain metal coatings such as platinum, gold, and copper. The study by Raju [7, 8] suggests that it is possible to achieve a work function as small as 1.75 eV by depositing an alloy of ratio 2Os:2Re:W on a B-type cathode to a thickness of about 0.3 micron.

22.3.2.1 Modern dispenser cathode

Modern dispenser cathode is made out of a porous tungsten pellet into which barium-calcium-aluminates are introduced through a process called impregnation at a temperature of about 1650°C in a dry hydrogen atmosphere. The impregnated mixture has different molar ratios of BaO:CaO:Al_2O_3 such as 4:1:1 (called S-type), 5:3:2 (called B-type), 6:1:2 and 3:1:1. At an operating temperature of about 1100°C, the impregnated mixture, that is present inside the pellet, reacts with the tungsten pore walls due to which free Ba/BaO is generated and smears the surface with a near monolayer. The generation of Ba/BaO occurs through the following reactions as given by Rittner et al. [9]:

$$5BaO + 2Al_2O_3 \rightarrow \frac{3}{2}Ba_3Al_2O_6 + \frac{1}{2}BaAl_2O_4 \tag{22.4}$$

$$\frac{2}{3}Ba_3Al_2O_6 + \frac{1}{3}W \rightarrow \frac{1}{3}BaWO_4 + \frac{1}{2}BaAl_2O_4 + Ba \tag{22.5}$$

$$\frac{2}{3}Ba_3Al_2O_6 \rightarrow \frac{1}{2}BaAl_2O_4 + BaO \tag{22.6}$$

Barium has two outermost electrons which are transferred to the tungsten to form a dipole monolayer. This formation of dipole monolayer of barium with tungsten reduces the work function drastically. Gibson and Thomas [10] showed, using electron energy loss spectroscopy (EELS), that barium and oxygen form a "near" coplanar structure.

During normal operation, there exists a near monolayer of barium and oxygen over the surface. The electropositive barium atoms and the electronegative oxygen atom form positive dipoles along with their image charges in the metal. The dipole field reduces

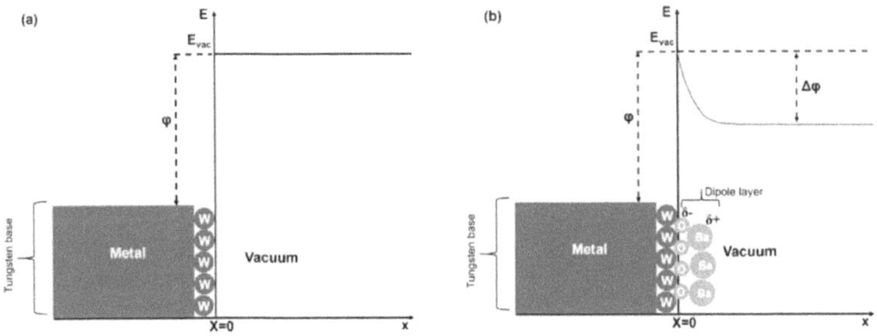

Figure 22.2 (a) Energy diagram for a pure metal surface, (b) Energy diagram with effect of dipole monolayer on the metal surface.

the potential at the surface and hence the work function of the surface is lowered by the adsorbed atoms [11]. This phenomenon can be explained as follows:

It is observed that electron emission can be slightly increased by applying some external electric field opposite to the cathode by electrical potential bias (or voltage). The effect of electron emission enhancement by applying the external potential can be modeled by a simple correction of the Richardson–Dushman equation which states that the emission current density mainly depends on two parameters *viz.* cathode temperature and cathode work function. In this present case, there is no change in cathode temperature, since only the electric field is changing. Therefore, the only possibility is slight changes in cathode work function which may be explained with the aid of Fig. 22.2. The energy level diagram of the metal surface without any applied field is shown in Fig. 22.2a. When an electric field, E_a, is applied (due to the dipole monolayer) to the cathode emission surface, a reduction of barrier is observed as shown in Fig. 22.2b. This field (E_a) reduces the barrier which an electron must overcome to emit from a metal surface [11].

The modified emission current density can be written as

$$J\,(\emptyset,\ T,\ E_{ext}) = A_0 T^2 \exp\left(-\frac{e}{kT}\left[\varphi - \Delta\emptyset\right]\right), \tag{22.7}$$

where

$$\Delta\varphi = \left(\frac{eE_{ext}}{4\pi\varepsilon_0}\right)^{1/2}. \tag{22.8}$$

E_{ext} is the externally applied electric field strength at the cathode, and ε_0 is the vacuum permittivity.

The resulting increase in emission current is known as the Schottky effect. Equation (22.7) is valid for electric field strengths less than about 10^8 V/m. The electron emission due to electric field strengths less than about 10^8 V/m is dominated by thermionic emission and this effect can be modeled by the Murphy–Good equation for thermo-field (T–F) emission [12]. At higher fields, i.e., electric field strengths of more than 10^8 V/m, the quantum tunneling effect begins to contribute, which is called field emission.

P. Zalm and van Stratum [5], using an electrostatic model, have explained that a coating of osmium or any other high-work-function material on a tungsten substrate results in lowering the effective work function to about 1.9 eV. Due to high-work-function-coated material, the dipole field at the surface increases and a large number of Ba atoms can be adsorbed. This argument is not applied to the platinum coating [13]. This is attributed to the atomic structure of the metal cathode surface in relation to adsorbed oxygen and barium atoms. This kind of coated cathode is called a modern M-type dispenser cathode, which can deliver about 30 A/cm^2 current density, and is presently using various kinds of high-power microwave sources.

22.4 Thermionic Cathode for Terahertz Devices

As we move GHz to THz frequency regime the dimensions of RF devices reduce drastically, hence the RF circuit size, beam tunnel dimension as well as beam size reduce accordingly. Generation of a required beam current using a conventional thermionic cathode requires large beam compression which becomes very difficult to design. Also, the integration and alignment of a large-size thermionic cathode with micro-fabricated circuits is another engineering challenge. An easy solution seems to use a small-size dispenser cathode operating at a higher temperature in order to obtain the required high current density. This proposition is not right as the life of a thermionic dispenser cathode diminishes exponentially with the operating temperature. Hence, the use of a conventional dispenser cathode at a high operating temperature is not a feasible solution due to its short life.

From the cathode front, there is a critical need for emission current density beyond 100 A/cm^2 [14] at a low operating temperature typically at 1100°C; and, from the beam point of view, out of many requirements, one of the critical needs for these THz VEDs is of current density electron beam >1000 A/cm^2. Another critical requirement is uniform emission from the cathode surface. However, producing such a high current density with a uniform emission from an electron source of miniature size is left as a challenge.

R. T. Longo [15] showed that during heating, barium is transported from the bulk to the surface through the pores, through diffusion and Knudsen flow, to form a Ba/BaO monolayer. Therefore, the parameters associated with the pore, *viz.* pore size, porosity, inter-pore connectivity, and pore distribution, play a critical role in the emission current, uniformity, and life.

The non-uniform pore size, shape, and its distribution across the surface leads to non-uniform Ba coverage resulting in a "patchy surface." A surface exhibiting wide work function distribution is said to be "patchy" as shown in Fig. 22.3b. The simulated I–V characteristics of two different cathodes having work function distributions narrower and wider, respectively, are shown in Fig. 22.3a. The cathode of the wider work function has a higher emission at higher anode voltages; however, the available emission at design voltage is low. This cathode is of no use when operated at low temperatures. Therefore, a direct control on pore distribution is an obvious solution.

The optimizations of the following parameters have a critical importance on the emission of the dispenser cathode. These parameters are graphically explained in Fig. 22.4.

(a) Porosity: The porosity has a bearing on the life of the cathode which is defined as the volumetric void space in the pellet. Cathode pellets having low porosity exhibit a low evaporation rate at the expense of barium coverage and hence low emission. On the other hand, for the case of high porosity, the coverage and

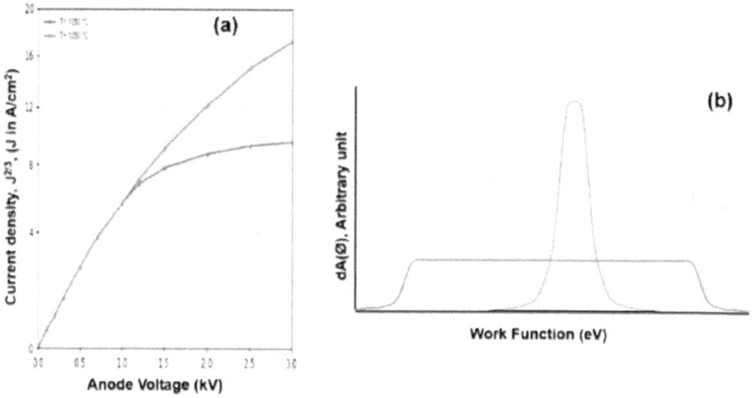

Figure 22.3 Comparison of the (a) I–V characteristics and (b) work function distribution at the same operating temperature of two different cathode.

evaporation increase, resulting in high emission at the expense of low life.

(b) Inter-pore connectivity: Another important parameter is pore-to-pore connectivity which is also called inter-pore connectivity. The supply of Ba/BaO from the bulk to the surface is affected for the pellet having a poor inter-pore connectivity, resulting in poor emission and short life.

(c) Pore distribution: Emission uniformity has been an important issue with dispenser cathodes [16] which is mainly determined by pore distribution. If the pores in the cathode button are distributed in random sizes and shapes across the surface, as shown in Fig. 22.4c, the result would be a non-uniform barium coverage, which leads to a large variation in work function over the surface. The emission from such an emitting surface would introduce noise at the output. Uniform barium coverage can be achieved through uniform pore arrangement, rendering a uniform work function distribution across the cathode surface.

(d) Pore geometry: Pore geometry such as pore size, the shape plays a critical role in the coverage of active material which directly controls the emission current density. For the too small pore size material, the barium supply from the bulk is restricted due to narrow paths. For very high pore size it will enhance the barium

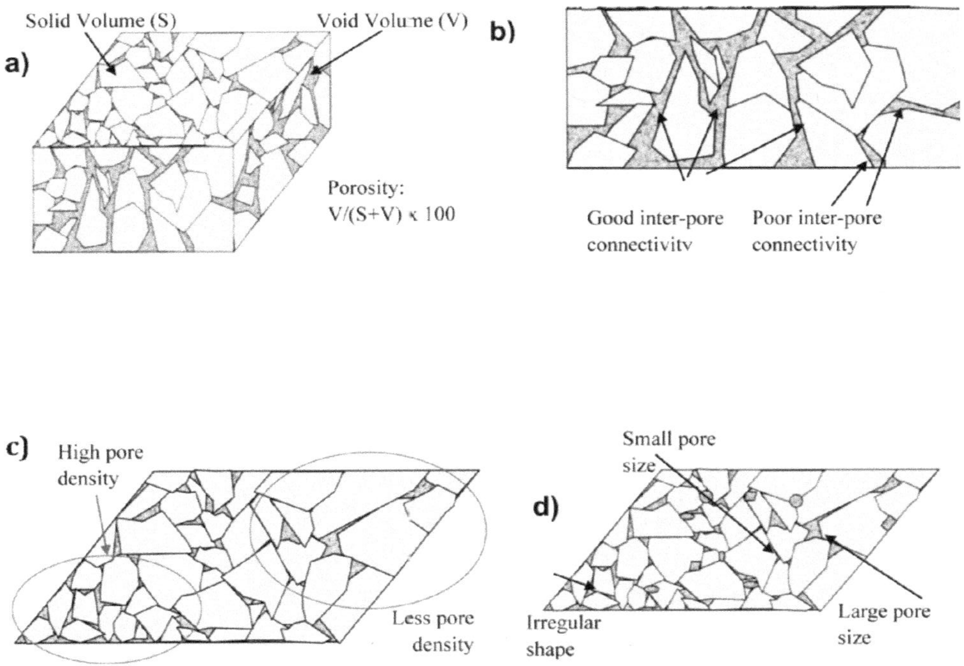

Figure 22.4 Schematic representation of: (a) porosity, (b) inter-pore connectivity, (c) pore distribution, and (d) pore geometry.

evaporation rate which will drastically reduce the cathode life. The optimum pore size and circular shape are effective and would result in a uniform coverage of active material.

22.4.1 CPD Cathode

Rutledge and Rittner [17] claim that the vapor transport of Ba/BaO from the bulk to the surface is through Knudsen flow. Upon reaching the surface, active materials migrate away from pore edges by diffusion. The pore parameters play a significant role in the Knudsen diffusion of barium/barium oxide (Ba/BaO) from the bulk to the surface and a subsequent radial spread over the surface. Researchers introduced a concept of a controlled porosity dispenser (CPD) cathode which offers a better control of pore size and their distribution across the emitting surface leading to enhanced emission uniformity. Initially, the CPD cathode was developed by creating or patterning holes over the surface through laser drilling, chemical etching, or lithography technique [18–21]. Later on, R. L. Ives et al. [22] developed controlled porosity reservoir (CPR) cathode out of a sintered bundle of W-wires.

CPR cathodes are fabricated out of tungsten wire by winding around a molybdenum material to form and sintering [22]. The sintered material results in a solid tungsten pellet having hexagonal arrays of pores throughout the structure. The emitting pellets are then formed by slicing the sintered material perpendicular to the pore axis, which produces a porous emission surface as shown in Fig. 22.5. The developed porous pellet integrated over an active material reservoir. During heating at an operating temperature, the active material comes out from the reservoir and sits on the emitting surface to form a dipole monolayer which reduces the work function and enhances emission current density.

A model was developed by AK Singh et al. [23] to determine the spread or the coverage (θ) of Ba/BaO on the surface. The aim of this modeling is to estimate the wire diameter which mainly determines the pore diameter and average pore separation. The diffusion length should be comparable to average pore separation to obtain a complete Ba coverage over the surface. R. T. Longo et al.

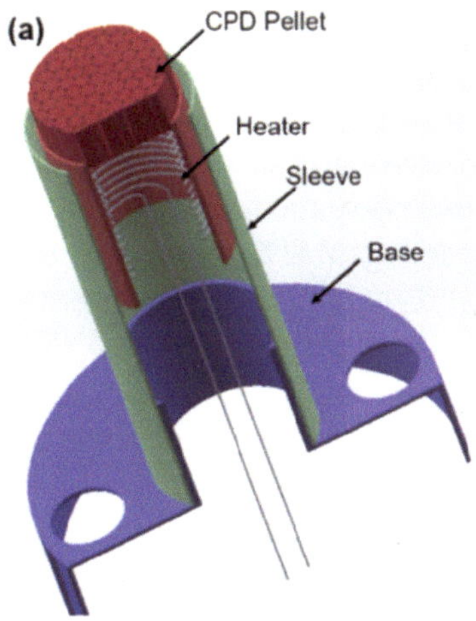

Figure 22.5 Schematic representation of: (a) CPD thermionic cathode, (b) Zoom view of CPD Pellet.

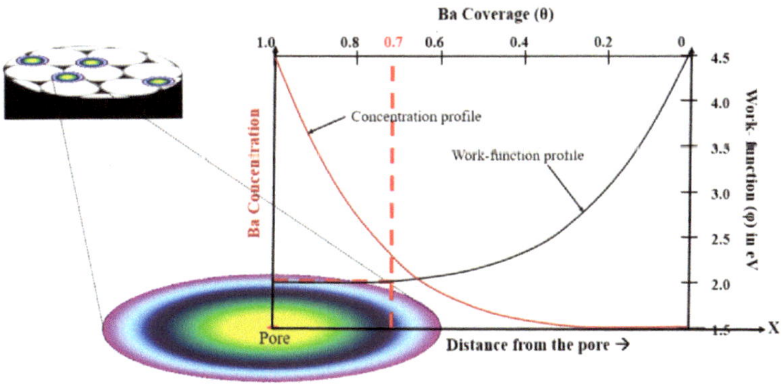

Figure 22.6 Barium coverage and work function distribution as a function of distance from the centre of pore.

[24] show that the work function reduces to its minimum value for a barium coverage of around 0.7 as shown in Fig. 22.6.

In this model, a pore is considered a source of Ba/BaO. During heating Ba/Bao comes out from the pore and covers the top surface. Hence barium coverage (θ) as well as work function (ϕ) are the functions of radial distance from the pore. Figure 22.6 depicts that as barium coverage reduces along radial distance, effective work function increases accordingly. Therefore, it is required to optimize the average pore separation and pore diameter to obtain a complete uniform barium coverage ($\theta = \sim 0.7$) over the surface which can deliver uniform and maximum emission current density. For this type of cathode pore separation and pore diameter depends on tungsten wire diameter. In this modeling, MATLAB software was used to obtain a pictorial representation of Ba/BaO spread over the surface. Three different bands of barium coverage, *viz*, $1.0 > \theta \geq 0.9$, $0.9 > \theta \geq 0.8$, and $0.8 > \theta \geq 0.7$, were chosen in this simulation. The pictorial results of barium coverage for different wire diameters are presented by AK Singh et al. [23].

Making a solid pellet out of optimized tungsten wire is a critical challenge which demands sintering the bundle of wires at an elevated temperature (2000–2200°C) with high pressure. The active sintering technique was adopted by A. K. Singh et al. to reduce the sintering temperature and duration substantially. Nickel was chosen as an active material which was coated on the surface of tungsten wire using the electro-polish technique. A specially designed spool was fabricated on which W-wire was wound using a transformer winding machine. The plating was carried out by dipping the whole assembly into a nickel-plating bath to obtain a coating thickness about 2–3 micron on the W-wire. After complete plating operation the bunch of wire was cut into pieces and wrapped with another tungsten wire to hold firmly. A specially designed molybdenum (Mo) compression fixture, consisting of four jaws in sectorial shape was developed for sintering wire bundles under high pressure. Sectorial jaws were used to compress the bundle in a radial direction by tightening the screws. Finally, the whole assembly was sintered in a dry hydrogen atmosphere at 1500°C for 10–15 min. After removing the sintered bundle, it was sliced to form pellets of 1 mm height. The excess nickel was then removed by firing sliced pellets at 1600°C under high vacuum conditions. After integrating with the heating assembly, the pellet was impregnated using barium-calcium-aluminate to form B-type cathode. The emission of CPD cathode is found to be more than twice that of a conventional thermionic dispenser cathode at the same temperatures [23].

The CPD cathode was characterized inside a UHV chamber in a closed-space diode configuration. The emission is found to be more than twice that of a conventional B-type cathode at the same temperatures as shown in Fig. 22.7. The sharp transition is observed from space charge region to temperature limited region which indicates that the uniformity of the coverage of active material as well as uniform emission current. The emission potential of the CPD cathode may be further improved by coating ternary alloy of 1W:2Re:2Os on the emitting surface.

The ternary alloy target pellet is made out of powders of W, Re, and Os in the ratio 1:2:2 by pressing in a die-punch. The pellet is then sintered around 1600°C in a dry hydrogen environment. Finally, the pellet is further sintered at 2200°C in the same environment followed by vacuum Argon arc melting at 3200°C temperature to form sigma phase alloy [25]. The target is fixed into a DC triode sputter coating system and the B-type cathode sample is kept below the target. The coating operation is carried out for a duration of 1 h 30 min to deposit a film of thickness of about 0.3 micron. During this time, the cathode is heated to ensure the adherence of coating is good and no hydrocarbons settle on the surface. Also, during heating at the operating temperature, the tungsten atoms from the base material diffuse into the coating material and Re and Os diffuse into the base which finally results in an equal proportion of W-R-Os. Auger spectroscopy analysis in [26] depicts that alloy coating results in a higher barium-to-oxygen ratio, which indicates that the barium is more in the metallic state. This, coupled with high Ba/Metal ratios results in high emission as compared to that of a B-type cathode.

The low work function together with good emission uniformity exhibited by CPD cathodes is proved to be an excellent candidate for use in vacuum sub-terahertz devices where moderate emission density ($\sim$60 A/cm^2) is required.

22.4.2 Nanoparticle-Based Cathode

Among the many varieties of impregnated cathode, the scandium oxide type cathodes developed by Figner et al. in 1967 [27], also known as pressed scandate cathode, made an important breakthrough. In 1977, Van Oostrom et al. [28] show that up to 10 A/cm^2 current density can be achieved at a lower operating temperature 950°C by adding Sc_2O_3. In 1979, Van Stratum et al.

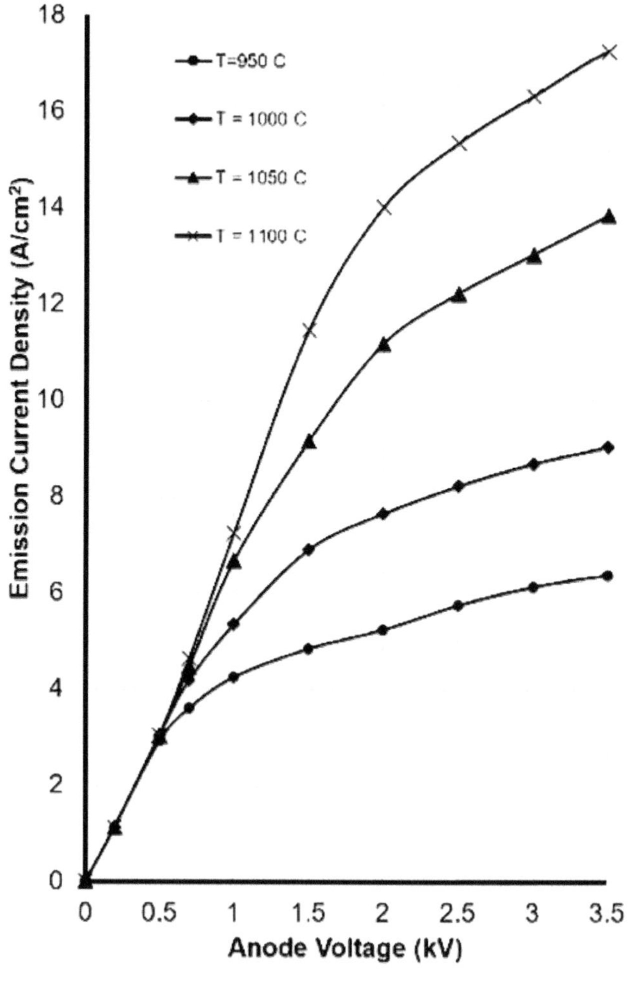

Figure 22.7 *I–V* plots of CPD cathode at different temperatures.

at Philips Research Laboratory [29] drew a comparable emission by adding Sc_2O_3 to the standard impregnate material. This type of cathode is known as impregnated scandate cathode. A mixture of tungsten and Sc_2O_3 called mixed matrix scandate cathodes was first introduced by Taguchi in 1984 [30]. Hasker et al. [31] improved this concept in 1990 by using tungsten and ScH_3 as a starting powder mixture and able to achieve emission current density about 100 A/cm^2 at the standard operating temperature. In 1986, Hasker et al. [32] also pioneered the top-layer scandate cathode. In this type of cathode, they coated a layer of W+ Sc_2O_3 mixed matrix of thickness about 5 μm on the top surface of porous W body impregnated with Ba-Ca-Aluminate. Next-generation top-layer scandate cathodes were introduced by S. Yamamoto et al. in 1988 [33] by sputter coating the W-base with W + Sc_2O_3. A combination of mixed matrix scandate cathode with lr, Os, Pt and Mo surface coating was published by S. Yamamoto [34] in 1989. There is significant progress in research in the area of high-current-density cathode due to the technological advancements over the last few decades. Researchers have improved the emission density and uniformity by adopting nanotechnology using a sol-gel method [35–37]. This technique enhanced the emission capability and improved the emission uniformity [38]. In 2007, Yiman Wang et al. [39] reported nanoparticle scandate cathode using sol-gel technique capable of emission current density of about 100 A/cm^2. The work by Young-Min Shin et al. [40] shows current density of up to 120 A/cm^2 can be achieved using the same technique. Due to nano-sized particles of Sc_2O_3, dispersed uniformly in near-atomic scale in the matrix, is the key factor for high emission and uniformity.

The emission properties are affected by the particle morphology, its size, and its distribution [37–39]. The diffusion of doped material from the bulk to the surface can be enhanced by using spherical-shaped particles, resulting in better emission [37]. Inter-pore connectivity and pore uniformity can be enhanced by using uniform grain size distribution particles which will enhance the life of the cathode. Finally, the emission uniformity of the cathode can be obtained by the homogeneous mixing of the doping material. Therefore, the synthesis of the spherical shape, a uniform grain size distribution, homogeneous mixing and a pure phase of scandia-doped tungsten nanoparticles are the critical requirements for the development of high-current-density cathodes for use in THz devices. Liquid-liquid doping using sol-gel technique may be the best solution for the above applications where tungsten and scandia will be completely dissolved into a solution.

A schematic flow diagram of the scandia-doped tungsten nanoparticle preparation is shown in Fig. 22.8. Tungsten oxide

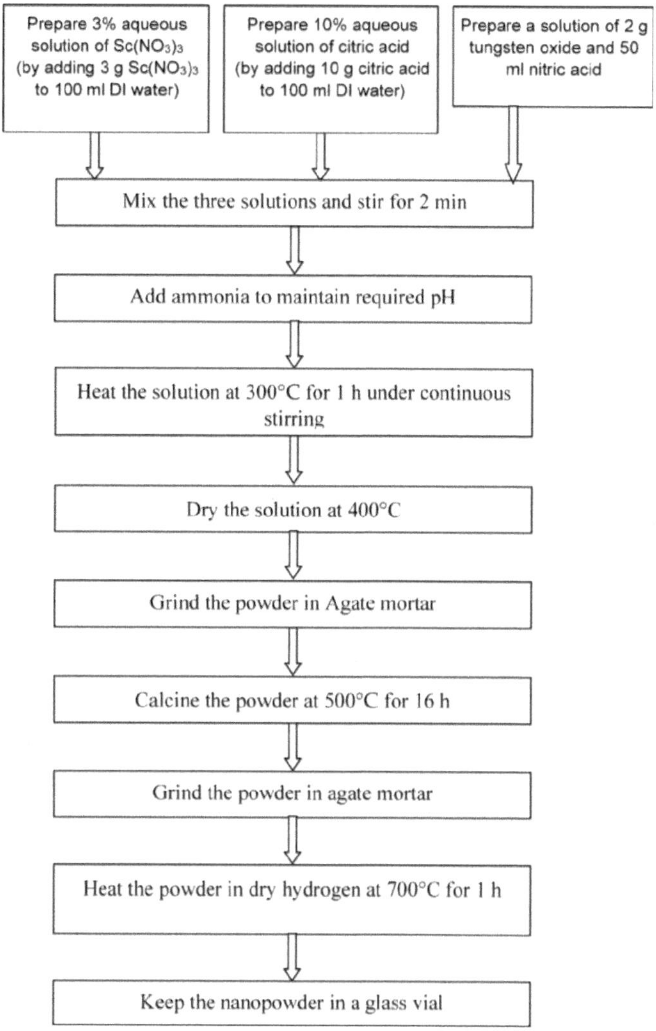

Figure 22.8 Synthesis flow diagram of scandia-doped tungsten nanoparticles.

powder of 99.9% purity with an average particle size of about 80 μm was used as the starting material. Scandia (Sc_2O_3)-doped tungsten powder was prepared using a sol-gel process by dissolving tungsten oxide in a mixture of nitric acid, citric acid, and ammonia at 300°C. An initial 3% aqueous solution of scandium nitrate ($Sc(NO_3)_3$) and an 8% aqueous solution of citric acid were added to 50 ml of nitric acid. Then, 2 gram of tungsten oxide was added to the solution, which was stirred continuously for 2 min. Ammonia solution was added slowly to maintain a required pH of the solution by monitoring with a pH meter. Stirring of the solution continued under hot conditions by maintaining the temperature at 300°C for 1 h to form a hydrolysis reaction. The solution was dried at 400°C for 30 min under continuous stirring. During this process, the sol (solution) becomes a gel. After the formation of the gel, it was allowed to further dry at the same temperature for a few hours to remove residual organics and water molecules. After that, the powder was collected and ground in an agate mortar. The powder was then calcined at 500°C for 16 h, after which it became a yellow tungsten trioxide nanopowder. Finally, scandia-doped tungsten nanopowder was produced by reducing the oxides under a dry hydrogen atmosphere at 700°C for 1 h in a furnace.

Another technological challenge is to develop solid pellets, out of scandia-doped tungsten nanoparticles, with high mechanical strength and 20% porosity. As the matrix (solid pellet) porosity is a function of sintering temperature, it is required to study the effect of sintering temperature on porosity. Experimental study reveals that at lower sintering temperatures, the particle shape remains spherical at the cost of lower mechanical strength. Whereas high-temperature sintering gives rise to distortion in shape, and further increase in temperature the circular shape is fully lost. The agglomeration of particles can be seen clearly, where the effective particle size increases with the sintering temperature and at the same time matrix porosity reduces. Therefore, optimization in sintering parameters such as pressure, temperature, and duration is essential to sinter the pellet at a relatively high temperature (>1500°C) to retain the compactness and avoid the crumpling of the pellet at the same time maintaining the particle shape with the required porosity. Subsequent pulse emission measurements after impregnating the pellet by barium, calcium, and aluminate show the emission capabilities more than 140 A/cm^2 as shown in Fig. 22.9.

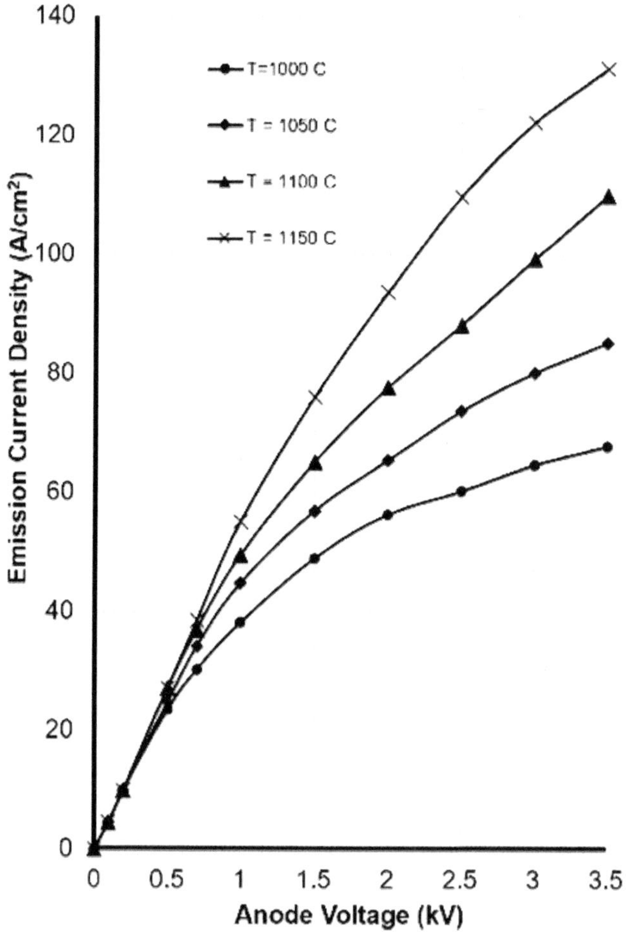

Figure 22.9 *I–V* characteristics of scandate cathode in diode.

However, in practice, there is a limit of maximum current drawing capabilities in CW even though the cathode is capable of delivering higher emission density. This limit is due to the electrical breakdown inside the vacuum. Conventionally the breakdown voltage in a vacuum environment between two electrodes is ~10 kV per mm. If the electrode is filled by high electron density the breakdown voltage will be reduced from 10 kV/mm. Breakdown field strength increases for a very short pulse voltage which can increase the limit of current drawing capabilities. In this experiment, the I–V characterization was carried out using a short pulse of pulse width 2 μs.

Y. M. Shin et al. [40] have developed terahertz sources using this kind of cathode. However, there is still skepticism about the commercial use of this kind of cathode. There are mainly two challenging issues which we need to solve while we move the microwave frequency range to the terahertz frequency range. As the operating wavelength is compatible with the dimensions of the high-frequency devices, the dimension of vacuum electron devices shrinks drastically as we move to the high-frequency region. The tentative size of vacuum terahertz devices reduces to less than 1 cm^3. As the device gets smaller and smaller, the beam tunnel becomes narrower and narrower. This is becoming increasingly difficult to engineer for the formation and transportation of high-current-density smaller electron beams through the beam tunnel. Also, the use of a thermionic cathode electron gun in terahertz vacuum electron devices offers serious issues.

The first issue is thermal loading while using a thermionic cathode which is working under very high operating temperatures. The induction of thermal energy from the cathode to the RF section may damage and de-tune the operating frequency. As a precaution, it is required to keep the thermionic cathodes at comparatively long working distances from the RF section to reduce thermal damage. This becomes a challenging issue to transport a high-current-density beam for a long distance while maintaining its laminarity, which is a necessary requirement to achieve the efficient beam–wave interaction required for THz generation.

Another challenging issue is creating a high vacuum through a narrow beam tunnel. For high-frequency devices, the beam tunnel becomes very narrow. Creating a high vacuum through a narrow beam tunnel becomes a critical challenge to engineers. Thermionic cathode easily gets poisoned in poor vacuum conditions, particularly the scandate cathode. Therefore, there are practical challenges to the commercial use of thermionic scandate cathode in terahertz region vacuum electron sources.

Field emission (FE) cold cathodes can overcome the limitations of the thermionic emitters. Since the field emitter cathode operates in cold conditions, there is no issue of thermal loading. The gap between the emitter and the RF section can be dramatically reduced by using a cold cathode which can relax the design difficulties. FE cathode can also work under moderate vacuum conditions which is another favorable advantage for application in high-frequency vacuum sources. Therefore, FE cathode may be a choice for the application in high-frequency devices; however, this kind of cathode has its limitations as it can be damaged easily when subjected to ion bombardment.

22.5 Field Emitter Cathode for Terahertz Devices

Emission of free electrons, due to very high applied electric fields ($>10^8$ V/m), from metals or semiconductors has attracted the attention of researchers for the past hundred years. J. E. Lilienfeld [41] was the first to demonstrate field emission from a pointed cold cathode in 1922. The theoretical foundation of the quantum mechanical tunneling emission current and field strength relation was first developed by Fowler and Nordheim in 1928 [42]. A systematic study of field emission has developed the technique field emission microscopy (FEM). Tungsten tips-based field electron microscope was first developed by E. W. Müller [43] in 1936 which is a landmark achievement for the development of practical devices using field emitter cathode. Recently, a detailed review was carried out by A. K. Singh et al. [44].

Recently, J. Zhao et al. [45], showed that field emitter cathode allows bunching to happen right from the cathode resulting in a high beam–wave interaction efficiency even for a short-length interaction structure. Thus, the use of FE cathodes in place of gridded thermionic cathodes can greatly enhance high-frequency and high-power capabilities. X. Yuan et al. [46] developed fully sealed CNT cold-cathode–based 0.22 THz, 500 mW gyrotron. 200 A/cm^2 current density obtained from vertically aligned graphene-based thin film by In-Keun Baek et al. [47]. They have realized this high current by the pre-mechanical shaping and post electrical conditioning of reduced graphene oxide (rGO) film emission. Steven B. Fairchild et al. [48]

demonstrated a high-emission-current-density cathode developed out of single bulk carbon nanotube (CNT) fibers. They have achieved a stable emission for 10 h at 9 mA at an applied field strength of 0.11 V/μm.

The criteria for the choice of a field emitter cathode for use in a high-power VED technology is based on the following:

- High electrical conductivity:
- High thermal conductivity:
- High chemical and mechanical stability: to enable large current handling capability,
- Stable surface conditions: (avoiding unwanted species adsorption and desorption) and sustaining the shape of the tip under high electric field (ion back bombardment) and high temperature.

Keeping in mind the above requirements of material properties, researchers have focused their interest on discovering new materials, such as nanowires made of silicon carbide (SiC), copper sulfide (Cu_2S), diamond films, nanocarbons (carbon nanotubes (CNTs), vertically aligned graphene), etc. In order to develop and design a field emitter-based electron gun one should know the basics of field emission theory. Various authors have presented field emission theory in detail. Here, an overview of the field emission mechanism is described.

22.5.1 Field Emission Theory

Emission current density from a metal due to an externally applied electric field was first modeled by Fowler–Nordheim (FN) in 1928, based on the quantum tunneling model. The Fowler–Nordheim equation predicts the emission current density from a planar surface of a bulk metal emitter by estimating the transport of electrons flux from bulk to the surface and the probability of electrons transmitted through the potential barrier at the emitted surface.

Fowler–Nordheim assumed that the emitter surface is perfectly planar, smooth, free of defects. Since the externally applied electric field cannot penetrate the metal body, the work function φ is assumed to be independent of the external electric field and uniform throughout the planner emitting surface. It is also assumed that the emitter follows the Sommerfeld model, i.e., the emitter consists of free electron gas and the electron emits from y–z plane towards x direction as shown in Fig. 22.10a.

Let us consider, E is the kinetic energy of an electron inside the metal and E_x is the kinetic energy of the electron along x direction. The electron will be seeing a barrier at the emitting surface having height $\varphi + \mu - E_x$. where, μ is the chemical potential of the electron. The potential energy of the electron will be as a function of its distance x from the surface of the metal. When an external electric field F is applied to the emitting surface of the metal emitter, the electron will be seeing a barrier thickness as $(\varphi + \mu - E_x)/Fe$, where, e is the electron charge. As a result, by the application of an external electric field, a triangular potential barrier will be created in the vacuum-metal interface as shown in Fig. 22.10b.

Let us consider a constant flux density of electrons N incident per unit area, per unit energy from metal to the emitting surface. Out of the incident electron flux density N a few electrons having transmission probability D are able to tunnel or transmit through the potential barrier $V(x)$ into the vacuum. These electrons will contribute to the field emission current density J. The rest of

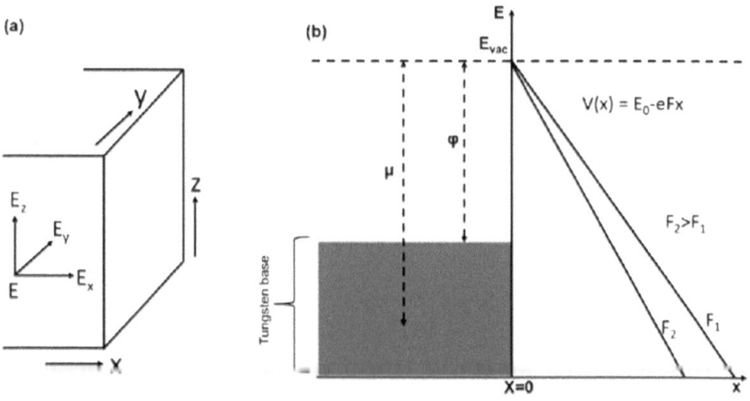

Figure 22.10 Schematic diagram: (a) kinetic energy components of electrons, (b) field emission model.

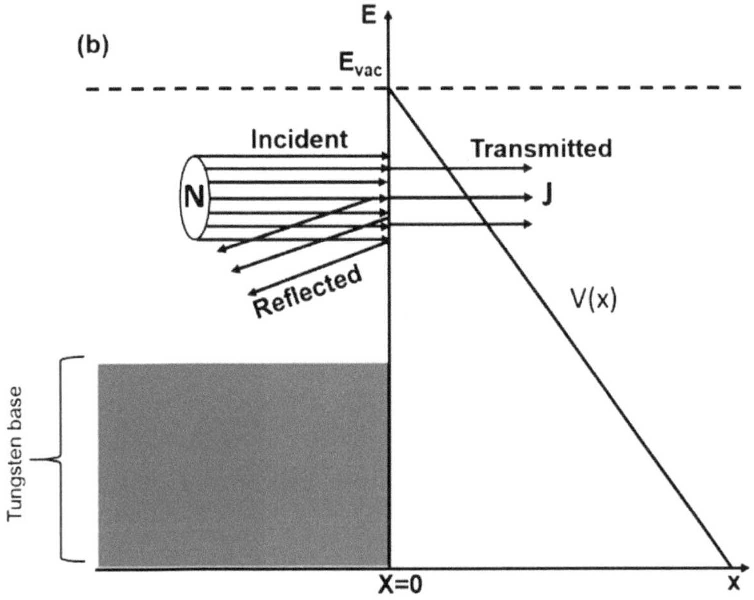

Figure 22.11 Tunnelling model of electron through potential barrier.

the electrons will be reflected back into the metal as shown in Fig. 22.11.

The emitted current density is given by the product of: (i) the number of electrons $N(E_x)$, which crosses a unit area paralleled to the emitting surface, per unit time towards the x direction. $N(E_x)$ is called the supply function, having kinetic energy in x direction between E_x and $E_x + dE_x$ and (ii) The probability of electrons which transmit or tunnel through the potential barrier $V(x)$ called the barrier-transmission probability $D(E_x)$.

The total emitted current density J_{FN} can be calculated by integrating over all normal energies as

$$J_{FN} = \int_0^{\infty} N(E_x) D(E_x) \, dE_x \tag{22.9}$$

The problem is how to calculate the supply function and the transmission coefficient.

22.5.1.1 Calculation of supply function and transmission coefficient:

Supply function $N(E_x)$ can be estimated from supply function density $n(E, E_x)$, as

$$N(E_x)\,dE_x = dE_x \int_{E_x}^{\infty} n(E, E_x)dE, \tag{22.10}$$

where, supply function density $n(E, E_x)$ is supply function per unit energy having total energy between E and $E + dE$. Therefore, the supply function density $n(E, E_x)$ can be the product of: (i) x-component current density (j_x) inside the bulk metal and (ii) Fermi energy $f(E)$ of emitter which is another assumptions of Fowler–Nordheim theory that the electrons obey Fermi–Dirac distribution function. A detailed calculation of supply function and transmission coefficient was carried out by [49]:

$$N(E_x) = \frac{em_e k_B T}{2\pi^2\hbar^3} ln\left[1 + \exp\left(-\frac{E_x - E_F}{k_B T}\right)\right] \tag{22.11}$$

$$D(E_x) = \exp\left(-\frac{4}{3}\frac{\sqrt{2qm^*}}{\hbar}\frac{\phi_B^{3/2}}{\phi_B/_L}\right) \tag{22.12}$$

Substituting the values of supply function and transmission probability from (22.11) and (Eq. 22.12) into Eq. (22.10) one can estimate field emission current density as

$$\begin{aligned} J_{FN} &= \int_0^{\infty} \frac{em_e k_B T}{2\pi^2\hbar^3} ln \\ &\times \left[1 + \exp\left(-\frac{E_x - E_F}{k_B T}\right)\right] \exp\left(-\frac{4}{3}\frac{\sqrt{2qm^*}}{\hbar}\frac{\phi_B^{3/2}}{\phi_{B/L}}\right) dE_x \\ J_{FN} &= \frac{em_e k_B T}{2\pi^2\hbar^3} \exp\left(-\frac{4}{3}\frac{\sqrt{2qm^*}}{\hbar}\frac{\phi_B^{3/2}}{\phi_B/_L}\right) \int_0^{\infty} ln \\ &\times \left[1 + \exp\left(-\frac{E_x - E_F}{k_B T}\right)\right] dE_x \end{aligned} \tag{22.13}$$

By integrating Eq. (22.13) one can get

$$J_{FN} = a_{FN}\frac{F^2}{\emptyset} exp\left(-\frac{b_{FN}\emptyset^{3/2}}{F}\right), \tag{22.14}$$

where,

$$a_{FN} = \frac{e^3}{16\pi^2\hbar} \approx 1.5414\,[\mu A][\text{eV}]/\text{V}^2$$

$$b_{FN} = \frac{4}{3}\frac{(2m_e)^{1/2}}{e\hbar} \approx 6.83089\ [\text{eV}]^{-3/2}\ [\text{V}]/\text{nm}$$

Over the last few decades, modern device fabrication technology has scaled down the emitter tip dimensions. The emission tip radius of curvature of modern field emitters typically ranges between 1–10 nm.

The Fowler–Nordheim theory was derived by assuming a flat emission surface. Therefore, a correction is required when an external field is applied in an extremely large curvature like a single carbon nanotube. The modified Fowler–Nordheim equation is as follows

$$J_{FN} = a_{FN}\frac{(\beta F)^2}{\emptyset}\exp\left(-\frac{b_{FN}\emptyset^{3/2}}{\beta F}\right), \tag{22.15}$$

where β is the correction term called field enhancement factor. The factor β depends on the emitting material geometry, the material crystallography, and the distance between the electrodes. β can be estimated from the F–N plot, using the following equation [50]:

$$\beta = \frac{b_{FN}\emptyset^{3/2}}{S}, \tag{22.16}$$

where S is the slope of the F–N plot.

A model was developed by Wang et al. [51] to estimate the field enhancement factor of carbon nanotube as follows:

$$\beta = \frac{h}{\rho} + 3.5 + A\left(\frac{h}{d}\right)^3, \tag{22.17}$$

where h is the height, ρ is the tip radius of the CNT, d is the separation between cathode and anode and A is a constant.

22.6 Conclusions and Future Prospects

The basic physics and parameters of electron emitters have been explained in a simple manner. The electron emitters (or cathode) are classified based on their electron emission mechanism namely thermionic emitters, field emitters, photo emitters, and secondary emitters. Out of these types of emitters, thermionic emitters and field emitters are the most suitable for applications in vacuum electron devices. A historical development along with the working mechanism of thermionic emitters has been discussed. Out of various types of thermionic emitters, CPR cathode, which was first developed by R. L. Ives et al. [22], seems to be a suitable candidate for VEDs from where a required current density around 50–60 A/cm^2 is possible. The modeling of CPD cathode, carried out by A. K. Singh et al. [23], gives a good insight into the emission mechanism and simple way of fabrication. Still there is a lot of scope for further work on this type of cathode such as: (a) better sintering process to make the cathode a better compact that can be used to make large-size cathodes and (b) rigorous analysis of active material coverage, etc., for a better understanding of emission mechanism. For sub-terahertz to terahertz range VEDs, where the required current density is more than 100 A/cm^2, nanoparticle-based scandate cathode may be one of the solutions. Although some practical device using nanoparticle-based scandate cathode has been demonstrated, still there is no commercial cathode available. Further work can be carried out on life estimation, poison study, patchy emission, etc.

A thermionic cathode requires heating at a high temperature in order to emit electrons which leads to a physically large device size. Also, a thermionic cathode cannot be switched on instantly, and electron beam modulation is usually carried out using additional structure. The field emitter cold cathode has the advantage of working under room temperature which reduces the physical size of the device. Also, it has fast switch-on time capabilities, which reduces the electron gun complexity. Therefore, a cold cathode is an ideal candidate for electron sources of terahertz region VEDs. Various practical devices, such as X-ray tubes, electric propulsion system, scanning electron microscope, field emitter display, etc., have been developed out of field emitter cold cathode. Some vacuum electron sources in gigahertz to terahertz ranges have also developed using cold cathode in laboratory demonstration. However, there is a further scope of work yet to be carried out on a cold cathode such as life estimation, emission current enhancement, mechanical strength enhancement, etc.

References

1. A. Wehnelt, Oxide coated cathode, *Annalen der Physik*, vol. 14, p. 425, 1904.
2. H. J. Lemmens, M. J. Jansen, and R. Loosjes, A new thermionic cathode for heavy loads, *Philips Tech. Rev.*, vol. 11, pp. 341–350, 1950.
3. R. C. Hughes, P. P. Coppola, and E. S. Rittner, US Patent 2 700 118, Nov. 1951.
4. R. C. Hughes, P. P. Coppola, and H. T. and Evans, Chemical reaction in barium oxide on tungsten emitters, *J. Appl. Phys.*, vol. 23, pp. 635–641, 1952.
5. P. Zalm, A. J. A. Van Stratum, Osmium dispenser cathodes, vol. 27, no. 3/4, pp. 69–75, 1966.
6. R. S. Raju, Studies on W-Re mixed-matrix cathodes, *IEEE Trans. Electron Devices*, vol. 56, no. 5, pp. 786–793, 2009.
7. R. S. Raju, Impregnated cathodes for use in high power microwave tubes, Ph.D. Thesis, Corpus Christi College, Cambridge University, February 1987.
8. R. S. Raju, C. E. and Maloney, Characterization of impregnated scandate using a semiconductor model, *IEEE Trans. Electron Devices*, vol. 41, pp. 2460–2465, Dec. 1994.
9. E. S. Rittner, R. H. Ahlert, W. C., and Ruttledge, Studies on mechanism of operation of L-cathode - I, *J. Appl. Phys.*, vol. 28, no. 2, Feb. 1957.
10. J. W. Gibson, and R. E. Thomas, *Appl. Surf. Sci.*, vol. 16, p. 163, 1983.
11. A. S. Gilmour, Jr. Klystrons, *Traveling Wave Tubes, Magnetrons, Crossed-Field Amplifiers, and Gyrotrons*, Artech House, Boston, London.
12. E. L. Murphy and R. H. Good, Thermionic emission, field emission, and the transition region, *Phys. Rev.*, vol. 102, no. 6, pp. 1464–1473, 1956.
13. H. B. Skinner, R. A. Tuck and P. J. Dobson, Theoretical models of dispenser cathode surface, *J. Phys. D: Appl. Phys.*, vol. 15, pp. 1519–1529, 1982.
14. J. Zhao, et al., Scandate dispenser cathode fabrication for a high-aspect-ratio high-current-density sheet beam electron gun, *IEEE Trans. Electron Devices*, vol. 59, no. 6, pp. 1792–1798, 2012.

15. R. T. Longo, Long life high current density cathodes, *Int. Elect. Defr. Meet. Tech. Digest*, IEEE, vol. 152, 1978.
16. R. K. Barik, Experimental study on nanoparticle-based high current density cathode for terahertz devices, Ph.D. dissertation, Department of Electrical Engineering and Computer Science, Seoul National University, Seoul, Korea, Aug. 2014.
17. E. S. Rittner, R. H. Ahlert, and W. C. Ruttledge, Studies on mechanism of operation of L-cathode - I, *J. Appl. Phys.*, vol. 28, no. 2, Feb. 1957.
18. R. E. Thomas and T. Pankey, Controlled-porosity dispenser cathode, U.S. Patent 4 101 800, Jul. 18, 1978.
19. R. E. Thomas and R. F. Greene, Controlled porosity sheet for thermionic dispenser cathode and method of manufacture, U.S. Patent 4 379 979, Apr. 12, 1983.
20. L. R. Falce and G. S. Breeze, Controlled porosity dispenser cathode, U.S. Patent 4 587 455, May 6, 1986.
21. R. F. Greene and R. E. Thomas, Method of manufacturing integral shadow gridded controlled porosity, dispenser cathodes, U.S. Patent 4 745 326, May 17, 1988.
22. R. L. Ives, L. R. Falce, S. Schwartzkopf, and R. Witherspoon, Controlled porosity cathodes from sintered tungsten wires, *IEEE Trans. Electron Devices*, vol. 52, no. 12, pp. 2800–2805, Dec. 2005.
23. A. K. Singh, M. Ravi, M. S. Bisht, R. K. Barik, S. K. Shukla, R. Prajesh, T. P. Singh, S. K. Saini, and R. S. Raju, Study and development of active sintered controlled porosity dispenser cathode, *IEEE Trans. Electron Devices*, vol. 62, no. 11, pp. 3837–3843, November 2015.
24. R. T. Longo, E. A. Alder, and L. R. Falce, Dispenser cathode life prediction model, *in Proc. IEDM*, pp. 318–321, 1984.
25. R. K. Barik, R. S. Raju, A. K. Tanwar, Supriyo Das and G. S. Park, High current density ternary-alloy-film dispenser cathode for terahertz vacuum devices, *in Proc. IEEE IVEC*, Feb. 21–24, 2011, Bangalore, India.
26. R. K. Barik, A. Bera, R. S. Raju, A. K. Tanwar, I. K. Baek, S. H. Min, O. J. Kwon, M. A. Sattorov, K. W. Lee, and G.-S. Park, Development of alloy-film coated dispenser cathode for terahertz vacuum electron devices application, *Appl. Surf. Sci.*, vol. 276, pp. 817–822, 2013.
27. A. Figner, A. Soloveichik, and I. Judinskaya, Metal porous body having pores filled with barium scandate, US Patent 3350170, Dec. 1967.
28. A. V. Stratum, J. V. Os, J. R. Blatter, and P. Zalm, Barium-aluminum-scandate dispenser cathode, US Patent 4007393, 1977.

29. A. V. Oostrom and L. Augustus, Activation and early life of a pressed barium scandate cathode, *Appl. Surf. Sci.*, vol. 2, pp. 173–186, 1979.

30. S. Taguchi, T. Aida, and S. Yamamoto, Investigation of Sc_2O_3 Mixed-matrix Ba-Ca aluminate-impregnated cathodes, *IEEE Trans. Electron Devices*, vol. 31, no. 7, pp. 900–903, 1984.

31. J. Hasker and J. E. Crombeen, Scandium supply after ion bombardment on scandate cathodes, *IEEE Trans. Electron Devices*, vol. 37, no. 12, pp. 2589–2594, 1990.

32. J. Hasker, J. V. Esdonk, and J. E. Crombeen, Properties and manufacture of top-layer scandate cathodes, *Appl. Surf. Sci.*, vol. 26, pp. 173–195, 1986.

33. S. Yamamoto, S. Sasaki, S. Taguchi, I. Watanabe, and N. Koganezawa, Application of an impregnated cathode coated with W-Sc_2O_3 to a high current density electron gun, *Appl. Surf. Sci.*, vol. 33–34, pp. 1200–1207, 1988.

34. S. Yamamoto, I. Watanabe, S. Taguchi, S. Sasaki, and T. Taguchi, Formation mechanism of a monoatomic order surface layer on a Sc-type impregnated cathode, *Jpn. J. Appl. Phys.*, vol. 28, pp. 490–494, 1989.

35. J. Wang, et al., A study of Scandia doped tungsten nano-powders, *J. Rare Earths*, vol. 25, pp. 194–198, 2007.

36. J. Wang, et al., A study of scandia and rhenium doped tungsten matrix dispenser cathode, *Solid State Sci.*, vol. 9, pp. 924–932, 2007.

37. Y. Wang, et al., Correlation between emission behavior and surface features of scandate cathodes, *IEEE Trans. Electron Devices*, vol. 56, no. 5, pp. 776–785, 2009.

38. J. Wang, et al., Sc_2O_3–W matrix impregnated cathode with spherical grains, *J. Phys. Chem. Solids*, vol. 69, pp. 2103–2108, 2008.

39. Y. Wang, J. Wang, W. Liu, K. Zhang, and J. Li, Development of high current-density cathodes with scandia-doped tungsten powders, *IEEE Trans. Electron Devices*, vol. 54, no. 5, 2007.

40. Y. M. Shin, J. Zhao, L. R. Barnett, and N. C. Lumann, Investigation of terahertz sheet beam traveling wave tube amplifier with nanocomposite cathode, *Phys. Plasmas*, vol. 17, pp. 123105-1–4, 2010.

41. J. E. Lilienfeld, Report on autoelectronic emission, 1922.

42. R. H. Fowler and L. Nordheim, Electron emission in intense electric fields, *Proc. R. Soc. Lond. A Math. Phys. Sci.*, vol. 119, no. 781, pp. 173–181, May 1928.

43. E. W. Muller, *Z. Tech. Phys.*, vol. 17, p. 412, 1936.

44. A. K. Singh, S. K. Shukla, M. Ravi, and R. K. Barik, A Review of Electron Emitters for High-Power and High-Frequency Vacuum Electron Devices, *IEEE Trans. Plasma Sci.*, vol. 48, no. 10, pp. 344–3454, 2020.
45. J. Zhao, et al., High current density and long-life nanocomposite scandate dispenser cathode fabrication, *IEEE Trans. Electron Devices*, vol. 58, no. 4, pp. 1221–1228, Apr. 2011.
46. X. Yuan, W. Zhu, Y. Zhang, N. Xu, Y. Yan, J. Wu, Y. Shen, J. Chen, J. She, and S. Deng, A fully-sealed carbon-nanotube cold-cathode terahertz gyrotron, *Sci. Rep.*, vol. 6, pp. 32936-1–9, doi: 10.1038/srep32936 2016.
47. I.-K. Baek, R. Bhattacharya, J. S. Lee, S. Kim, D. Hong, M. A. Sattorov, S.-H. Min, Y. H. Kim, and G.-S. Park, Uniform high current and current density field emission from the chiseled edge of a vertically aligned graphene-based thin film, *J. Electromagn. Waves Appl.*, vol. 31, pp. 2064–2073, 2017.
48. S. B. Fairchild, P. Zhang, J. Park, T. C. Back, D. Marincel, Z. Huang, and M. Pasquali, Carbon nanotube fiber field emission array cathodes, *IEEE Trans. Plasma Sci.*, vol. 47, no. 5, pp. 2032–2038, May 2019.
49. S.-D. Liang, *Quantum Tunneling and Field Electron Emission Theories*, World Scientific Publishing Company, pp. 157–207, 2014.
50. N. De Jonge, M. Allioux, M. Doytcheva, M. Kaiser, K. B. K. Teo, R. G. Lacerda, and W. I. Milne, Characterization of the field emission properties of individual thin carbon nanotubes, *Appl. Phys. Lett.*, vol. 85, pp. 1607–1609, 2004.
51. X. Q. Wang, M. Wang, P. M. He, Y. B. Xu, and Z. H. Li, Model calculation for the field enhancement factor of carbon nanotube, *J. Appl. Phys.*, vol. 96, no. 11, pp. 6752–755, 2004.

Chapter 23

Microfabrication Technologies

Colin D. Joye,[a] **Alan M. Cook,**[b] **and Diana Gamzina**[c]

[a] *SpaceX, Redmond, WA, USA*
[b] *U.S. Naval Research Laboratory, Washington, DC, USA*
[c] *SLAC National Accelerator Laboratory, California, USA*
colin.joye@spacex.com

23.1 Introduction

The method of fabrication to be used for creating vacuum electronic (VE) THz devices that require tight-tolerance features should not be left as an afterthought. At such scale-lengths as required for THz slow-wave devices, the fabrication method often drives what type of circuit can be readily produced, rather than being a mere means used to an arbitrary end. With that limitation in mind, it is important to identify what types of circuits may be achievable for a given fabrication technology.

23.1.1 Purpose/Objectives

The purpose of this chapter is to identify and quantify types of fabrication technologies that might be used to create THz structures.

Advances in Terahertz Source Technologies
Edited by Gun-Sik Park, Masahiko Tani, Jae-Sung Rieh, and Sang Yoon Park

ISBN 978-981-4968-89-8 (Hardcover), 978-1-003-45967-5 (eBook)
www.jennystanford.com

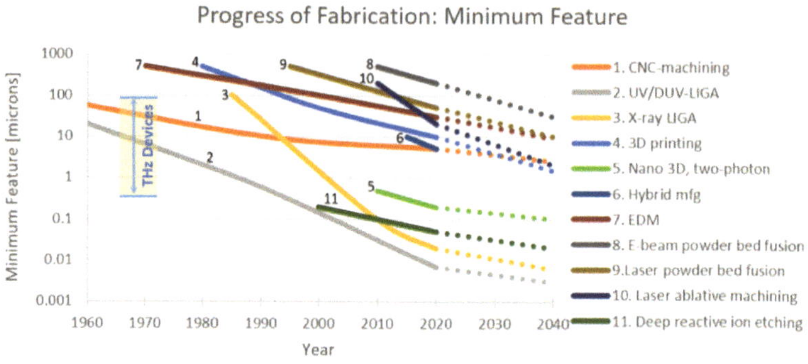

Figure 23.1 Illustration of the chronological progress of minimum feature sizes of various technologies, along with projections into the future (dotted lines). THz devices require minimum feature sizes approximately in the 0.5–100 micron range, depending on the target frequency. The data are found in the references [1–15].

In addition to exposing the latest developments and capabilities or various techniques, insights are provided that are specifically relevant to VE devices.

23.1.2 The State of Microfabrication Techniques

Microfabrication (MF) techniques have made enormous improvements since the year 2010. There has been much effort aimed at filling the "technology gap" between solid-state device fabrication techniques, covering size scales of nanometers to tens of microns, and conventional machining, covering tens of microns to meters. It is precisely in the range of 1–1000 microns that the THz device scales are of critical importance. Minimum feature sizes occupy the range of approximately 1–100 microns, depending on frequency, and limit the number of possible fabrication methodologies that can be applied (Fig. 23.1). In particular, for minimum feature sizes in the 1–10 micron range, there is a sort of technology gap—Certainly lithography, for example, is capable of the small-end of this fabrication range, but does not extend itself well to the large-end of the requirements, as typical THz circuits are 10–100 mm in length and can exceed 2 mm in thickness at the 100 GHz end of the spectrum. MEMS-compatible fabrication techniques such as

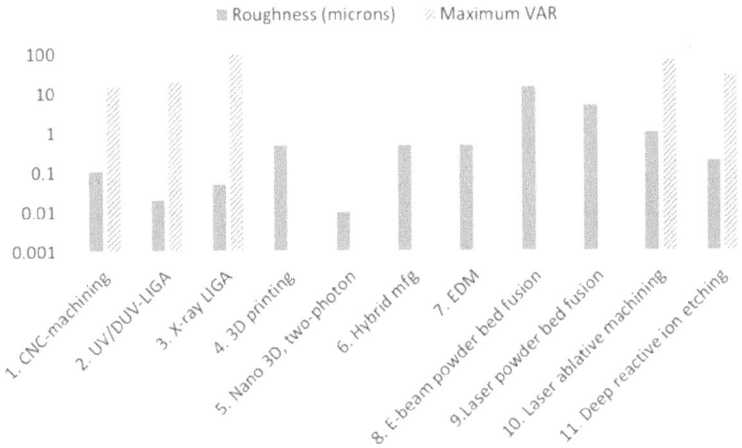

Figure 23.2 Current best-reported roughness estimates for various manufacturing techniques, along with the maximum vertical aspect ratio (VAR) achievable. Processes lacking a datum bar for Maximum VAR essentially have no fundamental limitation on VAR. Data are found in the references [16, 17] and the additional references from Fig. 23.1.

deep reactive ion etching (DRIE) are applicable, more or less, to the scales in question, but they are typically not relevant to THz device materials, such as copper.

23.1.3 Scope of Chapter, Scales

To provide examples of relevance, several prototypical slow-wave circuits are considered. Figure 23.3 shows three types of circuits with characteristic dimensions indicated, along with scaling curves and a subset of applicable fabrication methods. From this chart, it can be seen that to cover 100 GHz to 1 THz, the feature sizes must cover approximately 10 microns to 1 mm. Roughness is another issue entirely and must be orders of magnitude less than the relevant feature size at a given frequency in order to reduce millimeter-wave losses due to skin depth effects. Even though this chart is approximate, we get the sense that only the X-ray/UV-LIGA techniques are potentially applicable to the entirety of the 100 GHz to 1 THz range. LIGA is a German acronym, Lithographie, Galvanoformung, und Abformung, for a

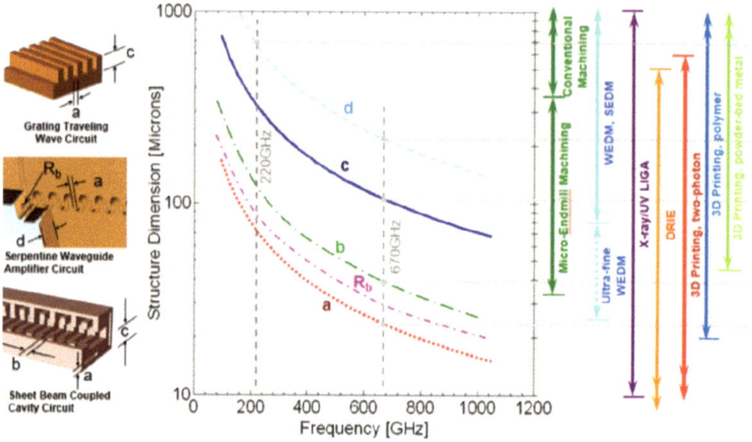

Figure 23.3 (left) Prototypical circuits to consider: (top) Grating. (middle) Serpentine waveguide. (bottom) Sheet beam coupled-cavity TWT. (center) Typical scaling curves for the dimensions indicated covering the THz regime, along with approximate applicability ranges for a subset of manufacturing methods (right).

process that combines a photo-generated mold or pattern with electroforming of metal. The "UV" in UV-LIGA stands for ultraviolet to differentiate it from X-ray LIGA, which uses X-rays as the photo-activator. At the lower end of the frequency spectrum in Fig. 23.3, the motor-driven methods dominate: computer numeric control (CNC) micromachining, electrical discharge machining (EDM), and conventional 3D printing. At the high end, the optical-based methods tend to prevail: UV-LIGA, two-photon 3D printing, and DRIE, the latter of which employs a photo-defined passivation mask.

Roughness is a very important feature to consider when evaluating fabrication techniques, and, unfortunately, since roughness comes in an incredibly wide range of characteristic scale sizes and geometries, the effects of roughness are relatively poorly understood in this 100 GHz to 1 THz range (Fig. 23.2).

Among important considerations for slow-wave circuit fabrication is the VAR required (Fig. 23.2). A low-VAR structure, say 1:1 ratio (a square) would be relatively easy to achieve compared to a 100:1 ratio—for example, walls that are 1 mm high but only 10 microns thick and separated by 10 micron gaps. High VAR is particularly important for low-beam-voltage devices, as the electron beam travels slowly and the interaction gaps are therefore required to be quite narrow. We now explore the tradeoffs in vertical aspect ratio, beam voltage, and operating frequency based on a prototypical structure, such as a serpentine waveguide TWT circuit:

Assuming an operating frequency, f_{op}, 25% above the cutoff frequency, f_c, and a waveguide long-dimension, a, then

$$a = (1.25)\frac{c}{2f_{op}}.$$

A practical upper limit assumption for the waveguide short dimension, b, which is also the beam-wave interaction gap, is $p/2$, where p is the serpentine waveguide half-period. For gap transit angle θ_g, the change in RF phase experienced by the electron as it traverses the interaction gap length $p/2$, and electron velocity v_e we have,

$$\frac{1}{p} = \frac{f_{op}}{v_e}\frac{\pi}{\theta_g}.$$

The VAR required for a given $\beta_z = v_e/c$ is simply,

$$\mathrm{VAR} = \frac{a}{\left(\frac{p}{2}\right)} = \frac{1.25}{\beta_z}\frac{\pi}{\theta_g},$$

where,

$$\beta_z = \sqrt{1 - \frac{1}{\left(1 + \frac{V_a e}{m_e c^2}\right)^2}},$$

where V_a is the accelerating voltage, e is the electron charge, m_e is the electron mass, and c is the speed of light. Figure 23.4 illustrates the dependence of VAR and the frequency-period product on transit angle and beam voltage. Clearly, the VAR goes very high, above 20, for accelerating voltages below about 5 kV for the 90° transit angle case. As the accelerating voltage goes lower, the frequency-period product must also decrease, implying that either the operating frequency must be reduced, or the period reduced, or both. Reduction in period means smaller features that are harder to fabricate and, ultimately, lower power handling capability.

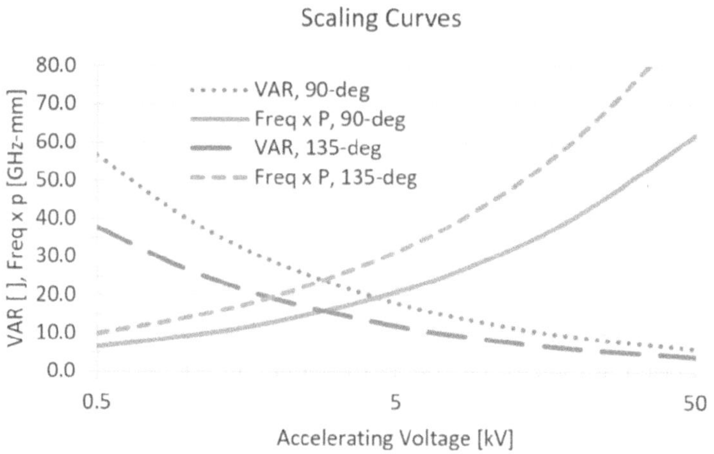

Figure 23.4 Scaling curves for vertical aspect ratio (VAR) and frequency-period product for 90° and 135° transit angles, assuming operating frequency is 25% above cutoff. The required vertical aspect ratios become extremely high for low accelerating voltages.

23.2 Microfabrication Materials

The pool of choices for microfabrication materials for THz circuits is, sadly, quite small. The key feature for successful circuit operation is high effective surface conductivity. It is possible to start with a high-conductivity material, such as copper, and make it less conductive by coatings, roughness, or sub-wavelength perforations, but it is generally not possible, by any of these means, to increase the effective conductivity better than copper, gold, silver, or aluminum, at least not over broad bandwidth. Therefore we tend to start with high-conductivity materials, or, where possible, use coatings of high-conductivity materials, in order to achieve the low losses necessary for efficient THz wave generation.

23.2.1 Copper

Copper is historically the go-to material for VE device fabrication owing to its high electrical conductivity, high thermal conductivity, high melting point, strong vacuum compatibility, ease of machining, ease of brazing and joining, relatively low cost (compared to silver or gold), and abundant availability in pure form. Oxygen-free high-conductivity (OFHC) copper is regarded as the "only" suitable material for VE device circuits.

Copper has plenty of drawbacks, however, mainly that it is very soft after brazing and therefore must be well protected from stresses. Also, some suppliers of copper may have copper tainted with oxygen (wrong grade), which looks and behaves exactly like OFHC copper until it is hydrogen brazed. When copper containing small amounts of oxygen (approximately 100 ppm or greater, by weight) it becomes embrittled in a hydrogen furnace and grows significantly due to the formation of steam as the hydrogen combines with entrapped oxygen. It is extremely difficult to de-oxygenate copper powder to less than 10 ppm at this time, and so 3D printing methods available as of this writing almost universally do not use truly OFHC copper. Copper powder with extremely low oxygen content sticks to itself and has been found difficult to use in powder-based 3D-printing applications [18].

Finally, the same properties that make copper useful for VE devices, namely high electrical and thermal conductivities, also make it very difficult to directly 3D print by laser and electron beam melting or sintering methods. The high thermal conductivity makes it problematic to obtain high 3D "voxel" resolution as the heat tends to spread outside of the local melt pool quickly, artificially enlarging the voxel region. The high electrical conductivity makes it resist absorption of infrared laser light, leading to low efficiency of energy transfer.

23.2.1.1 Electronic grade oxygen-free copper

Electronic grade oxygen free (pure) copper is generally the main type deemed suitable for VE devices. While grade C10100 OFHC copper is likely the most available, a stress-relieved grade, such as ASTM F68-16, often referred to as "F68," is preferred for THz devices because it has low residual stresses and, as a result, low levels of stress-induced deformation occur during fabrication. Annealed, pure copper should have a DC conductivity of at least 101% IACS, per ASTM F68 and B170. Copper that is very soft, however, becomes more difficult to machine using milling operations and can lead to excessive surface roughness and burrs. These burrs can usually be cleaned up by deburring milling operations and the use of diamond or coated carbide tooling.

23.2.1.2 Cupronickel

Alloys of copper and nickel are sometimes used, especially where greater post-braze strength is required, or where electromagnetic losses are not very important, such as for loads or horns. A wide spectrum of copper-to-nickel ratios is possible, but the loss of conductivity can be significant with moderate amounts of nickel, for example, 83% IACS conductivity is obtained for just 0.3% of nickel addition [19]. The more common 90-10 grades of cupronickel yield about 45% IACS conductivity, so this alloy is not commonly used in slow-wave circuits except as a lossy material that can be used to stabilize an amplifier by preventing spurious oscillations.

23.2.1.3 Glidcop®

Glidcop® is a dispersion-strengthened material consisting of nano-sized alumina mixed into the copper host. It provides a tremendous increase in copper hardness that persists even at the highest-temperature braze cycles, more so than alloys such as cupronickel or copper-zirconium. Small to modest amounts of alumina, such as grade Glidcop® AL-15 (0.3% alumina in OFHC copper) provide dramatic increases in hardness while having minimal effect on conductivity, 92% IACS [20] [21]. There have been reports that Glidcop® is more challenging to braze with certain braze materials owing to the alumina content. In these cases, a fresh skim-cut on the Glidcop® surface or a copper- or nickel-plating step can be used to enhance the brazing success. Since there is a high diffusion coefficient for silver-based brazing of Glidcop®, silver-based alloys should be carefully evaluated, or a coating of nickel on the Glidcop® will protect against diffusion [22].

23.2.1.4 Elkonite®

The trade-name Elkonite® applies to sintered refractory metal stock (most commonly tungsten, but also molybdenum or tantalum) that has been infiltrated with copper or even silver. Various grades are available, typically from 15–60% copper by weight (balance tungsten). For VE, a high-copper content would be beneficial (perhaps 80%), but the common grades, such as Elkonite® 1W3, contain only 44% copper and realize only 55% IACS conductivity, and are therefore not attractive for THz devices. Due to the refractory content, these materials can impart significantly more wear on cutting tools than pure copper.

23.2.2 Silver

If you need the highest conductivity possible at any expense, silver is a possible choice. With approximately 10% higher electrical and thermal conductivities than copper, it is likely to help only when the lowest possible losses can be tolerated, such as in high-Q resonant cavities. Silver comes with a lower melting point than copper, however, of about 962°C. Silver alloys have been successfully used in waveguides since the 1940s, but the price of silver recently has precluded its use except in the most demanding applications. Silver may have some unexplored benefits, however, such as ease of oxygen removal for use as a 3D-printed powder with low friction, but the high cost is likely to prevent its use except in the research environment. Additionally, it is important to consider the relatively high vapor pressure of silver at braze furnace temperatures: Approximately 10^{-2} Torr at 1020°C, versus 10^{-4} Torr for copper at the same temperature [23].

23.2.3 Aluminum

Aluminum alloys have been successfully used for waveguide components and other passive components (couplers, magic-tees, housings, etc.), especially if plated with gold for higher conductivity and for better environmental protection. While the vapor pressure of pure aluminum is only marginally higher than copper throughout the temperature spectrum [23], its melt point is considerably lower at approximately 660°C. Aluminum alloys almost universally include high vapor pressure materials such as zinc, manganese, and magnesium, and for that reason should be avoided in furnace and ultrahigh-vacuum applications.

23.2.4 Other Alloys

Copper-zirconium (e.g., Amzirc® [24], CW120C, C15000), copper-chromium (GRCop-84 [25]), and copper-chromium-zirconium have long been used in applications where high electrical conductivity, up to 95% IACS copper, and/or high thermal conductivity, combined with better mechanical strength are needed, such as for knife switches or resistance welding electrodes. Copper-zirconium has been used for VEDs [26], and features higher softening temperature of about 500°C, which is a typical bakeout (thermal processing) temperature, compared to pure copper. It also does not embrittle in a hydrogen braze environment owing to the gettering action of the zirconium. Precipitates tend to grow during repeated thermal braze cycles, continuously reducing the strengths of these alloys.

23.2.5 Lossy and Dielectric Materials

Multi-material microfabrication techniques are very attractive because they would enable a wide variety of devices to be easily fabricated. Controlled-loss materials, such as cupronickel, carbon, or lossy ceramic materials would be very useful to help stabilize THz amplifier circuits. Being able to lay them down at the same time as the low-loss circuit materials would be immensely useful.

Dielectric materials laid down with the circuit materials would also enable a wide range of interesting applications, but as the microfabrication of useful copper circuits is difficult enough in the THz range, it appears that the prospects of successfully adding lossy and/or dielectric fabrication steps is a while off yet.

23.3 Machines and Techniques

The first documented lathes begin to appear in the mid-13th century with the introduction of the belt drive and crank. The milling machine with a dedicated rotating spindle traces its roots back to the early 1800s. After improvements and developments, such as being able to machine six sides of a nut around 1830 (J. Nasmyth), the first milling machine was demonstrated in 1862 [27], having been designed by Joseph R. Brown, who later partnered with Lucian Sharpe to form Brown and Sharpe Manufacturing Company, which exists to this day. The first CNC mills were put into commercial operation in the early 1950s after development in the 1940s by John T. Parsons. CNC mills finally became standard equipment in machine shops around the 1980s, when they were able to be controlled by a desktop computer. In the late 1700s, Joseph Priestly discovered that spark discharges can erode metals—The nascent EDM technique. The first usable EDM machines emerged in the early 1950s. Around 1965, the first laser machine demonstrated the ability to drill holes in diamond substrates. The laser became more of a standard cutting tool in the 1970s. A lot of the early work and patents related to 3D printing originated in the early 1980s. 3D printers have now become so common that you can buy a 3D filament extrusion printer for your home use. So in many cases, there are decades from when a new invention comes out— In some cases hundreds of years— until it becomes widely used and commonplace. The point where micro manufacturing is today rests upon hundreds of years of developments in metalworking.

23.3.1 Subtractive Methods

Arguably the oldest and most established manufacturing technology is the subtractive manufacturing (SM) method. By this technique, material is removed from a larger block until it is the shape and size desired—From cutting a tree down in order to build a house, to nano-machining a highly precise micro-mirror, SM has become adept at doing it all. Subtractive CNC machining is only one example of SM—Other common methods include electrical discharge machining, laser cutting and ablation, water jet, and etching.

23.3.2 Additive Methods

The additive manufacturing (AM) concept is that material is added to, rather than subtracted from the base part. The widest-known modern commercial example is surely thermoplastic-extrusion 3D printing, also known as fused deposition modeling (FDM), where a part is normally built up directly onto a blank build plate.

An explosion of 3D-printing machines and techniques have flooded the marketplace since 2010. While many of these machine manufacturers have started to offer copper alloys as a standard material, they cannot yet process the oxygen-free grades. The most common and successful materials for 3D printing in direct metal are Inconel, nickel-based alloys, titanium, and even stainless steels. For metal-plated plastic, the challenge has been finding suitable chemistries to electrolessly plate metal onto polymeric forms. To get around the need for plating plastic, the inverse method allows a 3D-printed plastic mold to be electroformed with solid metal.

23.3.2.1 Direct AM

Direct 3D printing of arbitrary metal is a highly sought-after concept. Unfortunately, the same properties that make copper and silver attractive for their use in THz circuits, namely high electrical and thermal conductivity, also make fabrication by any direct route very difficult. Common direct fabrication methods are mostly powder-fusion based and include laser, electron beam, and binder jetting.

23.3.2.2 Indirect AM

Indirect additive manufacturing means that the ultimate material properties desired are not exhibited by the base AM material, but rather some other material that is coated, soaked in, sprayed, or otherwise imputed to the surface after the object has been manufactured to near-net shape using an AM process. Well-known examples include 3D-printed plastic that has been metalized to provide conductivity to the features [28]. Such resulting metalized plastic parts are low cost, lightweight, and can perform the same electrically as a solid part as long as the coating is several skin depths thick. Since the bulk material is still plastic, however, high average power handling capability is lost. Vacuum compatibility may be an issue for plated plastic, but the low weight of plated plastic makes this method attractive for non-vacuum, weight-critical uses, such as low-power airborne applications.

23.3.2.3 Inverse AM

It is also possible to create a temporary 3D-printed inverse mold around which the desired material is injected or applied, followed by removal of the mold. 3D-printed mold electroforming (3D-PriME) [29] is such a technique wherein copper is electroformed around a 3D-printed plastic mold, which is removed after the electroforming is completed. This is often an advantageous method for THz devices because only the vacuum region is 3D printed with metal being electroformed all around the mold. Downstream chemical etching (DCE) has been demonstrated to be effective for etching photoresist and 3D-printed polymers from a copper structure after electroforming [29, 30]. Investment casting would be considered inverse AM if the mold for the molten metal was to be made using AM methods. A similar technique to 3D-PriME using DRIE of silicon as a consumable mold for a nickel structure has been reported [31].

23.3.3 Hybrid Manufacturing

The combination of subtractive and additive machining has been termed "hybrid" manufacturing (HM) and generally applies to having both capabilities in a single machine such that the work piece can be added to and subtracted from multiple times in sequence without removal from the work table. The most common technique being offered at the time of this writing is simply making an AM tool that fits in the tool changer of a CNC milling machine. In other words, the CNC machine has both subtractive and additive tools available in its tool changer and can be programmed to add or subtract material while the substrate being operated on stays in the same fixture in the same machine (Fig. 23.5).

A major advantage of the HM process is that you can finish with the subtractive method in order to minimize roughness in a way that retains excellent surface conductivity.

On the other hand, the most successful materials for the AM-side of the HM process are low-thermal-conductivity materials for

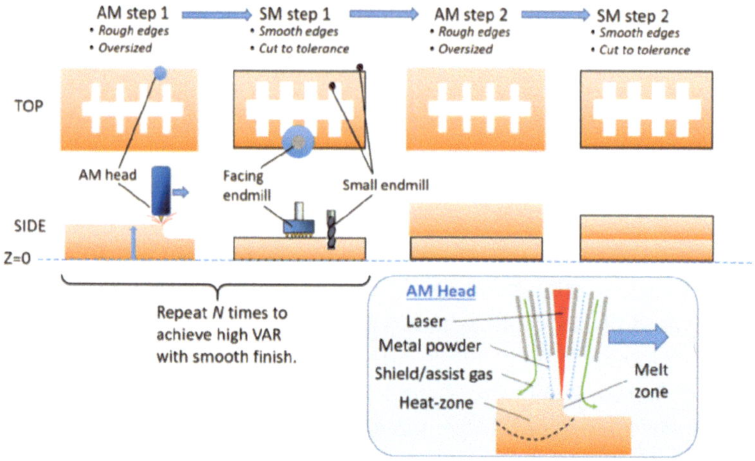

Figure 23.5 The hybrid manufacturing (HM) process consists of alternating between an additive process and a subtractive process. (Inset) Diagram of how a laser-based powder AM tool head works.

reasons stated previously. To further complicate the process, the SM operations are normally performed with copious amounts of coolant flowing in order to flush chips out and to keep the part cool. High-pressure cold air is a suitable alternative to flowing coolant. If coolant is used, the work piece's surfaces may need to be dried each time or even cleaned before the AM tool is put to use in order to promote adhesion of the AM material.

23.3.4 Multi-Material Manufacturing

The ideal system would be able to direct-print copper, lossy materials, and dielectrics all in one process. The main challenge, however, is that materials have different coefficients of thermal expansion (CTE), and processing for ultrahigh vacuum requires a 500°C bake out for efficient removal of water vapor adsorbed onto the surfaces. The CTE for silver is even worse (approximately 19.7×10^{-6} m/m-K at room temperature) than copper (17.7×10^{-6} m/m-K) or gold (14.2×10^{-6} m/m-K). Stainless steel is a remarkably close CTE match to copper even up to 1000°C and would be a likely candidate as a lossy material, at least for the vacuum-compatible grades, such as 316 and 304 L.

The ideal dielectric would be suitable as a vacuum window material and also have a close enough CTE match to copper or stainless steel that it would form a strong, leak-tight joint during bakeout. Modern glasses with high CTEs up to 20×10^{-6} m/m-K exist [32] that could be candidates for this type of multi-material integration, but a major challenge is getting the glass to wet to the base metal or vice versa.

We now study individual fabrication methods in detail.

23.3.4.1 Micro-CNC

CNC subtractive machining has made huge strides since even 2010. Movement capability down to nanometers is readily achievable now. Endmills down to 0.001" (25 microns) diameter are off-the-shelf, even made from diamond. Advances in the machining techniques have proven very helpful for THz VE devices. Constant-chip load cutting techniques, such as trochoidal cutting, produce noticeably cleaner finishes for THz devices than the traditional straight-line cuts or "G02/G03" circular arc cuts. The key is to keep constant loading on the tool so that the tool deflection is constant. Of course, zero tool deflection is always the goal, but as tool length increases, the deflection increases to the 3rd power, so high aspect ratio cuts, such as serpentine TWTs and other common THz slow-wave circuits always have to fight against tool deflection.

Tool deflection, δ, in the static case, can be estimated in millimeters by the following equation [33]:

$$\delta = C\frac{F}{E}\left(\frac{L_1^3}{D_1^4} + \frac{L_2^3 - L_1^3}{D_2^4}\right)^N,$$

where the constant C is equal to approximately 9.05, 8.30, or 7.93, and N is approximately 0.983, 0.981, or 0.980 for the case of 4-flute, 3-flute, or 2-flute cutters, respectively. In addition, F is the force applied (Newtons), E is the modulus of elasticity for the endmill material (GPa), L_1 and D_1 are the cutter length and diameter in mm, and L_2 and D_2 are the shank length and diameter in mm, respectively. Clearly, as the ratio of length to diameter of the cutting tool, L_1/D_1

increases, the deflection is in serious danger of becoming large due to the very high exponents of dependence. In fact, if the cutter diameter decreases by just 10%, the deflection increases by over 50%. Based on this analysis, a carbide tool with a cutting diameter of 0.5 mm and cutting length of 10 mm will deflect about 10 microns with a 0.05 N cutting force applied, so to keep the deflection to the 1 micron-level means the cutting force must be limited to 0.005 N instead, resulting in a much slower cut, or, alternately, many more passes that remove far less material with each pass. In either case, the result is far longer machining time for high L_1/D_1 ratios. In the dynamic case with high spindle speed, resonances can develop on the endmill that can significantly exceed this estimation of deflection.

For micro-CNC on copper THz circuits, flowing coolant (liquid or gas) is a must to remove chips that could get caught in the endmills during machining. Multiple finishing cuts are a must on high-aspect-ratio features in order to reduce tool deflection. Finishing passes for the purpose of deburring are often advisable in order to remove burrs from the circuits. Pure copper tends to "hang on" to burrs quite well, and some of the more tenacious burrs must be bent back and forth several times before breaking free. Alternatively, it is possible to chemically treat copper to remove burrs. In some cases, even on copper, tool wear can be a problem. Carbide or diamond tooling is preferred for these cases.

For any SM process, it is absolutely imperative that all chips, dust, slag, and other defects be fully removed from the structure by washing, pressurized air, ultrasonic agitation, and/or any other means possible before the circuit is brazed together. Figure 23.6 shows examples of metallic chips in a slow-wave circuit and waveguide captured by non-destructive 3D X-ray computed tomographic imaging.

Finally, the vertical aspect ratio of micromachining processes is limited by the deflection of long, thin endmills. For a prototypical slow-wave circuit, one can expect to achieve about 7:1 VAR, and maybe up to 15:1 in a research environment with the most careful and costly methods applied. A low-voltage, Ka-band TWT required 14:1 VAR, but it was made successfully in two halves of 7:1 VAR each [34, 35].

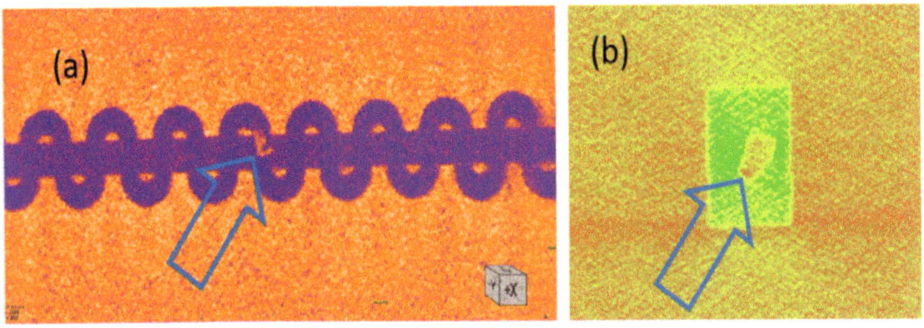

Figure 23.6 Non-destructive X-ray inspection reveals metallic chips in (a) a 235 GHz serpentine waveguide slow-wave circuit, (b) a waveguide stepped transition.

The limit for micro-CNC manufacture at this time seems to be in the 300 to 400 GHz range [16]. The reason for this limitation is tool vibration and size. The size of the tool becomes on the order of grain sizes of the material that the tools are made of, and the high-speed machining operation causes fractures to happen along the grain boundaries with high frequency. Additionally, with the smallest endmills, the depth-of-cut should be limited to 5 microns and the stepover approximately the same size [36]. It is also important that the tools be certified to be within a tight tolerance, typically $+/-3$ microns and the tool not loaded heavily on the final cuts [36]. If all of these recommendations are adhered to, CNC machining can produce an excellent product with a mirror-like finish.

23.3.4.2 Electron beam AM

Electron beam powder bed fusion (E-PBF) additive manufacturing typically uses a thin layer of metal powder, precisely laid down by a powder spreader, followed by melting or sintering by a moderate to high-power electron beam. The electron beam can be very quickly directed to anywhere on the build area, so "dithering"—rapidly moving the beam across many different spots - of the electron beam heating is commonly used to carefully control the melt pool sizes, prevent cracking, and increase precision.

A disadvantage of E-PBF is that it must be performed in a vacuum, which complicates the processing machinery dramatically and also disallows any type of reducing or inert atmosphere as a process possibility. This is especially problematic for copper, since elimination of all oxygen from pure copper results in a sticky powder that does not spread well. If hydrogen could be used in the chamber, it may be possible to use copper powder containing residual oxygen for ease of spreading, which could be scrubbed of oxygen during the dynamic melting process.

Recently, copper E-PBF technology has been developed to produce high-purity copper components suitable for ultrahigh-vacuum RF applications showing outgassing rates of printed materials equivalent to those of commercially available pure copper [37]. Surface roughness and feature size limitations remain to be a challenge for E-PBF. The spot size of the electron beam in commercially available systems is commonly limited to about 200 microns as a tradeoff between build time and resolution. As a result of this and other system parameters, surface roughness that can be achieved is limited to Ra of 25 microns and feature size to about a few millimeters. In addition, due to self-repulsive forces within the electron beam, there is a limit to how small the electron beam spot size can be for a fixed amount of current. However, particle size availability and thermal conduction typically limit the resolution to about 20 microns in copper. The smallest RF-relevant parts to date printed using E-PBF are at 100 GHz frequencies, as shown in Fig. 23.7d. Surface roughness achievable today causes high levels of loss and hence is unsuitable for manufacturing of most Terahertz systems [37, 38].

Figure 23.7a,b shows section views of traveling-wave circuit structures manufactured using the E-PBF process at X-band frequencies that could be utilized in particle accelerators and coupled-cavity traveling-wave tube amplifiers. The former has stringent surface roughness requirements not yet achievable via additive processes, while the latter can accommodate some degree of roughness without affecting performance significantly. Figure 23.7c demonstrates scalability of the process in the additive machining centers where multiple devices can be manufactured simultaneously without adding time to the process. This scalability significantly improves cost and lead times for vacuum electronic structures. Figure 23.7d shows a demonstration WR-10 waveguide manufactured via the E-PBF process to evaluate feature size, surface roughness,

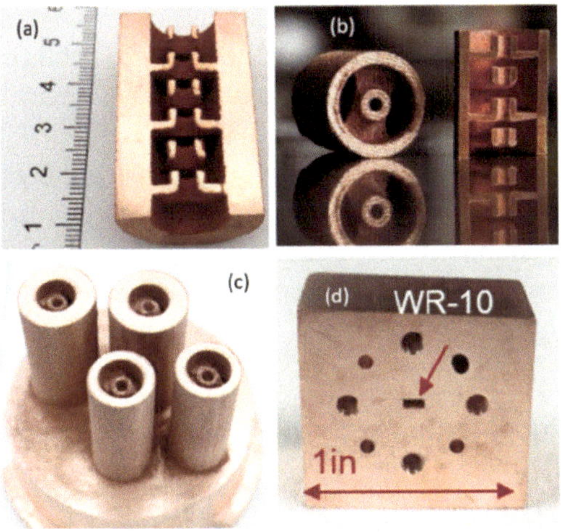

Figure 23.7 Shows photos of copper accelerator cavities created by E-PBF: (a) and (b) section views of traveling-wave circuit structures manufactured using an E-PBF process at X-band frequencies. (c) Four traveling-wave structures manufactured simultaneously on a 4 inch (100 mm) base. (d) A W-band (95 GHz) waveguide test part was created by E-PBF and then smoothed internally by magnetorheological polishing [39].

and tolerances, establishing approximate frequency limits for this technology for the time (circa 2018).

23.3.4.3 Laser powder bed fusion (L-PBF)

Like electron beam AM, L-PBF creates 3D objects by melting or sintering thin layers of metal powder precisely spread. Unlike E-PBF, however, the L-PBF can be done in inert or reducing atmospheres, which could be an advantage for easily-oxidized metals such as copper. Copper, however, does not readily absorb infrared laser light, so a few companies are tackling the use of green or blue-green (455 nm) lasers to significantly increase the power absorption in copper [40].

A major issue with copper and other high-thermal-conductivity materials is that the heat spreads very quickly, limiting the feature resolution that can be achieved, and diluting the concentration of power. This results in significant heating of the part, which in turn causes thermal expansion, which is another factor deleterious to the precision sought. CNC micromachines, by contrast, typically need to operate within a few degrees C or even less to achieve mirror-like finish [36], and the two-photon polymerization machines ask for under $+/-1°C$ stability [8].

Because the laser can be focused down quite tightly, small features down to about 10 microns in size can be readily made, if low-thermal-conductivity materials are used. Copper is entirely a different story and the resolution is again limited to about 20–30 microns, as in E-PBF.

23.3.4.4 Binder jetting

Like many of these AM techniques, binder jetting (BJ) relies on thin layers of spread metal powder to create 3D structures. BJ is a process akin to inkjet printing in that a binder "glue" is injected into the powder by a print head to create a 2D pattern. Then a new layer of metal is spread, followed by another round of binder printing. This part of the process is performed at room temperature, so high CTE materials do not move because they are not heating and cooling during the process.

Binder jetting produces only a weakly held powder—a "green body." The next step is to bake the green body to burn off the binder and begin to fuse the powder micro-spheres together. The part is then sintered at high temperature to densify it and to strengthen it. It is this sintering step where the problems happen, because the part necessarily must shrink during densification. Because the layers are stacked up in a different plane (Z) than the binder printing (X,Y)—in other words, all planes are not processed the same way due to mechanical properties and geometric constraints, the parts shrink anisotropically. This anisotropy makes compensating for the shrinkage difficult. In addition, because of the way the powder spheres sinter together, there are always very small channels that develop at the interfaces of the particles as they fuse together. These channels can cause vacuum leaks, both real and virtual, in the finished part.

The most successful materials for BJ, like other AM processes, tend to be low in thermal conductivity and electrical conductivity, so a higher-conductivity material is needed to coat the surfaces for use in THz devices. By performing only the minimal amount of sintering needed, the shrinkage can be minimized, but the density is low (typically around 60%), so there is a large pore volume to fill. Hence to restore conductivity, infiltration, a method of soaking up metal into the highly-porous "brown body" can be used. In addition to restoring conductivity, it also fills in all the space between the base metal particles and provides vacuum tightness. Infiltration is typically done in vacuum or in a reducing atmosphere at above the melt point of the infiltrate. For a sintered stainless steel base part, the copper can be infiltrated at around 1220°C in vacuum. The excess temperature above the melt point of copper allows the copper to wet the stainless steel surface. At this temperature, any cathode poisons and all water vapor will be baked off quite effectively, resulting in an inherently ultrahigh vacuum-compatible part. Additionally, infiltration performed in vacuum prevents gases from being trapped inside the pores, preventing virtual leaks—another benefit.

The key for making AM methods work for THz devices is limiting the surface roughness, and, unfortunately, powder-based AM methods result in high inherent roughness. Centripetal barrel polishing and magnetorheological polishing have been demonstrated to improve surface roughness to about Ra of 1–2 microns [39], but significant further development is needed to bring surface quality to nanoscale levels for demonstrating suitability to applications at 100 GHz and above. Figure 23.7d shows a W-band test waveguide that was magnetorheologically polished on the inside surfaces.

23.3.4.5 Electrical discharge machining

Wire electrical discharge machining (WEDM) and sinker EDM (also known as plunge EDM) both make use of an electric arc pulled between a wire (in the case of WEDM) or a mold (in the case of SEDM) and the work piece. Figure 23.8 illustrates the two methods. Both types are subtractive in nature. In WEDM, the wire is continuously spooled so that it does not also erode along with the work piece. In SEDM, the mold is made of an erosion-resistant material, such as graphite, and several molds are often required in

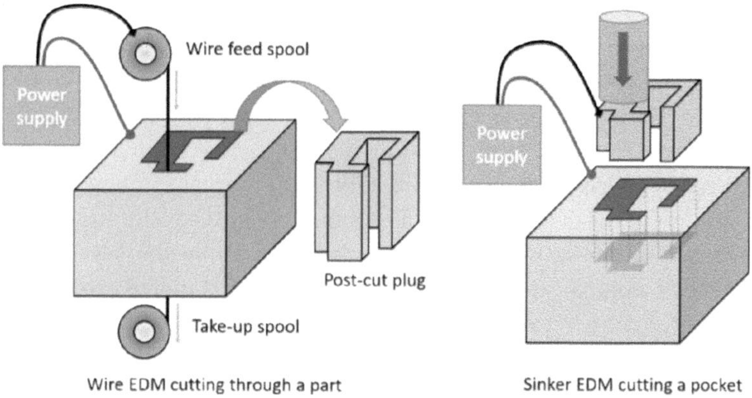

Figure 23.8 (Left) Diagram showing how wire EDM cuts through the entirety of a part using a thin, electrified wire. (Right) Diagram of sinker/plunge EDM where an electrified mold (often graphite) cuts a partial pocket into a block of metal.

order to finish the part to tight tolerance—akin to finishing passes in CNC machining.

The smallest wire presently available for WEDM is 0.0008 inch diameter (20 microns) and, due to the plasma sheath around the wire, cuts a path almost 0.0020 inch wide (50 microns). While WEDM is highly precise, its inherent requirement for the wire to cut entirely through the work piece limits it to 2D, essentially. Although a serpentine waveguide could be precisely cut out by WEDM, the two interdigitated portions of the serpentine need to maintain alignment to each other and then need to be bonded to top and bottom plates to complete a circuit. Typically, this serpentine circuit also needs to be split along the center plane in order to cut out a beam tunnel by CNC machining. The tolerances associated with the alignment of all these pieces makes WEDM not especially attractive for THz devices.

SEDM requires a sinker mold and therefore the precision depends not only on how accurate the molds are made, but also if the part being sinker EDM'd can tolerate the plasma sheath around the mold, which imparts a fillet-like profile on all edges. The use of finishing molds to reach the final shape means the tolerance on each mold must be excellent, and they must be extremely well aligned to each other and to the work piece. This process typically yields a part not better than +/−0.0005 inch (12 microns) in tolerance. But in SEDM, unlike WEDM, it is at least conceivable, if not awkward, to fabricate some types of 3D structures with a single mold operation.

23.3.4.6 Laser ablative machining (subtractive)

A new type of subtractive machining takes advantage of ultrashort pulse laser to ablate copper faster than the timescale on which heat transfer occurs. This method looks quite promising for machining copper without direct physical stresses and appears able to achieve high vertical aspect ratios needed in many slow-wave circuits.

The surface roughness, however, especially on the bottom surface perpendicular to the beam direction, is rather high at this time. Each laser pulse removes around 0.1 μo 1 μm of material, which defines the ultimate roughness [41, 42]. It is yet to be seen if further post-treatments, such as annealing or copper bright-dipping can reduce the roughness and make this process a winner for subtractive methods on copper at THz frequencies.

23.3.4.7 Lithography

Lithography is the process for using light, usually in the form of ultraviolet to X-rays, to create a positive or negative mold in a photoresist—A material that activates upon exposure to light or X-rays. Except for step "e," which is the subject of a patent, Fig. 23.9 shows a typical UV-LIGA process, which includes laying a photoresist (SU-8, KMPR, PMMA, etc.), and chemically altering it by UV light exposure followed by an electroforming step. While this process can be extremely accurate, down to wavelengths of light, it is inherently 2D. To create 3D features requires multiple layers, and even that has severe limitations. For example, the existing electroformed layer needs to be passivated somehow to prevent the activator (UV light, X-rays, etc.), on subsequent layers from causing unwanted structural changes to previous layers. Furthermore, substrate alignment is difficult to maintain because the substrate generally must be moved between many machines for processing. As an example, consider a UV-LIGA process with SU-8: The substrate will be in the UV exposure instrument, the SU-8 removal chemical bath, several cleaning steps,

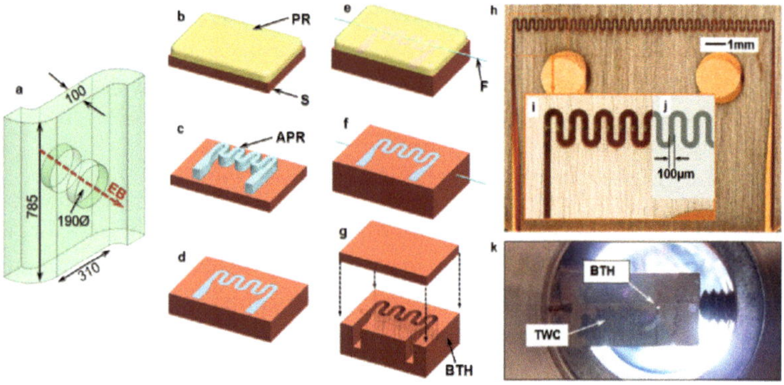

Figure 23.9 A hybrid LIGA process using an embedded filament in the LIGA process to create a TWT circuit. (a), View of the vacuum portion of a half-period of the circuit with location of electron beam EB; dimensions in micrometers. (b–g), Two-layer UV-LIGA microfabrication process using a polymer monofilament embedded in the photoresist (PR) to hold the size, shape, and location of the electron beam tunnel. APR: activated photoresist, F: filament, BTH: beam tunnel hole. (h-i), Photomicrographs of the completed all-copper serpentine waveguide traveling-wave circuit (TWC) with a 183 μm diameter steel gauge pin inserted into the beam tunnel. (j), Photo-mask pattern. (k), Photo of completed body with microfabricated circuit brazed in place.

an electroforming step in a chemical bath, a planarization step to cut the electroformed material back to a precise layer height, a possible passivation coating applied to underlying SU-8 and/or a hardbaking step, the new SU-8 application coating, the SU-8 softbaking step, repeat. Clearly, these steps are not amenable to integration in a single machine apparatus that can easily maintain alignment. Aides for alignment, such as substrate markers are often used, but these markers need to be either integrated into each layer, where they risk accumulated alignment errors, or need to be protected from damage during the other processing steps, which incurs yet more processing.

LIGA methods do not rely on particle sizes, as in powder-based AM methods, but rather on the wavelength of the activating light, so they are inherently much smoother in finish, resulting in inherently less roughness and therefore better potential for high surface conductivities than powder-based methods. They are also capable of high vertical aspect ratios of 20:1 or greater [43], although extraordinary development procedures may be required.

The embedded polymer filament patent invented at the U.S. Naval Research Laboratory [44, 45] is shown in Fig. 23.9 as a modified UV-LIGA process. In this process, a UV-transparent filament with an index of refraction similar to the photoresist is embedded into the photoresist prior to exposure. It then becomes captured in the photoresist and acts as a cylindrical mold for the shape of the electron beam tunnel as the electroforming steps are performed.

In X-ray LIGA, X-rays from a synchrotron are used to activate PMMA through a mask made from hi-Z materials. Because the X-ray wavelength is so short, the resolution is not usually limited by diffraction, as it is in UV-LIGA. Therefore, X-ray LIGA can be used to create extremely high VARs, in excess of 100:1 [46, 47]. However, due to the high cost of a synchrotron X-ray source, X-ray LIGA is possibly one of the more expensive processes listed here, and therefore is unlikely to garner plentiful commercial attention at this time, except for the most stringent, costly components.

In addition, because PMMA is a "positive" photoresist, one where the PMMA is destroyed by the X-rays, it is challenging to do multiple-layer processes, unless the mold features on the top layers are always larger or the same size as the underlying PMMA. Otherwise the X-rays will simply destroy all PMMA in its path. This behavior can be useful if multiple features are being cut into the bulk material from different angles in order to create a more complex structure, however. This positive resist behavior is different than a "negative" one, such as the UV-activated SU-8, where UV light acts to harden (activate) the photoresist, and subsequent exposure to UV has little effect on the previously-activated photoresist. Thus incidental exposure of previously-formed SU-8 layers will not be adversely affected by subsequent UV exposure while creating new layers.

More advanced LIGA processes are now in use that attempt to overcome the 2D limitation by simply stacking many layers on top of each other. Two examples that have been used for THz components are Collins Aerospace Z-Fab, and Nuvotronics PolyStrata [48]. It is possible to overlap dozens or even hundreds of fine layers together to create a tight-tolerance part. Some processes even allow for inclusion of a limited set of dielectric materials in limited geometries, which brings us closer to the dream of multi-material direct printing of 3D structures. As was noted before, LIGA is one fabrication process that purports to fulfill the needs of THz devices, and, while these processes are reputedly very expensive, they are absolutely within the feature range required. In addition, the roughness for these multi-layer processes is very low, owing in part to the fact that each time a new layer is added on to an old one, the old one must be planarized and polished first, leading to polished top and bottom walls. Since the sidewall roughness is optically defined, it is also in the 100 nm range or better, which is exactly what is needed for THz devices.

23.3.4.8 Deep reactive ion etching

This process, DRIE, borrowed from semiconductor processing, and now micro-electromechanical systems (MEMS) processing, takes silicon wafers and alternates between an etching process in reactive gas plasma and a passivation process. Very high vertical aspect ratios are achievable. Since this process does not work on metals yet, it would seem irrelevant to THz devices. But this technique was in fact used to create the first-ever 1.03 THz traveling-wave tube by plating the silicon microstructures with gold [49]. The authors also succeeded in demonstrating TWTs at 670 GHz [50], and 850 GHz [51]. Such a circuit cannot conduct heat away fast enough to operate at high average power since the bulk material is still silicon, but this DRIE process indeed demonstrated viability of a true 1 THz linear-beam TWT amplifier.

23.3.4.9 3D photopolymer printing

Distinct from thermoplastic FDM, 3D printing of photo-cured polymer plastics is the highest-resolution AM process available. This is due to the high resolution with which the liquid photopolymer resist material can be cured, which bears similarities to the optically defined precision achieved in photolithography. Stereolithography, Digital Light Processing (DLP), and two-photon polymerization (TPP) are the primary methods in this category [52]. Stereolithography and DLP rely on ultraviolet or high-visible range light to cure layers of photoresist using a focused laser or digital image projector, respectively. The DLP process has significantly faster build times because the projector cures an entire layer simultaneously, rather than raster-scanning the layer one pixel (voxel) at a time, as is the case in stereolithography and TPP. Minimum voxel sizes achieved in stereolithography and DLP are typically 25–35 μm, and as small as 15 μm in some machines [53].

TPP relies on a nonlinear optical process in which two or more photons in an ultrafast laser pulse simultaneously activate photo-initiators in the resist. Because of this highly-nonlinear thresholding phenomenon, the photopolymerization occurs over a very tiny volumetric space, with a voxel resolution down to 100 nm, or even less, which is far less than the diffraction limit of the 800 nm Ti:Sapphire lasers commonly used for the photo-initiation. The 3D printers by Nanoscribe GmbH [54], UpNano GmbH [55], among others, perform direct laser writing using two-photon interactions. Surface roughness can be as low as 10 nm [55]. Because these machines are designed for ultraprecision movements, and the write volumes are so tiny, it is very difficult to print large structures such as a 20 mm-long TWT circuit. Thus, use of the TPP method has-been mostly limited to nano-fabrication applications rather than THz devices [56]. However, full-size slow-wave circuit geometries at 1 THz in photoresist material have been printed by stitching together build areas and using sparse scanning techniques to reduce laser raster time [57].

As in the case of DRIE, a conductive metal plating on the polymer surface is required to produce an electromagnetic circuit by these methods. Electroplating and electroless plating methods have been developed to produce traveling-wave circuits and passive microwave components, which have been demonstrated in cold test in the upper-millimeter-wave range up to approximately 300 GHz using stereolithography and DLP [58–60]. Reported surface roughness in a DLP case was $R_a = 430$ nm [61].

23.3.5 Surface Treatments

Owing to the strong effect of roughness on performance at THz frequencies, low surface roughness is essentially a requirement

[62]. Surface treatments based on chemical, electromechanical, or mechanical means are available for some materials to reduce roughness. For example, the well-known copper bright dip solution can be used to effectively reduce roughness on pure copper [63]. But this process is extremely intense and can actually dissolve fine circuit features as if they were roughness, and the chemistry does not lend itself to dilution to help control the etch rates. Nevertheless, chemical means for polishing offer, perhaps, the best promise for increasing the effective electrical conductivity of THz slow-wave circuits.

Mechanical polishing of external surfaces is as old as the invention of the mirror. But polishing internal features mechanically is quite another challenge. In large cavities, it is possible to load polishing media inside the part and vibrate the part strongly. In small parts, this method tends to perhaps increase large-scale roughness while reducing small-scale roughness. Unfortunately, both scales matter for THz devices. One group employed magnetorheological polishing inside millimeter-wave cavity parts for surface roughness improvements [39], but this method is not very attractive for smoothing the inside of a complex and finely-structured slow-wave circuit, such as an extended interaction klystron (EIK) ladder, where it is difficult to conceive how the magnetic media could be walked between the rungs of the ladder in order to polish them. The risk of leaving any kind of particles inside a slow-wave circuit is not a risk that anyone in the production environment would want to take.

Electropolishing is a well-known process that is commonly used on UHV stainless steel components, surgical tools, food and pharmaceutical processing equipment, decorative applications, and the like. It works by immersing the part into an electrolyte solution, typically a sulfuric acid-centered solution, and applying a current. Micro-peaks on the surface are quickly removed, and flat areas are undamaged. The result is a highly-polished surface. In the case of UHV parts, the parts outgas less and have better ultimate pressures than those not electropolished [64].

While electropolishing can produce very good results on outer surfaces, even on complex shapes, it does not apply well to internal surfaces due to the need for a counter-electrode to maintain a constant spacing [65]. Therefore, for most slow-wave structures, it seems commercial electropolishing is probably not a viable option.

Laser surface polishing is a technique whereby the laser dwells on the surface just long enough to melt down the roughness peaks. While this has been successfully used on Inconel and other low-thermal-conductivity materials, it has not yet shown promise on copper [66].

For copper, perhaps one overlooked surface treatment, at least outside of the production environment, is to do nothing—nothing, except the normal brazing procedures that often exceed even 1000°C. At these temperatures, the remaining machining marks and roughness on copper essentially melt down as the copper grains grow, producing a highly smooth, mirror-like finish mottled with a spider web-like network of visible grain boundaries, as shown in Fig. 23.10. It is well known that both bulk thermal conductivity [67] and bulk electrical conductivity [68] improve with larger grain sizes and with annealing. For single-crystal, grain boundary-free

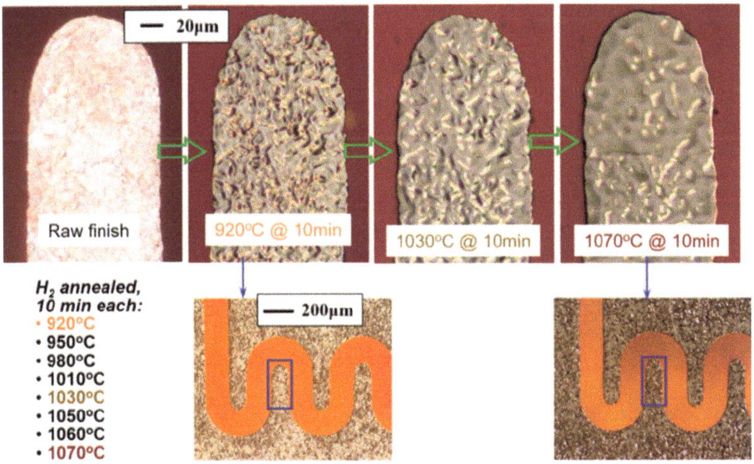

Figure 23.10 The effect of high-temperature annealing on a pure copper serpentine waveguide circuit at 220 GHz. The raw finish is electroformed copper followed by a molten salt bath step to remove SU-8 photoresist. The molten salt also etches the copper, leading to the initially rough surface finish. As the annealing temperature increases, as well as the number of annealing cycles, the copper becomes markedly smoother.

copper the conductivity reaches 113% IACS [69], which is above the conductivity of silver.

While many of these surface treatments apply to a part made of a single material, the same cannot be said for parts that have been plated or infiltrated. For example, a part binder-jetted in stainless steel using 30 micron powder and then infiltrated with copper, leaving at least 1 micron of pure copper on the THz circuit walls would have 90% of the conductivity of copper had it been made from solid copper at Ka-band due to the skin depth effect. However, the 30 micron stainless steel powder size making up the bulk of the part will impart hefty losses in the form of multi-micron Ra roughness, even after sintering. There is no obvious way to eliminate the roughness caused by the bulk stainless steel powder. So for plated or infiltrated parts, it is critically important to minimize the roughness produced by the base process.

23.4 Joining/Brazing

Due to the limitations of many of the fabrication techniques described above, it is usually necessary to join parts together by brazing, welding, bonding, soldering, or some other method. While ultrahigh-vacuum-compatible demountable flanges utilizing compressible copper gaskets, such as the Conflat® flange, are useful in research, they also tend to increase the size and weight of the end product, and/or limit the ability to bake out the device, and so are not really an option for the THz device production environment.

23.4.1 Brazing

Provided the process engineer has acquiesced to the necessity of brazing, one will find that fixturing is a constant challenge. There is always the need for several test brazes on a dummy part before performing the real braze, even for experienced technicians. This extra work raises the costs, the number of processing steps, the time-to-product, and also requires skilled technical labor. The main challenge is maintaining the correct alignment and correct pressure on the parts being brazed at the brazing temperature. In a common scenario, two parts are pinned for alignment and weighted for pressure. If these alignment pins are too loose, the top part may "skate" off of alignment as the braze foil turns to liquid. If the pins are too tight, the weight will fail to achieve the correct pressure as the upper half "hangs" on the pins. In some cases, choosing the right pin/hole size and number of pins is truly an art that requires experimentation. For copper-to-copper brazes, a suite of braze alloys (foils or wires usually) based on copper-gold alloys are typically used in sequence of descending liquidus point. Silver and silver alloys may also be used, provided the designer is aware that high-silver alloy content may dissolve into copper causing pits and fissures if too much is used. Silver also tends to travel around the surfaces of pure copper. Pure silver as a braze foil appears generally safe for foils of 0.001" (25 microns) or less in thickness.

23.4.2 Diffusion Bonding

Provided two parts have very flat faces that are in intimate contact with high pressure, two parts can be diffusion bonded together by simply raising the temperature near the melt point for an extended period of time [70]. This method has been successfully used for creating THz circuits, but has plenty of risks associated with the high pressures, which can cause deformation of the soft copper parts. Additionally, there is concern that pure copper will grow grain boundaries so large at temperatures near the melt point for extended periods of time that it may consume small features altogether.

23.4.3 Transient Liquid Phase Bonding

In transient liquid phase bonding, a filler is used that is only in liquid phase for a short amount of time as it dissolves into the host. A successful example is silver on copper [71]. A very thin silver layer is sandwiched between two copper parts and initially forms a eutectic alloy with a low melt point. As the silver continues to diffuse into the copper, the melt point raises, forming a solid copper joint. In the end, if the silver layer was very thin, this joint is essentially pure copper with an inconsequential silver impurity. Subsequent brazes or TLP bonds can be performed at the same temperature without affecting the first joint—Indeed such temperature excursions further enable the initial silver deposit to diffuse throughout the copper.

23.4.4 Laser welding

Laser welding is commonly used to weld vacuum flanges together, especially for electron gun-to-body joints. Such weld flanges are typically made from Inconel® or similar alloy for ease of welding and corrosion resistance. Inconel is rated for use in SLAC ultrahigh-vacuum accelerator systems [72]. It is convenient for joining two parts when you do not want to heat both parts up to braze temperatures for the sake of preserving the two parts being joined. It also gives a good "break point" in case future repairs or modifications are required. The downside, however, is that it assumes you already have a weld flange brazed on to both parts, and hence brazing is still required, typically.

23.5 Recommendations and Application to THz Devices

Up to about 400 GHz, micro-CNC machining seems the most reliable and cost-effective approach on copper. When micromachining THz circuits, use the trochoidal tool paths or other constant-chip-load cutting method combined with diamond-coated endmills and plenty of finishing passes to clean up burrs.

Above 200 GHz, lithography and electroforming seems to be a solid approach, if the 2D limitations do not pose a problem. For more 3D structure, the proprietary PolyStrata and Z-fab approaches seem to be well suited and properly developed for these frequency ranges, although they are likely to cost significantly more than CNC micromachining. X-ray LIGA is a viable option, provided one has access to a synchrotron light source.

At around 1 THz or above, TPP as a base method for 3D-printed mold electroforming (3D PriME) looks like a potent possibility, especially considering the likelihood for very low surface roughness. The major challenge for 3D PriME at THz frequencies, however, is that the electroformed copper needs to completely fill every tiny crevice. In addition, the completed polymer-free structure needs to be vacuum compatible, and, preferably, braze compatible too.

23.6 Discussion, Conclusion, and Outlook

The ideal manufacturing technique would be direct 3D copper printing capable of making very smooth surfaces and arbitrary structures (overhangs, cavities, etc.)—Better yet if lossy materials and dielectrics can be integrated. However, we are not there at present. Rather than simply pressing the "print" button in a CAD application, at this time there are still many processes to link together requiring much human intervention, such as processing the raw materials (de-oxygenating), optimizing printing parameters and working around limitations, de-powdering, removing support structures and bases, baking, sintering, infiltrating, polishing, plating/eletroforming, joining/brazing, etc. Putting these steps into a single, process-flexible machine is not impossible, it is just extremely expensive at this time. With the rapid advances in 3D fabrication technologies pushing toward faster build times, smaller minimum features sizes, and wider ranges of materials, it is likely that we can expect machines in the future that will be capable of creating THz devices with repeatability at lower cost than presently available.

Acknowledgments

C. D. Joye and A. M. Cook acknowledge the U.S. Office of Naval Research, who sponsored much of the research that led to this chapter. D. Gamzina acknowledges the DOE Office of Science High Energy Physics Early Career program.

References

1. D. E. Weisberg, Computer-Aided Design's Strong Roots at MIT, 2008. [Online]. Available: http://cadhistory.net/03%20MIT%20CAD%20 Roots%201945-1965.pdf. [Accessed 08 2021].

2. E. Brinksmeier, Ultraprecision Machining, *CIRP Encyclopedia of Production Engineering,* 2014.
3. M. P. Bohrer, Deep-UV and X-Ray Microlithography, AT&T Bell Laboratories, Murray Hill, NJ.
4. Newport Corp., Photolithography Overview, Newport Corp. [Online]. Available: https://www.newport.com/n/photolithography-overview. [Accessed 08 2021].
5. E. W. Becker, Production of Separation-Nozzle Systems, *Naturwissenschaften,* vol. 69, pp. 520–523, 1982.
6. 3Dsourced, The Complete History of 3D Printing from 1980 to 2021, 2021. [Online]. Available: https://www.3dsourced.com/guides/history-of-3d-printing/. [Accessed 11 08 2021].
7. Microlight-3D, Two-Photon Polymerization, Microlight-3D [Online]. Available: http://www.microlight.fr/TPP.html. [Accessed 08 2021].
8. Nanoscribe GmbH & Co. KG, Photonic Professional GT2, Nanoscribe GmbH & Co. KG [Online]. Available: https://www.nanoscribe.com/en/. [Accessed 08 2021].
9. Hybrid Manufacturing Technologies, Products, Hybrid Manufacturing, 2020. [Online]. Available: https://hybridmanutech.com/products/. [Accessed 08 2021].
10. DMG Mori, LASERTEC 65 3D DED Hybrid, DMG Mori [Online]. Available: https://us.dmgmori.com/products/machines/additive-manufacturing/powder-nozzle/lasertec-65-ded-hybrid. [Accessed 08 2021].
11. B. Rodgers, Buying a Wire EDM, Gardner Business Media, 2021. [Online]. Available: https://www.techspex.com/knowledgecenter/guide-to-buying-machine-tools/buying-a-wire-edm. [Accessed 08 2021].
12. Electrical Discharge Machining [Online]. Available: https://en.wikipedia.org/wiki/Electrical_discharge_machining. [Accessed 08 2021].
13. Digital Alloys [Online]. Available: https://www.digitalalloys.com/blog/powder-bed-fusion/. [Accessed 08 2021].
14. Promising Future Processing Technology, in *Advances in CMP Polishing Technologies,* Elsevier, Inc., pp. 229–295, 2012.
15. L. Gwennap, Apple's 5 Nanometer Chip Is Another Signpost That Moore's Law Is Running Out, Forbes, 12 10 2020. [Online]. Available: https://www.forbes.com/sites/linleygwennap/2020/10/12/apple-moores-law-is-running-out/. [Accessed 08 2021].

16. J. Feng, Y. Tang, D. Gamzina, X. Li, B. Popovic, M. Gonzalez, L. Himes, R. Barchfeld, H. Li, P. Pan, R. Letizia, C. Paoloni, and J. Neville C. Luhmann, Fabrication of a 0.346 THz BWO for Plasma Diagnostics, *IEEE Trans. Electron Devices,* vol. 65, no. 6, pp. 2156–2163, 2018.
17. L. A. Hof and J. A. Ziki, Micro-Hole Drilling on Glass Substrates—A Review, *Micromachines,* vol. 8, no. 53, https://doi.org/10.3390/mi8020053, 2017.
18. T. Horn and D. Gamzina, Copper Additive Manufacturing (short course), in *IEEE International Vacuum Electronics Conference,* Monterey, 2020.
19. Copper conductivity Materials Database, Copper Development Association [Online]. Available: https://copperalliance.org.uk/about-copper/conductivity-materials/. [Accessed 24 08 2021].
20. GlidCop AL-15 & GlidCop AL-25, J. I. Anthony & Co., 2005. [Online]. Available: http://www.jlanthonyandcompany.com/product/glidcop_al.html. [Accessed 08 2021].
21. GLIDCOP(r) Copper alloy system, Hoganas, 2021. [Online]. Available: https://www.hoganas.com/en/powder-technologies/glidcop/. [Accessed 08 2021].
22. F. Motschmann, Brazing and Vacuum Brazing at CERN, CERN [Online]. Available: https://indico.cern.ch/event/721360/contributions/2991902/attachments/1652143/2642948/Brazing_and_Vacuum_Brazing_Motschmann.pdf. [Accessed 24 08 2021].
23. R. E. Honig and D. A. Kramer, Vapor Pressure Data for the Solid and Liquid Elements, *RCA Rev.,* vol. 30, pp. 285–305, 1969.
24. A. C. Inc., AMZIRC: Oxygen Free Copper Plus Zirconium, *Alloy Digest,* vol. 33, no. 3, 1984.
25. D. L. Ellis, Precipitation-Strengthened High-Strength, High-Conductivity Cu-Cr-Nb Alloys Produced by Chill Blockmelt Spinning, Case Western Reserve University, Cleveland, OH (USA), 1989.
26. C. L. Kory, Simulation of TunneLadder Traveling-Wave Tube Cold-Test Characteristics: Implementation of the Three-Dimensional, Electromagnetic Circuit Analysis Code Micro-SOS, *Natl. Aeronautics Space Administration, Office Manag. Sci. Tech. Information Program,* vol. 3294, pp. 1–13, 1993.
27. [Online]. Available: https://www.metalworkingsuppliers.com/mw/milling-machine-history/. [Accessed 11 08 2021].
28. SWISSto12, 2020. [Online]. Available: https://swissto12.com/. [Accessed 08 2021].

29. A. M. Cook, C. D. Joye, R. L. Jaynes, and J. P. Calame, W-band TWT Circuit Fabricated by 3D-Printed Mold Electroforming, in *IEEE 19th International Vacuum Electronics Conference (IVEC)*, Monterey, CA, 2018.
30. R. Engelke, J. Mathuni, G. Ahrens, G. Gruetzner, M. Bednarzik, D. Schondelmaier, and B. Loechel, Investigations of SU-8 Removal from Metallic High Aspect Ratio Microstructures with a Novel Plasma Technique, *Microsyst. Technol.*, vol. 14, pp. 1607–1612, 2008.
31. C. H. Lee, P. G. Jung, S. M. Lee, S. H. Park, B. S. Shin, J.-H. Kim, K.-Y. Hwang, K. M. Kim, and J. S. Ko, Replication of Polyethylene Nano-Micro Hierarchical Structures Using Ultrasonic Forming, *J. Micromech. Microeng.*, vol. 20, p. 035018, 2010.
32. A. Hunger, G. Carl, A. Gebhardt, and C. Rüssel, Ultra-High Thermal Expansion Glass–Ceramics in the System $MgO/Al_2O_3/TiO_2/ZrO_2/SiO_2$ by Volume Crystallization of Cristobalite, *J. Non-Crystal. Solids,* vol. 354, pp. 5402–5407, 2008.
33. E. B. Kivanc and E. Budak, Development of Analytical Endmill Deflection and Dynamic Models, in *ASME International Mechanical Engineering Congress & Exposition*, Washington D.C., 2003.
34. A. N. Vlasov, J. C. Rodgers, R. L. Jaynes, C. D. Joye, J. A. Pasour, F. N. Wood, I. A. Chernyavskiy, S. J. Cooke, D. K. Abe, B. Levush, T. M. Antonsen, and D. Chernin, Design of a Low Voltage Folded Waveguide Four Beam Mini-TWT, in *IEEE 19th International Vacuum Electronics Conference (IVEC)*, Monterey, CA, 2018.
35. R. L. Jaynes, A. N. Vlasov, J. C. Rodgers, F. N. Wood, C. Joye, D. E. Pershing, A. M. Cook, and D. K. Abe, Fabrication and Testing of Ka-Band Multi-Beam TWT Circuits, in *IEEE 19th International Vacuum Electronics Conference (IVEC)*, Monterey, CA, 2018.
36. A. Turner, Achieving Mirror Surface Finishes, Commercial Micro Manufacturing, MST Global, 07 09 2020. [Online]. Available: http://www.cmmmagazine.com/cmm-articles/achieving-mirror-surface-finishes-by-milling/. [Accessed 08 2021].
37. C. Ledford, C. Rock, P. Carriere, P. Frigola, D. Gamzina, and T. Horn, Characteristics and Processing of Hydrogen-Treated Copper Powders for EB-PBF Additive Manufacturing, *MDPI Appl. Sci.*, vol. 9, no. 19, p. 3993, 2019.
38. G. Lanza, Outgassing of Electron Beam Printed Copper, in *IEEE International Vacuum Electrnoics Conference (IVEC)*, Netherlands, 2021.
39. I. Karakurt, K. Y. Ho, C. Ledford, D. Gamzina, T. Horn, N. C.Luhmann, and L. Lin, Development of a Magnetically Driven Abrasive Polishing Process for Tructures, *Procedia Manuf.*, vol. 26, pp. 798–805, 2018.

40. High Power Blue Laser - Nurubu [Online]. Available: https://nuburu.net/. [Accessed 08 2021].
41. C. Hart and S. Rajaraman, Low-Power, Multimodal Laser Micromachining of Materials for Applications in sub-5 μm Shadow Masks and sub-10 μm Interdigitated Electrodes (IDEs) Fabrication, *Micromachines,* vol. 11, no. 2, p. 178, 2020.
42. S. C. Christensen and C. P. Christensen, Laser Focus World, 1997. [Online]. Available: https://www.laserfocusworld.com/lasers-sources/article/16550905/small-excimer-lasers-create-tiny-pieces. [Accessed 13 08 2021].
43. D. Figura and J. Bartel, Fabrication of High Aspect Ratio SU-8 Structures Using UV Lithography and Megasonic-Enhanced Development, *Electrochem. Soc. Trans.,* vol. 25, no. 31, pp. 29–35, 2010.
44. C. D. Joye, J. P. Calame, K. T. Nguyen, and M. Garven, Microfabrication of Fine Electron Beam Tunnels using UV-LIGA and Embedded Polymer Monofilaments for Vacuum Electron Devices, *J. Micromech. Microeng.,* p. 015010ff, vol. 22, no. 1, 2012.
45. C. D. Joye, Microfabrication of Tunnels. USA Patent 9,777,384, 3 October 2017.
46. S. Lemke, P. Goettert, I. Rudolph, J. Goettert, and B. Löchel, Negative Resists for Ultra-Tall, *High Aspect Ratio Microstructures, in Proceedings of 37th International Conference on Micro and Nano Engineering (MNE), Berlin, Germany,* 2011.
47. A. L. Bogdanov and S. S. Peredkov, Use of SU-8 Photoresist for Very High Aspect Ratio X-Ray Lithography, *Microelectron. Eng.,* vol. 53, no. 1, pp. 493–496, 2000.
48. Nuvotronics, Inc., Nuvotronics - PolyStrata (R) Architecture, Nuvotronics, Inc. [Online]. Available: https://www.nuvotronics.com/ architecture.php. [Accessed 13 08 2021].
49. J. C. Tucek, M. A. Basten, D. A. Gallagher, and K. E. Kreischer, Operation of a Compact 1.03 THz Power Amplifier, in *IEEE 17th International Vacuum Electronics Conference (IVEC),* Monterey, CA, 2016.
50. J. C. Tucek, M. A. Basten, D. A. Gallagher, K. E. Kreischer, R. Lai, V. Radisic, K. Leong, and R. Mihailovich, A 100 mW, 0.670 THz Power Module, in *IEEE 13th International Conference on Vacuum Electronics (IVEC),* Monterey, CA, 2012.
51. J. C. Tucek, M. A. Basten, D. A. Gallagher, and K. E. Kreischer, Testing of a 0.850 THz Vacuum Electronic Power Amplifier, in *IEEE 14th International Vacuum Electronics Conference (IVEC),* Paris, France, 2013.

52. S. C. Ligon, R. Liska, J. Stampfl, M. Gurr, and R. Mulhaupt, Polymers for 3D Printing and Customized Additive Manufacturing, *Chem. Rev.,* vol. 117, pp. 10212–10290, 2017.
53. U. V. Andersen, D. B. Pedersen, H. N. Hansen, and J. S. Nielsen, In-Process 3D Geometry Reconstruction of Objects Produced by Direct Light Projection, *Int. J. Adv. Manuf. Technol.,* vol. 68, pp. 565–573, 2013.
54. Nanoscribe Products [Online]. Available: https://www.nanoscribe.com/en/products. [Accessed 13 08 2021].
55. UpNano GmbH [Online]. Available: https://www.upnano.at/nanoone/. [Accessed 13 08 2021].
56. L. C. Montemayor, L. R. Meza, and J. R. Greer, Design and Fabrication of Hollow Rigid Nanolattices via Two-Photon Lithography, *Adv. Eng. Mater.,* vol. 16, no. 2, pp. 184–189, 2013.
57. A. Cook, C. Joye, J. Calame, and F. Perkins, Three-Dimensional Printing and Electroforming for Millimeter-Wave Vacuum Electronics, in *2017 NRL Review, U.S. Naval Research Laboratory*, Washington, DC: DTIC accession number DSIAC-2190455, pp. 91–98, 2017.
58. A. M. Cook, C. D. Joye, and J. P. Calame, W-Band and D-Band Traveling-Wave Tube Circuits Fabricated by 3D Printing, *IEEE Access,* vol. 7, pp. 72561–72566, 2019.
59. X. Shang, P. Penchev, C. Guo, M. J. Lancaster, S. Dimov, Y. Dong, M. Favre, M. Billod, and E. de Rijk, W-Band Waveguide Filters Fabricated by Laser Micromachining and 3-D Printing, *IEEE Trans. Microwave Theory Tech.,* vol. 64, no. 8, pp. 2572–2580, 2016.
60. A. von Bieren, E. de Rijk, J.-P. Ansermet, and A. Macor, Monolithic Metal-Coated Plastic Components for mm-Wave Applications, in *39th International Conference on Infrared, Millimeter, and Terahertz Waves (IRMMW-THz)*, Tucson, AZ, USA, 2014.
61. J. Shen and D. S. Ricketts, Additive Manufacturing of Complex Millimeter-Wave Waveguide Structures Using Digital Light Processing, *IEEE Trans. Microwave Theory Tech.,* vol. 67, no. 3, pp. 883–895, 2019.
62. M. P. Kirley and J. H. Booske, Terahertz Conductivity of Copper Surfaces, *IEEE Trans. Terahertz Sci. Tech.,* vol. 5, no. 6, pp. 1012–1020, 2015.
63. M. Baldwin and J. Pope, Stanford Synchrotron Radiation Project User Specifications for Vacuum Systems and Components which Interface with the SPEAR Vacuum System, SLAC, Menlo Park, CA, 1973.
64. Y. Li and X. Liu, Vacuum Science and Technology for Accelerator Vacuum Systems, Cornell University, 2017. [Online]. Available: https://uspas.fnal.gov/materials/17UCDavis/Vacuum/USPAS%

202017%20Vacuum%20Session%205_4%20Fab_and_Clean.pdf. [Accessed 08 2021].

65. Z. Chaghazardi and R. Wuthrich, Electropolishing of Inside Surfaces of Stainless-Steel Tubing, in *The Electrochemical Society, MA2020-01*, Montreal, Canada, 2020.
66. Z. Fang, L. Lu, L. Chen, and Y. Guan, Laser Polishing of Additive Manufactured Superalloy, *Procedia CIRP*, vol. 71, pp. 150–154, 2018.
67. O. V. Gendelman, M. Shapiro, Y. Estrin, R. J. Hellmig, and S. Lekhtmakher, Grain Size Distribution and Heat Conductivity of Copper Processed by Equal Channel Angular Pressing, *Mater. Sci. Eng. A*, vol. 434, no. 1–2, pp. 88–94, 2006.
68. W. Yan, J. Chen, and X.-H. Fan, Effects of Grain Boundaries on Electrical Properties of Copper Wire, *Trans. Nonferrous Metals Soc. China*, vol. 13, pp. 1075–1079, 2003.
69. Y. C. Cho, S. Lee, M. Ajmal, W.-K. Kim, C. R. Cho, S.-Y. Jeong, J. H. Park, S. E. Park, S. Park, H.-K. Pak, and H. C. Kim, Copper Better than Silver: Electrical Resistivity of the Grain-Free, *ACS Crystal Growth Des.*, vol. 10, no. 6, pp. 2780–2784, 2010.
70. M. Samouhos, A. Peppas, P. Angelopoulos, and P. Tsakiridis, Optimization of Copper Thermocompression Diffusion Bonding under Vacuum: Microstructural and Mechanical Characteristics, *Metals*, vol. 9, p. 1044, 2019.
71. W. D. MacDonald and T. W. Eagar, Transient Liquid Phase Bonding, *Ann. Rev. Mater. Sci.*, vol. 22, pp. 23–46, 1992.
72. J. Khaw, Technical Specification for Vacuum Systems, OSTI.gov, Menlo Park, CA, 1987.

Index